ANNUAL REVIEW OF
ECOLOGY AND
SYSTEMATICS

EDITORIAL COMMITTEE (1994)

ANNUAL REVIEW OF ECOLOGY AND SYSTEMATICS

VOLUME 25, 1994

DAPHNE GAIL FAUTIN, *Editor*

University of Kansas

DOUGLAS J. FUTUYMA, *Associate Editor*

State University of New York at Stony Brook

FRANCES C. JAMES, *Associate Editor*

Florida State University

ANNUAL REVIEWS INC.   4139 EL CAMINO WAY   P.O. BOX 10139   PALO ALTO, CALIFORNIA 94303-0139

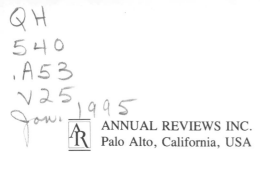

ANNUAL REVIEWS INC.
Palo Alto, California, USA

International Standard Serial Number: 0066–4162
International Standard Book Number: 0–8243–1425-5
Library of Congress Catalog Card Number: 71-135616

Annual Review and publication titles are registered trademarks of Annual Reviews Inc.

♾   The paper used in this publication meets the minimum requirements of
American National Standard for Information Sciences—Permanence of Paper
for Printed Library Materials, ANSI Z39.48-1984.

Annual Reviews Inc. and the Editors of its publications assume no responsibility for the
statements expressed by the contributors to this *Review*.

Typesetting by Kachina Typesetting Inc., Tempe, Arizona; John Olson, President;
Marty Mullins, Typesetting Coordinator; and by the Annual Reviews Inc. Editorial Staff

PRINTED AND BOUND IN THE UNITED STATES OF AMERICA

PREFACE

As regular readers of the *Annual Review of Ecology and Systematics* will have noticed, ARES's length has been increased. The additional pages allow us to accommodate more articles on what might be termed "applied ecology and systematics," such as forestry, fisheries, conservation biology, and similar areas. Intensifying scientific attention to these fields has resulted in increased research, new journals, and ultimately a need for reviews of the literature. Because these fields are linked to the more academic disciplines that have been the traditional purview of ARES, it is most appropriate that ARES embrace them.

Our desire to increase the number of chapters on applied topics reflects more than the magnitude of research effort in ecology and systematics. The growth of interest in these fields is a result of society's need for practical information and guidance. As revealed in many early ecology textbooks, ecological research prior to recent decades rarely mentioned humans, rightly recognizing activities of our species as often disruptive, but regarding them as somehow unnatural. Earlier integration of human activity into our studies might have prevented or forestalled some environmental situations that we now regard as crises. Next year's volume of ARES will have a topical section on environmental sustainability, in explicit recognition of the impact of humans on many of the issues of greatest concern to ARES readers.

The makeup of the current ARES editorial committee reflects the fields commonly published in ARES—fields in which ecological and systematic research has traditionally been done and students have traditionally been trained. Thus, for the immediate future, we must rely more than usual on ideas from the readers of ARES for topics and authors in the fields with which we are less familiar.

Chapters in this volume directly relevant to application of ecological and systematic knowledge are those by Schulze et al, who apply new technology to some pressing global issues, Chambers & MacMahon, who compare natural and managed systems, and Leigh Brown & Holmes on HIV. Holt & Lawton and Kuris & Lafferty, among others, explicitly relate principles they discuss to human concerns.

Linkages between and among chapters in this volume abound. Concern with evolution, the interface between ecology and systematics, is implicit in many chapters; it is explicit in by those by Moran on complex life cycles, Moreno on genetic architecture, Raff et al on metazoan phylogeny, and Vrijenhoek on unisexual fishes, and in that by Vermeij entitled "The Evolutionary Interaction among Species: Selection, Escalation, and Coevolution." Losos uses *Anolis* lizards and Frank & Leggett use fisheries as models and tests in ecology and evolution. Greenfield draws ecological comparisons in vocalizations of animals of various taxa, while Gerhardt deals with evolution of anuran vocalizations. The ecology of populations, with implications for evolution, is considered in several chapters. Hastings & Harrison analyze metapopulations in the terrestrial realm, emphasizing ecology; Palumbi examines genetic data for the marine realm, and Hoelzel focuses on cetaceans. These chapters connect with that by Kuris & Lafferty, who find evidence that interspecific competition structures populations of parasites, and with that by Holt & Lawton, who find populations structured in less direct ways, including by resource limitation, a phenomenon also addressed by Sterner & Hessen. Wootton deals with other indirect effects in community ecology, and Mitton considers population biology from a molecular perspective. Three chapters concern systematics of particular groups, two of them plants—legumes by Doyle, and Mesembryanthemacease by Ihlenfeldt. In their chapter on corals of the genus *Acropora,* Wallace & Willis describe how reproductive biology both explains and confounds systematics. Reproduction is the subject of chapters by Ketterson & Nolan (in birds) and by Wyatt & Broyles (in milkweeds).

This collection seems a rich harvest with which to celebrate ARES's twenty-fifth year of publication.

THE EDITORS

Annual Review of Ecology and Systematics
Volume 25 (1994)

CONTENTS

viii   CONTENTS (*continued*)

RELATED ARTICLES FROM OTHER *ANNUAL REVIEWS*

From the *Annual Review of Energy and the Environment,* Volume 19 (1994)

> *Trends in US Public Perceptions and Preferences on Energy and Environmental Policy:* BC Farhar
> *Water and Energy:* PH Gleick

From the *Annual Review of Entomology,* Volume 39 (1994)

> *Biology of Bolas Spiders,* KV Yeargan
> *Biology of Water Striders: Interactions Between Systematics and Ecology,* JR Spence and NM Andersen
> *Chemical Mimicry and Camouflage,* K Dettner and C Liepert
> *Insect Fauna of Coniferous Seed Cones: Diversity, Host-Plant Interactions, and Management,* JJ Turgeon, A Roques, and P de Groot
> *Phylogenetic Methods for Inferring the Evolutionary History and Processes of Change in Discretely Valued Characters,* DR Maddison
> *Selective Factors in the Evolution of Insect Wings,* JG Kingsolver and MAR Koehl

From the *Annual Review of Genetics,* Volume 28 (1994)

> *Evolution of* Hox *Genes,* FH Ruddle, JL Bartels, KL Bentley, C Kappen, MT Murtha, and JW Pendleton
> *Genetics and Phylogenetic Analyses of Endangered Species,* SJ O'Brien

From the *Annual Review of Phytopathology,* Volume 32 (1994)

> *Molecular Systematics and Population Biology of Rhizoctonia,* R Vilgalys and MA Cubeta
> *On Spread of Plant Disease: A Theory on Foci,* JC Zadoks and F den van Bosch

From the *Annual Review of Plant Physiology and Plant Molecular Biology,* Volume 45 (1994)

> *The Plant Mitochondrial Genome: Physical Structure, Information Content, RNA Editing, and Gene Migration to the Nucleus:* W Schuster and A Brennicke
> *Photoinhibition of Photosynthesis in Nature:* SP Long, S Humphries, and PG Falkowski

Annu. Rev. Ecol. Syst. 1994. 25:1–29

ALGAL NUTRIENT LIMITATION AND THE NUTRITION OF AQUATIC HERBIVORES

Robert W. Sterner

Department of Biology, Box 19498, The University of Texas at Arlington, Arlington, Texas 76019.[1]

Dag O. Hessen

University of Oslo, Post Office Box 1027, Blindern, N-0315 Oslo, Norway

KEY WORDS: nutrient limitation, mineral element limitation, stoichiometry, biogeochemical cycling, secondary production

Abstract

Organisms differ in the proportions of major elements that they contain, including N and P, which are known to be highly dynamic and potentially limiting to production of aquatic ecosystems. Such contrasting elemental composition between, for example, algae and herbivores, or between different herbivores, generates a suite of ecological predictions and opens up new dynamical possibilities. Here we review studies relating to the nutritional physiology of aquatic herbivores, especially freshwater pelagic species, and we relate element content to secondary production and nutrient recycling. A variety of evidence from many types of studies—physiological modelling, whole-ecosystem surveys, laboratory growth studies, etc—is assembled into an internally consistent picture of mineral limitation of aquatic herbivores. Herbivores with high nutrient demands (the best example is probably *Daphnia* and phosphorus) appear frequently to be limited not by the food quantity or energy available to them but by the quantity of mineral elements in their food.

[1]After June 1, 1994: Gray Freshwater Biological Institute, University of Minnesota, Box 100, County Roads 15 & 19, Navarre, Minnesota 55392

1

0066-4162/94/1120-0001$05.00

INTRODUCTION

Production of consumer biomass out of resources is similar to a complex chemical reaction in which resources are reactants and consumer biomass and wastes are products. Evolution favors organisms that perform this manufacturing process efficiently, optimizing the yield of the product, biomass, relative to the amount of reactant, resources, consumed in the reaction. However, as in all chemical reactions, ecological yields are dependent on the proportions of all reactants. Thus, stoichiometry can be an important rate-determining factor in ecological production dynamics.

All organisms have a set of nutritional requirements for metabolism and growth. When these required substances are balanced in optimal proportion, production efficiencies (production ÷ ingestion) for bulk food and all essential substances, including carbon or energy, will be equal. Imbalanced diets, on the other hand, result in less efficient use of substances that are overabundant, and highly efficient use of substances in short supply.

Diets are more likely to be balanced when their composition is close to the composition of the consuming organism. Thus, carnivores generally consume balanced diets while herbivores and especially detritivores may not. Accordingly, there are distinct differences in the assimilation efficiency of organisms with these three basic feeding modes (Figure 1). Assimilation efficiency of detritivores is low (mode = 0–20%), while that of carnivores is high (mode = 80–100%). Assimilation efficiency for herbivores is intermediate (mode = 60–80%), but it spans a wide range. Figure 1 argues that aquatic herbivores have a food base that is highly variable in quality.

The study of herbivore nutrition is well developed in several systems, for example, vertebrates and agriculturally important insects. About other systems

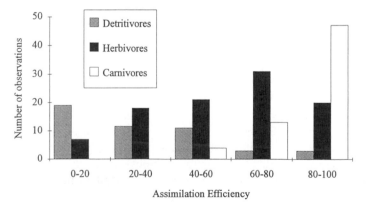

Figure 1 Frequencies of assimilation efficiencies for a wide variety of marine animals of three basic feeding modes. Redrawn from (167).

much less is known. Our goal in this contribution is to consider the nutrition of pelagic herbivores, which is not as well understood as that of some more economically important species. Some recently described features of the pelagic system add considerably to the integration of nutritional ecology with other branches of ecology, including biogeochemistry. Thus, our review focuses on those points. In addition, our review focuses on freshwater pelagic crustaceans, the group we know best; we have, however, sought to include representations of other habitats and groups as well.

OVERVIEW

In pelagic environments, unicellular or colonial autotrophs ("algae," which include cyanobacteria) are consumed by a variety of herbivores. Additionally, herbivore diets may include nonalgal components such as protozoans, detritus, and bacteria. Collectively, all such particles, which are caught on standard filters, are referred to as "seston." In freshwaters, the taxonomic diversity of herbivores is low; protozoans, rotifers, and two classes of crustaceans, copepods and cladocerans, account for virtually all herbivory. In lakes at most times, 5–10 species may account for 90% of the grazing pressure on the algal community; in contrast, marine systems are highly diverse.

Aquatic herbivores must obtain at least a very large fraction of their essential nutritive substances exclusively from their food. Some evidence suggests that soft-bodied marine herbivores may obtain quantitatively significant amounts of amino acids directly from the surrounding water (175), but the more impervious integument necessary for osmotic balance of freshwater organisms probably precludes such direct uptake. In *Daphnia,* for instance, direct uptake of orthophosphate can be demonstrated at high concentrations; at naturally occurring concentrations, however, direct uptake is negligible (108).

As in all animals, absolute dietary requirements of aquatic herbivores include not just energy, but also a wide variety of substances needed for structural growth. All animals require the macroelements H, C, N, O, Na, Mg, P, S, Cl, K, and Ca. In addition, approximately 14–19 essential trace elements have been described to date (141); some are essential only to certain species. Finally, requirements for biomolecules such as vitamins, amino acids, and fatty acids have been described for many animals, including pelagic herbivores (e.g. 15).

The biochemical discontinuity at the plant-animal interface is profound. Algal biomass consists mainly of common biomolecules such as carbohydrates, proteins, lipids, etc, and it can contain all the elements essential for herbivore nutrition. However, the concentrations of these substances may be very different in algal biomass compared to the concentrations optimal for the herbivores. In addition, these essential substances may occur in algal biomass within a matrix consisting of large amounts of nondigestible—or even digestion-impairing—material.

Our attention is concentrated on elements first examined by the oceanographer AC Redfield (112). Redfield found that samples of particulate matter from the offshore ocean consistently exhibited the ratio $C_{106}:N_{16}:P_1$, which is referred to as the "Redfield ratio" in his honor. (All ratios here will be atomic, not mass.) Redfield's early measurements have been confirmed by modern measurements (16, 41, 42, 45, 149). However, in spite of some claims to the contrary, not all organisms have identical composition. Thus, although from some analyses it is natural to wonder, "Why are the chemical compositions of living organisms so similar?" (86), we are more concerned here with their differences.

A major distinction between autotrophs and heterotrophs is in the degree to which cellular composition is held homeostatically constant despite different nutritional supplies (141). In autotrophs like algae, biochemical and elemental composition is very plastic. For instance, the elements N and P typically vary 10X relative to carbon in individual algal taxa variously limited by N and P. Interspecific variation among algae also is large (57, 152). Thus, the composition of herbivore diets is clearly a function of the species composition and growth conditions of its food. Animals, on the other hand, regulate their elemental content much more strictly. This difference in regulation between the trophic levels opens up a large set of interesting ecologically dynamic possibilities and feedbacks that this review describes.

The stoichiometric approach described in this review complements the much better developed "energetics" approach to animal growth. In energetics, the single currency, energy, is charted through acquisition, assimilation, physiological partitioning (e.g. growth vs reproduction), and outflow (e.g. respiration). Zooplankton were among the first organisms examined bioenergetically (117, 131), and the energetics of the freshwater cladoceran herbivore *Daphnia* has been extensively studied (37, 76–80, 82, 106). Provocative links between the energy (or carbon) balance of *Daphnia* and its life history characteristics have been identified. For example, the partitioning of assimilated energy as a function of body size (89, 91) indicates that the proportion of assimilate allocated to reproduction increases monotonically with body size and is independent of food quantity (90). Also, the energetic cost of producing molts may determine the largest attainable body size (90). Important links between physiology and life history evolution are now being uncovered.

The pros and cons of dealing with energy as a single "master" variable in ecological systems have been much debated (10, 23, 93). As we show in this review, a serious weakness of the energetics approach is its reliance on a single currency, for the growth of aquatic herbivores can be more sensitive to other aspects of their food, including minerals, than to the bulk food (carbon) or energy. We believe it proper to refer to such animals as being limited by minerals. Because somatic tissue, eggs, resting eggs, and molts have differing composition, their relative costs will likely vary with the identity of the most limiting currency in the animal's food. Eventually, we hope for an expansion of energetics studies to include other important dietary substances; this review lays the groundwork for integrating a physiology of multiple currencies into more traditional ecological views.

ALGAL NUTRIENT LIMITATION

Algal morphology and biochemistry are highly plastic. Due to this variability, good phycological practice is to subject an algal strain to many growth conditions by varying temperature, pH, nutrient level, etc, so the full phenotypic expression is seen. Likewise, algae in nature are subjected to a wide range of growth conditions; hence, planktonic herbivores must subsist on a resource base varying widely in terms of morphology and chemistry.

We focus on different levels of nutrient limitation of growth. Theoretical maximum growth rate under nutrient saturated conditions is termed μ_{max}. Realized, nutrient-limited growth rate is termed μ. The ratio μ/μ_{max} is a *relative growth rate* (RGR). The nutrient content of algal cells is referred to as the cell *quota*, symbolized by Q (units: moles/cell).

Physiology of Nutrient Content

Droop (27) first drew attention to the link between nutrient content and growth rate. Growth of cells limited by a chemical nutrient has reduced Q. Using the marine flagellate *Monochrysis lutheri* limited either by vitamin B_{12} or P, Droop showed that μ and Q of the limiting nutrient were related by

$$\mu = \mu' - (1 - 1/Q \cdot k_Q),$$

where μ' is at infinite Q and k_Q is the minimum cell quota found at zero μ. Thus, Droop found that the reciprocal of the cell quota was linearly related to growth rate. In some algae, this linear Droop relationship also holds when Q is substituted by the nutrient:carbon atomic ratio within the cells (134, 135, 141).

The Droop relationship is a special case in that it refers to a single substance within a single producer species limited by that element. More complexities are required to understand elemental content of mixed species in natural communities. For one, the parameters μ' and Q are species-specific and can be expected to vary for differing conditions of pH, light, etc. Also, the Droop relationship does not describe the content of nonlimiting elements (29). Rhee (116) examined both N and P contents of the chlorophyte *Scenedesmus* at a fixed growth rate in chemostats with an N:P supply ratio from 5 to 80.

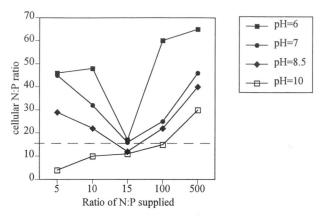

Figure 2   The ratio of cellular N:P on day 40 of a long-term, semicontinuous culture study of a natural phytoplankton assemblage from Lake Arlington, Texas. The algal community was cultured under a suite of N:P supply ratios and pH levels. Algal communities at pH = 10 had cellular N:P ratios increasing monotonically with increasing ratio of N:P supply, while at other pH levels, N-fixing cyanobacteria prospered at low ratio of N:P supply, and the ultimate outcome was a strongly P-limited (high cellular N:P ratio) community consisting almost entirely of cyanobacteria. At ratios of N:P supply greater than the Redfield ratio, the cellular N:P ratio increased with increasing ratio of N:P supply. Data previously unpublished. Horizontal dashed line = Redfield ratio.

Throughout this range, the cellular N:P ratio was equal to the supply ratio. Below a supply ratio of 30, cells accumulated excess N, and above this ratio, they accumulated excess P. Thus, the content of nutrients within algae has some relation to the allochthonous supply of nutrients.

In natural communities, shifts in species composition along such gradients of supply ratios occur (132, 158). To see how such species replacements influence cell quotas, a long-term culture study was undertaken (RW Sterner, unpublished). Gradients of the ratio of N:P supply (sensu 157) were established at four pH levels, which were maintained through daily titrations and by using organic buffers of appropriate pKa (38). At the end of the experiment (day 40), algal species composition reflected both the N:P supply ratio and the pH. At an N:P supply ratio less than the Redfield ratio, the three lowest pH cultures were dominated by N-fixing cyanobacteria, while nonfixing cyanobacteria and eucaryotes dominated the other cultures. The cellular N:P ratio (Figure 2) varied from less than 5 to more than 60. Unlike in Rhee's work (116), the cellular N:P ratio did not always mirror supply ratios. At low pH and low N:P ratio, cyanobacteria became P-limited and had high cellular N:P ratios. Similar observations have been made in whole-lake studies with low N:P supply ratios (124). Above the Redfield ratio, the cellular N:P ratio increased with increasing supply ratio at all pH values, but lower pH resulted in higher cellular ratios. We can see from this study that the elemental content of phytoplankton has a tendency to mirror the ratio of N:P supply, but the chemico-physical context (e.g. pH) or the species pool (e.g. presence/absence of N-fixers) can modify this expectation. These factors are among those creating highly variable nutritional qualities of food for pelagic herbivores.

In Situ Patterns in C:N:P Ratios

Culture studies help to elucidate physiological mechanisms. However, it is also necessary to know if the complexities of the natural environment disallow any of the culture trends. For example, algae may be forced to have very high C:P ratios in culture, but if growth rates in situ are always near maximum, such data would be irrelevant to natural communities. This point is especially important because of the already discussed constancy of C:N:P ratios in the particulate matter in the ocean.

In great contrast to the constancy of C:N:P ratios in offshore marine samples, the composition of seston in freshwater lakes is highly variable in space and time and often deviates greatly from Redfield proportions. Many studies on seston composition have appeared recently (2, 36, 55, 101, 154, 160, 163). Castle Lake, California, illustrates one pattern that seems general, namely, that large deviations from Redfield proportions can occur in surface waters but not near the thermocline (31). A study of Grosser Binnensee, Germany, a shallow, hypertrophic brackish lake, showed that departures from Redfield proportions were associated with periods of low RGR; both the C:P and C:N ratios were highest during periods when bioassays revealed significant nutrient limitation in algal growth (134). Joe Pool Lake, Texas, a warm monomictic reservoir (141, 142), and Funada-iki Pond, a shallow eutrophic pond in Japan (163), corroborated the association between RGR and large deviations from the Redfield ratio.

There is also a consistent trend with size. Fractionation of the seston of the eutrophic Lake Suwa and the mesotrophic reservoir Lake Okutama into four (Okutama) or five (Suwa) different size portions revealed a trend for increasing C:P and C:N ratios in larger size fractions compared to smaller ones (172). Such a pattern also occurred in the moderately eutrophic Schöhsee, Germany (133). In general, bacteria tend to have higher P and N contents, compared to C, than do the larger eucaryotic phytoplankton (155, 166). This makes bacteria potentially an important source of minerals for zooplankton (61).

An important review comparing the C:N:P ratio of seston in a wide diversity of lakes with marine and terrestrial organisms recently appeared (56). Frequency histograms of the C:N and C:P ratios in this data set of 51 lakes (Figure 3) illustrate that Redfield proportions are the exception, not the rule. Deviations from Redfield proportions relate to lake size and latitude (56), with smaller

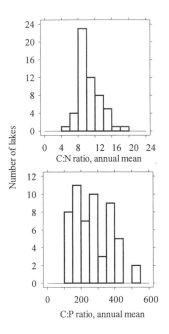

Figure 3 Frequency histogram of annual means for C:N (top) and C:P (bottom) ratios in 51 lakes worldwide surveyed by Hecky et al (56) (their Table 6).

lakes deviating more than larger lakes, and tropical lakes deviating more than high-latitude lakes. Lakes deviating most from Redfield proportions were shallow and had long residence times, characteristics suggesting more complete recycling of the particulate matter within the lake, in a way similar to the "endless summer" described by Kilham & Kilham (68).

Biochemical Changes

Among substances essential for crustacean herbivore growth is a set of biochemicals including but not limited to amino acids, fatty acids, and vitamins (15, 17, 21, 49), all of which can be supplied by phytoplankton to herbivores. However, the abundance of any one of these essential materials may be relatively low; thus, theoretically any one of these substances can be a limiting dietary factor. Considerable interspecific variation in algae exists in some biochemicals such as individual amino acids and fatty acids (97). Unlike the Droop relationship for nutrients, at present no general quantitative relationships exist between the content of biochemicals essential for herbivores and the algal growth rate. A few patterns suggested so far for algae are: protein content is reduced under N limitation (50); phospholipid content is reduced under P limitation (87); and total lipid content tends to be elevated under conditions of low RGR, no matter whether the limiting factor is N, P, or Si (126, 148). However, it is important not to treat these as general patterns because considerable interspecific variation in algal responses seems to be the rule, not the exception.

Some work in recent years has sought an understanding of the role of essential fatty acids in limiting herbivore growth. Interest in these compounds is particularly strong in aquaculture (13, 33, 35, 88, 98, 120, 125, 150, 176). For example, Ahlgren et al (1) concluded that the superiority of cryptomonad algae for zooplankton growth was due to their relatively high proportion of eicosapentaenoic and docosahexaenoic acids (both unsaturated fatty acids). In the absence of additional supportive evidence, it is premature to assign with confidence an advantage of particular algae or particular growth conditions to any one biochemical factor. We argue here from a variety of evidence for the likely importance of minerals in limiting zooplankton growth in nature, but we do recognize that particular biochemicals are probably also similarly important in different conditions.

Morphological Changes

Algal species vary in their resistance to grazing and therefore differ in quality for herbivores. Such resistance is largely attributed to morphological properties like size, shape, and cell-wall structure, but chemically based selection by copepods is also known (26). Several types of cell coverings influence edibility. Cyanobacteria, like heterotrophic gram negative bacteria, possess a peptidoglycan cell wall; assimilation of procaryotic food by pelagic grazers is highly variable but frequently low (22, 63, 65, 83, 104, 105, 164). Some chlorophyte and cyanobacterial species are surrounded by copious gelatinous sheaths that may also prevent assimilation and furthermore allow passage through the gut in viable condition (81, 110, 171).

Less well known is the fact that morphological features such as size and cell-wall thickness undergo substantial change with nutrient status, meaning that digestibility also can vary within single species. In one series of studies, P-starved cells of the green algae *Selenastrum capricornutum, Scenedesmus subspicatus,* and *Chlamydomonas reinhardtii* all exhibited increased cell-wall thickness, reducing their digestibility to the grazers *Daphnia magna* and *D. pulex* (168, 169). In contrast, N-limitation had only minor effects on cell morphology and digestibility. A mutant clone of *C. reinhardtii* without a cell wall was efficiently assimilated even when severely P-limited, implying that the cell wall was a digestive barrier. Accumulation of excess carbon as extracellular mucus in bacteria and some algae under nutrient deficiency (100, 137) may also block digestive enzymes (92).

HERBIVORE NUTRIENT BALANCE

Returning to our chemical reaction analogy, a mismatch in stoichiometry between the requirements of the consumer for maintenance, growth, and reproduction, and the composition of resources should affect yield and reproduction. Accordingly, it is important to understand the consumer requirements and then to determine how empirical patterns of pelagic herbivore nutrition relate to this analogy.

Again, we wish to expand upon energetics models. Empirically, foods of differing nutritive qualities can be brought within an energetics framework by the use of "net energy values" as developed for agriculture (127). For animals of given size, individual energy requirements for growth and maintenance have been determined empirically, and individual forages are equated to energetic equivalents. In cattle, for example, the net energy value of various foods is a function of percent roughage, percent digestible protein, etc (127). One problem with this approach is difficulty in accounting for complications such as "associative effects," in which the energetic value of a food depends on the rest of the animal's diet (8). Net energy values may help unite energetics of aquatic herbivores with more detailed studies of their nutrition, as has been done in other systems, but these values are at best post hoc, determining the energetic equivalent of given resources.

Interspecific Differences in Nutrient Content

Central to the stoichiometric approach is knowledge of the element content of the grazers. Individual species of freshwater crustacean herbivores possess a rather rigid stoichiometry, but those of different species can deviate considerably from each other (Figure 4) (3, 6, 59, 64, 145). A small amount of intraspecific variability is seen between juveniles and adults (7, 122); thus, homeostasis is not perfect. Cladocera do change their somatic ratios of carbohydrates, proteins and lipids, and thus also C:N:P ratios, but such intraspecific variability is small compared with interspecific variability and deviations between food and consumer stoichiometry.

The herbivorous cladocerans *Bosmina, Diaphanosoma, Holopedium,* and especially *Daphnia* possess relatively high specific P content (up to ca. 2% of dry weight) and thus have a low C:P ratio, while the carnivorous cladocerans *Polyphemus* and *Leptodora* have a P content of < 1% of dry weight (DW) and higher C:P ratios (Figure 4). Cladocerans are uncommon in marine habitats, but their few marine representatives have freshwater counterparts (Polyphemidae and Bosminidae) (Figure 4) that do not have very high P content. Freshwater copepods so far examined (*Heterocope* and *Acanthodiaptomus*) (Figure 4) have a low specific P content in the range of 0.4–0.8% P of DW,

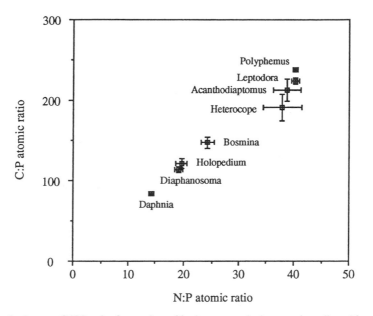

Figure 4 Average C:N:P ratios for a variety of freshwater zooplankton species collected from field samples. Horizontal and vertical bars represent approximate SDs from using first-order variance propagation on the ratio (element:DW) of two variables with uncorrelated errors (asymptotic standard deviation of ratios).

$$SD(z) = z * \sqrt{\frac{SD(x)^2}{x} + \frac{SD(y)^2}{y}}$$

Data from (3, 64).

similar to their marine counterparts (5). One survey (32) indicated that the N:P ratio of marine zooplankton is typically higher than that of freshwater zooplankton.

Differences in specific P content may be linked to the content of biochemicals containing P, such as nucleic acids, ATP-ADP-AMP, and phospholipids, or to minerals in the carapace. RNA constitutes 5–10% of DW in *Daphnia* (95), but only 2% in copepods (5). Assuming equimolar nucleotide makeup, RNA is 10% P by weight, meaning that RNA-P could account for as much as 1% of *Daphnia* DW, but only 0.2% of copepod DW. Thus, RNA alone can account for the entire difference in P content in these herbivores. RNA is apparently a major pool of P in *Daphnia,* but phospholipids could be a more important constituent of copepod P. Across a number of marine worms, phospholipids constitute 10–30% of DW; the proportion is 20–30% of DW in most copepod and euphausid species (115). Phospholipids may contain as much as

5% P, meaning that this could be a major pool of P in copepods. It is note-worthy, however, that a very large interspecific variability in phospholipids exists between arctic and antarctic species (115, 121); thus, copepods may exhibit much larger interspecific differences in their P content than has been measured so far.

Both cyclopoid and calanoid freshwater copepods have a somewhat higher N content than most cladocerans, with a range of 10–12% of DW, which is similar to that of medium-latitude marine copepods (5). With regard to the specific carbon content, both marine and freshwater copepods may show a pronounced variability. High-latitude marine (5, 64) and deep-water freshwater copepods (12) have more than 50% variability in their C:DW ratios. On the other hand, cladocerans and low-latitude copepods often have smaller lipid stores and a specific C content fluctuating closely around 45% of DW (3, 115). Such intraspecific variation implies that excess carbon can be allocated into storage compounds. Lipid stores in these copepods undergo strong seasonal cycles, and such pronounced changes in lipid:DW or lipid:protein ratios will invariably influence C:DW and C:N ratios (5, 151). Hence, there is a range in the degree of homeostatic regulation of C:N or C:P ratios among taxa of pelagic herbivores. The N:P ratio may nevertheless remain homeostatically controlled. For low- and medium-latitude copepods, with a less variable lipid content, most C:N ratios are clustered around 5 (5, 14).

In spite of the central importance of element content in the stoichiometric approach, and in spite of the relative ease with which such data can be collected, we lack knowledge of elemental content in many organisms. There are, for example, almost no data on C:N:P ratios in protozoans and rotifers in freshwater, and ratios in many marine plankton have not been reported.

Feeding

FEEDING RATES VS FOOD QUALITY   Within limits, feeding rate is under con-sumer control. The relationship between herbivore feeding rate and algal food quality enters strongly into consumer-resource dynamics and partially governs the physiological response of the consumer to its diet. For example, one *potential* response of an herbivore to mineral-deficient food would be a com-pensatory increase in the rate of food intake. Because slow-growing algae are generally mineral deficient, such a response would be highly destabilizing, increasing algal mortality at the same time that its growth rate slows.

The relationship between feeding rate and food quality is best understood in benthic invertebrates. A variety of models based on gut absorption, and varying in treatment of the costs of feeding and digestion (19, 109, 173), indicate that optimal feeding rates can either increase or decrease with decreas-ing food quality. In some models, a unimodal curve with maximal feeding at intermediate food quality is predicted. Effects on consumer-resource dynamics and stability of such complex relationships have not been determined, but need to be investigated.

Studies on the relationship between nutritional content of food and zooplankton feeding are needed. *Daphnia* exhibit decreased clearance rate in response to P- and N-limited algae (145, 147). This is a response potentially similar to that in the benthic worms (19, 109, 173), but the *Daphnia* experiments did not separate any effect of increased gut passage due to increased cell-wall thickness.

PARTICLE SELECTION   In nature, pelagic herbivores always co-occur with a diversity of potential foods; single collective descriptions such as the C:N:P ratio of seston collapse all this potentially interesting variation into a single value. The ability of herbivores to discriminate among such particles is highly variable (67).

Elegant and detailed experimental work indicates clearly that copepods can discriminate chemically among a diverse array of particles, selectively ingesting only a subset of all potential foods (25, 111, 159). When given a choice between algal cells of high nutrient content (high RGR) and low nutrient content (low RGR), both freshwater and marine copepods select the former in particle mixtures (11, 18). Thus, even if nutritionally poor particles are present in high concentration, highly selective species can avoid potential mineral limitation of growth by ingesting only high quality particles. Such selectivity carries costs, however (128), and these animals are less discriminating when high quality food is not plentiful (24). The complex neurological and morphological machinery necessary for particle selection suggests that there is a clear adaptive benefit to making such choices. Discriminating feeders have responded evolutionarily to a selective pressure of minimization of mineral limitation, suggesting that mineral limitation has been an important selective force.

However, some of the most successful pelagic herbivores are extremely poor at particle selection; large-bodied *Daphnia* are the best example. The lack of selection by this animal also carries significant costs, such as when unmanageable cyanobacterial filaments are common (40) or when large amounts of suspended clay occur (70–73). For nondiscriminating feeders, collective descriptions such as "seston" come close to indicating their actual diet.

Growth

If herbivores become mineral limited on low quality food, we would expect that herbivore growth per unit of food biomass would be lower on algae of low growth rate than on algae of high growth rate. Several studies are consistent with this expectation. Kiørboe (69) fed the marine copepod *Acartia tonsa* a single concentration (1.5 ppm) of the diatom *Thalassiosira weissflogii* with a widely varying N-limited RGR. The rate of ingestion of carbon by the copepod was constant across the entire range of RGR, but N ingestion increased linearly with increasing RGR due to increased N concentration in algae of higher μ. Egg production was closely related to N ingestion. Thus, the growth rate of the consumer population was highly variable even with a "single" food, and the dynamics were closely associated with the N content of the food. Similarly, *Daphnia* grown on N-deficient *Scenedesmus* have reduced growth rate compared to animals grown on that algae with higher RGR, according to Groeger et al (47). N-limited *Scenedesmus* had higher ratios of lipid:protein than N-sufficient cells; correspondingly, incorporation of lipid by *Daphnia* was higher on N-limited foods. These lipids were then allocated to eggs, and the resulting offspring had enhanced resistance to starvation. Responses to low-quality food apparently may be complex, e.g. lowered body growth rate and thus increased age at first reproduction, but enhanced survivorship of juveniles.

Lipids are a primary means of energy storage in pelagic herbivores. To achieve high body loads of storage lipid, animals must be in positive energy balance. Paradoxically, slow-growing *Daphnia* feeding on P-deficient food accumulate lipid to a considerable extent (144). Thus, positive energy balance does not equate to nutritional fitness. A large lipid load can also indicate a diet out of balance, with lipids in excess but one or more other substances in deficiency.

Toxicity in algae is well described; there are numerous taxa, especially among the cyanobacteria and dinoflagellates, that produce compounds that kill natural zooplankton herbivores as well as fortuitous "herbivores" such as cattle (34, 130). There must be some way to discriminate algae that are non-nutritious from those that are toxic. The operational definition of toxicity is that consumer mortality in the presence of a suspected toxic food is greater than it is in a food-less environment. Non-nutritious foods may also influence mortality, (e.g. 145), making this distinction less than clear-cut. Mitchell et al (96) found no evidence that N- or P-limited algae were toxic to *Daphnia*.

Correlative studies of animal growth and food nutrient content do not conclusively demonstrate that animal growth is mineral limited. Nutrient-limited algae exhibit many biochemical changes in addition to their reduction in mineral content. *Scenedesmus* of poor quality has low P, low N, low protein, low ash, and high carbohydrate contents (140). However, the correlation between animal response and mineral content in the food is very high (69, 140). Figure 5 shows the mass of 5-day-old *Daphnia obtusa* vs the P and N content in their food. With both food species on the graph, the P:C ratio in the algae is a good predictor of *Daphnia* growth, with noticeable reduction in food quality at a P:C ratio of about 0.005 (C:P = 200). The N:C ratio, on the other hand, shows more scatter, and whatever correlation there is may be due mostly

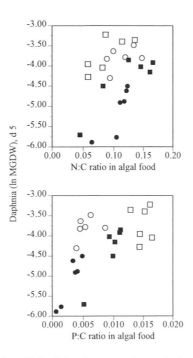

Figure 5 Dry mass of 5-day-old *Daphnia obtusa* reared on a single quantity (0.5 mg C/liter) of algal foods, but with the species and growth conditions of the algae varying: *open squares,* N-limited *Cyclotella; open circles,* P-limited *Cyclotella; closed squares,* N-limited *Scenedesmus, closed circles,* P-limited *Scenedesmus.* Horizontal axes indicate the elemental ratios in the algae. For experimental details for *Scenedesmus* studies, see (140); *Cyclotella* studies used similar methods (unpublished).

to the covariation between N and P in the algae. From such tight correlations, it must be concluded that the animals are responding directly either to the mineral content of the food or to some substance correlated with mineral content.

Note that the studies described in this section agree that algal food quality decreases with decreasing algal growth rate. Thus, the compounds controlling the nutritional basis for this response must be in lower concentration under nutrient limitation than under nutrient sufficiency. As described above, the Droop relationship for minerals is a quantitative description of the reduced concentration of minerals with decreasing RGR. Biochemicals are much more problematic here. Total lipid seems generally to increase with decreasing RGR, and fatty acid and amino acid (50, 94, 129, 148, 156) profiles are not clearly consistent with reductions in algal quality with decreasing growth rate: individual essential fatty acids are not clearly higher in algae of low RGR than in algae of higher RGR.

Growth and survivorship have different metabolic demands. The latter primarily requires energy while the former has structural requirements as well. Stoichiometry, as presented here, deals with yield in the sense of animal growth. If one were to factor out only effects on animal maintenance, different relationships might be predicted than those based on animal growth. A high demand for P in *Daphnia,* for example, is hypothesized to be due to body growth, in particular a high RNA content. P-limited foods that are poor for *Daphnia* growth are not poor for its maintenance (146), thus remaining consistent with stoichiometric principles.

Dietary Imbalance

C:element ratios are more homeostatic in animals than in plants. Animals must have means of chemically rearranging a C-rich food into a more suitable composition. The price of this rearrangement is reduced growth rate. A major but little-discussed difference between autotrophs and heterotrophs is in their ability to store essential substances for later use. Although metazoan animals store energy, they store little in the way of essential nutritional substrates. For example, individual amino acids must be in approximately balanced proportion in *every meal* for maximum utilization in humans (84). Although some mineral storage can occur in hard parts such as bones or shells, by and large, when an animal's diet has excess N or P, it cannot be used.

HOW TO COPE WITH EXCESS C   There are three potential physiological solutions to the problem of excess C in an herbivore's diet. First, C may not be assimilated across the gut wall. Second, if assimilated, excess C may be stored internally as carbon-rich compounds like starch or lipids, later to be used for metabolism or reproduction. Third, assimilated C may be disposed of through respiration or extracellular release of organic C-compounds.

To alter C assimilation relative to other substances, enzymatic activities must be modulated, e.g. phosphatase increased and lipase reduced in response to food of high P:C ratio. Marine copepods regulate gut enzyme activities in relation to their available food resources (51). The degree of specificity of individual enzyme induction, however, is not well understood (48, 52–54). Further experiments using herbivores feeding on foods of differing P:C or N:C ratios are needed before it will be possible to know the importance of this mechanism. Furthermore, the hypothesis of Reinfelder & Fisher (114)— that assimilation is closely related to the fraction of an element in the cytoplasm of the algal food cell—suggests that regulation at the gut is weak at best.

In homeostatic grazers, which do not store large amounts of lipids, excess C, like excess N or P, is not retained. Storage lipids in *Daphnia,* for example, are allocated into eggs every instar with little retained by the adult animal (46, 153). Lemcke & Lampert (85) and Elendt (28) recorded a drop of more than 50% in specific lipid content in *Daphnia* after severe starvation. Lemcke & Lampert (85) also recorded a decrease in C:DW ratio from 0.44 in well-fed to 0.38 in severely-starved individuals, corresponding to changes found by Hessen (58). At moderate starvation, the C:DW ratio changes were insignificant (85). Andersen & Hessen (3) were not able to induce any significant changes in C:DW, N:DW, or P:DW ratios by changing food conditions for the cladocerans *Daphnia longispina, Bosmina longispina,* and *Holopedium gibberum.*

Excess assimilated C might be respired or theoretically it might be excreted as complex C-rich macromolecules like polysaccharides, as in algae (100). Unknown is the extent to which the polysaccharide mantle in the freshwater cladoceran *Holopedium gibberum,* or extracellular mucus in marine gelatinous zooplankton, may be composed of "waste" carbon allocated into structural compounds. However, whether excess C is not assimilated or otherwise disposed of, the net outcome for food utilization would be the same: some proportion of available C would not be not allocated into somatic tissue or offspring. This would mean lower growth efficiency, reduced energy transfer across trophic levels, and reduced secondary production.

A final alternative is that animals simply are unable to cope with excess C, which inability materializes as a growth penalty. Enhanced respiration, for example, generates ATP, and unless this is metabolized through work done on the environment, it cannot easily be disposed of. The homeostatic animal suffers a growth and reproductive penalty through its inability to utilize a resource base very different from its needs, even if all essential substances are within that food.

POPULATION SCALE

So far, our attention has been at the organismal level and below. Mineral limitation has a number of important ramifications for higher levels of ecological organization, too.

Effects on Competition and Community Composition

Conventionally, competition between zooplankton herbivores has been thought to depend primarily upon food abundance or perhaps food quality (algal size, shape, or taxonomic composition). For example, *Diaptomus* was considered superior to *Daphnia* at limited food abundance because it had positive growth when food was very limited while *Daphnia* had negative growth (99). *Daphnia,*

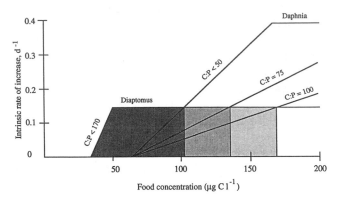

Figure 6 Hypothesized relations between algal food concentration (horizontal) and the rate of biomass gain (vertical) for an herbivore with low P demand (*Diaptomus*) and one with high P demand (*Daphnia*). The growth of the herbivore with high P demand is affected by the C:P ratio in the food; higher ratios increase the biomass of food necessary for given growth rate. Poor quality food expands the region in which *Diaptomus* is predicted to be competitively dominant.

however, was favored at great food abundance because it became satiated at higher food concentrations, giving it higher growth than the copepod at greater food abundance. In terms of quality, *Diaptomus* may be a more selective feeder than *Daphnia;* thus, it may do better when there is a high proportion of low-quality food (118).

Stoichiometry supplements these more traditional views. *Daphnia* has a relatively high P demand. Based on the recognized differences in C:P ratios between *Daphnia* and *Diaptomus, Diaptomus* should have a competitive advantage over *Daphnia* when food is P-deficient. From simple stoichiometric considerations, *Daphnia* can produce less than half of the biomass per unit P than does the copepod. The high RNA content of *Daphnia,* however, presumably allows for a higher growth rate than for *Diaptomus* when sufficient food is available. Thus a life history trade-off can be hypothesized between high growth with good food (the *Daphnia* strategy) and better growth on poor food (the *Diaptomus* strategy). Copepods with slightly higher N-demands than *Daphnia* (3) could theoretically face N- rather than P-limitation, but this is less likely due to the small amount of interspecific variation in herbivore C:N ratios and a general lack of high enough C:N ratios in algal foods. The competitive dominance of *Daphnia* at high food concentration would be reduced by increasing the C:P ratio in the algal food. With higher C:P ratios, more food is required for *Daphnia* growth to be greater than the growth of the copepod. With single-food species having defined element ratios, when food selectivity in copepods does not have to be accounted for, such shifts in competitive superiority along stoichiometry axes may be visualized as in Figure 6.

In nature, a calanoid would be less likely to face P-limitation in comparison with *Daphnia*, not only owing to the fact that its specific P-requirements are lower, but also because it could select the more nutritionally adequate food particles. If there are fundamental differences in element requirements, patterns of relative occurrence of these groups compared to food stoichiometry should be seen. Hessen (60) showed that while the abundance of cladocerans, and *Daphnia* spp. in particular, was well correlated with particulate P, the herbivorous calanoids were better correlated with particulate N.

ECOSYSTEM SCALE

Limitation of Secondary Production by Mineral Nutrients

When growth is determined by food abundance, not quality, herbivore growth should be adequately described by simple Monod-like saturation kinetics with carbon or biomass as the independent variable (79, 107, 123). In such situations, secondary production may be determined mainly by food quantity in terms of C. On the other hand, if food quality is low due to excess carbon, production per unit food C will be lower due to lowered growth efficiency, and some knowledge of production efficiency on that food is required.

Several similar models based on homeostasis and production efficiencies have recently been proposed (4, 60, 165); these aim to understand when the growth of grazing herbivores becomes limited by minerals. A general relationship based on mass balance is (60):

$$F(\alpha_C/\alpha_P) = Z,$$

where F = the C:P ratio of ingested food, Z = the C:P ratio of the animal, α_P = growth efficiency in terms of P, and α_C = growth efficiency in terms of C. Under C-limitation, grazers that utilize C maximally should be favored; let α_{C^*} denote this maximum. Likewise, under P limitation $\alpha_P \rightarrow \alpha_{P^*}$, and animals with higher α_{P^*} are favored. Along the transition from strict C to strict P limitation, the optimum strategy shifts suddenly between $\alpha_P/\alpha_{P^*} = 1$ and $\alpha_C/\alpha_{C^*} = 1$. Although respiration ensures that $\alpha_{C^*} < 1$, the extreme case of $\alpha_{P^*} = 1$ (i.e. no release of P) might be plausibly postulated, as assumed for example in the model of Olsen et al (102). However, such an optimum may not be reachable. Hessen & Andersen (62) assumed zero release only in the hypothetical case when food was totally depleted of P.

F^*, the threshold for transition between C and P limitation, is given by:

$F^* = Z\alpha_{P*}/\alpha_{C*}.$

The growth efficiencies of both C and P determine whether animal growth becomes P-limited. Smaller values of α_{C*} decrease its likelihood (increase the C:P threshold ratio), while smaller values of α_{P*} increase its likelihood (lower the C:P threshold ratio). As a first approximation for *Daphnia*, assuming α_{C*} = 0.6 and α_{P*} = 1.0, and setting the *Daphnia* C:P ratio to 83 (3), P-limitation is predicted at sestonic C:P ratios above 138. Above this threshold, α_C must become less than α_{C*}. Empirical evidence (Figure 5) indicates P limitation at a somewhat higher value, C:P ≈ 200. This empirical threshold suggests that α_{P*}/α_{P*} = 2.4, or that for α_{P*} = 1, α_{C*} = 0.42. Most of the 51 lakes surveyed by Heckey et al (56) fall above a C:P ratio of 200, suggesting that P-limitation in *Daphnia* should be common. Following the same lines of reasoning and based on the same assumptions, Hessen (60) estimated potential P-limitation for *Daphnia* in 23 out of 32 European lakes.

Because there also are strong seasonal fluctuations in elemental ratios within lakes and highly variable zooplankton communities, it is premature to make strict statements on the deviation between secondary production with or without P-limitation on the lake scale. In a typical temperate lake, with a C:P ratio close to 250 (56, 60, 161) (Figure 3), secondary production in terms of growth efficiency of carbon (α_C) of *Daphnia* would be 0.33 due to P-limitation, a reduction of ~ 20% compared to the theoretical maximum of 0.42. This is a minimum estimate, because α_P is set to 1.0.

Under the theoretical optima discussed above, an herbivore feeding on food with a C:P ratio higher than its threshold would have growth stimulated almost 1:1 by increasing P content in its food. Hence, we consider it to be limited by P. However, if these theoretical optima are not reached, such a tight correlation between growth and nutrient content will not be realized, as when, for example, the transition between α_{P*} and α_{C*} is not sudden. In addition, as previously stated, grazers face multiple nutritional challenges, and elemental P-limitation is but one of several possible constraints. Grazer deficiency of essential vitamins, amino acids, lipids, or fatty acids may also occur. From the perspective of secondary production, the outcome would be the same: growth efficiency in terms of C would decrease.

A nutritional imbalance will decrease growth efficiency for the herbivores, which should propagate through the food chain, giving reduced production at all trophic levels. While a strict element imbalance is less likely in carnivores, low food quality at the base of the food web may induce other qualitative constraints even in the carnivores. For example, rotifers feeding on algae that is low-quality in terms of fatty acid composition yield low growth in their fish predators for the same reason (174). Thus, for several reasons, the food quality of algae may severely affect secondary production even at distant trophic levels.

The stoichiometric view may also elucidate some major differences in trophic efficiency and energy transfer between herbivores and carnivores, and food quality may induce shifts in food web organization. Herbivores spend relatively little energy on food gathering; yet, not only do they face nutritional problems due to high contents of lignins, cellulose, or toxins, they also face a potentially severe imbalance in nutrient element supply. For these reasons, herbivores have low growth efficiency and low energy–transfer efficiency (Figure 1). Conversely, a carnivore may spend more energy on food gathering but may also be rewarded by a near optimal diet. From a purely stoichiometric point of view, it would be highly profitable to be a carnivore. For zooplankton communities, this implies that omnivores would be expected to prefer animal prey because the expenditure of capturing is counterbalanced by the gain from better quality. Also, strict herbivores like *Daphnia* would presumably benefit from including in their diet ciliates, which we assume are seldom or never mineral deficient. Long aquatic food chains may largely be sustained by extensive omnivory among planktonic predators (138), because even planktonic top predators like *Mysis* or large predatory copepods have frequent diet shifts and for periods also consume algae. Such dietary shifts in selective feeders may be related not only to prey abundance and vulnerability, but probably also to food quality. For selectively grazing marine copepods, such dietary shifts are common (39, 75) and seem to be a strategy to secure a balanced and optimal diet (74, 119).

Nutrient Cycling and Regeneration

The role of mineral-limited herbivores in pelagic nutrient cycling is different from that of herbivores feeding on nutrient-sufficient food. As a simple corollary to efficient retention of a mineral in short supply (an adaptive physiological trait, as argued above), nutrient release by such an herbivore will be greatly reduced relative to food intake. In addition, features such as changes in feeding rate or passage through the gut in viable condition further reduce nutrient release by the herbivore feeding on a mineral-deficient food. Thus, Olsen et al (102, 103) have shown greatly reduced P release with a rise in the C:P ratio of algal food.

A model based on herbivore homeostasis of the N:P ratio (139) makes several predictions about nutrient cycling: (i) homeostatic herbivores should accentuate the role of single elements in limiting algal growth (they do this by sequestering the single element most limiting to them and releasing relatively more of other elements); (ii) the ratio of N:P released by animals is negatively correlated with the N:P ratio of those animals themselves (thus, *Daphnia* should generally release compounds with a low N:P ratio, but copepods should do the reverse); and (iii) a plot of the ratio of N:P released by a homeostatic herbivore vs the N:P ratio in the algal food should be curvilinear but monotonically increasing. Although much work remains, experimental evidence for each of these points has appeared. Sommer (136) has shown that experimental microcosms consisting of *Daphnia* and P-limited algae tend toward one of two steady states: either high algae and low *Daphnia* with both consumer and algae having low growth rate, or low algae and high *Daphnia* with both having high growth rates (see also discussion in 141). Such dynamics are consistent with, but not identical to, prediction (i) above. Previous to the publication of the model being discussed, Elser et al (30) had shown that transitions between N- and P-limitation of algal growth in small lake systems occurred in a manner consistent with prediction (ii) above (see also 143). Finally, Urabe (162) has produced the most convincing application of the model to date, having shown a tight correspondence between the ratio of N:P released by zooplankton in a small pond and predictions generated by the model, including curvilinearity.

Although we have dealt exclusively with metazoan herbivores, it is worth mentioning the generality of these relationships. Goldman et al (43) found low regeneration of N and P in the protozoan *Paraphysomonas imperforata* feeding on N- and P-limited algal cells, respectively. Jürgens & Güde (66) demonstrated low P-regeneration by protozoans consuming P-deficient bacteria. Heterotrophic bacteria, subject to only small cellular changes in their C:N:P ratio, release N and P in similar ways from metabolizing dissolved substrates (44, 155). Higher trophic levels also seem to fall into this pattern. Excretion from dense schools of fish may substantially contribute to inorganic N and P in lakes (9). Braband et al (9) found a stable N:P ratio of 29 in excreta from roach feeding largely on bottom deposits low in P, while Dahl (20) recorded the very low average N:P ratio of less than 5 for fish feeding on zooplankton; these patterns are consistent with stoichiometric expectation if zooplankton have a lower N:P ratio than bottom sediments. Vanni (170) has calculated, using the homeostatic model, a widely varying ratio of N:P released by detritus-feeding fish (139).

CONCLUSIONS

We hope that we have demonstrated that mineral limitation of aquatic herbivores is likely to occur in nature. The conditions fostering mineral limitation include seston composition, herbivore demands, herbivore selectivity, food density, and doubtless several others too. We hope to have established the importance of an expanded bioenergetics of multiple currencies to physiological ecology. Energy alone does not explain all important aspects of growth dynamics. We hope also to have contributed to the unification of nutritional ecology and biogeochemistry by pointing out the clear stoichiometric relationships between these two fields.

When we view the herbivore as a complex adaptive system, seeking to maximize yield relative to food consumed, we seem able to make predictions about the role of those organisms in nutrient cycling and thus also about their positions in ecological food webs. Such a multivariate, "complementary" (113) model of ecosystems helps unite disparate fields far removed along the scale of biological organization. For example, should the RNA hypothesis discussed here hold up to further scrutiny, it will be possible to see a clear thread extending all the way from molecular biology to ecosystem function.

ACKNOWLEDGMENTS

RW Sterner acknowledges financial support from NSF grants BSR-9019722 and BSR-9119269.

Literature Cited

1. Ahlgren G, Lundstedt L, Brett M, Forsberg C. 1990. Lipid composition and food quality of some freshwater phytoplankton for cladoceran zooplankton. *J. Plankton Res.* 12:809–18
2. Aizaki M, Otsuki A. 1987. Characteristic of variations of C:N:P:Chl ratios of seston in eutrophic shallow Lake Kasumigaura, Japan. *J. Limnol.* 48:99–106
3. Andersen T, Hessen DO. 1991. Carbon, nitrogen, and phosphorus content of freshwater zooplankton. *Limnol. Oceanogr.* 36:807–14
4. Anderson TR. 1992. Modelling the influence of food C:N ratio, and respiration on growth and nitrogen excretion in marine zooplankton and bacteria. *J. Plankton Res.* 14:1645–71
5. Båmstedt U. 1986. Chemical composition and energy content. In *The Biological Chemistry of Marine Copepods,* ed. EDS Corner, SCM O'Hara, pp. 1–58. Oxford: Oxford Univ. Press
6. Baudouin MF, Ravera O. 1972. Weight, size and chemical composition of some freshwater zooplankton: *Daphnia hyalina* (Leydig). *Limnol. Oceanogr.* 17:645–49
7. Berberovic R. 1990. Elemental composition of two coexisting *Daphnia* species

during the seasonal course of population development in Lake Constance. *Oecologia* 84:340–50
8. Bjorndal KA. 1991. Diet mixing: non-additive interactions of diet items in an omnivorous freshwater turtle. *Ecology* 72:1234–41
9. Brabrand Å, Faafeng B, Nilssen JP. 1990. Relative importance of phosphorus supply to phytoplankton production: fish versus external loading. *Can. J. Fish. Aquat. Sci.* 47:364–72
10. Brown J. 1994. Organisms and species as complex adaptive systems: linking the biology of populations with the physics of ecosystems. In *Linking Species and Ecosystems,* ed. C Jones, J Lawton. New York: Chapman. In press
11. Butler NM, Suttle CA, Neill WE. 1989. Discrimination by freshwater zooplankton between single algal cells differing in nutritional status. *Oecologia* 78:368–72
12. Cabaletto JF, Vanderploeg HA, Gardner WS. 1989. Wax esters in two species of freshwater zooplankton. *Limnol. Oceanogr.* 34:785–89
13. Caric M, Sanko-Njire J, Skaramuca B. 1993. Dietary effects of different feeds on the biochemical composition of the

rotifer (*Brachionus plicatilis* Müller). *Aquaculture* 110:141–50

14. Checkley DMJ. 1985. Elemental and isotopic fractionation of carbon and nitrogen by marine, planktonic copepods and implications to the marine nitrogen cycle. *J. Plankton Res.* 7:553–68

15. Conklin DE, Provasoli L. 1977. Nutritional requirements of the water flea, *Moina macrocopa. Biol. Bull.* 152:337–50

16. Copin-Montegut C, Copin-Montegut G. 1983. Stoichiometry of carbon, nitrogen, and phosphorus in marine particulate matter. *Deep-Sea Res.* 30:31–46

17. Corner EDS, Conway CB. 1968. Biochemical studies on the production of marine zooplankton. *Biol. Rev.* 43:393–426

18. Cowles TJ, Olson RJ, Chisholm SW. 1989. Food selection by copepods: discrimination on the basis of food quality. *Mar. Biol.* 100:41–49

19. Dade WB, Jumars PA, Penry DL. 1990. Supply-side optimization: maximizing absorptive rates. In *Behavioural Mechanisms in Food Selection,* ed. RN Hughes, pp. 531–56. Berlin: Springer-Verlag

20. Dahl O. 1991. *Release of phosphorus and nitrogen from pelagic populations of roach* (Rutilis rutilus) *and bleach* (Alburnus alburnus). Master's thesis. Univ. Oslo, Oslo. 44 pp. (in Norwegian).

21. Dall W, Moriaty DJW. 1983. Functional aspects of nutrition and digestion. In *The Biology of Crustacea,* ed. LH Mantel, pp. 215–62. New York: Academic

22. de Bernardi R, Giussani G. 1990. Are blue-green algae a suitable food for zooplankton? An overview. *Hydrobiologia* 200/201:29–41

23. DeAngelis DL. 1994. Relationships between the energetics of species and the thermodynamics of ecosystems. In *Linking Species and Ecosystems,* ed. J Lawton, C Jones. New York: Chapman & Hall. In press

24. DeMott WR. 1989. Optimal foraging theory as a predictor of chemically mediated food selection by suspension-feeding copepods. *Limnol. Oceanogr.* 34:140–54

25. DeMott WR. 1990. Retention efficiency, perceptual bias, and active choice as mechanisms of food selection by suspension-feeding zooplankton. In *Behavioural Mechanisms in Food Selection,* ed. RN Hughes, pp. 569–94. Berlin: Springer-Verlag

26. DeMott WR, Moxter F. 1991. Foraging on cyanobacteria by copepods: responses to chemical defenses and resource abundance. *Ecology* 72:1820–34

27. Droop MR. 1974. The nutrient status of algal cells in continuous culture. *J. Mar. Biol. Assoc. UK* 54:825–55

28. Elendt B-P. 1989. Effects of starvation on growth, reproduction, survival and biochemical composition of *Daphnia magna. Arch. Hydrobiol.* 116:415–33

29. Elrifi IR, Turpin DH. 1985. Steady-state luxury consumption and the concept of optimum nutrient ratios: a study with phosphate and nitrate limited *Selenastrum minutum* (Chlorophyta). *J. Phycol.* 21:592–602

30. Elser JJ, Elser MM, MacKay NA, Carpenter SR. 1988. Zooplankton-mediated transitions between N- and P-limited algal growth. *Limnol. Oceanogr.* 33:1–14

31. Elser JJ, George NB. 1993. The stoichiometry of N and P in the pelagic zone of Castle Lake, California. *J. Plankton Res.* 15:977–92

32. Elser JJ, Hassett RP. 1994. A stoichiometric analysis of the zooplankton-phytoplankton interaction in marine and freshwater ecosystems. *Nature:* in press

33. Fernández-Reiriz MJ, Labarta U, Ferreiro MJ. 1993. Effects of commercial enrichment diets on the nutritional value of the rotifer (*Brachionus plicatilis*). *Aquaculture* 112:195–206

34. Forsyth DJ, Haney JF, James MR. 1992. Direct observation of toxic effects of cyanobacterial extracellular products on *Daphnia. Hydrobiologia* 228:151–55

35. Frolov AV, Pankov SL, Geradze KN, Pankova SA, Spektorova LV. 1991. Influence of the biochemical composition of food on the biochemical composition of the rotifer *Brachionus plicatilis. Aquaculture* 97:181–202

36. Gächter R, et al. 1985. Seasonal and vertical variation in the C:P ratio of suspended and settling seston of lakes. *Hydrobiologia* 128:193–200

37. Geller W. 1975. Die Nahrungsaufnahme von *Daphnia pulex* in Abhaengigkeit von der Futterkonzentration, der Temperatur, der Korpergrosse und dem Hungerzustand der Tiere. *Arch. Hydrobiol.* 48:47–107

38. Gensemer RW, Kilham SS. 1984. Growth rates of five freshwater algae in well-buffered acidic media. *Can. J. Fish. Aquat. Sci.* 41:1240–43

39. Gifford DJ, Dagg MJ. 1988. Feeding on the estuarine copepod *Acartia tonsa* Dana: carnivory vs. herbivory in natural microplankton assemblages. *Bull. Mar. Sci.* 43:458–68

40. Gliwicz ZM. 1990. *Daphnia* growth at different concentrations of blue-green filaments. *Arch. Hydrobiol.* 120:51–65
41. Goldman JC. 1984. Oceanic nutrient cycles. In *Flows of Energy and Materials in Marine Ecosystems Theory,* ed. MJ Fasham, pp. 137–70. New York: Plenum
42. Goldman JC. 1986. On phytoplankton growth rates and particulate C:N:P ratios at low light. *Limnol. Oceanogr.* 31: 1358–63
43. Goldman JC, Caron DA, Dennett MR. 1987. Nutrient cycling in a microflagellate food chain: IV. Phytoplankton-microflagellate interactions. *Mar. Ecol. Prog. Ser.* 38:75–87
44. Goldman JC, Caron DA, Dennett MR. 1987. Regulation of gross growth efficiency and ammonium regeneration in bacteria by substrate C:N ratio. *Limnol. Oceanogr.* 32:1239–52
45. Goldman JC, Peavey DG. 1979. Steady state growth and chemical composition of the marine chlorophyte *Dunaliella tertiolecta* in nitrogen-limited continuous cultures. *Appl. Environ. Microbiol.* 38:894–901
46. Goulden CE, Place AR. 1993. Lipid accumulation and allocation in daphnid cladocera. *Bull. Mar. Sci.* 53:106–14
47. Groeger AW, Schram MD, Marzolf GR. 1991. Influence of food quality on growth and reproduction in *Daphnia*. *Freshwater Biol.* 26:11–19
48. Harris RP, Samain JF, Moal J, Martin-Jezequel V, Poulet SA. 1986. Effects of algal diet on digestive enzyme activity in *Calanus helgolandicus*. *Mar. Biol.* 90:353–61
49. Harrison KE. 1990. The role of nutrition in maturation, reproduction and embryonic development of decapod crustaceans: A review. *J. Shellfish Res.* 9:1–28
50. Harrison PJ, Thompson PA, Calderwood GS. 1990. Effects of nutrient and light limitation on the biochemical composition of phytoplankton. *J. Applied Phycol.* 2:45–56
51. Hassett RP, Landry MR. 1983. Effects of food-level acclimation on digestive enzyme activities and feeding behavior of *Calanus pacificus*. *Mar. Biol.* 75:47–55
52. Hassett RP, Landry MR. 1990. Effects of diet and starvation on digestive enzyme activity and feeding behavior of the marine copepod *Calanus pacificus*. *J. Plankton Res.* 12:991–1010
53. Hassett RP, Landry MR. 1990. Seasonal change in feeding rate, digestive enzyme activity, and assimilation efficiency of *Calanus pacificus*. *Mar. Ecol. Prog. Ser.* 62:203–10
54. Head EJH, Conover RJ. 1983. Induction of digestive enzymes in *Calanus hyperboreus*. *Mar. Biol. Lett* 4:219–31
55. Healy FP, Hendzel LL. 1980. Physiological indicators of nutrient deficiency in lake phytoplankton. *Can. J. Fish. Aquat. Sci.* 37:442–53
56. Hecky RE, Campbell P, Hendzel LL. 1993. The stoichiometry of carbon, nitrogen, and phosphorus in particulate matter of lakes and oceans. *Limnol. Oceanogr.* 38:709–24
57. Hecky RE, Kilham P. 1988. Nutrient limitation of phytoplankton in freshwater and marine environments: A review of recent evidence on the effects of enrichment. *Limnol. Oceanogr.* 33:796–822
58. Hessen DO. 1989. Factors determining the nutritive status and production of zooplankton in a humic lake. *J. Plankton Res.* 11:649–64
59. Hessen DO. 1990. Carbon, nitrogen and phosphorus status in *Daphnia* at varying food conditions. *J. Plankton Res.* 12: 1239–49
60. Hessen DO. 1992. Nutrient element limitation of zooplankton production. *Am. Nat.* 140:799–814
61. Hessen DO, Andersen T. 1990. Bacteria as a source of phosphorus for zooplankton. *Hydrobiologia* 206:217–23
62. Hessen DO, Andersen T. 1992. The algae-grazer interface: feedback mechanisms linked to elemental ratios and nutrient cycling. *Ergebn. Limnol.* 35: 111–20
63. Hessen DO, Andersen T, Lyche A. 1989. Differential grazing and resource utilization of zooplankton in a humic lake. *Arch. Hydrobiol.* 114:321–47
64. Hessen DO, Lyche A. 1991. Inter- and intraspecific variations in zooplankton element composition. *Arch. Hydrobiol.* 121:355–63
65. Holm NP, Ganf GG, Shapiro J. 1983. Feeding and assimilation rates for *Daphnia pulex* fed *Aphanizomenon flos-aquae*. *Limnol. Oceanogr.* 28:677–87
66. Jürgens K, Güde H. 1990. Incorporation and release of phosphorus by planktonic bacteria and phagotrophic flagellates. *Mar. Ecol. Prog. Ser.* 59:271–84
67. Kerfoot WC, Kirk KL. 1991. Degree of taste discrimination among suspension-feeding cladocerans and copepods: implications for detritivory and herbivory. *Limnol. Oceanogr.* 36:1107–23
68. Kilham P, Kilham SS. 1990. Endless summer: internal loading processes

dominate nutrient cycling in tropical lakes. *Freshwater Biol.* 23:379–89

69. Kiørboe T. 1989. Phytoplankton growth rate and nitrogen content: implications for feeding and fecundity in a herbivorous copepod. *Mar. Ecol. Prog. Ser.* 55:229–34

70. Kirk K. 1991. Inorganic particles alter competition in grazing plankton: the role of selective feeding. *Ecology* 72:915–23

71. Kirk KL. 1991. Suspended clay reduces *Daphnia* feeding: behavioural mechanisms. *Freshwater Biol.* 25:357–65

72. Kirk KL. 1992. Effects of suspended clay on *Daphnia* body growth and fitness. *Freshwater Biol.* 28:103–9

73. Kirk KL, Gilbert JJ. 1990. Suspended clay and the population dynamics of planktonic rotifers and cladocerans. *Ecology* 71:1741–55

74. Kleppel GS. 1993. On the diets of calanoid copepods. *Mar. Ecol. Prog. Ser.* 99:183–95

75. Kleppel GS, Frazel D, Pieper RE, Holliday DV. 1988. Natural diets of zooplankton off southern California. *Mar. Ecol. Prog. Ser.* 49:231–41

76. Kooijman SALM. 1986. Population dynamics on the basis of budgets. In *The Dynamics of Physiologically Structured Populations,* ed. JAJ Metz, O Diekman, pp. 266–297. Berlin: Springer-Verlag

77. Lampert W. 1977. Studies on the carbon balance of *Daphnia pulex* as related to environmental conditions. I. Methodological problems of the use of ^{14}C for the measurement of carbon assimilation. *Arch. Hydrobiol. Suppl.* 48:287–309

78. Lampert W. 1977. Studies on the carbon balance of *Daphnia pulex* as related to environmental conditions. II. The dependence of carbon assimilation on animal size, temperature, food concentration and diet species. *Arch. Hydrobiol. Suppl.* 48:310–35

79. Lampert W. 1977. Studies on the carbon balance of *Daphnia pulex* as related to environmental conditions. III. Production and production efficiency. *Arch. Hydrobiol. Suppl.* 48:336–60

80. Lampert W. 1977. Studies on the carbon balance of *Daphnia pulex* de Geer as related to environmental conditions. IV. Determination of the "threshold" concentration as a factor controlling the abundance of zooplankton species. *Arch. Hydrobiol. Suppl.* 48:361–68

81. Lampert W. 1982. Further studies on the inhibitory effect of the toxic bluegreen *Microcystis aeroginosa* on the filtering rate of zooplankton. *Arch. Hydrobiol.* 95:207–20

82. Lampert W. 1984. The measurement of respiration. In *A Manual for the Assessment of Secondary Productivity in Fresh Waters,* ed. JA Downing, FH Rigler, pp. 413–468. Boston: Blackwell

83. Lampert W. 1987. Feeding and nutrition in *Daphnia.* In *Daphnia,* ed. RH Peters, R De Bernardi, pp. 143–92. Verbania Pallanza: Istituto Italiano di Idrobiologia

84. Lappe F. 1971. *Diet for a Small Planet.* New York: Ballantine

85. Lemcke HW, Lampert W. 1975. Veränderungen im Gewicht unter der chemischen Zusammensetzung von *Daphnia pulex* im Hunger. *Ergebn. Limnol.* 48:108–37

86. Li, Y-H. 1984. Why are the chemical compositions of living organisms so similar? *Schweiz. Z. Hydrobiol.* 46:177–84

87. Lombardi AT, Wangersky PJ. 1991. Influence of phosphorus and silicon on lipid class production by the marine diatom *Chaetoceros gracilis* grown in turbidostat culture. *Mar. Ecol. Prog. Ser.* 77:39–47

88. Lubzens E, Kolodny G, Perry B, Galai N, Sheshinski, R, Wax Y. 1990. Factors affecting survival of rotifers (*Brachionus plicatilis* O.F. Müller) at 4°C. *Aquaculture* 91:23–47

89. Lynch M. 1980. The evolution of cladoceran life histories. *Q. Rev. Biol.* 55:23–42

90. Lynch M. 1989. The life history consequences of resource depression in *Daphnia pulex. Ecology* 70:246–56

91. Lynch M, Weider LJ, Lampert W. 1986. Measurement of the carbon balance in *Daphnia. Limnol. Oceanogr.* 31:17–33

92. Malej A, Harris RP. 1993. Inhibition of copepod grazing by diatom exudates—a factor in the development of mucus aggregates. *Mar. Ecol. Prog. Ser.* 96:33–42

93. Mansson BA, McGlade JM. 1993. Ecology, thermodynamics and H.T. Odum's conjectures. *Oecologia* 93:582–96

94. Mayzaud P, Claustre H, Augier P. 1990. Effect of variable nutrient supply on fatty acid composition of phytoplankton grown in an enclosed experimental ecosystem. *Mar. Ecol. Prog. Ser.* 60:123–40

95. McKee M, Knowles CO. 1987. Levels of protein, RNA, DNA, glycogen and lipids during growth and development of *Daphnia magna* Straus (Crustacea: Cladocera). *Freshwater Biol.* 18:341–51

96. Mitchell SF, Trainor FR, Rich PH, Goulden CE. 1992. Growth of *Daphnia magna* in the laboratory in relation to the nutritional state of its food species,

Chlamydomonas reinhardtii. J. Plankton Res. 14:379–91

97. Moal J, Martin-Jezequel V, Harris RP, Samain J-F, Poulet SA. 1987. Interspecific and intraspecific variability of the chemical composition of marine phytoplankton. *Oceanologica Acta* 10:339–46

98. Mourente G, Rodriguez A, Tocher DR, Sargent JR. 1993. Effects of dietary docosahexaenoic acid (DHA 22/6n-3) on lipid and fatty acid compositions and growth in gilthead sea bream (*Sparus aurata* L) larvae during 1st feeding. *Aquaculture* 112:79–98

99. Muck P, Lampert W. 1984. An experimental study on the importance of food conditions for the relative abundance of calanoid copepods and cladocerans. 1. Comparative feeding studies with *Eudiaptomus gracilis* and *Daphnia longispina. Archiv für Hydrobiol. Suppl.* 66:157–79

100. Myklestad S. 1977. Production of carbohydrates by marine planktonic diatoms. II. Influence of the N/P ratio in the growth medium on the assimilation ratio, growth rate, and production of cellular and extracellular carbohydrates by *Chaetoceros affinis* var. *willey* (Gran) Hustvedt and *Skeletonema costatum* (Grev.)*Cleve. J. Exp. Mar. Biol. Ecol.* 29:161–79

101. Nakanishi M, Mitamura O, Matsubara T. 1990. Sestonic C:N:P ratios in the south basin of Lake Biwa with special attention to nutritional state of phytoplankton. *Jpn. J. Limnol.* 51:185–89

102. Olsen Y, Jensen A, Reinertsen H, Børsheim KY, Heldal, M, Langeland A. 1986. Dependence of the rate of release of phosphorus by zooplankton on the P:C ratio in the food supply, as calculated by a recycling model. *Limnol. Oceanogr.* 31:34–44

103. Olsen Y, Østgaard K. 1985. Estimating release rates of phosphorus from zooplankton: Model and experimental verification. *Limnol. Oceanogr.* 30: 844–52

104. Paerl H. 1988. Growth and reproductive strategies of freshwater blue-green algae (cyanobacteria). In *Growth and Reproductive Strategies of Freshwater Phytoplankton,* ed. CD Sandgren, pp. 261–315. New York: Cambridge Univ. Press

105. Paerl HW. 1988. Nuisance phytoplankton blooms in coastal, estuarine, and inland waters. *Limnol. Oceanogr.* 33: 823–47

106. Paloheimo JE, Crabtree SJ, Taylor WD. 1982. Growth model of *Daphnia. Can. J. Fish. Aquat. Sci.* 39:598–606

107. Peters RH. 1984. Methods for the study of feeding, filtering and assimilation by zooplankton. In *A Manual for the Assessment of Secondary Productivity in Fresh Water,* ed. JA Downing, FH Rigler, pp. 336–412. Oxford: Blackwell

108. Peters RH. 1987. Metabolism in *Daphnia.* In *Memorie dell'Istituto Italiano di Idrobiologia,* ed. RH Peters, R De Bernardi, pp. 193–243. Verbania Pallanza, Consiglio Nazionale Delle Richerche

109. Phillips NW. 1984. Compensatory intake can be consistent with optimal foraging models. *Am. Nat.* 123:867–72

110. Porter KG. 1973. Selective grazing and differential digestion of algae by zooplankton. *Nature* 244:179–80

111. Price HJ. 1988. Feeding mechanisms in marine and freshwater zooplankton. *Bull. Mar. Sci.* 43:327–43

112. Redfield AC. 1958. The biological control of chemical factors in the environment. *Am. Sci.* 46:205–21

113. Reiners WA. 1986. Complementary models for ecosystems. *Am. Nat.* 127: 59–73

114. Reinfelder JR, Fisher NS. 1991. The assimilation of elements by marine copepods. *Science* 251:794–96

115. Reinhardt SB, Van Vleet ES. 1986. Lipid composition of twenty-two species of Antarctic midwater species and fish. *Mar. Biol.* 91:149–59

116. Rhee G-Y. 1978. Effects of N:P atomic ratios and nitrate limitation on algal growth, cell composition and nitrate uptake. *Limnol. Oceanogr.* 23:10–25

117. Richman S. 1958. The transformation of energy by *Daphnia pulex. Ecol. Monogr.* 28:273–88

118. Richman S, Dodson SI. 1983. The effect of food quality on feeding and respiration by *Daphnia* and *Diaptomus. Limnol. Oceanogr.* 28:948–56

119. Roman MR. 1984. Utilization of detritus by the copepod *Acartia tonsa. Limnol. Oceanogr.* 29:949–59

120. Sakamoto M, Holland DL, Jones DA. 1982. Modification of the nutritional composition of *Artemia* by incorporation of polyunsaturated fatty acids using micro-encapsulated diets. *Aquaculture* 28:311–20

121. Sargent JR, Falk-Petersen S. 1988. The lipid biochemistry of calanoid copepods. *Hydrobiologia* 167/168:101–14

122. Sarnelle O. 1992. Contrasting effects of *Daphnia* on ratios of nitrogen to phosphorus in a eutrophic, hardwater lake. *Limnol. Oceanogr.* 37:1527–42

123. Scavia D, Lang GA, Kitchell JF. 1988. Dynamics of Lake Michigan plankton: a model evaluation of nutrient loading,

competition, and predation. *Can. J. Fish. Aquat. Sci.* 45:165–77

124. Schindler DW. 1977. Evolution of phosphorus limitation in lakes. *Science* 195:260–62

125. Shamsudin L. 1992. Lipid and fatty acid composition of microalgae used in Malaysian aquaculture as live food for the early stage of penaeid larvae. *J. Applied Phycol.* 4:371–78

126. Shifrin NS, Chisholm SW. 1981. Phytoplankton lipids: interspecific differences and effects of nitrate, silicate and light-dark cycles. *J. Phycol.* 17:374–84

127. Shirley RL. 1986. *Nitrogen and Energy Nutrition of Ruminants.* Orlando, Fla: Academic

128. Sierszen ME, Frost TM. 1992. Selectivity in suspension feeders: food quality and the cost of being selective. *Arch. Hydrobiol.* 123:257–73

129. Siron R, Giusti G, Berland B. 1989. Changes in the fatty acid composition of *Phaeodactylum tricornutum* and *Dunaliella tertiolecta* during growth and under phosphorus deficiency. *Mar. Ecol. Prog. Ser.* 55:95–100

130. Sivonen K, Namikoshi M, Evans WR, Carmichael WW, Sun F, et al. 1992. Isolation and characterization of a variety of microcystins from 7 strains of the cyanobacterial genus *Anabaena. Appl. Environ. Microbiol.* 58:2495–500

131. Slobodkin LB. 1954. Population dynamics of *Daphnia obtusa. Kurz. Ecol. Monogr.* 24:69–88

132. Smith VH. 1983. Low nitrogen to phosphorus ratios favor dominance by bluegreen algae in lake phytoplankton. *Science* 221:669–71

133. Sommer U. 1988. Does nutrient competition among phytoplankton occur in situ? *Int. Verein. theor. andgew. Limnol. Verh.* 23:707–12

134. Sommer U. 1989. Nutrient status and nutrient competition of phytoplankton in a shallow, hypertrophic lake. *Limnol. Oceanogr.* 34:1162–73

135. Sommer U. 1990. Phytoplankton nutrient competition—From laboratory to lake. In *Perspectives on Plant Competition,* ed. J Grace, D Tilman, pp. 193–213. New York: Academic Press

136. Sommer U. 1992. Phosphorus-limited Daphnia—intraspecific facilitation instead of competition. *Limnol. Oceanogr.* 37:966–73

137. Søndergaard M, Scierup HH. 1982. Release of extracellular organic carbon during a diatom bloom in Lake Mossø: molecular weight fractionation. *Freshwater Biol.* 12:313–20

138. Sprules WG, Bowerman JE. 1988. Om-nivory and food chain length in zooplankton food webs. *Ecology* 69:418–26

139. Sterner RW. 1990. N:P resupply by herbivores: zooplankton and the algal competitive arena. *Am. Nat.* 136:209–29

140. Sterner RW. 1993. *Daphnia* growth on varying quality of *Scenedesmus*: Mineral limitation of zooplankton. *Ecology* 74:2351–60

141. Sterner RW. 1994. Elemental stoichiometry of species in ecosystems. In *Linking Species and Ecosystems,* ed. C Jones, J Lawton. New York: Chapman & Hall. In press

142. Sterner RW. 1994. Seasonal and spatial patterns in macro and micro nutrient limitation in Joe Pool Lake, Texas. *Limnol. Oceanogr.* 39:545–50

143. Sterner RW, Elser JJ, Hessen DO. 1992. Stoichiometric relationships among producers and consumers in food webs. *Biogeochemistry* 17:49–67

144. Sterner RW, Hagemeier DD, Smith RF, Smith WL. 1992. Lipid-ovary indices in food-limited *Daphnia. J. Plankton Res.* 14:1449–60

145. Sterner RW, Hagemeier DD, Smith WL, Smith RF. 1993. Phytoplankton nutrient limitation and food quality for *Daphnia. Limnol. Oceanogr.* 38:857–71

146. Sterner RW, Robinson J. 1994. Thresholds for growth in *Daphnia magna* with high and low phosphorus diets. *Limnol. Oceanogr.* In press

147. Sterner RW, Smith RF. 1993. Clearance, ingestion and release of N and P by *Daphnia pulex* feeding on *Scenedesmus acutus* of varying quality. *Bull. Mar. Sci.* 53:228–39

148. Sukenik A, Wahnon R. 1991. Biochemical quality of marine unicellular algae with special emphasis on lipid composition I. *Isochrysis galbana. Aquaculture* 97:61–72

149. Takahashi T, Broecker W, Langer S. 1985. Redfield ratio based on chemical data from isopycnal surfaces. *J. Geophys. Res.* 90:6907–24

150. Tamaru CS, Murashige R, Lee CS, Ako H, Sato V. 1993. Rotifers fed various diets of bakers yeast and/or *Nannochloropsis oculata* and their effect on the growth and survival of striped mullet (*Mugil cephalus*) and milkfish (*Chanos chanos*) larvae. *Aquaculture* 110:361–72

151. Tande KS. 1982. Ecological investigations on the zooplankton community in Balsfjorden, northern Norway: generation cycles and variations in body weight and body content of carbon and nitrogen related to overwintering and reproduction in the copepod *Calanus finmarchi-*

cus (Gunnerus). *J. Exp. Mar. Biol. Ecol.* 62:129–42

152. Terry KL, Laws EA, Burns DJ. 1985. Growth rate variation in the N:P requirement ratio of phytoplankton. *J. Phycol.* 21:323–29

153. Tessier AJ, Goulden CE. 1982. Estimating food limitations in cladoceran populations. *Limnol. Oceanogr.* 27:707–17

154. Tezuka Y. 1985. The C:N:P ratios of seston in Lake Biwa as indicators of nutrient deficiency in phytoplankton and decomposition process of hypolimnetic particulate matter. *Japan J. Limnol.* 46:239–46

155. Tezuka Y. 1990. Bacterial regeneration of ammonia and phosphate as affected by the carbon:nitrogen ratio of organic substrates. *Microb. Ecol.* 19:227–38

156. Thompson PA, Harrison PJ, Whyte JNC. 1990. Influence of irradiance on the fatty acid composition of phytoplankton. *J. Phycol.* 26:278–88

157. Tilman D. 1982. *Resource Competition and Community Structure.* Princeton:Princeton Univ. Press

158. Tilman D, Kiesling R, Sterner R, Kilham SS, Johnson, FA. 1986. Green, blue-green and diatom algae: Taxonomic differences in competitive ability for phosphorus, silicon and nitrogen. *Arch. Hydrobiol.* 106:473–85

159. Turner JT. 1991. Zooplankton feeding ecology: do co-occurring copepods compete for the same food? *Rev. Aqat. Sci.* 5:101–95

160. Uehlinger U, Bloesch J. 1987. Variation in the C:P ratio of suspended and settling seston and its significance for P uptake calculations. *Freshwater Biol.* 17:99–108

161. Ulen B. 1978. Seston and sediment in Lake Norrviken. 1. Seston composition and sedimentation. *Schweiz. Z. Hydrobiol.* 40:263–86

162. Urabe J. 1994. N and P cycling coupled by grazers' activities: food quality and nutrient release by zooplankton. *Ecology* 74:2337–50

163. Urabe J. 1993. Seston stoichiometry and nutrient deficiency in a shallow eutrophic pond. *Arch. Hydrobiol.* 126:417–28

164. Urabe J, Watanabe Y. 1990. Difference in the bacterial utilization ability of four cladoceran plankton (Crustacea: Cladocera). *Nat. Hist. Res.* 1:85–92

165. Urabe J, Watanabe Y. 1992. Possibility of N or P limitation for planktonic cladocerans: an experimental test. *Limnol. Oceanogr.* 37:244–51

166. Vadstein O, Olsen Y. 1989. Chemical composition and phosphate uptake kinetics of limnetic bacterial communities cultured in chemostats under phosphorus limitation. *Limnol. Oceanogr.* 34:939–46

167. Valiela I. 1991. Ecology of water columns. In *Fundamentals of Aquatic Ecology*, ed. RSK Barnes, KH Mann, pp. 29–56. Oxford: Blackwell Sci. 2nd ed.

168. Van Donk E, Hessen DO. 1993. Grazing resistance in nutrient-stressed phytoplankton. *Oecologia* 93:508–11

169. Van Donk E, Hessen DO. 1994. Reduced digestability of UV-B stressed and nutrient limited algae by *Daphnia magna.* *Hydrobiologia.* In press

170. Vanni MJ. 1994. Nutrient transport and recycling by consumers in lake food webs: implications for algal communities. In *Food Webs: Integration of Pattern and Process*, ed. G Polis, K Winemiller. New York: Chapman & Hall. Submitted

171. Vanni MJ, Lampert W. 1992. Food quality effects on life history traits and fitness in the generalist herbivore *Daphnia*. *Oecologia* 92:48–57

172. Watanabe Y. 1989. C:N:P ratios of size-fractionated seston and planktonic organisms in various trophic levels. *Int. Verein. theor. angew. Limnol. Verh.* 24:195–99

173. Willows RI. 1992. Optimal digestive investment: a model for filter feeders experiencing variable diets. *Limnol. Oceanogr.* 37:829–47

174. Witt U, Quantz G, Kulmann D. 1984. Survival and growth of turbot larvae *Scophtalamus maximus* L. reared on different food organisms with special regard to long-chain polyunsaturated fatty acids. *Aquaculture Engineering* 3:177–90

175. Wright SH, Stephens GC. 1982. Transepidermal transport of amino acids in the nutrtion of marine invertebrates. In *The Environment of the Deep Sea*, ed. WG Ernst, JG Morin, pp. 302–23. New Jersey: Prentice-Hall

176. Yuacfera M, Lubián LM. 1990. Effects of microalgal diet on growth and development of invertebrates in marine aquaculture. In *Introduction to Applied Phycology*, ed. I. Akatsuka, pp. 209–227. The Hague, Netherlands: Academic Press.

Annu. Rev. Ecol. Syst. 1994. 25:31–44

GENETIC ARCHITECTURE, GENETIC BEHAVIOR, AND CHARACTER EVOLUTION

Gabriel Moreno

Department of Biology, College of Natural Sciences, University of Puerto Rico, San Juan, Puerto Rico 00931-3360

KEY WORDS: character evolution, epistasis, quantitative variation, genetic architecture.

Abstract

Nonadditive gene interaction or epistasis is a central component of hypotheses that postulate a creative role for small populations and founder events in character evolution and speciation. Because quantitative or small effect variation is considered to be the basis of evolutionary change, the claim that this variation shows mostly additive behavior has represented a serious obstacle for these hypotheses. The literature on classical phenotypic genetics as well as the contemporary counterpart of this work, developmental genetics, harbors a considerable wealth of information relevant to the issue of gene-character relationship. In particular, recent studies of gene interaction in *Drosophila* developmental genetics provide important insights into the structure and behavior of nonadditive gene action. It is proposed that the importance of these studies for population genetics lies in suggesting that the small-effect of at least some genetic variation can hide a potential for strong phenotypic effects which can be expressed through interlocus interactions within sets of functionally related loci. At the population level this genetic behavior implies that strong effects should become apparent in quantitative variation when the genotypic frequencies of the population are substantially disturbed. That at least some quantitative variation can show strong interaction effects with substantial disturbance of the population genotypic combinations is supported by results from quantitative genetic analyses of chromosome and inbred lines and long-term artificial selection on quantitative characters. An experiment comparing the genetic variance resulting from the same set of wild third-chro-

31

0066-4162/94/1120-0031$05.00

mosomes in a wild type and mutant backgrounds revealed substantial amounts of unexpressed strong effects for a typical quantitative character in a wild population of *Drosophila melanogaster*. The claim that quantitative variation will show strong epistatic behavior under population disturbance is of fundamental importance for theories that postulate a positive role for small population size and population bottlenecks in phenotypic evolution and speciation.

INTRODUCTION

The process of adaptation depends on how genes relate to the phenotype as much as it depends on natural selection. However, the study of the relationship between genes and the phenotype has received considerably less attention than the study of external ecological forces acting on organisms in evolutionary biology. Studies of genetic architecture have typically divided genetic factors into either major-effect mutations or small-effect polygenes. This categorical division has had deep consequences for evolutionary theory, in particular for hypotheses that postulate a creative role for nonadditive gene interactions in evolution. While we consider that gene interaction is a real genetic phenomenon, particularly among genes of discrete effects in the laboratory, we do not consider strong interaction effects to be an important attribute of polygenic variation (8, 12, 19, 30).

From the population genetics point of view, intrinsic properties like physiology and development may introduce nonlinearity in the relationship between genes and characters, resulting in nonadditive gene action (epistasis). One of the greatest contributions of Sewall Wright to evolutionary theory was to show how the biological properties of organisms manifested through nonadditive interactions among genes can play a creative role in evolution. Nonadditive gene interaction or epistasis is a central component of hypotheses that postulate a role for the intrinsic biological properties of organisms in phenotypic evolution (3, 4, 34, 52, 57).

Because quantitative or small effect variation is considered to be the basis of evolutionary change, the claim that this variation shows mostly additive behavior has represented a serious obstacle for hypotheses that postulate a creative evolutionary role for nonadditive genetic interactions (8). This belief, based mostly on analyses of variance components of quantitative characters, has reinforced the simpler picture of adaptation in which the biological properties of organisms do not play an active evolutionary role, and the belief has also discouraged the study of the relationship between genes and characters as an important component of evolutionary research. Determining to what extent nonadditive genetic behavior is an important component of the relation between genes and the phenotype, and in particular how it relates to small-effect variation, is therefore of fundamental importance to evolutionary theory and is critical for models of evolution that give a positive role to the intrinsic properties of organisms.

Research on the relationship between gene and character has typically occupied a marginal position in evolutionary research, although some historical periods have paid more attention to this issue than the current one. There is in fact a rich evolutionary tradition concerned with the relationship between genes and characters in the work of CH Waddington (54), II Schmalhausen (45), R Goldschmidt (16), and JM Rendel (40), among others, using *Drosophila* as a research organism. This review is in part an attempt to add and bring up to date concepts and experimental results of that tradition, namely, phenotypic evolutionary genetics.

The literature on classical phenotypic genetics as well as the contemporary counterpart of this work, developmental genetics, harbors a considerable wealth of information relevant to the issue of gene-character relationship. In particular, recent advances in *Drosophila* developmental genetics have important implications for our views of how genes relate to characters. In this paper I propose that recent studies of gene interaction in *Drosophila* provide important insights into the structure and behavior of nonadditive gene action; specifically, I suggest that the small-effect of at least some genetic variation may hide a potential for strong phenotypic effects that may be expressed through interactions within sets of functionally related loci. This genetic behavior may become important when genotypic frequencies are substantially disturbed in a population.

DEVELOPMENTAL-GENETIC BACKGROUND

Developmental geneticists working with *Drosophila* have been finding that phenotypic changes such as the addition of thoracic chaetae, wing veins, or homeotic transformations are under the control of sets of functionally interrelated loci (2, 9, 10, 14, 15, 37, 46). I hereafter use the term "developmental set" to refer to such groups of loci, following other authors (9, 46). Within these sets, mutations at the different loci produce similar phenotypic transformations (e.g. 9, 23, 46). For example, there are at least five loci that produce the homeotic transformation known as Polycomb (23, 28) and at least eight loci that produce gaps in all longitudinal wing veins (11).

The significance of these studies for population genetics lies not only in revealing that there is more developmental redundancy among major morphological loci than previously believed (55), but more importantly, they have shown that interactions at the phenotypic level within these sets of loci can be very strong. These studies have shown that mutations within these sets act as strong penetrance and dominance modifiers of each other. Mutations that have little or no phenotypic effect by themselves can in combination induce phenotypic transformations of magnitude similar to those typically produced by single mutations of large effect.

For example, recessive mutations at these loci often produce the same type of phenotypic transformation in double heterozygote combination that they produce as single homozygotes, although the penetrance of the transformation is typically lower in the former (1, 9, 20–22, 36, 46, 47, 53). A gap in wing vein L4 can be induced with penetrance of 47% if the mutations *Hairless* and *Abruptex* are combined as double heterozygotes, while each mutation as separate heterozygotes does not show this phenotype (21). Similarly, the mutations *trithorax, ash-1,* and *ash-2* behave as full recessives by themselves. However, the *trithorax* mutation induces a *trithorax* homeotic phenotype with 15.6% and 9.8% penetrance in double heterozygote combination with *ash-1* or *ash-2,* respectively. Furthermore, the triple heterozygote induces this transformation with 82.1% penetrance (46). The recessive mutations *comb gap* and *engrailed* produce vein gaps at low penetrance as double heterozygotes, increasing to 49% penetrance as triple heterozygotes with the recessive mutation *cubitus interruptus* (20, 22). Individuals heterozygous for a deficiency at either the second chromosome locus *daughterless* or a deficiency covering the *achaete-scute* complex are normal, while individuals doubly heterozygous for these mutations show missing macrochaetae (9).

The strongly nonadditive behavior that combinations of these loci show may be explained by the fact that the one-locus relationship between gene activity and phenotype is typically nonlinear, changes in activity of more than 50% being typically required for a significant phenotypic change (e.g. 18, 31). This nonlinear relationship results in the recessiveness of most deficiencies, amorphic, and hypomorphic mutations, i.e. mutations resulting in a 50% activity reduction or less at a locus, while activity changes of more than 50% typically produce a disproportionally larger phenotypic effect. (24, 56).

What studies of gene interaction show is that this nonlinear effect can also be observed when combining changes across loci. Activity changes at different loci of the same developmental set which by themselves have a small or no phenotypic effect, can have a large effect in combination. This behavior is illustrated in Figure 1. A gene-activity change δ has little or no phenotypic effect (P) because of the nonlinear relationship between gene activity and phenotype. However, if this change δ is combined with a change in activity β of about 50% at another locus in the same developmental set a large phenotypic change (P′) is produced. This genetic behavior is of extreme importance because it implies that the hidden molecular effect of some small-effect variation represents a potential for strong epistasis since it can be revealed by interaction with mutations at functionally related loci.

Such strong interactions are common among mutants of the developmental sets studied so far, i.e. those controlling homeotic transformations (23, 28, 46),

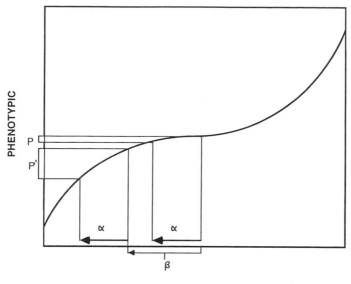

Figure 1   Relationship between gene dosage or activity and the phenotype for developementally related loci. A gene-activity change δ has little or no phenotypic effect (P) because of the nonlinear relationship between gene activity and the phenotype at one locus. However, if this change δ is combined with a change in activity of about 50% at another locus in the same developmental set, a large phenotypic change (P') is produced.

the development of the central nervous system (10), wing veins (11, 53), and bristles (9, 36).

Figure 2 shows an example of this genetic behavior in sternopleural bristle number in *Drosophila melanogaster,* a character traditionally used as a model quantitative character. The data show the phenotypic effect of seven recessive mutations that increase sternopleural bristle number when homozygous, as single heterozygotes and in double-heterozygote combinations with another mutation, *emc–* (a deficiency for the locus *emc*), with the same recessive effect. The phenotype of *emc* homozygotes is also shown for comparison. All of the seven mutations tested showed significant synergistic interactions with *emc–* as double-heterozygotes.

These results merely show that a similar increase in bristle number which requires a large change in activity at one locus (*emc–*) can be produced by combining functionally related mutations that have a small phenotypic effect by themselves. A 50% reduction in gene product at the locus *emc* produces only a slight change in the number of sternopleural chaetae. Similarly, individuals heterozygous for mutants at seven other loci known to affect chaeta

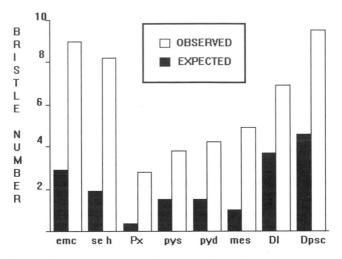

Figure 2 Number of extra sternopleural bristles above wild type for flies homozygous for *emc* and doubly heterozygous for a mutation at each of seven loci and a deficiency of the locus *emc*. White bars: observed score of the two mutations in combination. Black bars: additive expectation of the score of the two mutations taken separately. All interactions were significantly different from the additive expectation.

number show little or no effect on sternopleural chaetae. However, the number of sternopleural chaetae is synergistically increased either by combining the 50% dosage reduction at *emc* with (*a*) a further reduction at that locus (i.e. *emc/emc*), or with (*b*) a mutation in heterozygous state at any one of seven loci.

IMPLICATIONS FOR QUANTITATIVE VARIATION

The importance of this genetic behavior lies in suggesting that wild variation that may appear as small-effect variation may hide unexpressed molecular effects that can act as a potential source of strong phenotypic effects through gene-gene interaction. Physiological geneticists have long proposed that at least some of the standing quantitative variation in natural populations, rather than being a result of small molecular effects, may consist of mutations that have a considerable effect at the molecular level but show small phenotypic effects due to a nonlinear gene-phenotype relationship (e.g. 24, 27, 56). What developmental-genetic studies add to this idea is the observation that this molecular effect can be phenotypically expressed through interaction among mutations within sets of functionally related loci, as the results above demonstrate. The ubiquity of such strong synergistic interaction among morphological loci that recent developmental-genetic studies have revealed has not been a prediction of physiological-genetic models of metabolic control (5, 24, 27, 26).

The magnitude of the phenotypic effects that may be revealed through gene interaction in wild quantitative variation due to the presence of cryptic molecular effects is an empirical question currently being investigated. Populations may harbor mutations that in *homozygote* state show little or no phenotypic effect but that have a molecular effect of magnitude similar to the heterozygous mutations above. Therefore, a large phenotypic effect might be expressed if, for example, two of these mutations are combined as double homozygotes or in homozygote-heterozygote combination. It is not known whether mutations of this type are present in natural populations. Mutations that have little or no phenotypic effect in homozygous state would be difficult to detect. However, such variation should be revealed if mutations that change activity at other (functionally related) loci are introduced in the background.

REVEALING A POTENTIAL FOR STRONG EFFECTS IN QUANTITATIVE VARIATION

Figure 3 shows the results of an experiment that investigated this possibility by comparing the genetic variance produced by a set of ninety third chromosomes extracted from a wild population of *Drosophila melanogaster* in two genetic backgrounds, wild type and mutant (G. Moreno, in preparation). The mutant background was isogenic with the wild type background except that a mutation was introduced at an unlinked locus that mildly affected the mean of the character studied, sternopleural bristle number. This character is a typical quantitative character in that it shows mostly additive genetic behavior with little evidence of epistasis or major interaction effects in standard quantitative genetic analyses (12, 19, 48). The mutation chosen was a weak allele of the locus *Abruptex* decreasing total bristle number from about 17 to 15 in hemizygous males. By introducing a mutation at one locus, it is expected that the penetrance of mutations at functionally related loci that act as small-effect variation because of a nonlinear gene-character relationship will be increased, resulting in an increase in the genetic variance generated by the wild chromosomes. That is, it was hoped that the introduced mutation would act in a way analogous to change α in Figure 1 in revealing wild mutations acting as β. If on the other hand the nature of the polygenic variation in this character is truly additive, no increase in genetic variance is expected in the mutant background.

The results showed a two-fold increase in the genetic variance produced by the chromosomes in the mutant background ($T^1 = 4.47$, $P < 0.0001$) (7; see Table 1) supporting the idea that wild polygenic mutations in quantitative characters hide a significant amount of cryptic variability in the form of low penetrance mutations.

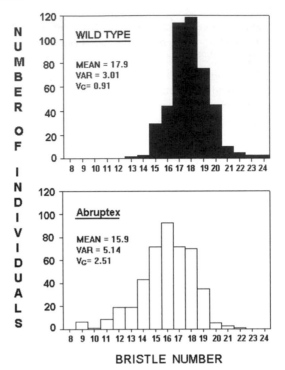

Figure 3   Histograms showing sternopleural bristle score of all flies for the wild third chromosomes tested in heterozygous state in both *Ax* and wild type background at 24°, with four flies per chromosome line. The mean, phenotypic variance (VAR), and genetic variance (VG) are shown for each case.

These results are important because they indicate that wild quantitative variation may hide a potential for stronger interaction effects than typically assumed. These cryptic interaction effects may become important when genotypic frequencies are substantially disturbed by selection, inbreeding, or population bottlenecks, as discussed below.

THE NATURE OF QUANTITATIVE VARIATION

How much of the small-effect variation in natural populations may be of this type is not known. The nature of quantitative loci and, thus, their behavior is at present an open question (12). The traditional distinction between small-effect polygenic variation and alleles of large effect assumes that the small effect of the former is relatively constant across genetic backgrounds. For brevity, I refer to this genetic behavior as Type 1. This idea is rooted in the belief that small-effect contributions from different loci to the same character

Table 1 Analysis of variance of the variability in sterno-pleural bristle number produced by a set of wild third chromosome lines tested in each of two genetic backgrounds: wild type and *Abruptex*.

Wild Type

Source	df	MS	Variance component	estimate
line	75	5.93	Var(line)	0.91[a]
rep(line)	76	2.29	Var(rep(line))	0.18
error	152	1.92	Var(error)	1.92

Abruptex

Source	df	MS	Variance component	estimate
line	75	13.14	Var(line)	2.51[a]
rep(line)	76	3.09	Var(rep(line))	0.44
error	152	2.22	Var(error)	2.22

[a] Between-line variances are significantly different at $p < 0.0001$ (see text).

are approximately additive, i.e., do not significantly alter each other's phenotypic effect.

This belief is based on quantitative genetic analyses of populations showing that interactions do not contribute substantially to the genetic variance (12, 19, 30). However, such results are observed in studies that typically leave wild genotypic frequencies relatively undisturbed. The genetic behavior described above (I refer to it as Type 2) is not incompatible with these results. Alleles that have a small effect because of the insensitivity of the wild type phenotype to genetic change will retain their small effect as long as dramatic changes do not occur at functionally related loci, for example, as a result of a substantial disturbance of genotypic frequencies. That is, this genetic behavior implies that strong epistasis will become apparent only when genotypic frequencies are substantially disturbed, increasing the frequency of combinations of alleles at two or more functionally related loci that express cryptic effects through interaction. Such alleles are expected to be present in populations at higher frequencies than single large-effect mutations given that they have little or no effect by themselves. Thus, any factors disturbing genotypic frequencies in a population, such as directional selection, inbreeding, or population bottlenecks, can potentially bring to sufficiently high frequency such combinations, resulting in the expression of strong effects.

This proposition is consistent with standard quantitative genetic studies finding little evidence of strong effects in populations in which an effort has been made to leave genotypic frequencies relatively undisturbed, while studies involving inbred lines detect considerable epistasis (sternopleural bristles: Refs. 3, 29, abdominal bristles: Refs. 3, 13, body size correlates: Refs. 25, 39, 42). Also, it is compatible with data showing that the scrambling of original genotypic frequencies by testing chromosomes in different combinations reveals a great deal of strong epistatic effects (42, 43).

Selection is also expected to reveal strong epistasis by changing genotypic frequencies. Artificial selection experiments show ample evidence of dramatic changes in the effect of quantitative genes induced by the change in genetic background during the course of selection (6, 13, 32, 33, 35, 38, 42, 44, 49, 50, 51, 58). For example, Spickett & Thoday (51) mapped three loci that accounted for 80% of the response to selection in sternopleural bristle number, two of which had a large effect on bristle number only when present in combination. Reeve & Robertson (39) found an increase in additive genetic variance in a *Drosophila melanogaster* strain selected for long wings; the effects of these chromosomes were found to be largely interdependent, showing abundant epistasis particularly when the chromosomes were tested in homozygous state (42).

Population bottlenecks are another way in which genotypic frequencies can be substantially disturbed, and thus, according to the present hypothesis, they are expected to induce epistatic interactions in quantitative variation. Epistasis has been proposed as an explanation for the documented increases in additive variance following a population bottleneck in several quantitative characters (2, 17). However, Bryant et al (2) only consider multiplicative gene action, and Goodnight (17) considers only additive × additive interactions, that is, interactions among the effects of genes that are expressed in the base population, as explanations for these results. The genetic behavior discussed here suggests that the most dramatic interactions will originate among alleles that in the base population remain cryptic because of a nonlinear gene-phenotype relationship. For example, mutations at two loci that have a small effect as single homozygotes may show a large effect in heterozygote-homozygote or double-homozygote combination. The other prevailing explanation of increases in additive genetic variance after bottlenecks is an increase in homozygosity of rare recessives (2, 41). The view of epistasis proposed here argues that, although some of the new additive variance can arise in this fashion, a significant portion of this new variance will come from interactions across loci of a developmental set that express otherwise recessive and low penetrance mutations.

PHENOTYPIC EPISTASIS AND FITNESS

It might be argued that interaction effects of the type described here may often result in reduced fitness, and thus this type of variation may not be important in character evolution. However, the possibility that these deleterious effects are much lower than those associated with a phenotypic change of equal magnitude produced by a single-locus mutation has not been investigated. In such cases, the pleiotropic fitness reduction associated with these interactions could be outweighed by favorable ecological selection for the new phenotype more often than in the case of single-locus mutations. It is also not known how much small-effect variation in natural populations is truly free of cryptic interaction effects of the type described here, as discussed above.

A related criticism is the idea that what ultimately determines the evolutionary trajectory is epistasis at the level of fitness, not at the level of morphology, and therefore nonadditive genetic behavior at the level discussed here is not what is important for character evolution. It is often argued by mathematical biologists that epistasis in morphology can be removed by studying evolution in the appropriate scale (e.g. 30). However, epistasis at the morphological level is important because it can affect the selection coefficients that alleles experience. If the character were transformed into a scale in which it behaved linearly, the selection coefficients would also change in this new scale. Epistasis at the level of morphology is therefore expected to affect evolutionary rates. The way gene effects behave at the level of the phenotype is therefore important for evolution. It is necessary to study this problem separate from the issue of fitness. Confounding these two issues can only lead to ignorance on how the relationship between genes and characters may affect evolution.

CONCLUSIONS

The hypothesis on genetic behavior espoused here suggests that quantitative variation harbors a potential for strong epistasis because the small phenotypic effect of this variation may often hide a larger molecular effect, and this larger effect can be revealed phenotypically through interactions within sets of functionally related loci. That at least some quantitative variation behaves this way is supported by results from analyses of inbred lines and long-term artificial selection on quantitative characters.

This view has profound implications for the mathematical modeling of evolution in quantitative characters by proposing that the evolutionary behavior of quantitative characters may be quite different from the simple additive models commonly assumed, particularly when gene frequencies are substantially disturbed, because the small additive effects in quantitative variation are a background-dependent property and not, as models typically assume, an inherent property of quantitative loci. This proposition implies that these models will tend to underestimate the evolutionary impact of epistasis and will fail to make realistic evolutionary predictions.

Finally, this genetic behavior also has important implications for hypotheses that postulate a positive role for small population size in discontinuous evolution. For example, it proposes the same evolutionary jumps that are typically attributed to a mutation of large effect may be triggered by a combination of mutations of small effect at a few loci in a developmental set. This suggests in turn that a polygenic architecture is not evidence against evolution by large steps.

ACKNOWLEDGMENTS

I thank Antonio Barbadilla, Daniel Dykhuizen, Walter Eanes, Douglas Futuyma, Lev Ginzburg, Joanne Labate, Richard Lewontin, Cedric Wesley, and Paul Wilson for the many discussions that helped clarify my ideas, as well as for their comments on the manuscript. Trudy Mackay and Bruce Walsh also provided valuable comments on the manuscript. I also thank Antonio García-Bellido and Luis García-Alonso for their encouragement in pursuing my interest in developmental genetics and for sharing with me information on the locus extramacrochaetae before it was published. I am grateful to the Drosophila Stock Centers at Bowling Green University and Indiana University for providing valuable assistance to my research. This work was supported by a National Science Foundation Dissertation Improvement Grant (BSR 8815268) and a National Science Foundation Postdoctoral Fellowship (RCD-9146323).

Literature Cited

1. Botas J, Moscoso del Prado J, García-Bellido A. 1982. Gene-dose titration analysis in the search for *trans*-regulatory genes in *Drosophila*. *EMBO J.* 1:307–10

2. Bryant EH, McCommas SA, Combs SM. 1986. The effect of an experimental bottleneck upon quantitative genetic variation in the housefly. *Genetics* 114: 1191–1211

3. Carson HL. 1975. The genetics of speciation at the diploid level. *Am. Nat.* 109:83–92

4. Carson HL, Templeton AR. 1984. Genetic revolutions in relation to speciation phenomena: the founding of new populations. *Annu. Rev. Ecol. Syst.* 15:97–131

5. Clark AG. 1990. Genetic components of variation in energy storage in *Drosophila melanogaster*. *Evolution* 44: 637–50

6. Clayton GA, Robertson A. 1957. An experimental check on quantitative genetical theory. II. The long term effects of selection. *J. Genet.* 55:152–70

7. Conover WJ. 1980. *Practical Nonparametric Statistics*. New York: Wiley. 2nd ed.

8. Crow JF. 1990. Sewall Wright's place in twentieth-century biology. *J. Hist. Biol.* 23:57–90

9. Dambly-Chaudière C, Ghysen A, Yan LY, Yan YN. 1988. The determination of sense organs in *Drosophila*: Interaction of *scute* with *daughterless*. *Roux's Arch. Devel. Biol.* 197:419–23

10. de la Concha A, Dietrich U, Weigel D, Campos-Ortega JA. 1988. Functional interactions of neurogenic genes of *Drosophila melanogaster*. *Genetics* 118: 499–508

11. Díaz-Benjumea J, González-Gaitán MAF, García-Bellido A. 1989. Devel-

opmental genetics of the wing vein pattern of *Drosophila. Genome* 31:612–19

12. Falconer DS. 1989. *Introduction to Quantitative Genetics.* London: Longman. 3rd ed.

13. Frankham R. 1969. Genetic analysis of two abdominal bristle lines. *Aust. J. Biol. Sci.* 22:1485–95

14. García-Alonso LA, García-Bellido A. 1988. Extramacrochaetae, a *trans*-acting gene of the *achaete-scute* complex of *Drosophila melanogaster* involved in cell communication. *Roux's Arch. Devel. Biol.* 197:328–38

15. García-Bellido A. 1986. Genetic analysis of morphogenesis. In *Genetics, Development, and Evolution,* ed. JP Gustafson, GL Stebbins, FJ Ayala, pp. 187–209. New York: Plenum

16. Goldschmidt R. 1938. *Physiological Genetics.* New York: McGraw-Hill

17. Goodnight CJ. 1988. Epistasis and the effects of founder events on the additive genetic variance. *Evolution* 42: 441–54

18. Hadorn E. 1961. *Developmental Genetics and Lethal Factors.* London: Methuen

19. Hill WG. 1982. Predictions of the response to artificial selection from new mutations. *Genet. Res.* 40:255–78

20. House VL. 1953. The interaction of mutants affecting venation in *Drosophila melanogaster* I. Interaction of *Hairless, engrailed,* and *cubitus interruptus. Genetics* 38:199–215

21. House VL. 1959. A comparison of gene expression of the *Hairless* and *Abruptex* loci in *Drosophila melanogaster. Anat. Rec.* 134:581–82

22. House VL. 1961. Mutant effects in multiple heterozygotes of recessive venation mutants in *Drosophila melanogaster. Genetics* 46:871

23. Jürgens G. 1985. Controlling the spatial expression of the bithorax complex in *Drosophila. Nature* 316:153–55

24. Kacser H, Burns JA. 1981. The molecular basis of dominance. *Genetics* 97: 639–66

25. Kearsey MJ, Kojima K. 1967. The genetic architecture of body weight and egg hatchability in *Drosophila melanogaster. Genetics* 56:23–37

26. Keightly PD. 1989. Models of quantitative variation in flux in metabolic pathways. *Genetics* 121:869–76

27. Keightly PD, Kacser H. 1987. Dominance, pleiotropy, and metabolic structure. *Genetics* 117:319–29

28. Kennison JA, Tamkun JW. 1988. Dosage-dependent modifiers of *Polycomb* and *Antennapedia* mutations in *Dro-*

sophila. Proc. Nat. Acad. Sci. USA 85: 8136–40

29. Kidwell JF. 1969. A chromosomal analysis of egg production and abdominal chaeta number in *Drosophila melanogaster. Can. J. Genet. Cytol.* 11: 547–57

30. Lande R. 1980. Genetic variation and phenotypic evolution during allopatric speciation. *Am. Nat.* 116:463–79

31. Lifschytz E, Green MM. 1979. Genetic identification of dominant overproducing mutations: The *Beadex* gene. *Molec. Gen. Genet.* 171:153–59

32. Mather K, Harrison BJ. 1949. The manifold effect of selection. Part I. *Heredity* 3:1–52

33. Mather K, Harrison BJ. 1949. The manifold effect of selection. Part II. *Heredity* 3:131–62

34. Mayr E. 1963. *Animal Species and Evolution.* Cambridge, Mass: Belknap/Harvard Univ. Press

35. Mohler JD. 1967. Some interactions of crossveinless-like genes in *Drosophila melanogaster. Genetics* 57:65–77

36. Moscoso del Prado J, García-Bellido A. 1984. Genetic regulation of the *achaete-scute* complex of *Drosophila melanogaster. Roux's Arch. Devel. Biol.* 193: 242–54

37. Neel JV. 1941. Studies on the interaction of mutations affecting the chaetae of *Drosophila melanogaster* I. The interaction of *hairy, polychaetoid,* and *Hairy wing. Genetics* 26:52–68

38. Ohh BK, Sheldon BL. 1970. Selection for dominance of *Hairy wing* in *Drosophila melanogaster.* I. Dominance at different levels of phenotypes. *Genetics* 66:527–40

39. Reeve ECR, Robertson FW. 1953. Studies in quantitative inheritance. II. Analysis of a strain of *Drosophila melanogaster* selected for long wings. *J. Genet.* 51:276–316

40. Rendel JM. 1967. *Canalisation and Gene Control.* London: Logos

41. Robertson A. 1952. The effect of inbreeding on the variation due to recessive genes. *Genetics* 37:189–207

42. Robertson FW. 1954. Studies in quantitative inheritance. V. Chromosome analysis in crosses between selected and unselected lines of different body size in *Drosophila melanogaster. J. Genet.* 52:494–520

43. Robertson FW. 1955. Selection response and the properties of genetic variation. *Cold Spring Harbor Symp. Quant. Biol.* 20:166–77

44. Scharloo W, Zweep A, Schuitema KA, Wijnstra WG. 1972. Stabilizing and dis-

ruptive selection on a mutant character in *Drosophila*. VI. Selection on sensitivity to temperature. *Genetics* 71:551–66

45. Schmalhausen II. 1949. *Factors in Evolution.* Chicago: Univ. Chicago Press

46. Shearn A. 1989. The *ash-1, ash-2* and *trithorax* genes of *Drosophila melanogaster* are functionally related. *Genetics* 121:517–25

47. Shepard SB, Broverman SA, Muskavitch MAT. 1989. A tripartite interaction among alleles of *Notch, Delta,* and *Enhancer of split* during imaginal development of *Drosophila melanogaster.* *Genetics* 122:429–38

48. Sheridan AK, Frankham R, Jones LP, Rathie KA, Barker, JSF. 1968. Partitioning of variance and estimation of genetic parameters for various bristle number characters of *Drosophila melanogaster. Theor. Appl. Genet.* 38:179–87

49. Shrimpton AE, Robertson A. 1988. The isolation of polygenic factors controlling bristle score in *Drosophila melanogaster.* I. Allocation of third chromosome sternopleural bristle effects to chromosome sections. *Genetics* 118:437–43

50. Spickett SG. 1963. Genetic and developmental studies of a quantitative character. *Nature* 199:870–73

51. Spickett SG, Thoday JM. 1966. Regular responses to selection. 3. Interaction between located polygenes. *Genet. Res.* 7:96–121

52. Templeton AR. 1982. Genetic architectures of speciation. In *Mechanisms of Speciation,* ed. C. Barigozzi, pp. 105–121. New York: Liss

53. Vässin H, Vielmetter J, Campos-Ortega JA. 1985. Genetic interactions in early neurogenesis of *Drosophila melanogaster. Neurogenetics* 2:291–308

54. Waddington CH. 1957. *The Strategy of the Genes.* London: Allen & Unwin

55. Wilkins AS. 1986. *Genetic Analysis of Animal Development.* New York: Wiley & Sons

56. Wright S. 1968. *Evolution and the Genetics of Populations. Genetic and Biometric Foundations.* Chicago: Univ. Chicago Press

57. Wright S. 1977. *Evolution and the Genetics of Populations. III. Experimental Results and Evolutionary Deductions.* Chicago: Univ. Chicago Press

58. Yoo BH. 1980. Long term selection on a quantitative character in large replicate populations of *Drosophila melanogaster.* II. Lethals and visible mutants with large effects. *Genet. Res.* 35:19–31

Annu. Rev. Ecol. Syst. 1994. 25:45–69

MOLECULAR APPROACHES TO POPULATION BIOLOGY

Jeffry B. Mitton

Department of Environmental, Population, and Organismic Biology, University of Colorado, Boulder, Colorado 80309

KEY WORDS: geographic variation, gene flow, protein polymorphisms, mtDNA, cpDNA, RFLP, hybridization

---

Abstract

Population biologists adopted protein electrophoresis to measure genetic variation in the 1960s and, more recently, have adopted a variety of techniques to detect variation in DNA sequences directly. In comparison to protein polymorphisms, DNA markers reveal more genetic variation, and provide population biologists with choices among sets of loci with different patterns of inheritance. Use of both protein polymorphisms and organellar genomes has dramatically increased the ability to detect patterns of mating in hybrid zones. The use of both proteins and DNA markers has also produced some intriguing puzzles; patterns of geographic variation and inferences of gene flow differ dramatically between these sets of loci, and nuclear markers may not show the correlations between heterozygosity and components of fitness that have been reported for protein polymorphisms.

INTRODUCTION

The study of population biology was profoundly changed (103) when Harris (69), Johnson et al (83), and Lewontin & Hubby (104) demonstrated that electrophoretic surveys of protein variation could be used to estimate the level of genetic variation within a species and to describe population structure. Numerous laboratories quickly adopted the techniques to study protein polymorphisms, and the tally of reports exceeded 1100 a decade ago (134). Now a second wave of innovation, more gradual than the first, has transformed the field again. This second revolution was the adoption of methods of extracting, cutting, amplifying, and detecting DNA. Many population biologists saw the

45

0066-4162/94/1120-0045$05.00

advantages of using mitochondrial DNA (mtDNA) to study populations of animals. This small molecule is inherited maternally, without recombination, and it generally accumulates mutations faster than nuclear loci, providing fine resolution within species and among closely related species. A second impetus to adopt DNA markers was the development of the polymerase chain reaction, which uses primers complementary to flanking regions to amplify a target sequence of DNA (4). Because the polymerase chain reaction works with a minute template of DNA, it permits the nondestructive sampling of populations and supports studies of DNA markers from dried or pickled museum specimens (see Glimpses into the Past).

This review points out new insights gained from the use of DNA markers, and compares and contrasts the perspectives gained from protein markers and DNA markers.

DNA MARKERS

The new techniques have spun off an array of acronyms that are now in common use. RFLPs (restriction fragment length polymorphisms) are fragments of DNA, produced by cutting with restriction enzymes, which differ in size due to the presence or absence of restriction enzyme sites among individuals. VNTRs (variable number of tandem repeats) are produced both by the presence or absence of restriction sites and by variable numbers of tandem repeats found commonly in minisatellite DNA (81, 82). The polymerase chain reaction (PCR) amplifies target DNA ranging in size from about 50 to 3000 base pairs, flanked by sites complementary to primers (155). Randomly amplified polymorphic DNAs (RAPDs) are produced by a single primer in a PCR amplification. RAPDs are dominant markers whose presence reflects priming sites flanking a segment of DNA suitable for amplification (30–3000 base pairs) or codominant markers revealing insertions or deletions between priming sites. CAPs are cleaved amplified polymorphic sequences—fragments of amplified DNA distinguished by the presence or absence of restriction sites.

Organellar RFLPs have the advantage of polarity over the alleles revealed by electrophoretic surveys of proteins. Migration rates of proteins in a supporting medium distinguish alternate alleles, but they provide no information on the relationships among alleles. In contrast, restriction sites can be mapped on the circular genomes of mitochondria and chloroplasts, and the evolutionary relationships among mtDNA or cpDNA haplotypes can be discerned. The phylogenetic relationships among mtDNA haplotypes yield intraspecific phylogenies revealing the matriarchal component of the genealogy among individuals or among populations. Organellar genealogies are now combined with patterns of geographic variation in the study of phylogeography, the principles and processes governing the geographic distributions of genealogical lineages (15, 16, 19).

Modes of Inheritance

One of the advantages of DNA markers over protein markers is a greater choice of modes of inheritance of markers. With only a few exceptions, protein markers are codominant, nuclear markers, and are biparentally inherited in sexual species. Many of the DNA markers have this pattern of expression and inheritance, but other patterns of inheritance are available as well. The mtDNA of animals is generally maternally inherited (16, 70, 127). In plants, mtDNA may be inherited biparentally, although most studies have revealed maternal inheritance. In most angiosperms, chloroplast DNA (cpDNA) is inherited maternally (139, 157), while paternal inheritance is reported in conifers. Both mtDNA and cpDNA are inherited without recombination, so that variation accumulates among clones by mutation, but the differences, once they arise, are not shuffled by recombination.

These patterns of inheritance are general, but exceptions have been reported. For example, mtDNA appears to be inherited biparentally near hybrid zones between the blue mussel, *Mytilus edulis,* and the Mediterranean mussel, *M. galloprovincialis,* and *M. trossulus* (77). And although mtDNA is usually in the size range of 16–18 kb in animals, exceptions have been reported; bark weevils of the genus *Pissodes* have mitochondrial genomes in the range of 30–36 kb (31), and the mtDNA of the deep-sea scallop, *Placopecten magellanicus,* ranges from 32.1 to 39.3 kb (162).

Although individual animals typically bear just one form of mtDNA, heteroplasmic individuals have been detected, and in some cases are common. For example, all 219 individuals examined from three species of bark weevils, *Pissodes strobi, P. nemorensis,* and *P. terminalis,* were heteroplasmic for two to five distinct size classes of mtDNA (31). Heteroplasmy has also been reported for *Drosophila melanogaster* (67) and *D. mauritiana* (163, 164), and the lizards *Cnemidophorus tesselatus* and *C. tigris* (46, 125, 126). Heteroplasmy is also relatively common in the mussels *Mytilus edulis, M. galloprovincialis* and *M. trossulus,* where they meet in hybrid zones (77). Of 150 mussels examined in hybrid zones, 72 were heteroplasmic for two mitochondrial genomes, 11 carried three genomes, and one carried 4 mitochondrial genomes. Heteroplasmy is also common near a hybrid zone between the crickets *Gryllus firmus* and *G. pennsylvanicus* (145). Sixty percent and 45% of *G. firmus* and *G. pennsylvanicus* were heteroplasmic, respectively, for the number of repeats of a 220 bp tandemly repeated segment.

Conifers of the family Pinaceae are unusual in that their cpDNA is inherited paternally (130–133, 171, 177, 178, 182), and their mtDNA is inherited maternally (132, 170, 176). However, both organellar genomes are inherited paternally in the coast redwood, *Sequoia sempervirens* (131), and in incense-cedar, *Calocedrus decurrens* (130). Exceptions to the general patterns of organellar inheritance are sufficiently common to justify studies of inheritance for DNA markers.

Mutation Rates

Mutation rates are heterogeneous among the components of the genome, and the pattern of variation differs between plants and animals. In vertebrates, the mutation rate of mtDNA exceeds that of the nuclear genome by a factor of at least three (32–34). In *Drosophila,* however, mutation rates of the mitochondrial and nuclear genomes are similar (142). In plants, the mutation rate of cpDNA is approximately three times the rate in mtDNA, and the mutation rate of nuclear DNA is five to ten times the mutation rate of mtDNA (187). The synonymous mutation rate of mammalian mtDNA is 100 times the rate in plant mtDNA (187). In animals, mtDNA evolves faster than the nuclear genome, but in plants, the nuclear genome is evolving fastest, followed by cpDNA, and then mtDNA.

Variability

Population genetics theory for neutral variation predicts that genetic variation will increase with population size (86, 87). As expected, allozyme variation increases with population size in animals (166), but the levels of variation are substantially below the theoretical expectations. Similarly, diversity of mtDNA increases with population size (14, 17), and once again, the levels of genetic variation are substantially below theoretical expectations. The departure from expectations is most often attributed to a violation of the assumption that the populations are at evolutionary equilibrium; if sufficient time has not elapsed since the establishment of the populations or since their last bottleneck, then genetic variation would not have risen to equilibrium expectations (17, 166).

Studies of mtDNA in deer mice, *Peromyscus,* first illustrated the utility of RFLPs for the study of natural populations (21). Six restriction enzymes revealed a total of 61 mtDNA haplotypes in *P. maniculatus* (98). MtDNA haplotypes defined five strongly differentiated geographic areas that did not correspond to recognized subspecies, or geographic variation of karyotypes, morphology, or allozyme polymorphisms (22). A summary of 13 studies of mtDNA (20) suggested that sequence divergence of 1% to 4% within species is common. The mtDNA sequences in *P. leucopus, P. maniculatus* and *P. polionotus* are differentiated from 13% to 20% (22). Thus, the number of mitochondrial haplotypes within populations and within species is quite high, and differentiation of congeneric species is substantial, providing many opportunities for the study of population biology and phylogeny (16, 108, 127).

One of the advantages of DNA markers over protein polymorphisms is the vastly higher number of markers that can be found. For example, surveys of 20–40 proteins have typically yielded 5–15 reliably scoreable polymorphic loci for studies of plant population biology. In contrast, 300 nuclear RFLPs have been mapped in loblolly pine, *Pinus taeda,* in the last several years (1, 48, 49). The advantage of DNA markers over protein polymorphisms is probably most extreme in species which, for one reason or another, maintain little protein polymorphism. For example, the haplo-diploidy of social insects reduces their genetic variation relative to diploid species. The level of protein polymorphism is low in honeybees, *Apis mellifera* (25, 101, 111, 140), but DNA variation appears to be abundant. Sixty-eight random primers revealed 90 differences between a drone and a queen (79). The inheritance of RAPDs produced by 13 of these primers yielded 20 presence/absence polymorphisms, 6 intensity polymorphisms, and 8 length polymorphisms in a pair cross. DNA from the M13 phage, when used as a probe, revealed DNA fingerprints in honeybees (29); in a sample of 23 haploid drones from a single queen, 21 distinct bands varied among drones, and only two drones had identical fingerprint profiles. A $(GATA)_4$ oligonucleotide probe revealed fingerprints with sufficient resolution to distinguish between super-sisters (sisters sired by the same haploid drone) and half-sisters (different fathers) (128) from a queen mated in a natural setting.

Animal mitochondrial DNA appears to be more variable than plant mitochondrial DNA. Perhaps the most extreme cases of mtDNA variability are found in menhaden *Brevoortia tyrannus* and *B. patrons* and the chuckwalla, *Sauromalus obesus* (18). When mtDNA from these species was digested with 18 endonucleases, 32 forms of mtDNA were found in a sample of 33 menhaden, and 35 forms of mtDNA were found in a sample of 51 chuckwallas. In contrast, mtDNA in plants appears to be relatively low. Rangewide surveys of mtDNA revealed only five forms of mtDNA in lodgepole pine, *Pinus contorta.* Only 10 forms of mtDNA were detected in a survey of 268 trees from 19 populations of Monterey pine, *Pinus radiata,* Bishop pine, *P. muricata,* and knobcone pine, *P. attenuata* (167). The relative degrees of variability in mtDNA in plants and animals reflect their relative mutation rates (187).

CpDNA, like mtDNA in plants, has relatively low levels of genetic variation within populations. For example, only three minor variants were found in a study of 100 individuals of the lupine, *Lupinus texensis* (23). A survey of 384 trees from 19 populations of Monterey pine, Bishop Pine, and knobcone pine revealed little or no cpDNA variation within or among populations of knobcone pine and Monterey pine (78). Bishop pine had some variation within populations, but more variation among populations. A total of 20 forms of cpDNA were detected in the 384 trees. A summary of the percent nucleotide changes per site in 46 studies of cpDNA yielded a mean of 0.07% at the intraspecific level, 0.8% at the interspecific level, and 3.4% at the intergeneric level (157).

In contrast, values of 1% to 8% were reported within species for the mtDNA of a diverse set of animals studied in the southeastern United States (14). Although variation in cpDNA is relatively low, it is sufficient to support studies of geographic variation in some species (110, 176, 177).

GLIMPSES INTO THE PAST

PCR primers can be designed from highly conserved sections of the genome, so that amplifications can be obtained for virtually any species, even when sequence data are lacking (89). Extraction techniques allow dried herbarium specimens and pickled museum specimens to provide the DNA needed for PCR (36), even from truly ancient material, such as mummified human remains, macrofossils from packrat middens, old bones, and insects trapped in amber (36, 38, 136–138). Amplified DNA has now been obtained from a bee trapped in amber for at least 25 million years, a weevil trapped in amber for more than 120 million years (37), partially fossilized remains of a magnolia leaf from the Miocene (64), from the extinct Quagga (76) and the extinct marsupial wolf (173), and Egyptian mummies (136–138). These studies have supported phylogenetic studies of magnolias (64), wolves (173), saber-toothed cats (80), and termites (47).

Museum specimens and contemporary population samples were combined in a study of geographic variation of the Panamint kangaroo rat, *Dipodomys panamintinus,* in California (174). The samples represented three subspecies with allopatric distributions, and contemporary samples were collected from the same localities from which the museum specimens had been collected more than 50 years previously. Based on a 225 bp portion of mtDNA near the D loop that was amplified and sequenced, 23 mitochondrial haplotypes were found in 106 individuals. The most divergent sequences differed at 12 positions; the mean pairwise difference among individuals at a locality was approximately six. The genetic distances and phylogenetic relationships among subspecies were similar in the contemporary samples and the samples stored in the museum. The frequencies of the mtDNA haplotypes had changed in the half-century between the collection of the samples, demonstrating that we have the capability to measure evolutionary change over decades or centuries.

THE NATURE OF CPDNA AND MTDNA VARIATION

DNA markers are now employed widely by population biologists, and for most applications, the variation is presumed to have no detectable effect upon physiological variation, but several reports suggest that variation in mitochondrial DNA is not always neutral in its effect upon physiology and components of fitness in animals.

Experimental removal and replacement of mitochondria from human tissue cultures suggested that variation in mitochondria has an impact upon cellular respiration (88). Mitochondria were removed from human (HeLa) tissue cells by chronic treatment with ethidium bromide. When mitochondria were reintroduced to the manipulated tissue culture, two measures of metabolism, respiration rate and time to doubling of the number of cells, returned to the values of the original HeLa tissue culture. This restoration of the measures of metabolism indicated that mitochondria could be extracted and later reintroduced without damaging the cell culture. However, when mitochondria from other individuals were introduced to the cell culture, the respiration rates of some of the new cultures were similar to the initial culture, while some were 30% higher, and others were 30% lower. Because the nuclear genome was not altered by either the elimination or reintroduction of mitochondria, this experiment demonstrated that variation in the mitochondrial genome can have a substantial influence upon respiration.

Experiments with laboratory populations of *Drosophila* (40, 57, 106) also suggest that variation in the mitochondrial genome influences components of fitness. Selection experiments revealed rapid changes in the frequencies of mtDNA haplotypes in *D. pseudoobscura* (106) and *D. subobscura* (57). Analyses of the segregation of a balancer chromosome identified strong interactions between the nuclear background and the cytoplasm of *D. melanogaster* (40). The interpretation of selection experiments involving mtDNA is difficult (40, 107, 161), but these first results justify further studies of the physiological and demographic consequences of mtDNA and cpDNA.

A study of direct competition between mitochondria revealed differences between species (112). Four replicate populations were started with *Drosophila melanogaster* eggs into which the mitochondria of *Drosophila mauritiana* had been injected. Within 30 generations, three of the four populations were fixed for the *mauritiana* mtDNA—the alien genome—and the fourth population had come to an equilibrium with 90% *mauritiana* and 10% *melanogaster* mtDNA. Unexpectedly, the alien mitochondrial genome had a selective advantage.

The common heteroplasmy in the crickets *Gryllus firmus* and *G. pennsylvanicus* (145) provided the opportunity to compare the frequency distributions of size classes of mtDNA in heteroplasmic females and their offspring (144). In each comparison, the offspring had higher frequencies of the smaller mtDNA, suggesting a replication advantage for smaller mtDNA molecules during development. Observations of the frequencies of size variants of mtDNA causing neuromuscular disease also suggest that smaller mtDNA molecules replicate faster than full-sized mtDNAs (178).

Studies of ongoing selection in composite crosses of barley, *Hordeum vulgare,* have revealed dramatic changes in fragments of DNA coding for ribosomal RNA that are consistent with strong selection (2, 153, 154, 194).

These diverse observations warn against sweeping generalizations concerning the intensity of natural selection influencing genes, genomes, or fragments of DNA. Some sequences must be subject to natural selection, others certainly are not. Because mtDNA and cpDNA are not shuffled by recombination, a single mutation can render the entire molecule either beneficial or deleterious. If selection acts upon organellar genomes, the time to fixation for beneficial genomes is much shorter than for neutral genomes (109, 172), and the amount of variability in the population is depressed at the time of fixation (172). Statistically, it makes most sense to assume that markers are selectively neutral, for that assumption allows the use of neutral models in population genetics to make predictions concerning the distribution of genotypes in space and in time. But we know little about the physiological consequences of variation in cpDNA and mtDNA; the assumption of neutrality should be made explicitly and deserves to be tested in the field.

EFFECTIVE POPULATION SIZE

The evolutionary dynamics of mtDNA and cpDNA differ from the evolutionary dynamics of nuclear genes (12, 13, 15, 16, 127, 157, 168, 176). The organellar genomes have no mechanism for recombination, so these molecules are inherited clonally; these genomes behave like a large, haploid, single locus. In addition, the effective population sizes of organellar genomes are smaller than those of nuclear genes (14, 15, 17, 26–28). In animals, the time to shared ancestry (172) of mtDNA is one fourth that of nuclear genes (15, 17). The decrement in the time to shared ancestry is attributable to a two-fold effect due to haploidy of the organellar genome and a second two-fold effect due to maternal inheritance in a dioecious population. The disparity in times to shared ancestry would be less in monecious species, such as ponderosa pine, for all mature individuals in the population have the capability to act as females. In monecious species, the times to shared ancestry in organellar genomes would be half those of nuclear genes. This reduction in the effective population size leads us to expect lower levels of genetic variation in organellar genomes in comparison to the nuclear genome.

DISCORDANT PATTERNS OF GEOGRAPHIC VARIATION

An unexpected revelation from molecular studies of geographic variation is that different sets of markers reveal dramatically different levels and patterns of geographic variation. Protein polymorphisms and DNA markers reveal discordant patterns of geographic variation in deer mice, horseshoe crabs, oysters, lodgepole pine, the closed-cone pines of California, and limber pine.

Allozyme polymorphisms and mtDNA reveal discordant patterns of genetic variation in the deer mouse, *Peromyscus maniculatus* (22, 97). Allelic frequencies at protein polymorphisms reveal slight-to-moderate population structure from central Mexico to central Canada, with values of F_{st} varying from 0.04 to 0.38 among six polymorphic loci. The same two alleles at aspartate aminotransferase segregated in all populations, with little variation in allelic frequencies. In contrast, mtDNA revealed five geographic groups distinguished by diagnostic haplotypes. For example, the mtDNA haplotypes in the eastern states and northern Michigan differ from the haplotypes in the central states by approximately 20 mutational steps. The shallow patterns of differentiation for allelic frequencies of protein polymorphisms do not correspond to the dramatic, historic population structure revealed by mtDNA haplotypes.

The horseshoe crab, *Limulus polyphemus,* first drew the attention of evolutionary biologists because it was considered to be a phylogenetic "relic" (159). The level of allozyme variation is moderate, and the degree of geographic variation is slight. I pooled allozyme data from samples from Woods Hole, MA, and Chincoteague, VA, and two populations from the panhandle of Florida to represent Atlantic and Gulf populations. F_{st}, which measures the allelic variation among populations, is 0.05 when calculated with these widely separated localities. *Nm*, the number of individuals moving between populations, was estimated from F_{st} by the equation of Wright (188):

$$F_{st} = 1/(1 + 4Nm).$$

From the degree of differentiation of allozyme loci, I estimated that approximately five individuals are exchanged between Atlantic and Gulf populations per generation. This is a relatively high level of gene flow, well above the level ($Nm \approx 1$) needed to prevent the differentiation of neutral alleles. However, a contrasting population structure was revealed by analyses of mtDNA (14, 15, 156). There is a striking discontinuity in northeastern Florida, marked by a 2% sequence divergence in mtDNA. This variation reveals that migration is not moving mtDNA across this recognized biogeographic boundary between warm-temperate and tropical marine faunas. The allozyme data suggest extensive gene flow, while the mtDNA data reveal a barrier to gene flow.

Patterns of geographic variation and inferences about the magnitude of gene flow also differ between protein polymorphisms and DNA markers in the American oyster, *Crassostrea virginica.* Allozyme frequencies in oysters exhibit very little variation from Maine to Texas, a geographic pattern initially interpreted as evidence for extensive gene flow among populations of oysters (35). From the original allozyme data, *Nm* was estimated to be approximately six (147). However, a survey of mtDNA variation revealed a striking discontinuity in mtDNA in northeastern Florida (147). A total of 82 mitochondrial haplotypes was revealed by 13 endonucleases in 212 oysters. But the mitochondrial haplotypes cluster neatly into two groups with mtDNA not being exchanged across a boundary in northeastern Florida. Clearly, the estimates of extensive gene flow from allozymes and very limited gene flow from mtDNA cannot both be correct. To determine which set of data was providing the more accurate inference of gene flow, data from four anonymous single-copy nuclear RFLPs were collected (84). These markers had the same pattern of geographic variation as the mtDNA. Thus, the differences between the patterns of geographic variation revealed by allozymes and mtDNA cannot be attributed simply to the maternal inheritance and the reduced N_e of mtDNA. The biparentally inherited nuclear RFLPs gave the same pattern of geographic variation as the mtDNA.

Similar comparisons of patterns of geographic variation can be made among the nuclear, chloroplast, and mitochondrial genomes of conifers. The conifers are wind pollinated, and cpDNA is inherited paternally and mtDNA is inherited maternally in the species examined below. Pollen may disperse hundreds of meters to many kilometers, while the distance of seed dispersal is generally similar to the height of the maternal parent. The nuclear, biparental markers have the paternal contribution moved by both pollen and seeds, while the maternal contribution is moved only by seeds. Because pollen moves further than seeds, cpDNA will get around much more than mtDNA, and therefore cpDNA should be less differentiated among localities than mtDNA. Thus, the potential for gene flow in conifers can be described as low for mtDNA, relatively high for nuclear markers, and highest for cpDNA.

Provenance studies and allozyme surveys appear to reveal discordant patterns of variation. Striking patterns of phenology, often associated with environmental variation, necessitate seed zones for the major timber species such as ponderosa pine, *Pinus ponderosa,* and Douglas-fir, *Pseudotsuga menziesii* (129). Common garden studies of ponderosa pine (151), Douglas-fir (150), lodgepole pine, *Pinus contorta* (149), and western larch, *Larix occidentalis* (148) revealed that survival and growth suffer dramatically if seeds are planted more than 250–300 meters higher or lower than the site from which they were collected (151). In contrast, differentiation of enzymes, at least measured with single loci, is typically slight (55, 66, 75, 105, 135, 189–193). We know that important patterns of phenological variation are genetically determined, but the intensity and patterning of phenological variation does not appear to match the pattern of enzyme variation, although multivariate analysis of enzyme variation improves the correspondence of the data sets (116, 181).

Three sets of genetic data—allozymes, cpDNA, and mtDNA—are available for comparison of patterns of geographic variation in the closed-cone pines of California. Allozyme surveys of genetic variation in knobcone pine, *Pinus attenuata,* Monterey pine, *P. radiata,* and Bishop pine, *P. muricata,* revealed the typical apportionment of genetic variation for pines—high amounts of variation within populations and relatively little differentiation among populations within a species. Millar et al (115) reported that 12–22% of the variation was among populations within a species. However, analyses of cpDNA (78) and mtDNA (167) yielded very different patterns of variation. CpDNA variation was revealed as RFLPs detected with probes from Douglas-fir. Neither knobcone pine or Monterey pine have much variation, either within or between populations. Bishop pine has little variation within populations, but marked differentiation [G_{st} = 87(±8%)] among populations. Similarly, while allozyme data reveal only 24% of the variation to be among species, the estimate from cpDNA is twice as great, 49%. MtDNA variation was revealed as RFLPs detected with a probe for the cytochrome oxidase-I gene amplified from knobcone pine with the polymerase chain reaction. Variability for mtDNA is low within populations, but G_{st} varied from 75% in Monterey pine to 96% in Bishop pine. Once again, DNA markers reveal greater variation among populations than do protein polymorphisms. Although cpDNA has the greatest potential for gene flow, and mtDNA has the lowest potential for gene flow, these genetic markers show similar degrees of differentiation.

The degrees of allozyme, cpDNA, and mtDNA differentiation appear to fit expectations based upon gene flow in lodgepole pine (53, 54, 183). Allozyme differentiation in lodgepole pine is typical of conifers; values of F_{st} estimated from allozymes rarely exceeded 6% (183) and are generally less than 10% in conifers (68). Two mtDNA polymorphisms were used to describe patterns of variation in 741 individuals throughout the ranges of jack pine and lodgepole pine. Probes for cytochrome oxidase I and II (COXI and COXII) identified mtDNA RFLPs from total DNA extracted from needle samples. A probe for the COXI gene revealed a diagnostic difference between species but almost no variation within populations or species. In contrast, a probe for the COXII gene revealed a small amount of variation within jack pine, but a high degree of variation within lodgepole pine. Large proportions of the variation within lodgepole pine were among populations within subspecies (F_{st} = 66%) and among subspecies (F_{st} = 31%). CpDNA variation at a single hot spot was revealed in digests from a single enzyme, probed with a 700-bp fragment isolated from lodgepole pine. Variation was diagnostic among species, but within species, variation was high within populations, with little differentiation among populations; values of θ were less than 0.05. Thus, differentiation of populations was strong for mtDNA, but slight for both allozymes and cpDNA.

Allozymes and DNA markers reveal contrasting patterns of genetic variation in limber pine, *Pinus flexilis*. Allozyme variation was used to describe variation within and among populations from lower tree line (1650 m) to upper tree line (3350 m) in Colorado (158). The average F_{st} for these loci was 0.02, indicating little differentiation among sites, and suggesting gene flow on the order of 10 migrants between populations per generation. However, this estimate of gene flow did not seem to be possible, for the pollination phenology of limber pine varies dramatically with elevation; most sites along the elevational transect differing by 400 m or more do not have overlapping pollination periods and therefore cannot exchange genes in a single generation. A preliminary survey of cpDNA RFLPs revealed striking differentiation between the elevational extremes of this transect. In limber pine, allozyme loci, pollination phenology, and rapid markers suggest very high, intermediate, and very low levels of gene flow, respectively.

These discordant patterns of variation present an intriguing challenge to population biologists. The disparity between the magnitudes of geographic variation revealed by provenance studies and allozyme surveys is not wholly unexpected. Lewontin (102) argued that it is generally easier to detect differences among population means than among allelic frequencies. While this matter of statistical sensitivity may contribute to the disparity between the two sets of observations, statistical sensitivity is probably not the most important explanation. From the available data, it appears that allozyme loci reveal only slight differences among samples, but DNA markers reveal dramatic differentiation of populations. The same statistical procedures are used to test the homogeneity of allozyme and DNA markers. The allozyme variation in deer mice, horseshoe crabs, oysters, California closed-cone pines, and jack, lodgepole, and limber pine all show a pattern of slight geographic variation and suggest high levels of gene flow, even among distant populations. This degree of gene flow is inconsistent with the provenance studies of conifers, with studies of pollination phenology in limber pine, and with much of the variation in DNA markers. The discrepancy is clearest in oysters, where mtDNA and nuclear RFLPs show the same pattern, sharply contrasting with the allozyme data. Similarly, the allozymes of the California closed-cone pines revealed much less differentiation than either mtDNA or cpDNA. While the relative degrees of differentiation of allozyme, mtDNA, and cpDNA markers in jack pine and lodgepole pine conform to expectations based upon gene flow, this particular pattern may depend upon the specific cpDNA marker as much as it does upon gene flow. The cpDNA marker is a single hot spot; a substantial proportion of the variants are produced by insertions and deletions. Dong & Wagner (54) pointed out that the degree of differentiation of organellar markers is dependent upon population size, gene flow, and mutation rate (28), and that the mutation rate of the cpDNA marker probably exceeds the rates characteristic of restriction sites. Thus, surveys of other cpDNA markers will determine if the patterns of geographic variation in jack and lodgepole pines differ from those of the California closed-cone pines. Clearly, for the majority of cases examined here, gene flow does not adequately predict the relative degrees of differentiation of allozyme, mtDNA, and cpDNA markers.

HETEROZYGOSITY AND COMPONENTS OF FITNESS

Studies that combine genetic and demographic data, and studies that combine genetic and physiological data, have revealed numerous reports of positive correlations between individual heterozygosity (124) and components of fitness (3, 99, 100, 118, 119, 122). For example, viability differs among allozyme genotypes in wild oats, *Avena barbata* (41), barley, *Hordeum vulgare* (42), annual ryegrass, *Lolium multiflorum* (117), the blue mussel, *Mytilus edulis* (93), the American oyster, *Crassostrea virginica* (160, 197), *Macoma balthica* (65), the killifish, *Fundulus heteroclitus* (118, 123), the guppy, *Poecilia reticulata* (24), the sulfur butterfly, *Colias eriphyle* (179, 180), and the palourde, *Ruditapes decussatus* (30). Growth rate is associated with allozyme genotype in the American oyster, *C. virginica* (198), the blue mussel, *M. edulis* (92), the tiger salamander, *Ambystoma tigrinum* (141), quaking aspen, *Populus tremuloides* (121), the coot clam, *Mulinia lateralis* (91), and white tailed deer, *Odocoileus virginianus* (43). Allozyme heterozygosity is negatively associated with oxygen consumption in *C. virginica* (94, 152), the snail, *Thais hemostomata* (58), the coot clam, *M. lateralis* (59), *M. edulis* (50, 51), the rainbow trout, *Salmo gairdneri* (45), and *A. tigrinum* (120). Rates of protein turnover and routine metabolic costs differ among allozyme genotypes in *M. edulis,* and this relationship may drive the correlations between allozyme heterozygosity and oxygen consumption, growth rate, and viability (72, 73). Empirical observations of positive correlations between heterozygosity and components of fitness are consistent with predictions from recent models in population genetics (60–63, 113, 114, 175, 185, 186). At first, these data were viewed as curiosities, suggestive of strong selection within a life cycle, but inconsistent with the convenient assumption of neutrality. These observations are numerous, have a broad systematic base (3, 117–119, 122, 195), and appear to be general, but certainly not universal.

Several hypotheses are consistent with the correlations between allozyme heterozygosity and components of fitness (118, 122, 195). The first hypothesis, overdominance, is that the correlations in fitness are attributable to the protein loci; kinetic differences among protein genotypes must influence physiology so that fitness advantages accrue to heterozygotes. A second hypothesis, associative overdominance, is that the protein loci are effectively neutral markers, in linkage disequilibrium with either balanced polymorphisms or deleterious alleles that have a direct effect upon components of fitness. A third hypothesis is that the protein loci are effectively neutral markers, reflecting differences in variation in the degree of inbreeding within the population. Inbred individuals have higher probabilities of homozygosity at all loci, regardless of their influence upon components of fitness. Additional hypotheses, including null alleles and genetic imprinting (39), are considered in detail in Zouros & Foltz

(195). Note that the first hypothesis attributes fitnesses to the protein polymorphisms, while the second and third hypotheses portray the protein markers as neutral markers that reflect fitness differentials at other loci. Occasionally, sufficient enzyme kinetic, physiological, and demographic data have been gathered to assign clear fitness differentials to protein polymorphisms (90, 95, 143), but in the vast majority of cases in which correlations between protein genotypes and fitness have been reported, the data are insufficient to convincingly reject all but one hypothesis, and so the question concerning the mechanism underlying the correlations between protein heterozygosity and fitness persists. However, a new experimental design promises to distinguish clearly between these hypotheses.

If the correlations between heterozygosity and components of fitness are attributable to linkage disequilibrium (hypothesis 2) or variation in the degree of inbreeding (hypothesis 3), then any genetic marker should reveal the associations with fitness that have been reported for protein polymorphisms. But if the associations are attributable to kinetic differences among protein genotypes, then heterozygosity at a different set of genetic markers, such as nuclear RFLPs, should not be correlated with components of fitness. Zouros & Pogson (196) employed this rationale in a study of the scallop, *Placopecten magellanicus*. Shell height was used to estimate growth rate, and heterozygosity was estimated with seven protein polymorphisms, two proteins of unknown function, five single-locus VNTRs, and three nuclear RFLPs. The correlation between heterozygosity at the seven enzyme loci and shell height was positive and significant ($r = 0.148$, $p < 0.05$), while the correlation between the other 10 markers and shell height was not significantly different from zero ($r = 0.032$, $p > 0.50$). Because the DNA markers did not reveal a correlation between heterozygosity and growth, the correlations appear not to be driven by either variation in inbreeding among individuals or linkage disequilibrium. I see no reason to hypothesize that the DNA markers would not reflect inbreeding, or that they are characterized by lower levels of linkage disequilibrium than loci coding for proteins. This experiment suggests that at least some of the fitness differentials in natural populations of scallops are attributable to genotypes at protein loci. This experiment needs to be replicated with other species, particularly those for which correlations between enzyme heterozygosity and components of fitness have been reported.

HYBRIDIZATION

Molecular genetic markers provide powerful tools for the study of hybridization. Molecular techniques have been used to document the number of hybrid origins of a species, to identify the pollen and seed parents of hybrids, and to document the level of back crossing. Molecular evidence has been used to strengthen putative cases of introgression, when morphological evidence was open to alternative interpretations. Nuclear loci allow the identification of hybrids and backcrosses, but the addition of organellar markers identifies the direction of crosses (71). The use of both nuclear and organellar markers allows the estimation of cytonuclear disequilibria (6, 10), a metric for the characterization of hybrid zones (5).

Using chloroplast markers, Soltis & Soltis (165) examined the hybrid origins of allopolyploid *Tragopogon mirus* and *T. miscellus*. They showed that *T. porifolius* was consistently the maternal parent, and *T. dubius* the paternal parent of *T. mirus,* despite morphological and cytological evidence for three separate hybrid origins of *T. mirus*. *T. miscellus* showed at least two separate origins, with reciprocal hybridization of *T. dubius* and *T. pratensis*. However, the common and widespread species, *T. dubius,* was generally the paternal parent of the hybrids, being the maternal parent in only two of fourteen populations of hybrid species. This suggests that pollen usually flows from common species to rarer species in hybridization. Similarly, in the Louisiana irises *Iris fulva* and *I. hexagona,* the direction of pollen flow in hybridization seems to be unidirectional, typically from *I. fulva* to *I. hexagona* (7–9). Individuals with a hybrid nuclear genotype (possessing diagnostic nuclear markers from both parents) show only the *I. hexagona* cpDNA genotypes.

The documentation of cases of introgression has also been improved by molecular techniques. It had been hypothesized (74) that introgression of *Helianthus debilis* ssp. *cucumerifolius* genes into *H. annuus* was the origin of *H. annuus* ssp. *texanus*. However, molecular data enabled a direct test of the hypothesis and clear documentation of introgression. Rieseberg et al (146) examined cpDNA and nuclear rDNA in sympatric populations of *H. annuus* ssp. *texanus* and *H. debilis* ssp. *cucumerifolius,* as well as in allopatric populations of *H. annuus*. Markers diagnostic to *H. debilis* ssp. *cucumerifolius* were found in 13 of 14 populations of *H. annuus* ssp. *texanus,* but not in allopatric populations of *H. annuus*. These data strongly supported Heiser's original (74) hypothesis of introgression in the formation of *H. annuus* ssp. *texanus*. Introgression of both maternally inherited cpDNA and nuclear rDNA markers was found. In a contrasting situation, a putative case of introgression between jack pine (*Pinus banksiana*) and lodgepole pine (*P. contorta*) (44) was not supported by evidence from paternally inherited cpDNA (177). A few trees morphologically identified as one species showed the cpDNA genotype of the other. However, no introgression was found outside the limited zone of sympatry, despite the inheritance of cpDNA through wind-dispersed pollen in these species.

Studies of introgression show that both the maternally inherited organellar genomes and nuclear genes can introgress into another species. In one of the first studies of introgression to use DNA markers, Ferris et al (56) demonstrated the presence of *Mus domesticus* mtDNA in populations of *M. musculus* in Scandinavia. In contrast, nuclear genes matched those of *M. musculus* in other parts of Europe. Thus, it was the maternally inherited genes that had introgressed. Behavioral differences between male treefrogs at their mating sites cause introgression of mtDNA from *Hyla cinerea* to *H. gratiosa* (96) and produce a substantial level of cytonuclear disequilibrium (11). At the mating ponds, *H. gratiosa* males call from the surface in open water, while *H. cinerea* males call from the shore. Females approaching the pond are eagerly grasped and amplexed by males. Female *H. gratiosa* need to evade the *H. cinerea* males at the shore to reach the *H. gratiosa* in open water. In a sample of 20 F_1 hybrids, all were produced by crosses of *H. gratiosa* females and *H. cinerea* males. Introgression of maternally inherited chloroplast DNA markers from maize, *Zea mays,* to *Zea perennis* has also been demonstrated (52). In a three-way hybrid zone involving *I. fulva, I. hexagona,* and *I. brevicaulis,* hybrids were found possessing markers diagnostic of all three species. However, cpDNA markers of *I. hexagona* were not observed (9). This suggests sympatric hybridization of *I. fulva* and *I. brevicaulis* with introgression of nuclear genes mediated by pollen flow from allopatric *I. hexagona.* Keim et al (85) analyzed a zone of hybridization between *Populus angustifolia* and *P. fremontii.* Of 24 hybrid trees, only F_1s and backcrosses to *P. angustifolia* were found, suggesting introgression of nuclear genes from *P. fremontii* to *P. angustifolia.*

A study of hybridization in white oaks revealed extensive introgression of the chloroplast genome and provided evidence that the nuclear genome introgressed less extensively than the chloroplast genome (184). This survey revealed three cases of local sharing among species of the chloroplast genome, with no morphological evidence of hybridization or introgression, suggesting a higher level of gene flow (for cpDNA) among sympatric species than among allopatric populations of the same species.

SUMMARY

The addition of DNA markers to the arsenal of techniques in population biology has provided considerably more flexibility and diversity than was previously available in electrophoretic surveys of protein variation. The numbers of markers for species such as honeybees and ponderosa pine has increased to more than one hundred. Furthermore, population biologists can now choose among markers with different modes of inheritance and rates of accumulation of mutations.

From the perspective of the population biologist, the DNA markers have not simply added more of the same information provided by protein polymorphisms. The uniparental inheritance of organellar genomes provides insight into historical events that shaped the population structure of today, and in some cases, the DNA markers provide interpretations that are inconsistent with data from protein polymorphisms. For example, surveys of cpDNA have revealed introgression in oaks that was not apparent in morphological characters or nuclear loci. Numerous empirical studies have reported positive correlations between fitness and individual heterozygosity for protein polymorphisms, but the first study employing both proteins and DNA markers reported, once again, positive correlations for proteins, but no association for the DNA markers. Furthermore, DNA and protein markers yield different estimates of the degree of geographic variation. In horseshoe crabs, oysters, Bishop pine, Monterey Pine, knobcone pine, lodgepole pine, and limber pine, DNA markers reveal much greater differentiation of populations than do protein polymorphisms. These results will call into question the estimates of gene flow inferred from F_{st}'s based upon protein polymorphisms. After considering alternate hypotheses for the discordant patterns of geographic variation in oysters, Karl & Avise (84) tentatively concluded that balancing selection acting upon enzyme loci caused this set of loci to diverge from the pattern of geographic variation exhibited by the mtDNA and nuclear RFLPs. Population biologists should not discard protein electrophoresis as they adopt DNA techniques, for the two sets of markers appear to reflect the predominance of different evolutionary forces.

Literature Cited

1. Ahuja MR, Devey ME, Groover AT, Jermstad KD, Neale DB. 1994. Mapped RFLP probes from loblolly pine can be used for restriction fragment length polymorphism mapping in other conifers. *Theor. Appl. Genet.* In press
2. Allard RW, Saghai-Maroof MA, Zhang Q, Jorgensen RA. 1990. Genetic and molecular organization of ribosomal DNA (rDNA) variants in wild and cultivated barley. *Genetics* 126:743–51
3. Allendorf FW, Leary RF. 1986. Heterozygosity and fitness in natural populations of animals. In *Conservation Biology: The Science of Scarcity and Diversity,* ed. ME Soulé, pp. 57–76. Sunderland, MA: Sinauer. 584 pp.
4. Arnheim N, White T, Rainey WE. 1990. Application of PCR: organismal and population biology. *BioSience* 40:174–82

5. Arnold J. 1993. Cytonuclear disequilibria in hybrid zones. *Annu. Rev. Ecol. Syst.* 24:521–54
6. Arnold J, Asmussen MA, Avise JC. 1988. An epistatic mating system model can produce permanent cytonuclear disequilibria in a hybrid zone. *Proc. Natl. Acad. Sci. USA* 85:1893–96
7. Arnold ML. 1992. Natural hybridization as an evolutionary process. *Annu. Rev. Ecol. Syst.* 23:237–61
8. Arnold ML, Bennett BD. 1993. Natural hybridization in Louisiana irises: genetic determinants. In *Hybrid Zones and the Evolutionary Process,* ed. RG Harrison, pp. 115–39. Oxford: Oxford Univ. Press
9. Arnold ML, Robinson JJ, Buckner CM, Bennett BD. 1992. Pollen dispersal and interspecific gene flow in Louisiana Irises. *Heredity* 68:399–404
10. Asmussen MA, Arnold J, Avise JC.

1987. Definition and properties of disequilibrium statistics for associations between nuclear and cytoplasmic genotypes. *Genetics* 115:755–68

11. Asmussen MA, Arnold J, Avise JC. 1989. The effects of assortative mating and migration on cytonuclear associations in hybrid zones. *Genetics* 122:923–34

12. Avise JC. 1986. Mitochondrial DNA and the evolutionary genetics of higher animals. *Philos. Trans. R. Soc. Lond. B.* 312:325–42

13. Avise JC. 1991. Ten unorthodox perspectives on evolution prompted by comparative population genetic findings on mitochondrial DNA. *Annu. Rev. Genet.* 25:45–69

14. Avise JC. 1992. Molecular population structure and the biogeographic history of a regional fauna: a case history with lessons for conservation biology. *Oikos* 63:62–76

15. Avise JC. 1994. *Molecular Markers, Natural History and Evolution.* New York: Chapman & Hall

16. Avise JC, Arnold J, Ball RM, Bermingham E, Lamb T, Neigel JE, et al. 1987. Intraspecific phylogeography: the mitochondrial DNA bridge between population genetics and systematics. *Annu. Rev. Ecol. Syst.* 18:489–522

17. Avise JC, Ball RM Jr, Arnold J. 1988. Current versus historical population sizes in vertebrate species with high gene flow: a comparison based on mitochondrial DNA lineages and inbreeding theory for neutral mutations. *Mol. Biol. Evol.* 5:331–44

18. Avise JC, Bowen BW, Lamb T. 1984. DNA fingerprints from hypervariable mitochondrial genotypes. *Mol. Biol. Evol.* 6:258–69

19. Avise JS, Helfman GS, Saunders NC, Hales LS. 1986. Mitochondrial DNA differentiation in North Atlantic eels: population genetic consequences of an unusual life history pattern. *Proc. Natl. Acad. Sci. USA* 83:4350–54

20. Avise JC, Lansman RA. 1983. Polymorphism of mitochondrial DNA in populations of higher animals. In *Evolution of Genes and Proteins,* ed. M Nei, RK Koehn, pp 165–90. Sunderland, MA: Sinauer

21. Avise JC, Lansman RA, Shade RO. 1979. The use of restriction endonucleases to measure mitochondrial DNA sequence relatedness in natural populations. I. Population structure and evolution in the genus *Peromyscus. Genetics* 92:279–95

22. Avise JC, Smith MH, Selander RK.

1979. Biochemical polymorphism and systematics in the genus *Peromyscus.* VII. Geographic differentiation in members of the *truei* and *maniculatus* species groups. *J. Mammal.* 60:177–92

23. Banks J, Birkey C Jr. 1985. Chloroplast DNA diversity is low in a wild plant, *Lupinus texensis. Proc. Natl. Acad. Sci. USA.* 82:6950–54

24. Beardmore JA, Shami SA. 1979. Heterozygosity and the optimum phenotype under stabilizing selection. *Aquilo. Ser. Zool.* 20:100–10

25. Berkelhamer RC. 1983. Intraspecific genetic variation and haplodiploidy, eusociality, and polygyny in the hymenoptera. *Evolution* 37:540–45

26. Birky CW Jr. 1978. Transmission genetics of mitochondria and chloroplasts. *Annu. Rev. Genet.* 12:471–512

27. Birky CW Jr. 1988. Evolution and variation in plant chloroplast and mitochondrial genomes. In *Plant Evolutionary Biology,* ed. LD Gottlieb, SK Jain, pp. 25–53. New York: Chapman & Hall

28. Birky CW Jr, Fuerst P, Maruyama T. 1989. Organelle gene diversity under migration mutation, and drift: equilibrium expectations, approach to equilibrium, effects of heteroplasmic cells, and comparison to nuclear genes. *Genetics* 121:613–27

29. Blanchetot A. 1991. Genetic relatedness in honeybees as established by DNA fingerprinting. *J. Hered.* 82:391–96

30. Borsa P, Jousselin Y, Delay B. 1992. Relationships between allozymic heterozygosity, body size, and survival to natural anoxic stress in the palourde, *Ruditapes decussatus* L. (Bivalvia: Veneridae). *J. Exp. Mar. Biol. Ecol.* 155:169–81

31. Boyce TM, Zwick ME, Aquadro CF. 1989. Mitochondrial DNA in the bark weevils: size, structure, and heteroplasmy. *Genetics* 123:825–36

32. Brown WM. 1983. Evolution of animal mitochondrial DNA. In *Evolution of Genes and Proteins,* ed. M Nei, RK Koehn, pp. 62–88. Sunderland, MA: Sinauer

33. Brown WM, George M Jr, Wilson AC. 1979. Rapid evolution of animal mitochondrial DNA. *Proc. Natl. Acad. Sci. USA* 76:1967–71

34. Brown WM, Prager EM, Wang A, Wilson AC. 1982. Mitochondrial DNA sequences of primates: tempo and mode of evolution. *J. Mol. Evol.* 18:225–39

35. Buroker NE. 1983. Population genetics of the American oyster *Crassostrea virginica* along the Atlantic coast and

the Gulf of Mexico. *Mar. Biol.* 75:99–112

36. Cano RJ, Poinar HN. 1993. Rapid isolation of DNA from fossil and museum specimens suitable for PCR. *BioTechniques* 15:433–35

37. Cano RJ, Poinar H, Pieniazek N, Acra A, Poinar GO Jr. 1993. Amplification and sequencing of DNA from a 120–135 million year old weevil. *Nature* 363: 536–39

38. Cano RJ, Poinar H, Poinar GO Jr. 1992. Isolation and partial characterization of DNA from the bee *Proplebeia dominicana* (Apidae: Hymenoptera) in 25–40 million year old amber. *Med. Sci. Res.* 20:249–51

39. Chakraborty R. 1989. Can molecular imprinting explain heterozygote deficiency and hybrid vigor? *Genetics* 122: 113–17

40. Clark AG. 1985. Natural selection with nuclear and cytoplasmic transmission. II. Tests with *Drosophila* from diverse populations. *Genetics* 111:97–112

41. Clegg MT, Allard RW. 1973. Viability versus fecundity selection in the slender wild oat, *Avena barbata* L. *Science* 181: 667–68

42. Clegg MT, Kahler AL, Allard RW. 1978. Estimation of life cycle components of selection in an experimental plant garden. *Genetics* 89:765–91

43. Cothran EG, Chesser R, Smith MH, Johns PE. 1983. Influences of genetic variability and maternal factors on fetal growth in white-tailed deer. *Evolution.* 37:282–91

44. Critchfield WB. 1985. *Can. J. For. Res.* 15:749–22

45. Danzmann RG, Ferguson MM, Allendorf FW. 1987. Heterozygosity and oxygen-consumption rates as predictors of growth and developmental rate in rainbow trout. *Physiol. Zool.* 60:211–20

46. Densmore LD, Wright JW, Brown WM. 1985. Length variation and heteroplasmy frequent in mitochondrial DNA from parthenogenetic and bisexual lizards (genus *Cnemidophorus*). *Genetics* 110:689–707

47. DeSalle RJ, Gatesy J, Wheeler W, Grimaldi D. 1992. DNA sequences from a fossil termite in Oligo-Miocene amber and their phylogenetic implications. *Science* 257:1933–36

48. Devey ME, Fiddler TA, Liu B-H, Knapp SJ, Neale DB. 1994. An RFLP linkage map for loblolly pine based on a three-generation outbred pedigree. *Theor. Appl. Genet.* In press

49. Devey ME, Jermstad KD, Tauer CG, Neale DB. 1991. Inheritance of RFLP loci in a loblolly pine three-generation pedigree. *Theor. Appl. Genet.* 83:238–42

50. Diehl WJ, Gaffney PM, Koehn RK. 1986. Physiological and genetic aspects of growth in the mussel *Mytilus edulis.* I. Oxygen consumption, growth, and weight loss. *Physiol. Zool.* 59:201–11

51. Diehl WJ, Gaffney PM, McDonald JH, Koehn RK. 1985. Relationship between weight standardized oxygen consumption and multiple-locus heterozygosity in the marine mussel *Mytilus edulis* L. (Mollusca). In *Proceedings of the 19th European Marine Biology Symposium,* ed. P Gibbs, pp. 531–36. Cambridge: Cambridge Univ. Press

52. Doebly J. 1989. Molecular evidence for a missing wild relative of maize and the introgression of its chloroplast genome into *Zea perennis. Evolution* 43: 1555–59

53. Dong J, Wagner DB. 1993. Taxonomic and population differentiation of mitochondrial DNA diversity in *Pinus banksiana* and *Pinus contorta. Theor. Appl. Genet.* 86:573–78

54. Dong J, Wagner DB. 1994. Paternally inherited chloroplast polymorphism in *Pinus*: estimation of diversity and population subdivision, and tests of disequilibrium with a maternally inherited mitochondrial polymorphism. *Genetics* 136:1187–94

55. El-Kassaby YA. 1991. Genetic variation within and among conifer populations: Review and evaluation of methods. In *Biochemical Markers in the Population Genetics of Forest Trees,* ed. S Fineschi, ME Malvolti, F Cannata, HH Hattemer, pp. 61–76. The Hague: SPB Acad. Publ.

56. Ferris SD, Sage RD, Huang C-M, Nielsen JT, Ritte U, Wilson AC. 1983. Flow of mitochondrial DNA across a species boundary. *Proc. Natl. Acad. Sci. USA.* 80:2290–94

57. Fos M, Dominguez A, Latorre A, Moya A. 1990. Mitochondrial DNA evolution in experimental populations of *Drosophila subobscura. Proc. Natl. Acad. Sci. USA* 87:4198–4201

58. Garton DW. 1984. Relationship between multiple locus heterozygosity and physiological energetics of growth in the estuarine gastropod *Thais haemastoma. Physiol. Zool.* 57:530–43

59. Garton DW, Koehn RK, Scott TM. 1984. Multiple-locus heterozygosity and the physiological energetics of growth in the coot clam, *Mulinia lateralis,* from a natural population. *Genetics* 108:445–55

60. Gillespie JH. 1978. A general model to account for enzyme variation in natural populations. V. The SAS-CFF model. *Theor. Popul. Biol.* 14:1–45

61. Gillespie JH. 1991. *The Causes of Molecular Evolution.* New York: Oxford Univ. Press

62. Ginzburg LR. 1979. Why are heterozygotes often superior in fitness? *Theor. Pop. Biol.* 15:264–67

63. Ginzburg LR. 1983. *Theory of Natural Selection and Population Growth.* Menlo Park, CA: Benjamin/Cummings

64. Golenberg EM, Giannasi DE, Clegg MT, Smiley CJ, Durbin M, et al. 1990. Chloroplast DNA sequence from Miocene Magnolia species. *Nature* 344: 656–58

65. Green RH, Singh SM, Hicks B, McCuaig JM. 1983. An arctic intertidal population of *Macoma balthica* (Mollusca, Pelecypoda): genotypic and phenotypic components of population structure. *Can. J. Fish. Aquat. Sci.* 4: 1360–71

66. Guries RP, Ledig FT. 1982. Genetic diversity and population structure in pitch pine (*Pinus rigida* Mill.). *Evolution* 36:387–99

67. Hale LR, Singh RS. 1986. Extensive variation and heteroplasmy in size of mitochondrial DNA among geographic populations of *Drosophila melanogaster. Proc. Natl. Acad. Sci. USA* 83:8813–17

68. Hamrick JL, Godt MJW. 1989. Allozyme diversity in plant species. In *Plant Population Genetics: Breeding and Genetic Resources,* ed. AHD Brown, MT Clegg, AL Kahler, BS Weir, pp. 43–63. Sunderland, MA: Sinauer

69. Harris H. 1966. Enzyme polymorphisms in man. *Proc. R. Soc. Lond. B* 164:298–310

70. Harrison RG. 1989. Animal mitochondrial DNA as a genetic marker in population and evolutionary biology. *Trends Ecol. Evol.* 4:6–11

71. Harrison RG. 1993. *Hybrid Zones and the Evolutionary Process.* New York: Oxford Univ. Press

72. Hawkins AJ, Bayne BL, Day AJ. 1986. Protein turnover, physiological energetics and heterozygosity in the blue mussel, *Mytilus edulis:* the basis of variable age-specific growth. *Proc. R. Soc. Lond. B* 229:161–76

73. Hawkins AJ, Bayne BL, Day AJ, Rusin J, Worrall CM. 1989. Genotype-dependent interrelations between energy metabolism, protein metabolism and fitness. In *Reproduction, Genetics and Distribution of Marine Organisms,* ed.

JS Ryland, PA Tyler, pp. 283–292. Predensborg, Denmark: Olsen & Olsen

74. Heiser CB. 1951. Hybridization in the annual sunflowers: *Helianthus annuus* X *H. debilis* var *cucumerifolius. Evolution* 5:42–51

75. Hiebert RD, Hamrick JL. 1983. Patterns and level of genetic variation in Great Basin bristlecone pine, *Pinus longaeva. Evolution* 37:302–10

76. Higuchi RB, Bowman B, Freiberger M, Ryder OA, Wilson AC. 1984. DNA sequences from the Quagga, an extinct member of the horse family. *Nature* 312:282–84

77. Hoeh WR, Blakley KH, Brown WM. 1991. Heteroplasmy suggests limited biparental inheritance of *Mytilus mitochondrial* DNA. *Science* 251:1488–90

78. Hong Y-P, Hipkins VD, Strauss SH. 1993. Chloroplast DNA diversity among trees, populations and species in the California closed-cone pines (*Pinus radiata, Pinus muricata,* and *Pinus attenuata.*) *Genetics* 135:1187–96

79. Hunt GJ, Page RE Jr. 1992. Patterns of inheritance with RAPD molecular markers reveal novel types of polymorphism in the honey bee. *Theor. Appl. Genet.* 85:15–20

80. Janczewski DN, Yuhki N, Gilbert DN, Jefferson GT, O'Brien SJ. 1992. Molecular phylogenetic inference from sabertoothed cat fossils of Rancho La Brea. *Proc. Natl. Acad. Sci. USA.* 89:9769–73

81. Jeffreys AJ, Wilson V, Thein SL. 1985. Hypervariable "minisatellite" regions in human DNA. *Nature* 314:67–73

82. Jeffreys AJ, Wilson V, Thein SL. 1985. Individual-specific "fingerprints" of human DNA. *Nature* 316:76–79

83. Johnson FM, Kanapi CG, Richardson RH, Wheeler MR, Stone WS. 1966. An operational classification of *Drosophila esterases* for species comparisons. *Univ. Texas Publ.* 6615:517–32

84. Karl SA, Avise JC. 1992. Balancing selection at allozyme loci in oysters: implications from nuclear RFLPs. *Science* 256:100–102

85. Keim P, Paige KN, Whitham TG, Lark KG. 1989. Genetic analysis of an interspecific hybrid swarm of *Populus:* Occurrence of unidirectional introgression. *Genetics* 123:557–65

86. Kimura M. 1983. *The Neutral Theory of Molecular Evolution.* Cambridge: Cambridge Univ. Press

87. Kimura M, Ohta T. 1971. *Theoretical Aspects of Population Genetics.* Princeton, NJ: Princeton Univ. Press

88. King MP, Attardi G. 1989. Human cells lacking mtDNA: Repopulation with ex-

ogenous mitochondria by complementation. *Science* 246:500–503

89. Kocher TD, Thomas WK, Meyer A, Edwards SV, Pääbo S, Villablanca FX, Wilson AC. 1989. Dynamics of mitochondrial evolution in animals: Amplification and sequencing with conserved primers. *Proc. Natl. Acad. Sci. USA* 86:6196–200

90. Koehn RK. 1987. The importance of genetics to physiological ecology. In *New Directions in Ecological Physiology,* ed. ME Feder, AF Bennett, WW Burggren, RB Huey, pp. 170–188. Cambridge, MA: Cambridge Univ. Press

91. Koehn RK, Diehl WJ, Scott TM. 1988. The differential contribution by individual enzymes of glycolysis and protein catabolism to the relationship between heterozygosity and growth rate in the coot clam, *Mulinia lateralis. Genetics* 118:121–30

92. Koehn RK, Gaffney PM. 1984. Genetic heterozygosity and growth rate in *Mytilus edulis Mar. Biol.* 82:1–7

93. Koehn RK, Milkman R, Mitton JB. 1976. Population genetics of marine pelecypods. IV. Selection, migration and genetic differentiation in the blue mussel *Mytilus edulis. Evolution* 30:2–32

94. Koehn RK, Shumway SE. 1982. A genetic/physiological explanation for differential growth rate among individuals of the American oyster, *Crassostrea virginica* (Gmelin). *Mar. Biol. Letters* 3:35–42

95. Koehn RK, Zera AJ, Hall JG. 1983. Enzyme polymorphism and natural selection. In *Evolution of Genes and Proteins,* ed. M Nei, RK Koehn, pp. 115–136. Sunderland, MA: Sinauer

96. Lamb T, Avise JC. 1986. Directional introgression of mitochondrial DNA in a hybrid population of tree frogs: the influence of mating behaviour. *Proc. Natl. Acad. Sci. USA* 83:2526–30

97. Lansman RA, Avise JC, Aquadro CF, Shapira JF, Daniel SW. 1983. Extensive genetic variation in mitochondrial DNAs among geographic populations of the deer mouse, *Peromyscus maniculatus. Evolution* 37:1–16

98. Lansman RA, Shade RO, Shapira JF, Avise JC. 1981. The use of restriction endonucleases to measure mitochondrial DNA sequence relatedness in natural populations. III. Techniques and potential applications. *J. Mol. Evol.* 17:214–26

99. Ledig FT. 1986. Heterozygosity, heterosis, and fitness in outbreeding plants. In *Conservation Biology: The Science of Scarcity and Diversity,* ed. ME

Soulea, pp. 77–104. Sunderlan, MA: Sinauer

100. Lesica P, Allendorf FW. 1992. Are small populations of plants worth preserving? *Conservation Biol.* 6:135–39

101. Lester LJ, Selander RK. 1979. Population genetics of haplodiploid insects. *Genetics* 92:1329–45

102. Lewontin RC. 1984. Detecting population differences in quantitative characters as opposed to gene frequencies. *Am. Nat.* 123:115–24

103. Lewontin RC. 1991. Twenty-five years ago in GENETICS. Electrophoresis in the development of evolutionary genetics: Milestone or millstone? *Genetics* 128:657–62

104. Lewontin RC, Hubby JL. 1966. A molecular approach to the study of genic heterozygosity in natural populations. II. Amount of variation and degree of heterozygosity in natural populations of *Drosophila pseudoobscura. Genetics* 54:595–609

105. Loveless MD, Hamrick JL. 1984. Ecological determinants of genetic structure in plant populations. *Annu. Rev. Ecol. Syst.* 15:65–95

106. MacRae AF, Anderson WW. 1988. Evidence for non-neutrality of mitochondrial DNA haplotypes in *Drosophila pseudoobscura. Genetics* 120:485–94

107. MacRae AF, Anderson WW. 1990. Can mating preferences explain changes in mtDNA haplotype frequencies? *Genetics* 124:999–1001

108. Maddison DR. 1991. African origin of human mitochondrial DNA reexamined. *Syst. Zool.* 40:355–63

109. Maruyama T, Birky CW Jr. 1991. Effect of periodic selection on gene diversity in organellar genomes and other systems without recombination. *Genetics* 127:449–51

110. Matos J. 1992. *Evolution within the* Pinus montezumae *complex of Mexico: population subdivision, hybridization, and taxonomy.* PhD Thesis. Washington Univ., St. Louis

111. Mestriner MA. 1969. Biochemical polymorphism in bees (*Apis mellifera ligustica*). *Nature* 233:188–89

112. Miki Y, Chigusa SI, Matsuura ET. 1989. Complete replacement of mitochondrial DNA in *Drosophila. Nature* 341:551–52

113. Milkman R. 1978. Selection differentials and selection coefficients. *Genetics* 88:391–403

114. Milkman R. 1982. Toward a unified selection theory. In *Perspectives on Evolution,* ed. R Milkman, pp. 105–118. Sunderland, MA: Sinauer

115. Millar CI, Strauss SH, Conkle MT,

Westfall R. 1988. Allozyme differentiation and biosystematics of the Californian closed-cone pines. *Syst. Bot.* 13: 351–70

116. Millar CI, Westfall RD. 1992. Allozyme markers in forest genetic conservation. *New Forests* 6:347–71. Also in *Population Genetics of Forest Trees*, ed. WT Adams, SH Strauss, DL Copes, AR Griffin, pp. 347–71. Dordrecht: Kluwer Acad.

117. Mitton JB. 1989. Physiological and demographic variation associated with allozyme variation. In *Isozymes in Plant Biology*, ed. D Soltis, P Soltis, pp. 127–145. Portland, OR: Dioscorides. 268 pp.

118. Mitton JB. 1993. Theory and data pertinent to the relationship between heterozygosity and fitness. In *The Natural History of Inbreeding and Outbreeding*, ed. N Thornhill, pp. 17–41. Chicago: Univ. Chicago Press

119. Mitton JB. 1993. Enzyme heterozygosity, metabolism, and developmental stability. *Genetica* 89:47–65

120. Mitton JB, Carey C, Kocher TD. 1986. The relation of enzyme heterozygosity to standard and active oxygen consumption and body size of tiger salamanders, *Ambystoma tigrinum*. *Physiol. Zool.* 59:574–82

121. Mitton JB, Grant MC. 1980. Observations on the ecology and evolution of quaking aspen, *Populus tremuloides*, in the Colorado Front Range. *Am. J. Bot.* 67:202–209

122. Mitton JB. Grant MC. 1984. Relationships among protein heterozygosity, growth rate, and developmental stability. *Annu. Rev. Ecol. Syst.* 15:479–99

123. Mitton JB, Koehn RK. 1975. Genetic organization and adaptive response of allozymes to ecological variables in *Fundulus heteroclitus*. *Genetics* 79:97–111

124. Mitton JB, Pierce BA. 1980. The distribution of individual heterozygosity in natural populations. *Genetics* 95:1043–54

125. Moritz C, Brown WM. 1986. Tandem duplication of D-loop and ribosomal RNA sequences in lizard mitochondrial DNA. *Science* 233:1425–27

126. Moritz C, Brown WM. 1987. Tandem duplications in animal mitochondrial DNAs: variation in incidence and gene content among lizards. *Proc. Natl. Acad. Sci. USA* 84:7183–87

127. Moritz C, Dowling TE, Brown WM. 1987. Evolution of animal mitochondrial DNA: Relevance for population biology and systematics. *Annu. Rev. Ecol. Syst.* 18:269–92

128. Moritz RFA, Meusel MS, Haberl M. 1991. Oligonucleotide DNA fingerprinting discriminates super-and half-sisters in honeybee colonies (*Apis mellifera* L.). *Naturwissenschaften* 78:422–24

129. Namkoong G, Kang HC, Brouard JS. 1988. *Tree Breeding: Principles and Strategies*. Berlin: Springer-Verlag

130. Neale DB, Marshall KA, Harry DE. 1991. Inheritance of chloroplast and mitochondrial DNA in incense-cedar (*Calocedrus decurrens*). *Can. J. For. Res.* 21:717–20

131. Neale DB, Marshall KA, Sederoff RR. 1989. Chloroplast and mitochondrial DNA are paternally inherited in *Sequoia sempervirens*. D. Don Endl. *Proc. Natl. Acad. Sci. USA* 86:9347–49

132. Neale DB, Sederoff RR. 1989. Paternal inheritance of chloroplast DNA and maternal inheritance of mitochondrial DNA in loblolly pine. *Theor. Appl. Genet.* 77:212–16

133. Neale DB, Wheeler NC, Allard RW. 1986. Paternal inheritance of chloroplast DNA in Douglas-fir. *Can. J. For. Res.* 16:1152–54

134. Nevo E, Beiles A, Ben-Shlomo R. 1984. The evolutionary significance of genetic diversity: Ecological, demographic and life history correlates. In *Lecture Notes in Biomathematics*, ed. GS Mani, 53:13–213. Lecture Notes in Biomath. Berlin: Springer-Verlag

135. O'Malley DM, Allendorf FW, Blake GM. 1979. Inheritance of isozyme variation and heterozygosity in *Pinus ponderosa*. *Biochem. Genet.* 17:233–50

136. Pääbo S. 1985. Molecular cloning of ancient Egyptian mummy DNA. *Nature* 314:644–45

137. Pääbo S. 1989. Ancient DNA: extraction, characterization, molecular cloning, and enzymatic amplification. *Proc. Natl. Acad. Sci. USA* 86:1939–43

138. Pääbo S, Higuchi RG, Wilson AC. 1989. Ancient DNA and the polymerase chain reaction. *J. Biol. Chem.* 264:9709–12

139. Palmer J, Osorio B, Aldrich J, Thompson W. 1987. Chloroplast DNA evolution among legumes: loss of a large inverted repeat occurred prior to other sequence rearrangements. *Curr. Genet.* 11:275–86

140. Pamilo P, Varvio-Aho S-L, Pekkarinen A. 1978. Low enzyme genetic variability as a consequence of haplodiploidy. *Hereditas* 88:93–99

141. Pierce BA, Mitton JB. 1982. Allozyme heterozygosity and growth in the tiger salamander, *Ambystoma tigrinum*. *J. Hered.* 73:250–53

142. Powell JR, Caccone A, Amato GD,

Yoon Y. 1986. Rates of nucleotide substitution in *Drosophila* mitochondrial DNA and nuclear DNA are similar. *Proc. Natl. Acad. Sci. USA* 83:9090–93

143. Powers DA, Smith M, Gonzalez-Villasenor I, DiMichele L, Crawford D, Bernardi G, Lauerman T. 1993. A multidisciplinary approach to the selection/neutralist controversy using the model teleost *Fundulus heteroclitus*. *Oxford Surveys in Evolutionary Biology*, ed. D Futuyma, J Antonovics, 9:82–157

144. Rand DM, Harrison RG. 1986. Mitochondrial DNA transmission genetics in crickets. *Genetics* 114:955–70

145. Rand DM, Harrison RG. 1989. Molecular population genetics of mtDNA size variation in crickets. *Genetics* 121:551–69

146. Rieseberg LH, Beckstrom-Sternberg S, Doan K. 1990. *Helianthus annuus* ssp. *texanus* has chloroplast DNA and nuclear ribosomal RNA genes of *Helianths debilis* ssp. *cucumerifolius*. *Proc. Natl. Acad. Sci. USA* 87:593–97

147. Reeb CA, Avise JC. 1990. A genetic discontinuity in a continuously distributed species: mitochondrial DNA in the American oyster, *Crassostrea virginica*. *Genetics* 124:397–406

148. Rehfeldt GE. 1982. Differentiation of *Larix occidentalis* populations from the northern Rocky Mountains. *Silvae Genetica* 31:13–19

149. Rehfeldt GE.1988. Ecological genetics of *Pinus contorta* from the Rocky Mountains (USA): a synthesis. *Silvae Genetica* 37:131–35

150. Rehfeldt GE. 1989. Ecological adaptations in Douglas-fir (*Pseudotsuga menziesii* var. *glauca*): a synthesis. *For. Ecol. Manage.* 28:203–15

151. Rehfeldt GE. 1993. Genetic variation in the *ponderosae* of the southwest. *Am. J. Bot.* 80:330–43

152. Rodhouse PG, Gaffney PM. 1984. Effect of heterozygosity on metabolism during starvation in the American oyster *Crassostrea virginica*. *Mar. Biol.* 80:179–88

153. Saghai-Maroof MA, Allard RW, Zhang Q. 1990. Genetic diversity and ecogeographical differentiation among ribosomal DNA alleles in wild and cultivated barley. *Proc. Natl. Acad. Sci. USA* 87:8486–90

154. Saghai-Maroof MA, Soliman KM, Jorgensen RA, Allard RW. 1984. Ribosomal DNA spacer-length polymorphisms in barley: Mendelian inheritance, chromosomal location, and population dynamics. *Proc. Natl. Acad. Sci. USA* 18:8014–18

155. Saikai RK, Gelfand DH, Stoffel S, Scharf SJ, Higuchi R, Horn GT, et al. 1988. Primer-directed enzymatic amplification of DNA with a thermostabile DNA polymerase. *Science* 239:487–91

156. Saunders NC, Kessler LG, Avise JC. 1986. Genetic variation and geographic differentiation in mitochondrial DNA of the horseshoe crab, *Limulus polyphemus*. *Genetics* 112:613–27

157. Schaal BA, O'Kane SL, Rogstad SH. 1991. DNA variation in plant populations. *Trends Ecol. Evol.* 6:329–33

158. Schuster WS, Alles DL, Mitton JB. 1989. Gene flow in limber pine: evidence from pollination phenology and genetic differentiation along an elevational transect. *Am. J. Bot.* 76:1395–403

159. Selander RK, Yang SY, Lewontin RC, Johnson WE. 1970. Genetic variation in the horseshoe crab (*Limulus polyphemus*) a phylogenetic "relic." *Evolution* 24:402–14

160. Singh SM. 1982. Enzyme heterozygosity associated with growth at different developmental stages in oysters. *Can. J. Genet. Cytol.* 24:451–58

161. Singh RS, Hale LR. 1990. Are mitochondrial DNA variants selectively nonneutral? *Genetics* 124:995–97

162. Snyder M, Fraser AR, LaRoche J, Gartner-Kepkay KE, Zouros E. 1987. Atypical mitochondrial DNA from the deep-sea scallop *Placopecten magellanicus*. *Proc. Natl. Acad. Sci. USA* 84:7595–99

163. Solignac MJ, Generemont J, Monnerot M, Mounolou JC. 1984. Genetics of mitochondria in *Drosophila*: inheritance in heteroplasmic strains of *D. mauritiana*. *Mol. Gen. Genet.* 197:183–88

164. Solignac MM, Monnerot M, Mounolou JC. 1983. Mitochondrial DNA heteroplasmy in *Drosophila mauritiana*. *Proc. Natl. Acad. Sci. USA.* 80:6942–46

165. Soltis DE, Soltis PS. 1989. Allopolyploid speciation in *Tragopogon*: insights from chloroplast DNA. *Am. J. Bot.* 76:1119–24

166. Soulé M. 1976. Allozyme variation: Its determinants in space and time. In *Molecular Evolution*, ed. FJ Ayala, pp. 60–77. Sunderland, MA: Sinauer

167. Strauss SH, Hong Y-P, Hipkins VD. 1993. High levels of population differentiation for mitochondrial DNA haplotypes in *Pinus radiata, muricata,* and *attenuata*. *Theor. Appl. Genet.* 85: 6065–71

168. Strauss SH, Neale DB, Wagner DB. 1989. Genetics of the chloroplast in conifers. *J. For.* 87:11–17

169. Deleted in proof

170. Sutton BCS, Flanagan DJ, Gawley JR, Newton CH, Lester DT, El-Kassaby YA. 1991. Inheritance of chloroplast and mitochondrial DNA in *Picea* and composition of hybrids from introgression zones. *Theor. Appl. Genet.* 82:242–48

171. Szmidt AE, Alden T, Hällgren J-E. 1987. Paternal inheritance of chloroplast DNA in *Larix. Plant Mol. Biol.* 9:59–64

172. Tajima F. 1983. Evolutionary relationships of DNA sequences in finite populations. *Genetics* 105:437–60

173. Thomas RH, Schaffner W, Wilson AC, Pääbo S. 1989. DNA phylogeny of the extinct marsupial wolf. *Nature* 340:465–67

174. Thomas WK, Pääbo S, Villiblanca FX, Wilson AC. 1990. Spatial and temporal continuity of kangaroo rat populations shown by sequencing mitochondrial DNA from museum specimens. *J. Mol. Evol.* 31:101–12

175. Turelli M, Ginzburg L. 1983. Should individual fitness increase with heterozygosity? *Genetics* 104:191–209

176. Wagner DB. 1992. Nuclear, chloroplast, and mitochondrial DNA polymorphisms as biochemical markers in population genetic analyses of forest trees. *New Forests* 6:373–90. Also in *Population Genetics of Forest Trees,* ed. WT Adams, SH Strauss, DL Copes, AR Griffin, pp. 373–390. Dordrecht: Kluwer Acad.

177. Wagner DB, Furnier GR, Saghai-Maroof MA, Williams SM, Dancik BP, Allard RW. 1987. Chloroplast DNA polymorphisms in lodgepole and jack pines and their hybrids. *Proc. Natl. Acad. Sci. USA* 84:2097–200

178. Wallace DC. 1989. Mitochondrial DNA mutations and neuromuscular disease. *Trends. Genet.* 5:9–13

179. Watt WB. 1983. Adaptation at specific loci. II. Demographic and biochemical elements in the maintenance of the *Colias* PGI polymorphism. *Genetics* 103:691–724

180. Watt WB, Cassin RC, Swan MS. 1983. Adaptation to a specific loci. III. Field behavior and survivorship differences among *Colias* PGI genotypes are predictable from in vitro biochemistry. *Genetics* 103:725–39

181. Westfall RD, Conkle MT. 1992. Allozyme markers in breeding zone designation. *New For.* 6:279–309. Also In *Population Genetics of Forest Trees,* ed. WT Adams, SH Strauss, DL Copes, AR Griffin, pp 279–309 (1992). Dordrecht: Kluwer Acad.

182. Whatley JM. 1982. Ultrastructure of plastid inheritance: green algae to angiosperms. *Biol. Rev.* 57:527–69

183. Wheeler NC, Guries RP. 1982. Population structure, genic diversity and morphological variation in *Pinus contorta* Dougl. *Can. J. For. Res.* 12:595–606

184. Whittemore AT, Schaal BA. 1991. Interspecific gene flow in sympatric oaks. *Proc. Natl. Acad. Sci. USA* 88:2540–44

185. Wills C. 1978. Rank-order selection is capable of maintaining all genetic polymorphisms. *Genetics* 89:403–14

186. Wills C. 1981. *Genetic Variability.* Oxford: Clarendon

187. Wolf KH, Li H-H, Sharp PM. 1987. Rates of nucleotide substitution vary greatly among plant mitochondrial, chloroplast, and nuclear DNAs. *Proc. Natl. Acad. Sci. USA* 88:9054–58

188. Wright S. 1931. Evolution in Mendelian populations. *Genetics* 16:97–159

189. Yeh FC. 1988. Isozyme variation of *Thuja plicata* (Cupressaceae) in British Columbia. *Biochem. Syst. Ecol.* 16:373–77

190. Yeh FC, El-Kassaby YA. 1980. Enzyme genetic variation in natural populations of Sitka spruce, (*Picea sitchensis* (Bong.) Carr.) I. Genetic variations patterns among trees from ten IUFRO provenances. *Can. J. For. Res.* 10:415–22

191. Yeh FC, Khalil MAK, El-Kassaby YA, Trust DC. 1986. Allozyme variation in *Picea mariana* from Newfoundland: genetic diversity, populations structure, and analysis of differentiation. *Can. J. For. Res.* 16:713–20

192. Yeh FC, Layton C. 1979. The organization of genetic variability in central and marginal populations of lodgepole pine, *Pinus contorta* ssp. *latifolia. Can. J. Genet. Cytol.* 21:487–503

193. Yeh FC, O'Malley D. 1980. Enzyme variations in natural populations of Douglas-fir *Pseudotsuga menziesii* (Mirb.) I. Genetic variation patterns in coastal populations. *Silvae Genetica* 29:83–92

194. Zhang Q, Saghi-Maroof MA, Allard RW. 1990. Effects on adaptedness of variations in ribosomal DNA copy number in populations of wild barley (*Hordeum vulgare* ssp. spontaneum). *Proc. Natl. Acad. Sci. USA* 87:8741–45

195. Zouros E, Foltz DW. 1987. The use of allelic isozyme variation for the study of heterosis. In *Isozymes: Current Topics in Biological and Medical Research,* ed. MC Rattazzi, JG Scandalios, GS Whitt, pp 2–59. Volume 13. New York: Alan R. Liss

196. Zouros E, Pogson GH. 1994. The present status of the relationship between

heterozygosity and heterosis. In *Genetics and Evolution of Aquatic Organisms*, ed. A Beaumont. London: Chapman & Hall. In press

197. Zouros E, Singh SM, Foltz DW, Mallet AL. 1983. Post-settlement viability in the American oyster (*Crassostrea virginica*): an overdominant phenotype. *Genet. Res. Camb.* 41:259–70

198. Zouros E, Singh SM, Miles HE. 1980. Growth rate in oysters: An overdominant phenotype and possible explanations. *Evolution* 34:856–67

Annu. Rev. Ecol. Syst. 1994. 25:71–96

UNISEXUAL FISH: Model Systems For Studying Ecology And Evolution

Robert C. Vrijenhoek

Center for Theoretical and Applied Genetics, Rutgers University New Brunswick, New Jersey 08903-0231

KEY WORDS: sex, asexuality, clones, coexistence, conservation

Abstract

Since their discovery in 1932, all-female "species" of fish have provided rich material for ecological and evolutionary studies. The significance of these rare organisms lies in the perspective they provide on what is considered normal (i.e. biparental sexuality). Study of these fish and their sexual relatives has contributed to our understanding of: (i) the origins and evolution of asexuality; (ii) the ecology of hybrids; (iii) genotypic and environmental effects on ecologically relevant traits; and (iv) the maintenance of sex in higher organisms. A consistent message emerging from these studies is the significance of genetic diversity for survival in spatially and temporally heterogeneous environments. Consequently, unisexual fish also serve as useful models for studying the role of genetic variation in the survival of small endangered populations. Finally, unisexual fish serve as genetically uniform systems for environmental and biomedical studies. The purpose of this review is to bring wider attention to the range of biological problems that can be attacked with these remarkable organisms.

INTRODUCTION

The first unisexual vertebrate discovered was the Amazon molly *Poecilia formosa,* named for the fabled tribe of all-female warriors (32). Based on morphological evidence, Carl and Laura Hubbs hypothesized that *P. formosa* was an interspecific hybrid. Their hypothesis was prophetic, because we sub-

0066-4162/94/1120-0071$05.00

sequently learned that all the known unisexual fish are hybrids, and probably all unisexual amphibia and squamate reptiles, as well (116). Recognition of a broader relationship between unisexuality, hybridization, and polyploidy led Schultz (83) to suggest that interspecific hybridization created an opportunity for emergence of nonrecombinant reproductive modes, which in turn were permissive to the evolution of polyploidy. We have since recognized 45 unisexual fish "biotypes," all of hybrid origin (116). Related biotypes typically differ in ploidy level and genomic composition. For example, Schultz (83) recognized three *Poeciliopsis* biotypes: diploid *P. monacha-lucida* (= $1n$ of *monacha* + $1n$ of *lucida*); and triploids *P. 2 monacha-lucida* (= $2n$ monacha + $1n$ *lucida*), and *P. monacha-2 lucida* (= $1n$ monacha + $2n$ *lucida*).[1] Genetically distinct clones of a particular biotype are assigned Roman numerals (e.g. *MML/I* and *MML/II*).

Some biotypes like *Poecilia formosa* were assigned formal species names before their hybrid composition was fully resolved. However, application of species names to unisexual organisms is problematic (47). Biotypes often contain multiple clones that had independent hybrid origins. Nevertheless, some taxonomists would assign species names to such biotypes, even though they are not monophyletic (13). Others would apply species names to each unisexual lineage that had an independent hybrid origin (21, 24). The potential for recognizing more unisexual "species" grows as we apply increasingly sensitive genetic tools; however, some techniques like mtDNA analysis or DNA fingerprinting do not necessarily reveal independent origins. It would be meaningless to name each newly discovered molecular variant. Overall, the hyphenated biotype designations recommended by Schultz (83) are more informative. Furthermore, including the hyphen disallows the use of these designations as formal species names and draws attention to the fact that these unisexuals are not biological species in the traditional sense. Echelle (21) provides a different view.

NONRECOMBINANT REPRODUCTIVE MODES

Among the vertebrates, true parthenogenesis occurs in unisexual lizards. Two, sperm-dependent, nonrecombinant modes of reproduction occur in fish and amphibia—gynogenesis and hybridogenesis (Figure 1).

Gynogenesis

Gynogenetic inheritance is strictly maternal. Sperm from males of a related sexual species are needed to activate embryogenesis in the unreduced ova

---

[1]For brevity, I frequently substitute abbreviations for the biotype names (*ML = P. monacha-lucida, MML = P. 2 monacha-lucida,* and *MLL = P. monacha-2 lucida*).

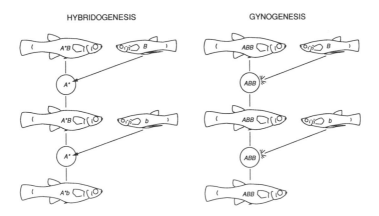

HYBRIDOGENESIS   GYNOGENESIS

Figure 1 Nonrecombinant reproductive modes in unisexual fish. During hybridogenesis, the hemiclonal A^* genome is transmitted to eggs. Although the paternal genome (B or b) is expressed, it is substituted in each generation. During gynogenesis, the entire genome is transmitted without recombination (e.g. the triploid genome ABB). Sperm is needed only to activate embryogenesis.

produced by gynogenetic females, but paternal genetic material is not incorporated or expressed. Cytogenetic details of gynogenesis have been examined in a variety of unisexual vertebrates. They have in common an endomitotic duplication of chromosomes just prior to a normal meiosis (17). Thus, crossing-over between replicated chromosomes is inconsequential, and inheritance is strictly clonal.

Hybridogenesis

Hybridogenetic females (A^*B) transmit a nonrecombinant A^* genome to each ovum; the B genome is discarded during oogenesis and replaced by mating with males of sexual host species BB (83). Paternal B genes are expressed in the hybrids but are not heritable. Only the A^* "hemiclonal" genome is transmitted between generations. Although hybridogenetic inheritance has been comprehensively investigated (1, 82, 115), cytogenetic details need further investigation. Apparently, A^* and B chromosome sets do not synapse and recombine in *Poeciliopsis,* because B chromosomes are isolated in a premeiotic cell division (11). Hybridogenesis also occurs in the European frog *Rana esculenta* (i.e. *ridibunda-lessonae*) (105) and the stick insect *Bacillus rossius-grandii* (55). All hybridogens share a process that discriminates between maternal and paternal chromosomes during oogenesis. How these chromosome sets are recognized in hybrid cytoplasm remains a mystery.

Parthenogens have fewer ecological constraints than sperm-dependent unisexuals. They can escape from competition with their sexual progenitors and play the role of fugitive species (48). In contrast, sperm-dependent gynogens and hybridogens must maintain intimate contact with a closely related sexual host. Despite the constraints of sperm-dependence, no parthenogenetic fish or amphibia are known. The requirements for initiation of embryogenesis must differ between the aquatic vertebrates (anamniotes) and lizards (amniotes) (104). Discovery of these initiating factors awaits investigation.

EVOLUTION OF UNISEXUAL FISH

Hybrid Origins

The hybrid origins of most unisexual vertebrates are well documented (18, and references therein). Interspecific hybridization often disrupts normal meiosis and leads to sterility. On rare occasions, novel cytological mechanisms that preclude synapsis between heterospecific chromosomes rescue egg production and lead to nonrecombinant reproduction (83, 129). Hybridization is the window that allows unisexuality to penetrate in vertebrates, but additional mechanisms can promote unisexuality in insects (98). The window for hybrid origins of unisexuality is narrow (59, 111). Sexual progenitors must be sufficiently different that meiosis is disrupted, and yet sufficiently similar that hybrids are viable and fertile.

No generalized level of interspecific divergence is associated with this window of opportunity. For example, genetic distances based on Nei's D (65) between the ancestors of *P. monacha-latidens, P. monacha-occidentalis,* and *P. monacha-lucida* are 0.50, 0.54, and 0.66, respectively (RC Vrijenhoek, unpublished). Hybrids between more closely related species like *P. monacha* X *P. viriosa* (D = 0.27) and *P. lucida* X *P. occidentalis* (D = 0.23) have normal meiosis (40). Yet, the presumed ancestors of *Menidia clarkhubbsi* (*M. berylinna-peninsulae*) are even more closely related (D = 0.14), and they produced a gynogen (22). Something other than genomic distance must be responsible for the meiotic disruptions that produced these unisexuals.

LABORATORY SYNTHESIS OF HYBRIDOGENS   Schultz (85, 86) was the first to synthesize a unisexual biotype in the laboratory. Crosses of *P. monacha* females X *P. lucida* males produced new *P. monacha-lucida* lineages that survive today. Reciprocal crosses produced no viable progeny. Schultz's experiments demonstrated that hybridogenesis is a spontaneous by-product of hybridization and not an evolved mechanism. Subsequent synthesis of 33 new *P. monacha-lucida* hemiclones corroborated this finding (124). These strains have been useful for studies of "frozen variation" (below). Hybridogenetic frogs and stick insects also have been synthesized in the laboratory (30, 55).

SYNTHESIS OF GYNOGENS   Cimino (11) used colcimide treatments to produce triploid embryos in *P. monacha-lucida* females, but unfortunately none were reared to maturity to test for gynogenesis. Attempts to synthesize gynogenetic *Poecilia formosa* through crosses of *P. mexicana* X *P. latipinna* produced male and female hybrids that either were sterile or had normal meiosis (102). Future attempts at synthesizing this biotype will require more detailed information on geographic variation in the putative ancestors (101, and BJ Turner, personal communication). To date, the only successful hybrid synthesis of a gynogen occurred with the homopteran insect genus *Mullerianella* (20).

Molecular Phylogenies

Studies of *Poeciliopsis* and lizards of the genus *Cnemidophorus* illustrate the power of comprehensive molecular studies to reveal unisexual origins. Allozymes can be used to identify the sexual ancestors, test for hybrid origins, establish clonal inheritance, and reveal clonal diversity. Analysis of mitochondrial DNA provides additional information because nuclear and cytoplasmic genes become permanently coupled once a unisexual lineage is formed (3). The rapid rate of mtDNA sequence evolution (5) also helps resolve recent evolutionary phenomena and reveal clonal diversity missed by allozyme studies (68).

POLYPHYLETIC AND PARAPHYLETIC ORIGINS   Matriarchal phylogenies of unisexual vertebrates can be revealed with analysis of mitochondrial DNA (3). Multiple hybrid events from a susceptible matriarchal lineage (i.e. paraphyletic origins) typify most unisexual lizards (*Cnemidophorus* and *Hemidactylus*), salamanders (*Ambystoma*), and the fish *Poecilia formosa* and *Poeciliopsis monacha-occidentalis*. Multiple hybrid events involving distinct matriarchal lineages (polyphyletic origins) typify the *Cnemidophorus tesselatus* complex, *Rana esculenta,* and the fish *Menidia clarkhubbsi* and *Poeciliopsis monacha-lucida*. Genetically diverse monophyletic unisexual lineages do not appear to be common.

The combination of allozyme and mtDNA data allowed discrimination among possible modes for triploid origins (Figure 2). Cuellar (15) espoused the spontaneous origins hypothesis. He argued that unisexual lizards arose from unreduced ova within a species, and that hybridization was a secondary process. If true, the homospecific *AA* genomes of an *AAB* triploid should be coupled with mtDNA from species *A*, and vice versa for *ABB*. Allozyme and mtDNA analysis of parthenogenetic *Cnemidophorus* refuted the spontaneous origins hypothesis (19, 61). Eight of ten biotypes did not exhibit the expected *AAB/mtA* coupling. Hybrid origins preceded triploidy.

Analyses of triploid *Poeciliopsis* also rejected the spontaneous origins hypothesis (70). Schultz (83) proposed that triploid *Poeciliopsis* arose by addition

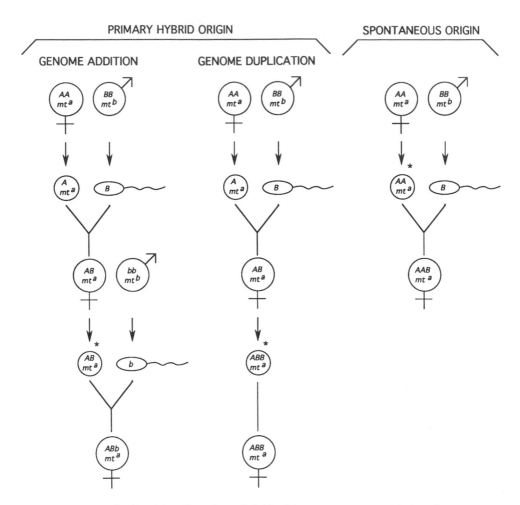

Figure 2 Triploid origins. The primary hybrid origin vs. spontaneous origin hypotheses are contrasted. The upper letters in each gamete or individual represent the nuclear genomes, and the lower letters represent the maternally transmitted mitochondrial (mt) DNA haplotype. Asterisks mark the point where reduction is suppressed. (Modified from Ref. 3).

of a third genome to diploid hybrids. Thus, an *AB* diploid that mates with species *BB* should produce triploids heterozygous for *B* alleles that segregate in the sexual ancestor (*AB* + b → ABb). However, duplication within an AB diploid should produce triploids that are homozygous for alleles at duplicated loci (*AB* → *ABB*). Heterozygosity for common ancestral alleles exists at homospecific loci in all clones of *P. monacha-2 lucida* and *P. 2 monacha-lucida* examined to date (70). The molecular data indicate that a single *ML* lineage produced both the *MML* and *MLL* triploids through genome addition. Allozyme studies of triploid *Poecilia formosa* also support the genome addition hypothesis (101a). The high diversity of triploid clones in the Rió Soto la Marina drainage of northeastern Mexico apparently results from additions of *mexicana* genomes to several diploid *Poecilia formosa* clones.

Longevity of Clones

Conservative interpretation of molecular data has led most students of unisexual vertebrates to suggest that clonal lineages arose recently (3). It is often impossible to determine whether novel alleles in unisexuals are a product of post-formational mutations and hence evolutionary longevity or just a consequence of incomplete sampling of the putative sexual ancestors. Nevertheless, two research groups independently reported that a mitochondrial lineage of unisexual salamanders of the genus *Ambystoma* might be four to five million years old (29, 94). However, these findings offer little insight regarding longevity of unisexual lineages. It is not surprising that a mitochondrial lineage might be old. Mitochondria have "existed with little recombination for 10^9 years" (52). The relevant issue is whether the nuclear genomes of unisexual *Ambystoma* have existed without recombination for such a long time. Thus, Spolsky et al (94) cautiously suggested that nuclear genome replacements from the sexual hosts might be important for the long-term survival of these unisexual *Ambystoma*. In at least one case (38), neither of the two sexual progenitors of nuclear genomes in an *Ambystoma* biotype was the mtDNA progenitor (e.g. *AAB/mt^c*. Clearly, nuclear replacements occur in *Ambystoma*.

The only comprehensive evidence for an old unisexual strain is provided by a monophyletic lineage of *Poeciliopsis monacha-occidentalis* (69). Multiple hybrid origins produced a diversity of *MO* hemiclones in southern Sonora. However, allozyme, tissue grafting, and mtDNA analyses identified a monophyletic lineage that colonized the northern rivers in a stepwise manner. Average mtDNA sequence divergence within the northern clade was 0.12%, and maximum divergence was 0.30%. Assuming mtDNA sequences diverge at the rate of 2% per million years (5), this lineage is probably between 60,000 and 150,000 years old. Maynard Smith (52) suggested that with respect to the longevity of most sexual species, this is "but an evening gone." Assuming that one unit of genetic distance represents approximately 18 MY of divergence (65), we can infer from allozyme studies that the sexual progenitors of unisexual *Poeciliopsis* have been in existence for from 4 to 12 MY (RC Vrijenhoek, unpublished). By these standards, our most ancient clone has led a relatively short existence. Notwithstanding, this age is relevant to the problem of "mutational meltdown" in clones (below).

The combined power of allozyme and mtDNA analysis for resolving questions about clonal origins and ages should draw attention to an unfortunate recent tendency to skip over old techniques in favor of newer, more fashionable ones. Phylogenies based strictly on single gene sequences risk confusing gene geneal-ogies with organismic phylogenies (2), and studies based only on mtDNA will fail to reveal hybrid origins and the longevity of nuclear genomic combinations.

THE PSEUDOGAMY PARADOX

Stable coexistence of unisexual sperm parasites and their sexual hosts is paradoxical (12). An all-female biotype should replace its host because sexual females must bear the cost of producing males. Yet, replacement of the sexual host also results in the demise of sperm-dependent unisexuals. Nevertheless, stable complexes involving pseudogamous biotypes exist in many taxa (97).[2]

Sperm Limitation

Moore & McKay (53, 58) described a frequency-dependent model for coex-istence in *Poeciliopsis*. Laboratory studies revealed that *P. lucida* males strongly prefer conspecific females over *P. monacha-lucida* as mates. Unisex-ual reproduction is sperm-limited in nature, but their mating success is fre-quency-dependent. When *P. lucida* is rare, males are solitary, and unisexuals fail to obtain sperm. As the sexuals increase in relative abundance, groups of males form dominance hierarchies. Then, unisexual mating success increases, because subordinate males participate in unisexual inseminations. However, a negative feedback exists between the relative abundance of unisexuals and their mating success. Computer simulations and analytical models revealed that this frequency-dependent process should result in equilibrium populations composed of 80 to 90% unisexual females (56, 58). Similarly detailed studies of behavior, population dynamics, and population modeling have not been undertaken with other unisexual fish.

Evolution of Sexual Mimicry

Wastage of sperm by promiscuous *P. lucida* males should favor evolution of improved mate discrimination, which, in turn, should favor unisexuals that match the behavior and morphology of sexual females (53). How does the excellent sexual mimicry of some *P. monacha-lucida* strains evolve? Genetic dissection of a Río Mocorito M^*L strain revealed that its M^* genome does not express *monacha* genital pigments, resulting in *lucida* mimicry (120). This mimicry was thought to result from silencing mutations in M^* genomes that control expression of *monacha*-like genital pigmentation. However, two lab-

---

[2]Hybridogens are not pseudogamous in the strict sense, but can be treated in the same theoretical context.

oratory synthesized *P. monacha-lucida* strains are good *lucida* mimics (43). Most of the synthetic strains are intermediate. Substantial variation in pigment expression can be "frozen" from the *P. monacha* gene pool. Frozen variation and interclonal selection may be sufficient to explain the evolution of sexual mimicry *P. monacha-lucida.*

Rare Clone Advantage

In the laboratory, some hemiclones enjoy a "rare clone advantage" with respect to mating success (34). Males of *P. lucida* learned to discriminate against a common hemiclone and favored mating with a newly introduced hemiclone. Thus, the probability of insemination in nature should be negatively correlated with clonal frequency, a testable prediction because these livebearing fish retain a record of their mating success in the number of embryos they carry relative to unfertilized eggs. If supported by field studies, the rare clone advantage hypothesis might help explain the paradox of coexistence among genetically similar clones.

Frequency-dependent sperm limitation was not observed in *Poecilia formosa* (3a, 31). However, the Moore/McKay model may not be necessary for stable coexistence in mixed reproductive complexes if niche overlap between the sexual host and sperm-dependent unisexual form is incomplete (36, 106, 107). Complete niche overlap with a sexual host is highly unlikely in unisexual vertebrates because the unisexuals are interspecific hybrids between species that might have very different niches (below).

Benefits to Sexual Males

It has been assumed that sexual males gain no benefit from mating with unisexual females, but a recent study of the sailfin molly, *Poecilia latipinna,* challenges that assumption (77). Apparently, males increase their attractiveness to conspecific females by "consorting" with unisexual *P. formosa.* It is unlikely that a similar advantage is gained by males of *Poeciliopsis,* because females do not seem to exhibit preferences with regard to males (53). However, males might benefit from the sexual experience gained by mating with unisexuals.

THE ECOLOGICAL NICHE OF HYBRID UNISEXUALS

Ecological Weeds

Several hypotheses exist for different ecological niches in unisexual-hybrids and their sexual progenitors. Wright & Lowe (132) argued that parthenogenetic *Cnemidophorus* are the equivalent of ecological weeds. True parthenogens should have a high rate of increase and exceptional colonization abilities, and thus, they can escape from competition with their sexual ancestors (48). Parthenogenetic lizards are most abundant in disturbed habitats and ecotones where the sexual progenitors are thought to be inferior competitors (15).

Intermediate Niche

Unisexual fish and amphibia are sperm-dependent and must coexist with a sexual host with which they are likely to share some ecological overlap. Moore (57) favored a variant of the weed hypothesis. He argued that hybrid unisexuals are inferior to the sexual progenitors in their respective niches but should be ideally suited for an intermediate niche. Thus, unisexuals should be most abundant in areas of overlap between geographical ranges of the sexual progenitors. Although consistent with broad geographical patterns of *P. monacha-occidentalis,* the intermediate niche hypothesis fails to explain local ecological distributions of *P. monacha, P. lucida,* and their diploid and triploid hybrids (75).

Heterosis and General Purpose Genotypes

The coupling between hybridization and unisexuality in the vertebrates suggested that unisexual organisms might be heterotic with respect to fitness (129). Schultz (84, 87, 88) suggested that unisexual heterosis might confer relatively high fitness in both parental niches plus the intermediate niche. Similarly, others have referred to unisexual organisms as "general purpose genotypes" with broad tolerance to temporal and spatial heterogeneity of the environment (67). Lynch (45) refined this hypothesis with respect to temporal variation—a general purpose genotype has the least variance in fitness over time and thus the highest geometrical mean fitness. Robust tolerance to extreme environmental conditions has been identified in two unisexual vertebrates. *Rana esculenta* appears to be more resistant to hypoxic stress than its sexual progenitors (100), and *P. monacha-lucida* has broader thermal tolerances than its progenitors. However, attributing enhanced tolerance to heterosis is questionable. Equally heterozygous *P. monacha-occidentalis* and *P. monacha-2 lucida* unisexuals did not show enhanced thermal tolerances relative to their sexual progenitors (6, 7). The broad thermal tolerance of some clones may be a product of multiple origins and interclonal selection. Studies of natural unisexuals cannot determine whether their characteristics are a spontaneous product of hybridity (i.e. heterosis) or a consequence of interclonal selection.

Examination of 33 laboratory synthesized *P. monacha-lucida* strains revealed no evidence for spontaneous heterosis (124). On average, the synthetic strains had lower survival and fertility, and higher frequencies of developmental defects than the sexual strains from which they derived. Only 14 of the 33 synthetic strains survived beyond the fourth laboratory generation. None of the synthetic strains exhibited enhanced fitness with respect to laboratory strains of *P. monacha* and *P. lucida.*

Variance in fitness among synthetic *P. monacha-lucida* strains results from the abilities of *monacha* genomes to combine with a *lucida* genome. However, natural *P. monacha-lucida* substitute a full range of *lucida* genomes produced by recombination. Successful hemiclones will be those that have the highest average fitness across these *lucida* combinations. Good combiners will be preserved as new hemiclones are frozen from the *P. monacha* gene pool. What has been attributed to heterosis in some hybridogens (6, 100) results from interclonal selection for the best combiners. Unlike the synthetic hemiclones, you do not see the poor combinations in nature; they are rapidly eliminated.

Frozen Niche-Variation

The experiments with synthetic hybridogens demonstrated that multiple hybrid origins can produce unisexuals with immense variation in ecologically relevant traits. In an under-exploited heterogeneous environment, interclonal selection should create a structured assemblage of clones that subdivide and efficiently exploit food and spatial resources (Figure 3) (107). Studies of unisexual *Poeciliopsis* revealed evidence for ecological diversification among coexisting clones. Clones of *P. 2 monacha-lucida* segregate along stream gradients (106). Clone *MML/II* is most abundant in highly productive downstream habitats and rare in less productive headwater pools. Clone *MML/I* occurs at a relatively low frequency and does not respond to the productivity gradient. Microhabitat preferences based on currents versus still-water pools also differ between these

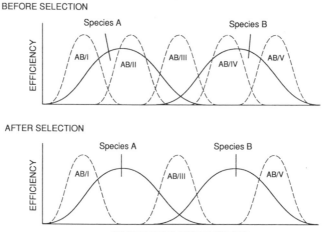

Figure 3   The Frozen Niche-Variation model. Before interclonal selection, multiple clones (*AB/I*, *AB/II*, ... *AB/V*) arising from independent hybridization events display a wide range of positions along a hypothetical resource axis. After selection, remaining clones exploit resources that are underutilized by the parental species.

clones and correspondingly affect their diets (75, 76). Strong dietary differences exist in the absence of spatial segregation of *P. monacha-lucida* hemiclones (123). In laboratory tests, hemiclone *ML/VIII* is a more effective predator than *ML/VII*. Examination of natural diets revealed that *ML/VIII* is a better insectivore, and *ML/VII* is predominantly a detritivore. Various clones of *Poeciliopsis* also differ in growth rate, fecundity, sexual aggressiveness, mate selection, and thermal tolerance (6, 7, 33, 34, 88). Similarly detailed ecological studies of other unisexual fish have not been performed. Examples of ecological differences between coexisting animal clones were reviewed elsewhere (113).

According to the "Frozen Niche-Variation" model, a unisexual population composed of multiple clones should use ecological resources more efficiently than a single clone. A survey of *Poeciliopsis* populations revealed that unisexual females typically outnumber sexual females when unisexual populations are multiclonal (107). In contrast, sexual females overwhelmingly predominate when unisexual populations are monoclonal. Differences frozen as a consequence of recurrent hybrid origins drive the ecological success of multiclonal assemblages. Genetically diverse unisexual populations have high local fitness and sufficient ecological breadth to displace the sexual ancestors from much of their ancestral niche.

Computer simulations of the frozen niche-variation model revealed that clonal invasion of a sexual population is nonrandom (122). New clones that freeze rare phenotypes are initially more successful than clones that freeze common phenotypes of the sexual ancestor. Clones invade from the ecological margins. This result is consistent with observations of food and spatial preferences of unisexual *Poeciliopsis*. For hybrid unisexuals, there are invasible margins to the ecological "right" and "left" of each sexual ancestor and in the space between them. Other models have shown that a diverse unisexual population should not completely displace its sexual ancestors unless the combined niche breadth of the multiclonal assemblage completely eclipses that of the sexuals (8).

Ecological Vacuums and Species Flocks

Why are the desert streams of northeastern Mexico home to unisexual *Poeciliopsis*? Asexual populations are more likely to be established in habitats that can be described as "ecological vacuums" or species depauparate environments with low interspecific competition (109). Frozen niche-variation will have its greatest effect in these places, creating new clones that can exploit under-utilized resources. Asexual lineages may arise as frequently in other regions, but species-packed environments may leave no room for additional subdivision.

The ecological diversification observed in unisexual fish populations provides a model for the evolution of species flocks. Differences among recently derived clones of *Poeciliopsis* reveal the potential for diversification in a sexual species. Cloning is a rapid way to fix divergent genotypes and increase niche breadth (74). Open recombination within a sexual species will retard diversification. Density- and frequency-dependent selection (i.e. soft selection) in which intermediate phenotypes have lower fitness can help to stabilize polymorphisms (131), but recombination recreates the intermediates. Ecological and behavioral factors that help to maintain genetic structure in a population can thwart the "tyranny of recombination" (73). For example, positive assortative mating based on habitat associations may lead to stable diversification and potentially to sympatric speciation in sexual species (49).

GENOTYPIC VARIANCE IN LIFE HISTORY TRAITS

Heritable variation is the raw material for evolution by natural selection. Nonetheless, determining genetic contributions to variation in ecologically relevant traits has been a difficult problem. To determine heritability in a sexual species, researchers must rely on correlations between known relatives or depend on controlled crosses. It is easier to dissect genotypic and environmental components of variance in clones. Variation within a clone is environmental, and the mean difference between clones raised in a common environment is genetic.

Synthetic Hemiclones

Poeciliopsis provide rich material for studies of genotypic variance in life-history traits. Under laboratory conditions, natural clones of *Poeciliopsis* differ substantially in growth rate and fecundity (88, 89). Were these differences frozen from the sexual gene pool or did they accumulate from post-formational mutations? This question motivated our efforts to synthesize *P. monacha-lucida* strains in the laboratory (126). The synthetic M^*L strains differed only with respect to their M^* genomes. We standardized the paternal genomes by mating each hemiclone with an isogenic *P. lucida* strain (L^i). All variation among the synthetic hemiclones could be attributed to the *monacha* genome frozen at the time of its origin.

Significant genetic variance in life-history traits was observed among 14 synthetic hemiclones (125, 126). The genetic component of variance was 47% for birth size, 42% for birth weight, and 15% for growth rate. In each case, more than half of this genetic variance could be attributed to differences between the haploid eggs produced by individual *P. monacha* females that founded the hemiclonal strains. Genetic differences in the timing of the first brood were largely a function of the river of origin of the *P. monacha* founder. Schultz (88) identified differences in brood size among synthetic *P. monacha-*

lucida strains. Ongoing studies with synthetic hemiclones are revealing genetic variance for predatory efficiency, innate predatory escape behavior of newborn progeny, and aggressiveness (NRW Lima, personal communication). The sexual ancestor of these synthetic hemiclones, *P. monacha,* obviously has substantial genetic variation for ecologically relevant traits. However, the results of these experiments cannot be used to estimate heritabilities. Phenotypic differences between hemiclones are products of whole genotypes (e.g. additive, dominance, and epistatic factors) interacting with the environment. Furthermore, some differences between hemiclones might reflect novel interactions between *monacha* and *lucida* genomes that could not be expressed in either sexual ancestor.

Reaction Norms

The adaptive value of phenotypic plasticity versus rigidity has been a controversial topic among students of life-history evolution (96). Unisexual fish provide simpler models for studying plasticity, because clonal genotypes can be replicated and raised in alternative environments. Only a few studies have examined phenotypic reaction norms in unisexual fish (89, 125). Genotype-by-environment interactions affect growth rates of clones reared under different diets and temperatures. Research in this area has only begun.

THE MAINTENANCE OF SEX

The paradox of sex has generated considerable debate (54, and references therein). An asexual lineage that avoids the costs of biparental sex should rapidly replace its sexual ancestors (50, 130). Yet, such replacements have not occurred in taxa that regularly give rise to asexual lineages. Even sperm-limited unisexual fish have not completely replaced their non-host sexual progenitors. Why does biparental sex predominate so overwhelmingly in the animal kingdom? The paradox of sex is based on a syllogism with two premises. The first premise assumes "all else is equal" between sexual and asexual lineages (50). The second premise assumes that there is a two-fold cost of biparental sex. Interested readers should consult comprehensive reviews of theories and evidence regarding the origins and maintenance of sex (4, 50, 54, 130). My purpose is to show how unisexual fish have served as models for evaluating the first assumption.

Ecological Equivalence?

Is it reasonable to assume that niche requirements, fertility, and survival are identical for sexual and asexual lineages? This assumption is undoubtedly wrong for unisexual vertebrates that are hybrids. However, if asexual lineages arise spontaneously within a sexual species, as in some insects (98), the sexual and asexual individuals may be similar. Nevertheless, computer simulations of clonal invasion revealed that clones are more successful initially if they freeze rare or marginal phenotypes (122). Regardless of their niche positions, character variances of sexual and asexual lineages will not be similar. An outcrossing sexual lineage is composed of an unlimited number of genotypes, and a clonal lineage is composed of one. The short-term benefits of sex are believed to result from this genotypic variance.

RESOURCE HETEROGENEITY MODELS   Ghiselin (25) suggested that a genetically diverse sexual population should be more efficient than a single genotype at extracting resources from a heterogeneous environment. This idea was fundamental to development of the Frozen Niche-Variation and Tangled Bank models (4, 107, 109). Variance in resource utilization can be partitioned into within- and between-phenotype components (74). If genotypic variance contributes significantly to the between-phenotype component, a sexual lineage should have greater niche breadth than a single clone.

Gynogenetic clones of *Poeciliopsis* exhibit reduced morphological variance when compared with sexual cognates (110). Hybridogens, which substitute paternal genomes, exhibit greater variance than gynogens, but less than sexuals. Although unisexual *Poeciliopsis* are highly heterozygous, their reduced phenotypic variance is not a product of enhanced developmental stability. Their ability to control random perturbations of development (fluctuating asymmetry) is no greater than that of the sexual cognates (118). Monoclonal *Cnemidophorus* populations also exhibit less character variance than multiclonal populations, which in turn are less variable than sexual populations (66). Comparisons of phenotypic variance in sexual and asexual taxa are complicated by cryptic clonal diversity that often exists in asexual populations. If within- and between-clone components of variance are not identified, the total variance of a multiclonal asexual population may match that of the sexual cognate (109).

Morphological variance is not easily related to niche breadth, however. Few attempts have been made to directly compare within- and between-genotype components of niche breadth in sexual and asexual taxa. The sexual lizard *Cnemidophorus tigris* and a coexisting clone of *C. sonorae* have similar within-individual components of dietary breadth, but sexuals have a wider between-individual component (9). *P. monacha,* the maternal ancestor of all *Poeciliopsis* clones, exhibits wider food diversity than individual clones, but within- and between-individual components of dietary breadth were not partitioned (76, 106, 123). A laboratory study also revealed that *P. monacha* has broader food use than related asexual lineages (121). In this limited environment, the between-individual component of resource use did not differ between cognates, but sexuals had a broader within-individual component. Additional field and experimental studies of niche breadth in well-defined sexual and asexual cognates are needed to resolve this issue. Nevertheless, given what we already know, I suspect it is unsafe to assume that individual sexual and asexual lineages have equivalent niche breadths.

THE RED QUEEN   According to the Red Queen hypothesis, genetic diversity produced by sexual reproduction is essential for species locked into coevolutionary struggles with biological enemies (4). Presumably, microparasites evolve rapidly to focus their attacks on the most common host phenotype. Rare host phenotypes have a temporary advantage that leads to frequency-dependent fitness and the maintenance of polymorphism (28). A common clone should have a higher parasite load than a rare clone, which in turn should have a higher load than a sexual cognate. Also, a common clone with no variance in immune response should have lower variance in number of parasites than its sexual cognate. Both predictions were supported by a study of *P. monacha* and two gynogenetic clones (44). Parthenogenetic lizards also appear to have higher parasite loads than sexual cognates (60). Ladle (39) criticized these studies because the hybrid nature of unisexual vertebrates might confer greater susceptibility to parasites. This argument is incorrect for *Poeciliopsis*. Rare clones were not more heavily parasitized than their sexual cognates, and parasite loads in the sexuals were variable, depending on whether or not the population was inbred (14). Frequency-dependent fitnesses resulting from Red Queen processes may provide the most powerful explanation for the maintenance of sex and genetic variability on ecological time-scales. Clearly, more field and experimental studies of parasite loads in sexual and asexual taxa are warranted.

Evolutionary Dead Ends

Asexual species are often considered evolutionary dead ends because of their presumed genetic inflexibility (16, 50, 128). Among vertebrates and insects, only 0.1% to 0.2% of species are strictly asexual (4). This rarity suggests a "mutation/selection-like" balance (111). New asexual lineages arise infrequently and go extinct rapidly. Extant asexual "species" are little more than scattered twigs at the tips of major phylogenetic branches (51). Except for bdelloid rotifers, asexual lineages have not speciated and diversified into rich asexual clades (95). Severe constraints on the origins of asexual lineages have been reviewed elsewhere (99, 111). Briefly, mutations or hybridizations that disrupt the meiotic process tend to cause sterility. Only rare accidents circumvent the recombinational process and simultaneously rescue egg production. Once formed, a new asexual lineage must find a suitable niche. Unoccupied niches are unlikely in species-packed environments but might exist in species-poor environments or ecological vacuums. Having passed these hurdles, a new asexual lineage will have trouble tracking environmental change and dealing with biotic challenges like parasites and competitors.

MULLER'S RATCHET   Genetic decay provides a final challenge to the persistence of clones. Muller (63) suggested that mutations would accumulate like a "ratchet mechanism" in asexual lineages. Recombination in sexual lineages produces offspring with higher and lower mutational loads than the parents, and purifying selection effectively maintains a low load. An asexual population cannot reduce its load below that of the "least loaded" clone. If by chance that clone is lost, the load has increased one step. Excluding back mutations, it cannot be reduced. For organisms like *Drosophila melanogaster,* as many as 35% of gametes may carry a new mutation with small viability effects (62). For organisms with larger genomes, the genomic mutation rate (U, a product of the per gene mutation rate and the number of genes per gamete) may be greater than 1. Assuming $U = 1$, the Poisson probability of producing an egg with no new mutations, e^{-U}, is 0.37. The rest of the eggs (63%) would have one or more new mutations, and most mutations are slightly deleterious (91). The probability of producing mutant-bearing eggs would be higher in permanently diploid ($e^{-2U} = 0.86$) and triploid asexual lineages ($e^{-3U} = 0.95$).

The ratchet mechanism of Muller is based on finite populations, independent gene effects, and weak selection. Random loss of the least loaded clone is a product of the number of gametes (N) and the intensity of selection (s) against deleterious mutations. Thus, the probability of ratchetting forward one notch equals $1 - Ne^{-U/s}$ (27). Polygenic mutations and genetic drift can result in "mutational meltdown" of small populations (10, 46). If accumulating mutations act synergistically, the ratchet will move forward very rapidly even in large asexual populations (37).

Hybridogenetic fish provided one of the few models for testing the mutational accumulation hypothesis. According to Lynch & Gabriel (46), only 10,000 to 100,000 generations are sufficient to cause the meltdown of clonal lineages. One *P. monacha-occidentalis* lineage has persisted at least that long, and it has accumulated several mutations in its *monacha* genome (69). A terminal branch of this lineage has a unique muscle-specific creatine kinase allozyme (Ck-A^c) and a silent carboxylesterase (Est-5°). The Est-5° allele produces a protein of normal size and charge that cross-reacts with antibody directed against "wild-type" Est-5 (93). A mutation must have altered its active site. A branch of this lineage is marked by a silent alcohol dehydrogenase (Adh^o). These mutations probably have no deleterious effects because they are permanently sheltered by wild-type alleles carried by the substitutable *occidentalis* genome.

Nei (64) predicted that mutations should accumulate in sheltered nonrecombinant genomes, even if the mutations confer a slight disadvantage in heterozygotes. His intent was to explain the reduction in size and gene content of Y-chromosomes in many unrelated taxa. A recent study of *Drosophila* revealed that only 35 generations were necessary to cause a significant accumulation of deleterious mutations in sheltered nonrecombinant chromosomes (72). Earlier studies of hybridogenetic *Poeciliopsis* provided the first evidence for mutational decay in sheltered genomes (41, 42). Fourteen natural hemiclones were genetically dissected to reveal the contents of their *monacha* genomes. Each strain was artificially inseminated with sperm from males of a *P. monacha* strain (M^+M^+). This cross places the hemiclonal M^* genome into a homospecific background (M^*M^+). After repeated attempts, only 7 of the 14 strains produced viable progeny. The M^* genomes of the nonproductive strains were altered to such a degree they were no longer compatible with a normal *P. monacha* genome. The ovaries of females from one strain contained advanced degenerating embryos. Six strains produced no discernible embryos, even though artificial insemination is highly effective with *P. monacha* sperm (84% success rate). The seven nonproductive strains carried developmental lethals that were recessive in the M^*L genetic background and dominant in the M^*M^+ background. These mutations were not frozen from the *P. monacha* gene pool, because dominant lethals should be exceedingly rare. The mutations accumulated subsequent to the formation of these hybridogenetic lineages.

The M^*M^+ progeny from several productive crosses exhibited gross developmental abnormalities (vertebral fusions, incomplete cranial caps, and fin creases), and juvenile mortality was high. Nevertheless two strains produced male M^*M^+ progeny that could be backcrossed to reveal recessive lethal genes. Mendelian segregation in a male that carries a single lethal gene in its hemiclonal genome (M^-) would deliver the lethal to 50% of its gametes. Backcross to the parental hemiclone would result in 50% mortality of progeny because they are homozygous for the lethal gene (M^-L X $M^-M^+ \rightarrow 0.5M^-M^- + 0.5M^-M^+$). If two lethal mutations exist, 75% of the progeny will die. If three exist, 88% will die, etc. A backcross to the *P. monacha* strain was used as a control for embryonic mortality. We estimated a minimum of two lethal equivalent mutations in hemiclone *ML/Va*, and four in *ML/IIIb* (a lethal equivalent is the expected number of lethals necessary to produce the observed embryonic mortality). We could not determine whether this mortality was due to a few discrete lethals or many diffusely spread semilethals. Notwithstanding, hemiclones *ML/Va* and *ML/IIIb* are probably among the least loaded hemiclones. At least they produced viable M^-M^+ progeny. Half of the hemiclonal genomes could not even be brought to that point. If most hybridogenetic strains are less than 100,000 years old, the extraordinary decay of these nuclear genomes requires explanation. Perhaps the ineffectiveness of purifying selection in sheltered genomes is sufficient to explain this decay rate, but other dysgenic processes may be involved. The mobilization of transposable elements is a likely candidate in such hybrids (35).

Experimental crosses with *Rana esculenta* also revealed deleterious recessive mutations in hemiclonal genomes (26). This problem should be examined in hemiclonal stick insects of the genus *Bacillus* (55). The genomes of gynogenetic and parthenogenetic species cannot be dissected to reveal mutational loads. However, comparisons of coding sequences and the load of transposable elements in sexual and asexual cognates might be informative.

The permanent association of divergent nuclear genomes in gynogenetic or parthenogenetic hybrids creates opportunities to study concerted evolution. In his review of the evolutionary genetics of *Poecilia formosa,* Turner (101) prophetically asked, "do selfish sequences of one of the parental species" take over' the genome?" The answer appears to be yes. Nuclear ribosomal DNAs (rDNAs) of several clones of the parthenogenetic lizard *Heteronotia binoei* express the rDNA pattern of only one of the two parental species, apparently a consequence of biased gene conversion (29a). Biased gene conversion and unequal crossing-over may drive such multi-copy genes to the homozygous state, subsequent to the hybrid origin of clonal lineasges. Clearly, this possibility needs to be explored in unisexual fish. It would be of great interest to compare gynogens that retain meiotic synapsis with hybridogens that apparently preclude synapsis and therefore offer no opportunity for unequal crossing-over.

CONSERVATION BIOLOGY

The foundations of conservation biology lie in theoretical ecology and population genetics, buttressed with observations of captively bred and naturally occurring small populations (23). Time for controlled studies typically does not exist when a crisis is declared and management plans are required for an endangered species. Studies of model organisms are needed to test many of the assumptions of conservation biology (92). Fish of the genus *Poeciliopsis* serve both as models and targets of species management plans. Although this work has been reviewed in greater detail (114), I focus on the role that unisexual organisms play in this model system for studying the ecological consequences of losing genetic diversity.

Endangered Populations

Arizona populations of *P. occidentalis* are federally listed as endangered (103). A survey of genetic diversity in remnant populations of this sexually reproducing fish revealed low variation in most Arizona populations and recommended that efforts to restore *P. occidentalis* throughout its former range use a genetically variable natural stock to replace the homozygous hatchery stock in use at that time (117). A subsequent laboratory analysis of fitness in these stocks revealed that the more heterozygous stock had higher survival, fecundity, growth rate, and developmental stability than did the homozygous strain (71). A genetically variable population with higher fitness should be a better choice for restocking a heterogeneous array of reclaimed habitats.

Unfortunately, such studies cannot ascertain whether heterozygosity at a few allozyme loci was involved (directly or indirectly) in fitness differences among these fish stocks, or whether it was just a spurious correlate. No historical data on inbreeding or population crashes existed for these populations. However, long-term studies (23 years, to date) of small, partially isolated populations of *Poeciliopsis monacha* provide significant insight. A benefit of studying *P. monacha* is that it coexists with gynogenetic clones that can be treated as genetic controls. The *monacha* complex lives in highly fluctuating desert streams of southern Sonora, Mexico. Although theoretical considerations lead us to believe that adaptively neutral variation should be lost under such conditions, *P. monacha* maintains relatively high levels of protein polymorphism (108). Balancing selection was demonstrated in this species, but the actual focus of selection (allozymes or associated linkage groups) was not determined (119).

Founder Events and Fitness Loss

Rapid loss of heterozygosity following a local extinction/recolonization event provided a unique opportunity to examine the fitness consequences of genetic diversity. Low heterozygosity in the founder population of *P. monacha* following recolonization was associated with reduced developmental stability, reduced tolerance to hypoxic stress, and increased parasite load relative to the local clones (44, 114, 118). Clones are protected against heterozygosity loss during extinction/recolonization events. Neighboring downstream *P. monacha* populations that retained heterozygosity during the study period did not exhibit similar reductions in relative fitness.

Fitness reduction in the founder population altered its dynamic relationship with coexisting gynogenetic clones. Prior to the extinction/recolonization event, the sexual population had genetic variability and constituted about 80% of the fish in the complex. After recolonization, the homozygous founder population constituted less than 5% of the fish, and that relationship persisted for five years, about 15 generations. Then, I introduced genetic variability into the founder populations with a small transplant of *P. monacha* females from a site less than 1 km downstream. In two years, the *P. monacha* population recovered its variability and again constituted about 80% of the fish. No comparable shifts in unisexual/sexual abundance or genetic diversity occurred at downstream sites.

Loss of variability during the founder event, and probably subsequent inbreeding, altered the ability of *P. monacha* to compete with the clones. If it weren't for sperm-dependence, *P. monacha* would have been driven to extinction in this headwater stream. It was temporarily extinct in some adjacent pools. The small transplant of genetic diversity into the founder population restored its ability to live and compete with the local gynogenetic clone (112). It also restored a lower parasite load relative to that of the local clone (44). The general significance of these findings should be apparent to conservationists dealing with habitat loss, population reductions, and the introduction of exotic parasites and competitors.

NEW DIRECTIONS

Much of what is known about carcinogenic substances in our environment is based on studies of mammalian models, primarily rodents. Increased attention is being paid to aquatic organisms as models for the study of water-borne carcinogens. Unfortunately, few genetically defined strains of aquatic vertebrates exist for such studies. Aquatic toxicology laboratories typically rely on the fathead minnow, *Pimephales promelus,* as a test organism, but defined strains are not available and geographical variation undoubtedly exists in this wide-spread species. Genetically defined strains of unisexual fish provide better models for such studies. Being viviparous, they are easy to raise, and isogenic strains are available without resorting to inbreeding and its potentially deleterious consequences. Chemically induced liver cancers have been studied in inbred sexual and hybrid unisexual strains of *Poeciliopsis* (78–81). Other studies of *Poeciliopsis* have explored the heat stress response at the molecular, cellular, and whole animal levels (90, 127). The stage has been set for using these well-characterized animals in environmental studies. Only time will tell if they are widely adopted.

ACKNOWLEDGEMENTS

This paper is dedicated to Professor RJ Schultz (University of Connecticut, Storrs) in the year of his retirement. The development of unisexual fish as model systems for ecological and evolutionary studies owes much of its success to his early discoveries, to his continuing enthusiasm for research, and to his dedication to the perpetuation and distribution of genetically defined fish strains as biological resources. I thank Bruce J Turner for his helpful comments and suggestions for the manuscript. Any omissions are my responsibility. This is contribution D-2-67175-6-94 of the New Jersey Agricultural Experiment Station and #94-19 of the Institute of Marine and Coastal Sciences, supported by State funds and NSF grants (BSR88-05351 and OCE93-02205).

Literature Cited

1. Angus RA, Schultz RJ. 1979. Clonal diversity in the unisexual fish *Poeciliopsis monacha-lucida*: a tissue graft analysis. *Evolution* 33:27–40
2. Avise JC. 1989. Gene trees and organismal histories: A phylogenetic approach to population biology. *Evolution* 43:1192–1208
3. Avise JC, Quattro JM, Vrijenhoek RC. 1992. Molecular clones within organismal clones: Mitochondrial DNA phylogenies and the evolutionary histories of unisexual vertebrates. In *Evolutionary Biology,* ed. M Hecht, B Wallace, R MacIntyre, 26:225–246. New York: Plenum
3a. Balsano JS, Randle EJ, Rach EM, Monaco PJ. 1985. Reproductive behavior and the maintenance of all-female *Poecilia. Poecilia. Env. Biol. Fish.* 12:251–63
4. Bell G. 1982. *The Masterpiece of Nature: the Evolution and Genetics of Sexuality.* Berkeley: Univ. Calif. Press
5. Brown WM, George M, Wilson AC. 1979. Rapid evolution of animal mitochondrial DNA. *Proc. Natl. Acad. Sci. USA* 76:1967–71
6. Bulger AJ, Schultz RJ. 1979. Heterosis and interclonal variation in thermal tolerance in unisexual fish. *Evolution* 33: 848–59
7. Bulger AJ, Schultz RJ. 1982. Origins of thermal adaptation in northern vs southern populations of a unisexual hybrid fish. *Evolution* 36:1041–50
8. Case ML, Taper TJ. 1986. On the coexistence and coevolution of asexual and sexual competitors. *Evolution* 40:366–87
9. Case T. 1990. Patterns of coexistence in sexual and asexual species of *Cnemidophorus* lizards. *Oecologia* 83: 220–27
10. Charlesworth D, Morgan MT, Charlesworth B. 1993. Mutation accumulation in finite populations. *J. Hered.* 84:321–25
11. Cimino MC. 1972. Egg production, polyploidization and evolution in a diploid all-female fish of the genus *Poeciliopsis. Evolution* 26:294–306
12. Clanton W. 1934. An unusual situation in the salamander *Ambystoma jeffer-*

sonianum (Green). *Occas. Pap. Mus. Zool. Mich.* 290:1–15
13. Cole CJ. 1985. Taxonomy of parthenogenetic species of hybrid origin. *Syst. Zool.* 34:359–63
14. Craddock C, Vrijenhoek RC, Lively CM. 1993. Fish hybrids and the Red Queen. *Trends Ecol. Evol.* 8:458
15. Cuellar O. 1977. Animal parthenogenesis. *Science* 197:837–43
16. Darlington CD. 1939. *The Evolution of Genetic Systems.* Cambridge, Eng: Cambridge Univ. Press
17. Dawley RM. 1989. An introduction to unisexual vertebrates. See Ref. 18, pp 1–18
18. Dawley RM, Bogart JP, ed. 1989. *Evolution and Ecology of Unisexual Vertebrates, Bulletin 466.* Albany: New York State Mus.
19. Densmore LD III, Moritz CC, Wright JW, Brown WM. 1989. Mitochondrial-DNA analyses and the origin and relative age of parthenogenetic lizards (genus *Cnemidophorus*). IV. Nine *sexlineatus*-group unisexuals. *Evolution* 43:969–83
20. Drosopoulis S. 1978. Laboratory synthesis of a pseudogamous triploid "species" of the genus *Mullerianella* (Homoptera, Delphacidae). *Evolution* 32:916–20
21. Echelle A. 1990. Nomenclature and non-Mendelian ("clonal") vertebrates. *Syst. Zool.* 39:70–78
22. Echelle AA, Echelle AF, Crozier CD. 1983. Evolution of an all-female fish, *Menidia clarkhubbsi* (Atherinidae). *Evolution* 37:722–84
23. Frankel OH, Soulé ME. 1981. *Conservation and Evolution.* Cambridge, UK: Cambridge Univ. Press
24. Frost DR, Wright JW. 1988. The taxonomy of uniparental species, with special reference to parthenogenetic *Cnemidophorus* (Squamata: Teiidae). *Syst. Zool.* 37:200–209
25. Ghiselin MT. 1974. *The Economy of Nature and the Evolution of Sex.* Berkeley: Univ. Calif. Press
26. Graf JD, Polls-Pelaz M. 1989. Evolutionary genetics of the *Rana esculenta* hybrid complex. See Ref. 18, pp 289–302

27. Haigh J. 1978. The accumulation of deleterious genes in a population: Muller's ratchet. *Theor. Popul. Biol.* 14: 251–57

28. Hamilton WD. 1982. Pathogens as causes of genetic diversity in their host populations. In *Population Biology of Infectious Diseases,* ed. RM Anderson, RM May, pp 269–296. New York: Springer

29. Hedges SB, Bogart JP, Maxson LR. 1992. Ancestry of unisexual salamanders. *Nature* 356:708–10

29a. Hillis DM, Moritz C, Porter CA, Baker RJ. 1991. Evidence for biased gene conversion in concerted evolution of ribosomal DNA. S 251:308–10

30. Hotz H, Mancino G, Bucci-Innocenti S, Ragghianti M, Berger L, et al. 1985. *Rana ridibunda* varies geographically in inducing clonal gametogenesis in interspecies hybrids. *J. Exper. Zool.* 236: 199–210

31. Hubbs C. 1964. Interactions between a bisexual fish species and its gynogenetic sexual parasite. *Univ. Tex. Mem. Mus. Bull.* 8:1–72

32. Hubbs CL, Hubbs LC. 1932. Apparent parthenogenesis in nature, in a form of fish of hybrid origin. *Science* 76:628–30

33. Keegan-Rogers V, Schultz RJ. 1984. Differences in courtship aggression among six clones of unisexual fish. *Anim. Behav.* 32:1040–44

34. Keegan-Rogers V, Schultz RJ. 1988. Sexual selection among clones of unisexual fish (*Poeciliopsis,* Poeciliidae): genetic factors and rare-female advantage. *Am. Nat.* 132:846–68

35. Kidwell MG, Kidwell JF, Sved JA. 1977. Hybrid dysgenesis in *Drosophila melanogaster* a syndrome of aberrant traits including mutation, sterility, and male recombination. *Genetics* 86:813–33

36. Kirkendall LR, Stenseth NC. 1987. Ecological and evolutionary stability of sperm-dependent parthenogenesis: effects of partial niche overlap between sexual and asexual females. *Evolution* 44:698–714

37. Kondrashov AS. 1988. Deleterious mutations and the evolution of sexual reproduction. *Nature* 336:435–40

38. Kraus F, Miyamoto MM. 1990. Mitochondrial genotype of a unisexual salamander of hybrid origin is unrelated to either of its nuclear haplotypes. *Proc. Natl. Acad. Sci. USA* 87:2235–38

39. Ladle R. 1992. Parasites and sex: catching the Red Queen. *Trends Ecol. Evol.* 7:405–8

40. Leslie JF. 1982. Linkage analysis of seventeen loci in the poeciliid fish genus *Poeciliopsis. J. Hered.* 73:19–23

41. Leslie JF, Vrijenhoek RC. 1978. Genetic dissection of clonally inherited genomes of *Poeciliopsis* I. Linkage analysis and preliminary assessment of deleterious gene loads. *Genetics* 90:801–11

42. Leslie JF, Vrijenhoek RC. 1980. Consideration of Muller's ratchet mechanism through studies of genetic linkage and genomic compatibilities in clonally reproducing *Poeciliopsis. Evolution* 34: 1105–15

43. Lima NRW. 1993. *Magnitude of genotypic diversity and phenotypic variability of morphological and behavioral characters in sexual and unisexual* Poeciliopsis *(Teleost: Poeciliidae).* PhD thesis. Federal Univ. São Carlos, São Carlos, Brazil

44. Lively CM, Craddock C, Vrijenhoek RC. 1990. The Red Queen hypothesis supported by parasitism in sexual and clonal fish. *Nature* 344:864–66

45. Lynch M. 1984. Destabilizing hybridization, general-purpose genotypes and geographical parthenogenesis. *Q. Rev. Biol.* 59:257–90

46. Lynch M, Gabriel W. 1990. Mutational load and the survival of small populations. *Evolution* 44:1725–37

47. Maslin TP. 1968. Taxonomic problems in parthenogenetic vertebrates. *Syst. Zool.* 17:219–31

48. Maslin TP. 1971. Parthenogenesis in reptiles. *Am. Zool.* 11:361–80

49. Maynard Smith J. 1966. Sympatric speciation. *Am. Nat.* 100:637–50

50. Maynard Smith J. 1978. *The Evolution of Sex.* Cambridge, UK: Cambridge Univ. Press

51. Maynard Smith J. 1986. Contemplating life without sex. *Nature* 324:300–1

52. Maynard Smith J. 1992. Age and the unisexual lineage. *Nature* 356:661–62

53. McKay FE. 1971. Behavioral aspects of population dynamics in unisexual-bisexual *Poeciliopsis* (Pisces: Poeciliidae). *Ecology* 52:778–90

54. Michod RM, Levin BR. 1988. *The Evolution of Sex: An Examination of Current Ideas.* Sunderland, Mass: Sinauer

55. Montovani B, Scali V. 1992. Hybridogenesis and androgenesis in the stick-insect *Bacillus rossius-grandii benazzii* (Insecta, Phasmatodea). *Evolution* 46:783–96

56. Moore WS. 1975. Stability of unisexual-bisexual populations of *Poeciliopsis* (Pisces: Poeciliidae). *Ecology* 56:791–808

57. Moore WS. 1984. Evolutionary ecology of unisexual fishes. In *Evolutionary Ge-*

netics of Fishes, ed. BJ Turner, pp 329–98. New York: Plenum

58. Moore WS, McKay FE. 1971. Coexistence in unisexual-bisexual species complexes of *Poeciliopsis* (Pisces: Poeciliidae). *Ecology* 52:791–99
59. Moritz C, Brown WM, Densmore LD, Wright JW, Vyas D, et al. 1989. Genetic diversity and the dynamics of hybrid parthenogenesis in *Cnemidophorus* (Teiidae) and Heteronotia (Gekonidae). See Ref. 18, pp 87–112
60. Moritz C, McCallum H, Donnellan S, Roberts JD. 1991. Parasite loads in parthenogenetic and sexual lizards (*Heteronotia binoei*): support for the Red Queen hypothesis. *Proc. R. Soc. Lond. B.* 224:145–49
61. Moritz CC, Wright JW, Brown WM. 1989. Mitochondrial-DNA analyses and the origin and relative age of parthenogenetic lizards (genus *Cnemidophorus*). III. *C. velox* and *C. exsanguis*. *Evolution* 43:958–68
62. Mukai T, Chigusa ST, Mettler LE, Crow JF. 1972. Mutation rate and dominance of genes affecting viability in *Drosophila melanogaster*. *Genetics* 72:335–55
63. Muller HJ. 1964. The relation of mutation to mutational advance. *Mutation Res.* 1:2–9
64. Nei M. 1970. Accumulation of nonfunctional genes in sheltered chromosomes. *Am. Nat.* 104:211–22
65. Nei M. 1972. Genetic distance between populations. *Am. Nat.* 106:283–92
66. Parker ED. 1979. Phenotypic consequences of parthenogenesis in *Cnemidophorus* lizards. I. Variability in parthenogenetic and sexual populations. *Evolution* 33:1150–66
67. Parker ED, Selander RK, Hudson RO, Lester LJ. 1977. Genetic diversity in colonizing parthenogenetic cockroaches. *Evolution* 31:836–42
68. Quattro JM, Avise JC, Vrijenhoek RC. 1991. Molecular evidence for multiple origins of hybridogenetic fish clones (Poeciliidae: *Poeciliopsis*). *Genetics* 127:391–98
69. Quattro JM, Avise JC, Vrijenhoek RC. 1992. An ancient clonal lineage in the fish genus *Poeciliopsis* (Atheriniformes: Poeciliidae). *Proc. Natl. Acad. Sci. USA* 89:348–52
70. Quattro JM, Avise JC, Vrijenhoek RC. 1992. Mode of origin and sources of genotypic diversity in triploid fish clones (*Poeciliopsis*: Poeciliidae). *Genetics* 130:621–28
71. Quattro JM, Vrijenhoek RC. 1989. Fitness differences among remnant populations of the Sonoran topminnow, *Poeciliopsis occidentalis*. *Science* 245:976–78
72. Rice WR. 1994. Degeneration of a non-recombining chromosome. *Science* 263:230–32
73. Rosenzweig ML. 1978. Competitive speciation. *Biol. J. Linn. Soc.* 10:275–89
74. Roughgarden J. 1972. Evolution of niche width. *Am. Nat.* 106:683–718
75. Schenck RA, Vrijenhoek RC. 1986. Spatial and temporal factors affecting coexistence among sexual and clonal forms of *Poeciliopsis*. *Evolution* 40:1060–70
76. Schenck RA, Vrijenhoek RC. 1989. Coexistence among sexual and asexual forms of *Poeciliopsis*: foraging behavior and microhabitat selection. See Ref. 18, pp 39–48
77. Schlupp I, Marler C, Ryan MJ. 1994. Benefit to male sailfin mollies of mating with heterospecific females. *Science* 263:373–74
78. Schultz ME, Kaplan LAE, Schultz RJ. 1989. Initiation of cell proliferation in livers of the viviparous fish *Poeciliopsis lucida* with 7,12-Dimethylbenz[a]anthracene. *Env. Res.* 48:248–54
79. Schultz ME, Schultz RJ. 1982. Diethylnitrosamine-induced hepatic tumors in wild vs. inbred strains of a viviparous fish. *J. Hered.* 73:43–48
80. Schultz ME, Schultz RJ. 1982. Induction of hepatic tumors with 7,12-dimethylbenz[a]anthracene in two species of viviparous fishes (genus *Poeciliopsis*). *Environ. Res.* 27:337–51
81. Schultz ME, Schultz RJ. 1985. Transplantable chemically-induced liver tumors in the viviparous fish *Poeciliopsis*. *Exp. Mol. Pathol.* 42:320–30
82. Schultz RJ. 1966. Hybridization experiments with an all-female fish of the genus *Poeciliopsis*. *Biol. Bull.* 130:415–29
83. Schultz RJ. 1969. Hybridization, unisexuality and polyploidy in the teleost *Poeciliopsis* (Poeciliidae) and other vertebrates. *Am. Nat.* 103:605–19
84. Schultz RJ. 1971. Special adaptive problems associated with unisexual fishes. *Am. Zool.* 11:351–60
85. Schultz RJ. 1973. Origin and synthesis of a unisexual fish. In *Genetic and Mutagenesis of Fish*, ed. JH Schroeder, pp 207–211. New York: Springer
86. Schultz RJ. 1973. Unisexual fish: Laboratory synthesis of a "species." *Science* 179:180–81
87. Schultz RJ. 1977. Evolution and ecology of unisexual fishes. *Evol. Biol.* 10:277–331
88. Schultz RJ. 1982. Competition and adaptation among diploid and polyploid

clones of unisexual fishes. In *Evolution and Genetics of Life Histories*, ed. H Dingle, JP Hegmann, pp 103–119. Berlin: Springer

89. Schultz RJ, Fielding E. 1989. Fixed genotypes in variable environments. See Ref. 18, pp 32–38

90. Schultz RJ, Kaplan LAE, Schultz ME. 1993. Heat induced liver cell proliferation in the livebearing fish *Poeciliopsis*. *Env. Biol. Fish.* 36:83–91

91. Simmons MJ, Crow JF. 1977. Mutations affecting fitness in *Drosophila* populations. *Annu. Rev. Genet* 11:49–78

92. Spielman D, Frankham R. 1993. Modeling problems in conservation genetics using captive *Drosophila* populations: improvement of reproductive fitness due to immigration of one individual into small partially inbred populations. *Zool. Biol.* 11:343–48

93. Spinella DG, Vrijenhoek RC. 1982. Genetic dissection of clonally inherited genomes of *Poeciliopsis*. II. Investigation of a silent carboxylesterase allele. *Genetics* 100:279–86

94. Spolsky CM, Phillips CA, Uzzell T. 1992. Antiquity of clonal salamander lineages revealed by mitochondrial DNA. *Nature* 356:706–8

95. Stanley SM. 1975. Clades versus clones in evolution: Why we have sex. *Science* 190:382–83

96. Stearns SC, Koella JC. 1986. The evolution of phenotypic plasticity in life history traits: Predictions of reaction norms for age and size at maturity. *Evolution* 40:893–913

97. Stenseth NC, Kirkendall LR, Moran N. 1985. On the evolution of pseudogamy. *Evolution* 39:294–307

98. Suomalainen E, Saura A, Lokki J. 1987. *Cytology and Evolution in Parthenogenesis*. Boca Raton, Fla: CRC

99. Templeton AR. 1982. The prophecies of parthenogenesis. In *Evolution and Genetics of Life Histories*, ed. H Dingle, JP Hegmann, pp. 75–102. Berlin: Springer

100. Tunner HG, Nopp H. 1979. Heterosis in the common European water frog. *Naturwissenschaften* 66:268–69

101. Turner BJ. 1982. The evolutionary genetics of a unisexual fish. In *Mechanisms of Speciation*, ed. C Barigozzi, pp 265–305. New York: Liss

101a. Turner BJ, Balsano JS, Monaco PJ, Rasch EM. 1983. Clonal diversity and evolutionary dynamics in a diploid-triploid breeding complex of unisexual fishes (*Poecilia*). *Evolution* 37:798–809

102. Turner BJ, Brett BL, Miller RR. 1980. Interspecific hybridization and the evo-

lutionary origin of a gynogenetic fish, *Poecilia formosa*. *Evolution* 34:917–22

103. US Fish and Wildlife Service. 1983. *Gila and Yaqui Topminnow Recovery Plan*. Albuquerque, NM: US Fish & Wildlife Serv.

104. Uzzell TM Jr. 1970. Meiotic mechanisms of naturally occurring unisexual vertebrates. *Am. Nat.* 104:433–45

105. Uzzell TM, Berger L. 1975. Electrophoretic phenotypes of *Rana lessonae*, *Rana ridibunda*, and their hybridogenetic associate *Rana esculenta*. *Proc. Acad. Nat. Sci. (Phila.)* 127:13–24

106. Vrijenhoek RC. 1978. Coexistence of clones in a heterogeneous environment. *Science* 199:549–52

107. Vrijenhoek RC. 1979. Factors affecting clonal diversity and coexistence. *Am. Zool.* 19:787–97

108. Vrijenhoek RC. 1979. Genetics of a sexually reproducing fish in a highly fluctuating environment. *Am. Nat.* 113: 17–29

109. Vrijenhoek RC. 1984. Ecological differentiation among clones: the frozen niche variation model. In *Population Biology and Evolution*, ed. K Wöhrmann, V Loeschcke, pp 217–231. Heidelberg: Springer

110. Vrijenhoek RC. 1984. The evolution of clonal diversity in *Poeciliopsis*. In *Evolutionary Genetics of Fishes*, ed. BJ Turner pp 399–429. New York: Plenum

111. Vrijenhoek RC. 1989. Genetic and ecological constraints on the origins and establishment of unisexual vertebrates. See Ref. 19, pp 24–31

112. Vrijenhoek RC. 1989. Genotypic diversity and coexistence among sexual and clonal forms of *Poeciliopsis*. In *Speciation and Its Consequences*, ed. D Otte, J Endler, pp. 386–400. Sunderland, Mass: Sinauer

113. Vrijenhoek RC. 1990. Genetic diversity and the ecology of asexual populations. In *Population Biology and Evolution*, ed. K Wöhrmann, S Jain, pp 175–197. Berlin: Springer

114. Vrijenhoek RC. 1994. Genetic diversity and fitness in small populations. In *Conservation Genetics*, ed. V Loeschcke, J Tomiuk, S Jain. Berlin: Birkhauser. In press

115. Vrijenhoek RC, Angus RA, Schultz RJ. 1978. Variation and clonal structure in a unisexual fish. *Am. Nat.* 112:41–55

116. Vrijenhoek RC, Dawley RM, Cole CJ, Bogart JP. 1989. A list of known unisexual vertebrates. See Ref. 18, pp 19–23

117. Vrijenhoek RC, Douglas ME, Meffe GK. 1985. Conservation genetics of en-

dangered fish populations in Arizona. *Science* 229:400–2

118. Vrijenhoek RC, Lerman S. 1982. Heterozygosity and developmental stability under sexual and asexual breeding systems. *Evolution* 36:768–76

119. Vrijenhoek RC, Pfeiler E, Wetherington J. 1992. Balancing selection in a desert stream-dwelling fish, *Poeciliopsis monacha*. *Evolution* 46:1642–57

120. Vrijenhoek RC, Schultz RJ. 1974. Evolution of a trihybrid unisexual fish (*Poeciliopsis*, Poeciliidae). *Evolution* 28:205–319

121. Weeks SC. 1991. *The short term advantage of sexual reproduction: theoretical and experimental tests.* PhD thesis. Rutgers Univ., NJ

122. Weeks SC. 1993. The effects of recurrent clonal formation on clonal invasion patterns and sexual persistence: A Monte Carlo simulation of the frozen niche variation model. *Am. Nat.* 141:409–27

123. Weeks SC, Gaggiotti OE, Spindler KP, Schenck RE, Vrijenhoek RC. 1992. Feeding behavior in sexual and clonal strains of *Poeciliopsis*. *Behav. Ecol. Sociobiol.* 30:1–6

124. Wetherington JD, Kotora KE, Vrijenhoek RC. 1987. A test of the spontaneous heterosis hypothesis for unisexual vertebrates. *Evolution* 41:721–31

125. Wetherington JD, Schenck RA, Vrijenhoek RC. 1989. Origins and ecological success of unisexual *Poeciliopsis*: the Frozen Niche Variation model. In *The Ecology and Evolution of Poeciliid Fishes,* ed. GA Meffe, FF Snelson Jr., pp 259–276. Englewood Cliffs, NJ: Prentice Hall

126. Wetherington JD, Weeks SC, Kotora KE, Vrijenhoek RC. 1989. Genotypic and environmental components of variation in growth and reproduction of fish hemiclones (*Poeciliopsis*: Poeciliidae). *Evolution* 43:635–45

127. White C, Hightower L, Schultz R. 1994. Variation in heat shock proteins among species of desert fishes (Poeciliidae, *Poeciliopsis*). *Mol. Biol. Evol.* 11:106–19

128. White MJD. 1973. *Animal Cytology and Evolution.* Cambridge: Cambridge Univ. Press. 3rd ed.

129. White MJD. 1978. *Modes of Speciation* San Francisco, Calif: Freeman

130. Williams GC. 1975. *Sex and Evolution.* Princeton, NJ: Princeton Univ. Press

131. Wilson DS. 1989. The diversification of single gene pools by density- and frequency-dependent selection. In *Speciation and its Consequences,* ed. D Otte, JA Endler, pp 366–385. Sunderland, Mass: Sinauer

132. Wright JW, Lowe CH. 1968. Weeds, polyploids, parthenogenesis and the geographical and ecological distribution of all-female species of *Cnemidophorus. Copeia* 1968:128–38

Annu. Rev. Ecol. Syst. 1994. 25:97–126

COOPERATION AND CONFLICT IN THE EVOLUTION OF SIGNAL INTERACTIONS

Michael D. Greenfield

Department of Entomology, University of Kansas, Lawrence, Kansas 66045

KEY WORDS: chorusing, rhythm, sexual communication, sexual competition, synchrony

---

Abstract

Various invertebrate and vertebrate species in which males produce acoustic or bioluminescent signals for long-range sexual advertisement exhibit collective patterns of temporal signal interactions. These patterns range from simple concentrations of signaling during a narrow diel interval to synchronous and alternating interactions entailing precisely timed phase relationships between neighboring individuals. Signals involved in synchrony and alternation are generally produced with rhythms that are under the control of central nervous oscillators. Neighboring individuals effect these interactions via mutual phase delays or phase advances of their oscillators or actual changes in the free-running periods of their oscillators. Both synchrony and alternation may represent adaptations to avoid spiteful behavior or to maximize the ability of a local group to attract females or evade natural enemies. Alternatively, these collective patterns may represent incidental outcomes of competition between males jamming each other's signals. The neural mechanisms that effect signal jamming can be selected for by critical psychophysical factors such as precedence effects. Additional competitive pressures that may generate synchrony, alternation, and other collective patterns of signal interaction include mutual assessment of rivals, evasion of detection by dominant individuals, disruption of communication within courting pairs, and narrowness of the time intervals during which receptive females are present.

INTRODUCTION

In many animal species in which males produce long-range sexual advertisement signals, population densities are often high enough for signaling neighbors to perceive one another. Under such conditions it is common for males to interact mutually by adjusting the timing of their signals. These interactions occur primarily in acoustic (3, 148) and bioluminescent signaling (26, 101), and they are best known among arthropods, anurans, and birds. Massive choruses of periodical cicadas in North America (4), dawn and evening choruses of various birds and other acoustic animals (81, 158), and synchronous flashing at firefly (lampyrid beetle) aggregation trees in Southeast Asia (15) count among the more impressive signal interactions, which one author (153, p. 331) has referred to as "great spectacles of the living world." Similar phenomena, perhaps less apparent to human observers, are found in numerous other species, and they involve varying degrees of temporal precision.

Because signal interactions are nearly always associated with individuals in aggregations, the question of their function in the context of social behavior arises (3, 17, 63, 112). Accordingly, the major thrust of this review is coverage of the various hypotheses and studies on the evolution of signal interactions, with emphases on the cooperative and competitive functions that these interactions may serve. To assess these hypotheses and data adequately, though, a proximate understanding of the neuroethological mechanisms controlling signal generation and perception is usually necessary: In many cases the collective pattern of signal interactions can be reduced to a summation of inter-individual stimuli and responses (3), and such processes are likely to be physiologically constrained in some fashion. Therefore, mechanisms are treated as well to afford a more thorough review and evaluation of the evolutionary issues. Fortunately, a wealth of information on controlling mechanisms exists, because most of the initial interest in signal interactions came from neuroethologists. Only recently, however, have explicit evolutionary questions concerning chorusing and related bioluminescent phenomena been posed. Nonetheless, some of these current studies have explicated interactive displays in surprising ways, and in doing so they have revealed critical aspects of sexual communication and sexual selection that would otherwise have remained unrecognized.

FORMATS OF INTERACTIVE DISPLAY

Levels of Temporal Precision

The intent of this section is not to provide a rigid framework with which to categorize various formats, but rather to introduce the assorted phenomena to be considered, arranged in a sequence that progresses toward increasing temporal precision of the interactions.

In a crude fashion, males in many species adjust their activity on a diel basis so that they signal at the same time as their neighbors. This type of collective display has been referred to as a "spree" (145), the temporal equivalent of a (spatial) lek. Such concentrations of signaling within a restricted time interval usually occur close to photoperiodic transitions: e.g. dawn and dusk choruses in birds (33, 81), cicadas (158), and acridid grasshoppers (58), calling during the beginning of the night in anurans and various orthopteran insects (92, 145), and twilight flashing in lampyrid beetles (87). Unlike all other signal interactions, these collective displays do not necessarily entail mutual perception of signals. The concurrent signaling that characterizes them could be effected by specific, invariant responses to photoperiodic or other environmental cues (32, 58, 87).

At a higher level of precision, acoustic signalers may engage in "unison bout singing." Here, individuals within a "hearing radius" all sing collectively for several seconds to a few minutes, remain silent for a variable interval, and then repeat the cycle many times during the diel activity period (2, 45, 55, 127, 152). In some cases, unison bout singing may involve more extensive groups of individuals (3) among which collective signaling is facilitated by a "chain-reaction effect." Within these singing bouts, "leaders," individuals who habitually sing first, can sometimes be distinguished (23, 55, 152). Comparable events involving chain reactions are known in bioluminescent signalers as well (14).

Among arthropods and anurans that produce rhythmic acoustic or bioluminescent signals, more specialized temporal interactions sometimes occur in which the signals of neighbors are related by a given phase angle (2, 3, 26, 49, 148). The most striking phenomena, generally termed *synchrony* and *alternation*, occur when phase angles approximate 0° and 180°, respectively. Intermediate phenomena and combinations of synchrony and alternation exist in some species. For example, the participants in a synchronous chorus may only partially overlap their songs (46, 131, 138), and alternating choruses may be punctuated by occasional occurrences of synchrony (2, 78, 130). As above, consistent leaders, individuals whose rhythm is slightly advanced relative to their neighbors, may exist among synchronizers (97, 124). Alternating species too may include distinctive individuals, those who tend to disrupt the regular pattern by bouts of rapid signaling (23, 79, 130, 157). Signal interactions can also involve the hierarchical nesting of one or more of the above levels within another; e.g. evening choruses and unison bouts may overlie synchrony or alternation (49).

I have chosen to limit this review to interactions involving only the long-range advertisement signals of conspecifics. Consequently, some well-known phenomena such as close-range male-female courtship dialogues found in avian dueting and in the flash communication of lampyrid beetles, heterospecific interactions, typically involving unilateral inhibition, and jamming avoidance responses of electrolocating gymnotiform fish are not considered here.

Communication Channels

Review of the literature suggests that while signal interactions are widespread among arthropod and vertebrate taxa, they are primarily restricted to certain channels: sound and light (bioluminescence). This may not reflect mere coincidence or sampling bias. Possibly, only acoustic and nocturnal bioluminescent signals are normally transmitted over sufficient distances that neighboring signalers can detect them (39). In fact, the use of alternative channels such as substrate vibration has been interpreted in some cases as an evolutionary shift to avoid heterospecific or conspecific "eavesdroppers" (9, 102). Even if intensity exceeds detection thresholds, signal:noise ratios and contrast in other communication channels, (diurnal) reflected light, for example, may not be high enough to select for signal interactions. That is, temporal adjustments by individual signalers may be quite inconspicuous, and conspecific and heterospecific receivers would remain uninfluenced by any alternating or synchronous event.

Additionally, sound and bioluminescence, unlike olfactory signals, are transmitted nearly instantaneously; they rise above perceptual threshold levels suddenly and do not fade out slowly upon termination; and their sources may be localized by a distant receiver (39). The last factor ensures that the signals of individuals retain some integrity and avoid being completely lost in a group's blended emission. Therefore, the fundamental reasons that might favor precise adjustments in signal timing relative to that of a particular neighbor(s) could only be compelling for signalers using these two channels.

RHYTHM GENERATION AND INTERACTIVE ALGORITHMS

Neural Oscillators

The precise timing and phase relationships in alternating and synchronous choruses and bioluminescent displays in arthropods and anurans occur in species that produce signals rhythmically. A variety of models have been proposed to account for the general regulation of timing in animals (e.g. see 54). Regulation of the rapid rhythms (period $< \sim 10$ sec) under consideration here, however, is probably best described by a neural oscillator model. This model assumes that a pacemaker in the central nervous system (CNS) is responsible for the regular periodicity of effector activity (26, 30, 78) and that the pacemaker continues whether or not the signaler perceives its own signal

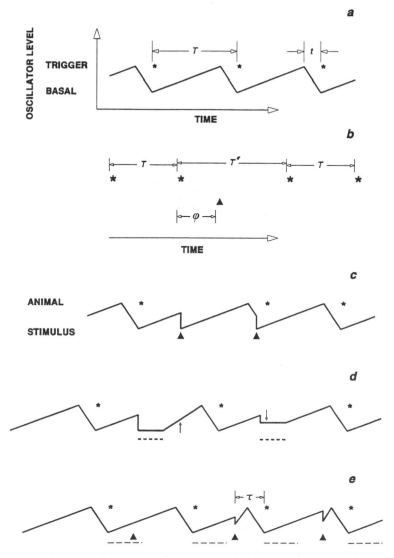

Figure 1 Timing of central nervous oscillators responsible for rhythmic signals in animals. (*a*) Free-running oscillator showing signal (*), signal period (*T*), effector delay (*t*), and basal and trigger levels of the oscillator. (*b*) Temporal relationships in a rhythmic signal regulated by a phase delay mechanism. ▲ is the stimulus, ϕ is the stimulus delay, (ϕ/T) • 360° is the stimulus phase, and ((T'-T)/T) • 360° is the response phase. (*c*) Phase delay resetting of oscillator by stimulus (adapted from ref. 18). (*d*) Inhibitory resetting of oscillator by a lengthy stimulus (adapted from ref. 62). After inhibition, return of oscillator to trigger level is faster due to either a steeper slope (↑) or incomplete resetting (↓). (*e*) Modification of phase-delay resetting common in anurans. A refractory period (- - - - -), during which the animal is immune to resetting, follows the signal. After the refractory period, the oscillator is reset by stimuli, but then returns very rapidly to the trigger level. τ is the rebound interval of the oscillator.

or even produces a signal. These criteria are generally met: Rhythmic effector activity can be elicited by arhythmic electrical stimulation of specific brain loci (74), perceptually deprived animals continue to signal rhythmically (18, 19), and harmonics may be conspicuous in the frequency distribution of signal periods of solo individuals (18). The harmonics are periods approximately n times the modal or preferred length (T), and they represent omission of effector output $n-1$ times in succession.

Under oscillator timing, signal onset necessarily occurs a short time interval (t; $t \ll T$) after the signal has been triggered by the CNS pacemaker (Figure 1a). Thus, t is an effector delay, and its length is constrained by the velocity of neural transmission and the duration of effector activation. Inferred measurements of t indicate values ranging from 50 to 200 ms (14, 16, 144). Rhythms tend to drift above and below the mean signaling rate to some extent. Such fluctuations may conceivably result from variation in T or t. A general statistical method for discerning the source of rhythmic fluctuation has been applied to lampyrid beetle flash rhythms, and the results implicated variation in T as the source (19). Spontaneous phase shifting, an abrupt change in the length of a single period, also occurs in the rhythms of individuals signaling in solo (147). Period lengthening may be more common here, possibly due to upper limits on CNS pacemaker rates that preclude excessive period shortening.

Homoepisodic vs Proepisodic Mechanisms

Various rhythmic arthropods and anurans interact with remarkable precision considering their high signaling rates. In acoustic Orthoptera and Hemiptera, synchrony may occur between neighbors calling as fast as 5 sec^{-1} (47, 50, 52). Because of the effector delay t discussed above, such interactions are all the more intriguing: How is it possible for individuals to adjust their timing given the constraints on effector response imposed by the CNS pacemaker?

Timing adjustments may be effected in either a homoepisodic or a proepisodic fashion (144). In the first case, an individual detects the signal of a neighbor and responds by producing its own signal. Thus, a synchronous event would be generated by a nearly immediate response to the concurrent signal of a neighbor(s). This sort of mechanism must be responsible for the synchronous onsets inherent in unison bout singing, because these bouts do not recur rhythmically. However, homoepisodic mechanisms cannot explain interactions involving many rhythmic signals: The time interval between the signal onsets of synchronizing neighbors is often shorter than a reasonable estimate for t (16, 144), and in acoustic signaling it may even be shorter than the duration of time required for sound to travel between the individuals.

Proepisodic mechanisms involve a response to a previous signal(s) of a neighbor that allows the focal male's signal to be produced at a given phase angle with respect to the neighbor's concurrent signal. Among synchronizing signalers, two primary types of proepisodic mechanisms have been identified: "phase delay synchrony" and "perfect synchrony" (14, 43). These mechanisms differ according to the degree of phase shifting, whether the full timing adjustment occurs immediately during one signal period or gradually over many, and whether the signaler adjusts its endogenous (free-running) rhythm. Both mechanisms were first described in synchronizing insects (66, 144), but recent findings and reinterpretations of earlier data indicate that modified versions also exist in alternating insects and in anurans.

Phase Delay Mechanisms

PHASE DELAY SYNCHRONY   Experiments with *Pteroptyx cribellata,* a Melanesian lampyrid which produces a 20-ms flash at 1 sec^{-1}, provided a detailed analysis of phase delay synchrony (18). A male presented with a single isolated flash stimulus, after one of his own signals, delayed his subsequent signal by a phase angle (response phase) equivalent to the phase angle (stimulus phase) between that first signal and the stimulus (Figure 1b). If the flash stimulus was presented immediately prior to the male's signal, however, this signal remained unaffected, but his next signal was advanced by an amount (response phase) equal to the stimulus phase. Presumably, the first signal had already been triggered prior to the stimulus.

The above responses may be summarized via a phase response curve (PRC) in which response phase is regressed against stimulus phase (62, 133, 144). PRCs obtained for *P. cribellata* have a slope ≈ 1 and pass through the origin (Figure 2a). A parsimonious explanation for this PRC is that (*a*) the CNS pacemaker ascends slowly from the basal to the trigger level, (*b*) the pacemaker descends rather steeply—but not instantaneously—to the basal level after being triggered, (*c*) certain external stimuli instantaneously reset the pacemaker to the basal level, and (*d*) the pacemaker resumes its free-running rhythm after being reset (Figure 1c). Thus, a stimulus such as the flash of a neighbor generates a positive phase shift (phase delay) if the stimulus phase is greater than 0° but less than approximately 315° (the transition phase) and a negative phase shift if between approximately −45° and 0° (after the CNS pacemaker has been triggered, but prior to effector output). If two neighboring males both signal at similar rates that fluctuate little, phase delays mutually align their rhythms within a single period and synchrony ensues thereafter. When the rates differ greatly (by $> (T{-}t)^{-1} : T^{-1}$; 1.14:1 in the above case), however, the faster individual signals with its free-running rhythm while the slower one is repeatedly reset to the basal level before its signals are triggered. At a given level of intra- and inter-individual variation in signaling rate, species with transition phases closer to 360° would be expected to synchronize less regularly and exhibit more frequent interruptions by alternation or solo signaling.

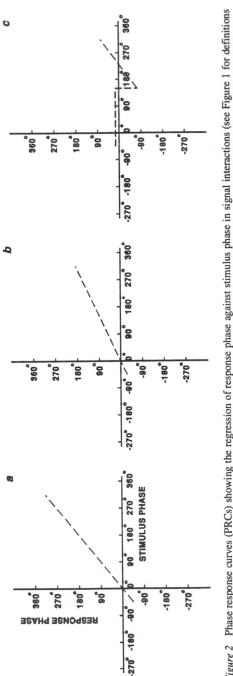

Figure 2   Phase response curves (PRCs) showing the regression of response phase against stimulus phase in signal interactions (see Figure 1 for definitions of terms and variables). (*a*) PRC found in synchrony generated by phase delay resetting (adapted from Refs. 14 and 665). PRC slope = 1 in the special case where $T = T + \phi$. (*b*) PRC found in alternation generated by phase delay resetting (adapted from Ref. 60). (*c*) PRC found in anurans (and possibly some insects) characterized by post-signal refractory periods followed by resetting with very rapid oscillator returns (adapted from Ref. 60; also see 88 and 91).

INHIBITORY RESETTING   Comparable studies on various acoustic insects and anurans have revealed that the above phase delay mechanism, with several modifications, is fairly widespread. This work has been augmented significantly by recent advances in digital technology that allow acoustic signals to be synthesized and/or extensively edited in the time domain and then broadcast to signaling individuals in controlled playback experiments. In several synchronizing species of acoustic Orthoptera such playback experiments have investigated responses to regular repetitions of call stimuli and single isolated stimuli, and resetting and PRCs with slopes between 0.8–1.0 and transition phases at 220°-300° have been inferred from the results (60, 62, 144). The species tested have signals of variable length, and this feature led to determination that a neighbor's signals, acting as external stimuli, not only reset a male's rhythm but may inhibit him as well. This modification has been termed "inhibitory resetting" (62).

   In inhibitory resetting an individual's pacemaker is reset to the basal level at the onset of the stimulus, and its pacemaker remains inhibited at this level until the signal terminates. At this juncture, the pacemaker returns to the trigger level but may do so more quickly than when the stimulus is absent (Figure 1d) (62). Rapid returns are implicated by PRCs with slopes < 1 (Figure 2b)—which have been observed in various synchronizing Orthoptera (60, 62, 144)—and they may represent an expression of the general neurophysiological phenomenon "post-inhibitory rebound" (see 130). Possibly, such rebound can occur because energy, as expended by effector output, is conserved during inhibition. A rapid return may be achieved by (a) an ascending slope from the basal to the trigger level that is steeper than in the free-running condition, and/or by (b) incomplete resetting to a state that is somewhat higher than the free-running basal level (Figure 1d).

PHASE SHIFT/SIGNAL LENGTH TRADE-OFF   Another modification of basic phase delay synchrony seen in acoustic insects involves changes in signal length that accompany phase shifts. Playback experiments with the snowy tree cricket, *Oecanthus fultoni* (Orthoptera: Gryllidae), a very precise synchronizer with a transition phase at 220°, showed that calls are lengthened during positive phase shifts and shortened during negative phase shifts (144). This relationship is apparently a trade-off, and as such it too may reflect energetic conservation or limitation. Trade-offs between signal rate and length are reported in several acoustic species (60, 61, 91, 130), and changes in *O. fultoni* signal length may represent a special case of a general phenomenon.

PHASE DELAY ALTERNATION   While early work recognized that alternation in acoustic insects and anurans was a form of interaction among rhythmic signalers, most of these studies otherwise considered alternation and synchrony as dissimilar events. Consequently, few attempts were made to explore the possibility that the two phenomena were generated by common mechanisms (for exceptions, see 77, 78, 91). However, recent findings and reanalyses of previously collected data suggest that modified versions of the phase delay model may regulate signal timing in many alternating species as well.

Via a series of playback experiments, it was determined that call timing in the alternating acridid grasshopper *Ligurotettix planum* is controlled by inhibitory resetting (60, 99). Analyses of responses in these tests yielded a PRC with a slope ≈ 0.6, which indicated a pacemaker that rebounded quite rapidly after inhibition when the stimulus occurred late in the insect's period. Two individuals with equivalent free-running rhythms who adhere to this particular mechanism would be predicted to alternate. If period lengths fluctuate stochastically, however, alternation would be punctuated by occurrences of synchrony when period lengths of each individual diverge from modal values in opposite directions. Synchrony would also occur if each individual happens to begin signaling, after a long silent interval, at the same time (133). This prediction—occasional punctuation by synchrony for one or two periods—occurs in *L. planum* and in the alternating tettigoniid *Pterophylla camellifolia* (3, 130). Reinterpretations of inhibitory responses to acoustic stimuli in *P. camellifolia* (130, 134) and in several other alternating orthopterans (25, 70, 76, 79, 80) indicate operation of an inhibitory resetting mechanism similar to that in *L. planum*. As in synchronizing orthopterans, a phase shift / signal length trade-off occurs in *P. camellifolia* (130).

The same modified phase delay model may also exist in many anurans, most of which, if they do engage in signal interactions, chorus in an alternating fashion (148). Various algorithms have been developed to model anuran chorusing (88, 91, 100, 160), but reexamination of data from playback experiments indicates that inhibitory resetting with a very rapid post-stimulus rebound can account for call timing in at least several species. For example, the leptodactylid frog *Leptodactylus albilabris,* which produces a 30-ms call at ~ 3.5 s^{-1} and both synchronizes and alternates with its neighbors, will respond to acoustic stimuli repeated 5–6 times its natural signaling rate by calling ~ 50 ms after every 4th–6th stimulus (100). Responses to slower stimulus rates are spread out over broader intervals, that still begin after 50-ms delays, but they occur after nearly every stimulus. In general, the frogs avoid calling during an interval from 20–50 ms after a stimulus. A "refractory period" (104) exists for ~ 150 ms following the animal's call during which a stimulus has no effect on the call period. This apparently complex set of responses can be predicted, however, via a simple inhibitory resetting model (Figure 1e) in which the pacemaker is immune to external influences during the refractory period but afterwards can be reset and then rebounds rapidly. *Eleutherodactylus coqui,* another leptodactylid that calls at a much slower modal rate ($\sim$ 0.45 sec^{-1}) and, in more typical anuran fashion, is a strict alternator with its (nearest) neighbors, displays similar responses (160) that are also predictable via the inhibitory resetting model in Figures 1e, 2c. In the hylid frog *Pseudacris streckeri* a similar model may apply; however, timing of both the first and second calls following a stimulus are influenced (88). Review of the PRCs of acoustic insects (Figure 2c), including both synchronizers and alternators (60, 62, 99, 144), suggests that brief post-call refractory periods may be present there as well.

PHASE-LOCKING IN SYNCHRONIZERS AND ALTERNATORS   The ability of *Leptodactylus albilabris* to time its calls regularly with respect to a rapidly delivered stimulus represents a form of entrainment or phase-locking. Such locking to stimuli has been reported in other alternating and synchronizing insects (25, 76, 133) and anurans (160), although none are capable of regular interaction with a stimulus repeated as frequently as six times the natural signaling rate. Recasting signal timing in terms of inhibitory-resetting models, however, removes some of the significance often accorded to phase-locking. These models indicate that a particular phase-locking regime can result incidentally from the appropriate combination of (animal) signaling and stimulus rates and that the exceptional 1:6 locking in *L. albilabris* is a simple effect of a very rapid rebound in the timing mechanism. That is, were the rebound slower, the frogs would have probably been repeatedly inhibited by the high stimulus rate.

COMMON MECHANISM FOR SYNCHRONY AND ALTERNATION   The value of the above interpretations is that they afford an opportunity to view a variety of signal interactions in a standard manner. This common perspective suggests that most cases of synchrony and alternation, rather than being generated by radically different mechanisms, actually represent the same basic responses which differ only in a time constant: Alternation occurs when the rebound interval (τ) following resetting (see Figure 1e) is relatively short compared with the modal signal period (T), and it appears to be restricted to species with slow signaling rates. Possibly, this restriction reflects a lower bound on τ that is physiologically constrained and common to all species regardless of T or signal length (x) or duty cycle (x/T). Such a bound would effect increased $\tau{:}T$ ratios at higher signaling rates. Eventually, τ would approach T, and regular alternation would be averted. This prediction is supported by the observation that, among insects, synchrony of acoustic (and bioluminescent) signals usually occurs in species with modal signaling rates exceeding 1 s^{-1} in solo, irrespective of the signal duty cycle (see 63). Conversely, most anuran signaling rates are slower than 1 s^{-1}, and alternation is the norm in this taxon (140, 148). Whereas these generalizations suggest that alternation and synchrony do have some taxonomic affinities, both interactions are known to occur in the same genus; e.g. *Platycleis* (Orthoptera: Tettigoniidae) (124). The finding that within *Platycleis* regular acoustic synchrony occurs only in *P. intermedia,* the species with the highest signaling rate (3.4 s^{-1}), lends further support to the notion that signal interaction format can often be predicted by solo rate.

Sismondo (133) addressed this issue from the perspective of PRCs, and he too demonstrated that synchrony and alternation can originate from a common mechanism, albeit not a pure phase delay one. He showed that a Southeast Asian tettigoniid, *Mecopoda* sp. S, that can both synchronize and alternate realized both these interactions when the slope of the PRC was lower than a critical value, but only synchrony if the slope was higher. A general model was then derived that predicted the set of conditions (stimulus and response phases) under which stable alternation (phase angle = 180°) could occur.

SYNCHRONY AND ALTERNATION AS EPIPHENOMENA   Implicit in the above generalization is the concept that when synchrony and alternation are generated by a phase delay mechanism, these interactions, no matter how precise, may be epiphenomena (62; also see 3). Regular synchrony and alternation result only when neighboring individuals happen to sustain similar signaling rates. In such circumstances, individuals in a synchronizing species usually signal at low absolute phase angles (0° to $(t/T) \cdot 360°$), but they occasionally drop out of the collective display or signal in solo for a period because of the inhibition that occurs when signaling rates diverge. Thus, synchrony is a default event. Individuals in an alternating species signal in the same two phase relationships; however, the proportions are reversed—synchrony being rare— because τ is $\ll T$. Specifically, some alternation is predicted whenever the PRC slope is $< [(T - 2t)/(T - t)]$, and at any slope below this critical value the likelihood of alternation would be proportional to $[(T - t)/T]$. In both synchrony and alternation, individuals do not signal during a "forbidden interval" ranging from t to $x + \tau$ after the onset of a neighbor's signal.

Although excitation in the form of increased rates during interactive signaling occurs in some species regulated by phase delay mechanisms (70, 77, 78), there are no firm indications in any of the species discussed above that individuals decrease their signaling rates to "accommodate" slower signalers while retaining a constant phase angle. Moreover, synchrony and alternation obviously depend on individuals signaling at consistent rates, but no data show that solo signalers reduce fluctuations in their rhythms when interacting with neighbors. Most of these natural signal interactions apparently relied on repeated resetting of the oscillator during each period rather than changes in the free-running rhythm, and 0° or 180° phase-locking to artificial stimuli only occurred if the stimuli were presented at particular rates.

Of special interest are several species of acoustic insects and anurans that exhibit both regular synchrony and alternation (78, 100, 133). These dualities, which may be produced by pacemaker time constants close to the bifurcation values that relegate signalers to one signal interaction format or the other, enlarge the probability that synchrony and alternation per se may not be adaptive in some cases. Rather, various of these collective displays may simply represent incidental by-products of phase delay mechanisms that were not selected for because of the synchrony or alternation generated.

Free-Running Oscillator Variation

While phase delay mechanisms can account for most rhythmic signal interactions, some cases of bioluminescent and acoustic synchrony are clearly controlled in other ways. Several species of Southeast Asian lampyrids (*Pteroptyx malaccae, Pteroptyx tener, Luciola pupilla*) time their flashes such that phase angles very close to 0° eventually result, and they do so by adjusting their oscillator periods (65). Because of this feature and the precision of the interaction, which may be partly derived from extremely stable solo rhythms, this type of proepisodic mechanism has been termed *perfect synchrony* (43). Perfect synchrony can occur even when neighbors do not have equivalent free-running rhythms.

Experiments with *P. malaccae* revealed a PRC with a slope only slightly greater than 0 (14, 65). By making small phase adjustments, commensurate with this PRC, over 20–40 successive periods, individuals who initially flashed at different solo rhythms converged on a single average rhythm and then remained in synchrony. After the interaction, the participants very slowly returned to their original solo rhythms. In this fashion, individual free-running rhythms can be adjusted by overall increases or decreases up to 15%. A coupled-oscillator model that allows free-running rhythms to adapt slowly to each other demonstrates theoretically this attainment of perfect synchrony (43).

Phase Advance Mechanisms

The rhythmic flash synchrony occasionally seen in dense populations of some North American lampyrids (*Photinus* spp.) is achieved with a homoepisodic mechanism in which an individual advances its phase upon perceiving a neighbor's signal (14, 21). This mechanism depends on the ability to signal shortly after (~ 200 ms) a stimulus that occurs during the latter portion of the signal period. The free-running oscillator is not affected here, and the normal rhythm resumes immediately after each advanced signal. Synchrony in species with slow signaling rates ($<< 1$ sec^{-1} in solo) may be sustained with this mechanism, since τ is not a factor. Possibly, the synchronous chorusing observed in periodical cicadas (*Magicicada cassini*; $T \approx 5$ sec (4) is accomplished in this fashion. A simple coupled-oscillator model showed that a population of signalers obeying this phase advance mechanism will tend toward a synchronized state and will do so even if individual free-running rhythms differ slightly (136).

COOPERATIVE FUNCTIONS

The pacemaker of the human heart includes approximately 10,000 cells whose electrical rhythms are synchronized (137). Because a normal heartbeat is dependent on such synchrony, this collective neural interaction is clearly adaptive. Necessity of neural synchrony was duly recognized, and an early coupled-oscillator model was developed to describe the cardiac pacemaker (115). Is it likewise possible that some collective signal interactions are adaptations as opposed to epiphenomena and that individuals actively cooperate in achieving and maintaining particular phase relationships?

Rhythmic Synchrony

RHYTHM PRESERVATION   In various rhythmically signaling species whose rates fluctuate negligibly, females only respond to male signals produced at the specific modal rate. This matching of rate and preference can hold even as rate changes with temperature in poikilothermic species (45). When female rate preference is specific as such, males who signal out of synchrony with their neighbors may prevent females from recognizing that the signaling rate of any individual male in the local group is correct (90, 144). Thus, a male signaling asynchronously would be committing a strongly "spiteful" act, behavior not expected because of evolutionary instability under most circumstances (86). Consequently, males may be subject to considerable selection pressure to cooperate by synchronizing, and some of the mechanisms discussed above may be adaptive.

Because receivers can more easily localize sources of bioluminescent signals than those of acoustic ones (39; but see 29), the rhythm preservation hypothesis may be a more likely factor in acoustic than bioluminescent synchrony. In fact, the function of precise synchrony in the gryllid *Oecanthus fultoni* has been explained as rhythm preservation, since temperature-specific signaling rates coupled with female rate preferences occur in this species (144). However, the predicted lack of responses by females presented with two or more asynchronous, but correct, calling rhythms has not been investigated. Adding to the uncertainty surrounding this hypothesis is the possibility of central neural separation of acoustic stimuli from two sources, an ability reported in another gryllid, *Teleogryllus oceanicus* (116). Therefore, it may be feasible, even in acoustic communication, for a female to recognize the signaling rate of a given individual amidst neighbors signaling at various phase angles, although the number of neighbors and their spatial separation could reduce this ability.

The rhythm preservation hypothesis may yet apply to bioluminescent signaling, but on an inter-group level. If signalers are tightly clustered due to any of various ecological factors promoting spatial aggregation (see 10, 82, 122), male groups may compete to attract females distributed between the groups, although individual males would, of course, compete for females already attracted to their own group. Thus, females presented with the choice of distant male groups in which individuals either synchronize their flashes or disregard phase may orient toward the former, if perceiving a correct rhythm is essential (see 90). Thereby, intra-group male cooperation via synchrony would be selected for, since it maximizes the group's female:male ratio and female encounter rates on a per male basis. This principle could also operate for acoustic signals.

MAXIMIZATION OF PEAK SIGNAL AMPLITUDE   Another way in which competition between male groups may conceivably generate synchrony derives from potential benefits of maximizing peak signal amplitude. Variations on this theme were originally proposed for lampyrids, and it was therefore termed the "beacon" effect (17). In either acoustic or bioluminescent signaling, if females are influenced more strongly by peak signal amplitude than by a time-averaged value, groups of males that synchronize would be more attractive than groups who do not. This argument does not rely on a per individual mating advantage of males signaling in larger groups per se, an advantage for which there is very little evidence among acoustic signalers [Morris et al (103) reported the only positive finding; see (1, 24, 37, 111, 129, 132, 139, and 146) for negative and equivocal results] and none among bioluminescent ones. Rather, the beacon effect as stated here only predicts that males who happen to be tightly clustered for any number of reasons should time their signals synchronously.

PERCEPTION OF FEMALE ANSWERS   Courtship dialogues between male and female signalers occur in some acoustic (71) and bioluminescent (20, 26) insects. Typically, the answers of receptive females to long-range sexual advertisements of males are low amplitude signals produced at a specific time following the male signal. Perception by males of these responses may depend on an absence of masking by the signals of other males (109, 112). Therefore, males may be selected to synchronize with their neighbors, because this would eliminate the potentially obfuscating noise from the signals of these other males.

AVOIDANCE OF NATURAL ENEMIES   Because male advertisement signals may be perceived by heterospecifics as well as by conspecific females, a male's risk of predation or parasitization is often increased during signaling (22, 24). Synchrony might reduce this risk because sound or light are thereby emitted from myriad directions at once. Several authors (63, 108, 144) have suggested that a natural enemy in the midst of such a deluge would be unable to localize the source of any one signal. If this is true, cooperation via synchrony would again be selected for quite strongly. This hypothesis is a short-term analogue of the argument advanced to explain the very long-term synchrony of emergence in periodical cicadas (73) and of parturition in wildebeests (44), though predator satiation, rather than perceptual confusion, is invoked in the latter cases.

Tuttle & Ryan (140) provided the only evidence supporting the natural enemy avoidance hypothesis, and it does not involve rhythmic signaling. They found that aggregated males in the hylid frog *Smilisca sila* sporadically produced synchronous calls at an average rate = 1.7 min^{-1}. Synchrony was attained via a homoepisodic mechanism (120), and playback experiments showed that phonotactic bats preying on the frogs were attracted more frequently to asynchronous than to synchronous calls.

The most likely candidates for which any of the cooperative synchrony models may be valid are those species in which the interaction is not facultative but occurs whenever neighboring signalers are sufficiently close. This condition tends to arise when synchrony is sustained by varying the free-running oscillator, by phase delay mechanisms in which the transition phase is low, and by some homoepisodic mechanisms. Unfortunately, data that could be used to test any of the above cooperative hypotheses for synchrony have been collected only rarely. Other than the experiments on *Smilisca sila*, no study has indicated definitively that synchrony is cooperative. Obviously, appropriate tests on perception and orientation in receivers—female, and male, conspecifics and natural enemies—are sorely needed.

Rhythmic Alternation

RECEIVER CONFUSION   Due to perceptual difficulties similar to those suggested to afflict natural enemies, female conspecifics too may be unable to locate the source of any one signal when all local signals are synchronized or overlap extensively in time. To prevent this situation, males may be selected to adjust their signaling rhythms such that alternation occurs. Among acoustic signalers this hypothesis has been tested in several anuran species, but supporting evidence has only been found in the centrolenellid frog *Centrolenella granulosa* (75; see 6, 114, 126, 129 for negative findings in other species). In acridid grasshoppers, which may not share with gryllids the ability to separate neurophysiologically two acoustic stimuli (see 142), reduced female phonotaxis to overlapped calls is known (98), and it may contribute, in part, to maintenance of signal alternation in *Ligurotettix planum* (60, 99).

If confusion by overlapped signals is a factor for both conspecific females and natural enemies, conflicting demands to both synchronize and alternate arise. Comparable dilemmas may be present in other situations, because it is conceivable that many combinations of factors could influence signal interactions. For example, competition between individual signalers, as shown in a later section, will often select for an outcome quite different from that predicted to result from competition between groups. Perhaps switching by the tettigoniid *Mecopoda* sp. S from synchrony to alternation (133) does represent an adaptive change in response to different influences: Factors operating specifically when neighbors are distant may select for alternation, and the resetting mechanism yielding a gradual PRC is therefore adopted when males move apart.

PRESERVATION OF SIGNAL COMPONENTS   Longer acoustic and bioluminescent signals are often temporally structured such that they comprise repeated components (and sub-components), commonly termed "pulses." Because features of individual components, such as pulse rate, may be more critical to females than features of entire signals (63, 108, 110), males may be selected to alternate because of limitations on the precision of their ability to adjust pulse timing. That is, were neighboring males to overlap signals, they would probably be unable to make the fine adjustments necessary to align temporally (synchronize) the pulses comprising their signals. Schwartz (126, 129) obtained support for this hypothesis in four-loudspeaker playback experiments conducted on several alternating anurans: Females were presented simultaneously with a pair of overlapping signals and a pair of nonoverlapping (alternated) ones. The signals comprising each pair were broadcast from loudspeakers separated by 180°; each of the four loudspeakers were separated by 90°. When pulses in the two overlapping signals were not synchronized, females oriented toward a signal from the other (alternated) pair. However, when the pulses in the overlapping pair of signals were synchronized, females no longer preferred signals from one pair of loudspeakers over the other. Therefore, female preference for alternated signals in the first experiment could not have resulted from an inability to localize overlapped calls.

The component preservation hypothesis is based on the assumption that receivers situated between two overlapped signals whose pulses are unsynchronized would perceive a pulse rate approximately twice the modal value (see 29 for this effect in bioluminescence) and remain nonresponsive. As such, component preservation is a fine-scale analogue of the rhythm preservation hypothesis, but it selects for signal alternation rather than synchrony.

MAXIMIZATION OF GROUP DUTY CYCLE   If females are attracted to a distant acoustic or bioluminescent signaling source based on an amplitude value integrated over a long period of time, competition between groups of males could select for alternation between signals of males within groups. By alternating signals, males would maximize the duty cycle and root-mean-square amplitude value of the group's collective emission. The latter effect results because the combined amplitude value of n synchronous, grouped signals, each of amplitude A, is $\leq n \cdot A$ (see 10).

Unison Bouts and Sprees

Competition between male groups for distant females may select for unison bouts and sprees due to reasons analogous to the peak amplitude and group duty cycle maximization hypotheses for rhythmic synchrony and alternation. This may be more likely for continuous signals than for rhythmically repeated ones, though. In the latter case, the collective timing effected by unison bouts and sprees may be too crude to generate the synchrony necessary for enhancing peak amplitude. As indicated in the previous section, however, the interval over which signal energy is integrated by a female receiver assessing amplitude may affect this issue critically. Unison bouts and sprees in rhythmically signaling species with longer integration time constants would be predicted to be more susceptible to this effect, because signals that do not overlap in time may yet be close enough that the peak amplitude assessed by females is enhanced.

COMPETITIVE SIGNALING

Sexual selection theory predicts that males may compete for females via signaling (151). The more obvious manifestations of such competition are the high amplitudes, long durations, and other extravagant features of many male sexual advertisement signals. Temporal interactions between the signals of neighbors may also represent such competition (3, 7, 63), but of a less direct nature.

Signal Jamming

COMPETITIVE SYNCHRONY   Competition and synchrony may appear to be incongruous activities. However, the frequency with which bouts of alternation and solo singing interrupt synchrony and the apparent failure of synchronizing individuals to slow their signaling rates while retaining constant phase angles suggest that many cases are not cooperative. As noted earlier, these non-cooperative cases may represent mere epiphenomena. Nonetheless, do situations in which synchrony arises by default reflect underlying competitive interactions?

PSYCHOPHYSICS   The attraction of females to male calls in certain orthopterans and anurans is influenced by a "precedence effect" in which preference is directed toward the first of two (or more) closely synchronized, but spatially separated signals (40, 62, 85, 98, 135, 152). In the tettigoniid *Neoconocephalus spiza*, the precedence effect will favor calls whose onsets are advanced as little as 13 ms in front of others (62). Preference for leaders will even override aspects of call energy such as duration and amplitude (62; also see 41), "non-arbitrary" features that may serve as female choice criteria in acoustic animals (see 68, 84). Consequently, males are strongly selected to adopt a timing mechanism that averts calling shortly after a neighbor and also relegates the neighbor to calling in this (following) role. The inhibitory resetting mechanism used by *N. spiza* meets these criteria, because a male adhering to it is inhibited from calling following a neighbor, yet during the next period he has a high probability of jamming the neighbor by calling in the leading role (or being the solo caller) owing to the rebound of his pacemaker following inhibition. If two males with similar modal calling rates both use inhibitory resetting, runs of synchrony are predicted to ensue. One male's signals will precede the other's by a short interval ($< t$) during these runs, with the leading role passing back and forth between the males. Synchrony results because τ is $\approx T$, and each male's pacemaker is triggered before the other male has called (62).

Computer simulation of the signaling controlled by inhibitory resetting showed that this mechanism is evolutionarily stable (ES) (see 93)—when compared with calling timed regardless of a neighbor's calls—provided that leading calls are more attractive than following ones (62). Thus, synchrony can be an ES outcome of a simple mechanism selected by the psychoacoustics of female choice. Different ES outcomes may be yielded by simulations of more complex situations: choruses in which three or more signalers are present, in which signalers can increase their calling rate for several successive periods, while maintaining the same long-term modal rate, or in which signalers may differ in available energy, which might lead to sustaining different maximum signaling rates or bout durations.

In *Neoconocephalus spiza*, the precedence effect does not result from any particular attractiveness of components of the call beginning (62). Rather, female preference may be directed toward the leading call because its sudden onset of sound, a critical feature, is not masked by the neighbor's call or occurs first and neurally inhibits the contralateral ear for ~ 100 ms (see 156 and forward masking in 133a). Such female choice could represent several types of indirect sexual selection (83). "Good genes" selection may apply if some males lead consistently and do so because of shorter τs which may reflect higher vigor. Alternatively, arbitrary (Fisherian) selection may apply, and it may have originated in a "sensory bias" (121) that simply renders leading sounds more easily localized or evaluated as louder. The sensory bias interpretation is suggested by the occurrence of precedence effects in diverse taxa (156), including various mammals (e.g. humans) in which the effect does not occur in the context of female choice. Possibly, the effects are generated by common ancestral elements of processing that are conserved in neural design (see 38). Precedence effects underline the importance of examining female choice via multichoice tests (see 36) and of considering temporal relationships between the alternative stimuli presented.

ALTERNATION   Precedence effects that select for inhibitory resetting mechanisms adaptive for evading jamming by neighbors may generate collective alternation if τ is relatively short. Whether the specific outcome is synchrony or alternation may be a mere artifact of the $\tau{:}T$ ratio, but alternation per se may be selected for in some species in which females cannot localize synchronized calls (98). That is, competitive and cooperative pressures may select for a mechanism including both a forbidden interval following a neighbor's calls and a short τ (or a long T) averting synchrony, respectively. As in synchrony, regular alternation, a one-for-one correspondence of calls between individuals, will only ensue when the chorus participants sustain equivalent signaling rates (see 61). That there are no indications of faster individuals in alternating choruses reducing their signaling rates to match slower neighbors while retaining constant phase angles, even in species where females cannot localize synchronized calls, suggests that occurrences of strictly regular alternation are artefacts of equally vigorous individuals.

Unlike synchrony, alternation would be impossible to maintain as population density increases, unless individuals decrease their signaling rates markedly or disregard the rules of signal timing (99). Several studies of acoustic insects and anurans have investigated this problem, and the typical solution found entails ignoring all but the nearest (loudest?) one or two neighbors (13, 99, 105, 106, 128). Thus, certain anomalous cases of simultaneous synchrony and alternation within choruses (e.g. 100) may be explained. Because nearby neighbors are likely to be a focal male's strongest competitors for females, such selective attention may be an adaptive response—in synchronizing as well as alternating species—to the predicament of being surrounded by multiple signalers. Finding that females too ignore more distant signalers when evaluating leader/follower roles or other attributes (a potential ability in Orthoptera given the existence of acoustic neurons capable of selective attention—117) would bolster this argument. Such support has been obtained in a tettigoniid (119) and a hylid frog (53), where playback experiments investigating the masking effects of chorus noise indicated that females did limit their choices as predicted.

Where local population density of signalers can be extremely high, as in anurans clustered at a breeding pond or occupying perches in a three-dimensional (arboreal) habitat, selective attention alone may not provide a male with calling time windows that are both nonoverlapping and outside forbidden intervals. Interactive algorithms in many anurans are characterized as phase delay mechanisms with very short τs. These mechanisms allow a male to insert his calls into the ends of short, unpredictable silent gaps in the chorus (105, 160). As such, they may be adaptations for averting production of overlapping (and following) calls while signaling in a dense aggregation.

The purported cases of jamming listed here all involve acoustic signaling. This may reflect constraints specific to the localization of acoustic signals discussed above. Perhaps, a leading bioluminescent flash does not mask the sudden onset of a following flash as perceived by a female receiver. Requisite experiments on precedence effects and female choice in bioluminescent signalers have not been done, however, and jamming interactions between long-range advertisement flashes cannot be ruled out.

Mutual Assessment

In his 1962 treatise on social behavior, Wynne-Edwards (155) suggested that animals chorus and exhibit other collective signaling displays to assess local density so that they can then regulate their population. While this sort of group-level interpretation has long since fallen into disfavor among behaviorists, various acoustic signal interactions are interpretable as adaptations that facilitate assessment on an individual level. This view may be particularly valid for alternated signals. It is indicated by the role of acoustic signaling in spacing (5, 12) and aggression (61, 143, 157) and by the difficulties that an animal may have in perceiving a neighbor's calls during its own signaling. Evidence for the latter comes both from behavioral experiments relying on the technique of interactive playback (see 34, 126) and from neurophysiological investigation.

INCOMPATIBILITY OF SIGNALING AND PERCEPTION   Sexually advertising males in many acoustic orthopteran and anuran species space themselves regularly or maintain minimum nearest-neighbor distances via mutual assessment of calls (5, 12, 105). In some acoustic species hearing is reduced during calling, a handicap due either to simple masking (69) or to neural (154) or biomechanical (107) devices. These devices may be adaptations that protect the animal's ears from high sound levels or that prevent self-excitation by territorial signals. Regardless of their origin, perceptual handicaps during signaling may select for avoidance of call overlap. Slower calling rates and/or interactive algorithms yielding alternation, such as phase delay mechanisms with low τ:T ratios, may be adaptations for accomplishing this and facilitating spacing. If a precedence effect in female choice is also present, this requirement for unimpaired assessment of neighbors should favor interactive mechanisms whose time constants yield alternation.

A corollary of the prediction that males avoid call overlap is that males in denser populations should signal more discontinuously to facilitate the monitoring of neighbors. Eiriksson (42) reported this effect in the acridid grasshopper *Omocestus viridulus,* and Dadour (35) found that the tettigoniid *Mygalopsis marki* widened gaps between its calls when intruders approached. Among *Neoconocephalus* tettigoniids, species that chirp, rather than buzz continuously, are those normally found in higher densities (57).

GRADED AGGRESSIVE SIGNALS   Circumvention of perceptual impairment may be especially important during production of graded aggressive signals. These signals are often evaluated precisely such that the individual who does not match its rival departs or assumes a subordinate role. Precise evaluation is believed to be essential because of an expected correlation between signal parameters and resource holding potential. Studies of acoustic insects (61, 157), birds (11), and mammals (31) show that graded aggressive signals are normally alternated between rivals. Alternation proceeds on a strict one-for-one basis until one individual begins to lag behind and then exits the encounter, presumably due to anticipated defeat if the contest were to continue and escalate. Failure to alternate could obviously lead to an inappropriate decision.

A group-level version of this phenomenon occurs in various mammals, and it may mediate assessment of social groups (94). Acoustic signals that are alternated between groups may indicate group size and resolve territorial boundaries accordingly, but the suggestion has been made that these choruses do not necessarily provide reliable information on the number of individuals present (67).

Evasion of Detection

Under certain competitive circumstances, nonjamming call overlap might actually be favored by the incompatibility of signaling and perception. A subordinate individual who synchronizes with a dominant neighbor may escape detection and probable eviction if he produces short signals that are completely overlapped by the longer signals of that neighbor (see 150, 159 on "blind spots" in electric and acoustic communication). Synchrony of this sort would therefore be a variant form of satellite behavior. Possibly, long and short, overlapped calls exhibited by synchronizing male neighbors in the tettigoniid *Neoconocephalus nebrascensis* (97) represent such evasion by satellites from dominants.

Courtship Disruption

In signaling systems that include male-female courtship dialogues, the answers of receptive females present numerous opportunities for males to jam or otherwise interfere with their male neighbors via signal interactions. For example, a male who synchronizes with his neighbor may be able to "interlope" in the courtship between a female and that neighbor (27, 28). A synchronizing male might also detect the responses of a distant female to a signaling male neighbor and then interlope in their courtship. Opportunities for interloping and detection exist because a synchronizing male's signals are timed correctly for interacting with females responding to neighbors. Otte & Smiley (112) and Buck (14) elaborate on diverse forms of these phenomena that may account for synchronous flashing in lampyrids. Analogous events involving the overlapping of calls may occur in orthopterans in which females respond acoustically (e.g. 51).

Coincidence with Female Activity Periods

SPREES   The concentration of signaling within a narrow diel interval may represent males taking advantage of times when environmental conditions are particularly favorable for signal transmission and reception (72, 158), when receptive females are most numerous (58, 145), or when other activities are not possible, yet energy that could be expended on signaling remains (96). Because the transmission of sound is particularly subject to wind, various authors have argued that dawn and dusk choruses—acoustic sprees—are timed to coincide with periods of calm (72, 158). Diel timing of signaling may also be influenced by biotic factors such as the activity schedules of natural enemies and of heterospecifics whose signals may interfere (56, 59, 119, 125). However, observations of males initiating signaling when presented with conspecific song (141) and of individual males elevating their levels of signaling when joined by others (131) suggest that sprees may be cooperative or competitive social phenomena.

Competitively, individual males may be selected to match or exceed the signaling output of their neighbors in order to remain attractive to females. Because of the energetic cost of male signaling (see 92, 118, 123, 149) and narrow diel activity periods of females, pressures on males to match or exceed their neighbors' signaling may translate into compressed signaling periods that coincide with female activity (48, 58). Competition, rather than atmospheric conditions, may also explain the tendency of sprees to occur at photoperiodic transitions. Receptive females are often "gated" such that they arrive at encounter sites at certain times (145). For example, in many nocturnal insects and frogs, females may mature and enter receptivity at any time over a 24-hr period, but they do not become active until evening. As predicted, male calling in most of these animals is concentrated during the initial half of the night. Analogously, morning choruses in diurnal acoustic insects such as acridids and cicadas may reflect the gating of receptive females after sunrise (58). Bimodal morning/evening choruses also occur in a few species (58, 158), and these may correspond with two daily gates for receptive females.

UNISON BOUTS AS EXPLODED SPREES   If the arrival and presence of receptive females extends over a long diel interval, males may have insufficient energy to signal for the duration of female presence. To allocate their available energy optimally, males may break their signaling into bouts so that the entire interval is covered effectively. This hypothesis appears to be a valid explanation for bout calling in the hylid frog *Hyla microcephala,* because males store insufficient glycogen to fuel continuous calling throughout the entire female activity period (127).

Given the possibility that energy limitation generates bout calling by individuals, unison bout calling per se may then arise due to any of the processes that create collective displays. In both acoustic insects and frogs, unison bouts are often initiated repeatedly by the same individuals (23, 55, 152). Bout leaders also tend to signal with the highest rates or longest calls. Surrounding males—who supposedly have less available energy—may signal following the bout leaders because the resulting unison bout affords an advantage to the group (and individual males) in attracting females or evading natural enemies. Followers may also evade detection by dominants and improve their chances of encountering females by calling at the same time as bout leaders. These complications may further explain why an early ES simulation of chorusing in the hylid frog *Hyla regilla* (113) did not fully resemble the unison bouts observed in this species.

SUMMARY

The investigation of signal interactions can open a window through which appreciation and understanding of the complex and subtle factors influencing the evolution of communication may be greatly enhanced. These interactions seem to be generated by a relatively limited number of neural mechanisms, yet the collective phenomena that we observe are diverse in format and may serve a variety of functions—or none at all! In many cases discerning these functions will require novel tests of how conspecific and heterospecific receivers perceive and evaluate signals in the field. Such tests promise to offer insight pertaining not only to signal interactions but to the fundamental roles played by phylogeny and current selective factors in constraining and shaping sexual signaling systems.

ACKNOWLEDGMENTS

A portion of the work reported in this review has been supported by NSF grant IBN-9196177, which is duly acknowledged. I am indebted to Bob Minckley, Stan Rand, the University of Kansas Bioacoustics Seminar (1991), the Smithsonian Tropical Research Institute Animal Behavior Group (1992), and the residents of the American Museum of Natural History Southwestern Research Station (1993) for valuable discussions, and to Jonathan Copeland, Bob Minckley, Josh Schwartz, and Enrico Sismondo for helpful criticisms leading to final preparation of the manuscript.

Literature Cited

1. Aiken RB. 1982. Effects of group density on call rate, phonokinesis, and mating success in *Palmacorixa nana* (Heteroptera: Corixidae). *Can. J. Zool.* 60:1665–72
2. Alexander RD. 1956. *A comparative study of sound production in insects, with special reference to the singing Orthoptera and Cicadidae of the eastern United States*. PhD thesis. Ohio State Univ., Columbus
3. Alexander RD. 1975. Natural selection and specialized chorusing behavior in acoustical insects. In *Insects, Science, and Society*, ed. D Pimentel, pp. 35–77. New York: Academic
4. Alexander RD, Moore TE. 1958. Studies on the acoustical behavior of seventeen-year cicadas (Homoptera: Cicadidae: *Magicicada*). *Ohio J. Sci.* 58:107–27
5. Arak A, Eiriksson T, Radesäter T. 1990. The adaptive significance of acoustic spacing in male bushcrickets *Tettigonia viridissima*: a perturbation experiment. *Behav. Ecol. Sociobiol.* 26:1–7
6. Backwell PRY, Passmore NI. 1991. Sonic complexity and mate localization in the leaf-folding frog, *Afrixalus delicatus*. *Herpetologica* 47:226–29
7. Bailey WJ. 1991. *Acoustic Behaviour of Arthropods*. London: Chapman & Hall
8. Bailey WJ, Rentz DCF. 1990. *The Tettigoniidae: Biology, Systematics, and Evolution*. Berlin: Springer-Verlag
9. Belwood JJ, Morris GK. 1987. Bat predation and its influence on calling behavior in Neotropical katydids. *Science* 238:64–70
10. Bradbury JW. 1981. The evolution of leks. In *Natural Selection and Social Behavior: Recent Research and New Theory*, ed. RD Alexander, DW Tinkle, pp. 138–69. New York: Chiron
11. Bremond J–C, Aubin T. 1992. Cadence d'emission du chant territorial du troglodyte (*Troglodytes*). *C.R. Acad. Sci. Paris. Ser. III.* 314:37–42
12. Brush JS, Gian VG, Greenfield MD.

1985. Phonotaxis and aggression in the coneheaded katydid *Neoconocephalus affinis*. *Physiol. Entomol.* 10:23–32
13. Brush JS, Narins PM. 1989. Chorus dynamics of a neotropical amphibian assemblage: comparison of computer simulation and natural behaviour. *Anim. Behav.* 37:33–44
14. Buck J. 1988. Synchronous rhythmic flashing in fireflies. II. *Q. Rev. Biol.* 63:265–89
15. Buck J, Buck E. 1966. Biology of synchronous flashing of fireflies. *Nature* 211:562–64
16. Buck J, Buck E. 1968. Mechanism of rhythmic synchronous flashing of fireflies. *Science* 159:1319–27
17. Buck J, Buck E. 1978. Toward a functional interpretation of synchronous flashing by fireflies. *Am. Nat.* 112:471–92
18. Buck J, Buck E, Case JF, Hanson FE. 1981. Control of flashing in fireflies. V. Pacemaker synchronization in *Pteroptyx cribellata*. *J. Comp. Physiol. A.* 144:287–98
19. Buck J, Buck E, Hanson FE, Case JF, Mets L, Atta GJ. 1981. Control of flashing in fireflies. IV. Free run pacemaking in synchronic *Pteroptyx*. *J. Comp. Physiol. A.* 144:277–86
20. Buck J, Case JF. 1986. Flash control and female dialog repertory in the firefly *Photinus greeni*. *Biol. Bull.* 170:176–97
21. Buck J, Hanson FE, Buck E, Case JF. 1982. Mechanism and function of synchronous flashing in the firefly *Photinus pyralis*. *Biol. Bull.* 163:398
22. Burk TE. 1982. Evolutionary significance of predation on sexually signalling males. *Fla. Entomol.* 65:90–104
23. Busnel M-C. 1967. Rivalité acoustique et hiérarchie chez l'ephippiger (Insect, Orthoptere, Tettigoniidea,). *Z. vergl. Physiol.* 54:232–45
24. Cade WH. 1981. Field cricket spacing and the phonotaxis of crickets and

parasitoid flies to clumped and isolated cricket songs. *Z. Tierpsychol.* 55:365–75

25. Cade WH, Otte D. 1982. Alternation calling and spacing patterns in the field cricket *Acanthogryllus fortipes* (Orthoptera; Gryllidae). *Can. J. Zool.* 60:2916–20

26. Carlson AD, Copeland J. 1985. Flash communication in fireflies. *Q. Rev. Biol.* 60:415–36

27. Carlson AD, Copeland J. 1988. Flash competition in male *Photinus macdermotti* fireflies. *Behav. Ecol. Sociobiol.* 22:271–76

28. Case JF. 1980. Courting behavior in a synchronously flashing, aggregative firefly, *Pteroptyx tener. Biol. Bull.* 159: 613–25

29. Case JF. 1984. Vision in mating behaviour of fireflies. See Ref. 89, pp. 195–222

30. Case JF, Buck J. 1963. Control of flashing in fireflies. II. Role of central nervous system. *Biol. Bull.* 125:234–50

31. Clutton-Brock TH, Albon SD. 1979. The roaring of red deer and the evolution of honest advertisement. *Behaviour* 69: 145–70

32. Crawford CS, Dadone MM. 1979. Onset of evening chorus in *Tibicen marginalis* (Homoptera: Cicadidae). *Env. Entomol.* 8:1157–60

33. Cuthill IC, Macdonald WA. 1990. Experimental manipulation of the dawn and dusk chorus in the blackbird *Turdus merula. Behav. Ecol. Sociobiol.* 26:209

34. Dabelsteen T. 1992. Interactive playback: a finely tuned response. See Ref. 94, pp. 97–109

35. Dadour IR. 1989. Temporal pattern changes in the calling song of the katydid *Mygalopsis marki* Bailey in response to conspecific song (Orthoptera: Tettigoniidae). *J. Insect Behav.* 2:199

36. Doherty JA. 1985. Phonotaxis in the cricket, *Gryllus bimaculatus* DeGeer: comparisons of choice and no-choice paradigms. *J. Comp. Physiol. A.* 157: 279–89

37. Doolan JM, MacNally RC. 1981. Spatial dynamics and breeding ecology in the cicada *Cystosoma saundersii*: the interaction between distributions of resources and intraspecific behavior. *J. Anim. Ecol.* 50:925–40

38. Dumont JPC, Robertson RM. 1986. Neuronal circuits: an evolutionary perspective. *Science* 233:849–53

39. Dusenbery DB. 1992. *Sensory Ecology. How Organisms Acquire and Respond to Information.* New York: Freeman

40. Dyson ML, Passmore NI. 1988. Two-choice phonotaxis in *Hyperolius marmoratus* (Anura: Hyperoliidae); the effect of temporal variation in presented stimuli. *Anim. Behav.* 36:648–52

41. Dyson ML, Passmore NI. 1988. The combined effects of intensity and the temporal relationship of stimuli on phonotaxis in female painted reed frogs *Hyperolius marmoratus. Anim. Behav.* 36:1555–56

42. Eiriksson T. 1992. Density dependent song duration in the (grasshopper *Omocestus viridulus. Behaviour* 122:121–32

43. Ermentrout B. 1991. An adaptive model for synchrony in the firefly *Pteroptyx malaccae. J. Math. Biol.* 29:571–85

44. Estes RD. 1966. Behaviour and natural history of the wildebeest (*Conochaetes taurinus* Burchell). *Nature* 212:999–1000

45. Ewing AW. 1989. *Arthropod Bioacoustics. Neurobiology and Behaviour.* Ithaca, NY: Cornell Univ. Press

46. Feaver M. 1977. *Aspects of the behavioral ecology of three species of Orchelimum (Orthoptera: Tettigoniidae).* PhD thesis. Univ. Mich., Ann Arbor

47. Finke C, Prager J. 1980. Pulse-train synchronous pair-stridulation by male *Sigara striata* (Heteroptera, Corixidae). *Experientia* 36:1172–73

48. Forrest TG. 1983. Calling songs and mate choice in mole crickets. See Ref. 64, pp. 185–204

49. Fulton BB. 1934. Rhythm, synchronism, and alternation in the stridulation of Orthoptera. *J. Elisha Mitchell Sci. Soc.* 50:263–67

50. Galliart PL, Shaw KC. 1991. Role of weight and acoustic parameters, including nature of chorusing, in the mating success of males of the katydid, *Amblycorypha parvipennis* (Orthoptera: Tettigoniidae). *Fla. Entomol.* 74:453–64

51. Galliart PL, Shaw KC. 1991. Effect of intermale distance and female presence on the nature of chorusing by paired *Amblycorypha parvipennis* (Orthoptera: Tettigoniidae) males. *Fla. Entomol.* 74: 559–68

52. Galliart PL, Shaw KC. 1992. The relation of male and female acoustic parameters to female phonotaxis in the katydid, (*Amblycorypha parvipennis. J. Orthop. Res.* 1:110–15

53. Gerhardt HC, Klump GM. 1988. Masking of acoustic signals by the chorus background noise in the green tree frog: a limitation on mate choice. *Anim. Behav.* 36:1247–49

54. Gibbon J, Church RM. 1992. Comparison of variance and covariance patterns in parallel and serial theories of timing. *J. Exp. Anal. Behav.* 57:393–406

55. Greenfield MD. 1983. Unsynchronized chorusing in the coneheaded katydid *Neoconocephalus affinis* (Beauvois). *Anim. Behav.* 31:102–12

56. Greenfield MD. 1988. Interspecific acoustic interactions among katydids (*Neoconocephalus*): inhibition-induced shifts in diel periodicity. *Anim. Behav.* 36:684–95

57. Greenfield MD. 1990. Evolution of acoustic communication in the genus *Neoconocephalus*: discontinuous songs, synchrony, and interspecific interactions. See Ref. 8, pp. 71–97

58. Greenfield MD. 1992. The evening chorus of the desert clicker, *Ligurotettix coquilletti* (Orthoptera: Acrididae): mating investment with delayed returns. *Ethology* 91:265–78

59. Greenfield MD. 1993. Inhibition of male calling by heterospecific signals. Artefact of chorusing or abstinence during suppression of female phonotaxis? *Naturwissenschaften* 80:570–53

60. Greenfield MD. 1994. Synchronous and alternating choruses in insects and anurans: common mechanisms and diverse functions. *Am. Zool.* 34: In press

61. Greenfield MD, Minckley RL. 1993. Acoustic dueling in tarbush grasshoppers: settlement of territorial contests via alternation of reliable signals. *Ethology* 95:302–26

62. Greenfield MD, Roizen I. 1993. Katydid synchronous chorusing is an evolutionarily stable outcome of female choice. *Nature* 364:618–20

63. Greenfield MD, Shaw KC. 1983. Adaptive significance of chorusing with special reference to the Orthoptera. See Ref. 64, pp. 1–27

64. Gwynne DT, Morris GK, ed. 1983. *Orthopteran Mating Systems: Sexual Competition in a Diverse Group of Insects.* Boulder, Colo: Westview

65. Hanson FE. 1978. Comparative studies of firefly pacemakers. *Fed. Proc.* 37: 2158–64

66. Hanson FE, Case JF, Buck E, Buck J. 1971. Synchrony and flash entrainment in a New Guinea firefly. *Science* 174: 161–64

67. Harrington FH. 1989. Chorus howling by wolves: acoustic structure, pack size and the beau geste effect. *Bioacoustics* 2:117–36

68. Hedrick AV. 1986. Female preferences for male calling bout duration in a field cricket. *Behav. Ecol. Sociobiol.* 19:73–77

69. Hedwig B. 1990. Modulation of auditory responsiveness in stridulating grasshoppers. *J. Comp. Physiol. A.* 167:847–56

70. Heiligenberg W. 1969. The effect of stimulus chirps on a cricket's chirping (*Acheta domesticus*). *Z. vergl. Physiol.* 65:70–97

71. Heller K-G, von Helversen D. 1986. Acoustic communication in phaneropterid bushcrickets: species-specific delay of female stridulatory response and matching male sensory time window. *Behav. Ecol. Sociobiol.* 18:189–98

72. Henwood K, Fabrick A. 1979. A quantitative analysis of the dawn chorus: temporal selection for communicatory optimization. *Am. Nat.* 114:260–74

73. Hoppensteadt FC, Keller JB. 1976. Synchronization of periodical cicada emergences. *Science* 194:335–37

74. Huber F. 1965. Brain controlled behaviour in orthopterans. In *The Physiology of the Insect Central Nervous System*, ed. JE Treherne, JWL Beament, pp. 233–46. New York: Academic

75. Ibáñez RD. 1993. Female phonotaxis and call overlap in the neotropical glassfrog *Centrolenella granulosa. Copeia* 1993:846–50

76. Jones MDR. 1963. Sound signals and alternation behaviour in *Pholidoptera. Nature* 199:928–29

77. Jones MDR. 1964. Inhibition and excitation in the acoustic behaviour of *Pholidoptera. Nature* 203:322–23

78. Jones MDR. 1966. The acoustic behaviour of the bush cricket *Pholidoptera griseoaptera.* 1. Alternation, synchronism, and rivalry between males. *J. Exp. Biol.* 45:15–30

79. Jones MDR. 1966. The acoustic behaviour of the bush cricket *Pholidoptera griseoaptera.* 2. Interaction with artificial sound signals. *J. Exp. Biol.* 45:31–44

80. Jones MDR. 1974. The effect of acoustic signals on the chirp rhythm in the bush cricket *Pholidoptera griseoaptera. J. Exp. Biol.* 61:345–55

81. Kacelnik A, Krebs JR. 1982. The dawn chorus in the great tit (*Parus major*): proximate and ultimate causes. *Behaviour* 83:287–309

82. Karban R. 1982. Increased reproductive success at high densities and predation satiation for periodic cicadas. *Ecology* 63:321–28

83. Kirkpatrick M, Ryan MJ. 1991. The paradox of the lek and the evolution of mating preferences. *Nature* 350:33–38

84. Klump GM, Gerhardt HC. 1987. Use of non-arbitrary acoustic criteria in mate choice by female gray tree frogs. *Nature* 326:286–88

85. Klump GM, Gerhardt HC. 1992. Mechanisms and function of call-timing in

male-male interactions in frogs. See Ref. 94, pp. 153–74

86. Knowlton N, Parker GA. 1979. An evolutionarily stable strategy approach to indiscriminate spite. *Nature* 279:419–21

87. Lall AB. 1993. Action spectra for the initiation of bioluminescent flashing activity in males of the twilight-active firefly *Photinus scintillans* (Coleoptera: Lampyridae). *J. Insect Physiol.* 39:123–27

88. Lemon RE, Struger J. 1980. Acoustic entrainment to randomly generated calls by the frog, *Hyla crucifer. J. Acoust. Soc. Am.* 67:2090–95

89. Lewis T, ed. 1984. *Insect Communication.* London: Academic

90. Lloyd JE. 1973. Model for the mating protocol of synchronously flashing fireflies. *Nature* 245:268–70

91. Loftus-Hills JJ. 1974. Analysis of an acoustic pacemaker in Strecker's chorus frog, *Pseudacris streckeri* (Anura: Hylidae). *J. Comp. Physiol.* 90:75–87

92. MacNally RC. 1984. On the reproductive energetics of chorusing males: costs and patterns of call production in two sympatric species of *Ranidella* (Anura). *Oikos* 42:82–91

93. Maynard Smith J, Price GR. 1973. The logic of animal conflict. *Nature* 246:15–18

94. McComb K. 1992. Playback as a tool for studying contests between social groups. See Ref. 94, pp. 111–19

95. McGregor PK, ed. 1992. *Playback and Studies of Animal Communication.* New York: Plenum

96. McNamara JM, Mace RH, Houston AI. 1987. Optimal daily routines of singing and foraging in a bird singing to attract a mate. *Behav. Ecol. Sociobiol.* 20:399–405

97. Meixner AJ, Shaw KC. 1986. Acoustic and associated behavior of the coneheaded katydid, *Neoconocephalus nebrascensis* (Orthoptera: Tettigoniidae). *Ann. Entomol. Soc. Am.* 79:554–65

98. Minckley RL, Greenfield MD. Psychoacoustics of female phonotaxis and the evolution of male signal interactions. *Ethol. Ecol. Evol.* Submitted

99. Minckley RL, Greenfield MD, Tourtellot MK. Chorus structure in tarbush grasshoppers: inhibition, selective phonoresponse, and signal competition. *Anim. Behav.* Submitted

100. Moore SW, Lewis ER, Narins PM, Lopez PT. 1989. The call-timing algorithm of the white-lipped frog, *Leptodactylus albilabris. J. Comp. Physiol. A.* 164:309–19

101. Morin JG. 1986. "Firefleas" of the sea: luminescent signaling in marine ostracode crustaceans. *Fla. Entomol.* 69:105–21

102. Morris GK. 1980. Calling display and mating behaviour of *Copiophora rhinoceros* (Orthoptera: Tettigoniidae). *Anim. Behav.* 28:42–51

103. Morris GK, Kerr GE, Fullard JH. 1978. Phonotactic preferences of female meadow katydids (Orthoptera: Tettigoniidae: *Conocephalus nigropleurum*). *Can. J. Zool.* 56:1479–87

104. Narins PM. 1982. Behavioral refractory period in neotropical treefrogs. *J. Comp. Physiol. A.* 148:337–44

105. Narins PM. 1992. Biological constraints on anuran acoustic communication: auditory capabilities of naturally behaving animals. In *The Evolutionary Biology of Hearing*, ed. DB Webster, RR Fay, AN Popper, pp. 439–54. Berlin: Springer–Verlag

106. Narins PM. 1992. Evolution of anuran chorus behavior: neural and behavioral constraints. *Am. Nat.* 139:S90–S104

107. Narins PM. 1992. Reduction of tympanic membrane displacement during vocalization of the arboreal frog, *Eleutherodactylus coqui. J. Acoust. Soc. Am.* 91:3551–57

108. Otte D. 1977. Communication in Orthoptera. In *How Animals Communicate*, ed. TA Sebeok, pp. 334–61. Bloomington: Indiana Univ. Press

109. Otte D. 1980. On theories of flash synchronization in fireflies. *Am. Nat.* 116:587–90

110. Otte D. 1992. Evolution of cricket songs. *J. Orthop. Res.* 1:25–49

111. Otte D, Loftus-Hills J. 1979. Chorusing in *Syrbula* (Orthoptera: Acrididae). Cooperation, interference competition, or concealment? *Ent. News.* 90:159–65

112. Otte D, Smiley J. 1977. Synchrony in Texas fireflies with a consideration of male interaction models. *Biol. Behav.* 2:143–58

113. Partridge BL, Krebs JR. 1978. Tree frog choruses: a mixed evolutionarily stable strategy? *Anim. Behav.* 26:959–60

114. Passmore NI, Telford SR. 1981. The effect of chorus organization on mate localization in the painted reed frog (*Hyperolius marmoratus*). *Behav. Ecol. Sociobiol.* 9:291–93

115. Peskin CS. 1975. *Mathematical Aspects of Heart Physiology.* New York: Courant Inst. Math. Sci. New York Univ.

116. Pollack GS. 1986. Discrimination of calling song models by the cricket, *Teleogryllus oceanicus*: the influence of sound direction on neural encoding of

the stimulus temporal pattern and on phonotactic behavior. *J. Comp. Physiol. A.* 158:549–61

117. Pollack GS. 1988. Selective attention in an insect auditory neuron. *J. Neurosci.* 8:2635–39

118. Prestwich KN, Walker TJ. 1981. Energetics of singing in crickets: effect of temperature in three trilling species (Orthoptera: Gryllidae). *J. Comp. Physiol. B.* 143:199–212

119. Römer H. 1993. Environmental and biological constraints for the evolution of long-range signalling and hearing in acoustic insects. *Philos. Trans. R. Soc. Lond. B.* 340:179–85

120. Ryan MJ. 1986. Synchronized calling in a treefrog (*Smilisca sila*). Short behavioral latencies and implications for neural pathways involved in call perception and production. *Brain Behav. Evol.* 29:196–206

121. Ryan MJ, Fox J, Wilczynski W, Rand AS. 1990. Sexual selection for sensory exploitation in the frog *Physalaemus pustulosus*. *Nature* 343:66–67

122. Ryan MJ, Tuttle MD, Taft LK. 1981. The costs and benefits of frog chorusing behavior. *Behav. Ecol. Sociobiol.* 8:273–87

123. Sakaluk SK, Snedden WA. 1990. Nightly calling durations of male sagebrush crickets, *Cyphoderris strepitans*: size, mating and seasonal effects. *Oikos* 57:153–60

124. Samways MJ. 1976. Song modification in the Orthoptera. I. Proclamation songs of *Platycleis* spp. (Tettigoniidae). *Physiol. Entomol.* 1:131–49

125. Schatral A. 1990. Interspecific acoustic behaviour among bushcrickets. See Ref. 8, pp. 152–65

126. Schwartz JJ. 1987. The function of call alternation in anuran amphibians: a test of three hypotheses. *Evolution* 41:461–71

127. Schwartz JJ. 1991. Why stop calling? A study of unison bout singing in a neotropical treefrog. *Anim. Behav.* 42:565–78

128. Schwartz JJ. 1993. Male calling behavior, female discrimination and acoustic interference in the Neotropical treefrog *Hyla microcephala* under realistic acoustic conditions. *Behav. Ecol. Sociobiol.* 32:401–14

129. Schwartz JJ. 1994. Male advertisement and female choice in frogs: findings and recent approaches to the study of communication in a dynamic acoustic environment. *Am. Zool.* 34: In press

130. Shaw KC. 1968. An analysis of the phonoresponse of males of the true ka-

tydid, *Pterophylla camellifolia* (Fabricius) (Orthoptera: Tettigoniidae). *Behaviour* 31:203–60

131. Shaw KC, Galliart PL, Smith B. 1990. The acoustic behavior of *Amblycorypha parvipennis* (Orthoptera: Tettigoniidae). *Ann. Entomol. Soc. Am.* 83:617–25

132. Shelly TE, Greenfield MD. 1991. Dominions and desert clickers (Orthoptera: Acrididae): influences of resources and male signaling on female settlement patterns. *Behav. Ecol. Sociobiol.* 28:133–40

133. Sismondo E. 1990. Synchronous, alternating, and phase-locked stridulation by a tropical katydid. *Science* 249:55–58

133a. Sobel EC, Tank DW. 1994. In vivo Ca^{2+} dynamics in a cricket auditory neuron: an example of chemical computation. *Science* 263:823–26

134. Soucek B. 1975. Model of alternating and aggressive communication with the example of katydid chirping. *J. Theor. Biol.* 52:399–417

135. Stiedl O. 1991. *Akusto-vibratorische verhaltensuntersuchungen an Ephippigerinen im Labor und im Biotop.* PhD thesis. Philipps Univ., Marburg, Germany

136. Strogatz SH, Mirollo RE. 1990. Synchronization of pulse coupled biological oscillators. *SIAM J. Appl. Math.* 50:1645–62

137. Strogatz SH, Stewart I. 1993. Coupled oscillators and biological synchronization. *Sci. Am.* 269(6):102–9

138. Sullivan BK. 1985. Male calling behavior in response to playback of conspecific advertisement calls in two bufonids. *J. Herpetol.* 19:78–83

139. Tejedo M. 1993. Do male natterjack toads join larger breeding choruses to increase mating success? *Copeia* 1993:75–80

140. Tuttle MD, Ryan MJ. 1982. The role of synchronized calling, ambient light, and ambient noise, in anti-bat-predator behavior of a treefrog. *Behav. Ecol. Sociobiol.* 11:125–31

141. Villet M. 1992. Responses of free-living cicadas (Homoptera: Cicadidae) to broadcasts of cicada songs. *J. Entomol. Soc. S. Afr.* 55:93–97

142. von Helversen D. 1984. Parallel processing in auditory pattern recognition and directional analysis by the grasshopper *Chorthippus biguttulus* L. (Acrididae). *J. Comp. Physiol. A.*154: 837–46

143. Wagner WE Jr. 1992. Deceptive or honest signalling of fighting ability? a test of alternative hypotheses for the function of changes in call dominant frequency by male cricket frogs. *Anim. Behav.* 44:449–62

144. Walker TJ. 1969. Acoustic synchrony: two mechanisms in the snowy tree cricket. *Science* 166:891–94
145. Walker TJ. 1983. Diel patterns of calling in nocturnal Orthoptera. See Ref. 64, pp. 45–72
146. Walker TJ. 1983. Mating modes and female choice in short-tailed crickets (*Anurogryllus arboreus*). See Ref. 64, pp. 240–67
147. Walker TJ. 1993. Phonotaxis in female *Ormia ochracea* (Diptera: Tachinidae), a parasitoid of field crickets. *J. Insect Behav.* 6:389–410
148. Wells KD. 1977. The social behaviour of anuran amphibians. *Anim. Behav.* 25:666–93
149. Wells KD, Taigen TL. 1989. Calling energetics of the Neotropical treefrog, *Hyla microcephala*. *Behav. Ecol. Sociobiol.* 25:13–22
150. Westby GWM. 1979. Electrical communication and jamming avoidance between resting *Gymnotus carapo*. *Behav. Ecol. Sociobiol.* 4:381–93
151. West-Eberhard MJ. 1984. Sexual selection, competitive communication and species-specific signals in insects. See Ref. 89, pp. 283–324
152. Whitney CL, Krebs JR. 1975. Mate se-
lection in Pacific treefrogs. *Nature* 255:325–26
153. Wilson EO. 1975. *Sociobiology.* Cambridge, Mass: Harvard Univ. Press
154. Wolf H, von Helversen O. 1986. "Switching–off" of an auditory interneuron during stridulation in the acridid grasshopper *Chorthippus biguttulus* L. *J. Comp. Physiol. A.* 158:861–71
155. Wynne-Edwards VC. 1962. *Animal Dispersion in Relation to Social Behaviour.* Edinburgh: Oliver & Boyd
156. Wyttenbach RA, Hoy RR. 1993. Demonstration of the precedence effect in an insect. *J. Acoust. Soc. Am.* 94:777–84
157. Young AJ. 1971. Studies on the acoustic behaviour of certain Orthoptera. *Anim. Behav.* 19:727–43
158. Young AM. 1981. Temporal selection for communicatory optimization: the dawn–dusk chorus as an adaptation in tropical cicadas. *Am. Nat.* 117:826–29
159. Zelick R. 1986. Jamming avoidance in electric fish and frogs: strategies of signal oscillator timing. *Brain Behav. Evol.* 28:60–69
160. Zelick R, Narins PM. 1985. Characterization of the advertisement call oscillator in the frog *Eleutherodactylus coqui*. *J. Comp. Physiol. A.* 156:223–29

Annu. Rev. Ecol. Syst. 1994. 25:127–65

EVOLUTIONARY BIOLOGY OF HUMAN IMMUNODEFICIENCY VIRUS

Andrew J. Leigh Brown

Centre for HIV Research, Division of Biological Sciences, University of Edinburgh, West Mains Road, Edinburgh EH9 3JN, Scotland

Edward C. Holmes

Department of Zoology, University of Oxford, South Parks Road, Oxford OX1 3PS, United Kingdom

KEYWORDS: virus-host interactions, lentivirus evolution, natural selection, HIV

Abstract

The human immunodeficiency viruses, HIV-1 and HIV-2, are members of a group of closely related viruses found in a number of different African primate species. More distantly related lentiviruses are found in several different mammalian orders. All are associated with long-term infections, but the outcome of infection ranges from a complete absence of symptoms to a rapidly developing immunodeficiency and death. While HIV-2 is probably directly related to a virus that is responsible for an asymptomatic infection in the Sooty Mangabey, no obvious candidate for the progenitor of HIV-1 has yet been found. Substantial genetic diversity is present in all immunodeficiency viruses, and phylogenetic analysis of HIV-1 sequences obtained from a wide range of geographic locations has revealed 5–7 groups of viral strains, all equally distant from each other. All groups have been found in Africa, but their distribution elsewhere reflects chance links between individuals at high risk of infection. In some areas large epidemics have spread through groups of such individuals to infect thousands within a few months, resulting in an increase in the global frequency of the particular strain responsible, without the occurrence of any significant diversification. In contrast, within infected patients, substantial

127

0066-4162/92/1120-0127$05.00

diversity is developed over the period of the infection, especially in regions of the *envelope* gene which are targets for immune recognition (frequency-dependent selection). However, this diversity appears to be reduced at transmission (stabilizing selection). Analysis of these different evolutionary forces gives insights into the development of drug resistance and to potentially protective immune responses which are of practical value, while providing novel observations on molecular evolution in real time.

INTRODUCTION—THE HIV PANDEMIC

By 1993 over 600,000 cases of AIDS had been diagnosed worldwide (175), and taking into account the incubation time of the disease and the probable extent of underreporting, it has been estimated that more than 10 million people could now be infected with HIV (21). These numbers are expected to continue to rise in the future, with a huge consequential cost in human suffering, serious economic loss, and even population decline in some nations in Africa (7). This situation is exacerbated by the fact that there is no agreed-upon view of how HIV causes AIDS despite the wealth of theories put forward (reviewed in 168), nor is there an effective vaccine available. Recently the long-term efficacy of AZT, the main antiviral drug in use against HIV, has also been questioned (1).

The international effort to develop effective vaccines and antiviral drugs is faced with a major problem in the sequence diversity exhibited by HIV both within and between infected individuals. This variation is a prominent feature of immunodeficiency viruses such as HIV and has been associated with the long-term maintenance of an infection in the presence of an active immune response, as well as with adaptation to a variety of cell types. A central task for biologists is therefore to understand the origin and maintenance of this variation. In particular, the analysis of HIV sequence data presents challenges for evolutionary biologists; methods of molecular evolutionary analysis, such as the reconstruction of phylogenetic trees and the estimation of the rates of nucleotide substitution, are an important means by which the origin of the virus and its interaction with the immune system can be understood. In this respect, HIV presents unique opportunities for the study of molecular evolution and also serves to illustrate the potential value of evolutionary methodology in providing an insight into the nature of infectious disease.

THE ORIGINS OF HIV

Introduction

The identification of a retrovirus as the etiological agent of AIDS within three years of the description of the disease was a major triumph of modern medical science (13). Like all retroviruses, HIV possesses a relatively simple structure consisting of two identical positive-sense single-stranded RNA molecules surrounded by a protein core, itself surrounded by a glycoprotein membrane (171). The RNA genome contains the three genes typical of retroviruses: the *gag* gene encoding the structural proteins of the viral core, comprising the matrix protein p17 (MA in the standardized retroviral terminology—98), the capsid protein p24 (CA) and the nucleocapsid proteins p6/p7 (NC); the *pol* gene which encodes the viral enzymes, the protease (PR) and the reverse transcriptase (RT), the structures of which have both been solved (85, 91) and the integrase (IN); and the *env* gene encoding gp120, the surface (SU), and gp41, the transmembrane (TM) envelope glycoproteins. HIV is also typical of retroviruses in its life-cycle, the most distinctive part of which is the production, by the viral enzyme reverse transcriptase, of a DNA copy of the RNA genome when the free virus enters a host cell (161). This DNA copy differs from the RNA template in that it possesses long terminal repeat (LTR) sequences at each end of the genome, which are produced during reverse transcription (162). The DNA copy of the virus is then integrated into the host genome where it resides as a provirus and is replicated, along with the host genome, by the cellular enzymes DNA polymerase I and RNA polymerase II. The latter enzyme transcribes the proviral HIV genome to generate viral mRNAs and the full-length genomic RNA molecule.

However, the structure of HIV differs dramatically from the C-type, or oncogenic retroviruses such as the ecotropic MLV and FeLV Type A (171) and the human T cell leukemia viruses (51). Most notably, the genome is substantially larger—about 9.5 kb vs about 7 kb for the C-type retroviruses of most animals. The additional length of HIV is associated with longer coding sequences, especially for *env*, and with a novel set of small regulatory genes, now known as *tat, rev, nef, vpu,* and *vif* (reviewed in 160). Of these new genes, specific functions have been assigned to the first two. The *tat* protein is responsible for activation of gene expression by interacting with a specific sequence (the TAR element) in the R region of the 5′ LTR, which is included in the transcript. Extensive investigations of its mode of action appear to indicate that TAT acts by binding to a sequence at the side of the loop formed by the TAR element in the newly transcribed mRNA and interacting with a cellular protein that appears to bind to the tip of the TAR loop to increase the rate of transcription (27, 28). The REV protein is responsible for determining the fate of transcripts; specifically, whether they enter the splicing pathway (the default mode) or whether they are protected from the nuclear processing machinery and exported into the cytoplasm as full-length genomic transcripts ready for packaging into viral particles. Again it is dependent on the presence of a specific sequence in the transcript—the REV response element (RRE), located at the 3′ end of the *env* gene (64). The presence of these regulatory proteins, along with the large variety of RNA transcripts that HIV produces and the ability to regulate its own gene expression, has been held to be adaptive in that it permits a flexible response to the cellular environment (116) and allows the infective state to be prolonged. Most recently it has been suggested that the *nef* gene may have the function of downregulating the cellular CD4 protein (63). Downregulation of their receptor is frequently observed in animal retroviruses, but the precise reason is not known.

Although HIV appeared to be significantly different from the C-type retroviruses found in other vertebrates, it shared certain features with visna, a virus associated with an epidemic of a wasting syndrome in Icelandic sheep in the late 1940s (147, 158) and a related virus found in goats (caprine arthritis-encephalitis virus, CAEV). Another virus which establishes a persistent cyclical infection, equine infectious anemia virus (EIAV) (17), had been linked with visna virus in a group known as the lentiviruses, to indicate their association with slowly progressive diseases. However, not until the identification of a second and clearly very distinct virus associated with AIDS in humans, HIV-2 (60), was it recognized that this was a major group of viruses with distinct properties.

The widespread nature of the lentiviruses among mammalian orders was confirmed by the publication in 1985 of news that rhesus monkeys in US primate colonies showing an AIDS-like disease were infected with a similar virus (30), and by the discovery of related viruses in domestic cats (feline immunodeficiency virus, FIV—122) and in the cow (bovine immunodeficiency-like virus, BIV—57, 58).

Whereas HIV, SIV in macaques, and FIV in domestic cats are associated with symptomatic, if chronic, infections, other viruses clearly related to HIV have now been discovered in a wide variety of different primates, where symptomatic infection appears to be the exception rather than the rule. In addition, antibodies that cross-react with FIV have been found in wild felid species worldwide, such as East African lions and Serengeti cheetahs, and related viruses have been isolated from North American pumas (123) with no associated pathological symptoms. Similarly, a number of different primate species are known to be infected with apparently species-specific immunodeficiency viruses. These viruses have been the subject of detailed phylogenetic studies which have clarified some important elements of their evolutionary history.

Primate Immunodeficiency Viruses

Although the first nonhuman primate in which an immunodeficiency virus was discovered was the rhesus macaque (*Macaca mulatta*; SIV_{mac}) (18), immunodeficiency viruses have so far been found to be endemic only in African cercopithecids. The monkey species infected include the sooty mangabey

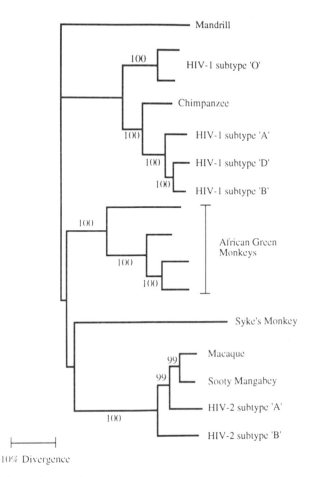

Figure 1 Phylogenetic tree of primate immunodeficiency viruses based on analysis of 2613 nucleotide sites in the *pol* gene. The analysis was carried out using the neighbor-joining method by Paul Sharp and colleagues (144), to whom we are grateful for permission to reproduce the tree. The five main lineages described to date are indicated by the brackets on the right.

(*Cercocebus atys*—SIV$_{sm}$) (66); the mandrill (*Papio sphinx*—SIV$_{mnd}$) (159); the Sykes monkey *Cercopithecus mitis albogularis* (SIV$_{syk}$) (67); and the African green monkey (*Cercopithecus aethiops*—SIV$_{agm}$) (4, 10, 50, 81). The last includes a cluster of closely related species, each with an apparently specific viral form, which are discussed in more detail below.

Phylogenetic analyses of these viruses, carried out using the nucleotide sequences of the relatively conserved *gag* and *pol* genes, have shown how the SIVs and HIVs are related (Figure 1). Remarkably, HIV-1 and the viruses from each of the four monkey species appear to be approximately equally distantly related to each other whatever method is used in the analyses. HIV-2, however, was found to be closely related to the immunodeficiency virus found in the rhesus macaque (18) and the sooty mangabey (68). SIV$_{mac}$ was isolated from captive rhesus monkeys in US primate centers in 1986 (18), but it was soon found that wild, Asian, rhesus macaques were not infected with the virus (35). Similar studies in wild-caught African primates in 1988 led to the discovery of the closely related virus SIV$_{sm}$ (68). Both monkeys were held in the same primate centers at the same time. Experiments have now shown that SIV from a sooty mangabey, which causes no disease in that species, will infect a rhesus macaque and often cause an AIDS-like disease, killing the animal within about 18 months (35). Epidemics of such diseases were recorded among macaques from some primate centers in the early 1980s (77, 99). The likely explanation is therefore that a transspecific infection occurred from a wild-derived sooty mangabey which then spread among captive macaques.

This leaves the question as to how HIV-2 comes to be so closely related. The answer is probably, also by a transspecific infection from the sooty mangabey. The evidence for this was initially circumstantial but suggestive. HIV-2 is the form of the virus found in West Africa and in people with West African connections. Sooty mangabeys are a West African species and are often found in close contact with humans, frequently being kept as pets. Bites from such animals are known to transmit viral infections, and this may be the main route of transmission among monkeys (M Gardner, personal communication). A recent detailed phylogenetic analysis of HIV-2 isolates from residents of rural Liberia has revealed sequences in infected humans that are more closely related to SIV$_{sm}$ than to other HIV-2 sequences (52). In the light of these observations, the simplest interpretation is that zoonotic transmissions occur continually at a low rate but are self-limiting. The reasons for the spread of one viral strain following the transspecific infection may relate to the epidemiological circumstances of the infected population, or possibly to some properties of that particular virus. However, it is not possible to link HIV-2 with any particular known variant of SIV$_{sm}$, and indeed, other HIV-2 isolates that are equally distinct have been described by other groups (39, 88, 89).

Where Did HIV-1 Come From?

The development of a plausible hypothesis to account for many features concerning the origin of the HIV-2 epidemic associated with West Africa has, however, contributed little to our understanding of the source of the global pandemic of HIV-1. The most direct evidence relating to the origin of HIV-1 came from the discovery of that 2 out of 83 wild-caught chimpanzees from Gabon carried antibodies that cross-reacted with HIV-1 (74). From one of these a distinct immunodeficiency virus was isolated that when sequenced grouped specifically with HIV-1. The distribution of this virus among wild chimpanzee populations is poorly known, and the infection appears to be very rare; only one other virus has been isolated from this source. Recently, however, some very distinct isolates of "HIV-1" have been obtained from the Cameroon (32). Provisional analysis indicates that these may be more divergent from other HIV-1s than SIV_{cpz} (116). Although contacts between humans and chimpanzees cannot therefore be ruled out as an hypothesis for the origin of HIV-1, along the lines proposed for HIV-2 and sooty mangabeys, the evidence is considerably weaker.

A more distant origin of HIV-1 has been proposed as a branch of the SIV_{agm} group (50). Phylogenetic analyses of the most complete dataset available at the time of writing show that whereas different SIV_{sm} sequences, like the different HIV-1s, cluster closely together on a tree of all primate immunodeficiency viruses, SIVs from the African green monkey differ very substantially from each other and are separated on the tree by relatively long branches, as can be seen in Figure 1.

The African green monkey represents a "superspecies" made up of at least four geographically isolated species: green monkeys (*Cercopithecus sabaeus*) from West Africa, grivets (*C. aethiops*) from the Horn of Africa, Ethiopia, and the Sudan, the tantalus monkey (*C. tantalus*) which is found in a belt extending across Africa from Uganda to Nigeria, and the vervets (*C. pygerythrus*) found over a wide range from Ethiopia southwards and extending throughout southern Africa and Angola. Each of the last three species is represented by more than one subspecies; over 10 are listed in the case of the vervet. The diversity shown by the SIVs from the superspecies as a whole (4, 10, 50, 81) led to suggestions that somewhere among the group a virus would be found that grouped with HIV-1. Instead, more recent analyses suggest that the diversity reflects phylogenetic relationships with the superspecies (4). An attempt is being made to test this hypothesis by investigating the molecular phylogenetics of green monkey mtDNA sequences (J Goudsmit, personal communication), and it will be interesting to learn more of the relationships among the group. As approximately 30% (or more) of indigenous African green monkeys are seropositive for SIV (121), it appears this virus has a long history in these species. While we cannot at present identify a particular nonhuman primate origin for HIV-1, it may lie among other species of the genus *Cercopithecus*, several more of which have antibodies that cross-react with SIV_{agm} (82).

The Age of HIV

A number of molecular evolutionary investigations have attempted to estimate how long immunodeficiency viruses have been associated with human populations. The basis for such attempts was, in several cases, the assumption that HIV sequence divergence is a continuing process and that the evolutionary divergence (i.e. number of nucleotide substitutions) between two strains isolated at different times can be used to estimate the rate of evolution. The likely time of origin is then inferred by extrapolation, and in several cases this was taken as the divergence of HIV-1 and -2. However, we have seen above that this is not relevant to the origin of HIV because it is now clear that the common ancestor of the two human viruses is to be found in a nonhuman primate. The divergence of HIV-1 and HIV-2 in fact appears to be close to the earliest node in primate lentiviral evolution (39).

Secondly, the dating procedure itself is highly suspect, as it is often assumed that HIV-1 followed a "progressive" or stem-like pattern, similar to that of influenza (15). In fact, the accumulation of sequence information on HIV-1 from different parts of the world and particularly Africa has shown that HIV-1 has undergone successive "star-like" radiations, with from five to seven distinct and equally distantly related groups detectable (see Global Variation of HIV-1). Such a lack of structure within the tree invalidates the assumptions being made in some dating procedures. Furthermore, in more than one case, the analysis was also confounded by comparing sequences from different branches of the radiation. Finally, the occurrence of multiple substitutions at single nucleotide sites and especially the preponderance of G to A transitions (see The Evolution of HIV Within an Individual), means that all approaches to inferring the time of origin for such highly divergent viruses are likely to be speculative, particularly when based on procedures, such as parsimony, which do not adequately correct for multiple nucleotide changes and differences in rates of evolution between lineages (150). The inherent difficulties in dating are indicated by the range of estimates obtained, from as little as forty to several thousand years for the divergence of HIV-1 and HIV-2 (100, 143, 150, 176).

Examination of the primate immunodeficiency tree (Figure 1) reveals an intriguing feature. All the major lineages of primate lentivirus appear to be related to each other by approximately equal distances. Similarly, the lentiviruses from different orders of mammals also appear to be related by approximately equal distances (57). Such observations have led to the radically different suggestion that immunodeficiency viruses have been present in mammals for millions of years and co-evolved with their hosts. The current picture of evolution of SIV_{agm} viruses described above is certainly in line with this, but there appears to be some difficulty in reconciling such a process with direct observations made on HIV sequence change within individual patients and infected communities (see below). Estimates of the error rate of HIV reverse transcriptase are on the order of 10^{-4} per site, per round of replication (136). We propose an alternative explanation which resolves these difficulties later in this chapter.

The Evolution of Virulence in Immunodeficiency Viruses

An important question arising from the comparison of primate lentiviruses outlined above is the mechanism by which viral adaptation to the host species evolves in the primate lentiviruses. Of particular interest in this respect is HIV-2, which has been associated with a lower pathogenicity than HIV-1. It is instructive in this case to refer to theoretical studies of host-parasite relationships. A commonly cited theory is that virulence is due to an unnatural relationship between virus and host, such as that caused by a transspecific infection, and that because infections often appear to be less virulent in their natural hosts, viruses should evolve to a state of avirulence. This view was based on the idea that viruses that inhabit their hosts for the greatest time are the most successful and that the death of host means the death of the virus.

However, this is a rather anthropomorphic view of viral evolution. The important parameter is not the duration of infection or even host mortality but rather the effective reproductive rate, R_e, defined as the average number of new infections generated by one infectious individual. Thus, a virus with a higher R_e will pass more of its progeny to future generations and consequently R_e should be maximized in evolution (6). Consequently it has been predicted that viruses with the highest probability of transmission will be those with intermediate rates of growth and levels of virulence. This evolutionary trade-off between host mortality and transmission probability is best illustrated in the natural "experiment" of myxoma virus released into wild rabbit populations in Australia and Europe. Here strains of intermediate fitness attained the highest frequency in the population; very virulent myxoma strains kill their hosts early and are consequently transmitted at lower rates between hosts than are strains of lower virulence (44, 45). In contrast, benign strains are cleared by the host immune system before they are transmitted.

It is unlikely that similar selective forces are the explanation for the apparent reduction in pathogenicity in HIV-2 as no one has claimed to have found evidence of the required dramatic decline in human population numbers, as was pointed out earlier (96). In contrast, the lower pathogenicity of HIV-2 may be due to functional differences. For example, the fact that the external glycoprotein of HIV-2 does not bind to the CD4 receptor with such high affinity as do many strains of HIV-1 could explain the lower pathogenicity; i.e. it could be due to incomplete adaptation to the host (79, 102, 113).

A recent analysis of rates of amino acid change in viruses with different pathogenicities (51) found that rates of amino acid change were two to three times higher in HIV-1 and SIV_{mac} (both pathogenic) than in the minimally pathogenic SIV_{agm} and SIV_{sm} viruses, as well as HIV-2, considered to be of intermediate pathogenicity. More specifically, a positive correlation was proposed between viral pathogenicity and the immunogenicity of the *env* (and particularly the surface glycoprotein gp120) and *gag* proteins. This suggests that whereas a weak immune response is able to "contain" the infection in African green monkeys and sooty mangabeys, a stronger immune response does not contain the infections in humans (with HIV-1) and macaques. It is therefore tempting to think that some SIV strains are inherently less virulent viruses. On the other hand, SIV_{mac} originated almost certainly from the transspecific infection of a rhesus macaque with SIV_{sm}(35). Either SIV_{sm} was inherently more pathogenic in the new host, a common occurrence among viruses (e.g. myxomatosis passing between rabbits of the genus *Sylvilagus* and *Oryctolagu*—44, 45), or more virulent variants were rapidly selected for as the virus passed from a "SIV-saturated" population into a "SIV-naive" one. It is interesting to note the presence of extreme variation in virulence within SIV_{sm}, represented by the strain SIV_{PBj14} (isolated from a pigtail macaque infected with a standard strain of SIV_{sm}) that kills macaques within days of infection. A limited number of nucleotide substitutions distinguish the lethal variant and the parent strain, and most of these have been found in the *env* gene (26).

Studies in areas of West Africa where HIV-2 is widespread and where HIV-1 has only recently entered the population have confirmed that progression to AIDS is significantly slower in HIV-2 infected patients, and that HIV-2 is much less easily transmitted (8). The frequency of mother-child transmission is about 10- to 20-fold higher in HIV-1-infected than HIV-2-infected mothers (43). Thus it is now clear that the two human immunodeficiency viruses differ fundamentally in virulence. Nevertheless, given the high mutation rate of HIV-1, it is likely that strains of low virulence currently exist. The study of individuals who possess these viruses, perhaps "long-term survivors" of HIV infection, is likely to be an important part of future HIV research. However, it is important to distinguish differences due to the viral strain from those due to host factors such as age (31) and HLA haplotype (153).

MOLECULAR EPIDEMIOLOGY OF HIV-1

Global Variation of HIV-1

As the first sequences of African isolates of HIV-1 were made available, it became clear that there was greater diversity among them than among those obtained from infected individuals in North America and Europe (3, 116). For this reason a simple African/Non-African subdivision is not sufficient to explain the worldwide diversity of HIV-1. Recent world-wide surveys of HIV-1 sequences have revealed a global pattern of geographical variation that is extremely complex.

From an analysis of complete *gag* and *env* sequences, Myers et al (116) and

Louwagie et al (103) suggest that there are about seven approximately equidistant subtypes of HIV-1. All seven subtypes have been found in Africa, but only a subset are present on each of the other continents. Subtype A is found in Central and West Africa and includes the well-characterized strains U455 (Uganda) and Z321 (Zaire). Subtype B is the predominant one found in North America and Europe and includes the first strain of HIV to be cultured, now known as HIV$_{LAI}$ (166). It appears to be present, but rare, in Central Africa. Although they are abundant in Brazil as well, some B subtype sequences there are quite distinct (132). Subtype C is found in South and Central Africa and recently has been found on the west coast of India, in Bombay and Goa, where it appears to have generated a major epidemic (36), while subtype D is found only in Central Africa. Subtype E was first found in the HIV epidemic in a northern part of Thailand among injecting drug users (107, 126), but related strains have since been identified from South and Central Africa. Whereas the grouping of viral strains into subtypes is, in general, similar whether sequences from the *gag* or *env* gene are used, the subtype E viruses have been found to group with A on the basis of the *gag* gene.

A sixth subtype "F" ("E" in Louwagie et al—103) has been identified which includes a second set of sequences described from Brazil (132), as well as a group from Romania (37) and Zaire. There are, in addition, sequences from Gabon that appear to fall into two more subtypes—G (which includes some more from Zaire) and H. Finally, at the time of writing, a last outlying sequence from Zaire may define a ninth subtype. In general, the *gag*- and *env*-based trees were highly congruent, with the exception of subtype E and a small number of D subtype viruses, including the Zairian isolate HIV$_{MAL}$ (3) whose *gag* gene sequence falls into subtype A. This suggests that recombination has occurred in the distant past in a patient infected with two divergent viruses. As well as this array of subtypes more or less equally distantly related, there are the sequences from the Cameroons that branch off outside SIV$_{cpz}$. Despite the fact that they have been assigned to a single subtype (referred to as "subtype O"), they are quite distant from each other as well (116).

WHAT CAN BE INFERRED FROM THE GLOBAL PATTERN OF DIVERSITY?   The HIV-1 subtypes differ at 10–13% of nucleotide sites and 20–25% of amino acids over the whole of *env*. This extent of diversity is not an inevitable consequence for an RNA virus. In particular, studies of the measles virus show that most of its genome is rather conserved, with six geographically distinct American strains isolated over a five year period, showing pairwise nucleotide distances of only 2.4% in the hemagglutinin protein gene (138), distances exceeded by HIV sequences found within a single patient. In an analysis of the VP1/2A region (222 bp/74 amino acids) of the genome from 59 strains of poliovirus type 1, another RNA virus, isolated over a 30-year period, Rico-

Hesse et al (134) found that they differed on average by approximately 1 amino acid substitution only from the Sabin vaccine strain, but they did differ at a high proportion of their synonymous sites. The diversity we have described in HIV would not arise inevitably from the high mutation rate associated with an RNA genome; instead it reflects the sum total of the evolutionary forces acting on the virus.

Several inferences about the evolutionary biology of HIV may be made from the information available about the distribution of the subtypes. First, different subtypes, which may differ substantially at the molecular level, can be found within the same geographical area, and especially in Central Africa. The fact that all the subtypes are found here, and that this therefore is the region of greatest genetic heterogeneity, may imply that the virus has existed here for the longest time. Secondly, the ability to partition the total spectrum of HIV diversity into a limited number of groups allows us to infer something about the spread of the virus. Eigen & Nieselt-Struwe (39), who undertook a phylogenetic analysis of a number of different isolates of HIV, considered the virus to be an example of a "constrained pandemic" where only a few genotypes coexist both geographically and temporally. This situation was contrasted to a phylogenetic tree of variants of type 1 poliovirus, where most branches give rise to descendants and where many genotypes appear to coexist. However, this interpretation of the phylogenetic tree structure in terms of epidemiological pattern was not based on an explicit evolutionary model. An alternative hypothesis is that the present structure of the global HIV tree with the remarkable expansion of lineages in some countries that are rare in the parent continent (Africa) is an example of the founder effect on a large scale; i.e. is a consequence of sampling. The following three cases serve as examples.

THAILAND   The need for a precise description of the genetic diversity of HIV within infected communities is apparent in an analysis of HIV in Thailand, a country that had an exponential increase in the frequency of HIV infections among high risk groups in the late 1980s and that is still experiencing a rapid increase in the number of infections. By mid-1991 approximately 0.5% of the population were estimated to be infected (107, 126). Two different subtypes (B and E) are present in this country; subtype E is predominant in North Thailand, while subtype B is the major one found in the central and southern regions, including Bangkok. No patients of dual infection have been found. Because of the very different global origins of the subtypes, it is likely that these viruses entered the Thai population independently. The two viruses are also associated with different risk groups: while 85% of the patients infected through sexual transmission possessed viruses of genotype E, only 24% of individuals infected through injecting drug use had viruses of this genotype. At present little is known of possible differences in efficiency of transmission by different routes between the different subtypes, but it is equally likely that the split arises from links between risk groups in different communities and is a reflection of demographic factors instead.

FINLAND   A similar picture of very different subtypes of HIV within a single country is provided by Finland (140). Despite the much lower prevalence of HIV in this country (compared to Thailand) four and possibly even five subtypes of the virus have been found there. Interestingly, the sequences obtained from individuals infected in nearby Estonia constitute an extremely homogeneous monophyletic group, which may reflect the close epidemiological relationships of individuals in a country where very few cases of HIV infection have been reported.

INDIA   Recent studies of high-risk groups in Bombay and Goa, on the western coast of the Indian subcontinent, have revealed another unexpected subtype. This recently infected group, in which HIV-1 is spreading rapidly, are infected with a strain belonging to subtype C. This was previously described from a single strain, HIV$_{NOF}$, found in South Africa, and the subtype has also been found at a low frequency in Zambia. Unusually, there is also a high prevalence of HIV-2 infected individuals, and several dual infections have been observed (36). Other centers of infection now appear to exist in India, and preliminary results suggest that, as for Thailand, these are characterized by viruses of different subtypes. The exact routes by which such an unusual combination of HIV strains reached Bombay/Goa remain unknown, but there is a possible association with long-established trading routes across the Indian Ocean.

In conclusion the pattern of distribution and the relative abundance of the different subtypes of HIV-1 appear to be primarily a consequence of demographic processes within the human population. There are significant differences in behavior, such as rates of sexual partner change, or frequency of sharing needles, which generate subpopulations that differ substantially with respect to their susceptibility to HIV infection (80). Studies of the dynamics of sexually transmitted diseases are now beginning to deal specifically with this issue (53). In our view these factors have had a major impact on the global distribution of the subtypes of HIV-1. The consequent level of variability means that any vaccine designed to eliminate a single specific strain may not be sufficient to protect all populations. Clearly a careful study of HIV diversity world-wide will be an important part of global intervention strategies.

Evolution of HIV Within a Community

The evolutionary analysis of HIV nucleotide sequence data can also provide information as to progression of the HIV epidemic within a single community. In particular, it is important to know how many HIV strains may have established the epidemic in a particular community, whether different risk (behavioral) groups possess characteristic strains, and how the distribution of these strains changes with time.

Studies on genetic variation of HIV in clustered outbreaks, several of which were associated with exposure of a number of patients to a common source of virus, such as infected blood or blood product, have revealed that there can be substantial similarities even in highly variable regions of the *env* gene (11, 22, 108). Such cases are highly informative because differences in the course of the disease cannot be ascribed to variation in the infecting virus strain. In Edinburgh, a number of hemophiliac patients, who had never been treated with commercial factor VIII, were found to have become infected in 1984. Analysis of their records revealed that they shared exposure to a single common batch of factor VIII which had been prepared in Scotland from plasma donated locally at a time when HIV infection was extremely rare (104). Sequences taken from some of these patients close to the time of the infection showed almost no divergence between patients, consistent with the infecting virus having derived from a single infected patient (177). Balfe et al (11) were therefore able to compare the rate of viral sequence change between different infected patients and to show that both the diversity and the rate of change varied substantially (11). In general the greater the diversity in the population, the more rapid the progress of disease. In these patients the presence of one particular HLA haplotype, A1 B8 DR3, was associated with rapid progression (153). Thus, host factors alone can be responsible for important differences in the infection and may also influence the development of diversity in the viral population.

The earliest stages of an explosive HIV epidemic appear to resemble the pattern seen in infection from a point source such as this. It was seen in Thailand and India (Bombay/Goa) that even when there is no known direct link between infected individuals, transmission appears to occur so rapidly that the virus does not diverge between individuals. In both Thailand and India, despite the highly divergent genotype of the virus found, virtually no inter-patient divergence was seen.

HIV EVOLUTION IN EDINBURGH   We have recently completed a detailed study of the molecular epidemiology of HIV-1 within Edinburgh (70). Although all the viruses found could be assigned to subtype B, phylogenetic analysis of the p17 (MA) region of the *gag* gene strongly suggested that distinct HIV strains circulate in this community, with different risk groups (hemophiliacs, heterosexuals, and injecting drug users) characterized by distinct viral variants. Specifically, when we sequenced the virus from all available hemophiliac patients who had been exposed to the common batch and become infected, a total of 14 patients (104), distinct *gag* sequence variants were found (Figure

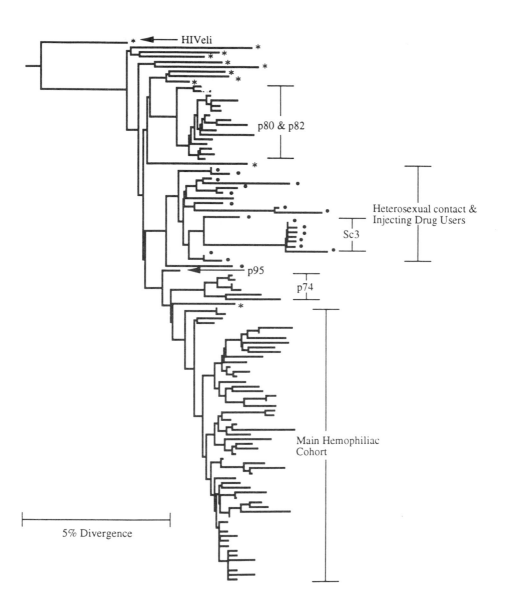

Figure 2 Neighbor-joining tree for 112 sequences of the MA (p17) coding region of the *gag* gene from 25 Edinburgh patients infected by three different routes (70). The sequences of 10 worldwide isolates of HIV-1 subtype B are included, along with the subtype D isolate HIVELI which roots the tree. Sequences from hemophiliac patients have no symbol. Sequences from patients infected by intravenous drug use or heterosexual contact are denoted by the • symbol and the sequences of worldwide isolates are identified by the symbol: *. p82 and p74 are hemophiliac patients who were exposed to the main implicated batch of factor VIII. p80 was not exposed to that batch but shared a different batch with p82. p95 was infected from commercial (US-derived) factor VIII. Patient Sc3 was infected by heterosexual contact.

2). Although the majority of patients still appear to have been infected by a single ("main") batch of factor VIII, one patient exposed to this batch (p82) became infected with a very different virus population. However, these sequences were very similar to those observed in another patient (p80) who had never been exposed to the main batch. Clearly there must have been more than one infected batch of factor VIII in use in Edinburgh in 1984.

Conversely, a single *gag* variant, different from those observed in the hemophiliacs, was found in the infected injecting drug-user (IDU) population in Edinburgh. This suggests that the HIV epidemic in this group may have been founded by a single HIV strain, a conclusion that is compatible with the rapid spread of the virus in this population, similar to the epidemic in Thailand (137). Significantly, this variant was also found in individuals thought to be infected through heterosexual contact, which suggests that the heterosexual epidemic may have been established by descendants of the variant that earlier infected the intravenous drug users, and that the latter may form a bridge into the heterosexual community.

The sequence of the p17/MA coding region appeared to be diversifying as the epidemic progressed, because sequences obtained from individuals infected during the recent stages of the HIV epidemic are often quite distinct (that is, connected by long branches on the phylogenetic tree) from sequences obtained from individuals infected at earlier time points. In contrast, little time-dependent evolution appears in the V3 loop of the *env* gene, a region of great immunological importance (see Sequential Studies). Here similar (and sometimes identical) sequences are found in unrelated individuals throughout the course of the epidemic, in marked contrast to the diversity in V3 often observed within individual patients (71). Studies on recent infections have suggested that only a subset of V3 sequences may be responsible for establishing many infections (84, 108, 173, 177, 179). As the V3 loop is likely to be an important component of any *env*-based vaccine, such a constraint against sequence change through time is a significant finding.

A similar observation was made by Kuiken et al (90) who examined the evolution of V3 sequences in Amsterdam from 1980 to 1991. The consensus V3 loop amino acid sequence taken at seroconversion (or a time near to seroconversion) from 74 individuals did not change over the course of the epidemic, although there were changes in the residues flanking the V3 loop. Furthermore, no risk group–specific changes could be identified at the amino acid level, although injecting drug users could be differentiated from homosexuals/hemophiliacs by two particular synonymous nucleotide substitutions.

HIV Forensics

In recent years there has been considerable interest in the use of molecular sequence data as a means to assess the likelihood of HIV transmission between infected individuals. While this subject, "HIV forensics," is still in its infancy, it shows a potentially important use of evolutionary methodology of HIV research. The most celebrated case began in 1990 when the Centers for Disease Control in Atlanta received a report of AIDS in a young woman with no recognized risk factors (designated patient A). About 24 months before this diagnosis, however, she had two teeth extracted by a dentist suffering from AIDS. Subsequent investigations identified six more former patients of the dentist (patients B to G) who were HIV positive.

To assess whether these patients were infected by the dentist, an analysis was made of their HIV sequences along with those of the dentist and 35 local controls (LCs) (125) (HIV infected individuals attending clinics within a 90 mile radius of the dental surgery). Three analyses were performed on a 350 bp long region from the *env* gene including the hypervariable V3 loop. The first involved a simple calculation of the genetic distances between all members of the study data set. An average distance of 4% (range 3.4 to 4.9%) was found between the dentist and five of the patients who were considered to have no behaviorial risk factors for HIV infection (patients A, B, C, E and G—now referred to as the "dental cohort"). These values are within the range of known HIV transmissions from a point source, as described above and are close to the levels of variation observed within single patients. In contrast, the average distance to the dentist for the LCs and the two patients who did have behaviorial risk factors (patients D and F) was 11%. A similarly high distance was found between the dental patients and the local controls.

The second piece of analysis involved the reconstruction of phylogenetic trees using the parsimony criterion. Although all the most parsimonious trees placed the dentist and the dental cohort closely together, it appeared that the grouping could not be supported by a "significant" number of bootstrap replications (only 79 out of 100 replications), although there is some debate as to at what level of support significance should be assigned (65). Finally, the authors performed a "signature pattern analysis" where eight (noncontiguous) amino acid residues common in the dentist's viral sequences, yet rare in the local control population, are used to constitute a signature sequence. The five dental cohort patients all had at least seven out of eight of these signature residues, whereas most of the local controls had only one or two of these residues (the maximum was five out of eight). This similarity in signatures was statistically significant.

Since the case of the Florida dentist occurred, a number of other attempts have been made to trace the source of HIV infections by the evolutionary analysis of viral sequence data. In the first, a region of the *pol* gene was used as evidence in a Swedish court to uphold the conviction of an individual for rape who knowingly infected a female patient with HIV (2). The rapist and the victim could be distinguished from a background population by a five amino acid signature sequence as well as in a phylogenetic analysis.

Another investigation involved an HIV-infected surgeon one of whose patients was found, on recall, to be a previously unknown seropositive (72). Significantly, the patient had also received a blood transfusion and the donor of one unit of the transfused blood had recently been found to be HIV positive, although it was not known whether this applied at the time of the donation. Phylogenetic trees were constructed on sequence data obtained from p17/MA coding sequences PCR-amplified from cells of the patient, the donor and plasma from the surgeon, as well as from a background of mainly North American sequences. All trees depicted the donor to be the source of the HIV infection in the patient. Furthermore, the patient-donor grouping was found to be significantly more likely than the patient-surgeon grouping in a maximum likelihood significance test, indicating that it was highly unlikely that the surgeon was the source of the HIV infection in the patient.

Although further cases are likely to arise in the future, a number of uncertainties still exist as to the strength of the conclusions that can be drawn on the basis of evolutionary analysis of viral sequence data. The key issues are the choice of a region of the HIV genome that best reflects evolutionary history and the accurate estimation of the probability that any sequence similarity observed is indicative of HIV transmission. These are obviously areas where the input from evolutionary biologists will be of great importance.

EVOLUTION OF HIV WITHIN AN INDIVIDUAL

The Quasispecies: Mutation and Selection

For many years we have known that many RNA viruses cannot be considered solely in the traditional taxonomic way that is appropriate for DNA viruses: as being composed of a limited number of strains, defined serologically. For example, from the earliest studies of HIV sequence variation, substantial sequence evolution during the course of an HIV infection was found such that infected individuals harbor a large population of viruses that may differ in both sequence and biological characteristics (48, 61, 139, 152).

Replication is an error-prone process in all retroviruses. These viruses are unusual because three enzymes are involved in replication: cellular DNA polymerase, cellular RNA polymerase, and the viral reverse transcriptase (24). Most studies of mutational frequency have concentrated on reverse transcriptase, particularly because of the lack of proof-reading activity associated with this enzyme. With mutation rates of the order of 10^{-4} per base pair, per replication (136), a population of viruses such as HIV-1 (genome size 10^4 nucleotides) will contain very few identical genomes.

A number of other mutational processes also appear to be important in the generation of sequence diversity in HIV. Hypermutation is a name given to the process whereby monotonous base substitutions, usually involving G to A transitions, occur during a single replication cycle, presumably because of a property of a particular reverse transcriptase (128, 163). Although many of the mutations generated by hypermutation will lead to viruses with defective genomes, recombination, another major source of diversity in immunodeficiency viruses (24), may lead to a proportion of these substitutions spreading through the viral population.

It is also evident, however, that natural selection plays an important part in the evolution of HIV, and especially in changing the antigenic properties of the virus. The central observation in this respect from studies of other lentiviruses is that serologically distinct viruses can be isolated later in the course of the infection that are not efficiently neutralized by the sera obtained shortly after inoculation, when the viruses are of the parental serotype (164). Thus, sequence variants can be serologically distinct from each other, and the differences between them have generally been localized to changes in the amino acid sequence of the envelope glycoprotein. In HIV, many amino acid substitutions have been observed in a putative loop structure within the third hypervariable (V3) region of the major envelope glycoprotein gp120. This region is the major immunodominant neutralizing epitope of HIV (59, 180), and mutations here are neutralization "escape" mutants. Despite the requirement for variability, V3 loop sequences, and especially the central GPG tripeptide, are relatively well conserved (94, 116) and can elicit antibodies capable of neutralizing a spectrum of viral isolates (169, 170). Furthermore, this region also encompasses an MHC–restricted determinant for cytotoxic T lymphocytes (CTLs) (155) and is part of the major determinant of macrophage and T cell tropism (See Genetic Variation, Cell Tropism, and Pathogenesis).

The interaction of mutation and selection gives rise to the quasispecies—a population of viral mutants with different replication rates (38, 117). A quasispecies population structure is commonly observed in RNA viruses (69), and the concept has been applied to HIV in recent years (109, 165), although it is often incorrectly used simply as a surrogate for genetic heterogeneity.

The basis of quasispecies theory is as follows. The virus is thought not to exist as a single (chemical) "species" but rather as a cloud of points within the space of all possible nucleotide sequences of that length. Each of the points within that cloud differs by one or more nucleotide substitutions from all others. However, there is a sequence that has the highest Darwinian fitness. This, the "master-sequence," may be present at very low frequency due to mutation pressure and so may be almost "notional," but there will be many sequences that are only one nucleotide (or "Hamming distance") away from it. In contrast to expectations from classical population genetics, the fitness of these sequences will be increased by their proximity in mutational distance to the fittest form, even if their own intrinsic fitness is significantly lower. The reason is simply that the frequency of mutations that will generate the "master-sequence" is not negligible. The population of sequences around the master-sequence will therefore replicate most rapidly. The "master-sequence" can be estimated by taking the most common nucleotide at each site—i.e. the consensus sequence. In the terms of population genetics, strong stabilizing selection acts as the countervailing force to the diversifying effect of the high mutation rate and constrains the spectrum of variation in the population.

This theory does not differ in fundamental respects from the core of population genetics theory as developed by Fisher, Haldane, and Wright, but rather in emphasis, considering in particular the consequences for a population of immense variability and very high mutation rates, and considering the genome as a whole. There have, for example, been earlier studies of the maximum size of an RNA genome, or genome segment (133). The most important distinction is that the frequency of a viral variant depends not only on its own replicative value but also on the likelihood with which it is produced by the erroneous replication of other variants. On the other hand, the concept of variable fitness is something that has not been a feature of quasispecies theory but is inherent to the population genetics of host-parasite interactions (23, 62, 105, 106, 142).

Evidently, the central evolutionary consequence of the quasispecies is that it allows selection from a pool of preexisting variants and provides potential for rapid adaptation to new environmental conditions, such as new antibody response or antiviral agents. Evidence for the existence and selection of such diversity is provided by the rapid resistance produced to antiviral therapy, particularly to the non-nucleoside drugs such as Nevirapine and inhibitors of HIV protease, against which resistant strains can appear in vitro (12, 40, 124). The antiretroviral drug in most use, azidothymidine (AZT), requires a total of 3–4 nucleotide substitutions from "wild type" to generate a fully resistant polymerase, but even this often occurs within six months of the start of treatment (14, 92, 93). The suggestion that the speed and reproducibility of the development of AZT resistance was due to the presence in HIV populations of naturally occurring variants has been raised by observations of some of the relevant mutations before exposure to AZT (112, 178).

Another important consequence of the quasispecies concept for those interested in understanding the evolutionary biology of HIV and especially through phylogenetic reconstruction is that, to date, all methods of phylogenetic reconstruction are most reliable when most substitutions are neutral (or nearly so), and thus it is difficult to assess the performance of such methods in the presence of natural selection, and particularly of convergent evolution (71). As well as natural selection, the bias in base composition and substitution seen in HIV, and particularly the strong tendency for G-to-A changes (114), is also likely to cause problems for methods of phylogenetic reconstruction and especially through the accumulation of multiple substitutions at single nucleotide sites.

Finally, it is important to note that the concept of a continuously diverging tree relating viral variants is not an evolutionary truism, but an assumption. The process of base substitution, in particular the concept that the frequency of a viral variant in the quasispecies depends partly on the erroneous replication of other variants, means that for some data sets a net, rather than a tree, may be the most realistic model for the evolution of immunodeficiency viruses.

In this context it is appropriate to point out that the classical debate over the relative contributions of genetic drift and natural selection to molecular evolution may by-pass the truly important features of viral evolution. Gojobori et al (56) conclude that the predominance of silent over replacement substitutions in certain genes of a number of different viruses, including HIV-1 is good evidence in support of the neutral theory. Coffin (24), on the other hand, suggests that no mutation in HIV may be neutral and that selective forces may be too small to measure. He further suggests that there is no such thing as a "rate of variation" because every base is subject to different selective forces, which themselves vary throughout the course of the infection, so that rates of substitution cannot be averaged across bases.

Sequential Studies—Three Phases of Viral Evolution

The process of viral evolution within a single patient is patently central to understanding the interaction between the virus and the human host. Studies of this process have understandably focused on those regions of the genome identified as important targets for the immune system. Until recently, the great majority concentrated just on the V3 region of gp120, sometimes extending to the other variable regions of *env*. However, studies have also focused on the regulatory genes *nef, rev* and *tat* (33, 109, 129), on *pol,* in the context of the development of AZT resistance (14, 83, 93, 151), and on *gag,* in investigations of escape from recognition by cytotoxic T lymphocytes (110, 130).

One of us earlier identified three distinct phases of HIV evolution within an infected individual (97). Phase I covers the short period from the time of infection with HIV until an antibody response develops, termed *seroconversion.* During this time, usually lasting seven to eight weeks, the virus replicates to high titre, often reaching 10^6 viral molecules per ml of plasma (29, 131). This phase is ended when the immune response, both CTL and humoral, develops fully and clears infectious virus from the peripheral circulation.

PHASE I   There have been a number of studies of viral population structure at this phase of infection, in patients infected through a variety of different routes. Zhang et al (177) examined the V3 and V4 sequence populations of individuals mostly infected through the parenteral route (i.e. hemophiliacs through transfusions of HIV-infected clotting factor), although some individuals infected through sexual contact were also included. The results were striking. No sequence variation was detected in the V3-V4 regions within any of the patients. Even more dramatic was that extremely similar (sometimes identical at the amino acid level) V3 loop regions were found in unrelated patients, despite the fact that the regions flanking the loop could be very different. Such a situation is in marked contrast to what appears to be happening in parts of the *gag* gene (the p17 region) in sequences taken at the same time. Although the *gag* gene is generally thought to evolve more slowly than *env,* sequence diversity was observed in a number of the patients. The pattern of this variation was also noteworthy in that in each patient there appeared to be a single sequence at high frequency and a number of variants that differed in a limited number of substitutions.

The extreme similarity in V3 sequences within and between patients at seroconversion has been confirmed in a number of studies (84, 108, 173, 179). Although Zhu et al (179) found less sequence similarity in the V3 region than some other studies, in vitro analyses showed all sequences to be of the macrophage tropic and NSI phenotype (see below). Furthermore, significantly less sequence diversity appeared in the envelope glycoprotein gp120 than in regions of the *gag* gene. Wolfs et al (173) examined sequence variation in V3 in seven presumed donor-recipient pairs of patients, mainly thought to be infected through sexual contact. Interestingly, the virus that dominated at seroconversion in the recipient was not always that observed at highest frequency in the donor. Whatever the sequence variability seen in the donor, viruses of the inferred macrophage tropic and NSI phenotype were found at highest frequency in the donor. The preferential presence of viruses of this phenotype at seroconversion has recently been confirmed in a large survey of infected individuals in Amsterdam (90). Similar features are also found in studies of HIV transmission from infected mothers to their children at some time in about 13–40% of such pregnancies. Scarlatti et al (1993) examined V3 sequences in a number of mother-to-child transmission cases. Although the authors concluded there were no obvious sequence similarities between the infected children, closer inspection suggests that the majority of the viruses had changes indicative of macrophage tropism and the NSI phenotype, again in agreement with studies of other modes of transmission. A similar observation was made by Wolinsky et al (174).

What processes can explain these patterns? The rapid increase in viral population size, coupled with the general sequence similarity (in the face of the sequence diversity often observed later in the infection) implies that there is selection acting on the viral population, either during the transmission of the virus from host to host or during the time between infection and seroconversion. Zhang et al (177) suggested the selection occurs at some time after infection and that a pool of viruses (although possibly of limited size) are transmitted. Very strong selection then acts on parts of the *env* gene, and most probably the V3 loop, during primary infection. The strength of this selection is such that it is detectable on genetically linked regions, causing a decline in sequence diversity across the *env* gene. The diversity sometimes seen in the p17 region of *gag* would then be explained by recombination between variants in the mixed pool that are transmitted. Although only a single V3 sequence appears to be abundant in early infection, more than one V3 sequence can be transmitted, as is shown in the case of the Florida dentist (see HIV Forensics), where two years after infection patient A was found to share two very different V3 sequences with the dentist. A similar observation has been observed in the context of a rape case in Sweden (2).

What is being selected? Most of the studies described above appear to generate sequences that possess amino acid residues associated with macrophage tropism and the NSI phenotype (see Genetic Variation, Cell Tropism and Pathogenesis), which suggests that macrophages are somehow critical for virus transmission or the establishment of infection. Selection for macrophage tropism could occur in a number of ways. Perhaps macrophages are more efficient in infection and replication in the absence of an immune response, or perhaps HIV is primarily transmitted in macrophages. It is also possible, however, that macrophages themselves are not the subject of selection but rather act to indicate some other feature of HIV infection, although it perhaps is noteworthy that the infection of macrophages is often associated with lentiviral disease (54).

PHASE II   Very different selective forces characterize the period after seroconversion. Phase II, which may last for many years while the patient is asymptomatic, is characterized by a continual viral replication (131, 178) and the presence of an active immune response, responsible for the dramatic drop in viral titre soon after seroconversion. Thus, the virus must be able to produce antigenically distinct mutants so as to evade the host immune response—a selective process confusingly termed "antigenic drift" following its use in studies on influenza virus (167).

Several more detailed studies of sequence variation in this phase of viral evolution have provided interesting insights into the nature of the selective forces. For such comparisons, nucleotide differences between sequences in protein coding regions can be usefully broken down into those that result in a change in an amino acid, known as *amino acid replacement substitutions,* and those that do not, which are termed *synonymous.* The proportion of replacement substitutions per *replacement* site (those nucleotide sites at which a replacement substitution is possible) is termed K_A and the proportion of synonymous substitutions per synonymous site is termed K_S. The ratio K_S/K_A is a useful indicator of selection on protein sequences, with high values indicating strong stabilizing selection for conservation of amino acid sequence change (directional selection) (75, 76). Simmonds et al (148) found evidence of the selective replacement of amino acids in an analysis of the extensive sequence variation, observed in the V3 and V4-V5 regions of the *env* gene from hemophiliacs infected from a single batch of factor VIII in Edinburgh. A K_S/K_A ratio of 0.67 was observed in V3 region, and a ratio of 1.24 in the CD4-binding site between V3 and V4. Similarly, Balfe et al (11) found a K_S/K_A ratio of 1.3 for the V4-V5 region and a ratio of 0.92 in V3 for sequences from the same patients. K_S/K_A ratios less than 1.0 are thought to be indicative of positive natural selection (75, 76). As the V3 region is the principal neutralization determinant of HIV and the V4-V5 hypervariable regions may contribute to conformational epitopes, many of these changes may be involved in the escape from immunological recognition. Balfe et al (11) estimated the mean rate of sequence change to be 0.4% per year in the V4-V5 region and 0.5% per year in V3 region.

The process of sequence evolution in the V4-V5 region in one of these patients was examined in more detail by Simmonds et al (149). Important differences were found between the free virus (RNA) population in the plasma and the lymphocyte-associated proviral (DNA) population in peripheral blood mononuclear cells (PBMCs). To summarize, V4 and V5 variants found in the plasma at seroconversion were not detected in a plasma sample taken three years later, yet they were still detectable as provirus in a contemporary PBMC sample. By the following year, a new population of variants was found in the plasma, and the PBMC sample contained variants found in the plasma in the previous year. Thus, it seems that variants first arise in the plasma, possibly following replication in lymph nodes or spleen, and then enter the PBMCs where they may reside for long periods of time. Conversely, it is true that much more rapid replacement of variants occurs in plasma, presumably because they are more directly exposed to the selective pressure exerted by the immune system.

Successive replacement of viral variants was also observed by Cichutek et al (22) who analyzed the V1 and V2 regions of the *env* gene from a hemophiliac infected from factor IX concentrate. Variants found 11 months after seroconversion were seen to replace those found 5 months after seroconversion, although seroconversion forms persisted in both these later samples. In this study 80% of all nucleotide sequence changes were nonsynonymous, again suggesting that positive natural selection was responsible for the fixation of these changes. Wolfs et al (172) examined sequence evolution in the V3 region from six children infected through a blood transfusion and the donor of the HIV-infected blood. The sequence distances to the donor and between the children increased with time although, interestingly, sequences identical to those found in the donor were detected in the two children with the fastest disease progression, suggesting that they might represent more virulent strains.

A higher rate of replacement substitution than silent substitution was observed in the V3 loop, again suggesting the action of positive natural selection. These rates were 9.5×10^{-3} and 11.4×10^{-3} for K_S and K_A, respectively. An increase in sequence diversity with time was also observed by Wolfs et al (173) in a study of V3 variation in two infected homosexuals. Samples were taken over a five-year period, starting from seroconversion, where the viral RNA population was found to be highly homogeneous. Substitutions that led to escape from antibody recognition could be assigned to specific amino acid changes in the V3 loop, although the precise mechanisms involved in the production of such escape mutants may be complex because neutralization-resistant viruses have been observed where the V3 loop remains completely unaltered (164). This suggests that it is not sufficient to analyze the V3 loop in isolation to understand how HIV-1 escapes neutralization. Escape mutants have also been assigned to regions of the *gag* gene where sequence variation leads to the loss of recognition of viral antigens by cytotoxic T lymphocytes (CTLs) (130).

The process of molecular evolution in phase II was examined in detail by Holmes et al (71). A phylogenetic analysis was performed on V3 sequence variants obtained over the course of seven years of infection within the plasma population of a single patient. Twenty-four different V3 loop amino acid sequences were observed over the course of the infection, beginning with a single sequence at seroconversion that was identical to the consensus for HIV-1 subtype B found worldwide (that is a sequence of inferred macrophage tropism, NSI phenotype). This sequence had disappeared from the plasma population by three years post-seroconversion, to be replaced by two distinct evolutionary lineages of variants which persisted for the remainder of the infection. A number of features of this phylogenetic pattern shed light on the underlying evolutionary process. First, the presence of two distinct lineages shows that, rather than the simple sequential replacement of one variant by its descendants, another, antigenically distinct, variant may take over. The coexistence of more than one lineage could raise the complexity of the dynamics of the viral population by allowing competition with each other. It also makes the estimation of evolutionary rates within a patient even more complex, as the nucleotide distances between populations sampled in different years were often very great. Such estimations which are not based on the phylogeny will be quite misleading. What is even more striking about these lineages is that their constituent viruses have residues associated with different phenotypes. Thus, the sequences assigned to one lineage were mainly inferred to be of the NSI phenotype (macrophage tropic) while the other lineage was mainly composed of viruses thought to be of the SI phenotype (T cell tropic) (see Genetic Variation, Cell Tropism and Pathogenesis).

Concurrent evolution of populations of viral sequences was also observed by Howell et al (73) who analyzed part of the V1 region of *env* in two samples separated by three months from a single patient. Two distinct populations of sequences were observed that differed in the presence/absence of an 18 bp duplication as well as a number of point mutations. These two populations evolved independently, although sequences with intermediate forms were also observed, suggesting that recombination was an important factor.

The second observation made by Holmes et al (71) was that there appeared to be a form of frequency dependent selection in the relationship between host and parasite. Specifically, variants that reached a high frequency in the plasma population in one year were generally eliminated by the following year, suggesting that the probability of neutralization of a particular viral variant depends upon its abundance. Furthermore, once a variant was eliminated from the plasma population, it was often replaced by a variant that differed by only a limited number of amino acids, usually only a single change. It is tempting to speculate that these represent escape mutants, and experiments are underway to test this hypothesis.

Sequential studies of other immunodeficiency viruses   Studies of SIV sequence evolution within individual infected monkeys have produced results comparable to those observed in HIV. Burns & Desrosiers (16) found that large amounts of sequence variation could be rapidly generated by infecting monkeys with a single cloned virus. Furthermore, 81% of nucleotide substitutions in *env* resulted in amino acid replacements with the average rate of replacement, which was 6.8×10^{-3}, some 4.3 times higher than the average synonymous rate (1.6×10^{-3}). This disparity was especially pronounced in the variable regions of the *env* gene, although, significantly, only the V5 variable region of SIV corresponded to a neutralization epitope of HIV and the V3 region of SIV was relatively well conserved. Intriguingly, the substitution rate in a monkey that remained asymptomatic was 1.6 times faster (1.12×10^{-2}) than in the monkey that progressed to simian AIDS (7.3×10^{-3}). In a similar study Baier et al (9) followed the evolution of the *env* gene in African green monkeys infected for both long (10 years) and short (15–20 months) periods. The V3 region was again found to be highly conserved, whereas the V1 and V2 regions were variable. Extensive variability was seen to evolve from a single genotype at a maximum rate of 7.7 mutations per 1000 bp per year, and 92% of all changes led to amino acid replacement. Finally, Almond et al (5) infected two cynomologus macaques with a mixed population of SIV. Sequence changes were found in specific regions of gp120—V1, V2, V4, and V5—but, significantly, not V3. Part of this diversity appears to be generated by recombination. Thus, one consistent feature of SIV studies, in contrast to what is observed in HIV, is that the region homologous to the V3 loop in SIV appears to be highly conserved across all isolates and within sequential studies of individual animals. Unfortunately, it is not known whether the V3 region of SIVs has a function in the determination of tropism, as proposed for HIV (see Genetic Variation, Cell Tropism and Pathogenesis).

Within the host, amino acid change in env is rapid; between hosts it is slower Although many of the above analyses are based around the relative numbers of synonymous and replacement substitutions, their uncritical acceptance may lead to startlingly misleading conclusions (115) about the nature and rate of molecular evolution. In a study of sequence variation in feline immunodeficiency virus (FIV), Rigby et al (135) found that estimates of K_A were often higher than those of K_S in regions of the *env* gene of FIV isolates taken from individual cats, or animals infected from the same source. In contrast, K_S was generally found to be higher than K_A in comparisons between more distantly related animals. This suggests that positive natural selection is responsible for the fixation of many mutants within individual animals, but this change is not progressive during the epidemic. Rather, the virus populations of different cats, or humans, evolve in overlapping regions of "sequence space." In consequence the effect of the within-patient positive selection is slowly eroded by silent changes that accumulate at greater evolutionary distances. This is precisely the opposite effect to that seen in nuclearly encoded genes where it is usual for synonymous substitutions to saturate at more distant comparisons.

Similar observations have been made on SIV populations in macaques infected with a molecular clone, and in longitudinal studies of HIV-infected patients. Therefore, observations of a K_S/K_A ratio close to 1.0 do not necessarily imply that the majority of substitutions are selectively neutral as was concluded from the first studies in this area, which were based entirely on comparisons between independent isolates (95). Furthermore, in FIV, as in HIV, identical amino acid changes occur on independent evolutionary lineages so that continual divergence does not occur at replacement sites. This further adds to the disparity between intra- and inter-individual estimates of K_S and K_A. A similar observation has also been made by Shpaer & Mullins (146).

Nowak & May (118) described how, in mathematical models of HIV and SIV evolution, different settings for parameters describing the interaction between the host immune system and the genetic characteristics of the pathogen can generate infections with a variety of different outcomes, all of which may be seen in lentiviruses in nature. They conclude that whether a particular lentivirus infection will be pathogenic depends on the settings of these parameters. A similar mathematical approach has been used to investigate the potential efficacy of vaccination against HIV (120). However, the nature of the interaction between host immune responses to viral sequence variants is not a simple example of adaptive evolution, a point discussed in greater detail below.

Other regions of the HIV genome do not appear to be subject to such strong selection pressures. Meyerhans et al (109) compared the sequence of the *tat* gene in virus populations from peripheral blood with those of virus cultures established from the same blood samples. They studied four sequential samples spanning a 2.5 year period from a single HIV-1 infected patient and found differences in the sequence compositions of the in vivo and in vitro populations such that there was a selection for minor forms of virus adapted to culturing conditions. There was no evidence of the selective replacement of one set of sequences by another in the in vivo samples. A similar observation was made by Delassus et al (34) who examined *nef* and LTR sequences from the same patient over a four-year period.

PHASE III   The third phase of within-patient HIV evolution was identified in the model of Nowak et al (119) as the time when the host immune response collapses and the patient progresses into AIDS. The aim of this model, which has a simple basis in quasispecies theory, is to explain the destruction of the immune system quantitatively in terms of a direct interaction between the virus and the $CD4^+$ T cell population. The central element in this model is the assumption of an asymmetric interaction between the viral quasispecies and the population of $CD4^+$ T cells (the central cellular targets for HIV infection), in that each $CD4^+$ cell is directed against a specific HIV antigen, but each virus strain can kill any CD4 cell. From such an assumption Nowak et al (119) predict the existence of an "antigenic diversity" threshold below which the immune system can regulate the viral population, above which the immune system collapses and the patient develops AIDS. It is predicted that when AIDS develops, and immune responses are lost, the variant with the highest replicative capacity will reach the highest frequency in the population and thus viral diversity will decline. However, we suggest that the population expansion at this stage will be less than in Phase I because of the much smaller number of uninfected $CD4^+$ T cells (97).

The experimental data do not at present give a clear picture of the relationship between viral genetic diversity and progression. McNearney et al (108) found only 0.01% amino acid and 0.02% nucleotide sequence differences in *env* were observed in two rapidly progressing neonatal transfusion recipients and their blood donor. No changes were observed in V3, nor were there any differences between isolates obtained from brain, lung, and blood. Other studies on this area are underway in a number of laboratories.

HIV-1 Genetic Variation, Cell Tropism and Pathogenesis

An important element in determining the pattern and process of HIV evolution within a single individual is the appearance of viruses that differ genetically in their cell tropism. Although the $CD4^+$ lymphocyte is the major HIV-infected population in PBMCs, cells of the monocyte-macrophage lineage appear to be a route of introduction of HIV into many tissues, including the CNS (87).

When HIV-1 is cultured from infected patients, it is commonly observed that the viral strains from different patients can differ in their rate of growth (46, 47), and in their ability to form syncytia, giant multinucleated cells that arise from gp120-CD4–mediated membrane fusion (156). While virus from most patients can be cultured in PBMCs (25), some strains ("SI" — syncytium-inducing) can form syncytia in transformed T cell lines, particularly the HTLV-I transformed lines C8166 and MT2. These strains, however, do not grow in macrophages, while those that do not form syncytia (or grow) in T cell lines ("NSI" — non–syncytium-inducing) can usually also infect macrophages. Thus, there appear to be two main tropisms: (i) viruses that can infect both macrophages and circulating peripheral blood T lymphocytes (macrophage-tropic strains), and (ii) viruses that can infect peripheral blood T lymphocytes and transformed T cell lines.

There have been intensive efforts toward the characterization of the genetic basis of differences in in vitro growth properties between strains of HIV-1. For example, *tat* gene activity influences growth rate, and naturally occurring variation in the level of *tat* gene activity has been demonstrated (109). However, variation in the *env* gene, and in particular the hypervariable V3 loop, appears to be primarily responsible for differences in pathogenicity and tropism (19, 20, 49, 78, 145). Specifically, substitutions of residues with acidic side chains at certain sites on the side of the V3 loop are associated with rapid growth and syncytium-induction. Milich et al (111) undertook an extensive analysis of the biological (phenotypic) properties of a large number of V3 loop sequences, with particular regard to the amino acid replacements implicated in the shift from NSI to SI viruses. They conclude that the transition is associated with the presence of nonconservative basic amino acid substitutions at four positions and most notably with a change from an acidic amino acid at position 25. As discussed above, macrophage tropic variants tend to appear early in the infection, and they possess V3 loop sequences similar to the consensus derived for subtype B, while T tropic variants tend to arise later in the infection and are more divergent in sequence.

Most importantly, many patients from whom such T cell tropic viral strains are isolated frequently have AIDS or develop AIDS within a short time (< 2 years) (86, 157), whereas those from whom only slow-growing, macrophage-tropic strains can be isolated have a better prognosis, in general. It has therefore been proposed that the development of SI variants is a determinant of AIDS. However, this possibility has to be reconciled with the fact that only NSI strains can ever be isolated from nearly 50% of patients with AIDS. The difficult question as to whether the appearance of SI variants is a cause or simply a consequence of the degradation of the immune system still remains to be answered.

Schuitemaker et al (141) conclude that monocytotropic HIV viruses are responsible for the persistence (and perhaps transmission) of HIV infection, while T cell tropic viruses are responsible for the development of AIDS. However, indirect evidence that macrophage tropism is central to the persistence of HIV infection comes from studies of HIV-infected chimpanzees. The HIV chimpanzee infection takes a very different course from that of humans. Specifically, neither plasma viremia, CNS disease, nor lymphatic hyperplasia are observed in HIV-infected chimpanzees (54). Both CNS and lymphatic hyperplasia in HIV-infected humans are characterized by the persistent infection of macrophages and dendritic cells. Significantly, HIV-1 appears to be unable to infect chimpanzee monocytes and macrophages, at least in vitro (54, 141), but the strains used to infect chimpanzees appear to be those such as HIV_{LAI} that are aggressively T cell tropic in human cells.

Finally, there has been much interest in HIV variants that specifically infect the brain. Analysis of brain-specific populations is especially important because HIV/AIDS often causes neurologic dysfunctions, such as encephalopathy; this may be due either to opportunistic infections or direct HIV invasion of the central nervous system. Epstein et al (42) examined variation in the V3 domain in brain and spleen of children with AIDS and found both tissue-specific and host-specific sequence variation, although the former had the strongest effect. A similar observation was made by Steuler et al (154) who looked at proviral gp41 sequences from blood and cerebrospinal fluid (CSF) from six individuals and found distinct populations in three of the patients and especially in an AIDS patient suffering from encephalopathy. Li et al (101) also examined the in vivo characteristics of HIV in the brain and found that it is characterized by the persistence of mixtures of fully competent and relatively homogeneous monocyte-macrophage tropic viruses.

Thus, HIV sequences obtained from the brain or CSF exhibit an increased tropism for macrophages, compared to isolates taken from blood. Furthermore, the brain may act as a reservoir that is poorly accessible to immune surveillance and viral clearance, and therefore antigenic variation may not be driven by immune selection. There is increasing evidence that virus replication and the accumulated burden of virus in lymphoid tissue may play an important and direct role in pathogenesis (41, 127) and notably in the brain where high levels of viral expression are associated with neurological dysfunction. In general, it seems likely that viral replication, burden, and variation are all important determinants of AIDS pathogenesis.

CONCLUSION—IMPLICATIONS FOR EVOLUTIONARY BIOLOGY

The wealth of data accumulated over the last few years has made the immunodeficiency viruses the most data-rich group of organisms for any evolutionary analyses. A number of specific selective forces have been identified. The effect of these agents can be investigated quantitatively in cell culture (both anti-viral drugs and neutralizing antibodies) and in the experimental models, mainly SIV and FIV, but increasingly now also HIV in the SCID-Hu mouse system. Integration of the observations with analysis continues in all the areas we have reviewed, but as yet there has been relatively little effort by evolutionary biologists to incorporate these conclusions into the body of evolutionary thought.

Apart from a lack of familiarity with the system, one possible reason for this lies in the fact that much effort is devoted in other systems on increasingly sophisticated analyses aimed at demonstrating selection acting at the molecular level. Such methodologies are clearly not relevant to the study of a viral population whose constitution is shaped almost completely by the complex mix of selective forces we have discussed. On the other hand, what is occupying those involved in such studies in the immunodeficiency viruses are the details of how these selective agents interact to determine the fitness of the population, a question that is barely accessible to investigation in anything other than a microbial system. Some aspects we have discussed, particularly the evolution of viruses with tropisms for different cell types, have echoes in Gillespie's "random environments" model (55). In other respects, particularly with regard to the speed of the adaptive response, quasi-species theory has made an important contribution, and work on the host-parasite interaction has led to vital insights relating to the interaction with the immune system. In such a complex and important area, these strands must be thoroughly integrated for the rapid advances in understanding, which are so desperately needed.

ACKNOWLEDGMENTS

Work in Andrew J. Leigh Brown's laboratory is supported by the Medical Research Council AIDS Directed Programme. EC Holmes is supported by Science and Engineering Research Council.

Literature Cited

1. Aboulker JP, Swart AM, Concorde Co-ordinating Committee. 1993. Preliminary analysis of the Concorde trial. *Lancet* 341:889–90

2. Albert J, Wahlberg J, Uhlen M. 1993. Forensic evidence by DNA sequencing. *Nature* 361:595–96

3. Alizon M, Wain-Hobson S, Montagnier L, Sonigo P. 1986. Genetic variability of the AIDS virus: nucleotide sequence analysis of two isolates from African patients. *Cell* 46:63–74

4. Allan JS, Short M, Taylor ME, Su S, Hirsch VM, et al. 1991. Species-specific

diversity among simian immunodeficiency viruses from African green monkeys. *J. Virol.* 65:2816–28

5. Almond N, Jenkins A, Heath AB, Taffs LF, Kitchin P. 1992. The genetic evolution of the envelope gene of simian immunodeficiency virus in cynomolgus macaques infected with a complex virus pool. *Virology* 191:996–1002

6. Anderson RM, May RM. 1982. Coevolution of hosts and parasites. *Parasitology* 85:411–26

7. Anderson RM, May RM, McLean AR. 1988. Possible demographic consequences of AIDS in developing countries. *Nature* 332:228–34

8. Andreasson P-A, Dias F, Naucler A, Andersson S, Biberfeld G. 1993. A prospective study of vertical transmission of HIV-2 in Bissau, Guinea-Bissau. *AIDS* 7:989–93

9. Baier M, Dittmar MT, Cichutek K, Kurth R. 1991. Development in vivo of genetic variability of simian immunodeficiency virus. *Proc. Natl. Acad. Sci. USA* 88:8126–30

10. Baier M, Garber C, Muller C, Cichutek K, Kurth R. 1990. Complete nucleotide sequence of a simian immunodeficiency virus from African green monkeys: a novel type of intragroup divergence. *Virology* 176:216–21

11. Balfe P, Simmonds P, Ludlam CA, Bishop JO, Leigh Brown AJ. 1990. Concurrent evolution of human immunodeficiency virus type 1 in patients infected from the same source: rate of sequence change and low frequency of inactivating mutations. *J. Virol.* 64:6221–33

12. Balzarini J, Karlsson A, Perez-Perez M-J, Camarasa M-J, et al. 1993. Treatment of human immunodeficiency virus type 1 (HIV-1)-infected cells with combinations of HIV-1-specific inhibitors results in a different resistance pattern than does treatment with single drug therapy. *J. Virol.* 67:5353–59

13. Barre Sinoussi F, Chermann JC, Rey F, Nugeyre MT, Chamaret S, et al. 1983. Isolation of a T-lymphotropic retrovirus from a patient at risk for acquired immune deficiency syndrome (AIDS). *Science* 220:868–71

14. Boucher CA, O'Sullivan E, Mulder JW, Ramautarsing C, Kellam P, et al. 1992. Ordered appearance of zidovudine resistance mutations during treatment of 18 human immunodeficiency virus-positive subjects. *J. Infect. Dis.* 165:105–10

15. Buonagurio DA, Nakada S, Parvin JD, Krystal M, Palese P, Fitch WM. 1986. Evolution of human influenza A viruses over 50 years: rapid, uniform rate of change in NS gene. *Science* 232:980–82

16. Burns DP, Desrosiers RC. 1991. Selection of genetic variants of simian immunodeficiency virus in persistently infected rhesus monkeys. *J. Virol.* 65:1843

17. Carpenter S, Evans LH, Sevoian M, Chesebro B. 1987. Role of the host immune response in selection of equine infectious anemia virus variants. *J. Virol.* 61:3783–89

18. Chakrabarti L, Guyader M, Alizon M, Daniel MD, Desrosiers RC, et al. 1987. Sequence of simian immunodeficiency virus from macaque and its relationship to other human and simian retroviruses. *Nature* 328:543–47

19. Cheng Mayer C, Quiroga M, Tung JW, Dina D, Levy JA. 1990. Viral determinants of human immunodeficiency virus type 1 T-cell or macrophage tropism, cytopathogenicity, and CD4 antigen modulation. *J. Virol.* 64:4390–98

20. Cheng Mayer C, Shioda T, Levy JA. 1991. Host range, replicative, and cytopathic properties of human immunodeficiency virus type 1 are determined by very few amino acid changes in tat and gp120. *J. Virol.* 65:6931–41

21. Chin J. 1991. Global estimates of HIV infections and AIDS cases: 1991. *AIDS* 5 (Suppl. 2):S57-S61

22. Cichutek K, Norley S, Linde R, Kreuz W, Gahr M, et al. 1991. Lack of HIV-1 V3 region sequence diversity in two haemophiliac patients infected with a putative biologic clone of HIV-1. *AIDS* 5:1185–87

23. Clarke B. 1976. The ecological genetics of host-parasite relationships. In *Genetic Aspects of Host-Parasite Relationships* (Symposia of the British Society for Parasitology, Vol. 14), ed. AER Taylor, R Muller, pp. 87–103. Oxford: Blackwell Sci.

24. Coffin JM. 1992. Genetic diversity and evolution of retroviruses. *Curr. Top. Microbiol. Immunol.* 176:143–64

25. Coombs RW, Collier AC, Allain J-P, Nikora B, Leuther M, et al. 1989. Plasma viremia in human immunodeficiency virus infection. *N. Engl. J. Med.* 321:1626–31

26. Courgnaud V, Laure F, Fultz PN, Montagnier L, Brechot C, Sonigo P. 1992. Genetic differences accounting for evolution and pathogenicity of simian immunodeficiency virus from a sooty mangabey monkey after cross-species

transmission to a pig-tailed macaque. *J. Virol.* 66:414–19

27. Cullen BR. 1986. Trans-activation of human immunodeficiency virus occurs via a bimodal mechanism. *Cell* 46:973–82

28. Cullen BR. 1991. Regulation of HIV-1 gene expression. *FASEB J.* 5:2361–68

29. Daar ES, Moudgil T, Meyer RD, Ho DD. 1991. Transient high levels of viremia in patients with primary human immunodeficiency virus type 1 infection. *N. Engl. J. Med.* 324:961–64

30. Daniel MD, Letvin NL, King NW, Kannagi M, Sehgal PK, et al. 1985. Isolation of a T-cell tropic HTLV-III-like retrovirus from macaques. *Science* 228:1201–4

31. Darby SC, Doll R, Thakrar B, Rizza CR, Cox DR. 1990. Time from infection with HIV to onset of AIDS in patients with haemophilia in the U.K. *Stat. Med.* 9:681–89

32. de Leys R, Vanderborght B, Haesevelde MV, Heyndrickx L, van Geel A, Wauters C, et al. 1990. Isolation and partial characterisation of an unusual human immunodeficiency retrovirus from two persons of west-central African origin. *J. Virol.* 64:1207–16

33. Delassus S, Cheynier R, Wain-Hobson S. 1992. Nonhomogeneous distribution of human immunodeficiency virus type 1 proviruses in the spleen. *J. Virol.* 66:5642–45

34. Delassus S, Meyerhans A, Cheynier R, Wain-Hobson S. 1992. Absence of selection of HIV-1 variants in vivo based on transcription/transactivation during progression to AIDS. *Virology* 188:811–18

35. Desrosiers RC. 1988. Simian immunodeficiency viruses. *Annu. Rev. Microbiol.* 42:607–25

36. Dietrich U, Grez M, von Briesen H, Panhans B, Geissendorfer M, et al. 1993. HIV-1 strains from India are highly divergent from prototypic African and US/European strains, but are linked to a South African isolate. *AIDS* 7:23–27

37. Dumitrescu O, Kalish ML, Kliks SC, Bandea CI, Levy JA. 1994. Characterization of human immunodeficiency virus type 1 isolates from children in Romania: identification of a new envelope subtype. *J. Infect. Dis.* 169:281–88

38. Eigen M, Biebricher CK. 1988. Sequence space and quasispecies distribution. In *RNA Genetics*, ed. E Domingo, JJ Holland, P Ahlquist, pp. 211–214. Boca Raton, Fla: CRC. 3rd ed.

39. Eigen M, Nieselt-Struwe K. 1990. How old is the immunodeficiency virus? *AIDS* 4 (suppl 1):S85-S93

40. El-Farrash MA, Kuroda MJ, Kitazaki T, Masuda T, Kato K, et al. 1994. Generation and characterisation of a human immunodeficiency virus type 1 (HIV-1) mutant resistant to an HIV-1 protease inhibitor. *J. Virol.* 68:233–39

41. Embretson J, Zupancic M, Ribas JL, Burke A, Racz P, et al. 1993. Massive covert infection of helper T lymphocytes and macrophages by HIV during the incubation period of AIDS. *Nature* 362:359–62

42. Epstein LG, Kuiken C, Blumberg BM, Hartman S, Sharer LR, et al. 1991. HIV-1 V3 domain variation in brain and spleen of children with AIDS: tissue-specific evolution within host-determined quasispecies. *Virology* 180:583–90

43. European Collaborative Study. 1992. Risk factors for mother-to-child transmission of HIV-1. *Lancet* 339:1007–12

44. Fenner F, Kerr PJ. 1994. Evolution of the poxviruses, including the coevolution of virus and host in myxomatosis. In *The Evolutionary Biology of Viruses*, ed. SS Morse, pp. 273–292. New York: Raven

45. Fenner F, Ratcliffe FN. 1965. *Myxomatosis.* Cambridge: Cambridge Univ. Press.

46. Fenyo EM, Albert J, Asjo B. 1989. Replicative capacity, cytopathic effect and cell tropism of HIV. *AIDS* 3(Suppl.) 1:S5–12

47. Fenyo EM, Morfeldt Manson L, Chiodi F, Lind B, von Gegerfelt A, et al. 1988. Distinct replicative and cytopathic characteristics of human immunodeficiency virus isolates. *J. Virol.* 62:4414–19

48. Fisher AG, Ensoli B, Looney D, Rose A, Gallo RC, et al. 1988. Biologically diverse molecular variants within a single HIV-1 isolate. *Nature* 334:444–47

49. Fouchier RA, Groenink M, Kootstra NA, Tersmette M, Huisman HG, et al. 1992. Phenotype-associated sequence variation in the third variable domain of the human immunodeficiency virus type 1 gp120 molecule. *J. Virol.* 66:3183–87

50. Fukasawa M, Miura T, Hasegawa A, Morikawa S, Tsujimoto H, et al. 1988. Sequence of simian immunodeficiency virus from African green monkey, a new member of the HIV/SIV group. *Nature* 333:457–61

51. Gallo RC, Kalyanaraman VS, Sarngadharan MG, Sliski A, Vonderheid

EC, et al. 1983. Association of the human type C retrovirus with a subset of adult T-cell cancers. *Cancer. Res.* 43:3892–99

52. Gao F, Yue L, White AT, Pappas PG, Barchue J, et al. 1992. Human infection by genetically diverse SIVsm-related HIV-2 in West Africa. *Nature* 358:495–99

53. Garnett GP, Swinton J. 1994. Dynamic simulation of sexual partner networks: which network properties are important in sexually transmitted disease epidemiology. In *Models for Infectious Human Diseases: Their Structure and Relation To Data*, ed. V Isuam, G Medley. Cambridge: Cambridge Univ. Press. In press

54. Gendelman HE, Ehrlich GD, Baca LM, Conley S, Ribas J, et al. 1991. The inability of human immunodeficiency virus to infect chimpanzee monocytes can be overcome by serial viral passage in vivo. *J. Virol.* 65:3853–63

55. Gillespie JH. 1992. *The Causes of Molecular Evolution.* Oxford: Oxford Univ. Press.

56. Gojobori T, Moriyama EN, Kimura M. 1990. Molecular clock of viral evolution and the neutral theory. *Proc. Natl. Acad. Sci. USA* 87:10015–18

57. Gonda MA. 1992. Bovine immunodeficiency virus. *AIDS* 6:759–76

58. Gonda MA, Braun MJ, Carter SG, Kost TA, Bess JW Jr, et al. 1987. Characterization and molecular cloning of a bovine lentivirus related to human immunodeficiency virus. *Nature* 330:388–91

59. Goudsmit J, Debouck C, Meloen RH, Smit L, Bakker M, et al. 1988. Human immunodeficiency virus type 1 neutralization epitope with conserved architecture elicits early type-specific antibodies in experimentally infected chimpanzees. *Proc. Natl. Acad. Sci. USA* 85:4478–82

60. Guyader M, Emerman M, Sonigo P, Clavel F, Montagnier L, Alizon M. 1987. Genome organization and transactivation of the human immunodeficiency virus type 2. *Nature* 326:662–69

61. Hahn BH, Shaw GM, Taylor ME, Redfield RR, Markham PD, et al. 1986. Genetic variation in HTLV-III/LAV over time in patients with AIDS or at risk for AIDS. *Science* 232:1548–53

62. Haldane JBS. 1949. Disease and evolution. *Ricerca scient. 19* (Suppl.):68–76

63. Harris M, Coates K. 1993. Identification of cellular proteins that bind to the human immunodeficiency virus type 1

nef gene product in vitro: a role for myristylation. *J. Gen. Virol.* 74:1581–89

64. Heaphy S, Dingwall C, Ernberg I, Gait MJ, Green SM, et al. 1990. HIV-1 regulator of virion expression (Rev) protein binds to an RNA stem-loop structure located within the Rev response element region. *Cell* 60:685–93

65. Hillis DM, Bull JJ. 1993. An empirical test of bootstrapping as a method for assessing confidence in phylogenetic analysis. *Syst. Biol.* 42:182–92

66. Hirsch VM, Dapolito G, McGann C, Olmsted RA, Purcell RH, et al. 1989. Molecular cloning of SIV from sooty mangabey monkeys. *J. Med. Primatol.* 18:279–85

67. Hirsch VM, Dapolito GA, Goldstein S, McClure HM, Emau P, et al. 1993. A distinct African lentivirus from Sykes' monkey. *J. Virol.* 67:1517–28

68. Hirsch VM, Olmsted RA, Murphey Corb M, Purcell RH, Johnson PR. 1989. An African primate lentivirus (SIVsm) closely related to HIV-2. *Nature* 339:389–92

69. Holland JJ, de la Torre JC, Steinhauer DA. 1992. RNA virus populations as quasispecies. *Curr. Top. Microbiol. Immunol.* 176:1–20

70. Holmes EC, Zhang LQ, Robertson P, Cleland A, Harvey E, et al. 1994. The molecular epidemiology of HIV-1 in Edinburgh, Scotland. *J. Infect. Dis.* In press

71. Holmes EC, Zhang LQ, Simmonds P, Ludlam CA, Leigh Brown AJ. 1992. Convergent and divergent sequence evolution in the surface envelope glycoprotein of human immunodeficiency virus type 1 within a single infected patient. *Proc. Natl. Acad. Sci. USA* 89:4835–39

72. Holmes EC, Zhang LQ, Simmonds P, Rogers AS, Leigh Brown AJ. 1993. Molecular investigation of human immunodeficiency virus (HIV) infection in a patient of an HIV-infected surgeon. *J. Infect. Dis.* 167:1411–14

73. Howell RM, Fitzgibbon JE, Noe M, Ren ZJ, Gocke DJ, Schwartzer TA, et al. 1991. In vivo sequence variation of the human immunodeficiency virus type 1 *env* gene: evidence for recombination among variants found in a single individual. *AIDS. Res. Hum. Retroviruses.* 7:869–76

74. Huet T, Cheynier R, Meyerhans A, Roelants G, Wain-Hobson S. 1990. Genetic organization of a chimpanzee lentivirus related to HIV-1. *Nature* 345:356–59

75. Hughes AL, Nei M. 1988. Pattern of nucleotide substitution at major histocompatibility complex class I loci reveals overdominant selection. *Nature* 335:367–70

76. Hughes AL, Nei M. 1989. Nucleotide substitution at major histocompatibility complex class II loci: evidence for overdominant selection. *Proc. Natl. Acad. Sci. USA* 86:958–62

77. Hunt RD, Blake BJ, Chalifoux LV, Sehgal PK, King NW, Letvin NL. 1983. Transmission of naturally occurring lymphoma in macaque monkeys. *Proc. Natl. Acad. Sci. USA* 80:5085–89

78. Hwang SS, Boyle TJ, Lyerly HK, Cullen BR. 1991. Identification of the envelope V3 loop as the primary determinant of cell tropism in HIV-1. *Science* 253:71–74

79. Ivey Hoyle M, Culp JS, Chaikin MA, Hellmig BD, Matthews TJ, et al. 1991. Envelope glycoproteins from biologically diverse isolates of immunodeficiency viruses have widely different affinities for CD4. *Proc. Natl. Acad. Sci. USA* 88:512–16

80. Johnson AM, Wadsworth J, Wellings K, Bradshaw S, Field J. 1992. Sexual lifestyles and HIV risk. *Nature* 360:410–12

81. Johnson PR, Fomsgaard A, Allan J, Gravell M, London WT, et al. 1990. Simian immunodeficiency viruses from African green monkeys display unusual genetic diversity. *J. Virol.* 64:1086–92

82. Johnson PR, Hirsch VM, Myers G. 1991. Genetic diversity and phylogeny of non-human primate lentiviruses. In *Annual Review of AIDS Research*, Vol. 1, ed. W Koff, pp. 47–62. New York: Marcel Dekker

83. Kellam P, Boucher CA, Larder BA. 1992. Fifth mutation in human immunodeficiency virus type 1 reverse transcriptase contributes to the development of high-level resistance to zidovudine. *Proc. Natl. Acad. Sci. USA* 89:1934–38

84. Kleim J-P, Ackerman A, Brackman HH, Gahr M, Schneweis KE. 1991. Epidemiologically closely related viruses from hemophilia B patients display high homology in two hypervariable regions of the HIV-1 *env* gene. *AIDS Res. Hum. Retroviruses* 7:417–21

85. Kohlstaedt LA, Wang J, Friedman JM, Rice PA, Steitz TA. 1992. Crystal structure at 3.5 A resolution of HIV-1 reverse transcriptase complexed with an inhibitor. *Science* 256:1783–90

86. Koot M, Vos AH, Keet RP, de Goede RE, Dercksen MW, et al. 1992. HIV-1 biological phenotype in long-term infected individuals evaluated with an MT-2 cocultivation assay. *AIDS* 6:49–54

87. Koyanagi Y, Miles S, Mitsuyasu RT, Merrill JE, Vinters HV, Chen IS. 1987. Dual infection of the central nervous system by AIDS viruses with distinct cellular tropisms. *Science* 236:819–22

88. Kreutz R, Dietrich U, Kuhnel H, Nieselt-Struwe K, Eigen M, Rubsamen Waigmann H. 1992. Analysis of the envelope region of the highly divergent HIV-2$_{ALT}$ isolate extends the known range of variability within the primate immunodeficiency viruses. *AIDS Res. Hum. Retroviruses* 8:1619–29

89. Kuhnel H, von Briesen H, Dietrich U, Adamski M, Mix D, et al. 1989. Molecular cloning of two west African human immunodeficiency virus type 2 isolates that replicate well in macrophages: a Gambian isolate, from a patient with neurologic acquired immunodeficiency syndrome, and a highly divergent Ghanian isolate. *Proc. Natl. Acad. Sci. USA* 86:2383–87

90. Kuiken CL, Zwart G, Baan E, Coutinho RA, van den Hoek JAR, Goudsmit J. 1993. Increasing antigenic and genetic diversity of the V3 variable domain of the human immunodeficiency virus envelope protein in the course of the AIDS epidemic. *Proc. Natl. Acad. Sci. USA* 90:9061–65

91. Lapatto R, Blundell T, Hemmings A, Overington J, Wilderspin A, Wood S, et al. 1989. X-ray analysis of HIV-1 proteinase at 2.7 A resolution confirms structural homology among retroviral enzymes. *Nature* 342:299–302

92. Larder BA, Darby G, Richman DD. 1989. HIV with reduced sensitivity to zidovudine (AZT) isolated during prolonged therapy. *Science.* 243:1731–34

93. Larder BA, Kemp SD. 1989. Multiple mutations in HIV-1 reverse transcriptase confer high-level resistance to zidovudine (AZT). *Science* 246:1155–58

94. LaRosa GJ, Davide JP, Weinhold K, Waterbury JA, Profy AT, Lewis JA, et al. 1990. Conserved sequence and structural elements in the HIV-1 principal neutralizing determinant. *Science* 249:932–35

95. Leigh Brown A, Monaghan P. 1988. Evolution of the structural proteins of human immunodeficiency virus: selective constraints on nucleotide substitution. *AIDS Res. Hum. Retroviruses* 4:399–407

96. Leigh Brown AJ. 1990. Evolutionary relationships of the human immunodeficiency viruses. *Trends Evol. Ecol.* 5:177–81

97. Leigh Brown AJ. 1991. Sequence variability in the human immunodeficiency viruses: pattern and process in viral evolution. In *AIDS '91 - A year in Review,* ed. MW Adler, JWM Gold, JA Levy. London: Curr. Sci. :1808–1809:535–42

98. Leis J, Baltimore D, Bishop JM, Coffin J, Fleissner E, et al. 1988. Standardized and simplified nomenclature for proteins common to all retroviruses. *J. Virol.* 62

99. Letvin NL, Eaton KA, Aldrich WR, Sehgal K, Blake BJ, et al. 1983. Acquired immunodeficiency syndrome in a colony of macaque monkeys. *Proc. Natl. Acad. Sci. USA* 80718–22

100. Li WH, Tanimura M, Sharp PM. 1988. Rates and dates of divergence between AIDS virus nucleotide sequences. *Mol. Biol. Evol.* 5:313–30

100a. Li W-H, Wu C-I, Luo C-C. 1985. A new method for estimating synonymous and nonsynonymous rates of nucleotide substitution considering the relative likelihood of nucleotide and codon changes. *Mol. Biol. Evol.* 2:150–74

101. Li Y, Hui H, Burgess CJ, Price RW, Sharp PM, et al. 1992. Complete nucleotide sequence, genome organization and biological properties of human immunodeficiency virus type 1 in vivo: evidence for limited defectiveness and complementation. *J. Virol.* 66:6587–600

102. Looney DJ, Hayashi S, Nicklas M, Redfield RR, Broder S, et al. 1990. Differences in the interaction of HIV-1 and HIV-2 with CD4. *J. Acquired Immun. Deficiency Syndrome* 3:649–57

103. Louwagie J, McCutchan FE, Peeters M, Brennan T, Sanders-Buell E, et al. 1993. Phylogenetic analysis of gag genes from 70 international HIV-1 isolates provides evidence for multiple genotypes. *AIDS* 7:769–80

104. Ludlam CA, Tucker J, Steel CM, Tedder RS, Cheingsong Popov R, et al. 1985. Human T-lymphotropic virus type III (HTLV-III) infection in seronegative haemophiliacs after transfusion of factor VIII. *Lancet* ii:233–236

105. May RM, Anderson RM. 1983. Epidemiology and genetics in the coevolution of parasites and hosts. *Proc. R. Soc. Lond. B* 219:281–313

106. May RM, Anderson RM. 1983. Coevolution of parasites and hosts. In *Coevolution,* ed. DJ Futuyma, M Slatkin, pp. 186–206. Sunderland, Mass: Sinauer

107. McCutchan FE, Hegerich PA, Brennan TP, Phanuphak P, Singharaj P, et al. 1992. Genetic variants of HIV-1 in Thailand. *AIDS Res. Hum. Retroviruses* 8:1887–95

108. McNearney T, Westervelt P, Thielan BJ, Trowbridge DB, Garcia J, et al. 1990. Limited sequence heterogeneity among biologically distinct human immunodeficiency virus type 1 isolates from individuals involved in a clustered infectious outbreak. *Proc. Natl. Acad. Sci. USA* 87:1917–21

109. Meyerhans A, Cheynier R, Albert J, Seth M, Kwok S, et al. 1989. Temporal fluctuations in HIV quasispecies in vivo are not reflected by sequential HIV isolations. *Cell* 58:901–10

110. Meyerhans A, Dadaglio G, Vartanian JP, Langlade Demoyen P, Frank R, et al. 1991. In vivo persistence of a HIV-1-encoded HLA-B27-restricted cytotoxic T lymphocyte epitope despite specific in vitro reactivity. *Eur. J. Immunol.* 21:2637–40

111. Milich L, Margolin B, Swanstrom R. 1993. V3 loop of the human immunodeficiency virus Type 1 env protein: interpreting sequence variability. *J. Virol.* 67:5623–34

112. Mohri H, Singh MK, Ching WT, Ho DD. 1993. Quantitation of zidovudine-resistant human immunodeficiency virus type 1 in the blood of treated and untreated patients. *Proc. Natl. Acad. Sci. USA* 90:25–29

113. Moore JP. 1990. Simple methods for monitoring HIV-1 and HIV-2 gp120 binding to soluble CD4 by enzyme-linked immunosorbent assay: HIV-2 has a 25-fold lower affinity than HIV-1 for soluble CD4. *AIDS* 4:297–305

114. Moriyama EN, Ina Y, Ikeo K, Shimizu N, Gojobori T. 1991. Mutation pattern of human immunodeficiency virus gene. *J. Mol. Evol.* 32:360–63

115. Myers G, Korber B. 1994. The future of human immunodeficiency virus. In *The Evolutionary Biology of Viruses,* ed. SS Morse, pp. 211–32. New York:Raven

116. Myers G, Korber B, Wain-Hobson S, Smith RF, Pavlakis GN. 1993. *Human retroviruses and AIDS 1991.* Los Alamos Natl. Lab.

117. Nowak MA. 1992. What is a quasispecies? *Trends Ecol. Evol.* 7:118–21

118. Nowak MA, May RM. 1992. Coexistence and competition in HIV infections. *J. Theor. Biol.* 159:329–42

119. Nowak MA, May RM, Anderson RM. 1990. The evolutionary dynamics of HIV-1 quasispecies and the development of immunodeficiency disease. *AIDS* 4:1095–103

120. Nowak MA, McLean AR. 1991. A mathematical model of vaccination against HIV to prevent the development of AIDS. *Proc. R. Soc. Lond. B* 246:141–46

121. Ohta Y, Masuda T, Tsujimoto H, Ishikawa K, Kodama T, et al. 1988. Isolation of simian immunodeficiency virus from African green monkeys and seroepidemiological survey of the virus in various non-human primates. *Int. J. Cancer* 41:115–22

122. Olmsted RA, Barnes AK, Yamamoto JK, Hirsch VM, Purcell RH, Johnson PR. 1989. Molecular cloning of feline immunodeficiency virus. *Proc. Natl. Acad. Sci. USA* 86:2448–52

123. Olmsted RA, Langley R, Roelke ME, Goeken RM, Adger-Johnson D, et al. 1992. Worldwide prevalence of lentivirus infection in wild feline species:epidemiologic and phylogenetic aspects. *J. Virol.* 66:6008–18

124. Otto MJ, Garber S, Winslow DL, Reid CD, Aldrich P, et al. 1993. In vitro isolation and identification of human immnodeficiency virus (HIV) variants with reduced sensitivity to C-2 symmetrical inhibitors of HIV type 1 protease. *Proc. Natl. Acad. Sci. USA* 90:7543–47

125. Ou CY, Ciesielski CA, Myers G, Bandea CI, Luo CC, et al. 1992. Molecular epidemiology of HIV transmission in a dental practice. *Science* 256:1165–71

126. Ou CY, Takebe Y, Weniger BG, Luo CC, Kalish ML, et al. 1993. Independent introduction of two major HIV-1 genotypes into distinct high-risk populations in Thailand. *Lancet* 341:1171–74

127. Pantaleo G, Graziosi C, Demarest JF, Butini L, Montroni M, et al. 1993. HIV infection is active and progressive in lymphoid tissue during the clinically latent stage of disease. *Nature* 362:355–58

128. Pathak VK, Temin HM. 1990. Broad spectrum of in vivo forward mutations, hypermutations and mutational hotspots in a retroviral shuttle vector after a single replication cycle: substitutions, frameshifts and hypermutations. *Proc. Natl. Acad. Sci. USA* 87:6019–23

129. Pedroza Martins L, Chenciner N, Asjo B, Meyerhans A, Wain-Hobson S. 1991. Independent fluctuation of human immunodeficiency virus type 1 rev and gp41 quasispecies in vivo. *J. Virol.* 65:4502–7

130. Phillips RE, Rowland Jones S, Nixon DF, Gotch FM, Edwards JP, et al. 1991. Human immunodeficiency virus genetic variation that can escape cytotoxic T cell recognition. *Nature* 354:453–59

131. Piatak MJr, Saag MS, Yang LC, Clark SJ, Kappes JC, et al. 1993. High levels of HIV-1 in plasma during all stages of infection determined by competitive PCR. *Science* 259:1749–54

132. Potts KE, Kalish ML, Lott T, Orloff G, Luo CC, et al. 1993. Genetic heterogeneity of the V3 region of the HIV-1 envelope glycoprotein in Brazil. *AIDS* 7:1191–97

133. Pressing J, Reaney DC. 1984. Divided genomes and intrinsic noise. *J. Mol. Evol.* 20:135–46

134. Rico-Hesse R, Pallansch MA, Nothay BK, Kew OM. 1987. Geographic distribution of wild poliovirus type 1 genotypes. *Virology* 160:311–22

135. Rigby M, Holmes EC, Pistello M, Mackay N, Leigh Brown AJ, Neil JC. 1993. Evolution of structural proteins of feline immunodeficiency virus; molecular epidemiology and evidence for selection. *J. Gen. Virol.* 74:425–36

136. Roberts JD, Bebenek K, Kunkel TA. 1988. The accuracy of reverse transcriptase from HIV-1. *Science* 242:1171–73

137. Robertson JR, Bucknall ABV, Welsby PD, Roberts JJK, Inglis JM, et al. 1986. Epidemic of AIDS related virus (HTLV-III/LAV) among intravenous drug abusers. *BMJ*. 292:527–29

138. Rota JS, Hummel KB, Rota PA, Bellini WJ. 1992. Genetic variability of the glycoprotein genes of current wild-type measles isolates. *Virology* 188:135–42

139. Saag MS, Hahn BH, Gibbons J, Li Y, Parks ES, et al. 1988. Extensive variation of human immunodeficiency virus type-1 in vivo. *Nature* 334:440–44

140. Salminen M, Nykanen A, Brummer-Korvenkontio H, Kantanen ML, Liitsola K, Leinikki P. 1993. Molecular epidemiology of HIV-1 based on phylogenetic analysis of in vivo gag p7/p9 direct sequences. *Virology* 195:185–94

141. Schuitemaker H, Koot M, Kootstra N, Dercksen MW, Goede RTD, et al. 1992. Biologial phenotype of human immunodeficiency virus type 1 clones at different stages of infection: progression of disease is associated with a shift from monocytotropic to T-cell-tropic virus populations. *J. Virol* 66:1354–60

142. Seger J. 1988. Dynamics of some simple host-parasite models with more than two genotypes in each species. *Phil. Trans. R. Soc. Lond. B* 319:541–55

143. Sharp PM, Li WH. 1988. Understanding the origin of AIDS viruses. *Nature* 336:315

144. Sharp PM, Robertson DL, Gao F, Hahn BH. 1994. Origins and diversity of human immunodeficiency viruses. *AIDS* 8(Suppl. 1): In press

145. Shioda T, Levy JA, Cheng Mayer C. 1991. Macrophage and T cell-line tropisms of HIV-1 are determined by spe-

cific regions of the envelope gp120 gene. *Nature* 349:167–69

146. Shpaer EG, Mullins JI. 1993. Rates of amino acid change in the envelope protein correlate with pathogenicity of primate lentiviruses. *J. Mol. Evol.* 37: 57–65

147. Sigurdsson B, Palsson PA, Grimsson H. 1957. Visna, a demyelinating transmissible disease of sheep. *J. Neuropathol. Exp. Neurol.* 16:389–403

148. Simmonds P, Balfe P, Ludlam CA, Bishop JO, Leigh Brown AJ. 1990. Analysis of sequence diversity in hypervariable regions of the external glycoprotein of human immunodeficiency virus type 1. *J. Virol.* 64:5840–50

149. Simmonds P, Zhang LQ, McOmish F, Balfe P, Ludlam CA, Leigh Brown AJ. 1991. Discontinuous sequence change of human immunodeficiency virus (HIV) type 1 env sequences in plasma viral and lymphocyte-associated proviral populations in vivo: implications for models of HIV pathogenesis. *J. Virol.* 65:6266–76

150. Smith TF, Srinivasan A, Schochetman G, Marcus M, Myers G. 1988. The phylogenetic history of immunodeficiency viruses. *Nature* 333:573–75

151. St Clair MH, Martin JL, Tudor Williams G, Bach MC, Vavro CL, King DM, et al. 1991. Resistance to ddI and sensitivity to AZT induced by a mutation in HIV-1 reverse transcriptase. *Science* 253:1557–59

152. Starcich BR, Hahn BH, Shaw GM, McNeely PD, Modrow S, et al. 1986. Identification and characterization of conserved and variable regions in the envelope gene of HTLV-III/LAV, the retrovirus of AIDS. *Cell* 45:637–48

153. Steel CM, Ludlam CA, Beatson D, Peutherer JF, Cuthbert RJG, et al. 1988. HLA haplotype A1 B8 DR3 as a risk factor for HIV-related disease. *Lancet* i:1185–88

154. Steuler H, Storch Hagenlocher B, Wildemann B. 1992. Distinct populations of human immunodeficiency virus type 1 in blood and cerebrospinal fluid. *AIDS. Res. Hum. Retroviruses.* 8:53–59

155. Takahashi H, Merli S, Putney SD, Houghten R, Moss B, et al. 1989. A single amino acid interchange yields reciprocal CTL specificities for HIV-1 gp160. *Science* 246:118–21

156. Tersmette M, Gruters RA, de Wolf F, de Goede RE, Lange JM, et al. 1989. Evidence for a role of virulent human immunodeficiency virus (HIV) variants in the pathogenesis of acquired immunodeficiency syndrome: studies on sequential HIV isolates. *J. Virol.* 63: 2118–25

157. Tersmette M, Lange JM, de Goede RE, de Wolf F, Eeftink Schattenkerk JK, et al. 1989. Association between biological properties of human immunodeficiency virus variants and risk for AIDS and AIDS mortality. *Lancet* 1:983–85

158. Thormar H. 1967. Cell-virus interactions in tissue cultures infected with visna and maedi viruses. *Curr. Top. Microbiol. Immunol.* 40:22–32

159. Tsujimoto H, Hasegawa A, Maki N, Fukasawa M, Miura T, Speidel S, et al. 1989. Sequence of a novel simian immunodeficiency virus from a wild-caught African mandrill. *Nature* 341: 539–41

160. Vaishnav YN, Wong Staal F. 1991. The biochemistry of AIDS. *Annu. Rev. Biochem.* 60:577–630

161. Varmus H. 1988. Retroviruses. *Science* 240:1427–35

162. Varmus H, Brown P. 1989. Retroviruses. In *Mobile DNA*, ed. DE Berg, MM Howe, pp. 53–109. Washington DC: Am. Soc. Microbiol.

163. Vartanian JP, Meyerhans A, Asjo B, Wain-Hobson S. 1991. Selection, recombination, and G——A hypermutation of human immunodeficiency virus type 1 genomes. *J. Virol.* 65: 1779–88

164. Wahlberg J, Albert J, Lundeberg J, von Gegerfelt A, Broliden K, et al. 1991. Analysis of the V3 loop in neutralization-resistant human immunodeficiency virus type 1 variants by direct solid-phase DNA sequencing. *AIDS Res. Hum. Retroviruses* 7:983–90

165. Wain-Hobson S. 1989. HIV genome variability in vivo. *AIDS* 3 (Suppl. 1): S13–S18

166. Wain-Hobson S, Vartanian JP, Henry M, Chenciner N, Cheynier R, et al. 1991. LAV revisited: origins of the early HIV-1 isolates from Institut Pasteur. *Science* 252:961–65

167. Webster RG, Laver WG, Air GM, Schild GC. 1982. Molecular mechanisms of variation in influenza viruses. *Nature* 296:115–21

168. Weiss RA. 1993. How does HIV cause AIDS? *Science* 260:1273–79

169. Weiss RA, Clapham PR, Cheingsong Popov R, Dalgliesh AG, Carne CA, et al. 1985. Neutralization of human T-lymphotropic virus type III by sera of AIDS and AIDS-risk patients. *Nature* 316:69–72

170. Weiss RA, Clapham PR, Weber JN, Dalgliesh AG, Lasky LA, Berman PW. 1986. Variable and conserved neutral-

ization antigens of human immunodeficiency virus. *Nature* 324:572–75

171. Weiss RA, Teich NN, Varmus HE, Coffin J. 1985. *RNA Tumor Viruses.* Cold Spring Harbor Lab. 2nd ed.

172. Wolfs TFW, de Jong JJ, Van den Berg H, Tijnagel JM, Krone WJ, Goudsmit J. 1990. Evolution of sequences encoding the principal neutralization epitope of human immunodeficiency virus 1 is host dependent, rapid, and continuous. *Proc. Natl. Acad. Sci. USA* 87:9938–42

173. Wolfs TFW, Zwart G, Bakker M, Goudsmit J. 1992. HIV-1 genomic RNA diversification following sexual and parenteral virus transmission. *Virology* 189:103–10

174. Wolinsky SM, Wike CM, Korber BTM, Hutto C, Parks WP, Rosenblum LA, et al. 1992. Selective transmission of human immunodeficiency virus type-1 variants from mothers to infants. *Science* 255:1134–37

175. World Health Organization. 1993. Statistics from the World Health Organization and the Centers for Disease Control and Prevention. *AIDS* 7:1287–91

176. Yokoyama S, Gojobori T. 1987. Molecular evolution and phylogeny of the human AIDS viruses LAV, HTLV-III, and ARV. *J. Mol. Evol.* 24:330–36

177. Zhang LQ, MacKenzie P, Cleland A, Holmes EC, Leigh Brown AJ, Simmonds P. 1993. Selection for specific sequences in the external envelope protein of human immunodeficiency virus type 1 upon primary infection. *J. Virol.* 67:3345–56

178. Zhang LQ, Simmonds P, Ludlam CA, Leigh Brown AJ. 1991. Detection, quantification and sequencing of HIV-1 from the plasma of seropositive individuals and from factor VIII concentrates. *AIDS* 5:675–81

179. Zhu T, Mo H, Wang N, Nam DS, Cao Y, Koup RA, et al. 1993. Genotypic and phenotypic characterization of HIV-1 in patients with primary infection. *Science* 261:1179–81

180. Zwart G, Langedijk H, van der Hoek L, de Jong JJ, Wolfs TF, et al. 1991. Immunodominance and antigenic variation of the principal neutralization domain of HIV-1. *Virology* 181:481–89

Annu. Rev. Ecol. Syst. 1994. 25:167–88

METAPOPULATION DYNAMICS AND GENETICS

Alan Hastings

Division of Environmental Studies, Center for Population Biology and Institute of Theoretical Dynamics, University of California, Davis, California 95616

Susan Harrison

Division of Environmental Studies and Center for Population Biology, University of California, Davis, California 95616

KEYWORDS: extinction, dispersal, persistence, colonization, gene flow

Abstract

Metapopulation dynamics as originally defined by Levins consists of the extinction and colonization of local populations. Theory suggests that these processes can profoundly affect demographic persistence, the coexistence of interacting species, genetic variation, and evolution. However, a review of empirical studies illustrates the limitations of the Levins definition and indicates the need for a more complex view of metapopulation dynamics. We describe a modeling approach that can incorporate a greater range of ecological and genetic processes within local populations. We discuss uses of this general formulation, its connections to other modeling frameworks, and directions for the future integration of ecological and genetic studies of metapopulations.

INTRODUCTION

The concept of a metapopulation can be traced to Nicholson & Bailey (111), who stated in 1935 (p. 590) that "A probable ultimate effect of increasing oscillation is the breaking up of the species-population into numerous small widely separated groups which wax and wane and then disappear, to be replaced by new groups in previously unoccupied situations." These authors

0066-4192/94/1120-0167$05.00

saw this as a resolution to the problem that the host-parasitoid models they produced were unstable.

Better-known antecedents include Wright's (172) shifting balance theory, Andrewartha & Birch's (5) observations on ephemeral insect populations, and den Boer's (27) ideas on the importance of asynchrony and dispersal. Levins (93) coined the term *metapopulation* to mean a "population of populations" existing in a balance between extinction and recolonization. This remains a common, often implicit definition of a metapopulation, although we use the term more broadly to connote any set of conspecific populations linked by dispersal.

The assumptions underlying most metapopulation models, and the general conclusions that result, are contained in Levins' (92, 93) original one-species model. This simple model assumes a very large number of habitat patches, so that stochasticity is unimportant, but not so large that density dependence arising from patch occupancy can be ignored. Patches are assumed to be equidistant from one another and identical in their underlying features. The population dynamics at the level of the patch are assumed to be so rapid that patches can be assigned to only two states, occupied or empty. The only biological processes operating are local extinction within a patch, and colonization of unoccupied patches by immigrants from occupied ones.

These assumptions lead to a simple model phrased in terms of the fraction p of patches occupied. Let e be a parameter measuring the rate of extinction within an occupied patch, and let m be a parameter describing the rate of colonization of unoccupied patches, which is proportional to the product of the fractions of occupied and unoccupied patches. Then the dynamics of p are given by the colonization rate minus the extinction rate, or

$$dp/dt = mp(1 - p) - ep \qquad\qquad 1.$$

The behavior of this model is quite simple: if the colonization rate is larger than the extinction rate, $m > e$, then there is a globally stable equilibrium with a fraction $p = 1 - e/m$ of the patches occupied. Otherwise, the only outcome is extinction at the metapopulation level.

Measurement of extinction and colonization rates is critical in relating this model to empirical studies. In fact, as a nondimensionalization would show, the two parameters enter into the dynamics only as the ratio e/m. Quantitatively, this ratio can be used to predict the fraction of occupied patches; however, to use this prediction empirically requires the potentially difficult task of identifying all available patches. The qualitative prediction, which is at the heart of the metapopulation concept, is that the system persists globally even though extinction is certain at the level of single patches.

Extensions of the Levins approach to interactions between species appeared shortly after Levins' original paper. So-called "patch" models examined metapopulations of either two competitors (89, 139), or a predator and prey (22, 58), in which one species is capable of causing the other's local extinction. These models demonstrated the possibility of "patchy coexistence," in which subdivision into asynchronous local populations can stabilize such a locally unstable interaction.

Yet as a review of empirical evidence will show, it is not clear that the dynamics of natural metapopulations are well described by Levins' and related models. We begin with studies of single species, turn to the cases of multiple species, and then review work on the genetic structure of metapopulations, based on the belief that further integration of ecological and genetic work will be fruitful. We then present a modeling approach that can overcome some of the limitations of the standard metapopulation concept and which is general enough to encompass all real metapopulations. This model is then tailored to some specific cases, to illustrate its uses in both ecological and genetic contexts. We also discuss the connections between this approach and other ways of modeling spatial structure.

ECOLOGICAL EVIDENCE

Single Species

Extinction and recolonization are sometimes measured directly, either by field observations (1, 46, 99, 117, 131, 144–146, 169) or using historical records (137, 152). Alternatively, these processes may be inferred from the spatial distribution of populations (35, 51, 56, 86, 114, 116, 143, 154, 157, 158), an approach that typically requires a complete regional survey of occupied and vacant habitat patches (plus certain assumptions about, e.g., the sources of colonists). Sometimes the failure of local populations to replace themselves by local recruitment is cited as evidence that a metapopulation exists (105, 147). Given the obvious difficulties of region-wide studies, it is not surprising that relatively few metapopulation studies have been done; nonetheless, several generalizations are possible (52, 53).

Many real metapopulations contain one or more local populations that virtually never go extinct (38, 56, 130, 131). Such "mainland and island" structure (52) can be created by natural differences in habitat size or quality, or even by human fragmentation of natural habitats. Because extinction and colonization affect only a subset of populations, they may have little effect on regional persistence. However, these processes do affect regional distribution. Small or isolated patches are less likely than large or central ones to support populations, whether there is a single mainland (35, 56, 116, 143, 157) or a constellation of large and small populations (81, 137, 153, 154). Geographic range limits may occur where habitat patches become too sparse (21, 137).

Species that occupy naturally patchy or transient habitats are often very good dispersers, so that most individuals visit more than one patch in their lifetime. Examples are insects on dung and carrion patches (71, 132, 133) or on weedy host plants (54, 54a), and amphibians in temporary ponds (36, 135). Such species form relatively unsubdivided, "patchy populations" (52), whose persistence is not highly sensitive to the precise spatial structure of the habitat. The same may be true when, for example, humans fragment a forest at a fine scale relative to the dispersal distances of birds that occupy it (53, 125).

Much interest lies in those intermediate cases where many but not all individuals occupy more than one habitat patch in their lifetimes (36, 50, 135, 137, 146). Local populations may or may not fluctuate independently of one another in such cases. However, the metapopulation may still be vulnerable to collapse should habitat patches become fewer or more isolated. An important area for investigation is the mortality associated with dispersal between patches. The greater this mortality, the more sensitive are the dynamics of such a metapopulation to the spatial structure of its habitat.

Very high rates of dispersal do not preclude an interesting interplay between local and regional dynamics. Marine invertebrates with planktonic larvae are generally thought to show little demographic or genetic subdivision over very large distances (but see Palumbi 1994, this volume). Yet even in models that assume a completely well-mixed regional propagule pool, spatial mosaics of different-aged populations may arise from the interaction between a variable propagule supply and local processes, such as competition for space (73, 126, 127) or Allee effects related to fertilization (94, 122). Regional persistence may depend on "source" populations that export larvae to much larger "sink" areas (73, 122, 126).

A final possibility is that either natural (16) or human-caused (166, 168) habitat fragmentation may result in local populations that are completely isolated from one another. To the extent that local extinctions occur, such "non-equilibrium metapopulations" (52) are bound for regional extinction. In the context of conservation, this implies that conspecific populations must sometimes be managed as separate entities rather than as a metapopulation.

Empirical evidence thus casts some doubt on the standard view of metapopulations, in which recolonization balances extinction. It may be inherently unlikely for real metapopulations to show enough subdivision so that populations can fluctuate and go extinct independently, yet enough connectedness that the metapopulation can persist, unless it contains some extinction-resistant "mainland" populations. A broader view of metapopulations seems called for, one that presumes neither independently fluctuating local populations, nor a necessary balance between extinction and colonization.

Species Interactions

Ideas about spatial structure and species interactions began with Hutchinson's (70) observations on fugitive competitors, and Huffaker's (68) experiments with predatory and prey mites on arrays of oranges. Though few studies have shown as clearly as Huffaker's that subdivision is stabilizing, a number have produced partial evidence (11, 134, 151). On arrays of plants in glasshouses, mite predators and prey show a potentially stabilizing degree of asynchrony (157a, 109, 128), and the same may be true in orchards (164). Prevalence of a fatal viral disease in terrestrial isopods decreases with patch isolation, again suggesting a link between patchiness and coexistence (39). Competing *Daphnia* species may exclude one another in experimental pools, but coexist in complexes of rockpools (12).

Other studies have shown the effects of subdivision on stability to be neutral or even negative. Isolating individual trees in cages did not alter the fluctuations of citrus scale and its parasitoid (108). The patchiness of ragwort (a weedy plant) had little effect on competition among its insect herbivores (54, 54a). Subdividing stands of goldenrod destabilized the interaction between ladybird beetles and aphids on these plants (77).

Perhaps reassuringly, the idea of patchy coexistence appears to be better supported by experiments on flightless organisms than by studies on more vagile species. However, because experiments typically control patch size, quality, and isolation, the relative importance of this mechanism in explaining coexistence in natural systems remains an open question. Like the classic, Levins notion of a metapopulation, patchy coexistence may require a fairly restrictive set of conditions, such as strong local interactions, an intermediate amount of dispersal, and no secure "mainlands" for either species.

Metapopulation (patch) and continuous (diffusion) models both predict that predator-prey coexistence requires low rates of dispersal by the predator. However, they make opposing predictions about the effects of high prey dispersal, which is stabilizing in the former (58) but can destabilize an equilibrium in the latter (89). To our knowledge no empirical studies have addressed this issue, e.g. by manipulating predator and prey dispersal differently. We discuss a theoretical resolution of this apparent paradox below.

GENETIC EVIDENCE

Wright (172, 174) attributed great significance to the evolutionary interplay between gene flow on the one hand, and local differentiation through founder effects, drift, and natural selection on the other. Recent theoretical work has furthered Wright's insight by showing that extinction and colonization may convert nonadditive into additive genetic variance, promoting a large role for small populations in adaptive evolution (37a,b, 161a, 169a).

Most evidence on genetic structure in natural metapopulations comes from electrophoretic studies. Allozyme frequencies from conspecific populations are used to calculate either Wright's F_{st}, which estimates their "average" differentiation, or the matrix of pairwise genetic distances between populations (9, 32,

43, 95, 141, 142). A few studies have measured spatial variation in quantitative genetic characters, a process that requires reciprocal transplant experiments (6). Molecular work on spatial structure in natural populations has just begun (30, 38a).

Allozyme studies show clearly that population subdivision promotes differentiation. Values of F_{st} are higher in subdivided than continuous habitats (7, 13, 19, 75, 79, 118, 119), higher in plants that disperse poorly than in those that disperse well (17, 43, 95), and higher in marine invertebrates without than those with planktonic larvae (7, 65, 69, 75, 167). Nonetheless, nonzero values of F_{st} have been found among populations linked by observed gene flow (13, 119, 135) and recolonization (99, 169). Patterns of correlation between genetic and geographic distances sometimes indicate limited ("stepping stone") gene flow between adjacent populations (13, 32, 65, 79, 119, 142).

The use of F_{st} to estimate average levels of gene flow requires the assumptions that gene flow is continuous and that genetic drift is the only cause of differentiation. This has proven to be problematic, because the assumptions of equilibrium and neutrality are often violated in natural systems, and because gene flow and recolonization may sometimes produce indistinguishable patterns (141, 142). Mitochondrial DNA lineages may prove to be more sensitive measures of infrequent gene flow (30), while disparities between patterns in mitochondrial and nuclear DNA may indicate population turnover (38a).

When populations experience different selective pressures, local adaptation and gene flow may interact to promote genetic variation in the metapopulation. Some evidence for this comes from coevolutionary studies. Karban (76) found fine-scale local adaptation of a flightless insect to clonal patches of its host plant. Wade (160) and Goodnight (37c) found enough between-population variation to permit local coadaptation in two competing species of flour beetles. Jarosz (74) and Antonovics (5a) found that local adaptation and gene flow determined genotype frequencies in a plant-pathogen system. Thompson (155) evaluates a variety of ways in which metapopulation structure influences the process and outcome of coevolution. Of course, coevolution and other forms of adaptation may also take place within (6) or among (31) single, isolated populations.

Extinction and colonization may produce a certain amount of genetic differentiation through the founder effect, if the groups that found new populations are sufficiently small and homogeneous (100, 102, 163). These conditions appear to be met in metapopulations of a milkweed beetle on patches of its host plant (99), and of fungus beetles on rotting logs (169). However, many species that undergo frequent population turnover are very good dispersers, and so their overall level of differentiation is still relatively low. Plants found in ephemeral habitats show low values of F_{st} (17, 37) and little correlation between genetic and geographic distances (8).

In the long term, extinction and recolonization should accelerate drift and the loss of heterozygosity, both within and among populations (102). Low levels of heterozygosity in some natural populations have been attributed to frequent turnover (159, 165); fluctuations in population size could also explain these patterns (138).

Interest in interdemic selection has motivated much of the work on metapopulation genetics. In laboratory studies, Wade (161) found that subdividing flour beetle populations for 10–15 generations could lead to heritable interdemic variation in fitness, and Wade & Goodnight (162) showed that these differences could be selected upon without depleting among-population variation. Not surprisingly, the evidence from natural systems is sparser. Newly founded plant populations may show higher levels of dispersal and outcrossing than do older ones (97, 113, 124), suggesting that turnover influences the evolution of life history traits; however, this is open to explanation in terms of individual selection.

In conservation biology, the effects of habitat fragmentation on population genetic structure are a major focus of concern (82, 85). Several studies indicate that recent fragmentation of habitat has increased the differentiation among local populations and/or reduced their internal genetic variation (88, 148, 149). Corridors or transplants among remnant populations are sometimes proposed to minimize inbreeding (14), although this may conflict with the goal of preserving differentiated populations (87, 104). In general, Lande (84) argues that the demographic consequences of fragmentation are likely to be more critical than the genetic ones.

METAPOPULATION MODELS

As we have seen, the empirical evidence casts doubt on the view of metapopulations that derives from the results of the simple Levins model. We therefore look at ways to extend and generalize this model. Rather than build up complex single-species models step by step, we move now to a framework that is broad enough to encompass any models that still make the assumption of ignoring spatial arrangement, while allowing very general assumptions about patch sizes and migration patterns (41, 49, 64, 106). The key step in formulating a general model is to recognize that the tools developed to look at structured populations for single species (106) can be used to look at structured models of metapopulations.

The i-State, p-State Model

We begin by describing the framework verbally to emphasize its generality. Our presentation follows that in Metz & Diekmann (106). The first step is to identify the potential states of the individual local populations. In Metz &

Diekmann's terms, these are the i-states, the states of individuals. The i-states could be population numbers or allele frequencies; in more complex models, the i-states could be population sizes of each of two species, or both allele frequencies and population sizes.

Processes within a single patch can cause the i-states just described to change smoothly with time, and can be described by differential equations. These equations for the continuous dynamics of the i-states will enter into the equations for the metapopulation model, in addition to the terms describing the jumps included in the classic formulation. The i-state could also include aspects that do not change such as the underlying size or quality of the habitat patch.

The model is then specified in terms of a "mass balance" equation for the p-state, the state of the population as a whole, or the fraction (more specifically, a density function for the fraction) of the patches that are in particular i-states. At this level, there are several processes affecting dynamics within a patch. One is the process included in the original Levins' formulation, namely, extinction. Related processes are sudden changes to or from other states. The final process is smooth changes within a patch due to processes such as population growth or natural selection.

In other words, the general model for changes within a patch is:

rate of change in fraction of populations in state i =   2.
rate of change due to smooth changes in the state within patches
$-$ (extinction rate + rate of sudden changes to other states)
+ contributions from sudden changes from other states.

Note that the terms on the right side of this equation represent changes in the fraction of patches in a particular state, not changes within a particular patch. In some cases, to this model we would need to append an equation describing the rate of founding of patches, i.e. colonization of empty habitat. Also note that the models such as Levins that only consider a finite number of patch types are special cases of this general model, where the terms for smooth changes in the state within a patch are absent; this corresponds to assuming that within-patch is much more rapid than population turnover.

We now look at how this general model is expressed in mathematical terms (106). For ease of exposition, we use a single variable, s, to describe the i-state. Let the dynamics of this state be given by $g(s, p(t\bullet))$, so that within a patch we are asssuming

$$\frac{ds}{dt} = g(s,p(t,\cdot)).$$ 3.

The dependence of the function g on the state of the metapopulation as a whole [implied by the notation $p(t\bullet)$] allows for the possibility of describing steady immigration (or gene flow) from other patches. Now we write the mathematical form of Equation 2, with the terms in the same order as Equation 2

$$\frac{\partial p\,(t,\,s)}{\partial t} = \frac{\partial g p\,(t,\,s)}{\partial s} - (\mu_e\,(s,\,p\,(t,\,\cdot),\,t) + \mu_j\,(s,\,p\,(t,\,\cdot),\,t\,))\,p\,(t,\,x)$$

$$+ \int \gamma\,(\sigma,\,s)\,\mu_j\,(\sigma,\,p\,(t,\,\cdot),\,t)\,p\,(t,\,\sigma)\,d\,\sigma, \qquad\qquad 4.$$

where $p(t,s)$ is a density function for the fraction of patches in state s at time t. The probability of extinction in a patch in state s is $\mu_e(s,p(t,\cdot),t)$. The probability of a sudden change from state s is $\mu_j(s,p(t,\cdot),t)$, and the probability that a patch leaving state σ ends up in state s is $\gamma(\sigma,s)$. To complete the model, we add a term for the colonization of empty patches,

$$g(0,\,p\,(t,\,\cdot))\,p\,(t,\,o)\,p\,(t,\,0) = \int_{0}^{\infty} M\,(\sigma,\,p\,(t,\,\cdot))\,p\,(t,\,\sigma)\,d\,\sigma, \qquad 5.$$

where the function M gives the rate of colonization of empty patches per patch of state s, which may depend on the state of the whole metapopulation. This framework, with the states described by a vector, and possibly with a vector of metapopulation variables p, is general enough to include any metapopulation description that ignores spatial arrangement of patches and assumes that processes at the metapopulation level are deterministic. In particular, this model may be used to describe systems where extinction is rare, yet metapopulation characteristics are important in determining overall dynamics.

This general framework, however, leads to a class of models that is typically too complex to analyze and is probably difficult to relate directly to empirical observations. So the difficult step is making appropriate simplifying assumptions that lead to tractable models (28). For example, one set of assumptions leads to the classic models used by population geneticists to explore population structure such as the island model of Wright (174). These models have been used to guide the estimation of the level of gene flow or dispersal among subpopulations (10). This general framework can be used to integrate genetic and ecological approaches, because the level of dispersal is a key determinant of the dynamics of spatially structured populations. We now turn to specific implementations, concentrating primarily on conclusions rather than on details of the models.

Ecological Models

SINGLE SPECIES MODELS   Hanski (44, 45) was among the first to extend Levins' approach by considering populations of several different sizes. His model, which incorporated two discrete population levels within a patch, was later shown by Gyllenberg & Hanski (41) to be a special case of a general model of the form of Equation 2 above. One important conclusion to emerge from such models is that immigrants from large patches to small patches can allow a species to persist if its overall abundance is high enough. Below a threshold level of total population size, the metapopulation collapses. This effect can cause the sudden extinction of a metapopulation as the extinction rate within patches is raised (60). Hanski (44, 45) also pointed out that models that incorporate different population levels and immigration into existing patches produce a positive correlation between local abundance and regional distribution (fraction of patches occupied).

Several papers (41, 50, 60, 64) have considered more general models that describe local population levels using a continuous variable, reflecting either underlying habitat differences or differences due to dynamics. These models lead to predictions of the equilibrium distributions of local population sizes. A major theme is that without immigration into existing populations there is a unique equilibrium for the metapopulation; with immigration into existing populations, multiple equilibria are possible. Similarly, only when immigration plays a significant role in local population dynamics is there a positive correlation between abundance and distribution (60); otherwise, the two are completely independent.

Models that are stochastic at the level of the metapopulation   All of the models discussed thus far make the assumption that the number of patches, or local populations, is so large that process at the level of the metapopulation can be viewed as deterministic. Obviously, when considering the extinction of real metapopulations, the effects of a finite number of patches and populations must be included. The simplest case is a system of just two populations. As shown by Quinn & Hastings (121), time to extinction is much longer for two small populations than for one large one, even when recolonization is ignored. Other studies have examined time to extinction in finite metapopulations in much more detail (55, 96). Spatial correlation among populations in their chances of extinction may or may not greatly reduce the survival time of a metapopulation (55).

COMPETITION MODELS   Early models analogous to Levins' first showed that two competitors could persist globally even if coexistence was impossible within a single patch (23a, 89, 139). Later models extended this result to cases of three (47) or large numbers (59) of competitors, and they examined the dependence of coexistence upon rates of disturbance. More recently, similar models building on (59) have shown that habitat removal or fragmentation could in some circumstances enhance diversity (110); that metapopulation dynamics may maintain diversity in plant communities (22a, 156); and that the coexistence of multiple diseases in hosts can be modeled as a metapopulation process (98). The latter illustrates the idea of diseases as metapopulations, with individual hosts as patches of habitat (4).

Few multispecies models have considered explicit population sizes, but

Hastings (61) determined a joint equilibrium distribution for two competing species with explicit local population sizes.

PREDATOR-PREY MODELS   The first models of predator-prey metapopulations (22, 58) demonstrated the possibility of global persistence despite local extinction, just as originally envisioned by Nicholson & Bailey and by Huffaker. More detailed models that included stochastic factors (40) provided a good fit to Huffaker's experimental results (68). Within the context of host-parsitoid metapopulation models, environmental variability combined with small enough migration rates has been shown to allow almost indefinite persistence (72, 123, 150)

More recently, these models have been elaborated and extended to more species. Prey dispersal in response to predators has been shown to be a potentially stabilizing influence (129). Metz & Diekmann (106) formulated and solved a model incorporating explicit local population sizes. Three-species models, derived from models with explicit consideration of population sizes (28), have been used to study the dynamics of food chains (129).

REACTION-DIFFUSION MODELS   Clearly, the spatial structure assumed in standard metapopulation models is highly idealized, so it is important to examine alternative ways to model spatial structure, such as movement as a diffusion process. Reaction-diffusion models in fact have a much longer history than metapopulation models and were first reviewed in these pages almost 20 years ago by Levin (90). Our goal here is to reconcile the conclusions of these models with those of the metapopulation models discussed above.

In verbal terms, the simplest reaction diffusion model approach states that:

rate of change of local population size =   6.
change due to local dynamics
+ change due to random exchange with other populations

This framework has much in common with the metapopulation model, such as the inclusion of local dynamics and movement. The differences are that metapopulation theory represents the world in terms of discrete patches, whereas reaction-diffusion theory can examine a continuous habitat; metapopulation theory allows the status of patches to change in discrete jumps (e.g. from occupied to empty), whereas reaction-diffusion theory portrays continuous changes in population density; and in the metapopulation framework, dispersal between patches is also typically a discrete process, while in the reaction-diffusion framework dispersal is a continuous process.

The simplest reaction-diffusion models, with no underlying spatial heterogeneity, show a global equilibrium identical to the equilibrium of the local dynamics. If the local equilibrium is unstable, subdivision and movement do not make it more stable (89). Moreover, an equilibrium that is stable without movement can become unstable after movement is included (89). This seems to contradict the predictions of the metapopulation model, in which movement can only be stabilizing.

This apparent paradox comes from the focus on equilibrium behavior of the reaction-diffusion models. Several recent models have shown that within the reaction-diffusion framework, a predator-prey system that would not persist in the absence of dispersal will persist with exchange (2, 112). However, persistence is not via a stable equilibrium, but through cyclic or more complex dynamics. Similarly, movement has been shown to replace chaotic dynamics in coupled single species models by much simpler dynamics (42, 62). In the case of two patches, the system often persists with a constant total population size, but with the populations in each patch alternating asynchronously between high and low levels. This is an exact analogue to the dynamics of classic metapopulation models. A related finding is that chaotic local dynamics can put local populations out of phase and thus allow a metapopulation to persist (3, 67).

These results are important because they show that the basic concept underlying metapopulation models, namely, that dispersal leads to a relatively constant global population size, can hold as well in models where local extinction does not occur. Such models may be closer to many natural metapopulations, where local extinctions may be the exception, and immigration into existing populations may play an important role. Metapopulation models may often be easier to analyze and understand than the nonequilibrium solutions of reaction-diffusion models.

Another reason for reconciling metapopulation and reaction-diffusion models is that the latter can be applied to questions that cannot be addressed with metapopulation models, such as rates of spatial spread of populations.

CELLULAR AUTOMATA   Spurred by increases in computational power, spatially explicit models such as cellular automata have become the subject of much investigation (23, 29). These models bridge the gap between metapopulation models and reaction-diffusion approaches by including spatial arrangement, yet allowing for stochasticity at the level of the local population. The basic approach is to represent space as a lattice of sites, each of which is described by one of a small number of states. At each discrete time step, the state of each site changes depending on its present state and those of neighboring sites. In the simplest case, the sites are empty or occupied, as in the Levins model. In fact, one reassuring result of the the analysis of cellular automata and related stochastic models is that their behavior is well described by Levins-type models (23, 29). In the study of stochastic analogues to cellular automata, the Levins model (or its analogues) is known as the mean field approximation. However, cellular automata or coupled lattice models typically show complex spatial and temporal dynamics not observed in analytic metapopulation models (63).

Recent cellular automata have examined predator-prey and host-parasitoid dynamics (24, 57, 170). Like metapopulation models, they have shown that dispersal allows global persistence despite local instability. Models of this form can provide insight into the interplay between spatial and temporal scales in ecology (33, 91). Cellular automata have also been applied to plant growth (20, 115) and competition (26, 107, 136). Conservation biologists increasingly are using spatially explicit simulation models to analyze the effects of habitat fragmentation on population viability (18, 82a 83a, 103, 120); however, some controversy surrounds the use of these models (28a, 53a).

Genetic Models

Analogues to virtually all of the ecological models discussed above have been used to investigate the role of metapopulation structure in maintaining genetic variability. These models have been extensively reviewed elsewhere (10, 66), and we primarily note that an integration of ecological and genetic approaches is still needed (113).

Genetic models usually focus on cases where local populations have fixed size and extinctions are ignored. The classic island model (171) has a geographic structure identical to the Levins model. It is simple enough that putting it into the i-state, p-state framework developed above seems forced, but this serves as a useful starting point to indicate how more realistic models may be produced. The state of each local population (i-state) is its inbreeding coefficient, which is the same in every local population because of the assumptions of the island model (10). Therefore the p-state, which is the distribution of inbreeding coefficients, F's, is just the same as each i-state. The result is that

$$F = \frac{(1 - m)^2}{2N_e - (1 - m)^2 (2N_e - 1)}$$

7.

$$\approx \frac{1}{4N_e m + 1}$$

8.

where m is the rate of migration (input into each island) per generation, N_e is the effective local population size, and the approximation is valid for small m. From either formula, one can estimate m, knowing N_e (or vice versa). The relationship among Wright's F-statistics has been extensively reviewed in the literature (10), which connects the theory with F_{st} discussed earlier.

That the island model actually produces a distribution of states is evident from its stochastic analogue (173) which includes drift in gene frequencies within a local population and migration between populations. Here the focus is on the allele frequency at a single locus, and the *i*-state is the frequency of the A allele. The *p*-state is the distribution of allele frequencies, which is a beta distribution at the steady state (25). However, this model still does not include extinctions or differences in population sizes.

Extinctions were included in a modification of the basic island model by Slatkin (140). When a population goes extinct in this model, the site is immediately recolonized by individuals chosen at random. Not surprisingly, turnover and gene flow both reduce *F*. In this model, if the extinction rate *e* is small enough, and *m* still smaller, one can replace *m* in Equation 8 by *m* + *e*/2.

Extinctions, but not variable population sizes, were included in Levins' (93) 1970 model focusing on allele frequencies. The technical difficulties in solving this model are formidable, but recent advances (106) may allow further progress. In a more realistic model that allows extinctions, population growth, and variable population sizes, the *p*-states might be still more complex. Rogers (124a) discusses some of the potential effects of realistic demography on F_{ST}.

There are also genetic models analogous to the reaction-diffusion and cellular automata formulations. The stepping stone models assume discrete demes, with exchange only between neighbors. This is similar to cellular automata, with a large number of local states. As in the ecological models, it is reassuring that general conclusions, such as the fact that populations are locally differentiated if $N_e m \ll 1$, hold for both the island model and the stepping stone model (78). Continuous space models also have a long history (109a), beginning with Fisher's (34) classic work on spatial spread.

CONCLUSIONS

Although orthodox metapopulation theory needs some rethinking, because the classic extinction-centered definition is too simplistic, its attention to turnover and connections between populations through dispersal clearly represents a perspective ecology must embrace. Theoretical extensions of the Levins framework, and comparisons with other modeling approaches, show that many conclusions are robust to changes in the assumptions that make the model more realistic. Subdivision and dispersal affect local population dynamics, genetic diversity, and the outcome of interactions between species, with or without extinction playing a central role.

An integration of genetic and ecological approaches will be essential for interpreting the structure of natural populations. It may be unrealistic to hope that genetic data can be used to determine the rates of either population turnover or gene flow, especially since the two may produce similar patterns. But gene frequency data may be used to identify the appropriate spatial scale for studying metapopulations. When values of F_{st}, or similar measures of differentiation, are plotted against the distance between sites, the distances at which F_{st} lies above zero but below its asymptote corresponds to the scale at which populations are separate yet linked by gene flow. Molecular techniques may provide more detailed insights into the rates and spatial patterns of dispersal.

From a theoretical point of view, there are two important directions for future work. First, the insights gained from studying natural populations can be used in simplifications of the general metapopulation model to produce analyses that yield biological insights. Second, further work on reconciling different approaches to modeling spatial structure will prove invaluable, because biological insights in different cases may be easiest to develop with a particular approach. For example, much of the discrepancy between reaction-diffusion and classical metapopulation models derives from the historical emphasis on stable equilibrium solutions when reaction-diffusion models are studied, to the exclusion of more complex dynamics. Closer inspection shows that dispersal can have many of the same effects in a reaction-diffusion framework as in a patch model framework. As we learn more about the connections between different theoretical approaches to metapopulations, we will need to pay more attention to how models differ in their data requirements.

To unite theory and empirical work, it will be necessary to parameterize relatively simple models with results from natural metapopulations. Important progress in this direction has been made by Hanski (50) whose spatially explicit simulation model (50, 50a) includes not only extinction and colonization, but also the effect of immigration on local population dynamics. Using data for the butterfly *Melitaea cinxia,* this model correctly predicts patterns of local population density and patch occupancy. In related work, Hanski (46, 46a) has developed a metapopulation model that can be parameterized using only presence and absence data.

ACKNOWLEDGMENTS

We thank Odo Diekmann, Martha Groom, Rick Grosberg, Peter Kareiva, Simon Levin, David McCauley, Chris Ray, John Thompson, Chris Thomas, and Michael Wade for comments. We acknowledge partial support from NIH Grant GM 32130 to Alan Hastings.

Literature Cited

1. Addicott JF. 1978. The population dynamics of aphids on fireweed: a comparison of local populations and metapopulations. *Can. J. Zool.* 56:2554–64
2. Adler F. 1993. Migration alone can produce persistence of host-parasitoid models. *Am. Nat.* 141:642–50
3. Allen JC, Schaffer WM, Rosko D. 1993. Chaos reduces species extinction by amplifying local population noise. *Nature* 364:229–32
4. Anderson RM, May RM. 1985. Vaccination and herd immunity to infectious diseases. *Nature* 318:323–29
5. Andrewartha HG, Birch LC. 1954. *The Distribution and Abundance of Animals.* Chicago: Univ. Chicago Press
5a. Antonovics JP, Thrall P, Jarousz A, Stratton D. 1994. Ecological genetics of metapopulations: the *Silene - Ustilago* plant-pathogen system. In *Ecological Genetics, ed. LA Real*, pp. 146–70. Princeton: Princeton Univ. Press
6. Argyres AZ, Schmitt JC. 1991. Microgeographic genetic structure of morphological and life-history traits in a natural population of *Impatiens capensis. Evolution* 45:178–89
7. Ayers DJ, Dufty S. 1994. Evidence for restricted gene flow in the viviparous coral *Seriatopora hystrix* on Australia's Great Barrier Reef. *Evolution*. In press
8. Barrett SCH, Husband BC. 1989. The genetics of plant migration and colonization. See Ref. 15, pp. 254–78
9. Barrowclough GF. 1983. Biochemical studies of the microevolutionary process. In *Perspectives in Ornithology*, ed. AH Brush, GA Clark, pp. 223–61. New York: Cambridge Univ. Press
10. Barton N, Clarke A. 1990. Population structure and process in evolution. In *Population Biology*, ed. K Wohrmann, SK Jain, pp. 115–73. New York: Springer-Verlag
11. Bengtsson J. 1991. Interspecific competition in metapopulations. *Biol. J. Linn. Soc.* 42:219–37
12. Bengtsson J. 1993. Interspecific competition and determinants of extinction in experimental populations of three rockpool *Daphnia* species. *Oikos* 67: 451–64
13. Berven KA, Grudzien TA. 1990. Dispersal in the wood frog (*Rana sylvatica*): implications for genetic population structure. *Evolution* 44:2047–56
14. Bleich VC, Wehausen JD, Holl SA. 1990. Desert-dwelling mountain sheep: conservation implications of a naturally fragmented distribution. *Cons. Biol.* 4: 383–90
15. Brown AHD, Clegg MT, Kahler AL, Weir BS. 1989. *Plant Population Genetics, Breeding, and Genetic Resources.* Sunderland, Mass: Sinauer
16. Brown JH. 1971. Mountaintop mammals: nonequilibrium insular biogeography. *Am. Nat.* 105:467–78
17. Broyles SB, Wyatt R. 1993. Allozyme diversity and genetic structure in southern Appalachian populations of the milkweed *Asclepias exaltata. Syst. Bot.* 18: 18–30
18. Burgman MA, Ferson S, Akcakaya HR. 1993. *Risk Assessment in Conservation Biology.* New York: Chapman & Hall
19. Burton RS, Feldman MW, Curtsinger JW. 1979. Population genetics of *Tigriopus californicus* (Copepoda: Harpacticoida): I. Population structure along the California coast. *Mar. Ecol. Prog. Ser.* 1:29–39
20. Cain ML, Pacala SW, Silander JA. 1991. Stochastic simulation of clonal growth in the tall goldenrod *Solidago altissima. Oecologia* 88:477–85
21. Carter RN, Prince SD. 1981. Epidemic models used to explain biogeographic limits. *Nature* 293:644–45
22. Caswell H. 1978. Predator-mediated coexistence: A nonequilibrium model. *Am. Nat.* 112:127–54
22a. Caswell H, Cohen JE. 1991. Disturbance, interspecific interaction and diversity in metapopulations. *Biol. J. Linn. Soc.* 42:193–218
23. Caswell H, Etter RJ. 1993. Ecological interactions in patchy environments: from patch-occupancy models to cellular automata. *Lect. Notes Biomath.* 96:93–109
23a. Cohen JC. 1970. A Markov contigency-table model for replicated Lotka-Volterra systems near equilibrium. *Am. Nat.* 104:547–60
24. Comins HN, Hassell MP, May RM. 1992. The spatial dynamics of host parasitoid systems. *J. Anim. Ecol.* 61: 735–48
25. Crow JF, Kimura M. 1956. Some genetic problems in natural populations. *Proc. Third Berk. Symp. Math. Stat. Prob.* 4:1–22
26. Czaran T, Bartha S. 1992. Spatiotemporal dynamic models of plant populations and communities. *Trends Ecol. Evol.* 7:38–42
27. den Boer PJ. 1968. Spreading of risk and stabilization of animal numbers. *Acta Biotheor* 18:165–94

28. Diekmann O, Metz JAJ, Sabelis MW. 1989. Reflections and calculations on a prey-predator-patch problem. *Act. Appl. Math.* 14:23–35

28a. Doak DF, Mills LS. 1994. A useful role for theory in conservation. *Ecology* 75: 615–26

29. Durrett R, Levin SA. 1994. Stochastic spatial models: a user's guide to ecological applications. *Philos. Trans. R. Soc. Lond. B* 343:329–50

30. Edwards SV. 1993. Long-distance gene flow in a cooperative breeder detected in genealogies of mitochondrial DNA sequences. *Proc. R. Soc. Lond. B* 252: 177–85

31. Ehrlich PR, White RR, Singer MC, McKechnie SW, Gilbert LE. 1975. Checkerspot butterflies: a historical perspective. *Science* 188:221–28

32. Epperson BK. 1989. Spatial patterns of genetic variation within plant populations. See Ref. 15, pp. 229–53

33. Fahrig L. 1992. Relative importance of spatial and temporal scales in a patchy environment. *Theor. Pop. Biol.* 41:300–14

34. Fisher RA. 1937. The wave of advance of advantageous genes. *Ann. Eugen.* 7: 355–69

35. Fritz RS. 1979. Consequences of insular population structure: distribution and extinction of spruce grouse populations. *Oecologia* 42:57–65

36. Gill DE. 1978. The metapopulation ecology of the red-spotted newt, *Notopthalmus viridescens* (Rafinesque). *Ecol. Mon.* 48:145–66

37. Godt MJW, Hamrick JL. 1991. Genetic variation in *Lathyrus latifolius* (Leguminosae). *Am. J. Bot.* 78:1163–71

37a. Goodnight CJ. 1987. On the effect of founder events on additive genetic variance. *Evolution* 41:80–91

37b. Goodnight CJ. 1988. Epistasis and the effect of founder events on additive genetic variance. *Evolution* 42:441–54

37c. Goodnight CJ. 1991. Intermixing ability in two-species communities of *Tribolium* flour beetles. *Am. Nat.* 138: 342–54

38. Gotelli NJ. 1991. Metapopulation models: the rescue effect, the propagule rain, and the core-satellite hypothesis. *Am. Nat.* 138:768–76

38a. Grant WS, Lesllie RW. 1993. Effect of metapopulation structure on nuclear and organellar DNA variability in semi-arid environments of southern Africa. *S. Afr. J. Sci.* 89:131–35

39. Grosholz ED. 1993. The influence of habitat heterogeneity on host-pathogen population dynamics. *Oecologia*. In press

40. Gurney WSC, Nisbet RM. 1978. Predator-prey fluctuations in patchy environments. *J. Theor. Biol.* 58:361–70

41. Gyllenberg M, Hanski I. 1992. Single-species metapopulation dynamics—a structured model. *Theor. Pop. Biol.* 42: 35–61

42. Gyllenberg M, Soderbackar G, Ericsson S. 1993. Does migration stabilize local population dynamics? Analysis of a discrete metapopulation model. *Math. Biosci.* 118:25–49

43. Hamrick JL, Godt MJW. 1989. Allozyme diversity in plant species. In Ref. 15, pp. 49–63

44. Hanski I. 1982. Dynamics of regional distribution: the core and satellite species hypothesis. *Oikos* 38:210–21

45. Hanski I. 1985. Single-species population dynamics may contribute to long-term rarity and commonness. *Ecology* 66:335–43

46. Hanski I. 1994. A practical model of metapopulation dynamics. *J. Anim. Ecol.* 63:151–62

46a. Hanski I. 1994. Patch occupancy dynamics in fragmented landscapes. *Trends Ecol. Evol.* 9:131–35

47. Hanski I, Ranta E. 1983. Coexistence in a patchy environment. *J. Anim. Ecol.* 52:263–79

48. Hanski I, Gilpin M. 1991. Metapopulation dynamics—brief history and conceptual domain. *Biol. J. Linn. Soc.* 42: 3–16

49. Hanski I, Gyllenberg M. 1993. Two general metapopulation models and the core-satellite species hypothesis. *Am. Nat.* 142:17–41

50. Hanski I, Kuussaari M, Nieminen M. 1994. Metapopulation structure and migration in the butterfly *Melitaea cinxia*. *Ecology*. In press

50a. Hanski I, Thomas CD. 1994. Metapopulation dynamics and conservation biology: a spatially explicit model applied to butterflies. *Biol. Conserv.* 68:167–80

51. Harrison S. 1989. Long-distance dispersal and colonization in the bay checkerspot butterfly. *Ecology* 70:1236–43

52. Harrison S. 1991. Local extinction in a metapopulation context: an empirical evaluation. *Biol. J. Linn. Soc.* 42:73–88

53. Harrison S. 1994. Metapopulations and conservation. In *Large-Scale Ecology and Conservation Biology*, ed. PJ Edwards, NR Webb, RM May, Oxford: Blackwell Sci. pp. 111–28.

53a. Harrison S, Doak DF, Stahl A. 1993. Spatial models and spotted owls: exploring some biological issues behind recent events. *Conserv. Biol.* 7:950–53

54. Harrison S, Thomas CD. 1991. Patchiness and dispersal in the insect community on ragwort (*Senecio jacobaea*). *Oikos* 62:5–12

54a. Harrison S, Thomas CD, Lewinsohn TM. 1994. Testing a metapopulation model of coexistence in the insect community on ragwort. *Am. Nat.* In press

55. Harrison S, Quinn JF. 1989. Correlated environments and the persistence of metapopulations. *Oikos* 56:293–98

56. Harrison S, Murphy DD, Ehrlich PR. 1988. Distribution of the bay checkerspot butterfly, *Euphydryas editha bayensis*: evidence for a metapopulation model. *Am. Nat.* 132:360–82

57. Hassell MP, Comins HN, May RM. 1991. Spatial structure and chaos in insect population dynamics. *Nature* 353: 255–58

58. Hastings A. 1977. Spatial heterogeneity and the stability of predator-prey systems. *Theor. Pop. Biol.* 12:37–48.

59. Hastings A. 1980. Disturbance, coexistence, history, and competition for space. *Theor. Pop. Biol.* 18:363–73

60. Hastings A. 1991. Structured models of metapopulation dynamics. *Biol. J. Linn. Soc.* 42:57–71

61. Hastings A. 1991. McKendrick von Foerster models for patch dynamics. *Lect. Notes Biomath.* 92:189–99

62. Hastings A. 1993. Complex interactions between dispersal and dynamics—lessons from coupled logistic equations. *Ecology* 74:1362–72

63. Hastings A, Higgins K. 1994. Persistence of transients in spatially structured ecological models. *Science.* 263:1133–36

64. Hastings A, Wolin CL. 1989. Within-patch dynamics in a metapopulation. *Ecology* 70:1261–66

65. Hellberg M. 1994. Relationships between inferred levels of gene flow and geographic distance in a philopatric coral, *Balanophyllia elegans*. *Evolution*. In press

66. Hedrick P. 1986. Genetic polymorphism in heterogeneous environments. *Annu. Rev. Ecol. Syst.* 17:535–66

67. Holt RD. 1994. Ecology at the mesoscale: the influence of regional processes on local communities. In *Species Diversity in Ecological Communities,* ed. RE Ricklefs, D Schluter, pp.77–88. Chicago: Univ. Chicago Press

68. Huffaker CB. 1958. Experimental studies on predation: dispersion factors and predator-prey oscillations. *Hilgardia* 27: 343–83

69. Hunt A. 1993. Effects of contrasting patterns of larval dispersal on the genetic

connectedness of local populations of two intertidal starfish, *Patiriella calcar* and *Patiriella exigua*. *J. Exp. Mar. Biol. Ecol.* 92:179–86

70. Hutchinson GE. 1951. Copepodology for the ornithologist. *Ecology* 32:571–77

71. Ives AR. 1988. Aggregation and the coexistence of competitors. *Ann. Zool. Fenn.* 25:75–88

72. Ives AR. 1992. Continuous-time models of host-parasitoid interactions. *Am. Nat.* 140:1–29

73. Iwasa Y, Roughgarden J. 1986. Interspecific competition among metapopulations with space-limited subpopulations. *Theor. Pop. Biol.* 30: 194–214

74. Jarosz AM, Burdon JJ. 1991. Host-pathogen interactions in natural populations of *Linum marginale* and *Melampsora lini*: II. Local and regional variation in patterns of resistance and racial structure. *Evolution* 45:1618–27

75. Johnson M, Black R. 1991. Genetic subdivision of the intertidal snail *Bembicum vittatum* (Gastropoda: Littorinidae) varies with habitat in the Houtman Abrolhos Islands, Western Australia. *Heredity* 67:205–13

76. Karban R. 1989. Fine-scale adaptation of thrips to individual host plants. *Nature* 340:60–61

77. Kareiva PM. 1987. Habitat fragmentation and the stability of predator-prey interactions. *Nature* 326:388–90

78. Kimura M, Maruyama T. 1971. Patterns of neutral variation in a geographically structured population. *Genet. Res.* 18: 125–31

79. King PS. 1987. Macro- and microgeographic structure of a spatially subdivided beetle species in nature. *Evolution* 41:401–16

80. Deleted in proof

81. Laan R, Verboom B. 1990. Effect of pool size and isolation on amphibian communities. *Biol. Conserv.* 54:251–62

82. Lacy RC. 1987. Loss of genetic diversity from managed populations: interacting effects of drift, mutation, immigration, selection and population subdivision. *Cons. Biol.* 1:143–58

82a. Lahaye WS, Gutierrez RJ, Akcakaya HR. 1994. Spotted owl metapopulation dynamics in Southern California. *J. Anim. Ecol.* 63: In press

83. Lamberson RH, McKelvey K, Noon BR, Voss C. 1992. A dynamic analysis of northern spotted owl viability in a fragmented forest landscape. *Cons. Biol.* 6: 505–12

83a. Lamberson RH, Noon BR, Voss C, McKelvey KS. 1994. Reserve design for territorial species: the effects of patch size and spacing on the viability of the northern spotted owl. *Conserv. Biol.* 8:185–95

84. Lande R. 1988. Genetics and demography in biological conservation. *Science* 241:1455–60

85. Lande R, Barrowclough GF. 1987. Effective population size, genetic variation, and their use in population management. In *Viable Populations for Conservation*, ed. ME Soule, pp. 87–124. Cambridge: Cambridge Univ. Press

86. Lawton JH, Woodroffe GL. 1991. Habitat and the distribution of water voles: why are there gaps in a species' range? *J. Anim. Ecol.* 60:79–91

87. Leary RF, Allendorf FW, Forbes SH. 1993. Conservation genetics of bull trout in the Columbia and Klamath River drainages. *Cons. Biol.* 7:856–65

88. Leberg PL. 1991. Influence of fragmentation and bottlenecks on genetic divergence of wild turkey populations. *Cons. Biol.* 5:522–30

89. Levin SA. 1974. Dispersion and population interactions. *Am. Nat.* 108:207–28

90. Levin SA. 1976. Population dynamic models in heterogeneous environments. *Annu. Rev. Ecol. Syst.* 7:287–311

91. Levin SA. 1992. The problem of pattern and scale in ecology. *Ecology.* 73:1943–67

92. Levins R. 1969. Some demographic and genetic consequences of environmental heterogeneity for biological control. *Bull. Entomol. Soc. Am.* 15:237–40

93. Levins R. 1970. Extinction. *Lect. Math. Life Sci.* 2:75–107

94. Levitan DR. 1991. Influence of body size and population density on fertilization success and reproductive output in a free-spawning invertebrate. *Biol. Bull.* 181:261–68

95. Loveless MD, Hamrick JL. 1984. Ecological determinants of genetic structure in plant populations. *Annu. Rev. Ecol. Syst.* 15:65–95

96. Mangel M, Tier C. 1993. Dynamics of metapopulations with demographic stochasticity and environmental catastrophes. *Theor. Pop. Biol.* 44:1–31

97. Manicacci D, Olivieri I, Perrot V, Atlan A, Gouyon PH, et al. 1992. Landscape ecology: population genetics at the metapopulation level. *Landscape Ecol.* 6:147–59

98. May RM, Nowak M. 1994. Superinfection, metapopulation dynamics, and the evolution of diversity. *J. Theor. Biol.* Submitted

99. McCauley DE. 1989. Extinction, colonization and population structure: a study of a milkweed beetle. *Am. Nat.* 134:365–76

100. McCauley DE. 1991. Genetic consequences of local population extinction and recolonization. *Trends Ecol. Evol.* 6:5–8

101. McCauley DE. 1991. The effect of host plant patch size variation on the population structure of a specialist herbivore insect, *Tetraopes tetraopthalmus. Evolution* 45:1675–84

102. McCauley DE. 1992. Genetic consequences of extinction and recolonization in fragmented habitats. In *Biotic Interactions and Global Change*, ed. PM Kareiva, JG Kingsolver, RB Huey, pp. 217–33. Sunderland, MA: Sinauer

103. McKelvey K, Noon BR, Lamberson RH. 1992. Conservation planning for species occupying fragmented landscapes: the case of the northern spotted owl. In *Biotic Interactions and Global Change*, ed. PM Kareiva, JG Kingsolver, RB Huey, pp. 424–50. Sunderland, MA: Sinauer

104. Meffe GK, Vrijenhoek RC. 1988. Conservation genetics and the management of desert fishes. *Cons. Biol.* 2:157–69

105. Menges ES. 1990. Population viability analysis for an endangered plant. *Cons. Biol.* 4:52–62

106. Metz JAJ, Diekmann O. 1986. *The Dynamics of Physiologically Structured Populations.* New York: Springer-Verlag

107. Moloney KA, Levin SA, Chiariello NR, Buttel L. 1992. Pattern and scale in a serpentine grassland. *Theor. Pop. Biol.* 41:257–76

108. Murdoch WW. 1994. Population regulation in theory and practice. *Ecology.* 75:271–87

109. Nachman G. 1991. An acarine predator-prey metapopulation inhabiting greenhouse cucumbers. *Biol. J. Linn. Soc.* 42:285–303

109a. Nagylaki T. 1989. The diffusion model for migration and selection. In *Some Mathematical Questions in Population Biology,* ed. A. Hastings, pp. 55–75. Providence: Am. Math. Soc.

110. Nee S, May RM. 1992. Dynamics of metapopulations: habitat destruction and competitive coexistence. *J. Anim. Ecol.* 61:37–40

111. Nicholson AJ, Bailey VA. 1935. The balance of animal populations. *Proc. Zool. Soc. Lond.* 3:551–98

112. Nisbet RM, Briggs CJ, Gurney WSC, Murdoch WW, Stewart-Oaten A. 1993. Two-patch metapopulation dynamics. *Lect. Notes Biomath.* 96:125–35

113. Olivieri I, Couvet D, Gouyon PH. 1990.

The genetics of transient populations—research at the metapopulation level. *Trends Ecol. Evol.* 5:207–10

114. Opdam P. 1990. Metapopulation theory and habitat fragmentation: a review of holarctic breeding bird studies. *Landscape Ecol.* 5:93–106

115. Pacala SW. 1986. Neighborhood models of plant population dynamics. 2. Multispecies models of annuals. *Theor. Pop. Biol.* 29:262–92

116. Peltonen A, Hanski I. 1991. Patterns of island occupancy explained by colonization and extinction rates in shrews. *Ecology.* 72:1698–708

117. Pokki J. 1981. Distribution, demography and dispersal of the field vole, *Microtus agrestis* (L.) in the Tvarminne archipelago, Finland. *Acta Zool. Fenn.* 164:1–48

118. Pounds A, Jackson JF. 1981. Riverine barriers and the differentiation of fence lizard populations. *Evolution* 35:516–28.

119. Preziosi R, Fairbairn DJ. 1992. Genetic population structure and levels of gene flow in the stream-dwelling water strider *Aquarius (Gerris) remigis* (Hemiptera: Gerridae). *Evolution* 46:430–44

120. Pulliam HR, Dunning JB Jr, Liu J. 1992. Population dynamics in complex landscapes: a case study. *Ecol. Appl.* 2:165–77

121. Quinn JF, Hastings A. 1987. Extinction in subdivided habitats. *Cons. Biol.* 1:198–208

122. Quinn JF, Wing SR, Botsford LW. 1993. Harvest refugia in marine invertebrate fisheries: models and applications to the red sea urchin, *Strongylocentrotus franciscanus*. *Am. Zool.* 33:111–37

123. Reeve JD. 1990. Stability, variability, and persistence in host-parasitoid systems. *Ecology* 71:422–28

124. Rice K, Jain SK. 1985. Plant population genetics and evolution in disturbed environments. In *The Ecology of Natural Disturbance and Patch Dynamics*, ed. STA Pickett, PS White, pp. 287–303. New York: Academic

124a. Rogers AR. 1990. Group selection by selective emigration: the effects of migration and kin structure. *Am. Nat.* 135:398–413

125. Rolstad, J. 1991. Consequences of forest fragmentation for the dynamics of bird populations: conceptual issues and the evidence. *Biol. J. Linn. Soc.* 42:149–63

126. Roughgarden J, Iwasa Y. 1986. Dynamics of a metapopulation with space-limited subpopulations. *Theor. Pop. Biol.* 29:235–61

127. Roughgarden J, Iwasa Y, Baxter C. 1985. Demographic theory for an open marine population with space-limited recruitment. *Ecology* 66:54–67

128. Sabelis MW, Laane WEM. 1986. Regional dynamics of spider-mite populations that become extinct locally from food source depletion and predation by phytoseiid mites (Acarina: Tetranychidae, Phytoseiidae). *Lect. Notes Biomath.* 68:345–76

129. Sabelis MW, Diekmann O, Jansen VAA. 1991. Metapopulation persistence despite local extinction–predator-prey patch models of the Lotka-Volterra type. *Biol. J. Linn. Soc.* 42:267–83

130. Schoener TW. 1991. Extinction and the nature of the metapopulation: a case system. *Acta Oecol.* 12:53–75

131. Schoener TW, Spiller DA. 1987. High population persistence in a system with high turnover. *Nature* 330:474–77

132. Shorrocks B, Rosewell J, Edwards K. 1990. Competition on a divided and ephemeral resource: testing the assumptions. I. Aggregation. *J. Anim. Ecol.* 59:977–1001

133. Shorrocks B, Rosewell J, Edwards K. 1990. Competition on a divided and ephemeral resource: testing the assumptions. II. Association. *J. Anim. Ecol.* 59:1003–17

134. Shorrocks B. 1991. Coexistence in a patchy environment. In *Living in a Patchy Environment*, ed. B Shorrocks, IR Swingland, pp. 91–106. Oxford, Eng: Oxford Univ. Press

135. Sinsch U. 1992. Structure and dynamics of a natterjack toad metapopulation (*Bufo calamita*). *Oecologia* 90:489–99

136. Silvertown J, Holtier S, Johnson J, Dale P. 1992. Cellular automaton models of interspecific competition for space—the effect of pattern on process. *J. Ecol.* 80:527–34

137. Sjogren P. 1991. Extinction and isolation gradients in metapopulations: the case of the pool frog (*Rana lessonae*). *Biol. J. Linn. Soc.* 42:135–47

138. Sjogren P. 1991. Genetic variation in relation to demography of peripheral pool frog populations (*Rana lessonae*). *Evol. Ecol.* 5:248–71

139. Slatkin M. 1974. Competition and regional coexistence. *Ecology* 55:128–34

140. Slatkin M. 1977. Gene flow and genetic drift in a species subject to frequent local extinction. *Theor. Pop. Biol.* 12:253–62

141. Slatkin M. 1985. Gene flow in natural populations. *Annu. Rev. Ecol. Syst.* 16:393–430

142. Slatkin M. 1993. Isolation by distance in equilibrium and nonequilibrium populations. *Evolution* 47:264–79

143. Smith AT. 1980. Temporal changes in insular populations of the pika (*Ochotona princeps*). *Ecology* 61:8–13

144. Solbreck C. 1991. Unusual weather and insect population dynamics: *Lygaeus equestris* during an extinction and recovery period. *Oikos* 60:343–50

145. Solbreck C, Sillen-Tullberg B. 1990. Population dynamics of a seed-feeding bug, *Lygaeus equestris*. I. Habitat patch structure and spatial dynamics. *Oikos* 58:199–209

146. Spight T. 1974. Sizes of populations of a marine snail. *Ecology* 55:712–29

147. Stacey PB, Taper M. 1992. Environmental variation and the persistence of small populations. *Ecol. Appl.* 2:18–29

148. Stangel PW, Lennartz M, Smith MH. 1992. Genetic variation and population structure of red-cockaded woodpeckers. *Cons. Biol.* 6:283–90.

149. Stiven AE, Bruce RC. 1988. Ecological genetics of the salamander *Desmognathus quadrimaculatus* from disturbed habitats in the southern Appalachian biosphere reserve center. *Cons. Biol.* 2:194–205

150. Taylor AD. 1988. Large-scale spatial structure and population dynamics in arthropod predator-prey systems. *Ann. Zool. Fenn.* 25:63–74

151. Taylor AD. 1991. Studying metapopulation effects in predator-prey systems. *Biol. J. Linn. Soc.* 42:305–23

152. Thomas CD, Harrison S. 1992. Spatial dynamics of a patchily-distributed butterfly species. *J. Anim. Ecol.* 61:437–46

153. Thomas CD, Jones TM. 1993. Partial recovery of a skipper butterfly (*Hesperia comma*) from population refuges: lessons for conservation in a fragmented landscape. *J. Anim. Ecol.* 62:472–81

154. Thomas CD, Thomas JA, Warren MS. 1992. Distributions of occupied and vacant butterfly habitats in fragmented landscapes. *Oecologia* 92:563–67

155. Thompson JN. 1994. *The Coevolutionary Process.* Chicago: Univ. Chicago Press

156. Tilman D. 1994. Competition and biodiversity in spatially structured habitats. *Ecology* 75:2–16

157. Toft CA, Schoener TW. 1983. Abundance and diversity of orb spiders on 106 Bahamanian islands: insular biogeography at an intermediate trophic level. *Oikos* 41:359–71

157a. van de Klashorst G, Redshaw JL, Sabelis MW, Lingeman R. 1992. A demonstration of asynchronous local cycles in an acarine predator-prey system. *Exp. Appl. Acar.* 14:185–99

158. Verboom J, Schotman A, Opdam P, Metz JAJ. 1991. European nuthatch metapopulations in a fragmented agricultural landscape. *Oikos* 61:149–56

159. Vrijenhoek RC. 1985. Animal population genetics and disturbance: the effects of local extinctions and recolonizations on heterozygosity and fitness. In *The Ecology of Natural Disturbance and Patch Dynamics,* ed. STA Pickett, PS White, pp. 265–85. New York: Academic

160. Wade MJ. 1990. Genotype-environment interaction for climate and competition in a natural population of flour beetles. *Evolution* 44:2004–11

161. Wade MJ. 1991. Genetic variance for rate of population increase in natural populations of flour beetles, *Tribolium* spp. *Evolution* 45:1574–84

161a. Wade MJ. 1992. Sewall Wright and the shifting balance theory. *Oxford Surveys Evol. Biol.* 8:35–62

162. Wade MJ, Goodnight CJ. 1991. Wright's shifting balance theory: an experimental study. *Science* 253:1015–18

163. Wade MJ, McCauley DE. 1988. Extinction and colonization: their effects on the genetic differentiation of local populations. *Evolution* 42:995–1005

164. Walde SJ, Nyrop JP, Hardman JM. 1992. Dynamics of *Panonychus ulmi* and *Typhlodromus pyri*: factors contributing to persistence. *Exp. Appl. Acar.* 14:261–91

165. Waller DM, O'Malley DM, Gawler SC. 1987. Genetic variation in the extreme endemic *Pedicularis furbishiae*. *Cons. Biol.* 4:335–40

166. Walters JR. 1991. Application of ecological principles to the management of endangered species: the case of the red-cockaded woodpecker. *Annu. Rev. Ecol. Syst.* 22:505–23

167. Ward RD. 1990. Biochemical genetic variation in the genus *Littorina* (Prosobranchia: Mollusca). *Hydrobiologia* 193:53–69

168. Welsh H. 1990. Relictual amphibians and old-growth forests. *Cons. Biol.* 3:309–19

169. Whitlock MC. 1992. Nonequilibrium population structure in forked fungus beetles: extinction, colonization, and genetic variation among populations. *Am. Nat.* 139:952–70

169a. Whitlock MC, Phillips PC, Wade MJ. 1994. Gene interaction affects the additive genetic variance in subdivided populations. *Evolution.* In press

170. Wilson WG, de Roos AM, McCauley E. 1993. Spatial instabilities within the diffusive Lotka-Volterra system—indi-

vidual-based simulation results. *Theor. Pop.* Biol.43:91–127

171. Wright S. 1931. Evolution in Mendelian populations. *Genetics* 16:97–159
172. Wright S. 1940. Breeding structure of populations in relation to speciation. *Am. Nat.* 74:232–48

173. Wright S. 1943. Isolation by distance. *Genetics* 28:114–38
174. Wright S. 1978. *Evolution and the Genetics of Populations. IV. Variability Within and Among Populations.* Chicago: Univ. Chicago Press

Annu. Rev. Ecol. Syst. 1994. 25:189–217

COMMUNITY STRUCTURE:
Larval Trematodes in Snail Hosts

Armand M. Kuris and Kevin D. Lafferty

Department of Biological Sciences and Marine Science Institute, University of California, Santa Barbara, California 93106

KEY WORDS: community structure, meta-analysis, recruitment heterogeneity, competition

Abstract

In species assemblages of larval trematodes in individual snail hosts, fewer multispecies infections are observed than might be expected by chance. Both interspecific competition and the isolating effect of heterogeneity in recruitment may explain this pattern of community structure. Here, we analyzed the expected and observed frequency of double infections, using data culled from 62 studies. Our analysis included 296,180 host snails. Of these, 62,942 were infected with one or more species of trematode (23% pooled over all studies, 24% average across studies). By incorporating information from subsamples, we were able to estimate the proposed isolating effect of heterogeneity in recruitment. Surprisingly, spatial and temporal heterogeneity as well as differential prevalence among host size classes typically led to intensification of interactions (average increases in interactions by +19%, +19%, and +23%, respectively), while partitioning among host species usually led to isolation of potential competitors (a −1% average decrease in interactions). We calculated the expected number of interspecific double infections by applying rules of independent assortment to the frequency of trematode species. The majority of the 14,333 expected interactions did not persist; only 4,346 double infections were actually observed (a 69% decrease, 62% average). Competition, via a variety of interspecific competitive mechanisms by dominant species, is the structuring process most consistent with this paucity of observed multispecies interactions. How important is competition? Overall, we estimated that 13% (10% average) of the trematode infections were lost to interspecific interactions. Subordinate species in particular suffered very high losses.

0066-4162/94/1120-0189$05.00

INTRODUCTION

In this review, we examine the efforts of ecologists and parasitologists to interpret structure in communities of larval trematodes. May (77) considered a group of species lacking statistical association to be "unstructured"; the more a community differs from a random association of species, the more "structured" it is. Likewise, forces "structure" a community if they cause an association of species to depart from a null model of species abundance and distribution. To determine if a given community is structured, it is necessary to test whether it is significantly different from a random assemblage of species. For example, certain species combinations may occur more or less frequently than expected by chance. A null model based on independent assortment can serve to construct "null communities" for comparison with observed communities. The degree to which a community departs from a null model represents a quantitative measure of community structure (42).

Despite this apparently straightforward analytical approach, considerable disagreement centers on whether certain communities are structured or just random assemblages of species (94, 98, 107). Although disturbance, physical stress, recruitment dynamics, predation, and competition all might alter the distribution and composition of species in a community, the role of interspecific competition is at the center of the community structure debate (15–17, 26, 43, 96). Recently, some ecologists, particularly those who work in marine systems, have found it useful to separate structuring forces that affect patterns of recruitment from those forces that occur after recruitment. Particularly interesting is how heterogeneity in recruitment can affect the importance of post-recruitment structuring forces (37, 47).

Larval trematodes in their first intermediate molluscan hosts provide useful systems to examine theories of community structure. Certain species of snails and small clams are host to a rather rich fauna (up to 20 species) of larval trematodes. As with all studies of parasite ecology, the hosts provide natural and discrete habitat units. These molluscs are also generally abundant and readily sampled in a blind fashion with respect to trematode infection status. Within each mollusc, the larval trematodes multiply asexually, often reaching about half the tissue weight of the parasitized snail. Trematodes usually castrate their snail hosts (50, 58, 59, 93, 116) by manipulating the endocrine control of the snail's reproductive system (55). Thus, with a few exceptions, larval trematodes treat their first intermediate hosts as a limiting resource; generally fully used by the asexual progeny of a single infecting parasite (usually a penetrating miracidium, sometimes an ingested egg).

A Brief History

Because of their obvious importance to medical and veterinary diseases, careful quantitative sampling of snail first intermediate hosts began in the early part of the twentieth century and often included samples from tropical regions. Sewell (97) recognized that, even when trematode prevalence (percentage of hosts infected) was high and snails were parasitized by a species-rich assemblage of trematodes, he infrequently observed interspecific double infections. The careful quantitative study of Dubois (29) emphasizes this point. Sewell proposed that a parasitized snail either lost its chemical attractiveness to other searching parasites, or infection altered its physiology to impede or prevent development of a later infection. Sewell felt that rare double infections were likely due to roughly simultaneous infection by two trematode species. Dubois concurred, assuming the absence of subsequent infections to be related to changes in immunity. In modern terminology, this would represent a form of concomitant immunity in which the presence of a parasitic infection induces a host defensive response against subsequent challenges. Dubois noted that there was no direct evidence, nor known mechanism, for hypotheses requiring biological antagonism, immunity, or incompatibility between two trematodes in a snail host. His extensive survey of Indian cercariae enabled Sewell (97) to suggest that only certain combinations of trematode species could coexist as double infections and that these double infections generally involved certain cercarial groups (furcocercariae, xiphidiocercariae, and monostome cercariae).

In contrast to Dubois and Sewell, Cort et al (19) reported a high frequency of multiple species infections in *Lymnaea* (= *Stagnicola*) *emarginata* from Douglas L., Michigan. They provided the first use of probability theory to compare the number of observed multiple infections (f_o) with the expected frequency, $f_e = (A \times B)/N$, where A and B are the number of observed infections of trematodes A and B, and N is the number of snails examined. If species A and B are not independent of each other, we should see a difference between the expected and observed numbers of double infections. Cort et al (19) noted that Dubois and Sewell collected their snails from many locations over a long time and suggested it was likely that prevalence was too low in most of their collections to produce multiple infections. They also strenuously rejected the postulated explanations (immunity, altered attractiveness, or interspecific trematode antagonism) due to lack of evidence and mechanisms. Cort et al (19) found many combinations to occur at frequencies similar to their random expectations but observed that several combinations occurred much less frequently than predicted (often not occurring at all). Although they rejected the hypothesis that certain types of trematodes were less likely to engage in multiple infections, they did not test the significance of differences between f_e and f_o.

Thus, work by the late 1930s had already framed the key issues in community ecology. Were communities of larval trematodes assembled at random? If not, what were the causes of nonrandom patterns of association?

The extensive experimental studies of Lie and colleagues revealed several mechanisms for the lack of double infections of certain interspecific combinations (reviewed by 12, 58, 66, 103–105). Using trematodes from the Malaysian snails *Lymnaea rubiginosa* and *Indoplanorbis exustus,* and the host of a human schistosome, *Biomphalaria glabrata,* they revealed the regular occurrence of predation by dominant species (with mouthed redial larval stages) on subordinate species. They also showed that certain species with only sporocyst larval stages (mouthless) were able indirectly to suppress the development of other subordinate species. In still other combinations, prior occupancy determined interspecific dominance. They were able to array the interspecific competitive abilities of these species in a largely linear dominance hierarchy (reviewed by 61, 66) (in our review, we refer to all interactions where one species negatively affects another species as *competition*). These interactions are comparable to intraguild predation as reviewed by Polis et al (82a). Other studies by Lie, Heyneman and co-workers (44, 68) established that prior infections altered the susceptibility of such snails to subsequent parasitization by other trematode species. However, rather than decreasing the likelihood of a second infection (interspecific heterologous immunity), these changes in snail defensive capabilities increase the likelihood of subsequent infections (44). Indeed, *Austrobilharzia terrigalensis* is an obligate secondary invader (113).

As an alternative to the interactive (competitive) hypothesis, Cort et al (19) noted that spatial and temporal factors certainly influenced the behavior of definitive hosts (vertebrates that harbor adult worms). They suggested this environmental heterogeneity would affect recruitment of trematodes to snail first intermediate hosts and would isolate trematodes from encounters as multiple infections. Others have supported (5, 54) or recently espoused this view (32, 34, 35, 99, 102, 104, 105).

Following the Cort's work with several freshwater snail-trematode systems in Michigan (19–21), reports of multiple infections from several hosts were reported over the next 50 years. Most studies concluded that double infections were rarer than expected in at least some pairs of species (5, 13, 18, 45, 56, 61, 64, 72, 74, 83, 85, 88, 102, 104, 111, 117). A few studies claimed the frequency of double infections to be similar to random expectations (23, 46, 54, 89, 99).

Sousa (102) reinvigorated the issue of community structure by acknowledging (with Kuris—61) that interspecific competition determined the outcome within the snail (infracommunity level). However, Sousa questioned whether this led to a significant impact on community structure at the snail population level (component community). By expanding the paradigm of intermediate disturbance developed to explain patterns of high diversity in communities of corals, tropical trees (14) and marine algae (100), Sousa predicted that if competition was important, maximum trematode diversity should occur at intermediate snail sizes. He recognized that, as a cohort of snails aged, the accumulating trematode infections would cause trematode species diversity to increase initially, and then interactions would cause the loss of subordinate species from older, larger snails with double infections. However, in the large majority of samples, in a system that is highly interactive within individual snails (*Cerithidea californica*), diversity did not significantly decline in large snails. Sousa (104) concluded that competition was not important for these assemblages and proposed that spatial (and temporal) heterogeneity of recruitment of trematodes to the snail population was the likely structuring force. Fernández & Esch (34, 35), Snyder & Esch (99), and Curtis & Hubbard (23) embraced Sousa's test; they also found no significant decline in trematode species diversity or richness in the largest size classes of snails. The most recent texts (32, 90) incorporate the paradigm that spatial and temporal heterogeneity in trematode recruitment to snails is a hypothesis, mutually exclusive to competition, that effectively accounts for the lack of observed multiple infections.

Recalling Robson & Williams' (88) discussion of the focal nature of transmission to snails, the structuring effect of heterogeneity results from the interplay of two independent and opposing factors. The first, isolation of species, occurs when the relative prevalence of each trematode species varies among subsamples. On the other hand, variation in the absolute total prevalence (all species combined) among subsamples intensifies the likelihood of double infections. Both factors act independently to determine whether species will interact more or less frequently than expected. If several species of vertebrate final hosts use the same site, or if one species of vertebrate acts as the final host for several species of trematodes, or if migratory behaviors of final hosts show similar patterns of seasonality, spatial and temporal heterogeneity may well concentrate trematode eggs at specific locations and times (38, 51, 74, 76, 82, 88, 119). These likely natural history patterns will intensify opportunities for multiple infections within snail first intermediate hosts (61, 64). Further, as discussed below, Sousa's (102) hyperbolic prediction will detect only competitive interactions so severe that they exclude subordinate species of trematodes from snail populations (61). It is a notably insensitive test for competition. The solution to these problems requires a method that analyzes effects of spatial and temporal (and other) sources of heterogeneity in conjunction with the competitive hypothesis. In other words, because heterogeneity in recruitment may *either* intensify *or* ameliorate competition, it is necessary to apportion the variance among recruitment and competition.

Recognizing that factors influencing recruitment of trematodes to snails must temporally precede interactions within the snails (47), Lafferty et al (64) developed methods to calculate preinteractive distributions of parasites among snails and to analyze available information on heterogeneous recruitment. They then assessed the impact of competition (or facilitation) and quantified the interaction between environmental heterogeneity and worm competition with respect to both magnitude and direction (isolation vs intensification for environmental heterogeneity, competition vs facilitation for worm interactions). In this review, we assume that no heterogeneity exists within a subsample. As we argue below, this is likely to offer a conservative estimate of the magnitude of competition.

In addition to the problem that interpretations of observed structure have floundered because they could not account for the effects of spatial and temporal heterogeneity in parasite recruitment. Researchers have often underreported structure itself for two reasons (64). First, all previous studies that used null models of independent assortment to calculate the expected frequency of double infections used observed frequencies to parameterize their null models (except 64). As is explained in more detail later, such an approach usually leads to an underestimation of the expected frequency of double infections. Second, statistical tests used to compare observed and expected values were often performed on data that were too finely subdivided to have sufficient statistical power to reject the null hypothesis of no structure.

Here, we apply the approach of Lafferty et al (64) to 62 data sets, compare the outcomes with other methods of analysis (generally published with particular data sets) and perform a meta-analysis to gauge the global importance of competition, facilitation and four sources of environmental heterogeneity (spatial, temporal, snail size, snail species) on community structure of larval trematodes in their first intermediate snail hosts. This approach yields results that, in several cases, differ from the interpretations of the original authors.

OUR APPROACH

We searched the larval trematode literature extensively for data sets suited for analysis, and we chose studies of natural, identified host populations that reported the total number of host snails dissected, the frequency of each trematode species, and the number of double, triple, and quadruple infections observed. Some data sets provided pooled data on multiple species infections but did not stratify them by potential sources of heterogeneity. In such cases, we used a weighted randomization procedure to distribute the multiple infections among subsamples. This approach was conservative with respect to both the competition and heterogeneity hypotheses, because we have defined community structure as the extent to which the distribution of species varies from a random assemblage. Multiple infections were usually so infrequent that we could assign them to any subsample without significantly affecting the outcome. Trematodes were often only partially identified (often to cercarial groups, e.g. echinostome, xiphidiocercariae, etc), and we included such studies treating the incomplete identifications as operational taxonomic units. This approach provided a conservative assessment of competitive interactions; the broader operational taxonomic units masked some potential interactions.

Estimating Prevalence

The application of the null model, $f_e = (A \times B)/N$, to the distribution of larval trematodes among snails is inadequate because individuals that infected a host but that competition later eliminated do not show up in samples and, therefore, are not entered into the model (64). This leads to an underestimate of the expected number of double infections. The magnitude of this error increases with the prevalence of dominant species in the assemblage. Correcting for this error requires parameterizing the null model with the prevalence of each species expected to have recruited before any interactions. Lafferty et al (64) provided the details of estimating the "pre-interactive" prevalence of all the species in an assemblage. In general, the prevalence of a species of parasite recruiting to a host population will be equal to the prevalence of that species observed among those hosts where no dominant parasite species are present. (See *Note Added in Proof A.*)

To generate the preinteractive prevalences of a trematode community, we postulated a dominance hierarchy based on evidence available for each host-parasite system, according to set rules. For well-studied systems (e.g. trematodes from *Cerithidea* spp., *Ilyanassa obsoleta, Helisoma anceps,* or *Lymnaea rubiginosa),* we made use of published dominance relationships based on laboratory experimentation, field mark recapture studies, or histological observations (25, 34, 61, 104). For more poorly known systems, we used taxonomic relationships and other more indirect assays to postulate dominance (61). Compared with a dominance hierarchy proposed by Fernández and Esch (34) based on experimental evidence for the trematodes of *Helisoma anceps,* a postulated dominance hierarchy based on the indirect rules of Kuris (61) for that system proved conservative, detecting a smaller competitive effect. We developed a conservative algorithm to construct dominance hierarchies in little-studied systems by postulating a dominance-subordinance relationship only for taxa with a consistent history of strong dominance (e.g. echinostomes, philophthalmids, heterophyids), weak dominance (e.g. notocotylids, schistosomes), or subordinance (e.g. xiphidiocercariae, strigeids). In the large majority of data sets, researchers determined the presence of trematode infections by crushing the snail hosts rather than merely observing shedding of cercariae into the water. Thus, most studies did not suffer from underestimation of prevalence due to reluctant shedding as demonstrated by Curtis & Hubbard (23). When several different studies of a host were available, we carefully reviewed the taxonomic literature to ensure that we consistently assigned species to the appropriate operational taxonomic units.

Structure

We could examine the effect of various potential sources of heterogeneity for a number of studies. Two studies reported subsamples by host sex, 9 by host size classes, 8 by host species, 31 by geographic locations (over a wide range of scales), and 19 by temporal variation (usually monthly, sometimes annually). To see whether the distribution of different trematode species among these subsamples affected the expected number of interactions (isolation of species or intensification of interactions), we used methods developed by Lafferty et al (64) to quantify the effects of spatial heterogeneity in recruitment. This required two steps: (i) applying the null model to the pooled subsamples yielded the expected number of double infections that would occur if recruitment was homogenous; (ii) applying the null model to each subsample separately and then summing the expected number of double infections across all subsamples yielded the expected number of double infections that occurred as a result of heterogeneity among subsamples. By comparing the expected number of double infections from steps 1 and 2, we estimated the effect of heterogeneity in recruitment on interactions. To make relative comparisons among studies, we then standardized this effect according to the equation (pooled − summed) / pooled. An alternative statistical analysis to ours would require (i) a heterogeneity chi-square of species by subsample (that omitted the number of uninfected hosts) to indicate the significance of isolation, and (ii) a heterogeneity chi-square of prevalence (of all species combined) by subsample. The drawback is that this alternative would not quantify the net effect of heterogeneity on the expected frequency of double infections.

For the 15 studies that reported more than one category of heterogeneity (e.g. both site and time), we assessed the effect of each type of heterogeneity independently from the other. As an example, for a study subdivided by site and date, we first calculated the expected prevalence of recruitment for each species. Then, to determine the effect of spatial heterogeneity, we pooled these values over all sites within each date. Next, we calculated the expected number of double infections for each date. Summing these values over all dates gave the expected number of double infections without spatial heterogeneity. Likewise, we determined the independent effect of temporal heterogeneity by pooling the expected prevalence of each species over all dates within a site, calculating the expected number of double infections for each site and then summing over all sites.

To quantify the effect of interspecific competition on the persistence of multispecies infections, we compared the expected number of double infections summed across subsamples with the number of double infections observed in nature. Because our null model predicted only double infections, we counted the few observed triple infections as three observed double infections (we counted a single quadruple infection as six observed double infections). Again, to make relative comparisons among studies, we standardized this effect according to the equation (observed − summed) / summed.

To quantify the importance of interspecific interactions for the entire assemblage, we estimated the proportion of the trematode individuals that were lost to competition. Clearly, if only a tiny fraction of the community interacts, effects of competition, though significant for interacting individuals, will have little consequence for the trematode community as a whole. Because interactions are more likely to negatively affect subordinate species, we made a more detailed assessment of the 25 studies that used littorine snails as hosts. Here, we classified species according to four levels of dominance. We then quantified, according to the level of dominance, the proportion of individuals lost from each species in each study.

Statistics

We statistically evaluated whether interactions or heterogeneity significantly structured a trematode assemblage in each study by calculating the confidence limits around the proportion of trematodes that were expected or observed to interact with another species (64). Specifically, we compared the proportion of interacting trematodes before (2 × pooled / number of trematodes) and after (2 × summed / number of trematodes) the effects of heterogeneity in recruitment. To assess the statistical significance of competition, we compared the expected (2 × summed / number of trematodes) and observed (2 × observed / number of trematodes) proportions of interacting trematodes. These statistics are conservative as they are not sensitive to structuring forces in a community that might act in opposite directions on different species (64). We excluded several data sets because they suffered from a combination of low sample size, low prevalence, or low species evenness in such a way that our statistical tests lacked the power to determine whether they were structured. As a rule of thumb, we required that studies have a sum of at least three expected double infections. A power analysis indicated that our statistical approach could always distinguish whether competition eliminated all double infections (64). We analyzed studies with fewer than three double infections separately to determine whether they were qualitatively different from the studies we included in our analysis.

We used meta-analyses (e.g. 43) with studies as independent replicates in a chi-square test to determine whether interactions and heterogeneity had significant effects on the structure of trematode communities. Since meta-analysis specifically includes studies that, by themselves, have low statistical power, we included the 15 studies that we excluded from our other analyses. However, it was necessary to pool information from these studies to accommodate the conditions for the chi-square analysis (>5 expected value). To test for an effect of interactions, we calculated the squared deviation from the expected value for each study as (observed − summed)2 / summed. We carried out an analogous approach for each of the four forms of heterogeneity except that here the squared deviation from the expected value was (summed − pooled)^2 / pooled.

RESULTS

We entered more than 300,000 snails, collected over eight decades, into our analysis. Considering that each host is a potential habitat for a trematode infra-community, this may be the most extensive community analysis for any system. We derived summary statistics (Table 1) by pooling the information from all studies as well as by averaging across studies. The Appendix presents

Table 1   Summary statistics for all studies combined[a].

Observed	pooled[b]	average[c]
Snails	296180	4356
Single infections	62942	926
Double infections	3871	57
Triple infections	155	2
Quadruple infections	1	0
Double interactions	4346	64
Prevalence	23%	24%
Number of trematodes	71153	1046
Trematodes interacting	12%	11%
Expected		
Summed double infections	14333	211
Number of trematodes	81621	1200
Trematodes interacting	35%	28%
Structuring changes		
Heterogeneity (net change in double infections)		
Spatial heterogeneity	+13%	+19%
Host size structure	+10%	+23%
Temporal heterogeneity	+10%	+19%
Host specificity	+2%	−1%
Change in interactions		
Double infections	−69%	−62%
Trematode abundance	−13%	−10%
Low power studies		
Change in double infections	−70%	−25%

[a] Unless noted, >3 expected double infections.
[b] Calculated by pooling values over all studies.
[c] Averaged across studies.

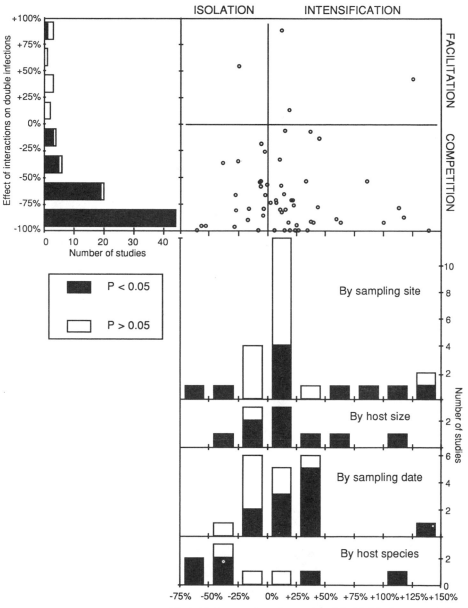

Figure 1 Effects of competition and heterogeneity on double infections. Scatter plot and histograms of the effects of heterogeneity and interactions on the expected and observed frequency of double infections. Each point represents a separate study. The effect of interactions on the number of double infections was calculated as (observed – sum of expected) / (sum of expected) while the effect of heterogeneity on the number of double infections was calculated as (sum of expected – expected of pooled) / (expected of pooled). A value of zero indicates no effect; –1 a complete loss of double infections. Filled bars represent studies in which a significant effect was detected; open bars represent studies where the effect was not significantly different from zero. Several studies were subdivided according to categories of sampling site, host size, sampling date and host species.

Table 2 Results from meta-analyses of the effect of heterogeneity and competition on the structure of trematode guilds.

Mechanism	chi-square	df[a]	p
Heterogeneity			
Sample site	617	23	<.001
Host size	184	8	<.001
Host species	241	8	<.001
Sample date	248	14	<.001
Competition	7,322	59	<.001

[a] The unit of replication is a study.

data from each study. Statistically significant instances of isolation and intensification occurred in all four types of subsamples (Figure 1). A meta-analysis for each type of heterogeneity resulted in strongly significant chi-square values (Table 2). Neither of the studies that investigated heterogeneity between host sexes (not shown in Figure 1) indicated an effect on the probability of interspecific interactions. Although differential use of host species had no consistent effect, spatial and temporal heterogeneity as well as differential prevalence among host size classes typically intensified the likelihood of interactions (Table 1).

Most expected interactions did not persist (Figure 1, Table 1). The studies that were not included in our analysis due to a lack of power (fewer than 3 expected double infections) indicated a similar effect (Table 1). In general,

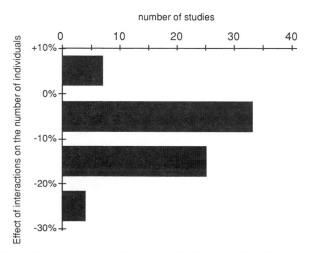

Figure 2 Effects of interactions on total trematode abundance for all studies.

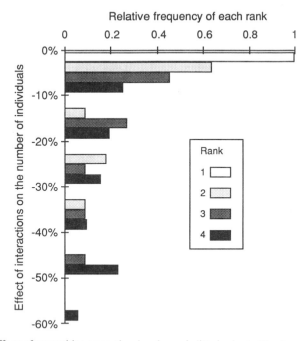

Figure 3 Effects of competition on species abundances in littorine hosts. The change in abundance (as a percentage) was calculated for each species in each study. Species were then grouped according to four ranks of competitive dominance (1 being most dominant). Each bar indicates the proportion of species (not pooled over studies) in a rank that suffer a certain percent loss to competition (e.g. the results show a higher proportion of subordinant than dominant ranking species suffer high losses).

interactions led to a dramatic decrease in the number of double infections. One significant case of facilitation (more double infections than expected) was evident. The meta-analysis for interaction also resulted in a strongly significant chi-square value (Table 2). There was no significant variation in the strength of competition among host species (ANOVA, $F = .987$, df = 7, 47, $p = .452$). The structuring effects of heterogeneity and competition were not associated (Figure 1, $r = 0.15$, $N = 59$, $p > 0.05$).

By comparing the total number of expected and observed trematodes, we estimated that competition eliminated 13% (pooled) or 10% (mean of studies) of the trematode individuals. Figure 2 indicates how this loss was distributed among the studies (several studies indicated a minor gain in individuals due to facilitation). Similarly, Figure 3 illustrates how, for the littorine studies, subordinate species suffered relatively high losses.

DISCUSSION

Our results did not support the recent paradigm for trematode communities. Isolation of species due to spatial and temporal heterogeneity does not explain the rarity of multispecies infections. The most frequent structuring forces in trematode communities involve a combination of intensification of interspecific interactions due to heterogeneity in recruitment followed by competitive interactions that greatly reduce the abundance of subordinate species.

Heterogeneity

Of greatest interest were those studies that discussed effects of heterogeneity and competition. Our study confirmed interpretations of researchers who found that size classes did not significantly structure recruitment and that competition significantly reduced negative interspecific interactions (40), differences in sites isolated potentially interacting species (76), and heterogeneity of recruitment over time and sites intensified the significant competitive interactions between larval trematodes (18, 82). Sousa (104, 105) characterized temporal and spatial heterogeneity exclusively as isolating forces, reducing the impact of competition between trematodes within snail hosts. Prevalences at his sites were significantly different (22% to 14%) and varied greatly between years (12% to 54% at one site, 5% to 26% at the other site; trends over time changing in opposite directions at the two sites). However, our analyses of these studies indicate that heterogeneity actually intensifies competition, because there were only minor variations in relative worm species abundance over space and time. Three studies by Esch and colleagues (34, 35, 118) all claimed (without providing analyses) marked isolating heterogeneity with respect to either or both space and time. Our statistical analyses detected no significant spatial or temporal heterogeneity in their data sets. These studies also argued that the great majority of pairs of trematodes did not significantly interact. In fact, although we find a strong net competitive effect in their studies, their analyses broke down the many sites and times so finely that they could not detect competition in their system. Thus, we reject their conclusions that the trematodes in *Helisoma anceps* host system are noninteractive and structured by spatial and temporal heterogeneity. We maintain the perspective that this system is highly interactive (competitive) and that variation in space and time neither isolates potential competitors nor intensifies their likely interactions.

Spatial heterogeneity is the form of variation in recruitment that we can most readily interpret. A higher proportion of trematodes will potentially experience interspecific interactions if heterogeneity concentrates their recruitment within a small region (e.g. 88), as opposed to being evenly spread over the host's distribution. An overlap in the habitat preferences of definitive hosts or spatial variation in transmission efficiency could produce such a pattern. Conversely, isolation in space will occur when vertebrate final hosts exhibit different habitat specificity.

Several mechanisms could foster intensification by host size class. Variation in parasite resistance with host age might produce intensification in certain snail size-classes. Increased resistance to larval trematode infections with increasing snail size has been well documented (reviewed by 70), and its genetic basis in *Biomphalaria glabrata* carefully described (71, 87). Ontogenetic changes in habitat use or behavior might also affect infection rates. The instantaneous risk of infection will likely be higher for large than for small snails if transmission occurs by way of the ingestion of trematode eggs (assuming large snails eat more) or because large snails present a large target for infective miracidial stages because these use a combination of random and chemosensory searching behavior (e.g. 75). In any case, if larger hosts are older, they will have been exposed to a greater cumulative risk of infection by several species, and nearly all studies show a monotonic increase in overall parasitic prevalence with increasing size of snails. We note that the relationship between age and growth in parasitized snails is complex and system specific. Some studies have experimentally shown that parasitized snails can grow faster than uninfected counterparts (81, 93, 116,), sometimes exhibiting gigantism (91). In other systems, trematodes strongly reduce the growth of parasitized snails (50, 60, 62, 101, 108). There have also been reports of high snail mortality rates associated with parasitism by larval trematodes (e.g. 19, 50, 54, 62, 92), and this may skew the size-prevalence relationship. Unless the sizes of the sexes are quite different, one would predict little effect of host sex as a source of heterogeneity, and our analysis revealed none for the two studies that reported infections by host sex.

The seasonal behavior of definitive hosts is the most likely factor responsible for temporal heterogeneity. When seasonality correlates positively among definitive hosts (e.g. migratory shore birds), recruitment will come in pulses and intensify interactions. Only when definitive hosts have opposing chronological patterns of activity will temporal heterogeneity act to isolate species. Such heterogeneity may not strongly alter the frequency of interactions because experimental studies have shown trematodes do not necessarily have to recruit at the same time to interact. Temporal heterogeneity may have subtler effects on structure, however. Prior residency is an important factor that determines dominance for some interactive pairs of trematodes (61, 69). Here, species that recruit early will have an advantage. Also, sampling during a period of pulsed recruitment (e.g. 46) should yield many double infections undergoing the process of competitive exclusion. Such communities may be highly interactive, yet yield a ratio of expected to observed double infections similar to less interactive systems. This was achieved experimentally by Heyneman & Umathevy (45) who added echinostome eggs to a pond where they had established a high prevalence of a subordinate species, *Schistosoma bovis* (46).

Long-lived snail hosts effectively integrate temporal variation in transmission. Thus, inter-trematode antagonisms will eventually happen. In contrast, short-lived hosts may either "miss" certain episodes of transmission, or experience relatively regular seasonal patterns of infection depending on the frequency, periodicity, and amplitude of trematode transmission and the strength of seasonality of snail population dynamics. Among the well-studied long-lived hosts that slowly accumulate parasites are *Cerithidea californica,* the littorines and *Ilyanassa obsoleta.* Here, in the largest size classes, prevalence often approaches 100% (13, 51, 61, 73, 76, 88, 102, 104, 115). Many freshwater species live only a few months and have strongly seasonal patterns of transmission (19, 34, 35, 99). It would be of interest to compare the temporal patterns and community structure of the larval trematodes of *Cerithidea californica* with *C. scalariformis* of the Atlantic coast of Florida as the latter lives one to two years (48) whereas the former lives at least 6–10 years (48, 84; AM Kuris, personal observation).

The trade-off between variation in host species suitability (how susceptible it is to infection) and specificity (the extent to which trematodes use it as an exclusive host) will modulate the effect of heterogeneity in recruitment among host species. The specificity of larval trematodes for certain snail hosts will cause heterogeneity in recruitment among snail species. Where some snail species serve a disproportionate role as hosts to several species, trematode species will interact more strongly. When each host species has its own unique assemblage of trematode species, isolation will occur.

Although a number of studies provide samples from different snail species, the definitive survey to analyze the effect of host specificity on larval trematode community structure has not yet been conducted. Sampling is very uneven, and well-known, heavily parasitized species, such as the littorines, attract repeated studies. Ecologists have tended to ignore species with few parasites, thus minimizing the impact of the isolation of highly host-specific species on community structure. Samples from taxonomic studies exhibit the reverse bias. Rare parasites often parasitize poorly studied host species and these can be over-represented in some surveys. We advocate a sampling procedure developed by Lafferty et al (64), who sampled snails at random until a previously specified number of *infected* snails were detected. This permits an analysis of community structure with sufficient statistical power to detect effects even when prevalence is uneven or low.

Despite sampling problems with surveys across host taxa, we can discuss at least the best studied marine systems involving mostly trematode parasites of shorebird hosts. First, a few species of generally abundant snails seem to channel a relatively large proportion of the available parasite species. These include *Ilyanassa obsoleta* on mud flats, *Cerithidea* spp. in salt marshes and mangroves, and *Littorina* spp. in rocky intertidal habitats. Second, a taxonomically very similar assemblage of larval trematodes occurs over great distances, and sometimes over related host species. For 15 studies of *Littorina* spp., at least 10 included *Cryptocolyle lingua, Renicola roscovita, cercaria lebouri* and a species of *Himasthla*. Four species of *Cerithidea* occur along the Atlantic coast of the United States, the Gulf of Mexico, the Pacific coast of the United States, and the Sea of Cortez. Yet all report *Parorchis acanthus,* three similar species of renicolids, a species of *Himasthla,* a species of *Austrobilharzia,* and three similar species of heterophyids (see also 7). This suggests a strongly historic component to these communities, perhaps involving co-accommodation and co-speciation. Third, the geographically widespread distribution of these assemblages suggests that a common system of dispersal and recruitment operates for each host-parasite system. Because the parasites often span several biotic provinces (defined by free-living animals), it suggests that, at least for marine species, vertebrate host behavior is a more powerful integrator of the parasite biotic provinces than are the ocean currents for their host snails.

Although heterogeneity in recruitment was shown here usually to be a weak structuring force relative to competitive interactions, several sources of heterogeneity could act in concert to alter the overall expected frequency of interactions. However, since heterogeneity generally intensified interactions, it is unlikely that unstudied aspects of heterogeneity in recruitment could provide a general alternative to competitive exclusion as an explanation for the low number of observed double infections. Further, one form of heterogeneity that we have not been able to assess is the repeatedly documented differential high susceptibility of previously infected snails (e.g. 44, 69). This source of heterogeneity could greatly intensify the expected frequency of multiple infections.

Competition

A few studies indicate facilitation, in which the presence of trematode infections increases the abundance of other species of trematodes. Most of these involve the only described obligate secondary invader, *Austrobilharzia terrigalensis* (113), which appears to use prior infections of other species in proportion to their availability (3). Our analysis provides an alternative conclusion for a number of studies. Several original studies reported a combination of negative, positive, or neutral associations between pairs of trematodes (5, 19, 21, 65, 88, 111, 115, 119). In all of these cases, the net effect of interactions significantly decreased the number of double infections. It was also common to find interactions reported as unimportant when, in fact, our analysis revealed significant negative associations. Some of these were due to problems with probability theory (35, 36, 54, 89); others occurred because double infections were numerous and the authors made no direct statistical comparison (20, 24,

109). Two studies (80, 112) claimed a significant effect, but our approach indicated that interspecific interactions did not significantly structure those communities. Both anomalies result from a lack of sufficient statistical power due to low prevalence or sample size. One study (51) emphasized the phenomenon of a priori infection increasing the susceptibility to a second infection, although our analysis indicated a net effect of competitive interactions rather than facilitation.

Four factors increase the frequency of competitive events: high species evenness, high prevalences of infection, intensification of interactions due to heterogeneity in recruitment, and dominance hierarchies. Clearly, competition eliminates the majority of interspecific interactions. How can such strong competition persist? Standard predictions made for competing populations in closed recruitment systems do not apply to larval trematodes; in the same way they fail to predict interactions in marine systems with open recruitment (37). Complex trematode life-cycles, when coupled with the dispersal capabilities of their definitive hosts, open the nature of trematode recruitment and make it possible for a number of fierce competitors to coexist in a rich, yet interactive assemblage.

As we mentioned previously, Sousa (102) predicted that if competition is an important structuring force at the component community level, the highest trematode diversity should occur in medium-sized snails (or populations with intermediate prevalence) because competitive exclusion will eliminate subordinate species from older (larger) snails (or from areas of high prevalence). Unfortunately, this prediction has limited generality. For any community, when *subordinate* individuals are common, their competitive exclusion will generate the pattern opposite to Sousa's prediction; diversity will increase. Therefore, a negative association between prevalence and diversity is likely to hold only in the limiting case in which competition leads to the total exclusion of some species (61). Such in fact occurred at 2 of Sousa's 38 sites. The hyperbolic association of worm diversity and snail size depends on simplifying assumptions that may be difficult to meet in snail-trematode systems. Variation in snail growth rates among populations (62, 102) or between infected and uninfected snails (60, 62, 81, 101, 102) may make size a poor indicator for the age of an infected snail. Hence, associations between trematode diversity and snail size are difficult to interpret. Sousa's hypothesis also predicts that in host populations with high prevalence, competition should be more intense and diversity should be low. This is difficult to assess because the opposing pattern, high diversity in areas of high prevalence, should occur if there is a positive association between the density of definitive hosts and the diversity of trematodes inhabiting those hosts. Even when the predicted hyperbolic diversity curve occurs, Fernández & Esch (35) pointed out that alternative explanations such as parasite-induced host mortality may obscure analyses of competitive effects. Finally, a general problem with predictions involving community diversity indices is that differences are difficult to detect (6). Therefore, absence of statistically significant changes in diversity indices may simply reflect the lack of power of this approach.

Although it is clear that interspecific competition reduces the number of interspecific interactions that can persist within a host (the infracommunity), does it affect the overall composition of the trematode assemblage that parasitizes a host population (the component community)? An interesting theoretical issue was raised by the claim that competition could be the predominant structuring force within the individual snails but not among the snails in a population (32, 35, 90, 102, 104). Is the system not additive? Can the extrinsic spatial and temporal heterogeneity effects be so large that they render intrinsic competitive effects unimportant at the level of the host population? Overall, our analysis indicates that infracommunity interactions are additive at the component community level because they must operate after potential isolating mechanisms occur at recruitment. "Important" is a loaded word, defined by the beholder. Our results show that competition significantly shapes the community beyond a simple reflection of recruitment. Its impact is most clearly important for subordinate species as their abundance is generally very much reduced by the impact of competitive dominants (Figure 3).

FUTURE WORK

Trematode Community Studies

Improved methods of analysis (64) have clearly altered the interpretation of past studies. Unfortunately, researchers usually did not design the stratification of samples to analyze heterogeneity in recruitment. Thus, the quality of our analysis reflects the quality of data that, often times, were not collected explicitly for our purpose. We hope that future studies will incorporate stratified sampling designs that balance the number of trematodes in each subsample. Also, it would be very instructive to investigate several levels of heterogeneity simultaneously to see how they interact to shape the number of expected interactions. Investigations into the effects of transmission to second intermediate and definitive hosts would also help determine whether heterogeneity transfers along the trematode life-cycle. Such hosts are often more vagile than snails and might act to homogenize effects of heterogeneity experienced by snails. (See *Note Added in Proof B.*) Finally, descriptive studies of larval trematode assemblages will continue to benefit from insights gained from experimental studies such as those pioneered by Sousa (102, 104), and there continues to be room for improved means of analysis.

Trematode Communities as Biomonitoring Tools

Gardner & Campbell (39) recently argued that parasites can act as probes for biodiversity because they track host food webs so broadly. A further use of parasites in biomonitoring studies stems from the similar hypothesis that parasites provide a representation of environmental quality and complexity. This is especially true for trematodes that usually have complex life-cycles involving trophic transmission. In a sense, trematode communities in snail hosts record the presence of definitive hosts and the abiotic conditions required for transmission in a particular wetland system. They also may indirectly represent the presence of second intermediate hosts. For example, Lafferty (63) noted fine-scale variation in prevalence between a population of snails from an undisturbed (25%) and a highly disturbed adjacent section (1%) of a salt marsh. Although some comparisons among sites could prove difficult to interpret, comparing changes in a location over time might be very instructive. Cort et al (18) did just this when they compared the larval trematodes from Douglas Lake in Michigan. Over a 20-year period, both prevalence and species diversity were greatly reduced (prevalence changed from 38–77% to 12–15%; species richness declined from 12–15 spp. to 3–4 spp.). They suggested that the increased number of summer cottages had led to a reduction in the number of vertebrates, especially shore birds, that visited the beaches. It would be interesting to see if this trend has continued over the past four decades. The considerable historical information that exists for trematode assemblages in several geographic areas provides ample opportunity for parallel comparisons.

Biological Control

Our findings that researchers have generally substantially underestimated competitive interactions in snail trematode systems, coupled with the demonstration that the most damaging human parasites (schistosomes) are largely competitive subordinates, should renew interest in the use of competitive dominants (notably echinostomes and cathaemasiids) as biological control agents. Many have suggested this approach (4, 10, 58, 66, 72, 80) and pilot studies have achieved success (reviewed by 12, 58, 66). However, biological control, along with other approaches to control of schistosomes by altering risk of transmission to humans, has lost favor largely because transmission models incorporate a low global prevalence of 1–2% (1). Our analysis shows that factors producing significant spatial and temporal heterogeneity can greatly elevate numbers of infected snails in likely foci of transmission. Thus, efforts at bio-control using trematode competitors should be explored, along with other efforts at local control of transmission to humans.

ACKNOWLEDGMENTS

We dedicate this review to Lie Kian Joe whose pioneering experiments revealed an unsuspected world of interactions. This analysis benefited greatly from discussions with A Bush, T Case, M Cody, S Cooper, G Esch, T Huspeni, W Murdoch, T Price, R Schmitt, W Sousa, T Stevens, A Stewart-Oaten, and R Warner. E Loker, W Wardle, and J McDermott graciously provided unpublished data sets for analysis. We made use of data sets published in German and Russian through the able translations of D Roberts and E Kogan.

NOTES ADDED IN PROOF

Despite the large number of studies encompassed in our analysis, we feel that the definitive study of trematode communities has yet to be done. Future analyses of the effects of recruitment and post-recruitment contributions to community structure should include evaluation of the impact of heterogeneity in snail densities at different sites. This can be incorporated by weighting samples proportional to density. We did not weight observed values according to sample size as in Lafferty et al. (64) because no other studies explicitly increased sample sizes from sites where prevalence was low. In data sets with multiple samples, large sample sizes contribute disproportionately to the pooled analysis. However, there was no discernible bias towards sites with either high or low prevalence. In some cases, the sample sizes may reflect abundance of snails. In a definitive study of community structure, this would be most appropriate. Analytical refinements are also needed to investigate the nature of changes in species composition over time. Changes as snails grow and as time passes may be due to either post-recruitment processes or to pulses in recruitment. The use of sentinel snail experiments would be very helpful to sort out these components. There is still much to learn.

APPENDIX

Appendix Data compiled from studies used in our analyses

| heterogeneity | Taxa[a] | S[b] | Total | Number of snails — Observed infections | | | Expected double | | p value | | Reference |
				Single	Double	Triple	Pool	Sum	Hetero.	Comp.	
none	bi	2	2011	576	4	0	—	6	—	.250	80
none	bu	3	628	377	0	0	—	89	—	<.001	28
none	ce	17	12995	7153	646	23	—	3392	—	<.001	74
none	ce	10	2908	433	12	0	—	21	—	.009	119
none	ce	7	1652	838	34	1	—	256	—	<.001	79
none	ce	15	305	140	11	1	—	51	—	<.001	d
none	ce	5	191	108	4	0	—	30	—	<.001	33
none	he	6	2000	406	0	0	—	22	—	<.001	118 A
none	he	8	806	518	0	0	—	60	—	<.001	22
none	he	5	556	207	0	0	—	12	—	<.001	40
none	il	8	5025	326	14	0	—	9	—	.068	112
none	il	6	379	162	134	65	—	346	—	.001	23
none	li	7	6843	2798	110	4	—	393	—	<.001	54 A
none	li	6	2690	1244	22	0	—	29	—	.070	54 B
none	ly	2	6281	502	1	0	—	10	—	<.001	4
none	ly	5	425	425	0	0	—	69	—	<.001	67
none	ly	4	649	103	0	0	—	5	—	.003	2
none	ly	9	323	163	2	0	—	31	—	<.001	27
none	ot	7	3817	2476	364	9	—	1578	—	<.001	53
none	ot	12	1165	570	44	4	—	193	—	<.001	114
none	ot	4	650	246	51	0	—	53	—	.780	46
none	ph	5	104	30	2	0	—	5	—	.090	8
sex	li	3	3049	753	20	0	18	19	0.830	.830	73
sex	li	11	838	707	7	0	101	102	0.806	<.001	13
site	bi	2	2255	862	34	0	80	75	0.366	<.001	41 A
site	bu	2	765	144	8	0	7	5	0.330	.167	41 B

site	ce	15	24252	3626	74	1	342	395	<.001	<.001	104
site	ce	9	849	493	7	0	88	104	0.008	<.001	64
site	ce	6	416	64	10	2	5	11	0.002	.150	7
site	he	8	3963	1209	11	0	83	80	0.576	<.001	34/35
site	he	6	2007	321	0	0	14	16	0.490	<.001	118 B
site	he	9	1179	958	79	0	333	241	<.001	<.001	20
site	li	6	6169	960	29	0	62	70	.210	<.001	115
site	li	6	3586	236	0	0	6	7	.517	<.001	9
site	li	8	2831	1230	48	3	340	681	<.001	<.001	38
site	li	8	2785	673	71	0	94	175	<.001	<.001	82
site	li	8	600	338	3	0	184	80	<.001	<.001	76 C
site	li	2	520	181	4	0	10	12	.490	.001	109
site	li-ot	3	417	106	0	0	9	9	.791	<.001	30
site	li	5	335	203	21	10^c	62	74	.006	<.001	51
site	ly	16	4795	2593	256	1	1055	1135	<.001	<.001	19
site	ot	3	8870	1024	155	0	75	84	.130	<.001	3
site	ot	5	4168	684	0	0	6	9	.232	.790	52
site	ot	4	1700	193	8	0	7	9	.543	<.001	57
site	ot	5	1244	803	237	0	310	293	.066	<.001	65
site	ph	7	5200	710	4	0	15	16	.866	<.001	95
site	ph	6	2491	762	9	0	77	126	<.001	<.001	21
site	bu	4	991	280	12	0	34	71	<.001	<.001	31
size	ce	9	970	320	7	0	41	66	<.001	<.001	e
size	he	8	4574	1374	11	0	65	54	.049	<.001	34/35
size	he	5	550	200	0	0	10	9	.683	<.001	40
size	li	4	5908	2194	145	0	516	634	<.001	<.001	88
size	li	6	1145	353	19	1	42	90	<.001	<.001	115
size	ot	6	3994	1656	192	0	266	294	.006	<.001	89
size	ot	5	1211	798	237	0	491	365	<.001	<.001	65
size	ph	6	2850	609	8	0	378	109	<.001	<.001	21
species	li-hy	10	4639	711	1	0	44	32	.004	<.001	86
species	li	8	2785	673	71	3	131	175	<.001	<.001	82
species	li	7	600	338	3	0	170	80	<.001	<.001	76 B

Appendix *(Continued)*

heterogeneity	Taxa[a]	S[b]	Total	Number of snails — Observed infections			Expected double		p value		Reference
				Single	Double	Triple	Pool	Sum	Hetero.	Comp.	
species	li	5	313	153	0	0	50	20	<.001	<.001	76 A
species	li-hy	24	42926	2561	25	0	64	40	<.001	<.001	11
species	ly-ot	41	2374	657	27	3	84	84	0.89	<.001	29
species	bi-bu	3	43526	2899	28	0	108	234	<.001	<.001	41 C
species	ot	4	248	57	11	0	8	10	.413	.640	36
time	bu	4	2045	437	3	0	22	16	.055	<.001	10
time	ce	15	24252	3626	74	1	356	395	<.001	<.001	104
time	he	8	4574	1374	11	0	56	54	.678	<.001	34/35
time	he	9	1179	958	79	0	247	241	.527	<.001	20
time	hy	13	15051	5249	428	7	1171	1100	<.001	<.001	111
time	il	6	14878	609	0	0	8	19	<.001	<.001	78
time	il	8	5677	1420	52	0	498	686	<.001	<.001	38
time	il	6	4314	254	0	0	5	6	.444	<.001	106
time	li	4	5876	2124	145	0	469	674	<.001	<.001	88
time	li	8	2785	673	71	3	186	175	.152	<.001	82
time	ly	16	4795	2593	256	10	1056	1135	<.001	<.001	19
time	ly	6	1639	571	195	15	199	288	<.001	<.001	5
time	ot	6	4920	2146	41	0	519	429	<.001	<.001	85
time	ot	7	1887	781	63	0	88	85	.678	.002	49
time	ot	9	304	94	0	0	11	14	.243	<.001	110
time	ph	6	1178	378	83	3	73	100	.001	.371	99
time	ph	6	697	356	5	0	83	104	<.001	<.001	21
time (mo)	ly	13	1697	678	53	2	171	209	<.001	<.001	18
time (yr)	ly	13	1697	678	53	2	195	209	.081	<.001	18

[a] Species codes are: bi = *Biomphalaria*, bu = *Bulinus*, ce = *Cerithidea*, he = *Helisoma*, hy = *Hydrobia*, il = *Illynassa*, li = *Littorina*, ly = *Lymnaea*, ot = other, ph = *Physa*.

[b] Species richness of larval trematodes.

[c] A quadruple infection was observed.

[d] Wardle, unpublished

[e] Kuris, unpublished

Literature Cited

1. Anderson RM, May RM. 1979. Prevalence of schistosome infections within molluscan populations: Observed patterns and theoretical predictions. *Parasitology* 79:63–94

2. Anteson RK. 1970. On the resistance of the snail, *Lymnaea catascopium pallida* (Adams) to concurrent infection with sporocysts of the strigeid trematodes, *Cotylurus flabelliformis* (Faust) and *Diplostomum flexicaudum* (Cort and Brooks). *Ann. Trop. Med. Parasitol.* 64:101–07

3. Appleton CC. 1983. Studies on *Austrobilharzia terrigalensis* in the Swan Estuary, Western Australia: frequency of infection in the intermediate host population. *Int. J. Parasitol.* 13:51–60

4. Boray JC. 1967. Host-parasite relationship between lymnaeid snails and *Fasciola hepatica. Proc. 3rd Int. Conf. World Assoc. Adv. Vet. Parasitol.* pp. 132–39

5. Bourns TKR. 1963. Larval trematodes parasitizing *Lymnaea stagnalis appressa* Say in Ontario with emphasis on multiple infections. *Can. J. Zool.* 41:937–41

6. Bouton CE, McPhereson BA, Weise AE. 1980. Parasitoids and competition. *Am. Nat.* 117:923–43

7. Bush AO, Heard RW, Overstreet RM. 1993. Intermediate hosts as source communities. *Can. J. Zool.* 71:1358–63

8. Byrd EE. 1940. Larval flukes from Tennessee. II. Studies on cercariae from *Physa gyrina* Say, with descriptions of two new species. *Rept. Reelfoot Lake Biol. Sta.* 4:124–31

9. Ching HL. 1962. Six larval trematodes from the snail, *Littorina scutulata* Gould of San Juan Island, U.S.A., and Vancouver, B.C. *Can. J. Zool.* 40:675–76

10. Chu KY, Dawood JK, Nabi HA. 1972. Seasonal abundance of trematode cercariae in *Bulinus truncatus* in a small focus of schistosomiasis in the Nile Delta. *Bull. WHO* 47:420–22

11. Chubrik GK. 1966. Fauna i ekologia lichinok trematod iz mollyuskov Barentsia i Belogo morei. *Akad. Nauk SSSR, Murmanskii Morskoi Biol. Inst. Trudy* 10:78–166 (in Russian)

12. Combes C. 1982. Trematodes: antagonism between species and sterilizing effects on snails in biological control. *Parasitology* 84:151–75

13. Combescot-Lang C. 1976. Étude des trématodes parasites de *Littorina saxatilis* (Olivi) et de leurs effets sur cet hôte. *Ann. Parasit. Hum. Comp.* 51:27–36

14. Connell JH. 1978. Diversity in tropical rain forests and coral reefs. *Science* 199:1302–10

15. Connell JH. 1980. Diversity and the coevolution of competitors, or the ghost of competition past. *Oikos* 35:131–38

16. Connell JH. 1983. On the prevalence and relative importance of interspecific competition: evidence from field experiments. *Am. Nat.* 122:661–96

17. Conner EF, Simberloff D. 1979. The assembly of species communities: chance or competition? *Ecology* 60:1132–40

18. Cort WW, Hussey KL, Ameel DJ. 1960. Seasonal fluctuations in larval trematode infections in *Stagnicola emarginata angulata* from Phragmites Flats on Douglas Lake. *Proc. Helm. Soc. Wash.* 27:11–13

19. Cort WW, McMullen DB, Brackett S. 1937. Ecological studies on the cercariae in *Stagnicola emarginata angulata* (Sowerby) in the Douglas Lake region, Michigan. *J. Parasitol.* 23:504–52

20. Cort WW, McMullen DB, Brackett S. 1939. A study of larval trematode infections in *Helisoma campanulatum smithii* (Baker) in the Douglas Lake Region, Michigan. *J. Parasitol.* 25:19–22

21. Cort WW, Olivier L, McMullen DB. 1941. Larval trematode infection in juveniles and adults of *Physa parkeri* Currier. *J. Parasitol.* 27:123–41

22. Crews A, Esch GW. 1986. Seasonal dynamics of *Halipegus occidualis* (Trematoda: Hemiuridae) in *Helisoma anceps* and its impact on fecundity of the snail host. *J. Parasitol.* 77:528–39

23. Curtis LA, Hubbard KM. 1990. Trematode infections in a gastropod host misrepresented by observing shed cercariae. *J. Exp. Mar. Biol. Ecol.* 143:131–37

24. Curtis LA, Hubbard KMK. 1993. Species relationships in a marine gastropod-

trematode ecological system. *Biol. Bull.* 184:25–35

25. DeCoursey PJ, Vernberg WB. 1974. Double infections of larval trematodes: competitive interactions. In *Symbiosis in the Sea,* ed. WB Vernberg pp. 93–109. Columbia, SC: Univ. S. Carolina Press

26. Diamond JM. 1975. Assembly of species communities. In *Ecology and Evolution of Communities,* ed. ML Cody, JM Diamond, pp. 342–444. Cambridge, Mass: Harvard Univ. Press

27. Donges J. 1972. Double infection experiments with echinostomatids (Trematoda) in *Lymnaea stagnalis* by implantation of rediae and exposure to miracidia. *Int. J. Parasitol.* 2:409–23

28. Dönges J. 1977. *Cercaria ogunis* n. sp. (Echinostomatidae) aus *Bulinus globosus* in Westafrika. *Z. Parasitenk.* 52:297–309

29. Dubois G. 1929. Les cercaires de la région de Neuchâtel. *Bull. Soc. Neuchâteloise Sci. Nat.* 53:3–177

30. Duerr FG. 1965. Survey of digenetic trematode parasitism in some prosobranch gastropods of the Cape Arago region, Oregon. *Veliger* 8:42

31. El-Gindy MS. 1965. Monthly prevalence rates of natural infection with *Schistosoma haematobium* cercariae in *Bulinus truncatus* in Central Iraq. *Bull. Endem. Dis.* 7:11–31

32. Esch GW, Fernández JC. 1993. *A Functional Biology of Parasitism: Ecological and Evolutionary Implications,* ed. P Calow. London: Chapman & Hall. 337 pp.

33. Epstein RA. 1972. *Larval trematodes of marine gastropods of Galveston Island, Texas.* MS thesis. Texas A&M Univ., College Station, Tex.

34. Fernández J, Esch GW. 1991a. Guild structure of larval trematodes in the snail *Helisoma anceps*: patterns and processes at the individual host level. *J. Parasitol.* 77:528–39

35. Fernández J, Esch GW. 1991b. The component community structure of larval trematodes in the pulmonate snail *Helisoma anceps. J. Parasitol.* 77:540–50

36. Flook JM, Ubelaker JE. 1972. A survey of metazoan parasites in unionid bivalves of Garza-Little Elm Reservoir, Denton County, Texas. *Tex. J. Sci.* 23:381–92

37. Gaines SD, Lafferty KD. 1994. Modeling the dynamics of marine species: the importance of incorporating larval dispersal. In *Ecology of Marine Invertebrate Larvae,* ed. L. McEdward. CRC. In press

38. Gambino JJ. 1959. The seasonal incidence of infection of the snail *Nassarius obsoletus* (Say) with larval trematodes. *J. Parasitol.* 45:440, 56

39. Gardner SL, Campbell ML. 1992. Parasites as probes for biodiversity. *J. Parasitol.* 78:596–600

40. Goater TM, Shostak JA, Williams JA, Esch GW. 1989. A mark-recapture study of trematode parasitism in overwintered *Helisoma anceps* (Pulmonata), with special reference to *Halipegus occidualis* (Hemiuridae). *J. Parasitol.* 75:553–60

41. Gordon RM, Davey TH, Peaston H. 1934. The transmission of human bilharziasis in Sierra Leone, with an account of the life-cycle of the schistosomes concerned, *S. mansoni* and *S. haematobium. Ann. Trop. Med. Parasitol.* 28:323–418

42. Grant P, Schluter D. 1984. Interspecific competition inferred from patterns of guild structure. In *Ecological Communities: Conceptual Issues and the Evidence,* ed. DR Strong, D Simberloff, LG Abele, AB Thistle, pp. 201–33. Princeton, NJ: Princeton Univ. Press,

43. Gurevitch J, Morrow L, Wallace A, Walsh J. 1992. A meta-analysis of competition in field experiments. *Am. Nat.* 140: 539–72

44. Heyneman DH, Lim K, Jeyarasasingam U. 1972. Antagonism of *Echinostoma liei* (Trematoda: Echinostomatidae) against the trematodes *Paryphostomum segregatum* and *Schistosoma mansoni. Parasitol.* 65:203–22

45. Heyneman D, Umathevy T. 1967. A field experiment to test the possibility of using double infections of host snails as a possible biological control of schistosomiasis. *Med. J. Malaya.* 21:373

46. Heyneman D, Umathevy T. 1968. Interaction of trematodes by predation within natural double infections in the host snail *Indoplanorbis exustus. Nature* 217:283–85

47. Holmes JC. 1987. The structure of helminth communities. *Int. J. Parasitol.* 17:203–8

48. Houbrick RS. 1984. Revision of higher taxa in genus *Cerithidea* (Mesogastropoda: Potamididae) based on comparative morphology and biological data. *Am. Malacol. Bull.* 2:1–20

49. Huehner MK. 1983. Aspidogastrid and digenetic trematode single and double infections in the gastropod, *Elimia livescens,* from the Upper Cuyahoga river. *Proc. Helm. Soc. Wash.* 54:200–03

50. Huxham M, Raffaelli D, Pike A. 1993. The influence of *Cryptocotyle lingua* (Digenea:Platyhelminthes) infections on

the survival and fecundity of *Littorina littorea* (Gastropoda:Prosobranchia); an ecological approach. *J. Exp. Mar. Biol. Ecol.* 168:223–38

51. Irwin SWB. 1983. Incidence of trematode parasites in two populations of *Littorina saxatilis* (Olivi) from the North Shore of Belfast Lough. *Ir. Nat. J.* 21:26–29

52. Ismail NS, Abdel-Hafez SK. 1987. Seasonal variation in infection rates of *Melanopsis praemorsa* (L. 1785) (Thiaridae) snails with larval trematodes in Azraq Oasis, Jordan. *Jpn. J. Parasitol.* 36:13–16

53. Ismail NS, Arif AMS. 1993. Population dynamics of *Melanoides tuberculata* (Thiaridae) snails in a desert spring, United Arab Emirates and infection with larval trematodes. *Hydrobiologia* 257: 57–64

54. James BL. 1969. The Digenea of the intertidal prosobranch, *Littorina saxatilis* (Olivi). *Z. Zool. Syst. Evol. Forsch.* 7:273–316

55. Jong-Brink M de, Elasaadany MM, Boer HH. 1988. *Trichobilharzia ocellata*: interference with the endocrine control of female reproduction of its host *Lymnaea stagnalis*. *Exp. Parasitol.* 68:93–98

56. Kendall SB. 1964. Some factors influencing the development and behaviour of trematodes in their molluscan hosts. In *Host-Parasite Relationships in Invertebrate Hosts*, ed. AE Taylor, pp. 51–73. Oxford: Blackwell Sci.

57. Køie M. 1969. On the endoparasites of *Buccinum undatum* L. with special reference to the trematodes. *Ophelia* 6: 251–79

58. Kuris AM. 1973. Biological control: Implications of the analogy between the trophic interactions of insect pest-parasitoid and snail-trematode systems. *Exp. Parasitol.* 33:365–79

59. Kuris AM. 1974. Trophic interactions: similarity of parasitic castrators to parasitoids. *Q. Rev. Biol.* 49:129–48

60. Kuris AM. 1980. Effect of exposure to *Echinostoma liei* miracidia on growth and survivorship of young *Biomphalaria glabrata* snails. *Int. J. Parasitol.* 10: 303–08

61. Kuris AM. 1990. Guild structure of larval trematodes in molluscan hosts: prevalence, dominance and significance of competition. In *Parasite Communities: Patterns and Processes*, ed. GW Esch, AO Bush, JM Aho, pp. 69–100. London: Chapman & Hall

62. Lafferty KD. 1993. Effects of parasitic castration on growth, reproduction and population dynamics of the marine snail *Cerithidea californica. Mar. Ecol. Prog. Ser.* 96:229–37

63. Lafferty KD. 1993. The marine snail, *Cerithidea californica*, matures at smaller sizes where parasitism is high. *Oikos* 68:3–11

64. Lafferty KD, Sammond D, Kuris AM. 1994. Analysis of larval trematode community structure: separating the roles of competition and spatial heterogeneity. *Ecology* 75: In press

65. Lauckner G. 1988. Larval trematodes in *Planaxis sulcatus* (Gastropoda, Planaxidae) from Heron Island, Great Barrier Reef. *Proc. 6th Int. Coral Reef Symp. Townsville, Australia* 3:171–76

66. Lie KJ. 1973. Larval trematode antagonism: Principles and possible application as a control method. *Exp. Parasitol.* 33:343–49

67. Lie KJ, Basch PF, Umathevy T. 1966. Studies on Echinostomatidae (Trematoda) in Malaya. XII. Antagonism between two species of echinostome trematodes in the same lymnaeid snail. *J. Parasitol.* 52:454–57

68. Lie KJ, Heyneman D, Jeong KH. 1976. Studies on resistance in snails. 7. Evidence of interference with the defense reaction in *Biomphalaria glabrata* by trematode larvae. *J. Parasitol.* 62:608–15

69. Lie KJ, Lim HK, Ow-Yang CK. 1973. Synergism and antagonism between two trematode species in the snail *Lymnaea rubiginosa. Int. J. Parasitol.* 3:719–33

70. Lim HK, Heyneman D. 1972. Intramolluscan inter-trematode antagonism: a review of factors influencing the host-parasite system and its possible role in biological control. *Adv. Parasitol.* 10:191–268

71. Loker ES, Bayne CJ 1986. Immunity to trematode larvae in the snail *Biomphalaria. Symp. Zool. Soc. Lond.* 56: 199–220

72. Loker ES, Moyo HG, Gardner SL. 1981. Trematode-gastropod associations in nine non-lacustrine habitats in the Mwanza region of Tanzania. *Parasitology* 83:381–99

73. Lysaght AM. 1941. The biology and trematode parasites of the gastropod *Littorina neritoides* (L.) on the Plymouth breakwater. *J. Mar. Biol. Assoc. U.K.* 25:41–67

74. Martin WE. 1955. Seasonal infections of the snail, *Cerithidea californica* Haldeman, with larval trematodes. *Essays Nat. Sci. Honor of Capt. A. Hancock.* pp. 203–10

75. Mason PR. 1977. Stimulation of the activity of *Schistosoma mansoni* mira-

cidia by snail-conditioned water. *Parasitology* 75:325–38

76. Matthews PM, Montgomery WI, Hanna REB. 1985. Infestation of littorinids by larval Digenea around a small fishing port. *Parasitology* 90:277–87

77. May RM. 1984. An overview: real and apparent patterns in community structure. In *Ecological Communities: Conceptual Issues and the Evidence,* ed. DR Strong, D Simberloff, LG Abele, AB Thistle, pp 3–18. Princeton, NJ: Princeton Univ. Press

78. McDaniel JS, Coggins JR. 1972. Seasonal larval trematode infection dynamics in *Nassarius obsoletus* (Say). *J. Elisha Mitchell Sci. Soc.* 88:55–57

79. McNeff LL. 1978. *Marine cercariae from* Cerithidea pliculosa *Menke from Dauphin Island, Alabama; life cycles of heterophyid and opisthorchiid Digenea from* Cerithidea Swainson *from the eastern Gulf of Mexico.* MA thesis. Univ. Alabama, Tuscaloosa

80. Nassi H. 1978. Données sur le cycle biologique de *Ribeiroia marini guadeloupensis* n. ssp., Trématode stérilisant *Biomphalaria glabrata* en Guadeloupe. Entretié du cycle en vue d'un contrôle éventuel des populations de Mollusques. *Acta Trop.* 35:41–56

81. Pan C. 1965. Studies on the host-parasite relationship between *Schistosoma mansoni* and the snail *Australorbis glabratus. Am. J. Trop. Med. Hyg.* 14:931–76

82. Pohley WJ. 1976. Relationships among three species of *Littorina* and their larval Digenea. *Mar. Biol.* 37:179–86

82a. Polis GA, Myers CA, Holt RD. 1989. The ecology and evolution of intraguild predation: potential competitors that eat each other. *Annu. Rev. Ecol. Syst.* 20:292–330

83. Porter A. 1938. The larval Trematoda found in certain South African Mollusca with special reference to schistosomiasis (bilharziasis). *Publ. S. Afr. Inst. Med. Res.* 42:1–492

84. Race MS. 1981. Field ecology and natural history of *Cerithidea californica* (Gastropoda: Prosobranchia) in San Francisco Bay. *Veliger* 24:18–27

85. Rankin JS. 1939. Ecological studies on larval trematodes from Western Massachusetts. *J. Parasitol.* 25:309–28

86. Rees FG. 1932. An investigation into the occurrence, structure, and life-histories of the trematode parasites of four species of *Lymnea* (*L. truncatula* (Müll.), *L. pereger* (Müll.), *L. palustris* (Müll.), and *L. stagnalis* (Linné)), and *Hydrobia jenkinsi* (Smith) in Glamorgan and Monmouth. *Proc. Zool. Soc. Lond.* 1932:1–32

87. Richards CS. 1984. Influence of snail age on genetic variations in susceptibility of *Biomphalaria glabrata* for infection with *Schistosoma mansoni. Malacologia* 25:493–502

88. Robson EM, Williams IC. 1970. Relationships of some species of Digenea with the marine prosobranch *Littorina littorea* (L.) I. The occurrence of larval Digenea in *L. littorea* on the North Yorkshire Coast. *J. Helminthol.* 44:153–68

89. Rohde K. 1981. Population dynamics of two snail species, *Planaxis sulcatus* and *Cerithium moniliferum,* and their trematode species at Heron Island, Great Barrier Reef. *Oecologia.* 49:344–52

90. Rohde K. 1993. *Ecology of Marine Parasites.* Wallingford, Eng: CAB Int. 298 pp. 2nd ed.

91. Rothschild M. 1936. Gigantism and variation in *Peringia ulvae* Pennant, 1777, caused by infection with larval trematodes. *J. Mar. Biol. Assoc. UK* 20:537–46

92. Rothschild M. 1938. Further observations on the effect of trematode parasites on *Peringia ulvae* (Pennant, 1777). *Novit. Zool.* 41:84–102

93. Rothschild M. 1941. Observations on the growth and trematode infections of *Peringia ulvae* (Pennant) 1777 in a pool in the Tamar Saltings, Pymouth. *Parasitology* 33:406–15

94. Sale PF. 1991. Reef fish communities: open non-equilibrial systems. In *The Ecology of Fishes on Coral Reefs,* ed. PF Sale, pp. 564–98. San Diego: Academic

95. Sankurathri CS, Holmes JC. 1976. Effects of thermal effluents on parasites and commensals of *Physa gyrina* Say (Mollusca: Gastropoda) and their interactions at Lake Wabamun, Alberta. *Can. J. Zool.* 54:1742–53

96. Schoener TW. 1983. Field experiments in interspecific competition. *Am. Nat.* 122:240–285

97. Sewell S. 1922. Cercariae Indicae. *Indian J. Med. Res.* 10:1–327

98. Sih A, Crowley P, McPeek M, Petranka J, Strohmeier K. 1985. Predation, competition and prey communities: a review of field experiments. *Annu. Rev. Ecol. Syst.* 16: 269–311

99. Snyder SD, Esch GW. 1993. Trematode community structure in the pulmonate snail *Physa gyrina. J. Parasitol.* 79:205–15

100. Sousa WP. 1979. Disturbance in marine intertidal boulder fields: the nonequilib-

rium maintenance of species diversity. *Ecology* 60:1225–39

101. Sousa WP. 1983. Host life history and the effect of parasitic castration on growth: a field study of *Cerithidea californica* Haldeman (Gastropoda: Prosobranchia) and its trematode parasites. *J. Exp. Mar. Biol. Ecol.* 73:273–96

102. Sousa WP. 1990. Spatial scale and the processes structuring a guild of larval trematode parasites. In *Parasite Communities: Patterns and Processes*, ed. GW Esch, AO Bush, JM Aho, pp. 41–67. London: Chapman & Hall

103. Sousa WP. 1992. Interspecific interactions among larval trematode parasites of freshwater and marine snails. *Am. Zool.* 32:583–92

104. Sousa WP. 1993. Interspecific antagonism and species coexistence in a diverse guild of larval trematode parasites. *Ecol. Monog.* 63:103–28

105. Sousa WP. 1994. Patterns and processes in communities of helminth parasites. *Trends Ecol. Evol.* 9:52–57

106. Stambaugh JE, McDermott JJ. 1969. The effects of trematode larvae on the locomotion of naturally infected *Nassarius obsoletus* (Gastropoda). *Proc. Pa. Acad. Sci.* 43:226–31

107. Strong DR, Simberloff D, Abele LG, Thistle AB. 1984. *Ecological Communities: Conceptual Issues and the Evidence.* Princeton, NJ: Princeton Univ. Press

108. Sturrock BM. 1966. The influence of infection with *Schistosoma mansoni* on the growth rate and reproduction of *Biomphalaria pfeifferi. Ann. Trop. Med. Parasitol.* 60:187–97

109. Threlfall W, Goudie RJ. 1977. Larval trematodes in the rough periwinkle, *Littorina saxatilis* (Olivi) from Newfoundland. *Proc. Helminth. Soc. Wash.* 44:229

110. Ullman H. 1954. Observations on a new cercaria developing in *Melanopsis praemorsa* in Israel. *Parasitology* 44:1–15

111. Vaes M. 1979. Multiple infection of *Hydrobia stagnorum* (Gmelin) with larval trematodes. *Ann. Parasitol.* 54:303–12

112. Vernberg WB, Vernberg FJ, Beckerdite FW. 1969. Larval trematodes: double infections in common mudflat snail. *Science* 164:1287–88

113. Walker JC. 1979. *Austrobilharzia terrigalensis:* a schistosome dominant in interspecific interactions with the molluscan host. *Int. J. Parasitol.* 9:137–40

114. Wardle WJ. 1974. *A survey of the occurrence, distribution and incidence of infection of helminth parasites of marine and estuarine mollusks from Galveston, Texas.* PhD thesis. Texas A&M Univ. College Station, Tex.

115. Werding B. 1969. Morphologie, Entwicklung and Ökologie digener Trematoden-Larven der Strandschnecke *Littorina littorea. Mar. Biol.* 3:306–33

116. Wesenberg-Lund C. 1934. Contributions to the development of the *Trematoda Digenea.* Part II. The biology of the freshwater cercariae in Danish freshwaters. *Mem. Acad. Roy. Sc. et Lett. Danemark, Sect. Sc.* 9 ser. 5:1–223

117. Wikgren BJ. 1956. Studies on Finnish larval flukes with a list of known Finnish adult flukes. *Acta Zool. Fenn.* 91:1–106

118. Williams JA, Esch GW. 1991. Infra- and component community dynamics in the pulmonate snail *Helisoma anceps,* with special emphasis on the hemiurid trematode *Halipegus occidualis. J. Parasitol.* 77:246–53

119. Yoshino TP. 1975. A seasonal and histological study of larval Digenea infecting *Cerithidea californica* (Gastropoda: Prosobranchia) from Goleta Slough, Santa Barbara County, Calif. *Veliger.* 18:156–61

Annu. Rev. Ecol. Syst. 1994. 25:219–36

THE EVOLUTIONARY INTERACTION AMONG SPECIES: Selection, Escalation, and Coevolution

Geerat J. Vermeij

Department of Geology and Center for Population Biology, University of California at Davis, Davis, California 95616

KEY WORDS: macroevolution, evolutionary trends, adaptation, Red Queen hypothesis, optimality theory

Abstract

The hypothesis of escalation states that enemies—competitors, predators, and dangerous prey—are the most important agents of natural selection among individual organisms, and that enemy-related adaptation and responses brought about long-term evolutionary trends in the morphology, behavior, and distribution of organisms over the course of the Phanerozoic. In contrast to this top-down view of the role of organisms in determining the directions of evolution, the hypothesis of coevolution holds that two interacting species or groups of species change in response to each other. I review and evaluate these hypotheses in the light of criticisms about the existence of evolutionary trends and the role of interactions of species in evolution.

Models describing the evolutionary effects organisms have on each other have been based largely on population dynamics and on cost-benefit analyses of the net outcome of interactions between species. Yet, the hypotheses of escalation and coevolution are statements about the nature, frequency, causes, and role of selection. Although these models have provided valuable insights and have forced some modifications in the hypotheses of escalation, studies seeking to distinguish between escalation and coevolution will require empirical observations and cost-benefit evaluations of the discrete events of interaction that collectively constitute organism-caused selection.

219

0066-4162/94/1120-0219$05.00

INTRODUCTION

In 1987, I published a book supporting and elaborating a hypothesis first clearly set forth by Darwin (19), which I called the hypothesis of escalation (79). This hypothesis states that enemies—predators, competitors, and dangerous prey— are the most important agents of selection among individual organisms, and that enemy-related adaptation brought about long-term evolutionary trends in the morphology, ecology, and behavior of organisms over the course of the Phanerozoic. With its emphasis on the evolutionary role of enemies, escalation embodies a "top-down" interpretation of the way evolution is affected by organisms themselves. I supported the hypothesis of escalation with paleontological data on the modes of life, environments, and functional designs of fossil organisms. Indeed, without such evidence from the fossil record, the hypothesis would not be testable and might never have been proposed in the first place.

A different conception of the role of organisms in evolution is that two interacting parties change in response to each other. This idea of coevolution, whose modern version was introduced by Ehrlich & Raven (24) in a landmark paper on the evolution of butterflies and plants, provided the impetus for Van Valen's (71) influential Red Queen hypothesis. According to this hypothesis, the probability of extinction of a population is approximately constant regardless of a taxon's age (see also 52, 56). Van Valen explained this constancy by noting that evolutionary changes in other species cause the environment for any given species to deteriorate unless the latter species compensates by evolving continuously. Organisms are therefore seen as major agents of extinction as well as of evolution.

Escalation, coevolution, and the Red Queen hypothesis have been criticized on several grounds. Gould (33) maintains that natural selection acting at the level of the individual organism has been of only incidental and fleeting significance in shaping the overall pattern of evolution, and that competition is not necessarily an important mechanism of natural selection. In Gould's view, the factors that determine speciation and extinction of lineages not only act independently of (or antagonistically to) natural selection, but in the long run they outweigh its evolutionary effect. A related claim is that evolutionary trends are at most short-term phenomena more often interpretable as changes in the variance of a trait than as changes in the mean or median value (34, 65). A true reading of the history of life might reveal few if any trends. Margulis (47) argues that innovation in evolution is mainly the result of symbioses arising from an integration of genomes with separate prior histories, and that natural selection among randomly generated mutants is at best a minor agency of evolutionary change. Still another viewpoint is that evolution is controlled largely by extrinsic change related to climate and tectonics rather than by organisms (11–13, 22, 37, 44).

Some of these differences in interpretation arise from ambiguities in the definition and recognition of trends. Others spring from doubts about the nature and importance of adaptation by natural selection. Still other criticisms flow from theoretical models of coevolution and escalation.

Scientific debates often end not in resolution but in confusion and indifference. Perhaps because of the appeal of the hypothetico-deductive method in science, possible explanations for patterns in nature are often portrayed as strict alternatives when they would be better regarded as complementary (see also 69). Critical assumptions remain hidden and unevaluated, and empirical observations are frequently neither sought nor considered. My purpose here is to evaluate models of coevolution, escalation, and related phenomena, and to redirect attention toward the study of the discrete events in the lives of individuals when selection due to other organisms takes place.

THE HYPOTHESIS OF ESCALATION

I begin with a summary of the hypothesis of escalation as presented in my 1987 book (79). The main features of the hypothesis are as follows.

1. Most resources needed by living things are either other organisms (prey and mates, for example) or are under the control of organisms (shelters, food, mates, information, nutrients, and energy). Therefore, acquisition and retention of resources by an individual organism requires that individual to prevent others from obtaining or monopolizing those resources (see also 71–73).

2. The survival and reproductive success of individuals depend on the ability of individuals to acquire and defend resources by competing with or eating other individuals. Insofar as traits related to this competition in the broad sense are heritable, competition for resources is an important means by which natural selection occurs.

3. Competition-related selection favors the evolution of several kinds of traits. (*a*) The rapid location and incorporation of resources and information enables individuals to deplete resources available to competitors and reduces exposure to enemies (see also 67). Enhanced sensory systems, high growth rates, and especially high metabolic rates make this possible. (*b*) Retention or appropriation of resources from other individuals is accomplished by resisting, interfering with, escaping from, or remaining undetected or unrecognized by would-be enemies. Armor, offensive weaponry, high locomotor performance, toxicity, crypsis, and intimate association with well-defended species are among the ways in which resources can be better retained, appropriated, and regulated.

4. Individuals often fail to acquire or retain resources during encounters with other individuals. Insofar as failure reduces the probability of survival or opportunities for reproduction, there is room for adaptive improvement. The

potential for improvement can be roughly gauged by the frequency and cost of failure.

5. The extent to which adaptation in any one direction can occur is limited by conflicting functional demands, that is, by trade-offs among incompatible requirements. These incompatibilities are most evident when populations are stable or in decline and when energy availability is low.

6. Circumstances that reduce functional trade-offs enable adaptive innovations to become established and escalation between species and their enemies to take place. These favorable circumstances include increased per-capita energy, that is, a higher metabolic rate, and expansion of populations in the presence of enemies. Such expansion is possible when habitable areas increase, when primary productivity rises, and when surviving populations recover after a major extinction event. Architectural innovations as well as symbioses may also eliminate functional trade-offs (see also 47, 60).

7. Extinctions accompany, and may often be caused by, reductions in primary productivity. They will tend preferentially to eliminate organisms with high metabolic demands as well as those with few energy reserves. Consequently, those species that are most functionally specialized, most highly escalated, and most energy-demanding are especially prone to extinction.

8. If natural selection due to enemies is an important cause of evolutionary change, there should be a long-term (but by no means a constant) trend for competition-related characteristics, which improves means of acquiring and controlling resources and information, to increase in expression and in frequency through time among functionally similar species inhabiting similar environments. Extinctions have temporarily interfered with escalation but have not eliminated the trend. This is because the diversity of life has generally increased through time (18, 39, 80).

9. Species unable to adapt to the increasing risks and hazards posed by enemies do not suffer extinction; instead they are restricted to environments where energy availability is low and where enemies are few in number and of small effect. These safe environments initially included the pelagic realm (see also 63), fresh water, the dry land, caves, the deep sea, and habitats within rocks, beneath the surface of sediments, or on or in the bodies of other organisms.

10. Escalation in the safe environments stimulates the recycling of nutrients and other resources and therefore increases the opportunity for further escalation in the biosphere as a whole. There is thus a strong positive feedback in the process of escalation (see also 25, 68).

11. Although the directions of evolution are determined largely by organisms, the timing of evolutionary events is dictated by extrinsic causes related to climate, sea level, tectonic movements, and mass extinctions.

THE RECOGNITION AND INTERPRETATION OF TRENDS

Almost all inferences about coevolution and escalation derive from the existence of trends in the expression of traits that function during interactions between species. Some criticisms of coevolution and escalation pertain to the reality of trends, which are defined as consistent statistical changes in the traits along a time axis (54).

With the general acceptance of cladistic methods by systematists, it has become commonplace to insist that evolutionary patterns, including trends, be looked for and verified only by analyzing ancestor-descendant relationships within monophyletic groups, or clades (see for example 34, 35, 54, 84, 85). Although many trends are indeed best sought in this way, others cannot in principle be detected within single clades and instead arise when ecologically and functionally comparable clades replace each other through time (53, 79).

All adaptive transformations occur within clades, but not all cases of branching or within-lineage evolution are expected to conform to the predictions of coevolution or escalation. Many, if not most, evolutionary changes involve an ecological or geographical shift rather than functional improvement relative to other species in the ancestral setting. This is nicely exemplified by lizards of the genus *Anolis* in the West Indies (46) and by intertidal gastropods of the genera *Littorina* and *Nucella* in the North Pacific and North Atlantic Oceans (16, 57). Consequently, it is generally inappropriate to test for trends with respect to traits involved in escalation or coevolution by tracking the mean value of such traits only within monophyletic groups. Only when descendant taxa are functionally and ecologically similar to ancestral ones would such within-clade tracking be suitable.

The most reliable method for detecting trends in escalation-related traits is to plot the expression of these traits among ecologically and functionally similar species through time. For example, one could test whether locomotor performance of large herbivorous tetrapods increased through time not by considering all such tetrapods, but by restricting the analysis to tetrapods of a given habitat such as open woodland. Failure to restrict the analysis could yield misleading results. Locomotor demands are quite different in open woodland as compared to deep forest or grassland environments, so that animals from these various settings are not functionally or ecologically comparable. For open woodland tetrapods, the analysis of the Mesozoic fossil record would concentrate mainly on dinosaurs, whereas the Cenozoic record is chiefly one of mammals and a few flightless birds. The important point is that, because adaptation is inextricably context-dependent, one cannot seek patterns in adaptation through time without taking the context into account.

Failure to incorporate the ecological context has led some critics to reject or downplay the role of interactions in selection, and to accord a minor role to natural selection in evolution. It is interesting that such points of view have been especially widely held by scholars living in cities. If one does not observe organisms functioning in nature, one is quite naturally inclined to discount the potential role that competition and selection play in daily life as well as in the longer-term dynamics of species, ecosystems, and the biosphere as a whole.

If biological interactions are important causes of selection, and if such selection is an important cause of evolutionary change (79), patterns in the expression of traits related to such interactions should be widespread. Whether these patterns arise from coevolution or escalation continues to be the subject of controversy. The remainder of this essay is devoted to a comparison and evaluation of models of coevolution and escalation, and to a discussion of how the two processes can be distinguished empirically.

MODELS OF COEVOLUTION AND ESCALATION

During its lifetime, any individual in nature interacts with members of many species, including its own. All these species are potentially important agents of selection. If two parties respond evolutionarily to changes in each other, the result is coevolution. This process thus implies reciprocal responses. If the responses involve only two species, there is said to be strict coevolution; if one or both interacting parties consist of more than one species, coevolution is said to be diffuse (30). Participants may be competitors, mutual beneficiaries, predator and prey, or host and guest. In coevolution, the two interacting parties may be each other's most important selective agents; that is, survival and reproduction of members of the interacting groups depend to a greater extent on interactions between the parties than on other potential sources of selection. In escalation, most evolution is caused by selection due to enemies—predators, competitors, parasites, and dangerous prey—that have the capacity to injure, kill, or depress the reproductive output of individuals. Escalation would be equivalent to coevolution if interacting parties are mutual enemies.

Most models that have been constructed to probe the effects species have on each other are variations on a coevolutionary framework. Although coevolution and escalation are ideas based on the roles and sources of natural selection, models of these processes have been built largely by ecologists more concerned with the population-level outcomes of selection than with the mechanics of the processes themselves.

The Red Queen Hypothesis and Models Derived From It

In 1973, Van Valen (71) published a compilation of paleontological data implying that the probability of extinction of a taxon is independent of the taxon's age, measured from the taxon's time of origin. From this point of departure, Van Valen developed the Red Queen hypothesis, which states that the environment of any given species deteriorates at a more or less constant rate. This deterioration, much of which is due to the evolution of co-occurring species, will eventually lead to extinction, unless the species in question adapts to counteract the environmental deterioration. Species must, in other words, be "running in place" (constantly evolving) just to keep up with the changes induced by evolution in their biological surroundings.

The Red Queen hypothesis is in many ways similar to the hypothesis of escalation, and it arises from a similar view of the importance of biological interactions during the course of evolution; but its two central predictions differ from those embodied in the hypothesis of escalation. These predictions are: (i) that there is continuous coevolution among interacting species, and (ii) that except during times of mass extinction, environmental deterioration due to organisms is the major cause of extinction. It is important to note that these predictions flow from theory, not from the data on extinction rates that prompted Van Valen to develop the theory. Indeed, the data are also compatible with interpretations for escalation. In my view of escalation, adaptation typically occurs only when populations of many species are able to expand simultaneously. At other times, there is a kind of evolutionary gridlock, or mutual stasis, because adaptive improvements in any one direction are prevented as organisms and the physical environment impose conflicting functional demands. The idea that organisms are important agents of extinction (50, 58, 61, 71, 89) is supported for island biotas exposed to continental invaders and perhaps for some instances among ants and large vertebrates, but not for most marine invertebrates, small vertebrates, or plants (79, 82).

Following pioneering work by Maynard Smith (49) and Stenseth & Maynard Smith (66), Rosenzweig and colleagues (59) modeled the evolution of interacting species by using assumptions collectively embodied in the so-called evolutionarily stable strategy (ESS). Traits are assumed to evolve toward a local optimum. In their model, Rosenzweig and colleagues (59) assumed that this optimal phenotype lies somewhere within the limits of the set of all possible phenotypes rather than at the extremes of the phenotypic range, because most phenotypes reflect adaptive compromises.

Although continuous evolution of the kind predicted by the Red Queen hypothesis is theoretically possible, a more likely outcome of the model by Rosenzweig and colleagues (59) is a mutual adaptational stalemate (60, 66). The only way to stimulate evolution among the interacting species is to introduce changes in the rules governing adaptive compromise. Such design changes are most likely when trade-offs among incompatible functions are reduced or eliminated (59, 60).

This can be achieved in several ways. One is by adaptive breakthroughs or "key innovations," changes in the developmental sequence (or rules of construction) that enable previously linked or covarying traits to vary independently (see also 10, 45, 62, 76). Another way is to enter into a mutually beneficial partnership with another organism (see also 47). Such partnerships enable animals to photosynthesize, vertebrates to digest fresh plant material, vascular plants to take up nutrients from the soil, hermit crabs to gain protection from sea anemones living on and increasing the size of their shell homes, and so on. Partnerships make for formidable competitors. Increases in metabolic rate provide still another means. With more energy available, fewer and less stringent compromises in energy allocation are necessary (79). Population expansion provides a fourth way to reduce the constraint of compromise. In expanding populations, costs associated with novel traits or novel combinations of traits are relatively low, because more individuals are able to survive than in populations that are stable or in decline.

The common element of these mechanisms is that the range of permissible phenotypes is increased. Moreover, all these mechanisms are subject to strong selection at the level of individual organisms. Margulis (47) has expressed the view that partnerships involving the genetic integration of two or more independently evolved organisms represent a departure from evolution by natural selection, but Maynard Smith (51) persuasively argues that such partnerships are subject to selection in the same way that other phenotypes are, provided that genetically intimate partnerships behave as units of evolution (that is, as entities characterized by multiplication, variation, and heredity).

The appearance of changes in design may be critical in propelling major episodes of escalation. For example, the evolution of predators may have led to the evolution of mineralized skeletons and of burrowers near the beginning of the Cambrian (81, 70). Partnerships between mycorrhizal fungi and vascular land plants may have set the stage for greater exploitation of the dry land by organisms beginning in the Devonian (64). Elsewhere I propose that these changes in design, which typically require high rates of supply of energy and nutrients, are most likely when extrinsic factors (especially the warming and nutrient enrichment associated with submarine volcanism and tectonically related phenomena) cause productivity to rise. The stimulating effects of extrinsic triggers are therefore greatly amplified through positive feedback by intrinsic improvement on per-capita performance.

An interesting case in which continuous evolution of the type predicted by the Red Queen hypothesis could occur is that of tightly coupled reciprocal coevolution involving chemical deterrents. Any new chemical variant may confer a survival advantage to a plant that is being exploited by a trophically specialized herbivore. If the new chemical is in some way harmful to the herbivore, any trait of the herbivore that reduces this harmful effect or puts the chemical to use as its own defense will be favored. Once reciprocal evolution has occurred, further alteration or abandonment of the chemical in the plant may begin the coevolutionary process anew. This kind of coevolution could occur continuously and, because there is no long-term genetic memory for previously abandoned defenses, need not result in a long-term directionality of adaptation. Rosenzweig and colleagues (59) point out, however, that the conditions required for this kind of aimless yet continuous evolution may be very seldom satisfied in nature.

The Life-Dinner Principle and Cost-Benefit Models

Dawkins & Krebs (20) took a cost-benefit approach to adaptation when they formulated the life-dinner principle to describe coevolution. They suggested that there is an inherent asymmetry in the evolutionary outcome for predator and prey in encounters between the two. Success for the prey means life (survival), whereas failure means death or, perhaps more commonly, injury. For the predator, success means another meal, whereas failure translates into postponing a meal. The stakes for the prey are thus usually higher than for the predator, although obvious exceptions come to mind. Failure for the predator might entail injury or, if the predator had not eaten for some time and expended much effort to acquire its prey, even death.

I took Dawkins & Krebs's (20) argument to imply that coevolution between predator and prey would be highly asymmetrical, with the prey responding more rapidly and more effectively to the predator than the predator does to the prey (77–79; see also 41 for an excellent example involving bivalved prey eaten by drilling naticid gastropods). For this reason, I accorded a minor role to coevolution between predator and prey, and I argued instead that both parties respond more effectively to their respective enemies than they do reciprocally (79).

Abrams (1–5) has taken issue with Dawkins & Krebs's (20) and my analyses. He modeled predator-prey coevolution by employing principles of cost-benefit analysis and population dynamics. For the prey, the per-capita rate of population growth was expressed in terms of the prey's food resources and in terms of the effect of the predator on the prey population density. The predator's per-capita population growth was expressed only in terms of the available prey resource. Out of 24 possible cases Abrams (1) examined, 20 show adaptive responses by the prey to improvements in prey capture by the predator; whereas 16 show responses by the predator to antipredatory adaptation in the prey. In 14 cases there is the potential for predator-prey coevolution. The outcomes depend on how improvements affect birth and death rates. These results showed that, although prey were more likely to respond to predators than predators were to prey, coevolution (a predator-prey arms race) is by no means inevitable when predators and prey interact. Abrams (1) also showed that the life-dinner principle would not apply when predators are rare and when their selective impact on the prey is small.

If selection by organisms depends on encounters between them, then one measure of the effectiveness of individuals and of their attributes during such encounters is the probability of success (77–79). From the prey's point of view, adaptive improvement means that the number of encounters during which the prey succeeds in not being eaten by the predator will increase relative to the total number of encounters with the predator. For the predator, improvement by the prey translates into a reduced proportion of successful attacks. I suggested further that improvement by the predator could be measured as an increase in the number of successful attacks on the prey relative to the total number of encounters (77–79).

Abrams (2), however, has convincingly argued that this increase does not necessarily imply improvement in the predator, and further points out that improvement by the predator often accompanies a decrease in the proportion of successful encounters. His argument may be summarized as follows. If failed encounters from the point of view of the predator are of low cost in energy or time and involve little risk to the predator, evolutionary changes that result in a higher absolute number of prey killed may be accompanied by a reduced success rate. As long as enough prey are killed, it may matter little to the predator's economic budget how many times the predator fails. Improvement, Abrams (2) and Kitchell (43) argue, should be measured in terms of absolute food intake. If predators do increase food intake while at the same time decreasing the proportion of successful attacks, both predator and prey will improve with respect to the predator-prey interaction. The result would therefore be what many observers would interpret as predator-prey coevolution in which both parties benefit.

Such positive feedback between interacting parties may be very common. For example, frequent failures by predatory crabs to break prey snail shells may be energetically cheap and take relatively little time. By increasing the rate of encounters without sacrificing efficiency, predators will increase their food intake and subject a greater proportion of the shell-bearing prey population to attacks. The larger the number of unsuccessful encounters from the point of view of the predator, the better adapted the prey is to the predator.

Interactions between drilling gastropod predators and bivalved prey provide the basis for coevolutionary models (21, 42, 43) whose results have been claimed to conflict with expectations derived from the hypothesis of escalation (43). These models assume a close and specialized link between predator and prey, as well as the existence of functional trade-offs among growth rate, reproductive output, and shell thickness (a measure of prey defense). Another central assumption is that each species maximizes energy intake. Evolutionary change is generated by the circumstance that energy maximization by interacting species is attempted but never fully realized.

The models indicate that patterns of energy allocation to growth and reproduction vary according to the intensity of predation. A low intensity of predation favors early reproduction by the prey and little allocation toward growth and shell thickening; whereas more intense predation favors an emphasis on rapid growth, shell thickening, and delayed reproduction (43). These predicted changes in energy allocation in the prey occur even if the predator's pattern of prey selection remains the same. Unidirectional response should therefore not be expected in predator-prey coevolution (43). The important point is that responses of interacting parties depend on the population densities and therefore on the frequency of interaction of the species involved (see also 5).

A good example of a complex response is offered by the evolution of cephalopods (15, 79, 86, 87). From the Late Cambrian (when cephalopods first appeared) to the Late Cretaceous, most cephalopods had an external shell, which functioned for protection as well as for buoyancy control. During this long interval, cephalopods exhibited many trends toward increased passive protection (greater sculpture, elaboration of the aperture, elaboration of septa separating internal shell chambers), as well as trends toward greater maneuverability and higher absolute speed. However, fundamental incompatibilities exist among buoyancy control, passive protection, and locomotor performance. In the Late Cretaceous, cephalopods had reached the limits of adaptational compromise in the functional design of the external shell as predators and competitors continued to become faster, more powerful, and more numerous. With further intensification of predation and competition, especially by fishes, only those cephalopod lineages that had abandoned the shell persisted and rediversified after the mass extinction of the end-Cretaceous. Without the shell, cephalopods were able to exploit high-energy modes of life, including rapid jet-propulsion and effective buoyancy control. Thus, although the rules of engagement may have remained broadly constant throughout the history of cephalopods, the increasing intensity of predation and competition eventually made previously workable adaptive solutions obsolete and disallowed the resurgence of long-established trends. The latter were replaced by trends emphasizing speed and maneuverability.

Because most predator-prey interactions take place in communities composed of many species, their coevolutionary nature is potentially influenced by interactions with these species. I argued (79) that predators are unlikely to respond uniquely to any one prey species because most predators have a catholic diet and tend not to be specialized to a single prey species (see also 26, 27, 40). Moreover, the composition of communities often changes dramatically through time even if individual species do not change significantly (8). Abrams (4), however, argues from models that alternative prey often magnify the response of a predator to its main prey. The predator's own predators have a similarly enhancing effect on the predator's evolutionary response to its principal prey (4).

Theory Versus Observation

Despite claims that the conclusions derived from the models of DeAngelis et al (21) and Kitchell (42, 43) conflict with the predictions of the hypothesis of escalation, these and other coevolutionary hypotheses as well as the escalation hypothesis incorporate ideas about functional trade-offs and positive feedback. Not surprisingly, therefore, observed patterns of evolution of interacting species are consistent with most if not all of the proposed models. The various models emphasize different aspects of interaction, but they are in most respects logically equivalent.

Conformity with reality does not, however, imply that the coevolutionary models are adequate or relevant. Their utility is compromised by unwarranted assumptions and by the even more fundamental problem that they do not consider the process of selection, which is after all supposed to be responsible for the evolutionary dynamics of interacting species. Abrams's (1–5) models, for example, incorporate the effects of predators on prey populations, but they do not take into account the effects of the predator's own enemies. The models by DeAngelis and his colleagues (21, 42, 43) assume a tight reciprocal linkage between predator and prey; yet, both predator and prey interact with a host of other species (17, 40), and in the system of drilling predators and bivalved prey for which these models were specifically designed, there is no evidence of reciprocal interaction. Instead, Kelley (41) finds in her exceptionally careful analysis of fossil evidence from the Miocene of Maryland that, although evolutionary changes in clams are interpretable as responses to drilling predators, the main changes in the predators are interpretable as defenses against the snails' own enemies. All the models treat selection very indirectly by incorporating its effects on population dynamics of interacting species.

An important question arising from the models, especially those of Abrams (1–5) and Kitchell (42, 43), is whether positive feedback between interacting populations is equivalent to coevolution (that is, reciprocal adaptive response). Is selection in favor of higher food intake by predators caused by the prey, or is it due to the predator's own enemies? The answer is likely to be complex. The evolutionary role of interactions among species depends on the costs and benefits involved. Minimal costs and minimal benefits will have little effect; high costs and large benefits make a big difference. For a very hungry predator, the cost of failure to secure a given prey may be very high; for a sated predator, the cost is apt to be low. Kitchell (43) is therefore correct in pointing out that, even if the pattern of selection (the traits of the winners as compared to the traits of the losers) remains unaltered, the effect of this selection depends on how costs and benefits change with such ecological variables as population density of predators relative to that of the prey.

For predatory mammals, fossil evidence points to a large role of competitors in the evolution of equipment such as claws, saber-like canines, and other weaponry for securing and gorging prey rapidly (7, 74, 75). West and her colleagues (88) argued that in the case of crabs and prey molluscs in Lake Tanganyika, the prey may be the more important agents of selection in view of the fact that crabs do not fight much among themselves. Whether this interpretation is correct, however, remains to be seen. Marine crabs with similarly enlarged claws frequently do fight and use their claws in defense against competitors and predators, as well as in overcoming the armor of prey molluscs.

Important as the insights offered by coevolutionary models are, I believe the models miss the essential distinction between coevolution and escalation. Populations of species may have positive effects on one another, but such effects do not imply reciprocal evolution. Population dynamics are not the same thing as selection. Given that coevolution and escalation are fundamentally about the nature, source, and intensity of selection, no models describing the population dynamics of interacting species can adequately describe selection-based processes such as coevolution and escalation.

COEVOLUTION OR ESCALATION: A DEBATE ABOUT SELECTION

A central issue in the debate about escalation and coevolution is the nature of selection. Most studies of adaptation and natural selection emphasize the net genetic or phenotypic outcome of the process rather than the way in which selection occurs. Yet, if we are to distinguish empirically between coevolution and escalation, it is essential to ascertain when selection occurs, how frequently selection takes place, and especially which agents or agencies are responsible.

My view is that selection is episodic. It occurs as events during which an individual organism has an encounter with an agency capable of influencing the individual's survival or reproduction. These events can be observed and counted, and their outcomes can be evaluated in terms of success or failure. A more difficult problem is to assess the cost of failure and the benefit of success. Some encounters may be inconsequential, whereas others mean the difference between death and survival or between reproductive failure or success. A coevolutionary interpretation implies that the selection that occurs during encounters between the two species (or groups of species) in question is more intense or more effective than is selection due to other agents. An interpretation of escalation, on the other hand, requires that selection due to enemies takes precedence over that effected by a food species, unless that food species is also a potential enemy.

As an example, consider the evolution of high speed in running. Speed in principle provides the obvious advantage of escape from predators; it also enables individuals to maintain larger territories and to arrive at food sources or mates before potential competitors do. If both predator and prey show increases in speed, a coevolutionary interpretation is tempting (see for example 7). A similar pattern could be obtained if both the prey and its predator were responding evolutionarily to their own enemies. The debate cannot be settled through modeling of predator-prey dynamics; a resolution depends entirely on empirical observations of both the prey and the predator. When do these animals use their high speed, and what are the outcomes of individual episodes of running? At least for lizards, the tentative answer to this question is that high speed is used mainly in escapes from predators (38).

Similar questions apply to the evolution of host-guest relationships. Small, poorly defended organisms may find substantial shelter from enemies by entering evolutionarily into intimate partnerships with larger, well-defended hosts. Although many of these partnerships are parasitic in nature, others have evolved into mutualisms, in which the guest provides competitive or defensive benefits to its host. These mutualisms imply an element of coevolution or mutual accommodation between host and guest, but the extent to which this partnership has evolved or become specialized may be determined strongly by selection imposed by enemies of both the host and the guest (for more or less similar views see 9, 14, 29). A better empirical understanding of the origin of intimate partnerships requires careful observations of encounters that the host and guest have with each other and with other species.

In order to make headway in the study of coevolution and escalation, we need to study the sources, frequencies, and cost-benefit effects of selection. This entails careful observation of encounters between species, together with an evaluation of the effects of such encounters on survival and reproduction.

Models should be constructed not only by expressing the net outcome of selection in terms of births and deaths or energy intake, but by evaluating systematically how encounters of various kinds affect opportunities for functional improvements among interacting species. It is important to express the encounter rate not merely in terms of population densities of the inter-acting species, but to scale these densities according to metabolic rates and, in the case of predators, according to handling time. High-energy species (those with high metabolic rates) are apt to encounter members of other species more frequently than do low-energy species (79). Predators that take a long time to find, pursue, and subdue prey will have a low effective encounter rate with prey even if their population density is high (3). Encounter rates, and therefore the opportunities for coevolutionary response or escalation, depend on population densities as well as on the metabolic properties of the species involved.

Because survival and reproduction require only that a given individual's genes make it more fit than its neighbors, models based on optimality theory should in my view be abandoned. Optimality in its many forms assumes that there exists a single "best" phenotype toward which selection "strives." It may be that individuals in some circumstances are extraordinarily well adapted, and that they therefore approach some engineering standard of perfection; but under other circumstances, individuals are sloppily constructed and would, on engineering grounds at least, seem to fall far short of the "best" or even a good design. Imperfection and errors abound in nature (28, 77). Optimality implicitly assumes that there are absolute standards of design, and that the further organisms are from meeting the specifications the faster evolution will proceed toward the optimum (49). The fact that degrees of specialization among ecologically similar species vary widely in the world today strongly implies to me that design standards in nature are relative, not absolute (83). Given that the correspondence between structure and function is often far from precise (6, 23, 32, 36), we should expect organisms to embody ad hoc and often rather clumsy solutions to functional demands, solutions that bear a deep stamp of history and ancestry. As long as they work, they will not be disadvantageous, and they will not be purged until a better solution comes along, usually in the bodies of individuals belonging to an entirely different evolutionary line. Even then, the tendency for incumbents to prevail may prevent superior solutions from gaining a foothold (60). Some unusual opportunity, such as an extinction or a circumstance favorable to population increase in many species simultaneously, may be necessary to permit more effective designs to become established.

CONCLUDING REMARKS

Debates about the role of selection in evolution and about the role of competition in selection are important. They should not be allowed to end in a tangle of unstated, unwarranted, or untested assumptions and partisan rhetoric. The modeling efforts by Abrams (1–5) and others (21, 42, 43, 49, 60) have yielded some important insights, but the roles of coevolution and escalation and the essential distinction between these two processes remain matters of empirical observation of living species, and detailed studies of interactions among fossil species over long intervals of time will go a long way toward understanding how interactions among species have affected the course of evolution.

Literature Cited

1. Abrams PA. 1986. Adaptive responses of predators to prey and prey to predators: the failure of the arms-race analogy. *Evolution* 40:1229–47
2. Abrams PA. 1989. The evolution of rates of successful and unsuccessful predation. *Evol. Ecol.* 3:157–71
3. Abrams PA. 1991. The evolution of anti-predator traits in prey in response to evolutionary change in the predators. *Oikos* 59:147–56
4. Abrams PA. 1991. The effects of interacting species on predator-prey coevolution. *Theor. Pop. Biol.* 39:241–62
5. Abrams PA. 1992. Predators that benefit prey and prey that harm predators: unusual effects of interacting foraging adaptations. *Am. Nat.* 140:573–600
6. Alexander RM. 1991. Apparent adaptation and actual performance. *Evol. Biol.* 25:357–73
7. Bakker RT. 1983. The deer flees, the wolf pursues: incongruencies in predator-prey coevolution. See Ref. 31, pp. 350–82
8. Bennett KD. 1990. Milankovitch cycles and their effects on species in ecological and evolutionary time. *Paleobiology* 16:11–21
9. Bernays E, Graham M. 1988. On the evolution of host specificity in phytophagous arthropods. *Ecology* 69:886–92
10. Bock WJ. 1959. Preadaptation and multiple evolutionary pathways. *Evolution* 13:194–211
11. Boucot AJ. 1975. *Evolution and Extinction Rate Controls*. Amsterdam: Elsevier
12. Boucot AJ. 1983. Does evolution take place in an ecological vacuum? *J. Paleont.* 57:1–30
13. Boucot AJ. 1985. Silurian-Early Devonian biogeography, provincialism, evolution and extinction. *Philos. Trans. R. Soc. Lond. (B)* 309:323–39
14. Brower LP. 1958. Bird predation and foodplant specificity in closely related procryptic insects. *Am. Nat.* 92:183–87
15. Chamberlain JA Jr. 1991. Cephalopod locomotor design and evolution: the constraints of jet propulsion. In *Biomechanics in Evolution*, ed. JMV Rayner, PJ Wootton, pp. 57–98. Cambridge: Cambridge Univ. Press
16. Collins T, Fraser K, Palmer AR, Vermeij GJ, Brown W. 1994. Evolutionary history of Northern Hemisphere *Nucella* (Gastropoda, Muricidae): Molecules, morphology, ecology and fossils. In review
17. Commito JA. 1987. *Polinices* predation patterns and *Mercenaria* morphology models. *Am. Nat.* 129:449–51
18. Cracraft J. 1985. Biological diversification and its causes. *Ann. Missouri Bot. Garden* 72:794–822
19. Darwin C. 1859. *The Origin of Species by Natural Selection or The Preservation of Favored Races in the Struggle for Life.* New York: Colliers
20. Dawkins R, Krebs JR. 1979. Arms races between and within species. *Proc. R. Soc. Lond. (B)* 205:489–511
21. DeAngelis DL, Kitchell JA, Post WM. 1985. The influence of naticid predation on evolutionary strategies of bivalve prey: conclusions from a model. *Am. Nat.* 126:817–42
22. Des Marais DJ, Strauss H, Summons RE, Hayes JM. 1992. Carbon isotope evidence for the stepwise oxidation of the Proterozoic environment. *Nature* 359:605–9
23. Dudley R, Gans C. 1991. A critique of symmorphosis and optimality models in physiology. *Physiol. Zool.* 64:627–37
24. Ehrlich PR, Raven PH. 1964. Butterflies and plants: a study in coevolution. *Evolution* 18:586–608
25. Fischer AG. 1984. Biological innovations and the sedimentary record. In *Patterns of Change in Earth Evolution,* ed. HD Holland, AF Trendall, pp. 145–57. Berlin: Springer
26. Fox LR. 1981. Defense and dynamics in plant-herbivore systems. *Am. Zool.* 21:853–64
27. Fox LR, Morrow PA. 1981. Specialization: species property or local phenomenon? *Science* 211:887–93
28. Frazzetta TH. 1970. From hopeful monsters to bolyerine snakes? *Am. Nat.* 104:55–71
29. Futuyma DJ. 1983. Evolutionary interactions among herbivorous insects and plants. See Ref. 31, pp. 207–31
30. Futuyma DJ, Slatkin M. 1983. Introduction. See Ref. 31, pp. 1–13
31. Futuyma DJ, Slatkin M. 1983. *Coevolution.* Sunderland, Mass: Sinauer
32. Garland T Jr, Huey RB. 1987. Testing symmorphosis: does structure match functional requirements? *Evolution* 41:1404–9
33. Gould SJ. 1985. The paradox of the first tier: an agenda for paleobiology. *Paleobiology* 11:2–12
34. Gould SJ. 1988. Trends as changes in variance: a new slant on progress and directionality in evolution. *J. Paleontol.* 62:319–29

35. Gould SJ. 1990. Speciation and sorting as the source of evolutionary trends, or "things are seldom what they seem." See Ref. 55, pp. 3–27

36. Gould SJ, Lewontin RC. 1979. The spandrels of San Marco and the panglossian paradigm: a critique of the adaptationist programme. *Proc. R. Soc. London (B)* 205:581–98

37. Guensburg TE, Sprinkle J. 1992. Rise of echinoderms in the Paleozoic evolutionary fauna: significance of paleoenvironmental controls. *Geology* 20:407–10

38. Hertz PE, Huey RB, Garland T Jr. 1988. Time budgets, thermoregulation, and maximal locomotor performance: are reptiles olympians or Boy Scouts? *Am. Zool.* 28:927–38

39. Hoffman A. 1989. *Arguments on Evolution: A Paleontologist's Perspective.* New York: Oxford Univ. Press

40. Howe HF. 1984. Constraints on the evolution of mutualisms. *Am. Nat.* 123:764–77

41. Kelley PH. 1992. Coevolutionary patterns of naticid gastropods of the Chesapeake Group: an example of coevolution? *J. Paleontol.* 66:794–800

42. Kitchell JA. 1986. The evolution of predator-prey behavior: naticid gastropods and their molluscan prey. In *Evolution of Animal Behavior: Paleontological and Field Approaches,* ed. MH Nitecki, JA Kitchell, pp. 88–110. New York: Oxford Univ. Press.

43. Kitchell JA. 1990. The reciprocal interaction of organism and effective environment: learning more about "and." In *Causes of Evolution: A Paleontological Perspective,* ed. RM Ross, WB Allmon, pp. 151–69. Chicago: Univ. Chicago Press

44. Knoll AH. 1992. The early evolution of eukaryotes: a geological perspective. *Science* 256:622–27

45. Lauder GV. 1981. Form and function: structural analysis in evolutionary morphology. *Paleobiology* 7:430–42

46. Losos JB. 1992. The evolution of convergent structure in Caribbean *Anolis* communities. *Syst. Biol.* 41:403–20

47. Margulis L. 1991. Symbiogenesis and symbionticism. See Ref. 48, pp. 1–14

48. Margulis L, Fester R. 1991. *Symbiosis as a Source of Evolutionary Innovation: Speciation and Morphogenesis.* Cambridge, Mass: MIT Press

49. Maynard Smith J. 1976. What determines the rate of evolution? *Am. Nat.* 110:331–38

50. Maynard Smith J. 1989. The causes of extinction. *Philos. Trans. R. Soc. London (B)* 325:241–52

51. Maynard Smith J. 1991. A Darwinian view of symbiosis. See Ref. 48, pp. 26–39

52. McCune AR. 1982. On the fallacy of constant extinction rates. *Evolution* 36:610–14

53. McKinney FK, Jackson JBC. 1991. *Bryozoan Evolution.* Chicago: Univ. Chicago Press. 2nd ed.

54. McKinney ML. 1990. Classifying and analyzing evolutionary trends. See Ref. 55, pp. 28–58

55. McNamara KJ. 1990. *Evolutionary Trends.* Tucson: Univ. Ariz. Press

56. Pearson PM. 1992. Survivorship analysis of fossil taxa when real-time extinction rates vary: the Paleogene planktonic Foraminifera. *Paleobiology* 18:115–31

57. Reid DG. 1990. A cladistic phylogeny of the genus *Littorina* (Gastropoda): implications for evolution of reproductive strategies and for classification. *Hydrobiologia* 193:1–19

58. Ricklefs RE, Cox GW. 1972. Taxon cycles in the West Indian avifauna. *Am. Nat.* 106:195–219

59. Rosenzweig ML, Brown JS, Vincent JL. 1987. Red Queens and ESS: the coevolution of evolutionary rates. *Evol. Ecol.* 1:59–94

60. Rosenzweig ML, McCord RD. 1991. Incumbent replacement: evidence of long-term evolutionary progress. *Paleobiology* 17:202–13

61. Roughgarden J. 1983. The theory of coevolution. See Ref. 31, pp. 33–64

62. Schaeffer B, Rosen DE. 1961. Major adaptive levels in the evolution of the actinopterygian feeding mechanism. *Am. Zool* 1:187–204

63. Signor PW, Vermeij GJ. 1994. The plankton and the benthos: origins and early history of an evolving relationship. *Paleobiology* 20: In press

64. Simon L, Bousquet J, Levesque RC, Lalonde M. 1993. Origin and diversification of endomycorrbizal fungi and coincidence with vascular land plants. *Nature* 363:67–69

65. Stanley SM. 1973. An explanation for Cope's Rule. *Evolution* 27:1–26

66. Stenseth NC, Maynard Smith J. 1984. Coevolution in ecosystems: Red Queen evolution or stasis? *Evolution* 38:870–80

67. Sterrer W. 1992. Prometheus and Proteus: the creative, unpredictable individual in evolution. *Evol. Cognition* 1:101–29

68. Thayer CW. 1983. Sediment-mediated biological disturbance and the evolution

of marine benthos. In *Biotic Interactions in Recent and Fossil Benthic Communities*, ed. MJ Tevesz, PL McCall, pp. 479–625. New York: Plenum

69. Turner JRG. 1983. Mimetic butterflies and punctuated equilibria: some old light on a new paradigm. *Biol. J. Linn. Soc.* 20:277–300

70. Valentine JW, Awramik SM, Signor PW, Sadler PM. 1991. The biological explosion at the Precambrian-Cambrian boundary. *Evol. Biol.* 25:279–356

71. Van Valen L. 1973. A new evolutionary law. *Evol. Theor.* 1:1–18

72. Van Valen L. 1976. Energy and evolution. *Evol. Theor.* 1:179–229

73. Van Valen L. 1983. How pervasive is coevolution? See Ref. 31, pp. 1–19

74. Van Valkenburgh B. 1991. Iterative evolution of hypercarnivory in canids (Mammalia: Carnivora): evolutionary interaction among sympatric predators. *Paleobiology* 17:340–62

75. Van Valkenburgh B, Hertel F. 1993. Tough times at La Brea: tooth breakage in large carnivores of the Late Pleistocene. *Science* 261:456–59

76. Vermeij GJ. 1973. Adaptation, versatility, and evolution. *Syst. Zool.* 22:466–77

77. Vermeij GJ. 1982. Unsuccessful predation and evolution. *Am. Nat.* 120:701–20

78. Vermeij GJ. 1983. Intimate associations and coevolution in the sea. See Ref. 31, pp. 311–27

79. Vermeij GJ. 1987. *Evolution and Escalation: An Ecological History of Life.* Princeton: Princeton Univ. Press

80. Vermeij GJ. 1989. Evolution in the long run. *Paleobiology* 15:199–203

81. Vermeij GJ. 1990 (1989). The origin of skeletons. *Palaios* 5:585–89

82. Vermeij GJ. 1991. When biotas meet: understanding biotic interchange. *Science* 253:1099–1104

83. Vermeij GJ. 1993. *A Natural History of Shells.* Princeton: Princeton Univ. Press

84. Vrba ES. 1980. Evolution, species and fossils: how does life evolve? *South Afr. T. Sci.* 76:61–84

85. Vrba ES. 1983. Macroevolutionary trends: new perspectives on the roles of adaptation and incidental effect. *Science* 221:387–89

86. Ward P. 1986. Cretaceous ammonite shell shapes. *Malacologia* 27:3–28

87. Wells MJ, O'Dor RK. 1991. Jet propulsion and the evolution of the cephalopods. *Bull. Mar. Sci.* 49:419–32

88. West K, Cohen A, Baron M. 1991. Morphology and behavior of crabs and gastropods from Lake Tanganyika, Africa: implications for lacustrine predator-prey coevolution. *Evolution* 45:589–607

89. Wilson EO. 1961. The nature of the taxon cycle in the Melanesian ant fauna. *Am. Nat.* 95:179–93

Annu. Rev. Ecol. Syst. 1994. 25:237–62

SYSTEMATICS OF THE CORAL GENUS *ACROPORA*: Implications of New Biological Findings for Species Concepts

C. C. Wallace

Museum of Tropical Queensland, 70-84 Flinders Street, Townsville 4810, Australia

B. L. Willis

Marine Biology Department, James Cook University, Townsville, 4811, Australia

KEY WORDS: systematics, corals, hybridization, biogeography

Abstract

The large coral genus *Acropora* occurs throughout the world's reefs and is potentially a model for evolution and development of modern reef faunas. New research including breeding trials and genetic analyses of sympatric populations of *Acropora* and other corals is suggesting misalignments of breeding, morphological and genetic boundaries such that species limits may be sometimes narrower, sometimes broader, than presently perceived. Ongoing biogeographic and phylogenetic analyses are reexamining coral species in space and time and generating hypotheses about the origination of species. Synthesis of new findings from these research areas with preliminary insights from molecular data on species boundaries and phylogenies is allowing assessment of the current taxonomic framework of *Acropora* and of the order Scleractinia. The tacit assumption that currently defined coral species encompass biological, evolutionary, and phylogenetic species concepts may be unfounded.

INTRODUCTION

> *The question is, how well do various discontinuities correspond: i.e. are the same sets of organisms delimited by discontinuities when we look at morphology, as when we look at ecology, or breeding?*
> Mishler & Donoghue 1982

237

It is widespread, abundant, and variable forms, such as these [Acropora], *which are the despair of old-fashioned systematists, but serve as stimulating problems for the modern naturalist....*

Haddon 1894

Systematic research is an interactive process in which taxa are defined or redefined by synthesis of all available information from biological, molecular, and other relevant areas of science (142). Systematics attempts to keep pace with developments in these fields so that the greatest possible accuracy of taxonomic interpretation will facilitate the greatest possible accuracy of research and experimental design. The taxonomist is charged with describing and naming species taxa according to the rules of nomenclature (86), while at the same time attempting to meet the requirements of a currently acceptable species concept. Both species descriptions and species concepts[1] constitute testable hypotheses, and both should be changing over time.

Times of intensified research, new directions, and new findings are important for assessment, development, and change of a taxonomic framework; such a time exists now for coral systematics. Following discovery of mass broadcast spawning in corals and overturning of the dogma of brooding as the dominant mode of coral sexual reproduction (6, 35, 36, 138), field-based taxonomic revisions (44, 108, 110–112, 117, 133, 144), and preliminary applications of molecular methodologies (4, 62–64, 89), the last decade has seen a long-awaited expansion of natural history research on corals as well as more novel experimental, breeding, genetic and biogeographic studies. The results of such studies must be continually assessed beside new developments relating to species concepts (7, 14, 18, 19, 56, 69, 71, 76, 103, 113), the process of speciation (13, 15, 22, 28), tests of species boundaries (1, 41, 76), and methods of phylogenetic and biogeographic interpretation (46, 70).

Large, diverse, and abundant genera have the greatest potential to provide new systematic information and yet are the potential source of greatest inaccuracies. We have chosen the coral genus *Acropora* as a model for evolution of modern reef faunas because it contains the greatest number of species, occurs in all tropical oceans, and dominates most reef habitats (29, 30, 55). *Acropora* has always carried problems of identification (84, 117, 122, 127), yet it is often the obvious choice for research. As the most documented participants in coral mass gamete-releasing events (6, 16, 24, 35, 39, 95, 96, 118, 138), its species provide much of the material for genetic, breeding, and other comparative and experimental studies capitalizing on the availability of gametes during the annual mass spawning of corals.

---

[1]For clarity of discussion, we reiterate Ereshefsky's distinction (23) between species taxa (groups of organisms), the species category (the class of all species taxa), and species concepts (philosophical constructs that attempt to define the species category).

Recent studies indicate some resolutions but also new taxonomic dilemmas within this genus. In this paper we review work in progress, particularly on *Acropora*, which is providing new data for testing species boundaries and species concepts in corals and which has the potential to lead to profound changes in the way we interpret the ecology and evolution of reefs and reef organisms. While a myriad of different kinds of research is contributing to the overall basis of coral systematics, we explore in particular:

1. breeding trials and their potential to question species boundaries and concepts using sympatric populations;
2. biogeographic analyses because of their potential to explore species boundaries using allopatric populations and thus to add another dimension— space—to considerations of species concepts;
3. phylogenetic analyses because of their emphasis on genealogy and the dimension of time and, in more practical terms, their potential for closer examination of characters and character polarity;
4. genetic studies and their potential to examine the molecular basis of inheritance and to be an independent assessment of species boundaries based on other criteria.

HYPOTHESIS-TESTING IN CORAL TAXONOMY

Many examples of hypothesis-testing in coral taxonomy have been presented and reviewed previously (45, 49, 53, 54, 136). We describe below some examples of experiments that followed publication of taxonomic revisions of the corals of eastern Australia (108–112). These revisions used an operational intraspecific category *ecomorph*, defined as having "intraspecific skeletal variations phenotypically and/or genotypically determined in response to specific ecological conditions" (108). To identify the source of variation would require hypothesis-testing on a species-by-species basis. Cases investigated to date suggest that the term *ecomorph* refers to many situations including inter- as well as as intraspecific variation and also that presently accepted coral species encompass a gradation of conditions ranging perhaps from "pseudo-sibling" species (49) to species that should be combined because their breeding ranges transcend morphological discontinuities.

In *Turbinaria mesenterina,* transplant experiments showed that "flat" and "convoluted" ecomorphs represent a phenotypic response to a depth-related factor, probably light intensity, whereas the same experimental approach, combined with genetic analysis, demonstrated that specific morphologies in *Pavona cactus* were associated with specific genotypes (135, 137). Physiological responses to light intensity, corroborated by morphometric revision, indicate that the *"mordax"* and *"pistillata"* ecomorphs of *Stylophora pistillata* should be recognized once again as separate species (27). An investigation of a novel breeding system, pseudogynodioecy, in the species *Galaxea fascicularis,* showed differences in spawning time which, followed by reexamination of morphologies, revealed that the existence of *G. alta* Nemenzo, 1980 on the Great Barrier Reef had been overlooked, although the two had not been synonymized (34).

Recognition of two morphologically close species *Goniastrea favulus* and *G. aspera* was validated by a finding of different breeding behaviors and gamete buoyancies in these two species (5). Similarly, recognition of two morphologically close species *Acropora (Isopora) palifera* and *A. (I.) cuneata* (both with considerable colony shape variability related to habitat—81, 112) is confirmed by genetic analysis and differences in timing of planular release (4, 52). Splitting of some species of *Platygyra* on the basis of morphological discontinuities (111), a vexed problem for some time (100, 133), is not justified according to genetic and breeding studies (67, 68). Breeding experiments reviewed below increase this list of tests of taxonomic hypotheses and indicate that the boundaries of species, assessed on the basis of discontinuities between morphological ranges, are sometimes too broad and sometimes not broad enough. Some breeding ranges may encompass morphological ranges far broader than those ever countenanced within the ecomorph concept.

THE SPECIAL CASE OF *ACROPORA*

Acropora was singled out by Wells (127, 128) as "the protean coral genus," a reference to the changeable god Proteus, because of its bewildering within-species variability coupled with a high degree of between-species similarity (84, 117, 122). *Acropora* provides an ideal subject for study of the nature and evolution of scleractinian reef coral species. With over 370 nominal species (112, 143) and around 150 valid species (129) even after extensive revision (112, 117), *Acropora* is by far the largest extant reef-building coral genus. It often provides the dominant coral cover of habitats or reefs in all three oceans (21, 59, 90). In the Indo-Pacific, its species are often: the dominant recruits to settlement plates (94, 119); the dominant prey item of *Acanthaster planci* and other corallivores (80, 125); the major contributors to mass spawning events (6, 16, 24, 35, 39, 95, 96, 118, 138); and structural components of reefal frameworks (e.g. 11, 17, 26, 55, 57, 59, 97). In the geological record, the genus first appears in the Eocene and is widely distributed by the Miocene (128). Records of up to 40 species of *Acropora* living sympatrically are not unusual (e.g. 21, 78, 121, 127). The genus thus provides opportunities for examining species concepts involving time and space, such as the evolutionary and phylogenetic concepts (19, 134), as well as the biological species concept that is strictly applicable only to sexual organisms living sympatrically at one point in time (14, 58).

The order Scleractinia (stony corals) has enjoyed a mostly stable classificatory arrangement since supra-specific revision based on skeletal characteristics, 50 years ago (105, 128). Within this scheme, *Acropora* stands out on two counts: the absence or extreme reduction of numerous plesiomorphic skeletal features such as dissepiments (layers of skeleton beneath the polyps) and columella (a skeletal formation below the center of a polyp), which provide diagnostic characters in other genera, and secondly, the synapomorphy of corallite dimorphism (121, 122). In no other coral genus is there an axial corallite (polyp skeleton) which forms the branch and buds off a second, radial type of corallite with differing morphology (with the exception of possibly nonhomologous axial growth in the ahermatype *Arcohelia*—130). The greater colony integration achieved through this polyp dimorphism has led to an enormous diversity of patterns of colony shape.

EXPLORING SPECIES BOUNDARIES AMONG SYMPATRIC POPULATIONS

Mating Systems in Mass-Spawning Corals: The Potential for Hybridization

From some points of view, hybridization studies are a Pandora's box releasing information challenging to notions of species boundaries, genetic cohesion, and the speciation process. The documented existence of natural hybrids in a broad spectrum of animals and plants (1, 87) has highlighted a role for hybridization in evolution (28, 37), and a need to consider hybrids in species concepts, reticulated phylogenies, and associated nomenclatural dilemmas (31, 41, 114). While most species concepts contain the idea of genetic cohesion, such cohesion is clearly subject to leakage in some cases through occasional or persistent hybridization leading to introgression, polyploidy, parthenogenesis, or diploid speciation, such that species boundaries are sometimes "semipermeable" (1).

In corals, since the discovery of mass spawning events (35), the potential for hybridization cannot be ignored (136). From many geographic locations, evidence exists for simultaneous or near-simultaneous (separated by hours or minutes) spawning of closely related species in *Acropora* and in other genera such as *Montipora, Platygyra, Favia,* and *Favites* (6, 16, 24, 35, 39, 95, 96, 118, 138). Mass gamete release by congeneric species represents an obvious opportunity for hybridization, as gametes released into the water column float to the sea surface where they reach high concentrations and can be formed into slicks which may last up to 24 hr (73, 141).

Most studies of hybrids start by detecting natural hybrids or hybrid zones

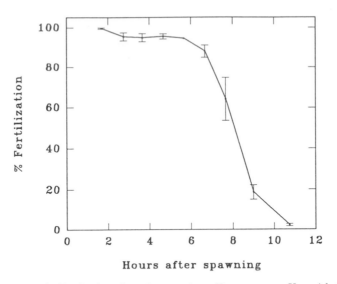

Figure 1 Percent fertilization in reciprocal crosses (eggs X sperm, sperm X eggs) between two colonies of *Acropora millepora*, showing effects of age on gamete viability (BL Willis, CK Wallace).

(38) and then explore their genetic composition, origin, and significance. There are few published suggestions that natural hybrids have been observed in corals (67, 117, p. 286), but innumerable publications refer to intergrading species or varieties and attempt to fit these into taxonomic and ecological context (e.g. 25, 68, 100, 108).

Among the species of *Acropora* involved in mass spawning, there are few obvious prezygotic mating barriers (20) or differences in recognition systems (76). All species of *A. (Acropora)* are hermaphroditic and produce positively buoyant bundles of eggs and sperm, which collect and then break apart at the sea surface (in contrast, species of the subgenus *A. (Isopora)* release brooded planula larvae; 2, 52). Gonads and gametes of different species are similar in size and structure (33, 48, 118, 120a), although it is possible that some morphometric differences in sperm ultrastructure may exist (33). Species specificity of egg/sperm recognition in *Acropora* and other coral genera is not as strong as that seen in most other marine invertebrate groups (R Miller, personal communication). A sperm attractant with incomplete species specificity has been isolated from the confamilial genus *Montipora* (12), and similar compounds that could be sperm attractants have been isolated but not tested in several *Acropora* species (B Bowden, personal communication). Many species spawn within one or two hours of each other (6)—adequate time for gamete viability, as eggs and sperm of *Acropora* are viable for up to 8 hr after release

(72; and a recent gamete viability study—Figure 1, cf 72). The obvious and unavoidable hypothesis is that species involved in mass congeneric spawning can interbreed (48, 85, 136, 139).

Breeding Trials

The potential for hybridization in mass-spawning corals has been tested repeatedly in a series of breeding trials on the Great Barrier Reef from 1989 to 1993 (139; B Stobart, KJ Miller, personal communication) and in Guam (85). Breeding trials on the Great Barrier Reef (298 trials, of which 250 were on *Acropora*) demonstrate overwhelmingly that many species of broadcast-spawning corals readily hybridize in controlled crosses, sometimes with very high fertilization success (140). In the same trials, examples of gamete incompatibility between colonies within presently defined coral species were also found, implying different mating types within populations.

Fertilization success within and between two species of *Acropora* illustrates the complexity found to be typical of scleractinian breeding systems (Figure 2). The species *A. millepora* and *A. pulchra* are morphologically distinct and belong to a group (112) that has tentatively been suggested to have natural hybrids (116, 117). In breeding trials between these two species, the following results have been documented (140).

HYBRIDIZATION BETWEEN MORPHOLOGICALLY DISTINCT SPECIES, INDICATING MISALIGNMENT OF MORPHOLOGICAL AND BREEDING BOUNDARIES   *Acropora pulchra* and *A. millepora* interbred with a mating success of 45%, just less than the mating success for within-species crosses of 56% and 48%, respectively (Figure 2), and hybrid offspring settled, grew, and survived for up to 15 months (BL Willis, CC Wallace, unpublished). A long-term grow-out program is attempting to establish sexual viability of these hybrids.

MATING INCOMPATIBILITIES BETWEEN COLONIES WITHIN PRESENT SPECIES, INDICATING BIOLOGICAL SPECIES WITH NARROW MORPHOLOGICAL BOUNDARIES   Whereas in hybrid crosses between *Acropora millepora* and *A. pulchra* the percentage of fertilization varied from 0% to 100% (Figure 2), the within-species crosses showed either no fertilization (less than 5%) or highly successful fertilization (i.e. 60–100%) (Figure 2). Further examination of the skeletal morphologies of the parents revealed that a different situation prevailed for each of the two species. In *Acropora millepora* two morphs could be consistently recognized ("thin branched" and "thick branched"), and these were not reproductively compatible. These probably conform respectively to the species *A. millepora* (Ehrenberg, 1834) and *A. squamosa* (Brook, 1893), previously combined on the basis of a graded series (117). In contrast, *A. pulchra* sometimes showed mating incompatibilities between morphologi-

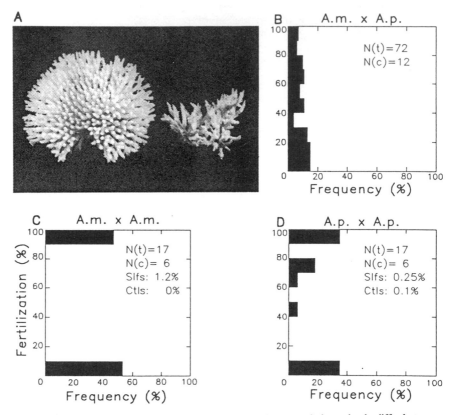

Figure 2 Breeding trials: (A) branching patterns and colony morphology clearly differ between *Acropora millepora (left)* and *A. pulchra (right)*; (B) percent fertilization for hybrid crosses between the two species; percent fertilization for within-species crosses of (C) *A. millepora* and (D) *A. pulchra* within-species crosses. *N(t)*: total number of trials (3 replicates per cross, except where some replicates were not recovered); *N(c)*: total number of colonies. (Redrawn from 140)

cally similar colonies. Because this species has a colony shape that is very conducive to cloning and because within-colony crossing (selfing) is negligible (Figure 3), it was first hypothesized that reproductively incompatible colonies were clonemates: However, genetic analysis disproved this (D Ayre, personal communication). The nature of the sexual barrier in this case is unknown.

BREEDING COMPATIBILITIES AMONG ACROPORA SPECIES   The *Acropora* (*Acropora*) species from the Great Barrier Reef are currently arranged in 14 species groups (112), which can be viewed as hypotheses of phylogenetic

	A. mil	A. pul	A. sel	A. ten	A. hya	A. cyt	A. nas	A. val	A. els	A. lon	A. for	A. nob
A. mil	49 (17)	45 (72)	40 (24)	0.2 (24)	0.6 (24)	1.5 (30)	0 (3)	0.2 (24)	0.5 (24)	0.2 (6)	2.2 (36)	0 (3)
A. pul		56 (17)			5.6 (12)	11 (23)	3.4 (6)		23 (12)	1.5 (24)	0.4 (12)	0.7 (6)
A. sel			94	1.4 (24)		12 (24)						
A. ten				98		0 (12)						
A. hy		Mean fertilization (%)			95 (12)	50 (48)	0 (6)	1.3 (24)	1.2 (24)	0.3 (12)		0 (6)
A. cyt						96 (18)	0.5 (12)		0.5 (12)	0.4 (24)		0.7 (12)
A. nas									0 (12)	0.7 (12)		0 (6)
A. val								93	6 (24)			
A. els									93	0.4 (48)		0.3 (12)
A. lon										92		0.1 (12)
A. for											99	
A. no												
Slf	1.2 (24)	5.2 (24)	0 (6)	0.5 (6)	0.6 (12)	0.9 (18)	0 (3)	13 (6)	4 (12)	1 (12)	0.5 (9)	0 (3)
Ctl	0 (30)	0.4 (24)	0 (6)	0 (6)	0.2 (12)	0.3 (18)	0 (3)	0 (6)	0.5 (12)	0 (12)	0.1 (9)	0 (3)

Legend (shading key, Mean fertilization %): 90 – 100; 40 – 60; 11 – 25; 1 – 10; < 1; (Not crossed)

Figure 3 Summary of mean fertilization success in crosses between 12 species of *Acropora*. Crosses along the diagonal represent within-species crosses. Double lines separate species into species groups (as per 112). Numbers in each square represent mean percent fertilization (central) and number of breeding trials performed (upper right). Abbreviations in column headings denote, in descending order: *A. millepora, A. pulchra, A. selago, A. tenuis, A. hyacinthus, A. cytherea, A. nasuta, A. valida, A. elseyi, A. longicyathus, A. formosa, A. nobilis,* self crosses (i.e. breeding trials involving eggs and sperm from the same colony), controls (eggs incubated in sperm-free sea water). (Redrawn from 140)

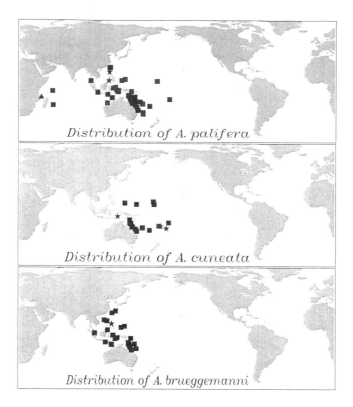

Figure 4 Distributions of Isoporan *Acropora* species based on specimen record (squares), type
localities of senior and junior synonyms (stars), and B. Riegel, personal communication (diamond).
(See discussion, p. 251)

relatedness (the hypothesis being that the closest relationships are between
species within each group and between adjacent groups) (124). In 250 crosses
involving 12 species of *Acropora* from 7 of these species groups, fertilization
occurred between 15 species pairs (Figure 3). Fertilization success was greatest
in hybrid crosses between morphologically similar species of *Acropora*, i.e.
the within-group pairs *A. cytherea/A. hyacinthus* and *A. pulchra/A. millepora*,
and the adjacent-group pair *A. selago/A. millepora.* Crosses between *A. elseyi*
and *A. pulchra, A. cytherea* and *A. selago,* and *A. cytherea* and *A. pulchra*
resulted in modest fertilization success (Figure 3), although species in these
pairs are separated by six, two, and one species group, respectively, in the
Veron and Wallace scheme (112). These findings suggest that hybridization
becomes less likely as morphological similarity decreases (and presumed phylogenetic distance increases) but see further discussions below.

Viability of Hybrids

Postzygotic viability of the offspring of hybrid crosses is not yet fully determined. Successful crosses develop into normal larvae and settle and survive at a rate similar to that of parental offspring (140; RC Babcock, KJ Miller, personal communication). Karyotypic analysis of an *Acropora pulchra/ millepora* cross shows that both parental and hybrid offspring have a chromosome number of 28 (JC Kenyon, personal communication), which is the common diploid number for corals studied so far (40, 48). However, unusual chromosome counts in some other species of *Acropora* suggest they may be polyploids generated from interspecific hybrization (JC Kenyon, personal communication). *Acropora* species take at least four years to reach reproductive maturity (118), and *Acropora* hybrids to date have been maintained for up to 15 months (140). Hybrid offspring of the brain corals *Platygyra daedalea* and *P. lamellina* survived and grew for three years (RC Babcock, KJ Miller, personal communication). These results indicate a good possibility that the laboratory-reared hybrids could survive to become adult corals, although their sexual viability is still undetermined. Although it seems likely that hybrid corals could be a significant component of adult coral assemblages, it is not yet known whether the requirement of the biologial species concept, that hybrid offspring be infertile, has been met.

MOLECULAR EVIDENCE OF GENE FLOW

Proliferation of molecular techniques is providing systematists with new data for evaluating species boundaries and genetic markers for interpreting processes of speciation (reviewed in 43). While analysis of biological macromolecules permits exploration of the molecular basis of inheritance, new sources of data may also pose new conflicts (42, 77)—for example when molecular evidence does not support taxonomic hypotheses constructed from morphological, distributional, and/or ecological analyses (88, 98, 102). There are many examples, particularly among vertebrates, in which the degree of morphological divergence between closely related species is either far greater or far less than would be predicted by the apparent degree of molecular divergence (115). If evolution at the morphological and molecular levels is uncoupled in corals, then our ability to formulate a species concept that fits the varied needs of taxonomists, ecologists, geneticists, and evolutionary biologists is greatly diminished. Here, we review molecular studies that are being used to explore species boundaries in scleractinian corals. The general expectation is that allozyme electrophoresis, molecular cytogenetics, restriction site analysis, and DNA/RNA sequencing will provide information appropriate for evaluating interspecies relationships (43). We concentrate on studies published since an earlier review (136).

MOLECULAR DATA SUPPORTING MORPHOLOGICAL EVIDENCE OF SPECIES BOUND-
ARIES   The few studies that have compared allozyme and morphometric data
for coral species have generally found that allozyme data support traditional
morphological interpretations of species boundaries. In some cases, insights
into genetic relatedness have suggested new taxonomic hypotheses that subsequent morphological or morphometric revision have corroborated. Within
the subgenus *Acropora* (*Isopora*), Ayre et al (1991) found that allozyme data
support the taxonomic distinctiveness of *Acropora palifera* and A. *cuneata,*
and provide a reliable means of separating the two species when morphological
criteria are ambiguous. Two species of *Montipora* (family Acroporidae), syn-
onomized on the basis of similarities in skeletal morphology (112), have
recently been separated because of fixed and frequency differences in alleles
at several loci (101), as well as breeding incompatibility in laboratory crosses
(140; B Stobart, personal communication) and differences in septal morphology (B Stobart, personal communication). Genetic, breeding, and morphological data were consistent across three populations. Differences in the degree
of genetic divergence detected by allozyme electrophoresis between morphs
of *Montastrea annularis* in the Panama (51) and Curacao (104) continue to
hamper decisions on species boundaries in this well-studied but taxonomically
difficult complex. Genetic, behavioral, growth rate, and oxygen isotope ratio
differences have been interpreted to suggest species-level distinctions between
the morphs in Panama (51), but genetic distances between the same morphs
in Curacao are more in accord with population-level differences (104).

Biochemical genetic studies of species within the genus *Porites* have now
been undertaken in a number of reef regions. Analysis of allozyme data confirm
the species status of five morphological species of *Porites* and two species of
Goniopora on the Great Barrier Reef (26a), although specimens of intermediate
morphology were not included in the analysis. Similarly, in a study of Caribbean and eastern Pacific species of *Porites* (126), electrophoretic groupings
agreed with morphological groupings. In a third study of the genus *Porites,*
however, allozyme data revealed the presence of cryptic species in Caribbean
populations (83). There is clearly a need for large-scale, regional biochemical
genetic studies of coral species to test the generality of findings described
above.

Preliminary data from restriction analysis of nucleic acids highlight the
potential usefulness of molecular techniques as taxonomic probes, particularly
when the function of a gene is known and related to a taxonomically significant
character. Although an unreplicated study using restriction analysis suggested
that the possession of one or two Taq I sites could be used to differentiate
between the morphologically distinct species A. *formosa* and A. *nobilis,* this
particular character was later found not to be unique and hence not useful as
a general taxonomic probe (64). Restriction analysis using "mini-collagen" (a

gene encoding a nematocyst protein) as a probe demonstrated that 13 of 14 colonies of *A. nasuta* had identical RFLP patterns. The morphology of the remaining colony indicated that it may have been misidentified (WG Wang, personal communication).

CONFLICTS IN DELIMITING SPECIES BOUNDARIES   In contrast to the general agreement between morphological and genetic data found in the above studies, boundaries defined by these two criteria apparently differ among species that hybridize experimentally (140; K. Miller, personal communication). In preliminary analyses, no fixed differences were found at 14 loci (of which 5 were polymorphic and 9 invariant) in each of two populations of *A. millepora* and *A. pulchra* (DJ Ayre, R Standish, BL Willis, unpublished data). *A. millepora* and *A. pulchra* are morphologically distinct (112, 117), but they readily hybridize in breeding trials (140). Although a lack of fixed differences is not conclusive evidence that gene flow is occurring, these same loci adequately separate other species of *Acropora* (4; DJ Ayre, personal communication). Similarly, a lack of fixed differences at 9 polypmorphic loci has been found for 7 morphospecies within the genus *Platygyra,* all of which interbreed in laboratory crosses (KJ Miller, personal communication). In these cases, breeding and genetic groups encompass two or more morphologically distinct groups. A large-scale electrophoretic survey of 10 coral species, including 7 species of *Acropora* in three sections of the Great Barrier Reef (DJ Ayre, TP Hughes, personal communication), will contribute substantially to present understanding of genetic differentiation within and among populations, and also between species of *Acropora*. Species that readily hybridize may represent cases where morphological and molecular evolution are uncoupled, although it is not clear what mechanisms maintain distinct morphological groupings in the face of gene flow.

BIOGEOGRAPHY AND PHYLOGENY: SPECIES IN SPACE AND TIME

Biogeography and phylogeny add the dimensions of space and time to considerations of species. Both are important for corals because of the dispersal potential of coral larvae (141) and the good fossil record of corals (e.g 105, 128). The evolutionary and phylogenetic concepts incorporate time into species concepts through considerations of lineage, the phylogenetic concept using character-based lineages as the basis of relationship assessment (7, 18, 71, 134). An evolutionary approach to classification has long been applied to corals at supraspecific levels (105, 128, and innumerable paleontological references). Through the contributions of paleontology, the concept of genealogical relationships, inherent in the evolutionary and phylogenetic species concepts, is also a strong component of coral systematics (8). More recently, the application of cladistic techniques to corals (10, 44, 124) is allowing biologists working with extant populations to contribute to the analysis of species lineages.

The processes involved in biogeographic and phylogenetic study should further the development of a species concept for corals. Collecting and identifying samples from new sites for distribution studies may lead to reassessment of species boundaries among allopatric populations and may highlight underexplored regions and habitats where searches for new species should be concentrated (44, 107, 120, 124, 129). Likewise, character selection and definition for phylogenetic analyses, and examination of character polarity within the analyses, contribute to the endeavor of examining species boundaries (e.g. 9, 10, 25). Equally as important, comparing distribution patterns with phylogenies, using cladistic biogeographic methods, may generate hypotheses about the nature and evolution of species which can then be tested using independent techniques such as genetic analysis.

Biogeographic Consistency of Morphological Species

Synonymization of many species names from historical, regional studies (44, 108, 112, 117, 133, 144) is a significant step forward for biogeographic studies, as previously, only generic distributions were available for hypothesis testing (44, 93, 99). Now, consistent morphological species of corals may be recognized in sympatric situations, as well as across regional scales (e.g. Great Barrier Reef, Japan) and throughout broad geographic ranges (e.g. the Indo-Pacific). The apparent morphological consistency of many species supports the use of a morphological species concept and suggests that distribution patterns can be used to formulate hypotheses about the origin of species. However, such hypotheses will become more powerful as the links between morphology and gene flow for coral species are clarified. In particular, the possibility that undetected sibling species exist (49, 50) or that several morphological species are united within single breeding units (140) needs to be clarified before the contributions of biogeographic studies can be fully assessed.

Biogeographic Studies Amplify Diversity

How diverse is the genus *Acropora*? Despite reduction of valid species names through recent revisions (112, 117), new species are being described from various parts of the Indo-Pacific, either from previously underexplored regions (e.g. Indian Ocean, Indonesia) or from habitats not previously known to contain *Acropora* (such as *Halimeda* banks and rubble slopes deeper than 25 m) (107, 120, 129). A large database at the Museum of Tropical Queensland, including actual specimens of *Acropora* and associated locality data from *Acropora* research world-wide (including 16, 30, 39, 107, 112, 120, 124, 140; B Reigl, personal communication; S Romano, personal communication; and approximately 100 unpublished collections), is providing distribution patterns as well as new information on morphological species. Some of the patterns are discussed below. (In *Acropora*, misidentifications and differences in the application of synonyms make it unwise to compile published records, thus we only include herein records from the specimen-based database plus type localities of senior and junior synonyms.)

Patterns of Species Distribution

A BIOGEOGRAPHIC CASE STUDY   The species *Acropora (Isopora) palifera, A. (I.) cuneata* and *A. (I.) brueggemanni* have been recognized as a subgenus (112), or, minus *A. brueggemanni,* as a genus (84). The *Acropora* distributions database shows that, while *A. palifera* has a broad Indo-Pacific distribution, *A. cuneata* and *A. breuggemanni* have separate but overlapping distributions, the former confined to the Pacific plate with little incursion into the Indo-Australian Arc, and the latter restricted to the central Indo-Pacific and Australian Plate, with a line of overlap on the western edge of the Pacific Plate (Figure 4; see Figure, p. 246). These species reproduce sexually by releasing brooded larvae that are ready to settle and thus are not subject to broad dispersal (2, 52). A vicariant event at the intersection of the Pacific and Australian plates may have been responsible for speciation in this group.

CLADISTIC BIOGEOGRAPHY   Cladistic biogeographic methods bring together genealogies and distribution patterns by examining the relationships among geographic areas in terms of the relationships within monophyletic groups of species occupying them. Very often it is found that numerous sets of organisms indicate the same set of area relationships, thus the earth's history can be examined for major events that could be responsible for disruptions in species ranges (46, 70). To apply these methods, complete monophyletic units must be used, thus not all phylogenies are useful. The phylogeny of only one monophyletic unit, the *A. selago* species group, has been available to date for *Acropora*. Cladistic-biogeographic analysis using this group generated an hypothesis of vicariant events occurring stepwise from west to east across the Indo-Pacific (124), an hypothesis corroborated by cladistic-biogeographic analysis of two other genera of corals (145, 146).

THE INDO-PACIFIC HIGH DIVERSITY REGION   The Indo-Australian Arc, embracing the region of Wallace's line, has long been known to be a location of extremely high biodiversity. For many terrestrial and freshwater groups, correlations of phylogenies and distribution patterns of species with the complex tectonic history of the region are explaining both the famous disjunct distribution patterns and the extraordinarily high biodiversity (65, 75, 131, 132). With notable exceptions (44), the corals of this region remain essentially unrevised and recorded only from type descriptions following nineteenth century expeditions. A database of Indonesian *Acropora* is being compiled (CC Wallace, JM Pandolfi, unpublished data), and species of *Acropora* are being found to have parapatric or stasipatric distributions meeting near the junction of the Indonesian Arc and the Pacific. Morphologically distinctive new species endemic to the Indonesian Arc or Indian Ocean-Indonesian region have also been found: for example, preliminary results suggest one species is endemic to the Banda Sea, another to the Lombok Straits, a third to Indonesia as well as the eastern Indian Ocean (120).

Centers of Diversity

The most frequently recorded feature of reef coral biogeography is that numbers and composition of coral genera decline in a concentric pattern from a high in the Indo-West Pacific, and also from smaller centers in the Western Indian Ocean and Caribbean (91, 92, 99). This distinctive pattern has influenced thinking about centers of diversity, dispersal, and origination of corals to the point where the centers of high diversity are commonly equated with origination, and the patterns with dispersal from a "center of origin" (60, 61, 93). Many hypotheses about coral distribution patterns (summarized in 93) are untestable (60, 61). Most hypotheses will become far more testable once species-level data become available (44, 93).

Taxonomic reviews of corals have concentrated on widespread species. For examining biogeographic patterns, widespread species are particularly uninformative, as they cause area assemblages to appear similar by masking narrower ranges of other species. When wide-ranging Indo-Pacific species are ignored, the patterns of distribution seen for *Acropora* species are strikingly different from generic distributions. Rather than following a concentric pattern around the center of diversity, some species are endemic to contained regions such as the Red Sea and the Caribbean; others are endemic to broad regions ranging to either side of the Indo-Pacific center of diversity, e.g Indian Ocean and western Indonesian Arc or Pacific Plate and Philippines Plate; and still others have unusual patterns, such as the subtropical Pacific and its boundaries, or are endemic to smaller areas in various parts of the Indo-Pacific (123–124; CC Wallace, JM Pandolfi, unpublished data). These species-distribution patterns, when linked more closely with information on species relatedness (e.g. by cladistic biogeography) and fossil species (BR Rosen, personal communication), will provide another way of examining the origination of species in *Acropora*.

Phylogenetic Studies on Acropora

The sheer size of the genus *Acropora* makes phylogenetic analysis cumbersome and the application of cladistic biogeography difficult (given the necessity to use monophyletic groups for area relationships). The proposed relationships among species groups of *Acropora* (112) are not yet tested by phylogenetic analysis (although such an analysis is in progress: CC Wallace, JM Pandolfi, personal communication). Although no formal phylogenetic hypothesis yet exists for the whole genus, species relationships have been examined in three kinds of subsets of species: (i) species group (124); (ii) a subset of species that have narrow ranges (123); and (iii) a subset of mass-spawning species from which molecular data were derived (62; WG Wang, personal communication).

PHYLOGENY AND THE SKELETON   In corals the skeleton has traditionally provided the basis of classification (105, 128). Given its reduced skeletal features, *Acropora* offers few morphological characters with unique or clearcut states (117, 122). Additional discrete characters can be detected by morphometric analysis, followed by discriminant analysis of branching patterns and corallite details (124; CC Wallace, JM Pandolfi, unpublished). Using morphometric as well as qualitative morphological characters gives only a slight improvement in phylogenetic resolution compared with using qualitative characters alone (124).

GENEALOGY AND THE SKELETON   While polarity of characters has not been reported, preliminary results (124; CC Wallace, JM Pandolfi, unpublished) indicate that the skeleton of *Acropora* may have developed from a primitive condition with few radial corallites relative to axials and dense elaborated microskeleton to forms with numerous, strongly differentiated radial corallites and simple, light microskeleton. This suggests a progression from slow growing colonies with low reproductive output (due to low polyp numbers) to rapidly growing colonies with vast reproductive output (120).

PHYLOGENY AND THE GENOTYPE   Studies of species relatedness using DNA-based techniques are at a very preliminary stage for corals. A significant proportion of what is known is based on unreplicated specimens and the choice of species has often depended on availability of material, e.g. the use of DNA extracted from sperm of species involved in mass spawning (62). We know of three research groups using molecular techniques to examine the phylogeny of *Acropora*. First, in a study that combined data from restriction analysis and DNA hybridization, McMillan et al (62) found relatedness groupings for single specimens of 12 species of *Acropora* that differed from Veron & Wallace's morphological groupings (112). Phylogenetic analysis of satellite DNA sequence data of repeat units from seven of these same specimens supported the molecular groupings (62). In particular, the relationship of the *A. formosa* and *A. tenuis* specimens with *A. millepora* and *A. pulchra* was the reverse of that expected from morphological considerations. Continuing studies are examining the phylogeny of the order Scleractinia and phylum Cnidaria (DJ Miller, personal communication). Secondly, the 16s ribosomal gene has been extracted from tissue samples of several cnidarians including two species of *Acropora* and is providing characters appropriate for cladistic analysis of genera and families. Thus data from analysis of this gene is also contributing to new assessments of the phylogeny of the order and phylum (89; SL Romano, personal communication). A third group has isolated a gene encoding a nematocyst protein (the "mini-collagen" gene) from sperm of *A. donei* (WG Wang, personal communication).

When the levels of variation in molecular sequence data within and between populations of a species are known, molecular data will contribute more to our understanding of the phylogeny of the *Acropora* and of the order. Also, as scleractinian genomes become better known, interpretation of genetic data will be facilitated by knowledge of the biological function of molecular sequences. Apart from a gene isolated from *Acropora tenuis* sperm, which is known to be a homeobox (developmental) gene (66), and the min-collagen gene (WG Wang, personal communication), the links between the molecular sequence data and the biology of corals are unknown. While a full application of genealogical analysis awaits the development of a more complete genetic library and in some cases more cost-effective techniques, the rapid progress in PCR and RAPD technology and a general interest in the genetics of corals, particularly *Acropora*, suggest this information will become available in the near future.

CONCLUSIONS

Significance of Hybrids to Species Concepts

Breeding experiments with corals, described above, demonstrate that the morphological discontinuities recognized as species boundaries by taxonomy are not always associated with barriers to interbreeding. Thus the "yardstick" of the biological species (reproductive isolation in sympatric populations) does not give unequivocal results in these cases. Once colonies have been bred out to adults whose morphologies can be examined, it will be possible to search for hybrids in the field. At the same time, as the genome becomes better understood, it should become possible to look for genetic indicators of the extent and role of hybridization within populations of these species. If hybridization is found to be a feature of coral communities, its implications will be profound. The species currently recognized by taxonomy have long been tacitly assumed to fit the biological species concept (106; review in 136). Although Mayr (58) was willing to dismiss hybrids as occasional "mistakes," widespread hybridization in corals would mean that different kinds of species (e.g. some which fit the biological species concept and some which do not) coexist sympatrically. Although hybridization is often considered to be a source of evolutionary novelty, providing the basis for rapid speciation, in certain conditions it may well have the opposite effect (28). In corals hybridization may be retarding divergence among evolutionary lineages and contributing to the high morphological variability characteristic of many species (140; see also 82 for an alternative mechanism for this variability).

The Role of Genetic Data in Assessing Hybrids

If, as is indicated by genetic results to date, at least two categories of species— biological species with fine-scale morphometric differences and consistent breeding incompatibilities versus polymorphic species (comprising two or more currently accepted species which are reproductively compatible)—exist side by side on modern reefs, we must contend with some difficult conclusions. First, there may have been at least two major modes of evolution leading to the diversity of species in this genus. Second, no single method of assessing species limits can be applied across the genus. Each species will have to be reassessed, on a case-by-case basis, and the rationale for restricting or expanding its limits fully documented. Both the concept of sibling species and that of broad morphological diversity within a single breeding boundary must be entertained as valid models. If hybridization and backcrossing occur frequently in coral populations, the ability of molecular data to elucidate phlyogenies will be severely challenged (28).

Phylogenetic Interpretations and the Species Concept

We must be careful about how we use characters to construct phylogenies, for example, in the way we use the criterion of hybridization. Sibling species, if shown to exist for corals, are by definition morphologically very similar but separated by a breeding boundary (58). Breeding tests have already shown examples of populations that have consistently discernible slight morphological differences but that are separated by breeding barriers (e.g. the *A. millepora* example given above and the example of two species discerned within *Montipora digitata* (140, B Stobbart, personal communication). In contrast to this, some corals from populations with discrete morphological characters interbreed in laboratory crosses. Thus the ability to interbreed (regarded by phylogeneticists as plesiomorphic; 113) may not be a suitable test for phylogenetic relatedness. Rather, it may indicate that some speciation in *Acropora* has occurred through allopatric divergence leading to species with different morphologies but no breeding barriers when reunited sympatrically (140). This is an extension of the hypothesis of Potts (82) that splitting and reuniting of gene pools has led to highly variable species in corals. In other cases, speciation may have occurred through minor changes in gamete recognition systems (74, 140).

The Search for Sibling Species

While we should continue the process of splitting species if sets of overlapping characteristics indicate the existence of "pseudo-sibling" species (49, 50), we should not fall into the trap of setting rules based on this for a new taxonomy for corals. The breeding barriers of known sibling species should be examined, because there are now strong indications that breeding barriers and species boundaries estimated from other criteria can be misaligned. Indeed we must remain open to the possibility that some species may have to be expanded to encompass discrete morphological boundaries within broader breeding barriers. Worse than this is the possibility that coral species defined by different characteristics (genetics, breeding barriers, morphology) may not coincide. The role of algal symbionts in affecting misalignments between morphological and other criteria is unexplored (8a). The tacit assumption of coral biologists that currently defined coral species fit the biological species concept, and that our classifications also express relatedness by descent, may be totally unsubstantiated (136).

Acropora—*Still the Protean Genus*

Acropora has always presented a parodox and can be expected to do so for some time. Just when we thought that some stability may have come into effect through revision of the taxonomy of the genus (112, 117), we must brace ourselves for major changes. New species probably await description on purely morphological grounds as areas of endemicity in the Indo-Australian Arc and new deep-water habitats are explored (120). Species currently in synonymy may have to be once again restored to species status as breeding barriers are discovered. Well-accepted species may have to be combined because of lack of breeding barriers, supported by genetic similarity. Sibling species (morphologically indistinguishable but with barriers to breeding) may be found to exist and have to be accommodated somehow in the systematic framework, as may be species with polyploid karyotypes and possibly also sterile hybrids. We may have to accept the use of different species concepts for different purposes.

ACKNOWLEDGEMENTS

This paper is dedicated to the memory of the great coral systematist John West Wells, who died on 12 January 1994. Research reported was funded by Australian Research Council grants to BL Willis, CC Wallace, and D Ayre; to CC Wallace; and to CC Wallace and JM Pandolfi. We thank our colleagues who reviewed this paper: PW Arnold, TP Hughes, B Scott, J Wolstenholme; and we also thank JD True, R. Willis, and J Wolstenholme for assistance with the manuscript.

Any *Annual Review* chapter, as well as any article cited in an *Annual Review* chapter, may be purchased from the Annual Reviews Preprints and Reprints service.
1-800-347-8007; 415-259-5017; email: arpr@class.org

Literature Cited

1. Arnold M. 1992. Natural hybridization as an evolutionary process. *Annu. Rev. Ecol. Syst.* 23:237–61
2. Atoda K. 1951. The larva and postlarval development of the reef-building corals. III. *Acropora brueggemanni* (Brook). *J. Morphol.* 89:1–15
3. Avise JC, Ball RM. 1990. Principles of genealogical concordance in species concepts and biological taxonomy. In *Oxford Surveys in Evolutionary Biology*, Vol. 7, ed. D Futuyama, J Antonovics, pp. 45–67. New York: Oxford Univ. Press
4. Ayre DJ, Veron JEN, Duffy SL. 1991. The corals *Acropora palifera* and *Acropora cuneata* are genetically and ecologically distinct. *Coral Reefs* 10:13–18
5. Babcock RC. 1984. Reproduction and distribution of two species of *Goniastrea* (Scleractinia) from the Great Barrier Reef province. *Coral Reefs* 21:87–95
6. Babcock RC, Bull GD, Harrison PL, Heyward AJ, et al. 1986. Synchronous spawnings of 105 species of scleractinian coral species on the Great Barrier Reef. *Mar. Biol.* 903:79–94
7. Baum D. 1992. Phylogenetic species concepts. *Trends Ecol. Evol.* 7:1–2
8. Budd AF. 1990. Longterm patterns of morphological variation within and between species of reef-corals and their relation to sexual reproduction. *Syst. Bot.* 15:150–65
8a. Buddemeier RW, Fautin DG. 1993. Coral bleaching as an adaptive mechanism: a testable hypothesis. *Bioscience* 43:320–26
9. Cairns SD. 1989. Discriminant analysis of Indo-West Pacific *Flabellum*. *Mem. Assoc. Australas. Palaeontols.* 8:61–68
10. Cairns SD, Macintyre IG. 1992. Phylogenetic implications of calcium carbonate minerology in the Stylasteridae (Cnidaria: Hydrozoa). *Palaios* 7:96–107
11. Chappel J. 1980. Coral morphology, diversity and reef growth. *Nature* 286: 249–52
12. Coll JC, Bowden BF, Meehan GV, Konig GM, et al. 1994. Chemical aspects of mass spawning in corals I. Sperm attractant molecules in the eggs of the scleractinian coral *Montipora digitata*. *Mar. Biol.* 118:177–82
13. Coyne JA. 1992. Genetics and speciation. *Nature* 355:511–15
14. Coyne JA, Orr HA, Futuyma DJ. 1988. Do we need a new species concept? *Syst. Zool.* 37:190–200
15. Cracraft J. 1983. Species concepts and speciation analysis. In *Current Ornithology*, Vol. 1, ed. R Johnston, pp. 159–87. New York: Plenum
16. Dai CF, Soong K, Fan TY. 1994. Sexual reproduction of corals in northern and southern Taiwan. *Proc. Int. Coral Reef Symp., 7th, Guam, 1992*, pp. 442–55
17. Davies PJ. 1983. Reef growth. In *Perspective on Coral Reefs*, ed. DJ Barnes, pp. Townsville: Aust. Inst. Mar. Sci.
18. de Queiroz K, Gauthier J. 1992. Phylogenetic taxonomy. *Annu. Rev. Ecol. Syst.* 23:449–80
19. de Queiroz K., Donoghue MJ. 1988. Phylogenetic systematics and the species problem. *Cladistics* 4:317–38
20. Dobzansky T. 1937. *Genetics and the Evolution of Species.* New York: Columbia Univ. Press. 1st ed.
21. Done TJ. 1982. Patterns in the distribution of coral communities across the central Great Barrier Reef. *Coral Reefs* 1:95–107
22. Endler JA. 1989. Conceptual and other problems in speciation. In *Speciation and Its Consequences*, ed. D Otte, JA Endler, pp. 625–648. Sunderland, Mass: Sinauer Associates. 679 pp.
23. Ereshefsky M. 1992. Introduction. In *The Units of Evolution*, ed. M

Ereshefsky, pp. 13–17. Cambridge, Mass: MIT Press. 405 pp.

24. Fadlallah YH, Lindo RH, Lennon DJ. 1992. Annual synchronous spawning event in *Acropora* sp. from the Arabian Gulf. *7th Int. Coral Reef Symp. Guam 1992.* Abstracts:29

25. Foster AB. 1984. The species concept in fossil hermatypic corals: a statistical approach. *Palaeontolographica Americana* 54:58–69

26. Frost SH. 1977. Miocene to Holocene evolution of Caribbean Province reef-building corals. *Proc. 3rd Int. Coral Reef Symp. Miami 1977,* pp. 353–359

26a. Garthwaite RL, Potts DC, Veron JEN, Done TJ. 1994. Electrophorectic identification of poritid species (Anthozoa: Scleractinia). *Coral Reefs* 13:49–56

27. Gattuso J-P, Pichon M, Jaubert J. 1991. Physiology and taxonomy of scleractinian corals: a case study in the genus *Stylophora. Coral Reefs* 9:173–82

28. Grant PR, Grant R. 1992. Hybridization of bird species. *Science* 256:193–97

29. Grigg RW. 1981. *Acropora* in Hawaii. 2. Zoogeography. *Pacific Sci.* 35:15–24

30. Grigg RW, Wells JW, Wallace CC. 1981. *Acropora* in Hawaii. I. History of the scientific record, systematics, and ecology. *Pacific Sci.* 35:1–13

31. Guiry MD. 1992. Species concepts in marine red algae. In *Progress in Phycological Research,* Vol. 8, ed. FE Round, DJ Chapman. pp. 251–78. Bristol: Biopress

32. Haddon AC. 1894. The genus *Madrepora.* (Review of Catalogue of the Madreporian corals in the British Museum (Natural History) by G.Brook. *Nature* (Suppl. Jan. 1894):9–10

33. Harrison PL. 1985. Sexual characteristics of Scleractinian corals: systematic and evolutionary implications. *Proc. Int. Coral Reef Congress, 5th, Tahiti, 1985.* 4:337–42

34. Harrison PL. 1988. Pseudo-gynodioecy: an unusual breeding system in the scleractinian coral *Galaxea fascicularis. Proc. Int. Coral Reef Symp., 6th, Townsville, 1988.* 2:699–705

35. Harrison PL, Babcock RC, Bull GD, Oliver JK, et al. 1984. Mass spawning in tropical reef corals. *Science* 223:1186–89

36. Harrison PL, Wallace CC. 1990. Reproduction, dispersal and recruitment of scleractinian corals. In *Ecosytems of the World,* No. 10 *Coral Reefs.* ed. Z Dubinsky, pp. 133–207. Amsterdam: Elsevier Sci. 550 pp.

37. Harrison RG. 1990. Hybrid zones: windows on the evolutionary process. In *Oxford Surveys in Evolutionary Biology,* Vol. 7, ed. J Futyama, J Antonovics. Oxford: Oxford Univ. Press. 314 pp

38. Harrison RG. 1993. Hybrids and hybrid zones: historical perspective. In *Hybrid Zones and the Evolutionary Process,* ed. RG Harrison, pp. 3–12. Oxford: Oxford Univ. Press. 364 pp

39. Hayashibara T, Shimoike K, Kimura T, Hosaka S, et al. 1993. Patterns of coral spawning at Akajima Island, Okinawa, Japan. *Mar. Ecol. Prog. Ser.* 101:253–62

40. Heyward A. 1985. Comparative coral karyotology. *Proc. 5th Int. Coral Reef Congress Tahiti, 1985.* 6:47–51

41. Highton R. 1990. Taxonomic treatment of genetically differentiated populations. *Herpetologia* 46:114–21

42. Hillis DM. 1987. Molecular versus morphological approaches to systematics. *Annu. Rev. Ecol. Syst.* 18:23–42

43. Hillis DM, Moritz C, eds. 1990. *Molecular Systematics.* Sunderland, Mass: Sinauer Assoc. 588 pp.

44. Hoeksema BW. 1989. Taxonomy, phylogeny and biogeography of mushroom corals (Scleractinia: Fungiidae). *Zoologische Verhandlingen, Leiden* 254:1–295

45. Hoeksema BW. 1989. Species assemblages and phenotypes of mushroom corals (Fungiidae) related to coral reef habitats in the Flores Sea. *Netherlands J. Sea Res.* 23:149–60

46. Humphries CJ, Parenti LR. 1986. Cladistic biogeography. *Oxford Monogr. Biogeography* No. 2. 98 pp.

47. Kenyon JC. 1992. Sexual reproduction in Hawaiian *Acropora. Coral Reefs* 11:37–43

48. Kenyon JC. 1994. Chromosome number in ten species of *Acropora. Proc. 7th Int. Coral Reef Symp. Guam 1992.* pp. 471–75

49. Knowlton N. 1993. Sibling species in the sea. *Annu. Rev. Ecol. Syst.* 24:189–216

50. Knowlton N, Jackson JBC. 1994. New taxonomy and niche partitioning on coral reefs: jack of all trades or master of some? *Trends Ecol. Evol.* 9:7–9

51. Knowlton N, Weil E, Weigt LA, Guzman HM. 1992. Sibling species in *Montastrea annularis,* coral bleaching, and the coral climate record. *Science* 255:330–33

52. Kojis BL. 1986. Sexual reproduction in *Acropora (Isopora)* species (Coelenterata: Scleractinia) I. *A. cuneata* and *A. palifera* on Heron Island reef, Great Barrier Reef. *Mar. Biol.* 91:291–309

53. Lang JC. 1976. Whatever works: the variable importance of skeletal and of non-skeletal characters in Scleractinian

taxonomy. *Palaeontographica Americana* 54:18–44

54. Lang JC, Chornesky EA. 1991. Competition between scleractinian reef corals—a review of mechanisms and effects. In *Ecosytems of the World*, No. 10 *Coral Reefs*, ed. Z Dubinsky, pp. 209–52. Amsterdam: Elsevier Sci. 550 pp.

55. Lewis JB. 1984. The *Acropora* inheritance: a reinterpretation of the development of fringing reefs in Barbados, West Indies. *Coral Reefs* 3:117–22

56. Loether R. 1990. Species and monophyletic taxa as individual substantial systems. In *The Plant Diversity of Malesia*, ed. P Baa, K Kalkman, R Geesink, pp. 371–78. Dordrecht, Netherlands: Kluwer Acad.

57. Macintyre IG. 1988. Modern coral reefs of Western Atlantic: new geological perspective. *Am. Assoc. Petroleum Geologists Bull.* 72(11):1360–69

58. Mayr E. 1992. A local flora and the biological species concept. *Am. J. Bot.* 79:222–38

59. Mayor AG. 1924. Growth-rate of Samoan corals. *Carnegie Inst. Wash. Pap. Dep. Mar. Biol.* 19:51–72, pl. 1–26

60. McCoy ED, Heck KL. 1976. Biogeography of corals, seagrasses and mangroves: an alternative to the center of origin concept. *Syst. Zool.* 25:201–10

61. McCoy ED, Heck KL. 1983. Centers of origin revisited. *Paleontology* 9:17–19

62. McMillan J, Mahony T, Veron JEN, Miller DJ. 1991. Nucleotide sequencing of highly repetitive DNA from seven species in the coral genus *Acropora* (Cnidaria: Scleractinia) implies a division contrary to morphological criteria. *Mar. Biol.*:110:323–27

63. McMillan J, Miller DJ. 1988. Nucleotide sequences of highly repetitive DNA from scleractinian corals. *Gene* 83:185–86

64. McMillan J, Miller DJ. 1990. Highly repeated DNA sequences in the scleractinian coral genus *Acropora*: evaluation of cloned repeats as taxonomic probes. *Mar. Biol.* 104:483–87

65. Michaux B. 1991. Distribution patterns and tectonic development in Indonesia: Wallace reinterpreted. *Aust. Syst. Bot.—Austral. Biogeography* 4:25–36

66. Miles A, Miller DJ. 1992. Genomes of diploblastic organisms contain homeoboxes: sequence of events, an even-*skipped* homologue from the cnidarian *Acropora formosa*. *Proc. R. Soc. Lond. B* 248:159–61

67. Miller KJ. 1994. Morphological varia-

tion in the scleractinian coral *Platygyra daedalea* (Ellis & Solander 1786) genetically or environmentally determined? *Proc. Int. Coral Reef Symp., 7th, Guam, 1992.* In press

68. Miller KJ. 1994. Morphological variation in the coral genus *Platygyra*: environmental influences and taxonomic implications. *Mar. Ecol. Prog. Ser.* pp. 550–56

69. Mishler BD, Donoghue MJ. 1982. Species concepts: a case for pluralism. *Syst. Zool.* 314:91–503.

70. Nelson G, Platnick NI. 1981. *Systematics and Biogeography, Cladistics and Vicariance.* New York: Columbia Univ. Press. 567 pp.

71. Nixon KC, Wheeler QD. 1990. An amplification of the phylogenetic species concept. *Cladistics* 6:211–23

72. Oliver JK, Babcock RC. 1992. Aspects of the fertilization ecology of broadcast spawning corals: sperm dilution effects and in situ measurements of fertilization. *Biol. Bull.* 183:409–17

73. Oliver JK, Willis BL. 1987. Coral spawn slicks in the Great Barrier Reef: preliminary observations. *Mar. Biol.* 94:521–29

74. Palumbi SR. 1992. Marine speciation on a small planet. *Trends Ecol. Evol.* 7:114–18

75. Parenti LR. 1991. Ocean basins and the biogeography of freshwater fishes. *Aust. Syst. Bot.–Austral. Biogeography* 4:137–49

76. Paterson HEH. 1985. The recognition concept of species. In *Species and Speciation. Transvaal Museum Monograph No. 4*, ed. ES Vrba, pp. 21–29. Pretoria: Transvaal Mus.

77. Patterson C. 1987. Introduction. In *Molecules and Morphology in Evolution: Conflict or Compromise?* ed. C Patterson, pp. 1–22. Cambridge: Cambridge Univ. Press

78. Pichon M. 1981. Dynamic aspects of coral reef benthic structures and zonation. *Proc. Int. Coral Reef Symp., 4th, Manila, 1981.* 1:581–94

79. Potts DC. 1976. Growth interactions among morphological variants of the coral *Acropora palifera*. In *Coelenterate Ecology and Behavior*, ed. GO Mackie, pp. 79–88. New York: Plenum

80. Potts DC. 1981. Crown-of-thorns starfish: man-induced pest or natural phenomenon? In *The Ecology of Pests: Some Australian Case Studies*, ed. RL Kitching, RE Jones, pp. 55–86. Melbourne: CSIRO

81. Potts DC. 1984. Natural selection in experimental populations of reef-build-

ing corals (Scleractinia). *Evolution* 38: 1059–78

82. Potts DC. 1985. Sea-level fluctuations and speciation in Scleractinia. *Proc. Int. Coral Reef Congress, 5th, Tahiti, 1985.* 4:127–32

83. Potts DC, Garthwaite RL, Budd AF. 1992. Speciation in the coral genus *Porites*. *7th Int. Coral Reef Symp., Guam Abstracts*:85

84. Randall R. 1981. Morphologic diversity in the Scleractinian genus *Acropora*. *Proc. Int. Coral Reef Symposium, 4th, Manila, 1981.* 2:157–64

85. Richmond RH. 1992. Fertilisation in corals: problems and puzzles. *Abstracts 7th Int. Coral Reef Symp., Guam*:89

86. Ride WDL, Sabrosky CW, Bernadi G, Melville RV. 1985. *International Code of Zoological Nomenclature.* Berkeley, Los Angeles: Univ. Calif. Press. 338 pp. 3rd ed.

87. Rieseberg LH, Wendel JF. 1993. Introgression and its consequences in plants. In *Hybrid Zones and the Evolutionary Process,* ed. RG Harrison, pp. 70–109. Oxford: Oxford Univ. Press. 364 pp.

88. Rieseberg LH, Carter R, Scott Z. 1990. Molecular tests of the hypothesized hybrid origin of two diploid *Helianthus* species (Asteraceae). *Evolution* 44: 1498–511

89. Romano SL, Palumbi SR. 1992. Molecular evolution of mitochondrial DNA sequences from scleractinian corals. *7th Int. Coral Reef Symp. Guam Abstracts*:91

90. Rosen BR. 1977. The depth distribution of hermatypic corals and its palaeontological significance. *Mem. Bur. Rech. Geol. Minier.* 89:507–17

91. Rosen BR. 1981. The tropical high diversity enigma—the coral's eye view. In *Chance, Change and Challenge: The Evolving Biosphere,* ed. PH Greenwood, PL Forey, pp. 103–29. Cambridge, Eng: British Mus. (Natural History) and Cambridge Univ. Press

92. Rosen BR. 1984. Reef coral biogeography and climate throughout the late Cainozoic: just islands in the sun or a critical pattern of islands? In *Fossils and Climate,* ed. P Brenchley, pp. 201–62. London: Wiley

93. Rosen BR. 1988. Progress, problems and patterns in the biogeography of reef corals and other tropical marine organisms. *Helgolander Meeresuntersuchengen* 42:269–301

94. Sammarco PW. 1983. Coral recruitment across the Central Great Barrier Reef: a preliminary report. In *Proc. Great Barrier Reef Conf.,* ed. JT Baker, R

Carter, PW Sammarco, K Stark, pp. 245–258. Townsville: James Cook Univ. Press

95. Shimada H, Yokochi H. 1992. Coral spawning and its synchrony at Iriomote Island, Southern Ryukus. *7th Int. Coral Reef Symp., Guam, 1992. Abstracts* pp. 95

96. Shimoike K, Hayashibara T, Kimura T, Omori M. 1994. Observations of split spawning in *Acropora* spp. at Akajima Island, Okinawa. *Proc. Int. Coral Reef Symp., 7th, Guam, 1992.* pp. 484–88

97. Smith SV, Harrison JT. 1977. Calcium carbonate production of the *mare incognitum,* the upper windward slope, at Enewetak Atoll. *Science* 197:556–59

98. Spooner DM, Sytsma KJ, Smith JF. 1991. A molecular reexamination of diploid hybrid speciation of *Solanum raphanifolium. Evolution* 45:757–64

99. Stehli FG, Wells JW. 1971. Diversity and age patterns in hermatypic corals. *Syst. Zool.* 20:115–26

100. Stephenson W, Wells JW. 1955. The corals of Low Isles, Queensland, August 1954. *Pap. Dep. Zoology Univ. Qld.* 1(4):1–59, 7 plates

101. Stobart B, Benzie JAH. 1994. Allozyme electrophoresis demonstrates that the scleractinian coral *Montipora digitata* is two species. *Mar. Biol.* 118:183–90

102. Sytsma KJ. 1989. DNA and morphology: Inference of plant phylogeny. *Trends Ecol. Evol.* 5:104–10

103. Templeton AR. 1989. The meaning of species and speciation: a genetic perspective. In *Speciation and Its Consequences,* ed. D Otte, JA Endler, pp. 3–27. Sunderland, Mass.: Sinaur Assoc. 679 pp

104. Van Veghel MLJ, Bak RPM. 1993. Intraspecific variation of a dominant Caribbean reef building coral, *Montastrea annularis:* genetic, behavioral and morphometric aspects. *Mar. Ecol. Prog. Ser.* 92:255–65

105. Vaughan TW, Wells JW. 1943. Revision of the suborders, families and genera of the Scleractinia. *Geol. Soc. America Spec. Pap.* 44. 363 pp, pl. 1–26

106. Veron JEN. 1981. The species concept in Scleractinia of eastern Australia. *Proc. Int. Coral Reef Symp., 4th, Manila 1981.* 2:183–86

107. Veron JEN. 1990. New Scleractinia from Japan and other Indo-West Pacific countries. *Galaxea* 9:95–173

108. Veron JEN, Pichon M. 1976. *Scleractinia of Eastern Australia I. Families Thamnasteriidae, Astrocoeniidae, Pocilloporidae. Aust. Inst. Mar. Science Monogr. Ser. 1.* 86 pp.

109. Veron JEN, Pichon M. 1979. *Scleractinia of Eastern Australia. Part 3. Families Agariciidae, Siderastreidae, Fungiidae, Oculinidae, Merulinidae, Mussidae, Pectiniidae, Caryophyliidae, Dendrophyliidae. Aust. Inst. Mar. Sci. Monogr. Ser. 4.* 459 pp.

110. Veron JEN, Pichon M. 1982. *Scleractinia of Eastern Australia Part 4. Family Poritidae. Aust. Inst. Mar. Sci. Monogr. Ser. 5.* 159 pp.

111. Veron JEN, Pichon M, Wijsman-Best M. 1977. *Scleractinia of Eastern Australia. Part 2. Families Faviidae, Trachyphylliidae. Aust. Inst. Mar. Sci. Monogr. Ser. 3.* 233 pp.

112. Veron JEN, Wallace CC. 1984. *Scleractinia of Eastern Australia. Part 5. Acroporidae. Aust. Inst. Mar. Sci. Monogr. Ser. 6.* 485 pp.

113. Vrana P, Wheeler W. 1992. Individual organisms as terminal entities: laying the species problem to rest. *Cladistics* 8:67–72

114. Wagner WH Jr. 1983. Reticulation: the recognition of hybrids and their role in cladistics and classification. In *Advances in Cladistics*, Vol. 2, ed. NI Platnick, VA Funk, pp. 64–79. New York: Columbia Univ. Press

115. Wake DB. 1981. The application of allozyme evidence to problems in the evolution of morphology. In *Evolution Today, Proc. 2nd Int. Cong. Syst. Evol. Biol.*, ed. GGE Scudder, JL Reveal, pp. 257–70

116. Wallace CC. 1974. A numerical study of a small group of *Acropora* specimens (Scleractinia, Acroporidae). *Mem. Qd. Mus.* 17:55–61

117. Wallace CC. 1978. The coral genus *Acropora* (Scleractinia: Astrocoeniina; Acroporidae) in the central and southern Great Barrier Reef. *Mem. Qd. Mus.* 18:273–319, pls 43–103

118. Wallace CC. 1985. Reproduction, recruitment and fragmentation in nine sympatric species of the coral genus *Acropora. Mar. Biol.* 88:217–33

119. Wallace CC. 1985. Seasonal peaks and annual fluctuations in the recruitment of juvenile scleractinian corals. *Mar. Ecol. Prog. Ser.* 21:289–98.

120. Wallace CC. 1994. New species and a new species group of the coral genus *Acropora* from Indo-Pacific locations. *Invertebrate Taxonomy* 8(4).

120a. Wallace CC, Christie C. 1992. Reproductive status of corals in December 1987. In *Reef Biology: A Survey of Elizabeth and Middleton Reefs, South Pacific*, pp. 61–68. Aust. Natl. Parks & Wildlife Serv., Canberra

121. Wallace C, Dale MB. 1979. An information analysis approach to zonation patterns of the coral genus *Acropora* on the outer reef slope. *Atoll Res. Bull.* 220:98–110

122. Wallace CC, Dallwitz MJ. 1982. Writing coral identification keys that work. *Proc. Int. Coral Reef Symp., 4th, Manila, 1981.* pp. 107–90

123. Wallace CC, Pandolfi JM. 1992. Historical patterns in the world-wide distribution of the coral genus *Acropora. 7th Int. Coral Reef Symp. Guam 1992* (Abstr.):106

124. Wallace CC, Pandolfi JM, Young A, Wolstenholme J. 1991. Indo-Pacific coral biogeography: a case study from the *Acropora selago* group. *Aust. Syst. Bot.–Austral. Biogeography* 4:199–210

125. Wallace CC, Watt A, Bull GD. 1986. Recruitment of juvenile corals onto coral tables preyed upon by *Acanthaster planci. Mar. Ecol. Prog. Ser.* 32:299–306

126. Weil E. 1992. Genetic and morphologic variation in Caribbean and Eastern Pacific *Porites. 7th Int. Coral Reef Symp., Guam* Abstracts:107

127. Wells JW. 1954. Recent corals of the Marshall Islands. *Prof. Pap. U.S. Geol. Surv.* 260-I:385–486, pls 94–187

128. Wells JW. 1956. Scleractinia. In *Treatise on Invertebrate Paleontology*, Part F. *Coelenterata*. ed. R Moore, pp. 328–444. Lawrence, Kans: Geol. Soc. Am. & Univ. Kansas Press

129. Wells JW. 1987. Notes on Indo-Pacific scleractinian corals II. A new species of *Acropora* from Australia. *Pacific Sci.* 39:338–9

130. Wells JW, Alderslade PN. 1979. The scleractinian coral *Arcohelia* living on the coastal shores of Queensland, Australia. *Rec. Aust. Mus.* 32:211–16

131. Whitmore TC. ed. 1981. *Wallace's Line and Plate Tectonics*. Oxford: Clarendon. 91 pp.

132. Whitmore TC. ed. 1987. *Biogeographical Evolution of the Malay Archipelago*. Oxford: Clarendon. 147 pp.

133. Wijsman-Best M. 1972. Systematics and ecology of New Caledonian Faviinae (Coelenterata–Scleractinia). *Bijdragen tot de Dierkunde* 42:1–90

134. Wiley EO. 1978. The evolutionary species concept reconsidered. *Syst. Zool.* 27:17–26

135. Willis BL. 1985. Phenotypic plasticity versus phenotypic stability in the reef corals *Turbinaria mesenterina* and *Pavona cactus. Proc. Int. Coral Reef Congress, 5th, Tahiti, 1985* 4:107–12

136. Willis BL. 1990. Species concepts in extant scleractinian corals: considera-

tions based on reproductive biology and genotypic population structures. *Syst. Bot.* 15:136–49

137. Willis BL, Ayre DJ. 1985. Asexual reproduction and genetic determination of growth form in the coral *Pavona cactus:* Biochemical genetic and immunogenetic evidence. *Oecologia* 65:516–25

138. Willis BL, Babcock RC, Harrison PL, Oliver JK, Wallace CC. 1985. Patterns in the mass spawning of corals on the Great Barrier Reef from 1981 to 1984. *Proc. Int. Coral Reef Congress, 5th, Tahiti, 1985.* 4:343–48

139. Willis BL, Babcock RC, Harrison PL, Wallace CC. 1992. Experimental evidence of hybridization in reef corals involved in mass spawning events. *7th Int. Coral Reef Symp. Guam* Abstracts 1992: 109

140. Willis BL, Babcock RC, Harrison PL, Wallace CC. 1994. Hybridization in mass spawning reef corals. Ms.

141. Willis BL, Oliver JK. 1990. Direct tracking of coral larvae: Implications for the dispersal of planktonic larvae in topographically complex environments. *Ophelia* 32:145–62

142. Wilson AO. 1985 Time to revive systematics. *Science* 230 (4731):1227–28

143. World Conservation Monitoring Centre. 1993. *Checklist of Fish and Invertebrates Listed in the CITES Appendices.* Joint Nature Conserv. Com., Peterborough. 171 pp.

144. Zlatarsky VN, Estacella NM. 1982. Les Scleractiniares de Cuba. *Sofia: Editions de l'Academie bulgare des Sciences.* 472 pp.

145. Pandolfi JM. 1992. Successive isolation rather than evolutionary centres for the origination of Indo-Pacific reef corals. *J. Biogeogr.* 92:593–609

146. Pandolfi JM. 1993. Tectonic history of Papua New Guinea and its significance for marine biogeography. *Proc. Int. Coral Symp. 7th,* Guam. In press

Annu. Rev. Ecol. Syst. 1994. 25:263–92

A DAY IN THE LIFE OF A SEED: Movements and Fates of Seeds and Their Implications for Natural and Managed Systems

Jeanne C. Chambers

USDA Forest Service, Intermountain Research Station, 920 Valley Road, Reno, Nevada 89512-2812

James A. MacMahon

Department of Biology and Ecology Center, Utah State University, Logan, Utah 84322-5305

KEY WORDS: seed fates, secondary dispersal, Phase II dispersal, abiotic and biotic influences, biome differences

---

Abstract

We develop a model that outlines the movements and fates of seeds after they leave the parent plant, and then we examine the relative influences of abiotic and biotic factors on those movements and fates. Phase I dispersal is movement of a seed from the parent to a surface, while Phase II dispersal includes subsequent horizontal or vertical movements. Although less studied, Phase II dispersal is more likely to account for the patterning of plants in communities and ecosystems and is the focus of this review. Abiotic factors influence Phase II dispersal—the distance and type of movement depend on seed morphology, surface attributes, and the nature of the physical forces. Biotic factors (animals) move seeds to new sites passively either on body surfaces or by ingestion, or actively by consuming fruits or hoarding seeds. Animals also influence the movements of seeds through digging and burrowing activities. Arrival at microsites suitable for germination and establishment is critical and is affected not only by abiotic and biotic factors but also by seed morphology and germination responses. We emphasize that seed banks are much more dynamic than they are usually portrayed. Although often poorly quantified, seed mortality can occur at any point in the model. Sufficient differences exist among biomes that certain generalizations can be made regarding seed dynamics. Knowledge of seed movements and fates is essential for ecosystem restoration and conservation efforts and for the control of alien species in all biomes.

> The consequence of all this activity of the animals and of the elements in transporting seeds is that almost every part of the earth's surface is filled with seeds or vivacious roots of seedlings of various kinds, and in some cases probably seeds are dug up from far below the surface which still retain their vitality. The very earth itself is a granary and a seminary, so that to some minds its surface is regarded as the cuticle of one great living creature.
>
> HD Thoreau (140, p. 151)

INTRODUCTION

Thoreau's view of the abundance of seeds in nature as forming a veritable granary is true in terms of the large number of seeds that often occur in nature, yet few of these "potential" plants survive to produce seeds themselves. This abundance of seeds and their importance in nature has spawned a plethora of studies that treat nearly every conceivable aspect of seed biology from the chemistry of individual seeds (56) to the accumulation of seeds in the seed pools of a variety of communities (41, 80). Few of these studies consider the day-to-day movements and fates of seeds in nature. In a very real sense we do not have a balance sheet, in space or time, that permits us to account for the seeds that a plant produces.

To organize our thinking we use a conceptual model (Figure 1) that outlines the pathways that seeds follow after leaving the parent plant, the states in which they reside, and some of the biotic and abiotic factors that influence them. We present a compilation of the extant literature to give a sense of the significance of the various factors that influence the movements and fates of seeds in nature. We then examine some basic differences among biomes and infer ways that knowledge of seed movements and fates can be used for the management of ecosystems and their components.

The Model

The model (Figure 1) begins with a cohort of potentially germinable seeds on a plant. Although we use the term seeds throughout, we are actually referring to diaspores, i.e. the seed and any investing structures. Seeds are treated as free-living, immature plants that are tracked until they germinate and become seedlings. Germinable seeds are subject to death from biotic or abiotic factors at any point in the model. Seed mortality resulting from consumption, weather, or other factors moves seeds to a sink because they have lost their potential to become plants. The movement of germinable seeds from the plant to a

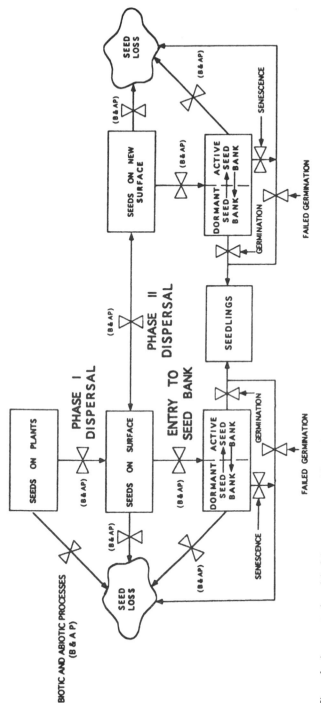

Figure 1   A conceptual model of the movements and fates of seeds. The germ of this model derives from consideration of Figure 4/1 of Harper (57), Figure 1 of Simpson et al. (131) and the dispersal phases of Watkinson (151). The model is explained in the Introduction.

surface is Phase I dispersal. We specifically use the terms "seeds on a surface" because many seeds arrive on surfaces other than soil, e.g. seeds of mistletoes, some bromeliads and orchids, are specifically adapted to disperse to tree branches or trunks (118). Phase I dispersal is the subject of a significant portion of the current seed literature. In part this emphasis is due to the fact that the movement of seeds from the plant to their first site of repose is amenable to direct observation and experimentation, including detailed aerodynamic studies and modeling (17, 49, 105). In contrast, we know much less about the fates of seeds once they land on a surface. Their size, mobility, and the fact that many are lost to animals or buried out of view makes them intractable objects of observation and experimentation. Consequently, the movements of seeds after initial dispersal from the parent plant are not well documented. In the model, Phase II dispersal includes both secondary horizontal and vertical movements of seeds. Thus, the model has an additional spatial component that is usually ignored. The point of entry into the seed bank depends on the plant species. Most seeds enter the seed bank after arrival on a surface, but in plant families characteristic of fire-prone habitats, e.g. Myrtaceae, Proteaceae, and Casuarinaceae, many species form aboveground seed banks (11). Seeds entering the seed bank may be in either an active or a dormant state. Physiologically active or nondormant seeds may germinate immediately, remain nondormant in the seed bank until the proper environmental conditions occur, or become dormant (8). We emphasize that seeds in seed banks are more dynamic than they are often portrayed. Seeds in seed banks may be moved by animals, wind, or other physical forces, lost to consumption by animals or attack by pathogens, change physiological status, or lose their germination potential because of senescence, but they are not static. As we detail below, the result of Phase I dispersal is often the arrival of a seed at a site close to the parent. The dispersal movements that account for the patterning of plants in communities and ecosystems is much more likely the result of Phase II than of Phase I dispersal, yet this is the area of our greatest knowledge gap. Although we examine all aspects of the model, our primary focus is on Phase II dispersal and its consequences.

PHASE I DISPERSAL

Phase I dispersal involves any mechanism by which a seed moves or is transported from the parent plant to a surface. Most plant seeds move only short distances from the parent (129). The resulting patterns of seed deposition are usually skewed, with a distribution represented either by a negative exponential function or by a curve that peaks a short distance from the plant and then shows a negative exponential decrease (154). Variation in this pattern results from factors such as habitat patchiness, seed vector behavior, or chance.

Details of the spatial result (seed shadows) of Phase I dispersal are well studied for a variety of individual species (155). Studies of Phase I dispersal for entire natural communities are much less common (20, 114).

Abiotic Influences

Abiotic dispersal may involve only gravity—seeds may simply fall beneath the parent. It may, however, involve specialized morphology for wind transport, such as samaras or plumes, or ballistic mechanisms in which seeds are ejected when hit by rain drops or when enveloping structures dry. When wind is the only agent, the height of plant, characteristics of surrounding vegetation, details of seed weight, size, and wind conditions during dispersal are normally sufficient data to predict patterns of seed deposition (49, 105). Recent models of wind-dispersed seeds have examined the effects of variations in meteorological conditions, seed mass, or form (49, 105). Models for other types of abiotic Phase I dispersal are more difficult to develop and are less common.

Biotic Influences

More attention has been devoted to biotic than abiotic dispersal, and several recent reviews discuss animal dispersal (62, 63, 136), frugivory (42, 70, 153), adhesion (133), and food hoarding and its dispersal consequences (113, 144). Biotic dispersal is described in terms of the method of seed acquisition and dispersal by the animal (136). Seed acquisition is categorized as passive when seeds or fruits are transported, by accident, on body surfaces (passive external) or consumed incidentally with other foods (passive internal). Active acquisition occurs when animals select seeds or fruits. Specific morphological characteristics of seeds and fruits facilitate the plant-animal interaction, whether during active or passive acquisition. Many seeds have adhesive properties such as hooks, barbs, or viscid surfaces for passive animal dispersal; these may result in longer dispersal than active dispersal by animals or wind dispersal (133). Actively acquired seeds are often attractive to animals because of fruit or seed color, odors, or the presence of a food reward such as investing pericarps or eliasomes on ant-dispersed seeds (136). The degree to which seeds are acquired while they are still attached to the parent plant or after they arrive on the soil surface depends on the life form and life history attributes of the plant species and on the behavioral characteristics of the animal. Animals that disperse seeds include birds, mammals, and ants as the main dispersers, but there are instances of dispersal by earthworms, beetles, Amazonian forest fish, tortoises, some herbivorous lizards, and even a frog (135, 136). Although not as common as for abiotic dispersal, models of biotic dispersal do exist. Murray (99) developed a simulation model that estimates reproductive output and relative "fitness" of neotropical gap-species from data on the seed shadows produced by different avian dispersers, germination requirements, and forest dynamics.

It is important that many studies fail to distinguish between Phase I and Phase II dispersal. Studies that quantify both Phase I and II dispersal and their relative effects on the various aspects of plant establishment are the most likely to lead to an understanding of plant population processes and community dynamics. Although many dispersal studies associate animals with plants as agents of dispersal only in the space dimension, many animal activities, e.g. hoarding, are equally important for dispersing plants in the time dimension—a function seldom quantified.

PHASE II DISPERSAL

Once a seed has arrived on a surface, it can remain where it initially came to rest, it can move to a new location (horizontal movement), or it can be incorporated into the soil (vertical movement). The probability of redistribution is determined by the nature of the abiotic or biotic factors that act on the seed, the characteristics of the site where the seed rests, and the interactions of the seed with abiotic or biotic factors.

Abiotic Influences

Relationships between a seed's physical dimensions, e.g. mass, length, width, and depth, and surface characteristics influence the horizontal and vertical movement of seeds after they have reached a surface. The type of movement and the distance moved depend on the nature of the physical forces. In steep terrain, gravity can move seeds downslope. The slope characteristics influence both the distance and direction of seed dispersal (152). Obviously, gravity facilitates entry into seed banks, but the amount of movement depends upon soil pore size and the physical dimensions and surface characteristics of the seeds (23). Wind contributes to the horizontal or surface movement of seeds in many environments, especially those where vegetation is sparse or low in structure (i.e. deserts and tundra). Wind is often a dominant dispersal agent for severely disturbed sites, e.g. Mount St. Helens (32). Wind has less effect on vertical movement of seeds, except for seed burial or reexposure along with wind-blown soil fines and litter (159). Horizontal transport of diaspores in rain wash can occur in any environment where the intensity and amount of precipitation is sufficient to result in overland flows, and such transport has been observed in a variety of ecosystems, including rain forests (122) and deserts (117). Precipitation also moves seeds vertically through the soil column (143). The duration and intensity of individual storm events and surface characteristics affect the vertical and horizontal movement of seeds. Several soil characteristics, including type, structure, and the amount of clay and colloid material, affect vertical movement of water, small soil particles, and, presumably, seeds. Precipitation and cryoturbation alter the structure of surface soils (128, 159) and, consequently, susceptibility of soils and seeds to later movement by wind or water. Cryoturbation alters the vertical distribution of seeds, moving seeds both upward and downward (143) and, in the case of solifluction lobes, may result in deep burial (94).

The type and intensity of the physical forces that act on seeds and the sites of deposition are largely determined by ecosystem characteristics. The vertical and horizontal structure of the vegetation, precipitation and temperature regimes, and the importance of wind vary significantly among ecosystems. Because of the importance of abiotic seed dispersal in systems with extreme environmental regimes, these environments have received the most study. In arid shrublands and woodlands where individual shrubs or trees are widely spaced, wind-blown or water-transported soil and litter accumulate under the long-lived shrubs or trees, resulting in a highly heterogeneous surface environment (88). Interspaces are often sparsely vegetated and characterized by a high percentage of bare ground. Wind velocities in interspaces are as much as four times greater than under shrubs (107). Consequently, interspaces serve as avenues of seed transport, with seed entrapment occurring primarily in soil cracks and crevices or under the litter-strewn canopies of shrubs or trees (74, 102, 117). The highest seed densities and greatest species richness in pinyon-juniper woodland are found at the interface between interspaces and the dense litter underneath the tree (79).

Within a given ecosystem, seeds dispersed from the parent plant can land within dense vegetation, on sites covered with litter, on exposed soils, on snow or ice, or even on other plants. The surfaces that seeds land upon are primary determinants of subsequent movement. Seeds that fall within dense vegetation generally have shorter secondary dispersal distances and are often more concentrated than seeds that land within open vegetation or on exposed soils (151). Within chalk grasslands, seed accumulation commonly occurs in patches of bryophytes (143). Plant litter also traps seeds. Seeds of *Bromus tectorum* are more likely to remain in place on natural or artificially littered microsites than on bare soil microsites (74).

Exposed soil can comprise most of the surface cover in extreme environments like deserts and tundras, in agronomic situations, or on sites disturbed by human activities. In many natural systems, soils exposed by small-scale disturbances serve as important sites of recruitment. Soil properties, climate, and disturbance characteristics determine the physical attributes and microtopography of exposed soils. In turn, these soil attributes influence both the horizontal and vertical movement of seeds. Wind-blown seeds often move farthest on smooth soils and remain in position, trapped in crevices, in rough soils (57, 102). A study of wind movement of four asymmetric samaras on four surface roughnesses (a smooth board, 250 μm, 50 μm and 2.0 mm particle size soil) showed that seeds remained in place longer on rougher surfaces because the threshold wind velocity was greater and the return time for that velocity was longer (69). Heavy seeds moved less frequently than light seeds. On exposed soils in a windy alpine environment, it was possible to develop predictive models of the vertical and horizontal movements of seeds with varying morphology over a range of soil particle sizes (23). In smaller particle size soils, smaller seeds tend to remain trapped in position and to reach greater depths in the soil column, while seeds that are longer or have higher eccentricity (length/width ratios) move horizontally over the soil and are not trapped. In larger particle size soils, both small and large seeds are trapped, but smaller seeds reach greater depths. For these alpine species, patterns of seedling emergence in the field could be explained largely by relationships between seed attributes and surface characteristics (22, 23). Depressions in the soil or obstructions serve as accumulation sites for seeds that move over the soil surface (117) by generating eddies or wind shadows and trapping seeds moved by overland flows. For artificially created depressions in Sonoran Desert soils, depression features such as perimeter, volume, depth, and surface area were significantly correlated ($R^2 = 0.9$ to 0.91) with longest seed dimension and density of seeds trapped in the depression (116). Obstructions or mounds frequently result from small mammal burrowing and tunneling. Such digging can expose surface and near-surface soils that act as catchments. In aspen woodlands, more seeds were found on pocket gopher mounds than in equivalent columns of adjacent topsoil (92).

Soil crusting and compaction occur on many types of exposed soils and can preclude seed entrapment and hinder seedling emergence (128). Soil crusts form when clods are broken down by raindrop impact and the dispersed finer particles are washed into the pores of the surface soil, resulting in a cemented seal (68). On crusted soils, seedling emergence (and presumably seed entrapment) most frequently occurs in soil cracks (34, 37, 58, 128). Similarly, for soils covered with cryptogamic crusts, seed entrapment and germination are highest in cracks in the crust (18). Soil compaction can result from vehicle passage and animal trampling as well as raindrop impact (59). Both seed incorporation and seedling emergence are lower on compacted than on noncompacted soils (128). Soil crusting, cracking, and compaction change over the growing season (159) owing to cyclical wetting and drying of soils and the effects of freezing and thawing.

Seeds of certain species can be wind-dispersed over snow-covered surfaces. Winter dispersal favors species whose inflorescences are located above the snow surface or that grow on exposed sites. Small, light seeds are transported the farthest. Sorting occurs, depending upon seed morphology and their associated aerodynamic properties, and distinct zones of debris accumulation, including seeds, occur on snowbeds in arctic and alpine tundras (47). For a gap colonizing tree, *Betula lenta*, secondary dispersal across a snow surface resulted in a distribution area 3.3 times greater than that covered by aerial dispersal alone (90). Seeds of *B. lenta* accumulated in shallow depressions in the snow at the bases of uprooted trees, resulting in aggregated seed distributions which increased the chances of successful regeneration.

Biotic Influences

Phase II dispersal by animals has multiple effects on seed fates. Seeds with morphological adaptations for abiotic or biotic primary dispersal or that apparently lack adaptations for dispersal can be secondarily dispersed by animals. Secondary dispersers often transport seeds farther than they are dispersed by primary mechanisms. For example, pine species that produce winged seeds typical of many wind-dispersed plants are often scatter-hoarded by rodents. In the Sierra Nevadas, animals had the capacity to harvest most of the naturally-produced seed of several pines (*Pinus jeffreyi, Pinus ponderosa,* and *Pinus contorta*) during the two-month period between seedfall and winter (146, 148). While wind-dispersed seeds of *Pinus jeffreyi* were distributed primarily within 12 m of source trees, chipmunk caches were found 2–69 m from trees (146). In some cases different animal species are responsible for Phase I and Phase II dispersal in the same plant species, and both animals may have mutualistic associations with the plant. In a Costa Rican Forest, ants harvest a high percentage of the seeds in bird defecations and cache them in their nests in partially decomposed twigs or deposit them on refuse piles (81). Similarly, for some species of figs a two-phase system exists that involves both birds and ants (71).

Many animals have direct or indirect effects on the vertical movement of seeds. Animal digging, burrowing, and tunneling can bury surface seeds or resurrect buried seeds. Large terrestrial mammals (armadillos, coatis, porcupines) and smaller animals (arthropods, caecilians, terrestrial crabs) dig holes that can influence seed movement (45). Estimates of the amount of soil moved by pocket gophers, through their tunnelling activities, range from 1 to 8.5 kg $m^{-2} yr^{-1}$ (1). This soil churning affects seed entrapment patterns, the microsites of establishment, and in the long term, plant community composition (66). Following the volcanic eruption of Mount St. Helens, the activities of surviving pocket gophers in bringing soil and propagules to the surface of ubiquitous and deep tephra deposits facilitated plant establishment (2).

Earthworms overturn large quantities of soil by burrowing and casting, while termites and ants overturn smaller quantities. Ingestion of seeds by earthworms is important in initial burial and in subsequent return of seeds to the surface in casts. On one temperate grassland area, 70% of the seedlings occurred on worm casts, although they comprised only 24% to 28% of the surface area (48). Similarly, in one year earthworms deposited on the soil surface a quantity of *Cerastium fontanum* seeds almost equal to the entire seed bank of the species

(139). Ingestion is dependent on seed size, and seeds in worm casts are smaller than those in the seed bank as a whole (139), indicating selective effects on seed bank dynamics.

Many animal species, especially those that function as Phase II dispersers, have associations with plants that are both antagonistic (granivory) and mutualistic (dispersal). Many of the seeds that are harvested are killed, but those that survive may exhibit substantial benefit (81, 144, 149). The tendency is often tc classify interactions between seeds and animals based on the most common outcome, even though the least common outcome may be of greatest importance to the plant (81). The outcomes of seed movement or handling by animals have been variously defined in terms of efficiency, quality, goodness, and reliability. Schupp (126) defined disperser effectiveness as a product of quantity or the number of seeds dispersed, and quality or the probability that a dispersed seed will produce a new reproductive adult. This approach requires that disperser effectiveness be defined in terms of the contribution that a disperser makes to plant fitness. This is appropriate for examining the fates of seeds, because it defines the best measure of disperser effectiveness as the number of new adult plants that result from the activities of a particular disperser relative to those that result from other dispersers or types of dispersal. Few studies have examined all of the components necessary to determine disperser effectiveness, although some come close (61, 78, 149).

The primary factors that influence disperser effectiveness include animal foraging behavior, seed availability, seed attributes, and seed location. The foraging behavior of animal species and the complement of animal species within a given area determine both the quantity and quality of seed dispersal. Plant species are often dispersed by a suite of unrelated animals that harvest varying numbers of seeds and place these seeds in quite different locations. Assemblages of dispersers may differ among locations, and both the likelihood of seed harvest and the fates of harvested seeds depend on the composition of the local fauna (14, 113, 126). In a chaparral community dominated by the shrub *Dendromecon rigida,* ant species differing in nesting and foraging behavior had varying effects on seed mortality, the microsites of germination, aggregation, and position on the landscape (14). In the Sonoran Desert, ants foraged only on the surface (85% of experimental seeds) while rodents harvested seeds from below the surface as well (96% of experimental seeds) (115).

Seed availability has important effects on the quality and quantity of seeds dispersed by animals. Within well-established populations of myrmecochorous plants with high seed densities, seed dispersal may be ant limited. Removal rates decrease over time, presumably because ants become satiated (132). Seed masting often increases the proportion of seeds that establish through predator satiation (28). In contrast, given low availability of seed, the animal disperser may consume a large portion of the seed crop, resulting in declines in plant populations (113). Seed availability may exhibit a high degree of stochastic variability, due to climatic and other environmental variability, and these fluctuations can be reflected in plant population densities and rates of expansion attributable to disperser activities (119). Rates of seed acquisition by animals for different plant species are often related to animal seed preferences. Most harvester ants exhibit preferences based on seed size, morphology, and availability (60). Seed use by ants in the Sonoran desert was related to seed size, with the smaller *Pheidole xerophila* specializing on small seeds, and the larger *Pogonomyrmex rugosus* preferring larger seeds (94). For *Pogonomyrmex occidentalis* in shrub steppe, relative seed abundance and size explained 28% of the variation in preference by ants (30). Decreases in the total potentially viable seed pool near the surface were estimated at 9% to 26%, while decreases in preferred species approached 100% (30).

Rodents usually exhibit seed preferences when presented with equally accessible seeds from several species (113). Factors influencing seed preference include the types of energy contained within seeds (carbohydrate vs protein), the presence of secondary compounds, and the ease of handling (144). Because energy gain usually increases with seed size, most rodents prefer large seeds as long as they are not too large to transport (115). Rodents discriminate between edible and inedible seeds using olfactory and, to a lesser degree, visual and tactile cues (144, 147), which result in greater harvesting of viable seeds.

Seed removal by secondary dispersers varies among the microsites in which seeds are located and among species (64). Selective harvest of clumped seeds by animals reduces the spatial patchiness of seeds in the soil (115) or redistributes the clumps through caching. Soil characteristics influence rodent foraging and, thus, seed distributions. Seed harvesting rates increase with seed density and soil density, and the rates decrease with soil particle size (112). Surface litter also influences the efficiency of seed harvesting (100), with the numbers harvested depending upon the seed species, the animal species, and perhaps overall seed availability (125).

The assumption that small size often allows seeds to "escape" seed harvesting by animals (67, 85) needs to be reevaluated in the context of Phase II dispersal. The probability of seed harvest is clearly related to the microsites of seed deposition and the size of the animal. If small seeds land on surfaces that promote incorporation into the soil or decrease seed-harvesting efficiency, then the probability of seed harvest decreases (113). However, if small seeds arrive on readily accessible surfaces or in identifiable depots (e.g. frugivore feces) they may be highly susceptible to harvest by small mammals or ants (14, 81, 113).

ENTRY INTO SEED BANK—SEED/MICROSITE INTERACTIONS

The potential benefits of seed dispersal to plant fitness depend on the microsites to which seeds are dispersed and the interactions of seeds with those microsites. Those microsites to which seeds are dispersed or that promote seed retention are not necessarily the best microsites for seed germination or seedling establishment (127).

Abiotic Factors

Soil surface attributes influence not only the sites of seed retention but also the microenvironment of germinating seeds and establishing seedlings. Larger soil particles increase the number of seeds trapped in soils (23) and result in longer residence times of large seeds and, consequently, higher seed germination (69). However, there is a soil particle size above which seedling establishment is compromised. Larger-particle-size soils seldom have the nutrient retention or water holding capacity of small-particle-size soils and may not provide the close root-soil contact needed for seedling growth and survival. In a seeding experiment on the pumice plains of Mount St. Helens, higher seedling emergence occurred in coarse pumice, while higher seedling survival was observed in fine pumice (158). Although not mentioned by the authors, higher seedling emergence on the coarse pumice was likely due to higher seed entrapment and retention.

Surface microtopography significantly affects the ultimate fates of seeds. Soil depressions or troughs tend to trap high numbers of seeds, and this "clumping" of seeds may result in higher seed predation (117). In arid environments, depressions can result in higher humidity and more favorable soil water relations for seeds and seedlings by trapping snow and accumulating precipitation. One-hundred times higher emergence of *Bromus tectorum* seedlings was reported for seeds sown in 9-mm soil pits than for seeds sown on bare soils (37). In humid environments, increased moisture may result in higher seed loss to pathogens.

Because seeds have many of the same physical attributes as "litter," abiotic forces deposit seeds in the same locations as plant litter (39). Animals often bury seeds in litter or under litter-covered surfaces. Relationships between seed morphology and the physical attributes of litter influence seed movement and burial on litter-covered surfaces. Litter facilitates the burial of seeds with hygroscopic awns in wind-blown arid environments (134) but retards burial of awned grass seeds such as *Aristida longiseta* (43). If extremely thick or coarse, litter can act as a physical barrier preventing seed penetration (55). Mortality results if germinating seedlings cannot emerge through the litter (150) or if the roots of seedlings cannot reach the soil (12, 43). In *Bromus tectorum,* an annual grass, variability in emergence timing may be due largely to variable penetration of both seeds and seedlings through the litter mat (86).

Biotic Factors

The role of animals in placing seeds in sites suitable for germination during Phase II dispersal is highly variable. Seed burial by animals may result in a more suitable physical environment for subsequent germination. However, the effects of seed burial on germination and establishment are highly species specific and are related to both soil conditions and depth of burial. In deciduous forests, seeds of some species such as oaks (50) may fail to germinate if they are not buried. In contrast, germination rates of white oak and pignut hickory are similar for seeds buried 2.5 cm deep or covered only with leaf litter (7). Soil water and temperature regimes influence both germination and decay of buried seeds. In arid areas, burial by animals may increase establishment by decreasing desiccation of the germinating seedling (113). Burial in cool dry soils may promote seed longevity, while burial in cool moist soils may result in natural stratification and increased seed germination. In warm humid areas, burial may increase the rate of seed decomposition and the probability of attack by pathogens.

Burial of seeds by food hoarders decreases the probability that seeds will be located and eaten by seed predators. Harvest rates of large seeds and nuts on the soil surface can be close to 100%, but rates may decrease as depth in the soil or litter increases (147). Not all secondary dispersal by animals results in seed burial. Ants often deposit seeds in refuse piles where they are susceptible to subsequent movement (14).

Animals may deposit seeds in nutrient rich environments, e.g. feces. In some ecosystems soil nutrients, including nitrogen and phosphorus, and soil aeration are significantly higher on ant refuse piles than on the surrounding area (9, 10). This may increase seed bank turnover rates by placing seeds in favorable germination sites and enhance seedling growth and survival (81). Ant nest sites are not always characterized by higher soil nutrients. In Australian sclerophyll vegetation, soil next to emerging seedlings of myrmecochores did not have higher levels of total N or available P than did soil around seedlings of nonmyrmecochors or soil not occupied by seedlings (120).

Seed Morphology and Seed Germination

Many seeds have specific morphological adaptations that influence seed movement into suitable germination microsites. Hygroscopic awns twist and untwist in response to changes in humidity, moving the diaspore over the soil surface and, given the proper surface, drilling the diaspore into the soil (108, 134). The pappuses of some species of Asteraceae collapse irreversibly or collapse and expand in response to humidity (128). This action also moves the diaspore over the soil surface and pushes it into the soil. Pappuses of other species remain rigid and maintain a constant angle between achene and soil, ensuring contact of the micropyle with the soil surface (128). Seeds of many species, including those with awns and pappuses, have antrorse (backwardly directed) bristles or barbs at the base of the seed that anchor the diaspore firmly in position. This prevents further movement and may produce a counter force to that of the radicle penetrating a compacted soil surface (109). In certain species the hairs on the hypocotyl swell in response to hydration and can raise seeds to a 30–45° angle with the soil surface, causing micropyle contact with the soil (52). The tips of the hairs become mucilaginous and, as they dry, bind to the soil, which helps the seedling to anchor and its roots to penetrate. Other species have mucilaginous seed coats that adhere to the soil surface when wetted (23). Morphological adaptations for primary dispersal do not preclude adaptations for movement on the soil surface. In *Erodium moschatum,* a geranium, seeds that moved on average 56 cm by explosive dispersal later moved another 7 cm over the soil surface through the activity of the hygroscopic awn (134).

Interactions of seeds with their environment can be highly specific, influencing not only the burial of individual seeds, but also the distributions of species. In southeastern Queensland, soils with a relatively high clay content had surfaces that were either loose and crumbly or that cracked on drying (110). Species with hygroscopically awned diaspores predominated, presumably because their diaspores encountered more suitable microsites. Soils with a high sand content had surfaces that were hard setting, had a tendency to form surface crusts, or were loose and single-grained. These surfaces were unsuitable for hygroscopically awned species and were dominated by unawned species.

Seed germination syndromes provide environmental cuing mechanisms that increase the probability of encountering conditions that are favorable for seedling growth and survival, essentially allowing seeds to disperse in time (3). Dormant seeds are those that will not germinate under normal environmental conditions and that must undergo afterripening or embryo maturation or be exposed to the proper environmental stimuli (8). Once seeds become active or nondormant, they must still experience the proper set of environmental conditions (light, temperature, and soil moisture regimes) to germinate. Seeds of many species, annuals and perennials, exhibit secondary dormancy or annual dormancy/nondormancy cycles that vary depending on species life histories and habitat characteristics (6, 8).

Because the seeds of most species require specific environmental conditions to become nondormant and then to germinate, the microsites of dispersal and the secondary movements of seeds are important determinants of seed germination. Small-scale variability in light, temperature, and moisture regimes exist within most communities. These environmental differences are accentuated by disturbance. Vegetation cover often results in lower soil temperatures, a lower red/far-red photon flux than sunlight and thus a low proportion of the active form of phytochrome, and in areas where growing-season precipitation is limited, lower soil moisture (22, 44). Higher soil temperatures, light fluxes, and soil moisture on disturbed soils or in open areas often favor seed germination (8). Although seldom studied, interactions among seed dispersal patterns and seed dormancy and germination undoubtedly influence seed turnover rates and the apparent spatial distribution of seeds in seed banks.

SEED LOSS

Abiotic Influences

Direct effects of abiotic factors on seed mortality include deep burial, crushing, abrasion, burning, water-logging, etc. As detailed below, the effects of abiotic factors on biotic interactions, senescence, and germination are probably more important for population and community dynamics.

Biotic Factors

Predation on fruits and seeds and its consequences for the plant community have been the subject of over a thousand studies and several reviews (28, 67, 85, 124). Recent papers emphasize the consequences of predation for plant population dynamics and plant community composition (65) and processes such as succession (31). The approach has shifted from investigating how many seeds are taken to determining the distributions and fates of those seeds that escape predation.

Seed predators and predation differ during the different life stages of plants. Predispersal predators are likely to be "small, sedentary, specialized feeders belonging to the insect orders Diptera, Lepidoptera, Coleoptera, Hymenoptera" (28). In addition to selective seed consumption by specialists, many seeds are lost by the consumption of flowers, seed heads, and fruits by large herbivores. Although Crawley (28) lists over 50 studies of predispersal seed predation, few of these demonstrate unambiguously the population or community consequences of predation (but see Louda (84, 85)).

Predation following Phase I dispersal is the common topic of seed predation studies. Here generalist and specialist vertebrates increase in importance. So long as the seed has not reached a safe site (often used synonymously with seed bank), it is subject to discovery and consumption. Even in the seed bank, seeds are not safe from pathogens and larger predators, especially vertebrates (some birds and mammals) that can find buried seeds. The consequences of predation range from effects on recruitment and distribution within a population of a single species to changes in composition of plant communities mediated directly by seed predation (85) or indirectly by changing the competitive balance between two seed consumers and subsequently between plants.

In contrast to animal predation, few ecological studies have addressed the effects of pathogens on seeds, especially those occurring in seed banks. Surface-contaminating fungi or bacteria can cause seed death directly through necrotic action or indirectly via production of toxic metabolic wastes (15, 16). In contrast, internally borne pathogens (some fungi and many viruses) often increase seed metabolic activity thus accelerating senescence (15, 16). In addition to seed death, fungal pathogens can decrease or stimulate germination (24) and may result in altered seedling survival following germination (29). Susceptibility to pathogen attack may be higher for physiologically active seeds than dormant seeds and may also increase as seeds age and membrane structure deteriorates (15, 16). Estimates of the effects of pathogens on soil seed dynamics are rare. In shrub steppe, overwinter seed decomposition and attack by fungi decreased the viable seed banks of common species by about 56% (29). In contrast, fungicide applications reduced seed loss of an invasive tropical shrub, *Mimosa pigra,* by only 10–16% over the seven month dry season, indicating that pathogens may have greater effects on germinating than on dormant seeds in this species (83).

Senescence

Environmental characteristics of the site and location in the seed bank determine the moisture and temperature regimes that seeds are exposed to and also can influence respiration and aging. Seed longevity is promoted under either extremely cold or dry conditions, and viability loss is increased under warm and moist conditions (98). The loss of seed viability over time can often be described with a negative exponential model, although a rectangular hyperbola sometimes provides a better fit to the data (98). While the relationship between seed longevity and temperature does not appear to vary among species, the relationship between longevity and soil moisture does (33).

The spatial and temporal effects of varying environmental conditions among the microsites of seed entrapment on seed longevity are largely unexplored. Most estimates of seed longevity come from studies in which seeds are buried under close-to-natural field conditions and retrieved over time (98). Because seeds are usually buried in soil under somewhat protected conditions, longevity of the seed population as a whole may be grossly overestimated. Such experiments often exclude seed harvesters and pathogens and thus fail to consider the effects of seed handling or pathogenic interactions on longevity. In addition, they restrict secondary seed movements and thus neglect the effects that exposure to varying environmental conditions may have on seed longevity.

Large differences in seed longevity exist among individual species in all ecosystems—tundra (19) to rain forest (45, 150). Two common generalizations regarding differences in species longevity are that there is a tendency for early seral or arable weed seeds to exhibit greater longevity than late seral species, and that small seed size is related to seed longevity. There is considerable evidence for the first generalization (51), but it is the second that is important for Phase II dispersal. An examination of nine species in the British flora showed that small seed size (mass) and low variance of seed or fruit dimensions (length, width, and depth) were related to greater seed longevity and either short-term (1–5 yr) or long-term (> 5 yr) persistence in the seed bank (138). Persistence of small and compact seeds in the seed bank is undoubtedly increased by a higher probability of seed burial. Short-lived seeds are usually larger and either flattened or elongate (138). Small and compact seeds exhibit higher soil incorporation than large seeds (23). Seed size is not a universal predictor of seed longevity, and seed morphological characteristics influence lifespans. While the large, hard-coated seeds of the Leguminosae are usually long-lived, the tiny, soft-coated seeds of *Salix* species live only a few weeks.

Failed Germination

Failed germination constitutes seed death soon after germination. We distinguish between failed germination and seedling death because seedling death is usually evaluated only after seedling emergence. This potentially ignores the death of numerous germinated seeds and underestimates seed bank losses. Antecedent conditions such as the effects of fungi (24) or handling by animal dispersers (147) increase mortality of germinating seeds. Consumption of germinating seeds by pathogens, predators, or granivores also reduces survivorship. Highly variable environmental conditions or unusual weather events can cause high mortality of germinating seeds (27). Unusual weather conditions can also "miscue" germination so that the timing of germination is inappropriate. Finally, seeds may arrive at microsites that provide the necessary conditions for germination but that are inadequate for growth and survival (127).

COMPARISONS AMONG BIOMES

Our presentation of the model provided examples from a variety of ecosystems to show the generality of our comments. This approach masks specific climatic or vegetational relationships that may exist and obscures some potentially interesting ecological differences. Here, we look for such general trends at the biome level of organization, of sufficient scale that the nuance of local variation can be ignored.

Desert

Among the world's deserts, a lack of specific adaptations for long-range dispersal is common, and species often have characteristics that hinder long-range dispersal (35). Many species have adaptations for short-range dispersal such as ballistic seeds. Wind dispersal varies in importance (35, 142). Abiotic redistribution of seeds on the soil surface by wind (74, 102) or overland flows (117) can be significant. In many deserts seed harvesting by ants and caching and hoarding by small mammals, particularly heteromyid rodents in North American deserts (113), influence seed distributions and plant establishment patterns. The importance of granivory by small mammals differs significantly among the world's deserts. Only in North America, Australia, and the Afro-Asian desert belt are there small bipedal desert mammals that specialize on seeds—other small desert mammals are opportunistic foragers on insects and vegetation (89). Predation by birds and ants is an important agent of mortality in all deserts. Additional seed losses due to fungi and other pathogens (29), as well as failed germination, can be significant.

Correspondence between the seed bank and current vegetation depends upon species composition. Few long-lived desert perennials have persistent seed banks, while persistence of annual species ranges from highly stable to transient (76). In the Sonoran and other hot deserts, high seasonal and annual variability in the seed bank indicates a lack of persistence or stability of the seed bank as a whole (76). Seed distributions in desert systems are highly clumped—a 10-fold variation in the numbers of seeds exists among various Sonoran desert microhabitats (117).

Tundra

In tundra ecosystems, seeds are often small, and few species invest in adaptations for seed dispersal. Adaptations for dispersal are primarily for wind, although alpine grasslands have small numbers of species with adaptations for ant, vertebrate, adhesive, or ballistic dispersal (157). Redistribution is strongly influenced by wind, with relationships between soil surface characteristics and seed attributes determining the horizontal and vertical movement of seeds (22, 23). Phase II seed dispersal by animals is rare. Seed-eating birds (horned larks and rosy finches) and insects (e.g. bruchid beetles) can be abundant in North American alpine tundra, and the potential for seed predation is high, although rates have not been measured (21). Low heat budgets and cold soils result in slow decomposition and may promote seed longevity relative to more temperate biomes (93). Deep burial of seeds can occur in solifluction lobes resulting in the long-term (300+ yr) preservation of genetic material (94).

Tundra floras are characterized by low numbers of species, and species identities in the seed bank and current vegetation are often similar (20, 93). Relative species abundances in the aboveground vegetation, seed rain, and seed bank depend upon species life histories and disturbance regimes (21). Both seed rain and seed banks are spatially and temporally variable, with seed banks exhibiting greater consistency than seed rain (20).

Grasslands

Important dispersal mechanisms in grasslands include wind, animals, and ballistic seeds. Seed production is often highly variable from year to year (121). Seed predation by ants, small mammals, or birds can be significant.

For many perennial grasslands, significant differences exist between the composition of the seed bank flora and the aboveground plant community. Depending upon the length of time since disturbance and the magnitude of disturbance, an abundance of both early seral natives and exotics may persist in the seed bank (121). Perennial species abundant in the standing vegetation may be absent from the seed bank, due to short-lived seeds (25) or heavy grazing and inflorescence removal (104). Highly variable seed production from year to year, combined with low seed longevity of late seral perennial grasses, can result in temporal variability in the seed bank. Spatial variability can occur as a result of individual species dispersal patterns interacting with small-scale topographic features (77).

Coniferous Forests

Coniferous forests tend to be characterized by trees that are dispersed either by wind or by cache-hoarding mammals and birds (40) or trees that exhibit no apparent adaptations for dispersal (157). Seed production often exhibits high periodicity (masting). Phase II dispersal of seeds adapted for Phase I dispersal by wind or animals can be significant (146). Early seral species exhibit a variety of dispersal mechanisms, but redistribution can be facilitated by rodents and ants. Both pre- and post-dispersal predation by insects can be high (4), and seed harvest by rodents can approach 100% (146, 148). Coniferous forests occur over a wide range of environments with varying rates of decomposition, but in warm or humid environments the potential for seed mortality due to pathogens is high.

Low correspondence between the composition of the seed bank and that of the vegetation exists (4). Many late seral species have low seed longevity, and early seral species are as abundant in the seed bank as late seral species in all stages of succession (4). Seed input and seed bank densities decline with time since disturbance, primarily due to differences in reproductive strategies among seral stages (73). Large-scale spatial variability can exist that is related to time since disturbance. Following disturbance, reestablishment of late seral trees may depend on residual plants within the area or chance seedling establishment (4).

Deciduous Forest

In deciduous forests, species are dispersed by wind or vertebrates, or they exhibit no specific dispersal mechanism (157). Masting by late seral species is common. Large-seeded species occur primarily in later seres because of the ability of seedlings to emerge through dense litter and to tolerate dense shade in mature forests (137). For certain forest herbs such as *Erythronium* and *Viola* species, Phase II dispersal is by ants (9). Rodents assume an important role as both predators and dispersers of many larger-seeded late seral species, as do some birds (146).

The composition of the seed bank in mature forests may be poorly correlated with that of the standing vegetation, and seeds of early seral species often have the highest abundance (111). Most seeds in persistent seed banks are small, possibly due to their ease in penetrating litter. The rarity of very large seeds in forest seed banks may relate to the selection value of masting and predator satiation, which are inconsistent with seed longevity (28, 130). Following disturbance, seed bank richness and density increase for about 30 to 100 years, after which both decline (101). Seed bank similarity to the standing vegetation is often highest in recently disturbed gaps or abandoned fields (111). Increased opening of gaps in the canopy as forests age (106) results in increased richness of seed banks of old growth forests and high spatial heterogeneity on the larger scale (111).

Rain Forest

In rain forests, birds and mammals are usually more important dispersal agents than wind (46). Brief longevity and rapid germination following dispersal are common in tropical rainforest seeds, but many species do exhibit delayed germination (45, 150). Plants often have a suite of dispersers that affect the germinability and longevity of seeds arriving at the soil surface (135). Important secondary dispersers include ants (81, 123), scatter-hoarding rodents (53), and dung beetles (36). Secondary seed movement can also be attributed to heavy rainfall or run-off. Seed longevity in tropical forests may be shorter than in more temperate areas due to warm, moist conditions and the abundance of pathogens and predators (45, 150).

Correspondence between the seed bank and aboveground vegetation in tropical forest is often related to successional stage. In the seed banks of both mature forests and regrowth sites, pioneer species usually dominate while primary species are poorly represented (45). However, forest seed banks are dominated by pioneer trees, while regrowth/farm sites are characterized by herbs. Older regrowth (30–50 yr) has higher numbers of trees and shrubs. The seed bank density of mature tropical forests is often lower than in young secondary regrowth and farms (45). Multiple factors influence the rate at which the seed bank returns to predisturbance size; these factors include degree of isolation, size, and severity of disturbance, and the regeneration strategies of colonizing species. Large-scale spatial variability in seed banks can be attributed to disturbance regime and gap characteristics. Smaller scale variability is high, and many species exhibit clumped distributions. This may be related to patterns of seed dispersal, abundance of predators, local edaphic conditions, or other factors (45).

MANAGEMENT IMPLICATIONS

Restoration and Conservation

Understanding the fates of seeds is a critical aspect of successfully restoring disturbed ecosystems or conserving endangered species. By using our knowledge of seed fates, it is possible to structure restoration or conservation processes to maximize establishment and persistence of desired species (87). The differences between restoration and conservation, in terms of seed dynamics, are a matter of scale; restoration reestablishes entire communities, while conservation maintains individual species. Because of this similarity, we consider restoration and conservation together.

Both restoration and conservation require knowledge of the influences on Phase II dispersal, seed germination, and seedling establishment. Because seed morphology and soil surface characteristics determine the vertical and horizontal movement of seeds in soils (23, 58, 110, 128), planting schemes can be devised or soil surfaces can be structured to maximize the entrapment and retention of seeds with varying morphology. Some seed attributes have fairly universal implications for establishment. Seeds that are small and lack morphological adaptations for dispersal are trapped after contacting the soil surface over a wide range of soil particle sizes (23). Thus, specialized soil surface treatments are unnecessary for entrapment, although the secondary erosion of seeds along with surface fines can be a problem. Because small seeds have low nutrient reserves, they must arrive at sites that are near the soil surface, but that have the necessary conditions for germination and establishment. Larger seeds are more likely to be moved over the soil surface after they arrive on the soil surface, and sites of deposition depend upon surface attributes (23). Because larger seeds have higher nutrient reserves, they can be seeded beneath the soil surface, placing the seeds in microsites where seed predation is less likely (113, 147), and where conditions for germination, such as higher soil moisture, are met.

Seed retention and seedling establishment of all types of seeds can be facilitated with several types of soil surface treatments. In areas where surface erosion due to wind or water is not a problem and there is adequate soil moisture for establishment, small seeds can be broadcast onto the soil surface and pressed into the soil surface either mechanically or by hand. In areas where erosion is a problem, surface mulches can hold both soil and seed in place. Organic and gravel mulches often have a similar function, and both trap naturally dispersed seeds. In northern climates, snow fences can trap wind-blown seeds and snow thus providing a more favorable environment for spring establishment. In arid areas, surface gouges trap seeds (115) and create moister microenvironments for germination and survival (159).

The importance of seed germination characteristics in restoration and con-servation efforts is becoming increasingly clear. Seeds of a variety of species from western North America exhibit large differences in germination response that vary among populations and that are habitat correlated (96, 97), empha-sizing the importance of using seeds from local or adapted populations. In Western Australia, seeds from local populations combined with detailed knowledge of seed germination responses are used to restore entire commu-nities following mining (11). These communities conform to the initial floris-tics model of succession (sensu Connell & Slatyer (26)), and the goal is to return as many species as possible in the first post-mining rainy season (11). Various seeding techniques are used that promote germination and establish-ment of the highly diverse species, including quickly returning topsoil with soil-borne seeds and appropriate mycorrhizal fungi, adding heat-shock respon-sive seed that have been collected by hand and pretreated by boiling, and mulching with the plant canopies of the original species to provide seed of serotinous species (11). In other highly diverse areas, such as the humid tropics, lack of information on seed germination responses impedes both restoration and conservation efforts (150).

Seed banks can be used to accelerate restoration or conservation of many species. As indicated above, direct replacement of topsoil that contains seeds of desired species and the appropriate fungal or bacterial symbionts promotes community or individual species reestablishment on disturbed sites (11). In some aquatic/semiaquatic systems, water-level management, augmentation of seed banks, and other techniques can be used to develop seed banks comprised of desirable species in the appropriate proportions (141). In some terrestrial systems, such as those disturbed by fire and grazing, management can produce desirable seed bank species mixes (72). Despite their importance for restoring or conserving certain life history groups, seed banks will not be useful tools for restoration or conservation of many late-successional species in closed habitats. These species seldom have long-lived seeds and, thus, do not form persistent seed banks (51, 111, 137). Because it is often difficult to manage seed banks for conservation purposes, the usual practice is to generate an artificial seed bank, i.e. a stored seed facility, to help preserve the biological and genetic diversity of certain species. Temperate species store best, and cryogenesis is an effective means of storage, if mutagenesis does not occur (13). Potential problems with this approach may include a reduction in overall genetic variation of the species and the introduction of genotypes with low relative fitness into the desired habitats (54).

In certain situations, it may be necessary to manage the animal component of the system to increase seedling establishment. In areas where seed predator densities are high, it may be possible to use economical, commercially available seed with high preference values, such as millet, to satiate the predators and minimize mortality of desired species (75). Seed-dispersing animals actively transport seeds into disturbed areas, thus expanding native populations into these areas (82). In areas with high densities of seed-dispersing animals such as heteromyid rodents, it may be possible to manage these animals actively to promote establishment of desirable native plant species. Bird dispersal to disturbed sites can be enhanced by bird-attracting structures such as snags or perches, increasing the abundance and diversity of bird-dispersed seeds (91). For plants that require dispersers but that do not form tightly coupled relationships with particular animals (many fleshy fruited species (156)), generalist animals within the area may be adequate dispersers. If a mutualism exists between a threatened plant and an animal, then both the specific animal population and the plant population must be managed.

Invasion of Undesirable Species

Invasion of new areas by alien plants can be caused by unusual long-distance migration events, either abiotic or biotic in nature. Alternatively, invasion may occur accidentally in impure seed crops, in soil or on nursery plants, on domestic livestock, or by allowing plants introduced as "useful" species to escape (5). Introductions are often short-lived if the plant species requires a disperser and no appropriate animal species is available. In contrast, some introduced species are relished by generalist, native animals that readily disperse the alien. The spread of the Brazilian pepper (*Schinus terebinthifolius*) in South Florida was facilitated by American robins (*Turdus migratorius*) and an introduced bird, the red-whiskered bulbul (*Pyononotus jocosus*) (38). Detailed understanding of the fates of seeds is very useful in the control of such aliens. Noble & Weiss (103) successfully modelled the movement of buried seeds (seed bank dynamics) of an invasive perennial (*Chrysanthemoides monilifera*) to determine the efficacy of biological control techniques that depended on pre-dispersal mortality of seeds. The model indicated that in addition to a pre-dispersal consumer, they needed a post-dispersal seed predator. This being unlikely, they suggested the possible use of an abiotic factor, fire, as a post-dispersal mortality agent. This was a good use of knowledge of the processes associated with seed fates.

SUMMARY

Phase I and Phase II dispersal are different phenomena, but workers have often failed to distinguish between them. Natural history type observations, many almost anecdotal, dominate the seed literature. Recent papers develop generalizations about the abiotic forces of dispersal, the influence of animal activities on plant populations and communities, and the effects of seed bank composition on subsequent vegetation and of vegetation on seed banks. Few studies quantitatively treat phase II dispersal, for either abiotic or biotic forces. Fewer studies treat the dynamics of seed banks, other than their composition. Virtually no study follows the fates of individual seeds. Instead, seed fates are determined from samples where the histories of individual seeds are not known (but see 74). General knowledge of the effects of abiotic factors on seed movements exists, but this type of information is uncommon and has seldom been used to explain plant establishment patterns and species distributions (but see 110). Some generalization can be made about the influences of particular animals on seed banks and the differential roles of animal-mediated dispersal as opposed to consumption. Multiple species interactions and interactions between abiotic factors and animals, likely to be extremely common in nature, are seldom documented and almost never quantified. The influence of redistribution of seeds via Phase II dispersal and the importance of seed morphology on both the spatial distributions of seeds in seed banks and on species turnover rates is becoming increasingly clear (e.g. 138, 139). Disturbance has long been recognized as an important force influencing seed bank dynamics, but the roles of seed morphology and of seed germination responses are still poorly understood. Seed loss from seed banks, especially due to pathogens, is undoubtedly significant but has received little study. Some generalization can be made about seed dynamics at the biome level. Knowledge of seed movements and fates is facilitating restoration and conservation efforts and has the potential for use in controlling the invasion of many undesirable species.

Fortunately, recent work is mechanistically oriented and treats a variety of processes and organisms simultaneously rather than cuing on a species or single species-species interactions. There is every indication that generalizations fitting the model we propose are possible and that they are important to both academic and applied ecology. Only after we accumulate more field observations and perform clever experiments can we understand what really happens during a day in the life of a seed.

ACKNOWLEDGMENTS

We thank Michael Kelrick for contributions to the formative stages of the manuscript and Doug Levey, Steve Vander Wall, Susan Meyer and Gene Schupp for constructive comments on the final versions.

Literature Cited

1. Andersen DC. 1987. Below-ground herbivory in natural communities: a review emphasizing fossorial animals. *Q. Rev. Biol.* 62:261–86

2. Andersen DC, MacMahon JA. 1985. Plant succession following the Mount St. Helens volcanic eruption: facilitation by a burrowing rodent, *Thomomys talpoides. Am. Midl. Nat.* 114: 62–69

3. Angevine MW, Chabot BF. 1979. Seed germination syndromes in higher plants. In *Topics in Plant Population Biology,* ed. DT Solbrig, S Jain, GB Johnson, PH Raven, pp. 188–206. New York: Columbia Univ. Press

4. Archibold OW. 1989. Seed banks and vegetation processes in coniferous forests. In *Ecology of Soil Seed Banks,* ed. MA Leck, VT Parker, RL Simpson, pp. 107–22. New York: Academic

5. Baker HG. 1986. Patterns of plant invasions in North America. In *Ecology of Biological Invasions of North American and Hawaii,* ed. HA Mooney, JA Drake, pp. 44–57. New York: Springer-Verlag

6. Banovetz SJ, Scheiner SM. 1994. Secondary seed dormancy in *Coreopsis lanceolata. Am. Midl. Nat.* 131: 75–83

7. Barnett RJ. 1977. The effect of burial by squirrels on germination and survival of oak and hickory nuts. *Am. Midl. Nat.* 98:319–33

8. Baskin JM, Baskin CC. 1989. Physiology of dormancy in relation to seed bank ecology. In *Ecology of Soil Seed Banks,* ed. MA Leck, VT Parker, RL Simpson, pp. 53–66. New York: Academic

9. Beattie, AJ. 1985. *The Evolutionary Ecology of Ant-Plant Mutualisms.* London: Cambridge Univ. Press. 182 pp.

10. Beattie AJ, Culver DC. 1983. The nest chemistry of two seed-dispersing ant species. *Oecologia* 56:99–103

11. Bell DT, Plummer JA, Taylor SK. 1993. Seed germination ecology in southwestern Western Australia. *Bot. Rev.* 59:24–73

12. Borchet MI, Davies FW, Michaelsen, J. 1989. Interactions of factors affecting seedling recruitment in blue oak (*Quercus douglasi*) in California. *Ecology* 70: 389–404

13. Brown AHD, Briggs JD. 1991. Sampling strategies for genetic variation in ex situ collections of endangered plant species. In *Genetics and Conservation of Rare Plants,* ed. DA Falk, KE Holsinger, pp. 99–119. New York: Oxford Univ. Press

14. Bullock SH. 1989. Life history and seed dispersal of the short-lived chaparral shrub *Dendromecon rigida* (Papaveraceae). *Am. J. Bot.* 76:1506–17

15. Burdon JJ. 1987. *Diseases in Plant Population Biology.* Cambridge, UK: Cambridge Univ. Press

16. Burdon JJ, Shattock RC. 1980. Disease in plant communities. *Appl. Biol.* 5:145–219

17. Burrows FM. 1986. The aerial motion of seeds, fruits, spores and pollen. In *Seed Dispersal,* ed. DR Murray, pp. 1–47. New York: Academic

18. Chartres CJ, Mucher HJ. 1989. The effects of fire on the surface properties and seed germination in two shallow monoliths from a rangeland soil subjected to simulated raindrop impact and water erosion. *Earth Proc. Landforms* 14:407–17

19. Chambers JC. 1989. Seed viability of alpine species: variability within and among years. *J. Range Manage.* 42:304–8

20. Chambers JC. 1993. Seed and vegetation dynamics in an alpine herb field: effects of disturbance type. *Can. J. Bot.* 71:471–85

21. Chambers JC. 1994. Disturbance, life history strategies and seed fates in alpine herbfield communities. *Am. J. Bot.* In press

22. Chambers JC, MacMahon JA, Brown RW. 1990. Alpine seedling establishment: the influence of disturbance type. *Ecology* 71:1323–41

23. Chambers JC, MacMahon JA, Haefner JH. 1991. Seed entrapment in alpine ecosystems: effects of soil particle size and diaspore morphology. *Ecology* 72: 1668–77

24. Clay K. 1987. Effects of fungal endophytes on the seed and seedling biology of *Lolium perenne* and *Festuca arundinacea. Oecologia* 73:358–62

25. Coffin DP, Lauenroth WK. 1989. Spatial and temporal variation of the seed bank of a semiarid grassland. *Am. J. Bot.* 76:53–58

26. Connell JH, Slatyer RO. 1977. Mechanisms of succession in natural communities and their role in community stability and organization. *Am. Nat.* 111: 1119–44

27. Cook RE. 1979. Patterns of juvenile mortality and recruitment in plants. In *Topics in Plant Population Biology,* ed.

DT Solbrig, S Jain, GB Johnson, PH Raven, pp. 207–23. New York: Columbia Univ. Press

28. Crawley MJ. 1992. Seed predators and plant population dynamics. In *The Ecology of Regeneration in Plant Communities,* ed. M Fenner, pp. 157–192. Wallingford, UK: CAB

29. Crist TO, Friese CF. 1993. The impact of fungi on soil seeds: implications for plants and granivores in a semiarid shrub-steppe. *Ecology* 74:2231–39

30. Crist TO, MacMahon JA. 1992. Harvester ant foraging and shrub-steppe seeds: interactions of seed resources and seed use. *Ecology* 73:1768–79

31. Davidson WD. 1993. The effects of herbivory and granivory on terrestrial plant succession. *Oikos* 68:25–35

32. Del Moral R, Bliss LC. 1993. Mechanisms of primary succession: insights resulting from the eruption of Mount St Helens. *Adv. Ecol. Res.* 24:1–66

33. Dickie JB, Ellis RH, Kraak HL, Ryder K, Tompsett PB. 1990. Temperature and seed storage longevity. *Ann. Bot.* 65: 197–204

34. Eckert RE, Jr, Peterson FF, Meurissee MS, Stephens JL. 1986. Effects of soil surface morphology on emergence and survival of seedlings in big sagebrush communities. *J. Range Manage.* 39: 414–21

35. Ellner S, Shmida A. 1981. Why are adaptations for long-range seed dispersal rare in desert plants? *Oecologia* 51:133–44

36. Estrada A, Coates-Estrada R. 1986. Frugivory in howling monkeys (*Alouatta palliata*) at Los Tuxtlas, Mexico: dispersal and fate of seeds. In *Frugivores and Seed Dispersal,* ed. A Estrada, TH Fleming, pp. 93–104. Dordrecht: Junk

37. Evans RA, Young JA. 1987. Seedbed microenvironment, seedling recruitment, and plant establishment on rangelands. In *Proc. Symp. Seed and Seedbed Ecology of Rangeland Plants,* ed. GW Frasier, RA Evans, pp. 212–20. Washington, DC: USDA Agr. Res. Serv. 311 pp.

38. Ewel JJ. 1986. Invasibility: lessons from South Florida. In *Ecology of Biological Invasions of North America and Hawaii,* ed. HA Mooney, JA Drake, pp. 214–30. New York: Springer-Verlag

39. Facelli JM, Pickett STA. 1991. Plant litter: its dynamics and effects on plant community structure. *Bot. Rev.* 57:1–32

40. Fenner M. 1985. *Seed Ecology.* New York: Chapman & Hall. 151 pp.

41. Fenner M, ed. 1992. Seeds. *The Ecology of Regeneration of Plant Communities.* Wallingford, UK: CAB. 373 pp.

42. Fleming TH. 1991. Fruiting plant-frugivore mutualism: the evolutionary theater and the ecological play. In *Plant-Animal Interactions,* ed. PW Price, TM Lewinsohn, GW Fernanades, WW Benson, pp. 119–44. New York: Wiley

43. Fowler NL. 1986. Microsite requirements for germination and establishment of three grass species. *Am. Midl. Nat.* 115:131–45

44. Franklin B. 1980. Phytochrome and seed germination. What's New? *Plant Physiol.* 11:28–32

45. Garwood NC. 1989. Tropical soil seed banks: a review. In *Ecology of Soil Seed Banks,* ed. MA Leck, VT Parker, RL Simpson, pp. 149–210. New York: Academic

46. Gentry AH. 1982. Patterns of neotropical species diversity. *Evol. Ecol.* 15:1–84

47. Glaser PH. 1981. Transport and deposition of leaves and seeds on tundra: a late-glacial analog. *Arc. Alp. Res.* 13: 173–82

48. Grant JD. 1983. The activities of earthworms and the fate of seeds. In *Earthworm Ecology,* ed. JE Satchell, pp. 107–22. London: Chapman & Hall

49. Greene DF, Johnson EA. 1993. Seed mass and dispersal capacity in wind-dispersed diaspores. *Oikos* 67:69–74

50. Griffin JR. 1971. Oak regeneration in the upper Carmel Valley, California. *Ecology* 52:862–68

51. Grime JP. 1979. *Plant Strategies and Vegetation Processes.* New York: Wiley. 222 pp.

52. Gutterman Y, Wiztum A, Evanari M. 1967. Seed dispersal and germination in *Blepharis persica. Israel J. Bot.* 16:213–34

53. Hallwachs W. 1986. Agoutis (*Dasyprocta punctata*), the inheritors of guapinol (Hymenaeacourbaril: Leguminosae). In *Frugivores and Seed Dispersal,* ed. A Estrada, TH Fleming, pp. 285–304. Dordrecht: Junk

54. Hamilton, MB. 1994. Ex situ conservation of wild plant species: time to reassess the genetic assumptions and implications of seed banks. *Conserv. Biol.* 8:39–49

55. Hamrick JL, Lee JM. 1987. Effects of soil surface topography and litter cover on germination, survival and growth of musk thistle. *Am. J. Bot.* 74:451–57

56. Harborne, JB. 1988. *Introduction to Ecological Chemistry.* New York: Academic. 356 pp.

57. Harper JL. 1977. *The Population Biol-*

ogy of Plants. New York: Academic. 892 pp.
58. Harper JL, Williams JT, Sagar GR. 1965. The behaviour of seeds in soil. I. The heterogeneity of soil surfaces and its role in determining the establishment of plants from seed. *J. Ecol.* 53:273–86
59. Hillel D. 1982. *Introduction to Soil Physics.* New York: Academic. 364 pp.
60. Hobbs RJ. 1985. Harvester ant foraging and plant species distributions in annual grassland. *Oecologia* 67:519–23
61. Holthuijzen AMA, Sharik TL, Fraser JD. 1987. Dispersal of eastern red cedar (*Juniperus virginiana*) into pastures: an overview. *Can. J. Bot.* 65:1092–95
62. Howe HF. 1986. Seed dispersal by fruit-eating birds and mammals. In *Seed Dispersal,* ed. DR Murray, pp. 123–90. New York: Academic
63. Howe HF, Smallwood J. 1982. Ecology of seed dispersal. *Annu. Rev. Ecol. Syst.* 13:201–28
64. Hughes L, Westoby M. 1990. Removal rates of seeds adapted for dispersal by ants. *Ecology* 71:138–48
65. Hulse EJ, Brown JH, Guo Q. 1993. Effects of kangaroo rat exclusion on vegetation structure and plant species diversity in the Chihuahuan Desert. *Oecologia* 95:520–24
66. Huntly N, Inouye R. 1988. Pocket gophers in ecosystems: patterns and mechanisms. *BioScience* 38:786–93
67. Janzen DH. 1971. Seed predation by animals. *Annu. Rev. Ecol. Syst.* 2:465–92
68. Jenny H. 1980. *The Soil Resource: Origin and Behavior. Ecological Studies 37.* New York: Springer-Verlag
69. Johnson EA, Fryer GI. 1992. Physical characterization of seed microsites—movement on the ground. *J. Ecol.* 80:823–36
70. Jordano P. 1992. Fruits and frugivory. In *Seeds: The Ecology of Regeneration in Plant Communities,* ed. M. Fenner, pp. 105–156. Wallingford, UK: CAB
71. Kaufmann S, McKey DB, Hossaert-McKey M, Horowitz CC. 1991. Adaptations for a two-phase seed dispersal system involving vertebrates and ants in a hemiepiphytic fig (*Ficus microcarpa:* Moraceae). *Am. J. Bot.* 78:971–77
72. Keddy PA, Wisheu IC, Shipley B, Gaudet C. 1989. Seed banks and vegetation management for conservation: toward predictive community ecology. In *Ecology of Soil Seed Banks,* ed. MA Leck, VT Parker, RL Simpson, pp. 347–66. New York: Academic
73. Kellman M. 1974. Preliminary seed budgets for two plant communities in coastal

British Columbia. *J. Biogeogr.* 1:1123–33
74. Kelrick MI. 1991. *Factors affecting seeds in a sagebrush-steppe ecosystem and implications for the dispersion of an annual plant species, cheatgrass* (Bromus tectorum). PhD thesis. Utah State Univ., Logan, Utah
75. Kelrick MI, MacMahon JA. 1985. Nutritional and physical attributes of seeds of some common sagebrush-steppe plants: some implications for ecological theory and management. *J. Range Manage.* 38:65–69
76. Kemp PR. 1989. Seed banks and vegetation processes in deserts. In *Ecology of Soil Seed Banks,* ed. MA Leck, VT Parker, RL Simpson, pp. 257–82. New York: Academic
77. Kinucan RJ, Smeins FE. 1992. Soil seed bank of a semiarid Texas grassland under three long-term (36-years) grazing regimes. *Am. Midl. Nat.* 128:11–21
78. Kjellsson G. 1991. Seed fate in an ant-dispersed sedge, *Carex pilulifera* L.: recruitment and seedling survival in tests of models for spatial dispersion. *Oecologia* 88:435–43
79. Koniak SD, Everett RL. 1983. Soil seed reserves in successional stages of pinyon woodland. *Am. Midl. Nat.* 108:295–303
80. Leck MA, Parker VT, Simpson RL, ed. 1989. *Ecology of Soil Seed Banks.* New York: Academic. 462 pp.
81. Levey DJ, Byrne MM. 1993. Complex ant-plant interactions: rain forest ants as secondary dispersers and post-dispersal seed predators. *Ecology* 74:1802–12
82. Longland WS. 1994. Seed use by desert granivores. In *Symposium on Ecology, Management, and Restoration of Intermountain Annual Rangelands.* Ogden, UT: USDA Forest Service. In press
83. Lonsdale WM. 1993. Losses from the seed bank of *Mimosa pigra:* soil microorganisms vs. temperature fluctuations. *J. Appl. Ecol.* 30:654–60
84. Louda SM. 1982. Distribution ecology: variation in plant recruitment over a gradient in relation to insect seed predation. *Ecol. Monogr.* 52:25–41
85. Louda SM. 1989. Predation in the dynamics of seed regeneration. In *Ecology of Soil Seed Banks,* ed. MA Leck, VT Parker, RL Simpson, pp. 25–51. New York: Academic
86. Mack RN, Pyke DA. 1984. The demography of *Bromus tectorum:* the role of microclimate, grazing and disease. *J. Ecol.* 72:731–49
87. MacMahon JA, Jordan WR. 1994. Ecological restoration. In *Principles of Conservation Biology,* ed. GK Meffe, R

Carroll, pp. 409–38. Sunderland, MA: Sinaur

88. MacMahon JA, Wagner FH. 1985. The Mojave, Sonoran and Chihuahuan deserts of North America. In *Hot Deserts and Arid Shrublands,* ed. M Evanari, I Noy-Meir, pp. 105–202. Amsterdam: Elsevier Sci.

89. Mares MA. 1993. Desert rodents, seed consumption, and convergence. *BioScience* 43:372–79

90. Matlack GR. 1989. Secondary dispersal of seed across snow in *Betula lenta,* a gap-colonizing tree species. *J. Ecol.* 77: 853–69

91. McClanahan TR, Wolfe RW. 1993. Accelerating forest succession in a fragmented landscape: the role of birds and perches. *Conserv. Biol.* 7:271–78

92. McDonough WT. 1974. Revegetation of gopher mounds on aspen range in Utah. *Great Basin Nat.* 34:267–75

93. McGraw JB, Vavrek MC. 1989. The role of buried viable seeds in arctic and alpine plant communities. In *Ecology of Soil Seed Banks,* ed, MA Leck, VT Parker, RL Simpson, pp. 91–106. New York: Academic

94. McGraw JB, Vavrek MC, Bennington CC. 1991. Ecological genetic variation in seed banks I. Establishment of a time transect. *J. Ecol.* 79:617–25

95. Melhop R, Scott NJ. 1983. Temporal patterns of seed use and availability in a guild of desert ants. *Ecol. Ent.* 8:69–85

96. Meyer SE. 1992. Habitat correlated variation in Firecracker Penstemon (*Penstemon eatonii* Gra: Scrophulariaceae) seed germination response. *Bull. Torr. Bot. Club.* 119:268–79

97. Meyer SE, Monsen SB. 1991. Habitat-correlated variation in Mountain Big Sagebrush (*Artemisia tridentata* spp. *vaseyana*) seed germination patterns. *Ecology* 72:739–42

98. Murdoch AJ, Ellis RH. 1992. Longevity, viability and dormancy. In *Seeds—The Ecology of Regeneration in Plant Communities,* ed. M Fenner, pp. 193–230. Wallingford, UK: CAB

99. Murray KG. 1988. Avian seed dispersal of three neotropical gap-dependent plants. *Ecol. Monogr.* 58:271–98

100. Myster RW, Pickett STA. 1993. Effects of litter, distance, density and vegetation patch type on postdispersal seed predation in old fields. *Oikos* 66:381–88

101. Nakagoshi N. 1985. Buried viable seeds in temperate forests. In *The Population Structure of Vegetation,* ed. J White, pp. 551–70. Dordrecht: Junk

102. Nelson JF, Chew RM. 1977. Factors affecting seed reserves in the Mojave

Desert ecosystem, Rock Valley, Nye County, Nevada. *Am. Midl. Nat.* 97: 300–20

103. Noble IR, Weiss PW. 1989. Movement and modelling of buried seed of the invasive perennial *Chyrsanthemoides monilifera* in coastal sand dunes and biological control. *Austr. J. Ecol.* 14:55–64

104. O'Connor TG, Pickett GA. 1992. The influence of grazing on seed production and seed banks of some African grasslands. *J. Appl. Ecol.* 29:247–60

105. Okubo A, Levin SA. 1989. A theoretical framework for data analysis of wind dispersal of seeds and pollen. *Ecology* 70:329–38

106. Oliver CD. 1981. Forest development in North America following major disturbances. *For. Ecol. Manage.* 3:153–68

107. Parmenter RR, MacMahon JA. 1983. Factors determining the abundance and distribution of rodents in a shrub-steppe ecosystem: the role of shrubs. *Oecologia* 59:145–56

108. Peart MH. 1979. Experiments on the biological significance of the morphology of seed-dispersal units in grasses. *J. Ecol.* 67:843–63

109. Peart MH. 1981. Further experiments on the biological significance of the morphology of seed-dispersal units in grasses. *J. Ecol.* 69:425–36

110. Peart MH, Clifford HT. 1987. The influence of diaspore morphology and soil-surface properties on the distribution of grasses. *J. Ecol.* 75:569–76

111. Pickett STA, McDonnell MJ. 1989. Seed bank dynamics in temperate deciduous forest. In *Ecology of Soil Seed Banks,* ed. MA Leck, VT Parker, RL Simpson, pp. 123–48. New York: Academic

112. Price MV, Heinz KM. 1984. Effects of body size, seed density, and soil characteristics on rates of seed harvest by heteromyid rodents. *Oecologia* 61:420–25

113. Price MV, Jenkins SH. 1986. Rodents as seed consumers and dispersers. In *Seed Dispersal,* ed. DR Murray, pp. 191–235. Sydney, Australia: Academic

114. Rabinowitz D, Rapp JK. 1980. Seed rain in a North American tall grass prairie. *J. Appl. Ecol.* 17:793–802

115. Reichman OJ. 1979. Desert granivore foraging and its impact on seed densities and distributions. *Ecology* 60: 1085–92

116. Reichman OJ. 1981. Factors influencing foraging desert rodents. In *Foraging Behaviour: Ecological, Ethnological and Psychological Approaches,* ed. A

Kamil, T Sargent, pp. 196–213. New York: STPM

117. Reichman OJ. 1984. Spatial and temporal variation of seed distributions in Sonoran desert soils. *J. Biogeogr.* 11:1–11

118. Reid N. 1991. Coevolution of mistletoes and frugivorous birds? *Aust. J. Ecol.* 16:457–69

119. Reynolds HG. 1958. The ecology of the Merriam kangaroo rat (*Dipodomys merriami* Mearns) on the grazing lands of Southern Arizona. *Ecol. Monogr.* 28: 111–27

120. Rice B, Westoby M. 1986. Evidence against the hypothesis that ant-dispersed seeds reach nutrient-enriched microsites. *Ecology* 67:1270–82

121. Rice KJ. 1989. Impacts of seed banks on grassland community structure and population dynamics. In *Ecology of Soil Seed Banks*, ed. MA Leck, VT Parker, RL Simpson, pp. 211–30. New York: Academic

122. Ridley HN. 1930. *The Dispersal of Plants Throughout the World*. Ashford, Kent, UK: Reeve. 745 pp.

123. Roberts JT, Heithaus ER. 1986. Ants rearrange the vertebrate-generated seed shadow of a neotropical fig tree. *Ecology* 67:1046–51

124. Sallabanks R, Courtney SP. 1992. Frugivory, seed predation, and insect vertebrate interactions. *Annu. Rev. Entomol.* 37:377–400

125. Schupp EW. 1988. Factors affecting post-dispersal seed survival in a tropical tree. *Oecologia* 76:525–30

126. Schupp EW. 1993. Quantity, quality and the effectiveness of seed dispersal by animals. *Vegetatio* 107/108:15–29

127. Schupp EW. 1994. Seed-seedling conflicts, habitat choice and patterns of plant recruitment. *Am. J. Bot.* In press

128. Sheldon JC. 1974. The behaviour of seeds in soil. III. The influence of seed morphology and the behaviour of seedlings on the establishment of plants from surface-lying seeds. *J. Ecol.* 62:47–66

129. Sheldon JC, Burrows FM. 1973. The dispersal effectiveness of the achene-pappus units of selected Compositae in steady winds with convection. *New Phytol.* 72:665–75

130. Silvertown JW. 1980. The evolutionary ecology of mast seeding in trees. *Biol. J. Linn. Soc.* 14:235–50

131. Simpson, RL, Leck MA, Parker, VT. 1989. Seed banks: general concepts and methodological issues. In *Ecology of Soil Seed Banks*, ed. MA Leck, VT Parker, RL Simpson, pp. 308. New York: Academic

132. Smith BH, Forman PD, Boyd, AE. 1989. Spatial patterns of seed dispersal and predation of two myrmecochorous forest herbs. *Ecology* 70:1649–56

133. Sorenson AE. 1986. Seed dispersal by adhesion. *Annu. Rev. Ecol. Syst.* 17:443–63

134. Stamp NE. 1989. Efficacy of explosive vs. hygroscopic seed dispersal by an annual grassland species. *Am. J. Bot.* 76:555–61

135. Stiles EW. 1989. Fruits, seeds, and dispersal agents. In *Plant-Animal Interactions*, ed. WG Abrahamson, pp. 87–122. New York: McGraw-Hill

136. Stiles EW. 1992. Animals as seed dispersers. In *Seeds. The Ecology of Regeneration in Plant Communities*, ed. M Fenner, pp. 105–56. Wallingford, UK: CAB

137. Thompson K. 1987. Seeds and seed banks. *New Phytol.* 106:23–34 (Suppl)

138. Thompson K, Band SR, Hodgson. 1993. Seed size and shape predict persistence in soil. *Func. Ecol.* 7:236–41

139. Thompson K, Green A, Jewels AM. 1994. Seeds in soil and worm casts from a neutral grassland. *Func. Ecol.* 8:29–35

140. Thoreau HD. 1993. *Faith in a Seed: The Dispersion of Seeds and Other Late Natural History Writings*. Washington, DC: Island Press/Shearwater Books. 283 pp.

141. van der Valk AG, Pederson RL. 1989. Seed banks and the management and restoration of natural vegetation. In *Ecology of Soil Seed Banks*, ed. MA Leck, VT Parker, RL Simpson, pp. 329–46. New York: Academic

142. Van Rooyen MW, Theron GK, Grobbelaar N. 1990. Life form and dispersal spectra of the flora of Namaqualand, South Africa. *J. Arid Environ.* 19: 133–45

143. van Tooren BF. 1988. The fate of seeds after dispersal in chalk grassland: the role of the bryophyte layer. *Oikos* 53:41–48

144. Vander Wall SB. 1990. *Food Hoarding in Animals*. Chicago: Univ. Chicago Press. 445 pp.

145. Vander Wall SB. 1991. Mechanisms of cache recovery by yellow pine chipmunks. *Anim. Behav.* 41:851–63

146. Vander Wall SB. 1992. The role of animals in dispersing a "wind-dispersed" pine. *Ecology* 73:614–21

147. Vander Wall SB. 1993. A model of caching depth: implications for scatter hoarders and plant dispersal. *Am. Nat.* 141:217–32

148. Vander Wall SB. 1994. Removal of

wind-dispersed pine seeds by ground-foraging vertebrates. *Oikos* 69:125–32

149. Vander Wall SB. 1994. Seed fate pathways of antelope bitterbrush: dispersal by seed-caching yellow pine chipmunks. *Ecology.* In press

150. Vázquez-Yanes C, Orozco-Segovia A. 1993. Patterns of seed longevity and germination in the tropical rainforest. *Annu. Rev. Ecol. Syst.* 24:69–88

151. Watkinson AR. 1978. The demography of a sand dune annual: *Vulpia fasciculata.* III. The dispersal of seeds. *J. Ecol.* 66:483–98

152. Westelaken IL, Maun MA. 1985. Spatial pattern and seed dispersal of *Lithospermum caroliniense* on Lake Huron sand dunes. *Can. J. Bot.* 63:125–32

153. Willson MF. 1986. Avian frugivory and seed dispersal in eastern North America. *Curr. Ornithol.* 3:223–79

154. Willson MF. 1992. The ecology of seed dispersal. In *Seeds: The Ecology of Regeneration in Plant Communities,* ed. M Fenner, pp. 61–86. Wallingford, UK: CAB

155. Willson MF. 1993a. Dispersal mode, seed shadows, and colonization patterns. *Vegetatio* 107/108:261–80

156. Willson MF. 1993b. Mammals as seed-dispersal mutualists in North America. *Oikos* 6:159–76

157. Willson MF, Rice BL, Westoby M. 1990. Seed dispersal spectra: a comparison of temperate plant communities. *J. Veg. Sci.* 1:547–62

158. Wood DM, Morris WF. 1990. Ecological constraints to seedling establishment on the Pumice Plains, Mount St. Helens, Washington. *Am. J. Bot.* 77:1411–18

159. Young JA, Evans RA, Palmquist D. 1990. Soil surface characteristics and emergence of big sagebrush seedlings. *J. Range Manage.* 43:358–67

Annu. Rev. Ecol. Syst. 1994. 25:293–324

THE EVOLUTION OF VOCALIZATION IN FROGS AND TOADS

H. Carl Gerhardt

Division of Biological Sciences, University of Missouri, Columbia, Columbia, Missouri 65211

KEY WORDS: sexual selection, energetic costs, mate choice, selective phonotaxis, sensory exploitation, reproductive character displacement

Abstract

The most commonly heard vocalizations of frogs are advertisement calls, which attract gravid females and mediate aggressive interactions between males. Frog vocalizations are energetically costly to produce, and body size often constrains the dominant frequency and intensity of vocalizations; propagation and degradation of these signals are affected by diverse physical and biotic factors. Behaviors and auditory mechanisms that mitigate these problems are discussed. With some exceptions, female preferences based on dominant frequency are intensity-dependent and mediate stabilizing selection within populations. Female preferences based on dynamic, gross-temporal properties typically mediate strong directional selection. The high values of these properties preferred by females increase a male's detectability in dense choruses and are a reliable predictor of his energetic investment in courtship. Female preferences based on fine-temporal properties (e.g. pulse rate) are often intensity-independent and usually mediate stabilizing selection within populations. The overall attractiveness of a signal depends on variation in more than one of these acoustic properties; their relative importance differs between species. Parsimony analysis supports the idea that auditory biases preceded the evolutionary appearance of call elements that enhance the attractiveness of advertisement calls in one species group of neotropical frogs. A more specific claim that the bias has not been modified by selection after the establishment of the new signal has little empirical support. Indeed, the selective consequences of positive phonotaxis to any "new" stimulus, whether or not there is a sensory

0066–4162/94/1120–0293$05.00

bias, must play a critical role in its establishment and maintenance as a mate-attraction signal and on the further evolution of the female preference. The hypothesis that present-day selective consequences of mate choice have also acted in the past evolution of call structure and preferences is supported by a few examples of reproductive character displacement. However, evolutionary divergence in signals and preferences will have multiple causes, most of which do not involve interactions between species. Phylogenetic analyses and studies of selection and other evolutionary forces in contemporary populations are complementary approaches to gaining insights about the evolution of frog vocalizations and animal communication in general.

INTRODUCTION

Vocalization dominates the reproductive behavior of many species of frogs and toads. In many species the courtship sequence is simple: the female moves to a calling male and initiates sexual contact (36, 57, 177, 178). Playbacks of recordings of male vocalizations or synthetic calls elicit phonotaxis by gravid females (57, 59, 103) and may cause cessation or increases in calling, changes in timing and type of call, and positive or negative phonotaxis by nearby males of the same or other species (69, 179). Because of their dual function, the loud, commonly heard vocalizations of most species are termed "advertisement calls" (177, 178). Acoustically mediated sexual behavior is similar in many insects (7, 42, 61a) and some songbirds (158).

Vocalizations may also function in the acquisition and defense of territories that may contain oviposition sites, resources for survival, space for unhindered courtship, or some combination of these factors (36, 80, 177, 178). Vocalization is used by amplexing males to discourage takeover attempts by other males (26). Male and female frogs may call in defense of diurnal retreats that are not used for reproduction (160, 177, 178), but most vocalization by anurans has the potential to affect mating success and hence is subject to sexual selection.

Close-range visual and tactile cues also play a role in sexual behavior of some species, such as dendobatid frogs that are diurnally active and whose males and females spend considerable time together before oviposition (178). Males engaged in active searching or satellite behavior probably rely on visual and tactile cues (177, 178). In some species there are short-range calls (72, 177–179) that may be used by either sex in both courtship and aggression. Vibratory or seismic stimuli produced when some species vocalize may also function as communication signals (113).

Speculation about the origin of frog vocalizations centers on the production of sounds that occur when air is forcibly ejected through the larynx during tactile stimulation or escape jumping (151). These sounds and most "release calls," which are given when males or unreceptive females are clasped, are typically short in duration, irregularly emitted, and highly variable in structure (151). Release calls nevertheless resemble more stereotyped, aggressive calls produced by males during close-range vocal exchanges or during physical combat (69, 179). There is often a structural gradation within a series of aggressive calls (179, 180), and there may also be a structural continuum between aggressive and advertisement calls (179, 180). Some species produce two-part calls, with one call-type directed at males and the other at females (179). In some species (e.g. *Hyla ebraccata, H. microcephala, Physalaemus pustulosus*) males append the second call-type one or more times when stimulated by a nearby rival (138, 179); in others (e.g. *Eleutherodactylus coqui, Geocrinia victoriana*), males normally produce two note-types, and the production of one call-type is modified or omitted in response to a rival's calls (94, 114).

This review focuses on the evolution of advertisement calls. After discussing basic morphological, physiological, and bioacoustical factors, I discuss primary sources of biotic selection, emphasizing the potential and demonstrated effects of female choice on male mating success. I briefly review a controversial subject: how female choice in the past may have shaped present-day signals and preferences of females.

NEUROENDOCRINE MECHANISMS AND DEVELOPMENT

Vocalization in most frogs depends on modifications of breathing behavior; one of the semi-independent brainstem oscillators identified by studying isolated brainstem preparations of the leopard frog (*Rana pipiens*) is homologous with the pulmonary respiration generator of other air-breathing vertebrates (151). An exception is underwater acoustic communication in pipids (188). There are extensive neural interconnections among neuroeffector circuits for vocalization, neuroendocrine centers, and auditory nuclei in the midbrain and forebrain (review in 185). The development and maintenance of vocalization is androgen-dependent (84); in *Xenopus laevis* there are critical times in the development of females during which exogenous androgen accomplishes partial masculinization, as judged by the acoustic structure of advertisement calls (175). The prevalence of vocalization in males relative to females is reflected in quantitative and qualitative sexual differences in laryngeal morphology, musculature controlling the larynx and its associated structures, and in the trunk muscles producing the airflow that drives laryngeal and other oscillators (84, 100, 101, 142, 152).

Most researchers tacitly assume that learning plays no role in either the development of vocalizations or female preferences. Males of the gray treefrogs (*Hyla chrysoscelis* and *H. versicolor*) were exposed throughout their development to sexual maturity to heterospecific calls; the calls they produced had the same species-specific pulse rate as did those of males exposed to conspecific calls (14). One artificially produced hybrid, *H. chrysoscelis* X *H. femoralis*, which was raised in acoustic isolation (31), produced calls with a structure intermediate between those of the parental species and similar to those of nonisolated hybrids and putative hybrids recorded in the field (49). Studies of deafened and acoustically isolated crickets provide stronger evidence that external acoustic models are irrelevant for the development of species-typical songs (review in 42). However, the possible role of acoustic experience in the development of structurally labile vocalizations (e.g. aggressive calls) has not been studied in species with prolonged breeding seasons and site fidelity. In some of these species, developing frogs have the chance to hear vocalization by adult conspecifics. Moreover, the ability of adult bullfrogs (*Rana catesbeiana*) to learn to distinguish between the advertisement calls of their territorial neighbors and those of unfamiliar males (27, 28) suggests that studies of the ontogeny of vocalization in this species may be worthwhile. Similarly, acoustic experience may affect development of vocalization in dendrobatid frogs, in which individuals also have relatively long-term associations (177, 178). Studies of the possible roles of experience, acoustic stimulation, and reproductive status on the selectivity of female anurans are completely lacking.

FACTORS AFFECTING COMMUNICATION DISTANCE

Selection will usually favor morphological characteristics and behaviors that increase the range of effective communication in the context of mate attraction. The maximum broadcast area of advertisement calls is limited by intrinsic factors such as morphology and energetic costs of signal production, and by direct, extrinsic limitations such as environmental attenuation and degradation of signals. Acoustically orienting predators and parasitoids represent more indirect evolutionary constraints on signaling behavior, which are not important for every population. Various combinations of these factors have been hypothesized to influence shifts from vocalization to alternative mate-acquisition tactics, such as searching (3, 177) and satellite behavior (3, 43, 45). However, few studies have measured the relative physiological condition of calling and silent males (e.g. 89), and none has compared the frequency of satellite behaviors in populations that differ in predator density.

Energetic and Morphological Constraints

Source amplitude is a major determinant of maximum broadcast range. Many anurans produce calls at peak amplitudes exceeding 100 dB sound pressure level (SPL—re 20 µPa) at a distance of 50 cm (52, 92, 118). Two interspecific comparisons revealed little correlation between body size and SPL (52, 118), whereas one interspecific (92) and two intraspecific (52, 70) comparisons found a positive relationship. Despite the abundance of data about call amplitude, only one study has experimentally estimated maximum communication distance. Gravid females of the barking treefrog, *Hyla gratiosa*, reliably showed phonotaxis to conspecific chorus sounds recorded 160 m from the chorus (66). The frogs did so when the amplitude in the frequency range corresponding to the low-frequency peak in the conspecific call (400–500 Hz) was 38–40 dB SPL, or about 6 dB lower than that of the chorus sound at 160 m. Thus, the estimated range of communication exceeds 160 m and could be as great as 320 m, depending on excess attenuation and other factors (66; see below).

Producing intense vocalizations is energetically costly (126, 127, 139, 165, 184). Aerobic metabolism supplies almost all of the immediate energy for sound production (e.g. 73, 126, 127, 165, 166), and metabolic rates observed in calling males exceed those of resting males by factors of 100–220% (126, 127, 184). Losses of body mass and condition occur in males of some species of frogs with prolonged breeding seasons (2, 3, 23, 98, 134), and calling rate is negatively correlated with growth rate in the carpenter frog, *R. virgatipes* (71).

Much of the variation in metabolic cost of calling is explained by variation in call duration and call rate, or a single combined measure—calling effort (183, 184). Species differences in energetic costs depend on body size and the type of call produced. For example, each pulse in the calls of the medium-sized gray treefrog *H. versicolor* requires a body wall contraction (184), resulting in much higher costs per call than in the smaller species *H. microcephala* and *Pseudacris crucifer* (184). Advertisement calls of the last two species appear to require only a single long contraction of the body wall; the rapidly repeated pulses of *H. microcephala* probably arise from passive amplitude modulation (101, 184).

Neither aerobic nor small anaerobic costs explain all of the energetic costs of calling (73, 127, 183). For example, in *H. versicolor,* males achieve the same calling effort and incur the same aerobic costs by producing either short calls at a high rate or long calls at a slow rate; however, males are much more likely to produce long calls at slow rates in competitive, dense choruses and to produce short calls at a high rate in sparse choruses (183). This context-dependent switch suggests that the production of long calls entails additional energetic costs that are not associated with short calls. The observation that males producing long calls vocalized for fewer hours during the night (183), and for fewer nights (164), than did males producing short calls further supports this hypothesis. Physiological explanations for these behavioral observations are not yet available. Wells & Taigen (184) discuss species differences in nightly and seasonal calling patterns related to costs of calling.

Sound frequency is the most important qualitative determinant of maximum communication distance. In general, an inverse relationship exists between frequency and excess attenuation, which is the drop in amplitude with distance in addition to that expected from spherical spreading from a point source (6 dB SPL per doubling of distance; 187). Many frogs approximate point sources, and some are omnidirectional radiators (52, 127). Although low-frequency sounds generally propagate more effectively than high-frequency sounds, the physics of sound production dictate that the most efficient coupling of sound with a medium occurs when the dimensions of the vibrating structure approximate the wavelengths of the sound being produced (126, 127, 139). Thus, frogs are generally constrained by their small body size to the production of relatively high-frequency sounds. Even so, mismatches between radiator dimensions and wavelength are probably the main cause of the low efficiency (acoustic power/metabolic power) of sound production in frogs and insects, which varies from less than 0.1% to 5% (126). A wide variety of behaviors ameliorate this problem, including use of elevated perches, plant leaves (as acoustic baffles), and burrows (as resonating chambers and exponential horns) (7, 8, 42, 81a, 126, 127, 136, 139, 181). Calling sites of the glass frog *Centrolenella fleischmanni* enhance signal propagation and increase mating success (181), and Australian burrowing frogs (*Heleioporus*) typically call from parts of burrows that optimize acoustic propagation of their calls (8). Whether males evaluate the acoustic properties of such sites before choosing them is a matter of speculation (181).

Dominant frequency is reasonably well correlated with body size in both intraspecific and interspecific comparisons (60, 122, 139, 190), suggesting that body size indirectly serves as a general constraint on the evolution of call frequency. This does not mean that sexual selection and other forces do not also act directly on call frequency. For example, relatively small but statistically significant geographical differences in mean dominant frequency were found, after the effects of body size were statistically removed, among 17 populations of the cricket frog (*Acris crepitans*) in Texas (150).

Frequency can be changed by altering the tension on the vocal cords or by changing the shape of arytenoid cartilages or their position relative to the vocal cords (101, 139, 142). The few species that produce substantially lower frequencies than might be predicted by their body size may have a fibrous mass associated with the vocal cords, thickened parts of vocal cords, or separate fibrous masses that vibrate in conjunction with or independently of the vocal cords (101, 139, 142). Individual frogs of two species alter call frequency to a relatively small (averages of 5% or less) but behaviorally effective extent in response to playbacks (97, 172, 173) or during vocal exchanges (172, 173).

Degradation of Signal Structure

The greater excess attenuation of high relative to low frequencies serves to filter broadband sounds. The relative decrease in high-frequency energy depends on habitat and the positions of the signaler and receiver (136, 187), with the most severe effect occurring when senders and receivers are both on a porous surface (46, 187). The spectrum will be distorted least when either the signaler or receiver is elevated and the other is at ground level. When both are elevated, some components of broadband signals will be attenuated and others enhanced, depending on their frequency, the distance between the animals, and the phase changes that occur when indirect soundwaves strike the substrate and interact with the waves that reach the receiver directly (187). Animals may use changes in the relative amplitudes of low- and high-frequency components to estimate the distance to the signaler (7, 136, 187).

Wiley & Richards (187) discuss two sources of environmental degradation of a signal's temporal structure: (a) reverberation, mainly caused by the redirection of sound waves from leaves, branches, and other objects; and (b) amplitude fluctuations, mainly caused by temporally unpredictable atmospheric turbulences. Reverberations result from multiple paths and variable delays of sound waves between the signaler and receiver. The spread of different arrival times at distant points from the signaler increases the apparent duration of sounds so that, for example, silent intervals between rapidly repeated sound pulses would be obscured. Amplitude fluctuations are caused by wind and rising pockets of warm air generated during daylight hours in open habitats. Because most frogs call at night, anuran signals are more likely to be affected by reverberation and scattering than by amplitude fluctuations unless there is a strong wind. Indeed, temperature inversions, in which cool air is trapped near the ground at night, can even provide a highly efficient, low-attenuation channel (wave guide) for sound propagation (187); one example of this phenomenon was documented in a study of the propagation of chorus sounds of the barking treefrog, *H. gratiosa* (66).

Ryan & colleagues (138, 139, 141, 149) have tested some predictions about environmental selection on the vocalizations of frogs. In the study of geographical variation in the calls of the cricket frog (*Acris*) mentioned above (150), habitat differences explained a considerable amount of variation in call frequency and some temporal properties. In another study (141), recordings of two subspecies of *Acris*, which differed in spectral and temporal properties, were played back in two different habitats. Signal degradation was estimated by cross-correlations between spectra (FFTs: fast-fourier transforms: amplitude vs frequency) of the signals rerecorded at various distances. As expected, degradation was more severe in a forested than in an open habitat; moreover, the calls of the subspecies *crepitans,* native to forested areas, suffered less degradation than those of the subspecies *blanchardi,* usually found in open areas. Although the methodology provides good estimates of spectral degradation, the phase information that is missing from FFTs is critical because, as indicated above, reverberant environments create multiple pathways and smearing of the temporal structure (139, 149). Moreover, temporal properties are important in female mate choice in this species (117). Thus, direct measurements of temporal degradation are highly desirable. For example, Ryan & Sullivan (149) measured changes in the depth of amplitude modulation with distance in a study of temporal degradation in the calls of two species of toads. Measurements of the decrease in call amplitude with distance as a function of call-type and habitat are also needed. The most direct approach for determining the biological significance of degradation of calls in various habitats would be to use female frogs themselves as judges. The temporal integrity of signals in the field has been assessed biologically in insects by monitoring the firing patterns of auditory neurons at various distances from singing males (136).

Masking Interference

Frog vocalizations have stereotyped properties and are usually produced repetitively for hours during a calling period. As in insects (136), such redundancy is one solution to the problem of signal degradation. However, redundancy also serves to counter the more severe problem of masking interference in breeding aggregations. Masking interference may be one of the many selective forces responsible for differentiation among species in a breeding community in calling site (92), breeding season (36, 139), or nightly calling periods (34). Species differences in dominant frequencies within breeding communities, together with the tuning of the auditory system to conspecific frequencies, may provide "private channels" for intraspecific communication (34, 115). However, hypothesized evolutionary responses to interspecific acoustic interference could also be generated by many other factors (69), and there have been few critical tests, such as comparisons of calls of groups of species in sympatry and allopatry (e.g. 35), or comparisons of observed patterns with null models. Moreover, the effectiveness of auditory tuning in reducing masking interference has been exaggerated by some authors (review in 69).

Calls of heterospecific males may produce some masking, but conspecific calls are a much more important source of interference; this interference can affect patterns of female choice. Moreover, some evolutionary changes (e.g. changes in call frequency) that may reduce interspecific masking are not options for reducing intraspecific masking in choruses. An extensive literature exists on the tactics used by insects and frogs to avoid or jam the signals of their neighbors (e.g. 42, 69, 74, 87, 115, 155, 179). Intraspecific and interspecific masking of a signal's acoustic fine-structure can affect female preferences

(69, 155, 156) and may help to explain density-dependent mate choice (e.g. 167). If the amplitude of an individual's call is not at least as great as that of the conspecific chorus background, then females of the green treefrog (*H. cinerea*) do not reliably find the source of individual calls (65). Directional hearing may alleviate the problem somewhat by allowing selective attention to signals of different frequencies arriving from a particular direction (the so-called "cocktail party effect"; 157), but in dense aggregations, females can probably only detect and process the calls of relatively few, nearby males (65, 136).

Predators and Parasites

Increasing a signal's range and conspicuousness may attract more predators and parasites. Acoustically orienting predators of frogs and toads include bats, owls, possums, and other frogs (82, 138). Blood-feeding dipterans (*Corethella* spp.) feed on treefrogs and are attracted to playbacks of their advertisement calls (105). Frogs may cease or reduce call rate when they detect a predator, call at lower rates on dark nights when predators are less detectable visually, or time very short calls to coincide closely with those of neighboring males (83, 138, 168).

FEMALE MATE CHOICE

Selective Phonotaxis

Some sounds readily elicit positive phonotaxis in isolation but have a low probability of doing so when other, more attractive signals are also present. For example, females of several species respond phonotactically to the advertisement calls of heterospecific males if conspecific calls are unavailable (6, 50, 57, 63, 69, 102, 148), and females may discriminate between two different kinds of heterospecific calls (6, 50) or between the calls of different hybrid individuals (49). Responses to these heterospecific or hybrid calls are usually rare if conspecific calls are presented simultaneously (49, 59, 60), but the addition of some elements of (allopatric) heterospecific calls may actually enhance the attractiveness of conspecific calls relative to unadorned conspecific calls (147, 148). Thus, the degree of selective responsiveness observed will depend critically on whether or not an animal is given a choice and, if so, on the array of signals available (see also 6, 60, 148 and below).

As illustrated in Figure 1 (a, b), the probability of observing positive phonotaxis to heterospecific calls may simply be related to the extent to which a signal excites the sensory system. Preferences based on such quantitative differences in stimulation can thus be abolished or even reversed by increasing the SPL of the heterospecific signal relative to the conspecific signal (e.g. 57).

By contrast, the structure of an audible heterospecific signal may be so different from a conspecific signal that it may not elicit positive phonotaxis at any SPL (Figure 1c), or it may elicit a different motor pattern, such as negative phonotaxis (Figure 1d). Here preferences or selective phonotaxis should be independent of the relative SPL of conspecific and heterospecific signals. Common examples are crickets and other insects, in which ultrasonic signals often elicit evasive behavior, whereas sounds of lower frequency with appropriate tem-

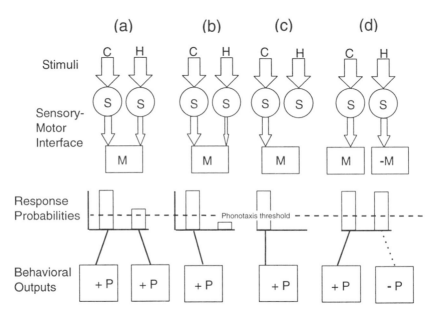

Figure 1 Diagrams illustrating hypothetical relationships between signals, the sensory-motor interface (S and M), and behavioral responses (+P or –P = positive or negative phonotaxis, respectively). The acoustic energy in C (conspecific) and in H (heterospecific) signals is equal. Each diagram compares the predicted output (probability of response) of the sensory-motor system in no-choice presentations. The relationship between sensory and motor systems is greatly oversimplified because the demarcation of the two systems is not always clear, and differential activation of a different subset of neuromuscular effectors in the same motor system may be used to generate positive or negative phonotaxis rather than there being a separate motor system for each (M,–M) for each (e.g. 124). (a) Activation of the sensory/motor interface elicits positive phonotaxis, but the probability of response is less when stimulated by a heterospecific signal than by a conspecific signal of the same sound pressure level because there is less stimulation of the sensory system. (b) If the heterospecific stimulus is even less effective than the conspecific one in stimulating the sensory system, the output of the sensory/motor system in response to H is inadequate to induce phonotaxis. However, increasing the intensity of H could do so. Thus, systems in (a) and (b) are conceptually the same in that the quantity of sensory stimulation determines the probability of positive phonotaxis. (c) H is audible but has no effect on the motor output regardless of intensity because the appropriate sensory/-motor connections are missing. (d) C and H do not differ in the degree to which they activate the sensory/motor system, and hence the probability of a behavioral response to either is the same; however, they produce different behavioral reactions to the sound.

poral patterns result in positive phonotaxis over a large range of SPL (7, 42, 81a, 123). Within a species, experience may play a role in category formation that also fits the pattern illustrated by Figure 1d. For example, territorial male bullfrogs respond to playbacks of the advertisement calls of familiar neighbors with counter-calling and to those of unfamiliar males with aggressive calls and positive phonotaxis (27, 28). I am more comfortable applying the term "recognition" to the processes responsible for the stimulus-response pattern exemplified by male bullfrogs than to those mediating simple discrimination defined merely by the presence or absence of phonotaxis (64a). However, because there are enormous difficulties in inferring underlying "decision rules" from the unconditioned behavioral responses of anurans, the terminology used to label patterns of selective responsiveness is a matter of personal preference as long as clear, operational definitions are provided (e.g. 148).

Female Selectivity and Acoustic Properties of Male Advertisement Calls

In early studies female frogs were offered choices between playbacks of recordings of males; preferences were observed for the advertisement calls of local, conspecific males over the aggressive calls of conspecific males, advertisement calls of distant populations of conspecific males, and the advertisement calls of heterospecific males and interspecific hybrids (e.g. 10, 49, 50, 57, 95, 117). Martof & Thompson (103) first used artificial sounds in two-stimulus choice tests with frogs. Subsequent studies confirmed their conclusions that synthetic sounds may omit certain acoustic properties typical of natural calls and still be as attractive as natural calls; the values of these irrelevant properties as well as those of some pertinent properties can be varied over a wide range without detracting from a signal's relative attractiveness (e.g. 59, 60, 131). Martof & Thompson's (103) conclusion that a call's overall effectiveness depends on multiple cues has been corroborated in experiments with insects, frogs, and birds (33, 38, 60, 61, 63, 112, 116).

Experiments using synthetic calls can identify acoustic differences that are responsible for various patterns of selective phonotaxis (59). Relating patterns of female preferences for a particular property to between-male variation in that property can provide a picture of the pattern and potential strength of female choice (60, 148). Extending the range of values beyond conspecific values in the population under consideration can reveal the existence of supernormal stimuli, hidden preferences or sensory biases, and the bases for population-based preferences and discrimination against heterospecific calls (4, 5, 54, 55, 59, 60, 63, 86, 103, 131, 140, 145, 153, 154, 162). The majority of studies just cited explored the significance of both intra- and interspecific variation in call properties, thus countering concerns (e.g. 148) that researchers tend to focus exclusively on either sexual selection or species recognition.

Preferences Based on Differences in Call Frequency and Spectral Patterns

Experiments using synthetic calls usually show that females prefer calls with frequencies near the population mean over calls with frequencies at one, the other, or both ends of the distribution of call frequency (32, 51, 56–58, 63) or that females prefer low- over high-frequency calls (38, 39, 112, 144, 145). However, the number and spacing of sound sources playing back calls of different frequencies reduced female selectivity in green treefrogs (58) and painted reed frogs (*Hyperolius marmoratus*)(39). Preferences for low-frequency calls over high-frequency calls were reversed when the call rate of the high-frequency alternative (112) was increased or the relative timing altered so that high-frequency calls led low-frequency ones (38).

Female preferences based on differences in frequency usually depend on the relative sound pressure level (SPL) of the alternatives. Lowering the SPL of a synthetic call with a preferred frequency by just 3–6 dB is usually sufficient to abolish the preference (32, 51, 58, 63, 112, 145). Two exceptions are green treefrogs (58), which, depending on absolute playback levels, preferred calls of 0.9 kHz over calls of 0.6 and 1.2 kHz that were 6–12 dB higher in SPL, and cricket frogs, in which females from one population preferred calls of 3.2 kHz with an SPL that was 6 dB less than an alternative of 3.5 kHz (145). In the green treefrog, greater frequency selectivity occurs at low SPL rather than at high SPL for frequency differences in the low-frequency range; the reverse is true for discrimination in the high-frequency range (58). In species producing calls with bimodal spectra, the presence and relative amplitudes of the two spectral peaks are important properties (e.g. 51, 55, 154). In the green treefrog, deficiencies in high-frequency energy relative to low-frequency energy are more important at high SPL than at low SPL (55).

Some of these patterns correlate well with the physiology of the peripheral auditory system. Anurans have two auditory organs (20, 44, 55, 58, 90, 91, 189). The basilar papilla is simpler in anatomy, has fewer hair cells, an overall higher threshold, and is tuned to higher frequencies than the amphibian papilla, which is probably capable of intensity-independent frequency resolution by a place mechanism in some species (90, 91). Interspecific variation in the extent and complexity of the caudal extension of the amphibian papilla has been the subject of speculation about sensory constraints on acoustic communication and speciation rates (90, 91, 139). If there is a single emphasized frequency or frequency band in the calls of a species, the basilar papilla is usually broadly tuned to the same frequency range (189). Part of the amphibian papilla's region of sensitivity may correspond to a low-frequency spectral peak in species with advertisement calls with bimodal spectra (20, 44, 189), but whether or not such energy is present in the call, the low-frequency sensitivity of this organ may play a role in predator detection (20).

Tuning curves of auditory neurons and audiograms are not necessarily accurate predictors of female preferences because both provide estimates of auditory threshold as a function of frequency. Most frogs undoubtedly make choices between calls at signal and noise levels that are well above threshold (20, 21, 30), and the auditory system is notoriously nonlinear (21, 189). At low SPL, the frequencies that elicit the greatest neural response (spike rate, magnitude of multi-unit activity) correspond closely to best frequencies, i.e. the frequencies eliciting the lowest threshold response; at high SPLs, the most effective frequencies may be substantially different from such best frequencies (21; JJ Schwartz, HC Gerhardt, unpublished data).

Frequency tuning and female preferences may be size-dependent (139, 186, 189). Because females are, on average, larger than males, this can generate sexual differences in tuning. For example, females in one population of spring peepers (*P. crucifer*) are tuned to lower frequencies than males and have higher sensitivity to conspecific calls than do males (189). A behavioral study of spring peepers from Missouri found that females preferred calls near the population mean over calls at the low-frequency end of the distribution, and there was no preference for average over high-frequency calls (32). Neuro-physiological studies failed to demonstrate a sex difference in tuning, but the sample size was small (10 males and 6 females), and the individuals studied did not differ greatly in body size (30). In cricket frogs (*Acris*), females are tuned to lower frequencies than males, and variation in body size may offer a partial explanation for some population-based preferences (145). In the túngara frog (*Physalaemus pustulosus*), another species in which females are significantly larger than males (138), females are tuned to, and prefer dominant frequencies in, "chucks" that are below the population mean (147). The claim that there are no statistically significant sexual differences in tuning is based on a sample of audiograms from five individuals of unspecified sex (143); many more data are required to verify the lack of size(sex)-dependent tuning in this species.

Preferences Based on Differences in Fine-Temporal Properties

Preferences based on differences in fine-temporal structure (pulse rate and shape) of advertisement calls are consistently biased toward values near the mean in local populations and are strongly intensity-independent (e.g. 59, 60, 63; see 33, 42 for comparable results from insects). Females may show a greater extent of intensity-independent discrimination against pulse rates at one end of the distribution than against pulse rates at the other end (57, 63). For example, females of *H. chrysoscelis* more strongly reject alternatives with lower-than-average pulse rates than they do those with higher-than-average pulse rates; females of *H. versicolor* show the opposite pattern (57, 63).

Fine-temporal properties also include the waveform periodicity evident in calls with harmonic spectra and rapidly amplitude-modulated (AM) wave-forms. Females of the green treefrog generally prefer synthetic calls with the periodicity typical of conspecific calls (approximately 300-Hz) over alterna-tives of much lower and higher periodicity, provided that the periodicity is clearly evident in the overall amplitude-time envelope rather than just in the fine details of the waveform (54, 159). These preferences mediate discrimi-nation of conspecific advertisement from aggressive calls, which have a strong, 50 Hz AM superimposed on the basic 300-Hz periodicity. Females choose unmodulated calls over calls with just two cycles of 50 Hz AM, and females also discriminate between pairs of calls that differ by just one cycle of 50-Hz AM (53). These results suggest that, in addition to mediating preferences for advertisement over aggressive calls, the temporal resolution ability of females would permit discrimination among the graded series of signals produced by some species of anurans (179, 180, 182). Neurophysi-ological correlates of fine-temporal preferences are described for several species (20, 44, 137, 174).

Preferences Based on Gross-Temporal Properties and Call Amplitude

In every species tested, females prefer calls played back at high rates over low call-rate alternatives; most species with pulsed calls preferred long over short calls (59, 60, 119, 164). Female green treefrogs prefer calls with a rate that was about 15% higher than that of an alternative; they also choose calls with a rate that is twice that of the mean rate in the population but do not approach calls with rates four times that of the mean, even in no-choice situations (58). Females of the gray treefrog (*H. versicolor*) prefer calls of supernormal dura-tion (about twice the mean duration) (86), which are sometimes produced once or twice by a male as a female approaches closely (43).

In *H. microcephala,* preferences for a stimulus with a call rate twice that of an alternative are abolished by a 3-dB reduction in the relative SPL of the fast-rate call and reversed by a 6-dB reduction (156). In *H. versicolor* and *H. chrysoscelis,* however, female preferences persist when the SPLs of fast or long calls are reduced by 6 dB relative to slower (50%) or shorter (50%) alternatives (60). These preferences persist when the 6-dB dif-ference is created by placing the speaker that is playing back fast-rate or long calls twice as far from the release point of the female as the speakers playing back the slower or shorter alternatives (HC Gerhardt, ML Dyson, SD Tanner, unpublished data).

Preferences for long calls in both species of gray treefrogs are not explained simply by total time of acoustic stimulation. Females preferred long calls played back at a slow call rate to short calls played back at a high rate even though the total acoustic on-time of the two alternatives was the same (86). Moreover, preferences for long calls in *H. versicolor* persisted when short calls were played back continuously and long-call playback was interrupted; many females that were in the process of approaching the source of short calls when long-call playback resumed reversed their direction and returned to the source of long calls (HC Gerhardt, ML Dyson, SD Tanner, unpublished data).

Call amplitude is strongly influenced by the environment, and the amplitude experienced by a female depends on the distance to the calling male. However, there is also significant intraspecific variation in call amplitude (52, 70, Ref. 42 for insect studies), and females typically choose the louder of two calls (43, 59, 60, 144). Experiments with *Bufo calamita* (2) suggest that females do not use differences in the sound pressure gradient, the rate of change in amplitude with distance (higher for closer sound sources than more distant ones) to determine the relative amplitudes of calls at their source. Different patterns of preferences based on, or influenced by, differences in call amplitude have been labeled as "active choice" and "passive attraction" (3, 4), but these terms have often led to the confounding of proximate and ultimate causation (140, 163).

Temperature Effects on Call Properties and Female Preferences

Cooling female green treefrogs from about 24°C to about 18°C reverses their normal preference for calls with a low-frequency peak of 0.9 kHz over calls of 0.5 kHz (67); temperature-dependent shifts in auditory tuning also occur in this and other anurans (20, 161). The change in preference is not matched by a change in call frequency when males are cooled over the same temperature range (67). The shift in the preference at low temperature may result in preferences for calls of sympatric barking treefrogs over those of conspecific males. This temperature effect also results in a sensory bias that should favor larger-than-average conspecific males at cooler-than-average temperatures. In other species, changes in temperature produce relatively small changes in dominant frequency (152, 174).

Temperature-dependent shifts in preferences (and neural correlates thereof; e.g. 137) for particular pulse rates are matched by concomitant shifts in the pulse rate of male calls (63). Changes in call rate and duration are more weakly correlated with temperature than is pulse rate (60), but preferences for joint variation in pulse rate and call rate change in the expected direction with changes in temperature (63). Such temperature coupling as well as the genetic coupling described for some interspecific hybrids (31, 33) is consistent with the hypothesis that common mechanisms, or perhaps even common genes, regulate and coordinate signal production and recognition in animal communication systems (e.g. 33). However, alternative explanations for these results are compelling (18, 31, 33, 139), and there could still be common genetic control of physiologically separate networks for call structure and preferences (18). The few genetic studies of other taxa with designs appropriate to test the hypothesis of genetic coupling have had negative or equivocal results (18).

Female Preferences, Call Variation, and Male Mating Success

Preferences for mean values over values at one end of the natural distribution of a call property, for values at one end of the distribution over values at the other end, or for values other than those at or near the mean in the population all constitute directional selection (144, 148). Even if female preferences for particular values of a call character would not affect the relative mating success of any male in the current population, preferences that result in discrimination for, or against, values outside the current range of variation could mediate directional selection on new mutants (148).

The expected efficacy of selection based on any pattern of preference will be determined mainly by the minimum difference in an acoustic property that elicits the preference relative to the between-male variability in that property in the population. Spectral and fine-temporal properties are usually static within calling bouts of individual males, and more importantly, between-male variance in these properties tends to be limited in most populations even though this is not a necessary consequence of individual stereotypy (60). Gross temporal properties are dynamic properties that often change considerably within a single calling bout; nevertheless, the repeatability (between-male variation/ total variation) of some dynamic properties may be just as high as that of static properties (60, 164). Because the range of between-male variation in mean values of dynamic properties is much greater than that of static properties (60), we might expect to observe more positive correlations between male mating success and dynamic properties than static properties. Females often show preferences between stimuli that differ by as little as 15% in either a dynamic or static property (59, 60), and the greater the extent of between-male variation in a property, the more likely a female is to encounter a small group of males in which there is a difference of 15% in that property. Moreover, the high values of dynamic properties (call rate and duration) preferred by females should increase the detectability of such signals in dense choruses.

This prediction is supported by field studies that examined male mating success over a substantial part of at least one breeding season in species in which females usually initiate matings (22, 23, 40, 111, 119, 164; additional references in 60, 62). A positive relationship was found between mating success and call rate in 6 of 7 field studies, but only 4 of 20 studies documented a large-male mating advantage in species in which call frequency is correlated with body size. Two of these 20 studies also investigated the possibility of direct correlations between mating success and call frequency in three species; no significant relationship was observed (review in 62). None of the studies above cited reported size-assortative mating, but two others did so (12, 134).

Mate choice based on call frequency could be obscured by weak correlations between body size and frequency or by large variation in the operational sex ratio throughout the breeding season (e.g. 110). The strength of mate choice based on frequency and call rate could be overestimated if these properties covary with each other or with call amplitude. Indeed, future studies should use multivariate procedures to examine the relationship between mating success and call properties.

Evolutionary Response to Female Choice

Even weak selection can have important effects over evolutionary time, but if the additive genetic variance for a trait is low, then even strong selection cannot cause rapid evolutionary change. We can conclude little about the relative importance of female choice on static and dynamic call properties over evolutionary time, because we lack estimates of the additive genetic variance of call properties in anurans. However, the majority of studies of insects have documented substantial amounts of genetic variability in properties under strong directional selection by mate choice (e.g. 19, 29, 42, 77). Indeed, the morphological, physiological, and biotic factors discussed above are probably more important constraints on the evolution of vocalizations, in the form of counter-selective forces, than is the extent of genetic variability in call properties subject to female choice.

EVOLUTION OF FEMALE PREFERENCES

The evolution of female preferences in species in which the female does not benefit directly from her choice has generated enormous controversy (85, 104). On the one hand, models of indirect selection (Fisher-effect and good-genes models) cannot easily be distinguished because both are expected to generate assortative mating that should, in turn, produce genetic correlations between the male signal and female preference. Moreover, both models predict that in the early stages of the co-evolution of trait and preference, the male trait may be an accurate indicator of viability fitness. Studies of taxa other than anurans provide the best empirical evidence for indirect selection (e.g. 9, 109, 117a). On the other hand, there is widespread agreement that direct selection on female preferences, which can also generate genetic correlations, is likely to swamp indirect effects, and recent theoretical and empirical studies have emphasized sources of direct selection that previously may have been overlooked (e.g. 78, 128, 132). For example, preferences for attributes of sounds that indicate their proximity and make them easy to locate could have evolved because they allow females to minimize energetic costs and predation risks (85, 104, 132).

Direct selection may be the best explanation for the evolution or maintenance of preferences in three species of anurans that result in size-assortative mating (12, 134) or large-male advantage (138). When the relative sizes of males and females in pairs were manipulated to be different from the usual pattern, fertilization success was decreased (12, 134, 138; but see 88). Females were sometimes observed by-passing nearby males to mate with distant ones (e.g. 12, 138). These extra movements almost certainly increase predation risks in some populations, where predators that rely on movement to detect prey (e.g. snakes and other frogs) are common. Another potential agent of direct selection is male parental care. Using egg survivorship as a measure of male parental ability, Marquez (99) found that male midwife toads with high mating success were not better parents than males with low mating success.

Preferences that reduce hybridization have been classified as targets of direct selection (60, 85) but are sometimes better viewed as examples of indirect selection. Fertilization success in crosses between many pairs of anuran species is often comparable to that of homospecific matings (108), and the deleterious genetic consequences of mate choice are expressed in the ecological, behavioral, or reproductive inferiority of hybrid offspring (49, 68).

There is no unequivocal evidence for the evolution of intraspecific preferences by the good genes mechanism in anurans. Male size and age are poor predictors of offspring quality (81). Field studies have shown that mating success was usually correlated with high call rates, which are energetically demanding; these are possible examples of viability (fitness) assessment, as discussed by Maynard Smith (104). Even though some of the greater mating success of males that produce long- or high-rate calls is probably attributable to the increased detectability of such signals in dense choruses (58, 60, 179), females of *H. versicolor* preferred long, low-rate calls over short, high-rate calls, which were adjusted to have the same ratio of sound to silence (86). The relative detectability of these alternatives should not be affected by varying levels of background noise. However, the relationship between call rate or duration and fertilization success has not been tested in this or any other species, so that we do not know if direct selection acts on these preferences. Demonstrating differences in heritable variation in offspring viability as a consequence of mating decisions is an even more difficult task. Call duration and rate are not reliable indicators of parasite load (76), as predicted by the parasite-indicator model (85).

As emphasized by Kirkpatrick & Ryan (85), demonstration of present-day correlations between a preference, a male trait, and direct or indirect fitness benefits does not prove that such relationships have existed throughout the evolutionary history of the species in question. However, comparative studies using phylogenetically informative characters are also problematic and controversial (e.g. 125). Such studies generate hypotheses about the order of the evolutionary appearance of signals and preferences, but they cannot identify the evolutionary forces that give rise to the preference, nor can they predict the evolution of the preference after the male trait appears, acquires significance as a signal, and thus becomes subject to selection generated by the consequences of mate choice (e.g. 125, 131). The two approaches—studies of current selective consequences and comparative and phylogenetic analyses—are complementary (106, 107, 147), and they both generate hypotheses about past evolutionary history that depend on assumptions that are probably impossible to validate or refute (147). Studies of contemporary selection on signals and preferences can generate hypotheses about the future evolution of these processes that are, in principle, testable in some systems (e.g. insects, fish). Moreover, contemporary studies can provide important insights about proximate mechanisms that can retard or accelerate evolutionary responses to selection (e.g. 106).

Variability in Female Preferences

Models of the evolution of female preferences assume that there is intrapopulational variation in preference (e.g. 85). Geographical differences in preference in anurans (e.g. 61, 145) and experimental evidence from other taxa (review in 15a) provide support for this assumption. However, multiple tests of female anurans have seldom demonstrated patterns of preference consistent with phenotypic variation among females from the same population (4, 60, 61, 112). That is, when a sample of females as a whole did not prefer one stimulus over its alternative, individual females were also inconsistent in choosing one or the other of the sounds (see 15 for a comparable example from insects). Additional multiple tests (to generate a score for each female) or new experimental designs may be required to estimate the extent of phenotypic variation in preference for particular call properties within populations. If these efforts are successful, it will be worthwhile to conduct the breeding and artificial selection experiments required to estimate directly levels of additive genetic variation.

Sensory Exploitation

The sensory exploitation hypothesis states that sensory biases currently affecting female choice may evolve before the appearance of male signals; such biases probably functioned originally in some other context such as predator or prey detection, or in mate choice for a signal that was subsequently lost

(85, 140, 143, 146–148). Endler (41), who emphasizes the role of the relation-ship between signal structure, perception, and environmental variables, views sensory exploitation as a subset of evolutionary processes he terms "sensory drive." The existence of preferences for novel stimuli or values of signal properties not currently expressed in a species or population is well docu-mented in both the acoustic and visual modalities (3, 5, 140, 146–148). Arak & Enquist (5) argue that these preferences are natural consequences of the imperfection of mechanisms of signal recognition.

Studies of the *Physalaemus pustulosus* species group provide the best evi-dence for sensory exploitation in anurans (131, 140, 146–148). The addition of several sonic elements, such as chucks, squawks, and doublet-whines, en-hances the attractiveness of whines (frequency modulated sweeps of several harmonics) that are necessary and sufficient for female attraction. These ad-ditional elements appear to have arisen independently in several species in the group. The addition of three chucks from a *P. pustulosus* to a whine of *P. coloradorum* (which does not produce chucks) enhanced its attractiveness to females of *P. coloradorum* relative to a whine without chucks (147). In *P. pustulosus* addition of a prefix typical of another species (*P. pustulatus*) or the playback of doublet whines, which are sometimes produced by males of *P. coloradorum*, enhanced the relative attractiveness of whines (147). Curiously, females of *P. coloradorum* did not prefer the doublet calls typical of conspe-cific males as did females of *P. pustulosus* (147). In *P. pustulosus* some elements of the whine are required to elicit phonotaxis, some are dispensable, and some enhance its attractiveness even though they are not required to elicit phonotaxis (131).

These results are consistent with the sensory exploitation hypothesis: pref-erences for various acoustic appendages exist in two species in which such signals are absent. The range of effective signals both in frequency and form (white noise is as effective as chucks) suggests that these biases are probably general in nature and can be mediated by either of the two auditory inner ear organs (131, 146). The effectiveness of chucks for *P. coloradorum* was antic-ipated by comparisons of audiograms of this species with those of *P. pustulosus* (143). Both showed a general region of enhanced frequency sensitivity—cor-responding to the tuning of the basilar papilla—in the part of the spectrum where most of the energy in chucks occurs.

Although Ryan & colleagues (140, 146) concede that preferences (sensory biases) may have evolved further after the appearance of the male trait in other systems (e.g. repertoire size in songbirds; 157a), they argue that there has been no further evolution of the sensory bias in *P. pustulosus* after the evolutionary appearance of the chuck. This more specific hypothesis assumes that the preference for chucks of lower-than-average frequency found in females of *P. pustulosus* in Panama has been conserved from the ancestral state; direct selection in the form of increased fertilization success of females that mate with males producing lower-than-average frequency chucks could maintain the bias (143) but presumably has not altered it. The data to support this argument consist of the "statistically indistinguishable" (147, p. 190) tuning properties of *P. pustulosus* and *P. coloradorum,* which was based on a comparison of audiograms of only five individuals of the first species and six of the second (143). Moreover, there were unspecified numbers of males and females in each of these small samples. Given the uncertainties of extrapolating from audiograms to frequency preferences and the possibility of sexual dimorphism in tuning, the hypothesis that there has been no evolution of the frequency preference for lower-than-average chucks should be tested behaviorally. Specifically, females of *P. coloradorum* and other species without chucks should be tested with chucks of low and high frequency. If the same proportion of females prefer low-frequency chucks as in *P. pustulosus* (147), then the hypothesis would be supported. As in any test of a null hypothesis, the magnitude of a difference in preference (or tuning) that could falsify the hypothesis should be specified, and enough data should be collected to permit detection of such a difference.

As the cartoons of Figure 1 illustrate, the probability of observing a response depends not only on a signal's effectiveness in stimulating the sensory system, but also on whether and how that stimulation translates into such behavior. We might predict, for example, that the evolutionary changes in the sensory-motor interface required for a sensory bias to serve a new function in mate choice might be more easily accomplished if its former function was to detect prey rather than predators. Obviously, positive phonotaxis would be likely in the first case, and negative phonotaxis, in the second. I think that the conversion of any stimulus, whether or not it taps some sensory bias, into a mate-attraction signal must depend critically on the selective consequences of the response— just as in the analogous process of reinforcing a learned response to a previously irrelevant stimulus. Herrnstein (78a) makes the same point in discussing discrimination and categorization of visual stimuli in animals. Other authors (125, 131a) emphasize the importance of selection in their discussions of the sensory exploitation hypothesis, and even Ryan & Keddy-Hector (144, p. S24) suggest that selection can, in general, play a role in the establishment of preexisting biases.

In any event, a transition from ultrasonic sensitivity, which has evolved numerous times in insects (42, 48), from it usual role in bat avoidance (42, 123, 124) to a mate-attraction function has apparently taken place in some groups of insects, including many species of katydids (7). Behavioral studies have also shown that in the cricket, *Teleogryllus oceanicus,* the high frequencies that normally elicit evasive behavior can also elicit positive orientation towards a stimulus with an appropriate temporal pattern (124). In the

Australian whistling moth, *Hecatesia thyridion,* the attraction of females to ultrasonic signals produced by conspecific males may have been facilitated by a switch to diurnal activity, which eliminated the need to detect nocturnal bats (48). Comparative studies of wide-ranging species in areas with different predators that emit or detect acoustic signals may yield important insights into the evolution of female preferences and animal communication in general.

SEXUAL SELECTION, SPECIES RECOGNITION, AND GEOGRAPHICAL VARIATION

Signal divergence (10, 24, 57, 117, 150) and strong female selectivity vis-á-vis intraspecific variation in conspecific calls may exist in areas where other species with similar calls are absent (e.g. 57, 61, 138, 145). Nevertheless, the usual pattern of homospecific pairing observed within assemblages of broadly sympatric species is usually interpreted as evidence for "species recognition" or "species isolation." Littlejohn (93) prefers the term "homogamy" (positive assortative mating) because it is a more general term that skirts perennial debates about species concepts (75, 121) and avoids the difficulty of applying species labels to individuals in hybrid zones (75, 93). However, because hybrid swarms consist of many individuals of diverse genetic makeup, defining sets of "like" individuals is often arbitrary (75). I prefer the term "mate choice" because it does not specify arbitrary classes of individuals and carries no implications about cognitive processes. Indeed, there is a system in which females are expected to prefer the calls of another sympatric taxon. In the hybridogenetic system of the European water frogs (1), females of *Rana esculenta,* which originated as a hybrid between *R. lessonae* and *R. ridibunda,* eliminate the *lessonae* genome premeiotically, and thus require matings with males of *R. lessonae* to ensure that zygotes regain the lost genome. Homospecific matings rarely produce viable offspring. Females of *R. esculenta* preferentially mate with males of *R. lessonae* rather than conspecific males, but the roles of species-specific calls and selective phonotaxis in mediating such preferences require additional study (1).

Although females select among conspecific calls in areas of allopatry, the adverse consequences of hybridization in sympatric areas may constrain directional female choice for values of one or more call properties that might be favored in areas of allopatry (57). For example, females of the green treefrog (*Hyla cinerea*) from eastern Georgia prefer calls with frequencies at or near the mean in local populations; these preferences reduce the chances of mating with males of two sympatric species: squirrel treefrogs (*H. squirella*), which produce calls of higher frequency; and barking treefrogs (*H. gratiosa*), which produce calls of lower frequency. The preference is sufficiently narrow that both low- and high-frequency calls of some conspecific males would also be less attractive than calls of about average frequency (51, 58, 60). Discrimination against the two kinds of heterospecific males is a current consequence of the preference and a current constraint on directional selection for dominant frequency in sympatric areas. Females that fail to discriminate against calls of an extreme type might mate with a male of another species (57).

The stabilizing pattern of frequency preferences in green treefrogs could have evolved because of the benefits derived from intraspecific mate choice, avoiding hybridization, or both. One way to examine this question is to compare the selectivity of females from areas of sympatry with that of females from areas of allopatry (see below). Stabilizing patterns of preference also exist in the barking treefrog in eastern Georgia (56, 57), where there is no risk of mating with another species with calls of lower frequency. Thus, hybridization does not constrain female choice for lower-than-average call frequencies in these populations, nor was hybridization in the past likely to have been responsible for the evolution of discrimination against low-frequency calls. Even in populations of green treefrogs, sympatric with barking and squirrel treefrogs, preferences for other, independently varying properties of the call such as call rate mediate strong directional female choice (57, 58). The main point of Gerhardt (57) was not that a requirement for species recognition is a general constraint on (directional) female choice, but that interspecific interactions in the context of mate-choice potentially influence the evolution of particular properties of calls and preferences—a hypothesis that contradicts Paterson's (120) position that such interspecific interactions (e.g. reinforcement) never play such a role. This perspective was misinterpreted by Ryan & Rand (148; see also 140, 150), who cite Gerhardt (57), along with Paterson (121), as taking the position that "if there is species recognition there can not be sexual selection (148, p. 647)."

Reproductive Character Displacement

Geographical variation in the structure of vocalizations (10, 36, 47, 57, 61, 69, 92, 95, 96, 117, 150) and female preferences (61, 95, 145) are commonly observed in anurans. Because the forces promoting geographical variation are multiple and undoubtedly change in their relative importance through the evolutionary history of a lineage, explanations for patterns of geographical variation are especially susceptible to the confounding of cause and effect. The interplay of ecological, genetic, and behavioral factors that may determine the outcome of interactions between partially differentiated populations is the subject of several reviews (69, 75, 93), and the relative importance of these same forces in speciation is a particularly controversial topic (e.g. 75, 79, 120, 121).

As discussed above, the adverse consequences of hybridization represent a form of selection on female preferences and male call structure. Patterns expected from reproductive character displacement or reinforcement are greater female selectivity in areas of sympatry than in areas of allopatry, divergence in the call structure of one or both taxa in areas of sympatry, or both. The best examples for call structure include members of the *Litoria ewingi* complex (92, 95), *Pseudacris triseriata feriarum* and *P. n. nigrita* in Georgia and Florida (47), and *Gastrophryne carolinensis* and *G. olivacea* in Texas (96). No evidence for character displacement was found in the cricket frogs *Acris gryllus* and *A. crepitans* (117), nor in the subspecies of the latter species (150). Rather, most of the clinal variation in call frequency in *A. crepitans* was attributed to a pleiotropic consequence of a clinal increase in body size due to selection for resistance to desiccation in the more arid western part of the range (117). Within Texas, however, some nonclinal variation was attributed to habitat effects and clinal variation, to gene flow (150). A gradual, east-to-west increase in pulse rate occurs in the gray treefrog (*H. chrysoscelis*)(57). A small amount of nonclinal variation in this property may be due to sympatry with *H. versicolor* in some areas (130), but much more of the interpopulational differentiation is associated with the distribution of genetic lineages defined by mtDNA sequence divergence and chromosome markers (129; HC Gerhardt et al, in preparation).

Waage (171) hypothesized that character displacement of signal structure need not occur if the signals are discriminably different when sympatry is reestablished; selection would then act primarily on female selectivity. A study (61) of *H. chrysoscelis* supports this hypothesis. The relative importance of pulse rate, a species-specific character, and call duration, which mediates strong intraspecific directional selection, depended on whether females came from areas of sympatry with *H. versicolor* or areas of remote allopatry (61). More specifically, females from sympatric areas were much more likely than females from allopatric areas to choose a short call with a pulse rate typical of local, conspecific males over a long call with an inappropriate pulse rate. Additional studies of geographical variation in female selectivity are badly needed, and in the *H. versicolor* complex, additional comparisons of sympatric and allopatric populations of the same genetic lineage are in progress.

POLYPLOID SPECIATION AND THE STRUCTURE OF VOCALIZATIONS

Polyploid speciation is uncommon but widespread among anuran families (169). Some species, notably in the genus *Xenopus* (169), arose through hybridization (allopolyploidy); the call structure of such hybrid species would be expected to be distinguishable from that of the putative parental species. Indeed, in the hybridogenetic water frogs, the pulse rates of the hybrid, *Rana esculenta*, are intermediate with respect to the parental forms, *R. lessonae* and *R. ridibunda* (13). However, most polyploid species are thought to have arisen via autopolyploidy (169) possibly through a triploid intermediate (11). Bogart & Wasserman (11; see also 130) speculated that polyploidy affects pulse rate through changes in cellular dimensions or tissue mass or density (see also 130); this hypothesis was suggested by the observation that the blood cells of the polyploid species *H. versicolor* and *Odontophrynus americanus* are larger and the pulse rates of male advertisement calls are slower than those of their diploid counterparts. New experimental evidence from artificially produced tetraploids of *Hyla japonica*, supports this hypothesis: Mean pulse rate in tetraploids was about 20% lower than that of diploid controls (170). In the *H. versicolor* complex, autopolyploidy should result at least in call differentiation sufficient to serve as the basis for selection for assortative mating by ploidy, even if the ploidy change does not generate call differences of the magnitude observed at present. Otherwise, we would have to hypothesize at least three mutational events in the same direction to account for the fact that males of three lineages of *H. versicolor*, which arose independently from two of more diploid lineages (129), have very similar pulse rates (HC Gerhardt, MB Ptacek, unpublished data).

CONCLUSIONS

Two major philosophical themes emerge from this review. First, studies of proximate mechanisms and of the relevant physical aspects of the environment have much to contribute to our understanding of the evolutionary processes affecting frog vocalizations and female choice. Second, there are multiple factors influencing the evolution of communication systems, and we should therefore keep our minds open to a diversity of theoretical explanations and approaches to gathering empirical data.

ACKNOWLEDGMENTS

I thank B Buchanan, S Tanner, M Ryan, M Kirkpatrick, K Wells, L Dugatkin, G Watson, and especially C Murphy for comments and criticisms. My research has been generously supported by the National Science Foundation and National Institutes of Mental Health.

Literature Cited

1. Abt G, Reyer U-J. 1993. Mate choice and fitness in a hybrid frog: *Rana esculenta* females prefer *Rana lessonae* males over their own. *Behav. Ecol. Sociobiol.* 32:221–28
2. Arak A. 1983. Sexual selection by male-male competition in natterjack toad choruses. *Nature* 306:261–62
3. Arak A. 1983. Male-male competition and mate choice in anuran amphibians. In *Mate Choice*, ed. P Bateson, 8:67–107. Cambridge: Cambridge Univ. Press. 462 pp.
4. Arak A. 1988. Female mate selection in the natterjack toad: active choice or passive attraction? *Behav. Ecol. Sociobiol.* 22:317–27
5. Arak A, Enquist M. 1993. Hidden preferences and the evolution of signals. *Philos. Trans. R. Soc. Lond. B* 340:207–13
6. Backwell PRY, Jennions MD. 1993. Mate choice in the Neotropical frog, *Hyla ebraccata*: sexual selection, mate recognition and signal selection. *Anim. Behav.* 45:1248–50
7. Bailey WJ. 1991. *Acoustic Behaviour of Insects: An Evolutionary Perspective.* London/New York/Tokyo/Melbourne/Madras: Chapman & Hall. 225 pp.
8. Bailey WJ, Roberts JD. 1981. The bioacoustics of the burrowing frog *Heleioporus* (Leptodactylidae). *J. Nat. Hist.* 15:259–88
9. Bakker TCM. 1993. Positive genetic correlation between female preference and preferred male ornament in stickelbacks. *Nature* 363:255–57
10. Blair WF. 1974. Character displacement in frogs. *Am. Zool.* 14:1119–25
11. Bogart JP, Wasserman AO. 1972. Diploid-polyploid cryptic species pairs: a possible clue to evolution by polyploidization in anuran amphibians. *Cytogenetics* 11:7–24
12. Bourne GR. 1993. Proximate costs and benefits of mate acquisition at leks of the frog *Ololygon rubra. Anim. Behav.* 45:1051–59
13. Brzoska J. 1982. Vocal response of male European water frogs (*Rana esculenta* complex) to mating and territorial calls. *Behav. Processes* 7:37–47
14. Burger J. 1980. *The effects of acoustic experience on the calls of gray treefrogs.* MA thesis, Univ. Missouri, Columbia. 111 pp.
15. Butlin RK. 1993. The variability of mating signals and preferences in the brown planthopper, *Nilaparva lugens* (Homoptera: Delphacidae). *J. Insect Behav.* 6:125–40
15a. Butlin RK. 1994. Genetic variation in mating signals and responses. In *Speciation and the Recognition Concept: Theory and Application,* ed. DM Lambert, HC Spencer. Baltimore & London: Johns Hopkins Univ. Press. In press
16. Deleted in press
17. Deleted in press
18. Butlin RK, Ritchie MG. 1989. Genetic coupling in mate recognition systems: What is the evidence? *Biol. J. Linn. Soc.* 37:237–46
19. Cade WH. 1984. Genetic variation underlying sexual behavior and reproduction. *Am. Zool.* 24:355–66
20. Capranica RR, Moffat AJM. 1983. Neurobehavioral correlates of sound communication in anurans. In *Advances in Vertebrate Neuroethology,* ed. J-P Ewert, RR Capranica, DJ Ingle, 36:701–30. New York: Plenum. 1238 pp.
21. Capranica RR. 1992. Untuning of the tuning curve: is it time? *Semin. Neurosci.* 4:401–8
22. Cherry MI. 1992. Sexual selection in the leopard toad, *Bufo pardalis. Behaviour* 120:164–76
23. Cherry MI. 1993. Sexual selection in the raucous toad, *Bufo rangeri. Anim. Behav.* 45:359–73
24. Claridge MF, Morgan JC. 1993. Geographical variation in acoustic signals of the planthopper, *Nilaparvata bakeri* (Muir), in Asia: species recognition and sexual selection. *Biol. J. Linnean Soc.* 48:267–81
25. Crossley SA. 1975. Changes in mating behavior produced by selection for ethological isolation between ebony and vestigial mutants of *Drosophila melanogaster. Evolution* 28:631–47
26. Davies NB, Halliday TR. 1978. Deep croaks and fighting assessment in toads *bufo. Nature* 274:683–85
27. Davis MS. 1987. *Neighbor recognition by sound in the North American bullfrog,* Rana catesbeiana. PhD thesis. Univ. Missouri, Columbia. 70 pp.
28. Davis MS. 1988. Acoustically mediated neighbor recognition in the North American bullfrog, *Rana catesbeiana. Behav. Ecol. Sociobiol.* 21:185–90
29. De Winter AJ. 1992. The genetic basis and evolution of acoustic mate recognition signals in a *Ribautodelphax* planthopper (Homoptera, Delphacidae) 1. The female call. *J. Evol. Biol.* 5:249–65
30. Diekamp BM, Gerhardt HC. 1992. Mid-

brain auditory sensitivity in the spring peeper (*Pseudacris crucifer*): correlations with behavioral studies. *J. Comp. Physiol.* A 171:245–50

31. Doherty JA, Gerhardt HC. 1984. Acoustic communication in hybrid treefrogs: sound production by males and selective phonotaxis of females. *J. Comp. Physiol.* 154:319–30

32. Doherty JA, Gerhardt HC. 1984. Evolutionary and neurobiological implications of selective phonotaxis in the spring peeper (*Hyla crucifer*). *Anim. Behav.* 32:875–81

33. Doherty JA, Hoy RR. 1985. Communication in insects. III. the auditory behavior of crickets: some views of genetic coupling, song recognition, and predator detection. *Q. Rev. Biol.* 60: 457–72

34. Drewry GE, Rand AS. 1983. Characteristics of an acoustic community: Puerto Rican frogs of the genus *Eleutherodactylus. Copeia* 1983:941–53

35. Duellman WE, Pyles RA 1983. Acoustic resource partitioning in anuran communities. *Copeia* 1983:639–49

36. Duellman WE, Trueb L. 1986. *Biology of Amphibians.* New York: McGraw-Hill. 670 pp.

37. Deleted in proof

38. Dyson ML, Passmore NI. 1988. Two-choice phonotaxis in *Hyperolius marmoratus* (Anura: Hyperoliidae): the effect of temporal variation in presented stimuli. *Anim. Behav.* 36:648–52

39. Dyson ML, Passmore NI. 1992. Effect of intermale spacing on female frequency preferences in the painted reed frog. *Copeia* 1992:1111–14

40. Dyson ML, Passmore NI, Bishop PJ, Henzi SP. 1992. Male behavior and correlates of mating success in a natural population of African painted reed frogs (*Hyperolius marmoratus*). *Herpetologica* 48:232–42

41. Endler JA. 1993. Some general comments on the evolution and design of animal communication systems. *Philos. Trans. R. Soc. Lond.* 340:215–25

42. Ewing AW. 1989. *Arthroprod Bioacoustics: Neurobiology and Behavior.* Ithaca, NY: Comstock/Cornell. 260 pp.

43. Fellers GM. 1979. Aggression, territoriality and mating behavior in North American treefrogs. *Anim. Behav.* 27: 107–19

44. Feng AS, Hall JC, Gooler DM. 1990. Neural basis of sound pattern recognition in anurans. *Prog. Neurobiol.* 34: 313–29

45. Forester DM, Lykens DV. 1986. Significance of satellite males in a population of spring peepers (*Hyla crucifer*). *Copeia* 1986:719–24

46. Forrest TG, Green DM. 1991. Power output and efficiency of sound production by crickets. *Behav. Ecol.* 2:327–38

47. Fouquette MJ. 1975. Speciation in chorus frogs. I. Reproductive character displacement in the *Pseudacris nigrita* complex. *Syst. Zool.* 24:16–23

48. Fullard JH, Yack JE. 1993. The evolutionary biology of insect hearing. *Trends Ecol. Evol.* 8:248–52

49. Gerhardt HC 1974. Vocalizations of some hybrid treefrogs:acoustic and behavioral analyses. *Behaviour* 49:130–51

50. Gerhardt HC. 1974. Behavioral isolation of the treefrog *Hyla cinerea* and *Hyla andersonii. Am. Midl. Natur.* 91:424–33

51. Gerhardt HC. 1974. The significance of some spectral features in mating call recognition in the green treefrog (*Hyla cinerea*). *J. Exp. Biol.* 61:229–41

52. Gerhardt HC. 1975. Sound pressure levels and radiation patterns of the vocalizations of some North American frogs and toads. *J. Comp. Physiol.* A 102:1–12

53. Gerhardt HC. 1978. Discrimination of intermediate sounds in a synthetic call continuum by female green tree frogs. *Science* 199:1089–91

54. Gerhardt HC. 1978. Mating call recognition in the green treefrog (*Hyla cinerea*): the significance of some fine-temporal properties. *J. Exp. Biol.* 74:59–73

55. Gerhardt HC. 1981. Mating call recognition in the green treefrog (*Hyla cinerea*): importance of two frequency bands as a function of sound pressure level. *J. Comp. Physiol.* 144:9–16

56. Gerhardt HC. 1981. Mating call recognition in the barking treefrog (*Hyla gratiosa*): responses to synthetic calls and comparisons with the green treefrog (*Hyla cinerea*). *J. Comp. Physiol.* A 144:17–25

57. Gerhardt HC. 1982. Sound pattern recognition in some North American treefrogs (Anura: Hylidae): implications for mate choice. *Am. Zool.* 22:581–95

58. Gerhardt HC. 1987. Evolutionary and neurobiological implications of selective phonotaxis in the green treefrog (*Hyla cinerea*). *Anim. Behav.* 35:1479–89

59. Gerhardt HC. 1988. Acoustic properties used in call recognition by frogs and toads. In *The Evolution of the Amphibian Auditory System,* ed. B Fritzsch, T Hethington, MJ Ryan, W Wilczynski, W Walkowiak, 21:455–83. New York: Wiley

60. Gerhardt HC. 1991. Female mate choice

in treefrogs: static and dynamic acoustic criteria. *Anim. Behav.* 42:615–35

61. Gerhardt HC. 1994. Reproductive character displacement on female mate choice in the grey treefrog *Hyla chrysoscelis. Anim. Behav.* 47:959–69

61a. Gerhardt HC. 1994. Selective phonotaxis to long-range acoustic signals in insects and anurans. *Am. Zool.* In press

62. Gerhardt HC, Daniel RE, Perrill SA, Schramm S. 1987. Mating behaviour and male mating success in the green treefrog. *Anim. Behav.* 35:1490–503

63. Gerhardt HC, Doherty JA. 1988. Acoustic communication in the gray treefrog, *Hyla versicolor*: evolutionary and neurobiological implications. *J. Comp. Physiol. A* 162:261–78

64a. Gerhardt HC, Dyson ML, Tanners, Murphy CG. 1994. Female treefrogs do not avoid heterospecific calls as they approach conspecific calls; implications for mechanisms of mate choice. *Anim. Behav.* 47:1323–32

65. Gerhardt HC, Klump GM. 1988. Masking of acoustic signals by the chorus background noise in the green treefrog: a limitation on mate choice. *Anim. Behav.* 36:1247–49

66. Gerhardt HC, Klump GM. 1998. Phonotactic responses and selectivity of barking treefrogs (*Hyla gratiosa*) to chorus sounds. *J. Comp. Physiol. A* 163: 795–802

67. Gerhardt HC, Mudry KM. 1980. Temperature effects on frequency preferences and mating call frequencies in the green treefrog, *Hyla cinerea* (Anura: Hylidae). *J. Comp. Physiol.* 137:1–6

68. Gerhardt HC, Ptacek MB, Barnett L, Torke K.1994. Hybridization in the diploid-tetraploid treefrogs *Hyla chrysoscelis* and *Hyla versicolor. Copeia* 1994:51–9

69. Gerhardt HC, Schwartz JJ. 1994. Interspecific interactions in anuran courtship. In *Amphibian Biology,* Vol 2: *Social Communication,* ed. H Heatwole, BK Sullivan. Sydney: Surrey Beattley & Sons. In press

70. Given MF. 1987. Vocalizations and acoustic interactions of the carpenter frog, *Rana virgatipes. Herpetologica* 43: 467–81

71. Given MF. 1988. Growth rate and the cost of calling activity in male carpenter frogs, *Rana virgatipes. Behav. Ecol. Sociobiol.* 22:153–60

72. Given MF. 1993. Male response to female vocalizations in the carpenter frog, *Rana virgatipes. Anim. Behav.* 46:1139–49

73. Grafe TU, Schmuck R, Linsenmair KE.

1992. Reproductive energetics of the African reed frogs, *Hyperolius viridiflavus* and *Hyperolius marmoratus. Physiol. Zool.* 65:153–71

74. Greenfield MD. 1994. Evolution of alternating and synchronous chorusing: cooperation or conflict? *Am Zool.* In press

75. Harrison RG. 1990. Hybrid zones: windows on evolutionary processes. In *Oxford Surveys in Evolutionary Biology,* ed. D Futuyma, J Antonovics, 7:69–128. Oxford: Oxford Univ. Press

76. Hausfater G, Gerhardt HC, Klump GM. 1990. Parasites and mate choice in gray treefrogs, *Hyla versicolor. Am. Zool.* 30:299–311

77. Hedrick AV. 1988. Female choice and the heritability of attractive male traits: an empirical study. *Am. Nat.* 132:267–76

78. Hedrick AV, Dill L. 1993. Mate choice by female crickets is influenced by predation risk. *Anim. Behav.* 46:193–96

78a. Herrnstein RJ. 1990. Levels of stimulus control: a functional approach. *Cognition* 37:133–66

79. Howard DS. 1993. Reinforcement: origin, dynamics, and fate of an evolutionary hypothesis. In *Hybrid Zones and the Evolutionary Process,* ed. RG Harrison, pp. 46–69

80. Howard RD. 1978. The evolution of mating strategies in bullfrogs, *Rana catesbeiana. Evolution* 32:850–71

81. Howard RD, Whiteman HH, Schueller TI. 1994. Sexual selection in American toads: a test of good genes hypotheses. *Evolution.* In press

81a. Huber F, Moore T, Loher W, eds. 1989. *Cricket Behavior and Neurobiology.* New York: Cornell Univ. Press. 565 pp.

82. Jaeger RG. 1976. A possible prey-call window in the anuran auditory system. *Copeia* 1976:833–4

83. Jennions MD, Backwell PYR. 1992. Chorus size influences on the anti-predator response of a Neotropical frog. *Anim. Behav.* 44:990–92

84. Kelley DB, Gorlick DL. 1990. Sexual selection and the nervous system. *Bioscience* 40:275–83

85. Kirkpatrick M, Ryan MJ. 1991. The evolution of mating preferences and the paradox of the lek. *Nature* 350:33–38

86. Klump GM, Gerhardt HC. 1987. Use of non-arbitrary acoustic criteria in mate choice by female gray treefrogs. *Nature* 326:286–88

87. Klump GM, Gerhardt HC. 1992. Mechanisms and function of call-timing in male-male interactions in frogs. In *Playback and Studies of Animal Communi-*

cation, ed. PK McGregor, pp. 153–74. New York/London: Plenum

88. Kruse KC. 1981. Mating success, fertilization potential, and male body size in the American toad (*Bufo americanus*). *Herpetologica* 37:228–33

89. Lance SL, Wells KD. 1993. Are spring peeper satellite males physiologically inferior to calling males? *Copeia* 1993: 1162–65

90. Lewis ER, Lombard ER. 1988. The amphibian inner ear. In *The Evolution of the Amphibian Auditory System,* ed. B Fritzsch, T Hethington, MJ Ryan, W Wilczynski, W Walkowiak, 5:93–124. New York: Wiley. 705 pp.

91. Lewis ER, Hecht EL, Narins PM. 1992. Diversity of form in the amphibian papilla of Puerto Rican frogs. *J. Comp. Physiol. A* 171:421–35

92. Littlejohn MJ. 1977. Long-range acoustic communication in anurans: an integrated and evolutionary approach. In *The Reproductive Biology of Amphibians,* ed. DH Taylor, SI Guttman, 8:263–94. New York/London: Plenum. 475 pp.

93. Littlejohn MJ. 1993. Homogamy and speciation: a reappraisal. In *Oxford Surveys of Evolutionary Biology,* ed. D Futuyma, J. Antonovics, 9:135–64. Oxford: Oxford Univ. Press

94. Littlejohn MJ, Harrison PA. 1985. The functional significance of the diphasic advertisement call of *Geocrinia victoriana* (Anura: Leptodactylidae). *Behav. Ecol. Sociobiol.* 16:363–73

95. Littlejohn MJ, Loftus-Hills JJ. 1968. An experimental evaluation of premating isolation in the *Hyla ewingi* complex (Anura:Hylidae). *Evolution* 22:659–63

96. Loftus-Hills JJ, Littlejohn MJ. 1992. Reinforcement and reproductive character displacement in *Gastrophryne carolinensis* and *G. olivacea* (Anura: Microhylidae): a reexamination. *Evolution* 46:896–906

97. Lopez PT, Narins PM, Lewis ER, Moore SW. 1988. Acoustically-induced call modification in the white-lipped frog, *Leptodactylus albilabris. Anim Behav.* 36:1295–308

98. Mac Nally RC. 1981. On the reproductive energetics of chorusing males: energy depletion profiles, restoration and growth in two sympatric species of *Ranidella* (Anura). *Oecologia* 51:181–88

99. Marquez R. 1993. Male reproductive success in two midwife toads, *Alytes obsteticans* and *A. cisternasii. Behav. Ecol. Sociobiol.* 32:283–91

100. Marsh RL, Taigen TL. 1987. Properties enhancing aerobic capacity of calling muscles in gray treefrogs, *Hyla versicolor. Am. J. Physiol.* 252:R786–93

101. Martin WF. 1972. Evolution of vocalizations in the genus *Bufo.* In *Evolution in the Genus* Bufo, ed. WF Blair, 15:279–309. Austin: Univ. Texas Press

102. Martof BS. 1961. Vocalization as an isolating mechanism in frogs. *Am. Midl. Natur.* 71:118–26

103. Martof BS, Thompson EF. 1964. A behavioral analysis of the mating call of the chorus frog, *Pseudacris triseriata. Am. Midl. Natur.* 71:198–209

104. Maynard Smith J. 1991. Theories of sexual selection. *Trends Ecol. Evol.* 6: 146–51

105. McKeever S, French FE. 1991. *Corethrella* (Diptera: Corethellidae) of eastern North America: laboratory life history and responses to anuran calls. *Ann. Entomol. Soc. Am.* 84:493–97

106. McKitrick MC 1993. Phylogenetic constraints in evolutionary theory: Has it any explanatory power? *Annu. Rev. Ecol. Syst.* 24:307–30

107. McLennan DA. 1991. Integrating phylogeny and experimental ethology—from pattern to process. *Evolution* 45: 1773–89

108. Mecham JS. 1965. Genetic relationships and reproductive isolation in southeastern frogs of the genera *Pseudacris* and *Hyla. Am. Midl. Natur.* 74:269–308

109. Møller AP. 1992. Female swallow preference for symmetric male sexual ornaments. *Nature* 357:238–40

110. Morris MR. 1989. Female choice of large males in the treefrog *Hyla chrysoscelis:* the importance of identifying the scale of choice. *Behav. Ecol. Sociobiol.* 25:275–81

111. Morris MR. 1991. Female choice of large males in the treefrog *Hyla ebraccata. J. Zool. Lond.* 223:371–78

112. Morris MR, Yoon SL. 1989. A mechanism for female choice of large males in the treefrog *Hyla chrysoscelis. Behav. Ecol. Sociobiol.* 25:65–71

113. Narins PM. 1990. Seismic communication in anuran amphibians. *Bioscience* 40:268–74

114. Narins PM, Capranica RR. 1978. Communicative significance of the two-note call of the treefrog *Eleutherodactylus coqui. J. Comp. Physiol. A* 127:1–9

115. Narins PM, Zelick R. 1988. The effects of noise on auditory processing and behavior in amphibians. In *The Evolution of the Amphibian Auditory System,* ed. B Fritzsch, W Wilczynski, MJ Ryan, T Hetherington, W Walkowiak, pp. 511–536. New York: Wiley

116. Nelson DA. 1988. Feature weighting in

species song recognition by the field sparrow (*Spozella pusilla*). *Behaviour* 106:158–81

117. Nevo E, Capranica RR. 1985. Evolutionary origin of ethological isolation in cricket frogs, *Acris. Evol. Biol.* 19:147–214

117a. Norris K. 1993. Heritable variation in a plumage indicator of viability in male great tits *Parus major. Nature* 362:537–39

118. Passmore NI. 1981. Sound levels of mating calls of some African frogs. *Herpetologica* 37:166–71

119. Passmore NI, Bishop PJ, Caithness N. 1992. Calling behaviour influences mating success in male painted reed frogs, *Hyperolius marmoratus. Ethology* 92:227–41

120. Paterson HEH. 1978. More evidence against speciation by reinforcement. *S. Afr. J. Sci.* 74:369–71

121. Paterson HEH. 1985. The recognition concept of species. In *Species and Speciation,* ed. E Vrba, pp. 21–29. Pretoria: Transvaal Mus. Monogr. No. 4

122. Penna M, Veloso A. 1990. Vocal diversity in frogs of the South American temperate forest. *J. Herpetol.* 24:23–33

123. Pollack GS, El-Feghaly E. 1993. Calling song recognition in the cricket *Teleogryllus oceanicus:* comparison of the effects of stimulus intensity and sound spectrum on selectivity for temporal pattern. *J. Comp. Physiol. A* 171:759–65

124. Pollack GS, Hoy RR. 1989. Evasive acoustic behavior and its neurobiological basis. In *Cricket Behavior and Neurobiology,* ed. F Huber, T Moore, W Loher, 11:340–63. New York: Cornell Univ. Press. 565 pp.

125. Pomiankowsky A. 1994. Swordplay and sensory bias. *Nature* 368:494–95

126. Prestwich KN. 1994. Comparative energetics of calling in insects and anurans. *Am. Zool.* In press

127. Prestwich KN, Brugger KE, Topping M. 1989. Energy and communication in three species of hylid frogs: power output and efficiency. *J. Exp. Biol.* 144:53–80

128. Price T, Schluter D, Heckman NE. 1993. Sexual selection when the female directly benefits. *Biol. J. Linnean Soc.* 48:187–211.

129. Ptacek MB, Gerhardt HC, Sage RD. 1994. Speciation by polyploidy in treefrogs: multiple origins of the tetraploid, *Hyla versicolor. Evolution,* In press

130. Ralin DB. 1977. Evolutionary aspects of mating call variation in a diploid-tetraploid species complex of treefrogs (Anura). *Evolution* 31:721–36

131. Rand AS, Ryan MJ, Wilczynski W. 1992. Signal redundancy and receiver permissiveness in acoustic mate recognition by the tungara frog, *Physalaemus pustulosus. Am. Zool.* 32:81–90

131a. Reeve HK, Sherman PW. 1993. Adaptation and the goals of evolutionary research. *Q. Rev. Biol.* 68:1–32

132. Reynolds JD, Gross MR. 1990. Costs and benefits of female mate choice: is there a lek paradox? *Am. Nat.* 136:230–43

133. Ritchie MG. 1992. Setbacks in the search for mate-preference genes. *Trends Evol. Ecol.* 7:328–29

134. Robertson JGM. 1986. Female choice, male strategies and the role of vocalization in the Australian frog *Uperoleia rugosa. Anim. Behav.* 34:773–84

135. Robertson JGM. 1990. Female choice increases fertilisation success in the Australian frog, *Uperoleia laevigata. Anim. Behav.* 39:639–45

136. Römer H. 1992. Ecological constraints for the evolution of hearing and sound communication in insects. In *The Evolutionary Biology of Hearing,* ed. DB Webster, RR Fay, AN Popper, pp. 79–93. New York, Berlin, Heidelberg: Springer Verlag

137. Rose GJ, Brenowitz EA, Capranica RR. 1985. Species specificity and temperature dependency of temporal processing by the auditory midbrain of two species of treefrogs. *J. Comp. Physiol. A* 157:763–69

138. Ryan MJ. 1985. *The Túngara Frog: A Study in Sexual Selection and Communication.* Chicago: Univ. Chicago Press. 230 pp.

139. Ryan MJ. 1988. Constraints and patterns in the evolution of anuran acoustic communication. In *The Evolution of the Amphibian Auditory System.* ed. B Fritzsch, T Hetherington, MJ Ryan, W Wilczynski, W Walkowiak, 28:637–77. New York: Wiley

140. Ryan MJ. 1990. Sexual selection, sensory systems and sensory exploitation. In *Oxford Surveys of Evolutionary Biology,* ed. D Futuyma, 7:157–95. Oxford: Univ. Oxford Press

141. Ryan MJ, Cocroft RB, Wilczynski W. 1990. The role of environmental selection in intraspecific divergence of mate recognition signals in the cricket frog, *Acris crepitans. Evolution* 44:1869–72

142. Ryan MJ, Drewes RC. 1990. Vocal morphology of the *Physalaemus pustulosus* species group (Leptodactylidae): morphological response to sexual selection

for complex calls. *Biol. J. Linnean Soc.* 40:37–52

143. Ryan MJ, Fox JH, Wilczynski W, Rand AS. 1990. Sexual selection for sensory exploitation in the frog *Physalaemus pustulosus. Nature* 343:66–67

144. Ryan MJ, Keddy-Hector A. 1992. Directional patterns of female mate choice and the role of sensory biases. *Am. Nat.* 139:S4-S35

145. Ryan MJ, Perrill SA, Wilczynski W. 1992. Auditory tuning and call frequency predict population-based mating preferences in the cricket frog, *Acris crepitans. Am. Nat.* 139:1370–83

146. Ryan MJ, Rand AS. 1990. The sensory basis of sexual selection for complex calls in the túngara frog, *Physalaemus pustulosus* (sexual selection for sensory exploitation). *Evolution* 44:305–14

147. Ryan MJ, Rand AS. 1993. Sexual selection and signal evolution: the ghost of biases past. *Philos. Trans. R. Soc. Lond. B* 340:187–95

148. Ryan MJ, Rand AS. 1993. Species recognition and sexual selection as a unitary problem in animal communication. *Evolution* 47:647–57

149. Ryan MJ, Sullivan BK. 1989. Transmission effects on temporal structure in the advertisement calls of two toads, *Bufo woodhousei* and *Bufo valliceps. Ethology* 80:182–89

150. Ryan MJ, Wilczynski W. 1991. Evolution of intraspecific variation in the advertisement call of a cricket frog (*Acris crepitans*, Hylidae). *Biol. J. Linnean Soc.* 44:249–71

151. Schmidt RS. 1991. Neural correlates of frog calling: production by two semi-independent generators. *Behav. Brain Res.* 50:17–30

152. Schneider H. 1988. Peripheral and central mechanisms of vocalization. In *The Evolution of the Amphibian Auditory System,* ed. B Fritzsch, T Hetherington, MJ Ryan, W Wilczynski, W Walkowiak, 52:537–58. New York: Wiley

153. Schwartz JJ. 1986. Male call behavior and female choice in a neotropical frog. *Ethology* 73:116–27

154. Schwartz JJ. 1987. The importance of spectral and temporal properties in species and call recognition in a neotropical treefrog with a complex vocal repertoire. *Anim. Behav.* 35:340–47

155. Schwartz JJ. 1987. The function of call alternation in anuran amphibians: a test of three hypotheses. *Evolution* 41:461–71

156. Schwartz JJ. 1993. Male calling behavior, female discrimination and acoustic interference in the neotropical treefrog *Hyla microcephala* under realistic acoustic conditions. *Behav. Ecol. Sociobiol.* 32:401–14

157. Schwartz JJ, Gerhardt HC. 1989. Spatially mediated release from auditory masking in an anuran amphibian. *J. Comp. Physiol. A* 166:37–41

157a. Searcy WA. 1992. Song repertoire and mate choice in birds. *Am. Zool.* 32:71–80

158. Searcy WA, Andersson M. 1986. Sexual selection and the evolution of song. *Annu. Rev. Ecol. Syst.* 17:507–33

159. Simmons AM, Buxbaum RC, Mirin MP. 1993. Perception of complex sounds by the green treefrog, *Hyla cinerea:* envelope and fine-structure cues. *J. Comp. Physiol. A* 173:321–27

160. Stewart MM, Rand AS. 1991. Vocalizations and the defense of retreat sites by male and female frogs, *Eleutherodactylus coqui. Copeia* 1991: 1013–24

161. Stieber IB, Narins PM. 1990. Temperature-dependence of auditory nerve response properties in the frog. *Hearing Res.* 46:63–82

162. Straughan IR. 1975. An analysis of the mechanisms of mating call discrimination in the frogs *Hyla regilla* and *Hyla cadaverina. Copeia* 1975:415–24

163. Sullivan BK. 1989. Passive and active female choice: a comment. *Anim. Behav.* 37:692–94

164. Sullivan BK, Hinshaw SH. 1992. Female choice and selection on male calling behaviour in the grey treefrog *Hyla versicolor. Anim. Behav.* 44:733–44

165. Taigen TL, Wells KD. 1985. Energetics of vocalization by an anuran amphibian (*Hyla versicolor*). *J. Comp. Physiol. B* 155:163–70

166. Taigen TL, Wells KD, Marsh RL. 1985. The enzymatic basis of high metabolic rates in calling frogs. *Physiol. Zool.* 58:719–26

167. Telford SR, Dyson ML, Passmore NI. 1989. Mate choice only occurs in small choruses of painted reed frogs (*Hyperolius marmoratus*). *Bioacoustics* 2:47–53

168. Tuttle MD, Ryan MJ. 1982. The role of synchronized calling, ambient light, and ambient noise, in anti-bat-predator behavior of a treefrog. *Behav. Ecol. Sociobiol.* 11:125–31

169. Tymowska J. 1991. Polyploidy and cytogenetic variation in frogs of the genus *Xenopus.* In *Amphibian Cytogenetics and Evolution,* ed. DS Green, SK Sessions, pp. 259–97. San Diego: Academic

170. Ueda H. 1993. Mating calls of autotriploid and autotetraploid males in *Hyla*

japonica. Sci. Rep. Lab. Amphibian Biol., Hiroshima Univ. 12:177–89

171. Waage JK. 1979. Reproductive character displacement in *Calopteryx* (Odonata: Calopterygidae). *Evolution* 33:104–16

172. Wagner WE Jr. 1989. Fighting, assessment, and frequency alternation in Blanchard's cricket frog. *Behav. Ecol. Sociobiol.* 25:429–36

173. Wagner WE. 1989. Graded aggressive signals in Blanchard's cricket frog: vocal responses to opponent proximity and size. *Anim. Behav.* 38:1025–38

174. Walkowiak W. 1988. Neuroethology of anuran call recognition. In *The Evolution of the Amphibian Auditory System,* ed. B Fritzsch, W Wilczynski, MJ Ryan, T Hetherington, W Walkowiak, 22:485–509. New York: Wiley

175. Watson JT, Kelley DB. 1992. Testicular masculinization of vocal behavior in juvenile female *Xenopus laevis* reveals sensitive periods for song duration, rate, and frequency spectra. *J. Comp. Physiol. A* 171:343–50

176. Deleted in proof

177. Wells KD. 1977. The social behaviour of anuran amphibians. *Anim. Behav.* 25:666–93

178. Wells KD. 1977. The courtship of frogs. In *The Reproductive Biology of Amphibians,* ed. DH Taylor, SI Guttman, 7:233–62. New York/London: Plenum. 475 pp.

179. Wells KD. 1988. The effects of social interactions on anuran vocal behavior. In *The Evolution of the Amphibian Auditory System,* ed. B Fritzsch, W Wilczynski, MJ Ryan, T Hetherington, W Walkowiak, 20:433–54. New York: Wiley

180. Wells KD. 1989. Vocal communication in a neotropical treefrog, *Hyla ebraccata:* responses of males to graded aggressive calls. *Copeia* 1989:461–66

181. Wells KD, Schwartz JJ. 1982. The effect of vegetation on the propagation of calls in the neotropical frog *Centrolenella fleischmanni. Herpetologica* 38:449–55

182. Wells KD, Schwartz JJ. 1984. Vocal communication in a neotropical treefrog, *Hyla ebraccata:* aggressive calls. *Behaviour* 91:128–45

183. Wells KD, Taigen TL. 1986. The effect of social interactions on calling energetics in the gray treefrog (*Hyla versicolor*). *Behav. Ecol. Sociobiol.* 19:9–18

184. Wells KD, Taigen TL. 1989. Calling energetics of a neotropical treefrog, *Hyla microcephala. Behav. Ecol. Sociobiol.* 25:13–22

185. Wilczynski W, Allison JD, Marler CA. 1993. Sensory pathways linking social and environmental cues to endocrine control regions of amphibian forebrains. *Brain Behav. Evol.* 42:252–64

186. Wilczynski W, McClelland BE, Rand AS. 1993. Acoustic, auditory, and morphological divergence in three species of neotropical frog. *J. Comp. Physiol. A* 172:425–38

187. Wiley RH, Richards DG. 1978. Physical constraints on acoustic communication in the atmosphere:implications for the evolution of animal vocalizations. *Behav. Ecol. Sociobiol.* 3:69–94

188. Yager DD. 1992. A unique sound production mechanism in the pipid anuran *Xenopus borealis. Zool. J. Linnean Soc.* 104:351–75

189. Zakon HH, Wilczynski W. 1988. The physiology of the anuran eighth nerve. In *The Evolution of the Amphibian Auditory System,* ed. B Fritzsch, W Wilczynski, MJ Ryan, T Hetherington, W Walkowiak, 6:25–155. New York: Wiley. 705 pp.

190. Zimmerman BL. 1983. A comparison of structural features of calls of open and forest habitat frog species in the central Amazon. *Herpetologica* 39:235–46

Annu. Rev. Ecol. Syst. 1994. 25:325–49

PHYLOGENY OF THE LEGUME FAMILY: An Approach to Understanding the Origins of Nodulation

Jeff J. Doyle

L. H. Bailey Hortorium, 466 Mann Library Building, Cornell University, Ithaca, New York 14853

KEY WORDS: Leguminosae, phylogeny, nodulation, nodulin, multigene families

Abstract

Members of Leguminosae (legumes), the third largest family of flowering plants, are cosmopolitan in distribution, diverse in form, and of considerable ecological, agricultural, and scientific importance. Objective phylogeny reconstruction at all taxonomic levels is in the process of reshaping the taxonomy of the family as well as providing new hypotheses concerning the affinities of the family with other angiosperms. Cladistic analyses of morphological and DNA variation for the entire family are in relatively good agreement and echo long-held beliefs, based on more intuitive methods, that many recognized taxa are unnatural. Of the three subfamilies, Mimosoideae and Papilionoideae are most probably monophyletic, while Caesalpinioideae, as suspected, is a paraphyletic grade of basal elements. Phylogenetic hypotheses are being used to address a diversity of questions including biogeography, evolution of pollination systems, origins of economically important species, and genome evolution. Phylogenetic considerations suggest that the ability of legumes to fix atmospheric nitrogen in symbiosis with soil bacteria (nodulation) has arisen several times in the family. Nodules appear to be structurally homologous across the family. The orthology relationships and expression patterns of genes expressed solely or predominantly in the nodule (nodulins) may eventually provide additional criteria for elucidating homologies.

325

0066-4162/94/1120-0325$05.00

INTRODUCTION

The legume family (Leguminosae or Fabaceae) is the third largest family of flowering plants, with approximately 650 genera and nearly 20,000 species. Cosmopolitan in its distribution, its species range from dominant tropical canopy trees to tiny alpine annual herbs. The Leguminosae contains many taxa of agricultural or other economic importance, notably various "beans." Pea (*Pisum*), common bean (*Phaseolus*), and soybean (*Glycine*) have provided model systems for genetics and plant molecular biology whose importance is rivalled or surpassed only by maize and, more recently, *Arabidopsis*. Ecologically as well as agriculturally, legumes owe much of their importance to their ability to fix atmospheric nitrogen in partnership with various soil bacteria ("rhizobia") housed in novel structures called nodules, found on roots or in some cases stems.

Studies of legume phylogeny and evolution by an active international scientific community have been enhanced and to some extent coordinated during the last two decades through a series of International Legume Conferences, jointly sponsored by the Royal Botanic Gardens, Kew, and the Missouri Botanical Garden, St. Louis. These have resulted in a series of publications on various aspects of legume biology, including developmental biology, nitrogen metabolism, systematics, and the fossil record (39, 51, 73, 92, 95, 116–118). Progress in reconstructing phylogenetic relationships at higher taxonomic levels in the family is the subject of one of the most recent Conference volumes (21), and primarily this work is reviewed here. Achieving a phylogenetic perspective is a critical first step in addressing diverse evolutionary questions, a point increasingly appreciated throughout "comparative" biology. Some such questions in the legume family are reviewed, particularly the origin(s) and evolution of the nitrogen fixing symbiosis and the plant genes involved in it.

TAXONOMY AND PHYLOGENY: THE FOUNDATION

Legumes are generally treated as a single family with three subfamilies, although some systems recognize three families. The defining character of the family is the fruit that gives the family both its formal and informal names. The unicarpellate ovary matures to produce this legume, typically dehiscent along both sutures. Although some theories derive Leguminosae from multicarpellate apocarpous taxa, the few multicarpellate Leguminosae appear to be advanced rather than basal in the family (38). Many modifications to the fruit occur, presumably as adaptations to different dispersal mechanisms, and these have often been given undue weight in taxonomic schemes. Thus, for example, indehiscent fruits with wings or fleshy exocarps appear to have arisen inde-

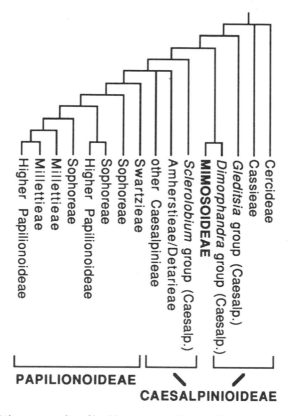

Figure 1   Cladistic representation of intuitive concepts of relationships among legume subfamilies (bold), tribes, and generic groups (Caesalp. = Caesalpinieae) (50, 93, 94, 96).

pendently in several different groups. The bilaterally symmetrical flower commonly associated with the family is not found throughout its members, but rather occurs only in the bulk of the largest of its three subfamilies, the Papilionoideae (typical "beans": ca. 450 genera, 12,000 spp.). Even in this subfamily, this morphology represents the derived, albeit most common condition. Quite different floral morphologies occur in the remaining subfamilies. Members of Mimosoideae (e.g. *Mimosa, Acacia:* ca. 65 genera, 3000 spp.) typically have inflorescences with aggregated, individually inconspicuous, radially symmetrical flowers in which anthers, rather than petals, are the attractive organs. Both Papilionoideae and Mimosoideae are thought to be natural, monophyletic groups (96). In contrast, the third subfamily, Caesalpinioideae (ca. 150 genera, 2500 spp.), has been considered paraphyletic. It is thought to comprise the basal elements of the family, including the sister groups of Mimosoideae and Papilionoideae as well as other taxa not involved in this "main radiation" of the family (96; Figure 1). Not surprisingly, caesalpinioid species display a variety of floral types, from radially symmetrical, often showy flowers in many taxa, to highly reduced, unisexual flowers, to zygomorphic morphologies presumably convergent with the papilionoid form.

Problems of paraphyly and polyphyly are not confined to the subfamily level. Previous tribal revisions (95) corrected some of the more egregiously unnatural groupings, but a number of tribes admittedly remained grades rather than clades. This was particularly true of putative basal elements of Caesalpinioideae and Papilionoideae. Objective phylogenetic analyses of major groupings of legumes have appeared only within the last decade, and in general they confirm earlier intuitive hypotheses, though there have been some surprises. Such cladistic analyses were initially based on morphological data, but most recently much information has come from molecular sources.

Affinities of Leguminosae

The same distinctiveness that suggests monophyly for the legumes also obscures their relationships with other plant families, which has led to considerable disagreement about their taxonomic placement (23, 28, 121). Several cladistic analyses of the growing sequence database for the chloroplast gene ribulose bisphosphate carboxylase/oxygenase (*rbc*L) bear on the issue of legume relationships (19, 76, 77). Cladograms from these studies surprisingly place Polygalaceae as sister group to a monophyletic Leguminosae. In studies of rosid families (76, 77) and in unpublished cladistic analyses (MD Chase, unpublished information), other, newly added, sequences join these two families in a "legume" clade. These include members of Surianaceae as well as taxa with questionable affinities, which, however, have been placed in such families as Rosaceae, Sapindaceae, or Surianaceae. Some families considered close to Leguminosae (e.g. Chrysobalanaceae, Connaraceae) are not too distant from this clade but are no closer than many other rosid or even some nonrosid taxa (e.g. various "higher" Hamamelidae). The bulk of Sapindaceae, a family considered particularly close by Thorne (121), are still more distant. The emerging pattern includes elements of both the Thorne (121) and Cronquist (23) systems, as well as relationships not predicted by any of the current classification systems (e.g. Polygalaceae sister to Leguminosae, Connaraceae more distant). Confirmation with other sources of data, both molecular and nonmolecular, are needed, but the *rbc*L hypotheses open some interesting questions about legume character evolution. In particular, the Polygalaceae, with zygomorphic flowers having a grossly papilionoid aspect, make a provocative outgroup for the legumes.

Relationships within Leguminosae: Subfamilial and Tribal-Level Studies

Phylogenetic analysis of *rbc*L sequences within the family thus far has concentrated on Caesalpinioideae and basal Papilionoideae (5, 33) or even more narrowly on the tribe Genisteae of Papilionoideae (56, 134). A preliminary cladistic analysis of the entire family, mainly using nonmolecular data, has been conducted by Chappill (17), while a second nonmolecular cladistic analysis (125) focuses on basal legumes and emphasizes floral ontogenetic characters. Despite differences in sampling, topologies of both the nonmolecular Chappill study and the broader *rbc*L analysis have many similarities (Figure 2). Topologies of the study emphasizing ontogenetic characters are rather different, but all three analyses suggest a paraphyletic Caesalpinioideae and monophyletic Mimosoideae and Papilionoideae. The very base of the family is unresolved in strict consensus trees from the Chappill study. The caesalpinioid tribe Cercideae is basal in current *rbc*L trees and at the base of the large caesalpinioid/mimosoid clade of Chappill. A basal or near basal position of Cercideae is perhaps consistent with its general distinctiveness and agrees

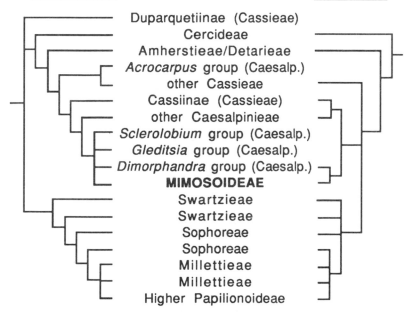

Figure 2   Summary of results from cladistic analyses for Leguminosae using either mostly nonmolecular (17) or *rbc*L sequence (5, 33, 56, 134) data. Taxa above Mimosoideae are Caesalpinioideae; below are Papilionoideae.

with its placement outside the "main radiation" of the family (96). However, Cercideae is the only tribe of Caesalpinioideae for which reliable fossil evidence is not present by the Middle Eocene, by which time all three subfamilies as well as many modern genera are represented (50). In contrast, the earliest putative legume wood fossils are from the Upper Cretaceous, and they resemble modern Cassieae (Caesalpinioideae) (131). Several caesalpinioid genera with reduced flowers that have been considered most primitive in the family (96) are not basal in either the rbcL or Chappill studies, nor are they present in the early fossil record, leading to the suggestion that reduced flowers are unlikely to be a plesiomorphic feature in the family.

Tribes Cassieae and Caesalpinioideae are paraphyletic or polyphyletic in all analyses, while the merger of the caesalpinioid tribes Amherstieae and Detarieae, suggested in a study focusing on their genera (10), is supported by the Chappill study. Caesalpinieae have been considered to represent the basal elements of a number of natural groups in the family (Figure 1), anticipating the findings of these cladistic analyses (Figure 2) as well as studies within the *Caesalpinia* group (67). Within Mimosoideae, the Chappill analysis suggests that none of the currently recognized mimosoid tribes are monophyletic, a finding supported by a separate morphological cladistic analysis of Ingeae and Acacieae (18). Like Mimosoideae, Papilionoideae is monophyletic in all three analyses. The papilionoid clade includes at or near its base members of the tribe Swartzieae, a tribe of controversial status that historically has been placed either in Papilionoideae or Caesalpinioideae. Sophoreae join Swartzieae as basal elements of Papilionoideae. Neither tribe is monophyletic, both as predicted (94) and as shown in recent morphological cladistic analyses (49).

Within Papilionoideae, a 50 kb chloroplast DNA inversion (88) appears to be a good synapomorphy for a "higher papilionoid" clade that includes most Sophoreae and Swartzieae (33, 37). Of the large-scale analyses only that of Chappill includes many higher papilionoids, and its results disagree on a number of points either with intuitive schemes or with some molecular results. The loss of one copy of the ca. 25 kb chloroplast DNA inverted repeat (IR) has been suggested as a synapomorphy for a major component of Papilionoideae, the "temperate herbaceous" tribes that include peas, lentils, and clovers, plus some members of the woody, largely tropical tribe Millettieae (63, 68). The Chappill topologies, in contrast to preliminary rbcL trees, are more traditional in that they optimize the IR loss as homoplasious, requiring a separate loss in Millettieae and a likely regain of the IR in Loteae/Coronilleae.

Cladistic studies at the tribal level in higher Papilionoideae have suggested that a number of tribes are not natural. Molecular phylogenetic studies within the temperate herbaceous tribes, for example, suggest that Galegeae is a paraphyletic assemblage representing basal elements of several other tribes (105). Molecular systematic studies of Phaseoleae and its putative allies suggest that several of its subtribes may be natural but nest Desmodieae and some (but not all) Millettieae well within this largest tribe of the subfamily (14, 33, 35). The broader cladistic analysis of Chappill (17) also places Desmodieae within Phaseoleae.

The Australian and South African tribes of the "Genistoid alliance" comprise a clade in the Chappill trees, placed rather surprisingly as a highly derived group. Australian Mirbelieae has been the focus of considerable cladistic and revisionary effort in recent years (22), while much work on the South African tribes by van Wyk and colleagues (128) has suggested a number of taxonomic changes that are necessary in order to produce monophyletic groupings. These results also suggest that there is no direct relationship between African and Australian groups, and that the Australian tribes are unrelated to the American Brongniartieae.

Recent cladistic analyses have produced objective phylogenetic hypotheses for other papilionoid tribes such as Indigofereae (108). A series of morphological and molecular studies by Lavin (62) has clarified relationships within Robinieae (including Sesbanieae) and has also strengthened the contention that this tribe is derived from ancestors that would be classified in the modern tribe Millettieae, a tribe whose paraphyly or polyphyly has been demonstrated in a number of cladistic analyses (17, 35, 45, 62, 63). This supports the view that Millettieae is a grade of generally woody, tropical genera one level advanced over the grade represented by Sophoreae (93).

Reliance on convenient "key" characters has in large part been responsible for creating unnatural groupings in Papilionoideae. Examples include free stamens in Sophoreae, the combination of twining habit with trifoliolate leaves in Phaseoleae, and indehiscent fruits in Dalbergieae. While these shortcomings have in some cases been recognized for years, it is only now that cladistic analyses, often using molecular data, are providing the means for rectifying them in an objective manner.

APPLICATIONS OF PHYLOGENETIC HYPOTHESES

Phylogenies have been used to address a number of evolutionary questions in the family at various taxonomic levels. For example, Lavin & Luckow (64) have utilized cladistic analyses of Robinieae and Mimoseae to test biogeographic hypotheses. Their findings that South American taxa are relatively derived are consistent with a "boreotropical" floristic hypothesis and suggest that reconsideration of some biogeographic assumptions is warranted, particularly for legumes.

The diversity of pollination systems in the family (3) provides numerous groups in which to study shifts in pollinator preference from a phylogenetic perspective. In one of her series of elegant studies of floral ontogeny in the family, Tucker (124) notes that many legume taxa have evolved bird pollination independently, and she suggests that many of the morphological characters involved in this syndrome arise as relatively late modifications in development. Phylogenetic studies at lower taxonomic levels provide the foundation for studying such shifts in closely related groups where the underlying ontogenetic framework is similar. Recent cladistic studies in the Australian papilionoid tribe Mirbelieae (22) suggest independent shifts from bee pollination to bird pollination within one generic clade, while phylogenies reconstructed from both molecular and morphological characters support independent origins of hummingbird pollination and concomitant parallel floral evolution within *Erythrina* (Papilionoideae: Phaseoleae) (12, 13).

Other phylogenetic studies, often using restriction site variation in the chloroplast genome, have focused on the evolution of genera containing cultivated legumes, such as *Glycine* (36), *Vigna* (126), *Phaseolus* and allies (27), *Sphenostylis* (97), and *Pisum* (87). Cytological evolution has been addressed indirectly in a series of papers on the huge papilionoid genus *Astragalus* (reviewed in 105), where evidence suggests a single origin of aneuploid species in the New World. In contrast, multiple origins of polyploidy have been documented for wild species of *Glycine* using DNA variation (34).

The evolution of both organellar and nuclear genome organization in the family has also been studied. The structure of the legume chloroplast genome has provided several characters of phylogenetic utility (15, 33, 37, 63, 88), and phylogenies in turn can ultimately provide insights on the forces shaping chloroplast genome evolution. The legumes include a number of cases of apparent gene transfer to the nucleus and loss from the chloroplast and mitochondrial genomes (30, 37, 44, 83). Studies of the transfer of *cox*II from the mitochondrion (83), and of *rpl*22 from the chloroplast (44) have involved the use of both organismal and gene phylogenies. Transfers can greatly predate loss from the organellar genome, and the eventual stochastic loss of redundant organellar genes provides a molecular example of Hecht & Edwards' (48) warnings concerning the fallibility of "loss" characters in phylogeny reconstruction (37). Structural evolution of the much larger nuclear genomes of legumes is also being explored. Though as yet linkage maps are published for only a handful of taxa, notably economically important species of papilionoid genera (11, 82, 84, 127, 132), it is clear that within at least some papilionoid tribes there is considerable conservation of linkage groups. For example, approximately 40% of the known linkage map of *Lens* (lentil) shares its gene order with that of another member of the tribe Vicieae, *Pisum* (132). Knowledge of papilionoid phylogeny will be important in choosing "bridging" taxa to make connections with the linkage maps of more distantly related taxa such as Phaseoleae.

The Origin and Evolution of Symbiotic Nitrogen Fixation in Leguminosae

A hallmark of the legume family is the ability of many of its members to enter into symbiotic relationship with any of several genera of nitrogen-fixing soil bacteria (collectively termed "rhizobia"). This process continues to be the subject of numerous reviews and conferences (40, 86, 112, 113), but it has not been discussed in the context of the developing phylogenetic hypotheses for the legumes. The origin or origins of this complex symbiosis remain a mystery. Although over 90% of Papilionoideae and Mimosoideae presumably are capable of forming symbioses with rhizobia, very few members of Caesalpinioideae do so (25, 112). The most parsimonious optimization of the character "nodulation" on a summary cladogram for the family (Figure 3) suggests that the syndrome has arisen independently at least three times, once each in lineages giving rise to Mimosoideae and Papilionoideae, and a third in the genus *Chamaecrista* (Cassieae: Caesalpinioideae).

Neither symbiotic nitrogen fixation nor even rhizobial nodulation are unique to Leguminosae. *Parasponia,* a member of the elm family (Ulmaceae), also nodulates in partnership with rhizobia and indeed can be induced to nodulate with strains that typically nodulate legumes (7). Genera from eight other dicot families (Betulaceae, Casuarinaceae, Coriariaceae, Datiscaceae, Elaeagnaceae, Myricaceae, Rhamnaceae, Rosaceae) nodulate with *Frankia,* an "actinomycete," while *Gunnera* (Gunneraceae) has a symbiotic relationship with cyanobacteria (4, 112). In conventional taxonomic systems such as that of Cronquist (23), these groups appear quite unrelated, but recent molecular systematic studies (19, 72, 76, 77; D Soltis, P Soltis, D Morgan, S Swenson, P Martin, B

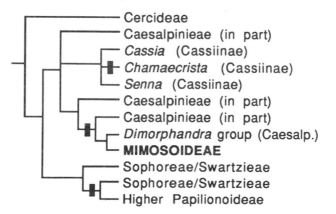

Figure 3 Summary cladogram of Leguminosae with most parsimonious optimization of the nodulation syndrome (black rectangles).

Mullin, J Dowd, unpublished information) have suggested that all symbiotic partners except *Gunnera* belong in a single large clade. Such taxa are the minority within the clade and often within their own families. This, coupled with differences both in the bacterial taxa involved in the relationship and a number of ontogenetic and structural differences in nodules, suggests that the syndrome has arisen more than once. However, it is intriguing to speculate that the progenitor of this entire clade uniquely possessed the necessary components to permit nodulation to evolve in its descendants (D Soltis, P Soltis, D Morgan, S Swenson, P Martin, B Mullin, J Dowd, unpublished). The nature of this apparent "apomorphic tendency" (104) is unknown.

The production of a nodule is energetically expensive, and nodules may not be formed even in legume species that commonly do nodulate, if a readily accessible nitrogen source is available, or, conversely, under adverse environmental conditions (111). Models involving numerous independent losses must therefore be considered along with the more parsimonious parallel gain model. If nodulation has indeed arisen more than once, there could be evidence of homoplasy in the nodule itself at some level, from gross morphology to molecular aspects of nodule development and details of biochemistry, or perhaps in the phylogenies of their symbionts. The prospects for such a "character analysis" of nodulation are reviewed below.

NODULE DEVELOPMENT AND MORPHOLOGY   The nitrogen-fixing symbiosis is a tightly choreographed series of interactions between the two partners that involve host flavonoids and bacterial exopolysaccharides (40, 112). The result of this interaction is the production of a novel plant organ, the nodule, on the root or stem. Nodules vary considerably in shape, and may be either determinate or indeterminate in their growth patterns. Nodule morphology is known to be under the control of the host, rather than the bacterial symbiont, because infection of different legume (or even nonlegume) species with the same bacterial strain results in nodules typical of the particular host species (7, 112). Although nodules commonly occur on legume roots, their anatomy is stem-like, with a peripheral rather than central vascular system (110, 112). This is true of nodules of all legume taxa and distinguishes them both from actinomycetous nodules and from the rhizobial nodules of *Parasponia*.

Corby (20) has classified the various types of legume nodules and has noted some correspondence between nodule morphology and taxonomy. However, viewed in phylogenetic context, it is clear that nodule morphology is an extremely labile character. In several cases more than one nodule type occurs within a tribe, genus, or species or even in an individual (20). At the level of morphology and development, the predominant nodule type in *Chamaecrista*, Mimosoideae, and basal Papilionoideae is the large, indeterminate caesalpinioid type. Infection occurs through the formation of "infection threads" and the bacterial partner ("bacteroid") is surrounded by a peribacteroid membrane (PBM), an essential "novel organelle" (130) produced and maintained by the host plant. Nodules in different species contain quite different proportions of infected vs. uninfected cells, and although in most legume species bacteria are released into the cytoplasm of infected cells by rupture of infection threads, in other cases, infection threads are persistent, with no bacterial release (26). Such persistent infection threads are confined to Caesalpinioideae and some basal Papilionoideae, while *Chamaecrista* is polymorphic for this character (78). If presence of persistent infection threads is plesiomorphic, as seems likely, then the apomorphic condition of release into host cells has evolved independently at least three times in the family. Patterns of morphological and anatomical variability thus suggest that all legume nodules are homologous, but that additional transformations have occurred, sometimes in parallel, in different lineages of the family.

BACTERIAL SYMBIONTS   The legume-rhizobium relationship is perhaps the best-studied of all symbioses, and much is now known about phylogeny of rhizobia (135, 136). If the evolution of nodulation has involved diversifying coevolution (120), symbiont phylogenies may provide information on the homology of nodulation in different taxa. However, in a review of the symbiosis, Young & Johnson (136) conclude that there is little correlation between bacterial and plant phylogenies, and the acquisition of new data on legume phylogeny has done nothing to change this conclusion. The bacterial symbionts are themselves taxonomically diverse, representing several deep branches of the alpha group of the Proteobacteria (purple bacteria). Two of the most prevalent symbiotic bacterial genera, *Bradyrhizobium* and *Rhizobium,* have ribosomal RNA genes that are as dissimilar from one another as those of plants and animals, yet both form effective symbioses with species of the single genus *Glycine. Bradyrhizobium* species also nodulate the ulmaceous genus *Parasponia,* while some strains of a single *Rhizobium* species may have very broad host ranges within Leguminosae and may nodulate species in different subfamilies (110). Bacterial relationships therefore do not appear to provide criteria for hypothesizing homologies of nodules.

NODULIN GENES   "Nodulin" is a widely used term for gene products specific to nodules, found nowhere else in the plant (66, 102). The existence of genes whose products perform unique functions is not unlikely, because the nodule itself is a novel structure. On the other hand, according to Sprent (110), "There is little reason to suspect that nodulating legumes possess unique genes ready to be switched on when appropriate rhizobial strains are around." Nap & Bisseling (79) suggest that "the genetic resources to evolve a symbiosis with *Rhizobium* may be general in all dicotyledonous plants" and that "most nodulin

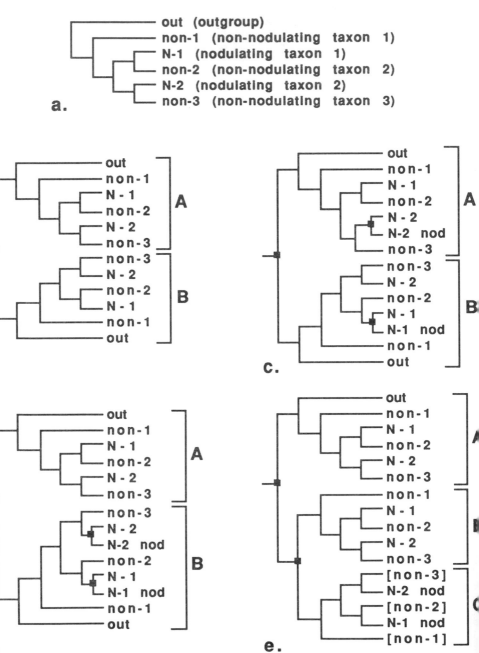

genes originate from genes already functioning in the plant." Nevertheless, these and other authors suggest that the evolution of various nodulin genes may have involved exon shuffling, duplication, or recombination. Such genes could still be legitimately called "novel" and might provide clues to the homologies of nodules in different taxa.

Wholly novel nodulin genes, perhaps transferred horizontally or assembled by exon shuffling, would presumably not occur in most non-nodulating taxa except perhaps in the immediate sister groups of nodulating species. The finding of mutually exclusive suites of such novel genes in different lineages of nodulating taxa might suggest multiple origins of nodulation. Alternatively, the origin of the nodulation syndrome could have involved duplications in one or more different gene families, producing a new gene encoding a nodulin isozyme. If duplications have occurred in the course of independent origins of nodulation, the gene tree for such a multigene family should differ from that reconstructed if only a single origin has occurred, even allowing for complete loss of nodulin orthologues (41) in taxa that have lost the ability to nodulate (Figure 4). Finally, existing genes could have been recruited to perform novel functions in the new context of the nodule. Only if paralogous, and not orthologous genes were recruited to be nodulins in two different plant lineages (Figure 5) could gene trees suggest independent recruitment and hence independent origins of nodulation in the two groups. The information necessary to construct the gene trees necessary for performing such phylogenetic tests of homology is not yet available even for most of the better-known nodulin genes. Moreover, these simple scenarios may be complicated by expression shifts, recombination, or concerted evolution, further reducing the number of genes that could be used for testing nodule homology.

Early Nodulins (ENODs)   There are a number of genes expressed during the initial stages of nodulation whose products are thought to perform such functions as preparation for growth of the infection thread, maintenance of oxygen diffusion barriers, metal ion transport across the peribacteroid membrane, or nodule morphogenesis. As many as 50% of ENOD gene products are proline-rich proteins (e.g. ENOD2, MsENOD10, ENOD12) (90) chemically related to those that are common structural components of plant cell walls. Some (ENOD5, GmENOD55, N#36) are related to arabinogalactan proteins such as

---

Figure 4   Origins of nodulins by gene duplication. Duplication events are shown by black squares. In a two gene family the species tree topology (a) is also the topology of each set (A or B) of orthologous sequences (b). Duplication during independent origin of nodulation in taxa N-1 and N-2 may involve either paralogous (c) or orthologous (d) genes. In (e), duplication has occurred during a unique origin of nodulation. Genes of the C clade are nodulins in nodulating taxa but are presumably superfluous in non-nodulating taxa and may be present as pseudogenes or be lost entirely.

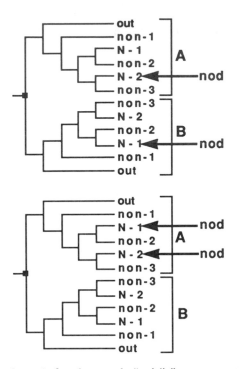

Figure 5 Recruitment (arrows) of paralogues to be "nodulin" genes suggests independent origins of nodulation (top). Recruitment of orthologues (bottom) is equally consistent with single or multiple origins.

gums and exudates (24, 58, 79, 90), while one (ENOD8) is similar in sequence to nonproline-rich domains of anther-specific proteins of *Arabidopsis* and *Brassica* (29). Several ENOD genes are known to be expressed outside the nodule, generally in stem or floral tissue (42, 53, 58, 106). This suggests that ENODs have been recruited from other functions and are not therefore nodulins in the strict sense of the word (42). Several ENODs are encoded by multigene families, some of which may be quite complex and include non-nodulins (1, 55, 70, 79, 101, 107); orthologies have not been described.

ENOD genes have not been described from any nonlegume nodulating species (4, 7) and may thus provide additional criteria for postulating non-homology between rhizobial nodules of Leguminosae and *Parasponia*. The observation that transcripts of several different ENOD genes are detected in stems is interesting in this regard. Because *Parasponia* nodules have a root-like anatomy rather than the stem-like anatomy of legume nodules, it might be expected that genes recruited for its nodules would normally be expressed in root tissue, rather than stem tissue. In that case, its ENOD analogues may be "root" genes, unlike the "stem" ENOD genes. Within Leguminosae, ENOD genes have yet to be sought in the genomes of Mimosoideae or *Chamaecrista*, so this criterion of nodule homology may also be applicable within the family. However, nodules of Caesalpinioideae and Mimosoideae are also stem-like in their anatomy, so if ENOD genes represent essential nodule functions they are perhaps also likely to be "stem" genes.

Peribacteroid Membrane Proteins   The peribacteroid membrane is essential for maintaining bacteria in the host and requires the synthesis of many components by the host plant, and perhaps also by the bacteria (130). A number of proteins of unknown function associated with the PBM in soybean belong to an extended multigene family, some of which are among the most actively transcribed genes in the nodule. The evolution of this family is thought to involve gene and domain duplications, incorporation of "foreign" segments, and gene conversion, coupled with functional divergence (46, 54, 81, 99, 130). Although the duplication and apparently rapid divergence of this family are considered to be relatively recent, the family as a whole is of ancient enough origin that it does occur in other Phaseoleae (16). Evolution by internal duplications, as suggested for some members of the N20/44 family, is proposed for nodulins of other legumes (57, 98, 129), but whether the genes involved are thought to be homologous is not stated.

Another soybean PBM nodulin, N26, is thought to be involved in dicarboxylic acid transport across the PBM (85). In contrast to the N20/44 family, N26 belongs to a class of membrane proteins that includes such diverse members as vertebrate eye lens protein, *Drosophila* neurogenic protein, plant tonoplast integral membrane protein, a pea turgor-regulated protein, and bacterial glycerol facilitator (75, 103). Gene trees for this family suggest that the six major membrane-spanning components are divided into two large domains, presumably derived by an internal duplication (89). Results of experiments with transgenic plants have been cited to suggest that nodule specificity of N26 expression has involved the evolution of species-specific repressors in soybean roots during the recruitment of this gene (75). N26 orthologues have not been reported in taxa other than *Glycine*.

Leghemoglobins (Lb)   These quintessential late nodulins are responsible for the pinkish color of sectioned nodules exposed to air. The evolutionary origin of Lb has intrigued biologists for some time (reviewed in 2). For years it was thought that these were the only globins to occur outside the animal kingdom, which led to the speculation that their presence in legumes was due to horizontal gene transfer. However, more recently it has been shown that globins and the genes that encode them are present in nonleguminous plants, including both nodulating and nonnodulating species. Globin genes have also been described from bacteria, and as it now seems that they are a ubiquitous feature of living organisms, horizontal gene transfer need not be postulated to account for their occurrence in plants. In the nodule, Lb is thought to facilitate oxygen diffusion for rhizobial respiration while maintaining an environment in which oxygen concentration is low enough not to inhibit oxygen-sensitive bacterial nitrogenases (112, 130). Lb may also play a role in peribacteriod membrane degradation during nodule senescence (52). The function of plant globins outside the nodule is less well understood, but they may serve in oxygen sensing or in oxygen transport (2). Leghemoglobins are likely not nodulins in the narrow sense but may have been recruited for nodulation in addition to their more obscure roles.

Leghemoglobins are a multigene family in legumes (6, 9, 60, 65, 74). In alfalfa there are six different classes of transcripts belonging to two major groups (6, 69). In soybean, like alfalfa a polyploid, characterization of the family has included the demonstration of homeologous loci, as well as pseudogenes (65). Leghemoglobin genes appear to undergo considerable, but not complete concerted evolution (Figure 6). The two major groups described from *Medicago sativa* are clearly paralogous, and the single described *Pisum*

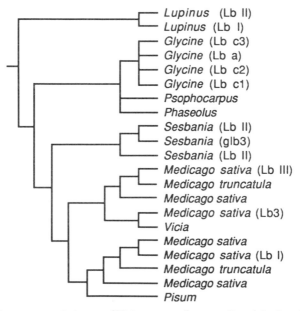

Figure 6   Strict consensus cladogram of 78 shortest trees from equally weighted parsimony analysis of leghemoglobin amino acid sequences (JJ Doyle, unpublished data). Trees were rooted with Ulmaceae and Casuarina sequences.

and *Vicia* proteins join different clades despite the close relationship between these two genera. A paradox has been noted concerning regulation of leghemoglobin genes in different species. Despite the extensive degree of divergence within the *Medicago* gene family, all of its members appear to be expressed simultaneously and identically in the nodule (6, 43). In contrast, the more homogeneous family of *Glycine* is expressed differentially, with different genes activated sequentially and to different levels over a several-day period (71). These differences in leghemoglobin expression pattern may be causally related to the different types of nodules in the two genera, though the mechanisms or functional constraints involved are unknown (43). These and other regulatory differences (109) may prove interesting as a source of characters within Papilionoideae, but it does not appear likely that leghemoglobin genes will provide useful criteria for establishing homologies of nodules among the subfamilies.

Metabolic Nodulins   In addition to demands imposed by maintenance and growth of the nodule itself, the bacteroids require a carbon source for growth and nitrogen fixation. Moreover, the ammonia produced by the bacteria must be incorporated into compounds for transport to other parts of the plant. Thus it is not surprising that various genes involved in carbon or nitrogen metabolism should be highly expressed in the nodule. Some of these genes encoding metabolic nodulins are the sole members of multigene families showing nodule-enhanced or nodule-specific expression, which in theory should permit phylogenetic tests for multiple nodule origins.

Glutamine synthetase (GS), the best studied metabolic nodulin, catalyzes the formation of glutamine from ammonia produced by bacterial nitrogen fixation, beginning the mobilization and transport of nitrogen in the plant. The subunits of octameric GS are generally encoded by a multigene family in plants, the evolutionary history of which has been studied in both plants and animals (31, 91). In flowering plants, the nuclear gene encoding the chloroplast-specific isozyme is clearly paralogous to other members of the family, apparently having arisen by an ancient duplication (31). Phylogenetic analyses that include sequences published since the 1991 study (31) of the gene family do group all legume cytosolic genes separately from those of tobacco or grasses (Figure 7). This pattern suggests either low rates of concerted evolution or expansion of the gene family within taxonomic groups. If the latter hypothesis is correct, it may be possible to test some of the hypotheses of nodulin origin described above.

However, GS expression patterns in legumes are very complex. The "nodulin" GS of *Phaseolus* is also expressed in other tissues, including developing cotyledons (8), while a second GS isozyme gene is also transcribed at low levels in nodules. Similarly, *Glycine* nodules contain both a nodule-en-

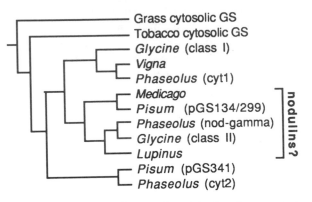

Figure 7 Single most parsimonious tree from equally weighted parsimony analysis of nucleotide sequence data for cytosolic glutamine synthetase coding regions (JJ Doyle, unpublished data). Trees were rooted with chloroplast GS sequences.

hanced GS and a GS isozyme encoded by a gene that is expressed in other tissues (100). In *Medicago,* the nodule-enhanced GS gene is also expressed at lower levels in other tissues, while another cytosolic GS gene, expressed most strongly in cotyledons and roots, is expressed to a lesser degree in nodules (114). In pea there is no nodulin, but transcription of two root cytosolic GS genes increases 10- to 20-fold in nodules (122, 131). The gene tree for legume GS contains three clades of putative orthologues, one of which includes published genes best fitting the definition of nodulins (Figure 7). It is possible that this clade could be due to a duplication event unique to nodulating Papilionoideae, but additional sampling clearly is needed.

Other metabolic nodulins, for example, NADH-dependent glutamate synthase and sucrose synthase, are represented by single genes currently described only from one or two species (47, 59, 101). Additional metabolic enzymes may also be considered nodulins (42, 101, 123), but whether the genes encoding any of these metabolic proteins are uniquely expressed in nodules remains to be seen. Most appear to be nodule-enhanced, rather than truly nodulins. For example, members of Phaseoleae transport nitrogen as ureide compounds rather than the more common amides, and nodule-specific uricase genes have been described in both *Glycine* and *Phaseolus* (80, 100). Although nodule uricase genes probably evolved totally separately from the genes encoding root uricase activity (79), n-uricase is also expressed at low levels in uninfected soybean roots (61), and its protein has been detected in cotyledons (119). Recruitment of existing genes during the evolution of nodulation seems the most likely scenario for most metabolic nodulins.

Summary: Nodulin Origins and Evolution as Homology Criteria.

Speculation about the origin and evolution of nodulins is common in the molecular biology literature, but few studies have been rigorously phylogenetic. Some nodulin genes, notably those encoding PBM nodulins, do appear to be "novel," perhaps formed by recombination and/or duplications of whole genes or particular domains within genes. Many more appear to have been recruited to perform their traditional tasks in a new compartment and context. For such genes, the key to understanding their evolution may lie outside of their coding regions, in the regulatory sequences that may define them as "nodulins." The presence of shared promoter sequences in the upstream regions of several different nodulin genes (130) suggests that recruitment could have occurred by addition of novel controlling elements to diverse existing genes.

The above review makes clear that a considerable amount of new data will be necessary to utilize nodulin genes as criteria for testing homology of nodules in different plant groups. Too few nodulins have yet been described to permit critical comparisons. That the nodulins of economically less important Caesalpinioideae and Mimosoideae have not been studied is perhaps not surprising. But in many cases the same nodulins have yet to be described in more than one or two of the major papilionoid "model systems." This may be because many nodulins are taxon-specific, or because rapid evolution has obscured homologies. Immunological results have been cited to suggest that nodulins may be specific to particular types of nodules (115), but this approach suffers from its inability to identify apomorphic causes of non-cross reactivity. Phylogenetic analyses of gene sequences are much needed.

CONCLUSIONS

Considerable progress has recently been made in resolving long-standing phylogenetic questions in the Leguminosae. As is true elsewhere in plant systematics (32), the data have often been molecular, but it has been the application of rigorous objective approaches to phylogeny reconstruction that has provided the foundation for this success. The generally good agreement of these results with more intuitive interpretations of relationships in the family is a tribute to the quality of taxonomic insight beginning with George Bentham's work in the mid-1800s.

The availability of rigorous phylogenetic hypotheses permits diverse evolutionary questions to be addressed in their proper historical context. Current phylogenetic hypotheses for the family suggest that nodulation has arisen more than once in the family, but structural criteria suggest that nodules are homologous throughout the family. Additional criteria may become available as the molecular developmental genetics of nodulation become better understood. Reciprocally illuminating phylogenies of genes and organismal taxa can and should make major contributions to this process.

ACKNOWLEDGMENTS

I thank Janet Sprent, Roger Polhill, Doug and Pam Soltis, Mike Crisp, Doyle McKey, Mark Chase, David Morgan, and Shirley Tucker for providing unpublished material, encouragement, and/or helpful comments, and Jane Doyle for reading (too) many drafts and for her general support. I thank the National Science Foundation DEB Systematic Biology Program for supporting US research on legume systematics, notably recent grants for the Third International Legume Conference (BSR-9200904) and for my own current research (BSR-9107480).

Literature Cited

1. Allison LA, Kiss GB, Bauer P, Poiret M, Pierre M, et al. 1993. Identification of two alfalfa early nodulin genes with homology to members of the pea ENOD-12 gene family. *Plant Mol. Biol.* 21:375–80

2. Appleby CA, Dennis ES, Peacock WJ. 1990. A primaeval origin for plant and animal haemoglobins? *Aust. Syst. Bot.* 3:81–89

3. Arroyo MTK. 1981. Breeding systems and pollination biology in Leguminosae. See Ref. 95, pp. 723–69

4. Baker DD, Mullin BC. 1992. Actinorhizal symbioses. See Ref. 113, pp. 259–92

5. Ballenger JA, Dickson EE, Meyer M, Doyle JJ. 1993. DNA sequence data and phylogeny of Leguminosae. *Am. J. Bot.* 80(6):130–31 (Abstr.)

6. Barker DG, Gallusci P, Lullien V, Khan H, Ghérardi M, Huguet T. 1988. Identification of two groups of leghemoglobin genes in alfalfa (*Medicago sativa*) and a study of their expression during root nodule development. *Plant Mol. Biol.* 11:761–72

7. Becking J-H. 1992. The *Rhizobium* symbiosis of the nonlegume *Parasponia*. See Ref. 113, pp. 497–559

8. Bennett MJ, Lightfoot DA, Cullinore JV. 1989. cDNA sequence and differential expression of the gene encoding the glutamine synthetase gamma polypeptide of *Phaseolus vulgaris L. Plant Mol. Biol.* 12:553–65

9. Bojsen K, Abildsten D, Jensen EO, Paludan K, Marcker KA. 1983. The chromosomal arrangement of six soybean leghemoglobin genes. *EMBO J.* 2:1165–68

10. Breteler FJ. 1994. The boundary between Amherstieae and Detarieae (Caesalp.). See Ref. 21

11. Brummer EC, Bouton JH, Kochert G. 1993. Development of an RFLP map in diploid alfalfa. *Theor. Appl. Genet.* 86: 329–32

12. Bruneau A. 1993. *Systematics of Erythrina (Leguminosae: Phaseoleae) and implications for the evolution of pollination systems.* PhD thesis. Cornell Univ., Ithaca, NY. 259 pp.

13. Bruneau A, Doyle JJ. 1993. Cladistic analysis of chloroplast DNA restriction site characters in *Erythrina* (Leguminosae: Phaseoleae). *Syst. Bot.* 18:229–47

14. Bruneau A, Doyle JJ, Doyle JL. 1984. Phylogenetic relationships in Phaseoleae: Evidence from chloroplast DNA restriction site characters. See Ref. 21

15. Bruneau A, Doyle JJ, Palmer JD. 1990. A chloroplast DNA structural mutation as a subtribal character in the Phaseoleae (Leguminosae). *Syst. Bot.* 15:378–86

16. Campos F, Padilla J, Vazquez M, Ortega

JL, Enriquez C, Sanchez F. 1987. Expression of nodule-specific genes in *Phaseolus vulgaris* L. *Plant Mol. Biol.* 921–32

17. Chappill JA. 1994. Cladistic analysis of the Leguminosae: The development of an explicit hypothesis. See Ref. 21

18. Chappill JA, Maslin BR. 1994. A phylogenetic assessment of tribe Acacieae. See Ref. 21

19. Chase MW, Soltis DE, Olmstead RG, Morgan D, Les DH, et al. 1993. Phylogenetics of seed plants: An analysis of nucleotide sequences from the plastid gene *rbc*L. *Ann. Missouri Bot. Gard.* 80:528–80

20. Corby HDL. 1988. Types of rhizobial nodules and their distribution among the Leguminosae. *Kirkia* 13:53–123

21. Crisp MD, Doyle JJ, eds. 1994. *Advances in Legume Systematics.* Part 7: *Phylogeny.* Kew: Royal Botanic Gardens. In press

22. Crisp MD, Weston PH. 1994. Mirbelieae. See Ref. 21

23. Cronquist A. 1981. *An Integrated System of Classification of Flowering Plants.* New York: Columbia Univ. Press. 1262 pp.

24. de Blank C, Mylona P, Yang W-C, Katinakis P, Bisseling T, Franssen H. 1993. Characterization of the soybean early nodulin cDNA clone GmENOD55. *Plant Mol. Biol.* 22:1167–71

25. de Faria SM, Lewis GP, Sprent JI, Sutherland JM. 1989. Occurrence of nodulation in the Leguminosae. *New Phytol.* 111:607–19

26. de Faria SM, McInroy SG, Sprent JI. 1987. The occurrence of infected cells with persistent infection threads in legume root nodules. *Can. J. Bot.* 65:553–58

27. Delgado-Salinas A, Bruneau A, Doyle JJ. 1993. Chloroplast DNA phylogenetic studies in New World Phaseolinae (Leguminosae: Papilionoideae: Phaseoleae). *Syst. Bot.* 18:6–17

28. Dickison W. 1981. The evolutionary relationships of the Leguminosae. See Ref. 95, pp. 35–54

29. Dickstein R, Prusty R, Peng T, Ngo W, Smith ME. 1993. ENOD8, a novel early nodule-specific gene, is expressed in empty alfalfa nodules. *Mol. Plant-Microbe Interact.* 6:715–21

30. Downie SR, Palmer JD. 1992. Use of chloroplast DNA rearrangements in reconstructing plant phylogeny. In *Molecular Systematics of Plants,* ed. PS Soltis, DE Soltis, JJ Doyle, pp. 14–35. New York: Chapman & Hall. 434 pp.

31. Doyle JJ. 1991. Evolution of higher plant glutamine synthetase genes: tissue specificity as a criterion for predicting orthology. *Mol. Biol. Evol.* 8:366–77

32. Doyle JJ. 1993. DNA, phylogeny, and the flowering of plant systematics. *BioScience* 43:380–89

33. Doyle JJ. 1994. DNA data and legume phylogeny: A progress report. See Ref. 21

34. Doyle JJ, Doyle JL. 1990. Analysis of a polyploid complex in *Glycine* with chloroplast and nuclear DNA. *Aust. Syst. Bot.* 3:125–36

35. Doyle JJ, Doyle JL. 1993. Chloroplast DNA phylogeny of the papilionoid legume tribe Phaseoleae. *Syst. Bot.* 18:309–27

36. Doyle JJ, Doyle JL, Brown AHD. 1990. A chloroplast DNA phylogeny of the wild perennial relatives of soybean (*Glycine* subgenus *Glycine*): congruence with morphological and crossing groups. *Evolution* 44:371–89

37. Doyle JJ, Doyle JL, Palmer JD. 1993. Structural changes in the organellar genomes of legumes and their potential as phylogenetic markers. *Am. J. Bot.* 80 (6):147 (Abstr.)

38. Elias TS. 1981. Mimosoideae. See Ref. 95, pp. 143–151

39. Ferguson IK, Tucker SC, eds. 1994. *Advances in Legume Systematics.* Part 6: *Structural Botany.* In press

40. Fisher RF, Long SR. 1992. Rhizobium-plant signal exchange. *Nature* 357:655–59

41. Fitch W. 1970. Distinguishing homologous from analogous proteins. *Syst. Zool.* 19:99–113

42. Franssen HJ, Nap J-P, Bisseling T. 1992. Nodulins in root nodule development. See Ref. 113, pp. 598–624

43. Gallusci P, Dedieu A, Journet EP, Huguet T, Barker DG. 1991. Synchronous expression of leghemoglobin genes in *Medicago truncatula* during nitrogen-fixing root nodule development and response to exogenously supplied nitrate. *Plant Mol. Biol.* 17:335–50

44. Gantt JS, Baldauf SL, Calie PJ, Weeden NF, Palmer JD. 1991. Transfer of rpl22 to the nucleus greatly preceded its loss from the chloroplast and involved the gain of an intron. *EMBO J.* 10:3073–8

45. Geesink R. 1984. Scala Millettiearum. *Leiden Bot. Series,* 8:1–131. Leiden: EJ Brill

46. Gottlob-McHugh SG, Johnson DA. 1991. Detection of a subfamily of genes within the soybean nodulin A multigene family. *Can. J. Bot.* 69:2663–69

47. Gregerson RG, Miller SS, Twary SN, Gantt JS, Vance CP. 1993. Molecular

characterization of NADH-dependent glutamate synthase from alfalfa nodules. *Plant Cell* 5:215–26

48. Hecht MK, Edwards JL. 1976. The determination of parallel or monophyletic relationships: the proteid salamanders—a test case. *Am. Nat.* 110:653–77

49. Herendeen PS. 1994. Phylogenetic relationships of the tribe Swartzieae. See Ref. 21

50. Herendeen PS, Crepet WL, Dilcher DL. 1992. The fossil record of the Leguminosae: Phylogenetic and biogeographic implications. See Ref. 51, pp. 303–16

51. Herendeen PS, Dilcher DL, eds. 1992. *Advances in Legume Systematics.* Part 4. Kew: Royal Botanic Gardens. 326 pp.

52. Herrada G, Puppo A. Moreau S, Day DA, Rigaud J. 1993. How is leghemoglobin involved in peribacteroid membrane degradation during nodule senescence? *FEBS Lett.* 326:33–38

53. Hirsch M, Asad S, Fang Y, Wycoff K, Löbler M. 1993. Molecular interactions during nodule development. See Ref. 86, pp. 291–96

54. Jacobs FA, Zhang M, Fortin MG, Verma DPS. 1987. Several nodulins of soybean share structural domains but differ in their subcellular locations. *Nucleic Acids Res.* 15:1271–80

55. Kardailsky I, Yang W-C, Zalensky A, van Kammen I, Bisseling T. 1993. The pea late nodulin gene PsNOD6 is homologous to the early nodulin genes PsENOD3/14 and is expressed after the leghaemoglobin genes. *Plant Mol. Biol.* 23:1029–37

56. Kass E, Wink M. 1992. *rbc*L sequences from lupins and other legume species. *Plant Mol. Evol. Newslett.* 2:21–26

57. Kiss GB, Vincze E, Végh Z, Tóth G, Sooas J. 1990. Identification and cDNA cloning of a new nodule-specific gene, Nms-25 (nodulin-25) of *Medicago sativa. Plant Mol. Biol.* 14:467–75

58. Kouchi H, Hata S. 1993. Isolation and characterization of novel nodulin cDNAs representing genes expressed at early stages of soybean nodule development. *Mol. Gen. Genet.* 238:106–19

59. Kuester H, Fruehling M, Perlick AM, Pühler A. 1993. The sucrose synthase gene is predominantly expressed in the root nodule tissue of *Vicia faba. Mol. Plant-Microbe Interact.* 6:507–14

60. Kuhse J, Pühler, A. 1987. Leghemoglobins in *Vicia. Plant Sci.* 49: 137–43

61. Larsen K, Jochimsen B. 1986. Expression of nodule-specific uricase in soybean callus tissue is regulated by oxygen. *EMBO J.* 5:15–19

62. Lavin M. 1994. Tribe Robinieae and allies: model groups for assessing Early Tertiary northern latitude diversification of tropical legumes. See Ref. 21

63. Lavin M, Doyle JJ, Palmer JD. 1990. Systematic and evolutionary significance of the loss of the large chloroplast DNA inverted repeat in the family Leguminosae. *Evolution* 44:390–402

64. Lavin M, Luckow M. 1993. Origins and relationships of tropical North America in the context of the boreotropics hypothesis. *Am. J. Bot.* 80:1–14

65. Lee JS, Verma DPS. 1984. Structure and chromosomal arrangement of leghemoglobin genes in kidney bean suggest divergence in soybean leghemoglobin gene loci following tetraploidization. *EMBO J.* 3:2745–52

66. Legocki RP, Verma DPS. 1980. Identification of "nodule-specific" host proteins (nodulins) involved in the development of *Rhizobium*-legume symbiosis. *Cell* 20:153–63

67. Lewis GP, Schrire BD. 1994. A reappraisal of the *Caesalpinia* group (Caesalpinioideae: Caesalpinieae) using phylogenetic analysis. See Ref. 21

68. Liston A. 1994. Use of the polymerase chain reaction to survey for the loss of the inverted repeat in the legume chloroplast genome. See Ref. 21

69. Löbler M, Hirsch AM. 1992. An alfalfa (*Medicago sativa* L.) cDNA encoding an acidic leghemoglobin MsLb3. *Plant Mol. Biol.* 20:733–36

70. Löbler M, Hirsch AM. 1993. A gene that encodes a proline-rich nodulin with limited homology to PsENOD12 is expressed in the invasion zone of *Rhizobium meliloti*-induced alfalfa root nodules. *Plant Physiol.* 103:21–30

71. Marcker A, Lund M, Jensen EO, Marker KA. 1984. Transcription of the soybean leghemoglobin genes during nodule development. *EMBO J.* 3:1691–95

72. Martin PG, Dowd JM. 1990. A protein sequence study of the dicotyledons and its relevance to the evolution of the legumes and nitrogen fixation. *Aust. Syst. Bot.* 3:91–100

73. McKey D, Sprent JI. 1994. *Advances in Legume Systematics.* Part 5: *Nitrogen Economy.* Kew: Royal Botanic Gardens. In press

74. Metz BA, Welters P, Hoffman HJ, Jensen EO, Schell J, de Bruijn FJ. 1988. Primary structure and promoter analysis of leghemoglobin genes of the stem-nodulated tropical legume *Sesbania rostrata:* conserved coding sequences,

cis-elements, and trans-acting factors. *Mol. Gen. Genet.* 214:181–91

75. Miao GH, Hong Z, Verma DPS. 1992. Topology and phosphorylation of soybean nodulin-26, an intrinsic protein of the peribacteroid membrane. *J. Cell Biol.* 118:481–90

76. Morgan DR, Soltis DE. 1993. Phylogenetic relationships among members of Saxifragaceae sensu lato based on *rbc*L sequence data. *Ann. Missouri Bot. Gard.* 80:631–60

77. Morgan DR, Soltis DE, Robertson KR. 1994. Systematic and evolutionary implications of *rbc*L sequence variation in Rosaceae. *Am. J. Bot.* In press

78. Naisbitt T, James EK, Sprent JI. 1992. The evolutionary significance of the legume genus *Chamaecrista* as determined by nodule structure. *New Phytol.* 122:487–92

79. Nap J-P, Bisseling N. 1990. The roots of nodulins. *Physiol. Plant.* 79:407–14

80. Nguyen T, Zelechowska, MG, Foster V, Bergmann H, Verma DPS. 1985. Primary structure of the soybean nodulin-35 gene encoding nodule specific uricase II localized in peroxisomes of uninfected cells of soybean. *Proc. Natl. Acad. Sci. USA* 82:5040–44

81. Nirunsuksiri W, Sengupta-Gopalan C. 1990. Characterization of a novel nodulin gene in soybean that shares sequence similarity to the gene for nodulin-24. *Plant Mol. Biol.* 16:835–50

82. Nodari RO, Tsai SM, Gilbertson RL, Gepts P. 1993. Towards an integrated linkage map of common bean. 2. Development of an RFLP-based linkage map. *Theor. Appl. Genet.* 85:513–20

83. Nugent JM, Palmer JD. 1991. RNA-mediated transfer of the gene *cox*II from the mitochondrion to the nucleus during flowering plant evolution. *Cell* 66:473–82

84. O'Brien SJ, ed. 1993. *Genetic Maps: Locus Maps of Complex Genomes.* Cold Spring Harbor, NY: Cold Spring Harbor. 6th ed.

85. Ouyang LJ, Whelan J, Weaver CD, Roberts DM, Day DA. 1991. Protein phosphorylation stimulates the rate of malate uptake across the peribacteroid membrane of soybean nodules. *FEBS Lett.* 293:188–90

86. Palacios R, Mora J, Newton WE, eds. 1993. *New Horizons in Nitrogen Fixation.* Boston: Kluwer Academic. 788 pp.

87. Palmer JD, Jorgensen RA, Thompson WF. 1985. Chloroplast DNA variation and evolution in *Pisum:* patterns of change and phylogenetic analysis. *Genetics* 109:195–213

88. Palmer JD, Osorio B, Thompson WF. 1988. Evolutionary significance of inversions in legume chloroplast DNAs. *Curr. Genet.* 14:65–74

89. Pao GM, Wu LF, Johnson KD, Hofte H, Chrispeels MJ, et al. 1991. Evolution of the MIP family of integral membrane transport proteins. *Mol. Microbiol.* 5:33–37

90. Perlick AM, Pühler A. 1993. A survey of transcripts expressed specifically in root nodules of broadbean (*Vicia faba* L.). *Plant Mol. Biol.* 22:957–70

91. Pesole G, Bozzetti MP, Lanave C, Preparata G, Saccone C. 1991. Glutamine synthetase evolution: a good molecular clock. *Proc. Natl. Acad. Sci. USA* 88: 522–26

92. Pickersgill B, ed. 1994. *Advances in Legume Systematics.* Part 8: *Evolution of Cultivated Legumes.* Kew: Royal Botanic Gardens. In press

93. Polhill RM. 1981. Papilionoideae. See Ref. 95, pp. 191–208

94. Polhill RM. 1981. Sophoreae. See Ref. 95, pp. 213–30

95. Polhill RM, Raven PR, eds. 1981. *Advances in Legume Systematics.* Parts 1, 2. Kew: Royal Botanic Gardens. 1049 pp.

96. Polhill RM, Raven PR, Stirton CH. 1981. Evolution and systematics of the Leguminosae. See Ref. 95, pp. 1–26

97. Potter D, Doyle JJ. 1992. Origins of African yam bean (*Sphenostylis stenocarpa* (Hochst, ex A. Rich.) Harms): Evidence from morphology, isozymes, chloroplast DNA and linguistics. *Econ. Bot.* 46:276–93

98. Rice SJ, Grant MR, Reynolds PHS, Farnden KJF. 1993. DNA sequence of nodulin-45 from *Lupinus angustifolius. Plant Sci.* 90:155–56

99. Richter HE, Sandal NN, Marker KA, Sengupta-Gopalan C. 1991. Characterization and genomic organization of a highly expressed late nodulin gene subfamily in soybeans. *Mol. Gene. Genet.* 220:445–52

100. Roche D, Temple SJ, Sengupta-Gopalan C. 1993. Two classes of differentially regulated glutamine synthetase genes are expressed in the soybean nodule: a nodule specific class and a constitutively expressed class. *Plant Mol. Biol.* 22: 971–83

101. Sanchez F, Campos F, Padilla J, Bonneville, J-M, Enriquez C, et al. 1987. Purification, cDNA cloning and developmental expression of the nodule-specific uricase from *Phaseolus vulgaris* L. *Plant Physiol.* 84:1143–47

102. Sanchez F, Padilla JE, Pérez H, Lara

M. 1991. Control of nodulin genes in root-nodule development and metabolism. *Annu. Rev. Plant Physiol. Plant Mol. Biol.* 42:507–28

103. Sandal NN, Marcker KA. 1988. Soybean nodulin 26 is homologous to the major intrinsic protein of the bovine lens fiber membrane. *Nucleic Acids Res.* 16:9347

104. Sanderson MJ. 1991. In search of homoplastic tendencies: statistical inference of topological patterns in homoplasy. *Evolution* 45:351–58

105. Sanderson MJ, Liston A. 1994. Molecular phylogenetic systematics of Galegeae, with special reference to *Astragalus.* See Ref. 21

106. Scheres B, van de Wiel C, Zalensky A, Horvath B, Spaink H, et al. 1990. The ENOD12 gene product is involved in the infection process during the pea-*Rhizobium* interaction. *Cell* 60:281–94

107. Scheres B, van Engelen F, van der Knaap E, van de Wiel C, van Kammen A, et al. 1990. Sequential induction of nodulin gene expression in the developing pea nodule. *Plant Cell* 2:687–700

108. Schrire BD 1994. Evolution of the tribe Indigofereae (Leguminosae: Papilionoideae). See Ref. 21

109. She Q, Lauridsen P, Stougaard J, Marker KA. 1993. Minimal enhancer elements of the leghemoglobin *lba* and *lbc3* gene promoters from *Glycine max* L. have different properties. *Plant Mol. Biol.* 22:945–56

110. Sprent JI. 1989. Which steps are essential for the formation of functional legume nodules? *New Phytol.* 111:129–53

111. Sprent JI. 1994. Nitrogen acquisition systems in the Leguminosae. See Ref. 73

112. Sprent JI, Sprent P. 1990. *Nitrogen Fixing Organisms*. New York: Chapman & Hall. 256 pp.

113. Stacey G, Burris RH, Evans HJ, eds. 1992. *Biological Nitrogen Fixation*. New York: Chapman & Hall. 944 pp.

114. Stanford AC, Larsen K, Barker DG, Cullinore JV. 1993. Differential expression within the glutamine synthetase gene family of the model legume *Medicago truncatula*. *Plant Physiol.* 103:73–81

115. Steginck SJ, Vaughn KC. 1990. Immunotaxonomy of nodule-specific proteins. *Cytobios* 61:7–19

116. Stirton CH, ed. 1987. *Advances in Legume Systematics*. Part 3. Kew: Royal Botanic Gardens. 466 pp.

117. Stirton CH, Zarucchi JL, eds. 1989. *Advances in Legume Biology*. St. Louis: Missouri Botanic Garden. 842 pp.

118. Summerfield RG, Bunting AS, eds. 1980. *Advances in Legume Science.* Kew: Royal Botanic Gardens. 667 pp.

119. Tajima S, Ito H, Tanaka K, Nanakado T, Sugimoto A, et al. 1991. Soybean cotyledons contain a uricase that cross-reacts with antibodies raised against the nodule uricase nod-35. *Plant Cell Physiol.* 32:1307–12

120. Thompson JN. 1989. Concepts of co-evolution. *Trends Ecol. Evol.* 4:179–83

121. Thorne RF. 1992. Classification and geography of flowering plants. *Bot. Rev.* 58:225–348

122. Tingey SV, Walker EL, Coruzzi GM. 1987. Glutamine synthetase genes of pea encode distinct polypeptides which are differentially expressed in leaves, roots and nodules. *EMBO J.* 6:1–9

123. Tsai F-Y, Coruzzi GM. 1990. Nodule-enhanced asparagine synthetase in pea. *EMBO J.* 9:323–32

124. Tucker SC. 1993. Floral ontogeny in Sophoreae (Leguminosae: Papilionoideae). I. *Myroxylon (Myroxylon* group) and *Castanospermum (Angylocalyx* group). *Am. J. Bot.* 80:65–75

125. Tucker SC, Douglas AW. 1994. Ontogenetic evidence and phylogenetic relationships among basal taxa of legumes. See Ref. 39

126. Vaillancourt RE, Weeden NF, Bruneau A, Doyle JJ. 1993. Chloroplast DNA phylogeny of Old World Vigna (Leguminosae). *Syst. Bot.* 18:642–51

127. Vallejos CE, Sakiyama NS, Chase CD. 1992. A molecular marker-based linkage map of *Phaseolus vulgaris* L. *Genetics* 131:733–40

128. van Wyk B-E, Schutte AL. 1994. Phylogenetic relationships in the tribes Podalyrieae, Liparieae and Crotalarieae. See Ref. 21

129. Végh Z, Vincze E, Kadirov R, Tóth G, Kiss GB. 1990. The nucleotide sequence of a nodule-specific gene, Nms-25 of *Medicago sativa:* its primary evolution via exon-shuffling and retrotransposon-mediated DNA rearrangements. *Plant Mol. Biol.* 15:295–306

130. Verma DPS, Hu C-A, Zhang M. 1992. Root nodule development: origin, function and regulation of nodulin genes. *Physiol. Plant.* 85:253–65

131. Walker EL, Coruzzi GM. 1989. Developmentally regulated expression of the gene family for cytosolic glutamine synthetase in *Pisum sativum. Plant Physiol.* 91:702–8

132. Weeden NF, Muehlbauer FJ, Ladizinsky G. 1992. Extensive conservation of linkage relationships between pea and lentil genetic maps. *J. Hered.* 83:123–29

133. Wheeler E, Baas P. 1992. Fossil wood

of the Leguminosae: A case study in xylem evolution and ecological anatomy. See Ref. 51, pp. 281–301

134. Wink M, Käss E, Kaufmann M. 1993. Molecular versus chemical taxonomy. *Soc. for Med. Plant Res., 41st Congress, Düsseldorf,* ed. A Nahrstedt, pp. 17–18

135. Young JPW. 1992. Phylogenetic classification of nitrogen-fixing organisms. See Ref. 113, pp. 43–86

136. Young JPW, Johnston AWB. 1989. The evolution of specificity in the legume-rhizobium symbiosis. *Trends Ecol. Evol.* 4:341–49

Annu. Rev. Ecol. Syst. 1994. 25:351–75

USING DNA SEQUENCES TO UNRAVEL THE CAMBRIAN RADIATION OF THE ANIMAL PHYLA

Rudolf A. Raff

Institute for Molecular and Cellular Biology, and Department of Biology, Indiana University, Bloomington, Indiana 47405

Charles R. Marshall

Department of Earth and Space Sciences, and Molecular Biology Institute, University of California, Los Angeles, Los Angeles, California 90024-1567

James M. Turbeville

Department of Biology, University of Michigan, Ann Arbor, Michigan 48109

KEY WORDS: metazoan phylogeny, Cambrian radiation, molecular systematics, ribosomal DNA sequences, animal phyla

---

Abstract

Most animal phyla appeared in the Cambrian radiation, and many have a rich fossil record. The phylogenetic relationships among phyla are, however, still poorly understood. Although systematics based on morphological characters has been revitalized by cladistic methods, incongruent results have been obtained for most relationships. Gene sequence data offer the potential for adding substantial new characters for evaluating phylum level relationships. Molecular studies of the Cambrian radiation are just beginning, and only a few of the potential characters available from genes have been examined. The achievements and controversies between molecular studies are reviewed and compared to corresponding studies based on morphological data. Because molecular systematics is intensely dependent upon the methods used to infer trees, important methodological considerations that affect the studies of phylum-level

351

0066-4162/94/1120-0351$05.00

molecular systematics are discussed. The analysis of available gene sequence data has resolved conflicts between morphological interpretations. However, molecular systematics has not yet produced phylogenetic trees of broad phylum relationships more robust than those based on morphology.

INTRODUCTION

Despite a century of work on metazoan phylum-level phylogeny using anatomical and embryological data, it has not been possible to infer a well-supported metazoan phylogeny. Doing so has become even more important with the growing realization that most phyla arose during or just before the start of the Tommotian (23), which was formerly set at the base of the Cambrian but is now placed as Middle Early Cambrian, 530 million years ago (16). The fossil record records little of the evolutionary events leading up to the Cambrian radiation. To understand the pattern of that radiation, the phylogenetic relationships of the phyla must be understood. Such an understanding is also necessary to comprehend the patterns of evolution of genes and developmental processes among the phyla.

Phyla are disjunct in their features, and shared-derived features linking phyla are rare. We may have exhausted most of the informative morphological features, and interpretation of characters is not always an unequivocal matter. Systematists may interpret the same morphological character in diametrically different ways [examples of the effects of differing character interpretation/selection are found in recent cladistic analyses of metazoan relationships (9, 27, 98)]. Because phylogenetically informative features are few, mistakes will have severe effects on the outcome. Character conflicts are inevitable. When they occur, there is no way a priori to resolve character conflicts. One has to depend on parsimony, with plausible characters given equal weight (e.g. 28).

Gene sequence data offer a wealth of new phylogenetically informative characters. An average animal genome contains one to a few billion base pairs, far more potential characters than we will ever access. Here we review the ways that gene sequence data have been applied to unraveling phylum level evolutionary relationships. The field is still young, and the results limited and imperfect. Nonetheless, a number of important insights have been gained. These are discussed, as are some of the methodological issues that affect our ability to infer phylogenies from gene sequences.

METHODOLOGICAL ISSUES

Given a set of nucleotide sequences, the recovery of accurate phylogenies may be confounded by two major factors: incorrect alignments, and/or use of inappropriate tree recovery algorithms. Each is considered below. An equally important question is whether specific nodes in a phylogeny that was derived from an accurate alignment and appropriate tree recovery algorithm are significantly supported, an issue that is also reviewed.

Alignment

No general purpose algorithm provides reliable alignments. For 18S rRNA studies, most alignments have been done by eye (36, 110, 111), often with reference to secondary structure models (46, 47). Incorrect alignments can lead to incorrect trees and, for sequences that are especially difficult to align, the order in which sequences are added to the alignment can affect the final topology of the tree (65, 66). The most common way of dealing with alignment problems is to eliminate the difficult-to-align regions (e.g. 36, 110, 111), even though this reduces the data available.

Tree Recovery Algorithms and Consistency

A tree recovery algorithm is consistent if the probability that it yields the correct tree increases with increasing data (34). A method is inconsistent when the chance of recovering an incorrect tree increases with the size of the data set. Inconsistency generally results from an inability to account properly for multiple substitutions. Maximum parsimony may be inconsistent if there is great rate disparity between branches (32). An analysis of the "zone of inconsistency" for 16 tree recovery algorithms for the special case of four taxa with unequal rates of evolution showed that, even for artificial sequences generated with overly simple models of sequence evolution, all standard methods for inferring trees (maximum parsimony, Lake's evolutionary parsimony, neighbor joining, UPGMA, and least square methods) may be inconsistent (55). However, methodological advances such as a 12 parameter correction for multiple hits (66) will increase the range of data for which consistent methods of analysis exist.

The problem of consistency is often couched in terms of unequal rate effects, but a better rubric is "long branches attract." This is important because inconsistency may occur when long branches (usually terminal branches) are in proximity to short branches (typically internodes) even for sequences that have evolved according to a molecular clock (50). For example, maximum parsimony may be inconsistent with five taxa having equal rates of change (50, 126) and six taxa even with equal and slow rates of evolution (50). Unfortunately, the condition of adjacent long terminal branches and short internodes is common in key evolutionary events such as the Cambrian radiation of metazoan phyla. Breaking long branches by adding new taxa improves the chances of attaining consistency (50). However, it is generally unknown how many, and where, long branches should be broken to gain consistency (83). Disturbingly, all studies of inconsistency explored to date involve simple models of sequence evolution (e.g. all sites evolve at the same rate); in the real world, cases of inconsistency may be far more prevalent (126).

Distance methods are also inconsistent if rates of substitution vary (76), or if the distances have not been accurately corrected for multiple substitutions (55). Maximum likelihood is less susceptible to the difficulties associated with the attraction of long branches than is parsimony because a unique match of the same nucleotide between two species is down-weighted if the sequences are highly divergent (i.e. the long branches are "taken into account"—104). However, maximum likelihood has been used little in the analysis of large metazoan data sets due to its cost in computer time, and the requirement that a particular model of sequence evolution must be specified. If the model does not accurately reflect the mode of sequence evolution, maximum likelihood methods may also be inconsistent.

Although the attraction of long branches has been the primary focus of the studies of inconsistency, related sources of inconsistency include site-to-site variation in the rates of nucleotide substitution (66, 76), as well as substitution biases (73) and different GC compositions (68, 101). A dynamically weighted parsimony approach, designed to remove the effects of site-to-site rate variation and substitution biases, improved the topology of an 18S rRNA tetrapod phylogeny (judged on morphological grounds) (121). However, the selection of the weighting parameters is arbitrary and not rooted in biology. A distance method designed to deal with site-to-site variation in rates has been developed that employs different corrections for multiple substitutions according to the degree of variability observed at a site (113). If one uses morphology as the criterion for judging the quality of molecular trees, then these corrections significantly improved the resolution of the relationships within phyla.

Lake's invariant technique, called evolutionary parsimony (63, 104), was designed to prevent long branchs attracting; it has been applied to metazoan phylum relationships (64) but may be inconsistent (55). Evolutionary parsimony also fails to converge on the correct tree as the rate disparity between taxa increases (55, 116). The recently developed Hadamard approach to phylogenetic reconstruction (49, 51, 83, 84) can eliminate homoplasy if the sequences have evolved according to the Kimura three-parameter model (60). However, it is unlikely to yield significantly better phylum-level phylogenies than other methods because molecules such as 18S rRNA, with its variably conserved paired and unpaired regions, probably did not evolve according to such a simple model. The power of the approach may still be useful if regions of the molecule with similar rates can be identified and analyzed in isolation.

Identifying Significantly Supported Nodes

Even with a consistent method, the best tree need not be the correct tree; all that consistency guarantees is that with increasing data the chance of recovering the correct tree increases. Random error still may result in the wrong topology, especially for small data sets or for data sets with limited numbers of substitutions. This issue of signal-to-noise ratio is essentially a question of convergence (102), and one must ask how much information is needed to guarantee at some significance level that all branches are correct (49)? The problem of identifying significantly supported nodes is usually expressed in terms of whether hierarchical information is contained within the data, or how much better optimal trees are than nearly optimal trees.

Several approaches assess the significance of the shortest tree or of particular nodes within a tree. For determining whether a data set has significantly more hierarchical information than expected from random sequences, one can use the skewness of the tree length distribution (53, 102) or a randomization test (6, 31). However, these tests do not identify where in the tree the hierarchical information is contained (53). For assessing the strengths of individual nodes on a tree one uses bootstrapping (33, 35) or a randomization test (30). A weakness common to all these methods is that they measure only the relative strengths of the conflicting signals within a data set; they cannot distinguish between signal due to history (and therefore of phylogenetic value) from signal due to confounding effects such as the attraction of long branches, substitution biases, etc (e.g. see 73, 102). Convergence may occur with inconsistent as well as with consistent methods (102). With maximum parsimony, distance methods, and maximum likelihood (67), it is difficult to determine how much better the optimal tree is than sub-optimal trees. An approach with heuristic value is to examine a set of the next-to-shortest trees (17, 25, 99).

Robustness of Molecular Phylogenies

Given the difficulties associated with alignment and with establishing the conditions of consistency and convergence, it is clear that molecular phylogenies should not be accepted uncritically as accurate representations of the degree of relatedness between organisms. However, despite the formidable methodological problems that still confront us (alignment, long branches attracting, accounting for site-to-site rate variation, and identifying insignificantly supported nodes), it is striking that most nodes on most molecular trees are generally concordant with other data. Phylogenetically accurate information can be extracted from molecules, even if it is difficult to prove the case in a completely rigorous fashion. We also are beginning to gain sufficient understanding of the conditions under which molecular phylogenies will be inaccurate that in specific instances we can rule out the possibility of artifacts caused by alignment errors, long branches attracting, or compositional or substitution biases (e.g. 83).

FIRST RESULTS

Molecular phylogeny of the metazoan phyla became feasible a few years ago, when RNA sequencing methods made possible the generation of large amounts of data by direct sequencing of the small ribosomal subunit RNA (78). This molecule was particularly suitable for deep-ranging phylogenetic studies. The gene is present in all organisms; it occurs in only one version in animals; it contains a significant number of bases (about 1800); it evolves slowly; and it appears not to have been subject to horizontal gene transfer. In addition, direct sequencing of 18S ribosomal RNAs could be done because of the high concentration of rRNAs in cells. With the advent of polymerase chain reaction (PCR) techniques, DNA sequencing has since superseded RNA sequencing in rapidity and accuracy. However, RNA sequencing is historically important for making large-scale phylogenetic sequencing possible.

The initial study of Field et al (36) sequenced ribosomal RNAs from 20 classes in 10 phyla. That substantial metazoan database did not resolve the old phylogenetic controversies, but the results were important. As expected, cnidarians were the animals deepest in the tree. In fact, with the Fitch-Margoliash distance method used, they joined the fungi (36). This result was consistent with an early origin of diploblastic grade animals, but disturbing in indicating a polyphyletic origin for animals. Later analyses with more complete data and other inference methods show that Cnidaria branched early but are part of a monophyletic metazoan clade (37, 64, 80, 115). Second, in accord with traditional morphological hypotheses (8, 56, 57), platyhelminths were the sister group of the coelomate phyla.

The results obtained with the coelomate phyla provided important insights but proved to be of low resolution. Using a distance method, Field et al (36) distinguished four major coelomate animal radiations—arthropods, coelomate protostomes, chordates, and echinoderms—but could not resolve the branching order among these groups. Reanalysis of their data by parsimony (80) and evolutionary parsimony (64) revealed more structure. New ambiguities were introduced by Lake's analysis (64), which, for instance, gave a paraphyletic Arthropoda.

The analyses of further and more complete sequences have produced a more coherent 18S rDNA sequence tree for the metazoan phyla. Thus, the tree, inferred by Turbeville et al (110) shows several important features of the metazoan radiation. The noncoelomate bilaterian platyhelminths branch is the sister group of the coelomates. The coelomates are monophyletic, with distinct deuterostome and protostome clades. These results are consistent with the most generally accepted view of metazoan phylogeny (56). The short internal branches separating phyla are consistent with a rapid radiation of metazoan lineages. That may reflect the events of the Cambrian radiation, but we have no clock to correlate the timing of the gene radiation with the appearance of disparate morphologies in the early Cambrian.

THE BASE OF THE METAZOAN TREE

Animals have traditionally been shown as originating from an unknown protist root. Both 18S rDNA sequences and protein sequences indicate that the protists comprise a large number of disparate phyla (44, 45, 48, 100, 115). To determine the sister group of the animals has required the resolution of the relationship of metazoans to the other two great multicellular eukaryotic kingdoms, the fungi and plants. Because the divergence of animals, plants, and fungi was apparently close in time, early studies produced conflicting results. For example, inferences from some studies of rRNA and protein sequences suggested that plants and animals are sister groups (26, 42, 114). Other studies of ribosomal RNA sequences led to the inference of a close fungal-animal linkage (45, 52, 115). With larger databases and improved analyses, recent reassessments from several gene sequences are consistent with fungi and animals as sister groups [elongation factor 1a] (10, 45) ; [catalase] (77); and [18s rDNA sequences] (115). The congruence of results from three genes is impressive.

Sponges

Sponges have a cellular grade of organization, with no discrete tissues or organs, and only a few cell types. It has been questioned whether they are really metazoans or are "parazoans," with a separate origin from true animals. Sponges possess a specialized collared monociliated cell type, the choanocyte, that generates water currents through the pores of the animal. There is an obvious similarity to a group of colonial protists known as choanoflagellates that possess a similar monociliated cell with a collar of microvilli. An 18S ribosomal DNA maximum likelihood tree shows that sponges occupied the base of a monophyletic metazoan tree, and that choanoflagellates were the sister group of the metazoans including the sponges (115). The fungi emerged as the sister group of a choanoflagellates plus metazoan clade. Cells similar to choanocytes are present in other metazoans, probably representing a metazoan plesiomorphy (87). If sponges had independently evolved multicellularity, we would not expect them to share basic features of animal cell-cell interactions in development. The complex extracellular matrix of metazoans, composed of collagen, proteoglycans, adhesive glycoproteins, fibronectin, and the integrin link to the intracellular cytoskeleton, is present in sponges and unites sponges with eumetazoans (74).

Diploblastic Animals

In the tree of Wainwright et al (115), the diploblastic clades (ctenophores, cnidarians plus placozoans) branch off above the sponges and form sister groups to the triploblastic, bilaterian phyla. This result is reasonably consistent with cladistic analyses of morphological features (8, 98). The simplest diploblastic animal, *Trichoplax adherens,* has been placed in its own phylum, the Placozoa. It has only a few thousand cells organized into a dorsal sheet of flagellated cells, a more columnar ventral sheet of flagellated cells, and few mesenchymal cells in an extracellular fluid filling the space between the cell sheets. Grell (40) has suggested that *Trichoplax* is a simple diploblastic animal, and that the dorsal and ventral cell sheets are homologous to the ectoderm and endoderm of cnidarians. The 18S rDNA sequence tree bears him out, with *Trichoplax* falling into the tree with cnidarians and ctenophores (115), although it lacks major metazoan apomorphies such as muscle and nerve cells and thus conflicts with a recent cladistic analysis of morphology (8).

Trees derived from a 450-base region of 28S rDNAs have yielded a diphyletic Metazoa, in which both diploblasts (cnidarians, ctenophores, placozoans, and sponges) and triploblasts are monophyletic but split from each other by intervening plants (20). Adoutte & Philippe (2) have rethought the possibility of a diphyletic Metazoa, considering new 18S rDNA trees using both distance and parsimony methods. Their distance analyses yielded a diphyletic Metazoa, whereas parsimony yielded either metazoan monophyly or diphyly depending on the diploblast species sampled. They suggest that the controversy is not yet solved. However, the large number of genes and morphological features shared by diploblastic and triploblastic metazoans (18, 97) strongly supports metazoan monophyly.

THE RADIATION OF THE BILATERIA

The origin of the Bilateria marks a radical departure in organization of animal bodyplans from the diploblasts; it had to have occurred prior to the Cambrian radiation. The phylogenetic placement of the few living phyla—the platyhelminths, nemertines, and according to some workers, the molluscs that have been considered as being of acoelomic, triploblastic grade—are controversial. Monophyly of the acoelomate condition itself has been questioned (88).

Platyhelminths

Given the key position of flatworms in metazoan phylogenetics, their placement is of crucial importance, but on present data this should be treated with caution. Cladistic analysis of morphological features of the platyhelminths suggests they are monophyletic (29). Analyses of 18S RNA sequence data also indicates monophyly (14, 90). One neighbor-joining study (59) suggests that flatworms are paraphyletic. However, in all analyses with partial sequences (and some complete sequences—see 123), they diverge prior to the rest of the bilaterian animals. In analyses with complete sequences, flatworms are sometimes placed in a clade with nematodes as the sister group of the protostome coelomates (JM Turbeville, unpublished observation). Ax (7) has tied a phylum of very small "worms," the gnathostomulids, to the flatworms on the basis of shared morphological features, but this link has been questioned (103). No molecular data are available.

Are the flatworms the last remnant of a radiation of acoelomate phyla early in metazoan history, or have the acoelomates always had a low disparity? If so, perhaps the evolution of a coelom made possible the exploitation of most metazoan niches, triggering the evolution of the coelomate phylum bodyplans in the Cambrian radiation. Flatworms may offer a view of the primitive bilaterian bodyplan at the base of the radiation. Investigations of homeobox-containing genes of flatworms are just beginning and offer an important opening into understanding the evolution of the body axis in bilaterian animals (12).

Nemertines

The predominantly marine nemertine worms possess a complete digestive tract and a closed circulatory system. In development, they exhibit spiralian cleavage and are protostomatous. The traditional view (based predominantly on symplesiomorphic characters) is that nemertines are acoelomate in body organization and are most closely related to flatworms (110). However, unlike flatworms, which have no body cavities, nemertines have two kinds of cavities, a rhyncocoel, housing an eversible proboscis with which nemertines capture prey, and a system of cell-lined cavities referred to traditionally as a blood vascular system. The interpretation of these cavities has been controversial. The minority view (109) is that the body cavities of nemertines are coelom homologs. Turbeville (108) investigated these spaces ultrastructurally and argued, on the basis of the classic criteria of homology, that the cavities were coelom homologs. They correspond in position (lateral), cytology (lined by a continuous epithelium), and formation (splitting of mesodermal bands) to coeloms of spiralian coelomates. Invertebrate blood vessels are typically lined by an extracellular matrix or, in some taxa, with a discontinuous cell lining (108). To address the nemertine controversy with new data, Turbeville et al (110) determined 1300 bp of a nemertine 18S rRNA. Analysis with parsimony and other methods placed nemertines within the coelomate protostome clade, refuting the view that nemertines are the sister group of the platyhelminths. Thus, the interpretation of the body cavities as coelomic spaces appears to be correct.

Molluscs

Molluscs pose a similar problem. The percardial cavity of molluscs has been proposed as the coelom homolog because it is lined by mesoderm, and it is associated with the gonads (18), but that suggestion has been debated. Based on an interpretation that molluscs lack a coelom and on the mode of movement of some molluscs on a slime trail produced by the flat foot (21), an acoelomic radiation has been suggested (see 18). Molluscs develop in a way very similar to annelids, but their embryology does not allow an unequivocal decision on whether molluscs are coelomates. In a cladistic analysis of anatomical, embryological, and molecular features, Scheltema (96) has suggested that molluscs are the sister group of the unsegmented sipunculans and so places this clade as the sister group of the annelids within the coelomate protostomes. The 18S rRNA sequence results put molluscs strongly into the protostome coelomates (36, 64, 80), again congruent with one of the two conflicting morphologically derived hypotheses (i.e. that molluscs are coelomates). This result is supported by elongation factor 1a sequences (61). Annelids, pogonophorans, and molluscs form a eucoelomate protostome sister group to the arthropods, consistent with the 18S rRNA result (61). This placement of molluscs means that the molluscan pericardial cavity is probably derived from a coelom. The segmentation of annelids may have been acquired after the divergence of that lineage from a stem lineage shared with molluscs. Alternatively, the ancestors of the molluscs may well have exhibited body segmentation that was later lost. Whether or not segmentation is primitive to this group is crucial in deciding whether segmentation of annelids is homologous to that of arthropods.

THE PSEUDOCOELOMATES

The pseudocoelomates (also called aschelminths) include almost a third of all living phyla, the rotifers, gnathostomulids, priapulids, gastrotrichs, kinorhynchs, nematodes, nematomorphs, acanthocephalans, and loriciferans [a phylum only described in 1983 (62)]. Only the priapulids have a meaningful fossil record. Thus, the role of most of these phyla in the Cambrian radiation is completely obscure. The nematodes are the most diverse group, and make up a sizable portion of the world's biomass. Some are parasitic. The free-living soil nematode *Caenorhabditis elegans* is now one of the major model systems for studies in developmental genetics. Its phylogenetic position among the major metazoan phyla is thus important. Traditionally, pseudocoelomates have been united by possession of a body cavity that lacks a peritoneal lining and is hypothesized to be a persistent blastocoel (57).

However, detailed studies of these body cavities have revealed that they are not persistent blastocoels, and that there is variability in body cavity organization (93). For example, the body cavity of some nematodes is anatomically and ontogenetically similar to a eucoelom, whereas some nematomorphs and gastrotrichs are essentially acoelomate. These observations suggest that a "pseudocoelom" is unlikely to represent a homologous feature of these taxa.

Based on comparative studies of the body cavities among bilaterian adults and larvae, it has been suggested that these taxa could have arisen independently from eucoelomates by progenesis (89). In a cladistic analysis, Lorenzen (69) concluded that there are no identifiable morphological synapomorphies supporting monophyly of these taxa. A cladistic analysis by Schram (98) by contrast finds support for a monophyly of aschelminthes (inclusive of chaetognaths!). However, the characters (including a pseudocoelom) supporting the clade exhibit very low similarity and their certainty is tenuous at best. In the analysis of Eernisse et al (27), which included only nematodes, kinorhynchs, and priapulids as representatives of the pseudocoelomates, monophyly was not supported. The consensus tree resulting from a reanalysis of Schram's data (27) indicated that pseudocoelomates formed the primitive bilaterian taxon. In the consensus tree of Eernisse et al (27), nematodes and kinorhynchs were more closely related to arthropods than to other spiralians. The position of priapulids is unresolved: in one tree it was the outgroup of a (nematode (kinorynch (tardigrade, arthropod))) clade. In two other trees it was the sister taxon to the above clade plus a spiralian clade.

Analyses of morphological characters do support monophyly of certain taxa (69, 98). The putative synapomorphies of the free-living rotifers and the parasitic acanthocephalans are a syncitial epidermis containing an intracellular "cuticle," and a system of tubules formed by the invagination of the apical epidermal cell membrane. A pair of anteriorly situated sacs (lemnisci) associated with the protrusible anteriors of bdelloid rotifers and acanthocephalans supports monophyly of these taxa (69). Nematodes and the parasitic nematomorphs share a similar cuticle and possess only longitudinal body wall muscles, suggesting that they are sister taxa (69, but see 93). Other features suggest that the nematomorphs might fall into a nematode clade (69). Putative synapomorphies of gastrotrichs and nematodes include a unique bilateral cleavage, and similarly organized myoepithelial pharynges and muscle innervation patterns (92), and thus suggest gastrotrichs as the sister group of a nematode plus nematomorph clade (69).

The priapulids were an important component of the Cambrian Burgess Shale fauna (120). They thus are a major group for reconstructing the events of the Cambrian radiation. Neither anatomical nor embryological features have provided reliable clues to their placement among the Bilateria (9, 27, 98).

The pseudocoelomates pose a challenge to the use of gene sequence data in high level metazoan systematics, because their morphology has provided few clues to their times of origin or phylogenetic relationships. Although complete and partial 18S rDNA and partial 28S rDNA sequences (39) of nematodes have been included in analyses of metazoan phylogeny, the results have been conflicting (81). Complete 18S rDNA sequences of an additional nematode (Ascaris), a nematomorph (Gordius), and a rotifer (Brachionus) have recently been obtained (JM Turbeville, RA Raff, unpublished), as has an acanthocephalan sequence (105). Our preliminary analyses with these sequences find no support for pseudocoelomate monophyly, but they do support a sister group relationship of rotifers and acanthocephalans (see above). The pattern of relationships observed when two poriferans and three cnidarians were used as outgroups with parsimony analysis was (Porifera(Cnidaria ((Nematomorpha(((Rotifera, Acanthocephala)((Nematoda, Chaetognatha)-Platyhelminthes)) (Mollusca, Arthropoda)))Deuterostomia))). This tree also joins the parasitic flatworms to the longest branched taxa, the chaetognath and nematodes. Of course, thorough analyses with additional pseudocoelomate taxa will allow a more reliable assessment of their placement among the other phyla. The preliminary data at least suggest a closer relationship of these taxa to spiralian or protostome phyla. Their depth in the tree suggests that the pseudocoelomates were an early part of the metazoan radiation.

THE EUCOELOM, PROTOSTOMES, AND LOPHOPHORATES

As noted above, most 18S rDNA trees are consistent with a monophyletic Metazoa and with morphology. The sponges form the sister group of all other metazoans. Diploblastic phyla branch next as the sister group of the Bilateria. Finally, the platyhelminths, as acoelomic triploblastic bilaterians, branch as the sister group of all eucoelomate bilaterians in most analyses. Neither molecular data nor the fossil record tells us precisely when the coelom originated and the radiation of the eucoelomate phyla began. These events had to have occurred by 530 million years ago, prior to the first abundant shelly fossils at the base of the Tommotian, marking the minimum age of the metazoan radiation. Possible diploblastic metazoans are recorded in the Ediacara fauna (about 560 to 580 million years), as are the first probable bilaterian trace fossils (23). The appearance of trace fossils in rocks deposited just prior to the Cambrian indicates the origin of the coelomic body cavity that allowed efficient burrowing by large animals.

Not all cladistic analyses of morphology agree on the monophyly of eucoelomic phyla (18, 27). The 18S rRNA sequence trees are largely consistent with coelomate monophyly, but trees are sensitive to the outgroups used. The primary classical division of coelomates is into protostome and deuterostome superphyla on the basis of developmental features (18, 56). Some (but not all) recent molecular phylogenies (e.g. 110, 113, 115; JM Turbeville, RA Raff, unpublished) are consistent with this division. Major problems remain for molecular systematics in addressing coelomate monophyly and relationships between and within coelomate superphyla.

Coelomate Protostomes

An unexpected feature pointed out by Field et al (36) is that protostome coelomate phyla, the annelids, molluscs, pogonophorans, sipunculids, nemertines, and brachiopods form a clade. This grouping suggested an unsuspected major radiation of a large coelomate protostome clade after the initial coelomate radiation. The spectacular radiation of coelomate protostome phyla probably followed not long after the origin of coelomates, but long enough to have left a strong signal in the 18S rDNA gene. Arthropods and annelids did not form an articulate clade. Arthropods were placed as the sister group of the coelomate protostome clade that includes the annelids, and thus they are no closer to the segmented annelids than they are to the unsegmented molluscs [this result is consistent with Eernisse et al's (27) reanalysis of Schram's morphological character matrix, but not with other cladistic analyses of morphological characters (18, 27, 118)]. If correct, the placement requires that segmentation is primitive in the ancestor of the arthropod-plus-protostome coelomate clade (and was lost by many of these taxa), or that annelids and arthropods evolved segmentation independently. The two scenarios hold different consequences for the pattern of the Cambrian radiation and for those seeking common molecular mechanisms for segmentation. That nemertines, brachiopods, and molluscs should be component phyla of the protostome coelomate clade may appear surprising. However, in each of the cases conflicting hypotheses based on morphology exist. The molecular data in all three cases support one of the contending morphological hypotheses.

Lophophorates

Lophophorates (phoronids, brachiopods, and bryozoans) share a ring of hollow tentacles surrounding the mouth (the anus is situated outside the ring), and they have been considered a monophyletic clade, although lophophorate monophyly has been disputed (8, 18). Hyman (56) considered them as related to protostomes. Other workers have evaluated their morphological features as indicating a membership in the deuterostomes (18), as the sister group of the deuterostomes (8), or as intermediates between deuterostomes and protostomes (see 122). The tripartite coelom and mesocoelic tentacles are considered to unite lophophorates with deuterostomes. Radial embryonic cell cleavage and the origin of coelomic pouches in development from the sides of the archenteron is consistent with a deuterostome affinity [however, radial cleavage is probably a symplesiomorphy of the Bilateria (8)]. The mode of coelom formation in bryozoans is unclear. The larval mouth is formed in the protostome manner from or near the blastopore in phoronids and some brachiopods (38, 75).

All analyses of a brachiopod partial 18S rRNA sequence support their placement among coelomate protostomes (36, 64, 80). More extensive molecular analyses of lophophorates are apparently forthcoming (43), and these indicate that lophophorates are most closely related to protostomes. A larger sampling of taxa will allow a test of lophophorate monophyly. The discrepency between brachiopod 18S rRNA and morphology suggests other possibilities. One is that brachiopods really are deuterostomes, but that during their history, a horizontal transfer of ribosomal genes from a protostome has occurred, and the deuterostome sequence was eliminated. That is unlikely, but readily tested by determining the relationships of other gene sequences that would reveal the deuterostome genetic background. Another interesting possibility is that brachiopods are protostomes by ancestry as indicated by the 18S rRNA sequence, but that their deuterostome-like developmental features have resulted from convergent evolution. If true, that would both provide a caution about the phylogenetic "value" of developmental features and support the idea that early development can exhibit a very high evolutionary flexibility (125).

The distribution and sequence of the oxygen transport protein hemerythrin provides additional molecular support for a coelomate protostome clade. This protein is found only in priapulids, sipunculans, lingulid brachiopods, and some annelids (24, 91). This distribution pattern supports the hypothesis that brachiopods are coelomate protostomes and also suggests that the priapulids, which are generally classified as pseudocoelomates, may belong to this group as well.

ARTHROPODS

Monophyly vs Polyphyly

There are more kinds of arthropods than any other animals, and they present some of the most complex and interesting phylogenetic problems. Unlike annelids and other coelomates, they are covered with a hard cuticle. Arthropodization has resulted in a profound reorganization of bodyplan. Annelids (the most often proposed sister group) possess a hydrostatic skeleton in which the body wall musculature works on the fluid-filled coelom. This allows the animal to utilize its segments to expand or contract parts of the body to burrow in sediments. Arthropods have no hydrostatic skeleton and have evolved jointed appendages. In more primitive arthropods, the segments and appendages are uniform, but one of the major trends in arthropod evolution has been tagmosis, the morphological differentiation of appendages and segments for specialized functions. Arthropods not only share a plan of external and internal body segmentation, they add segments in a manner similar to that of annelids via growth and patterning of a posterior growth zone, the teloblast. The arthropod nervous system is similar in layout to that of annelids, with a dorsal brain with connections surrounding the esophagus. The paired main nerve trunk runs down the ventral side of the body. The coelom has been drastically reduced, but arthropods possess a haemocoel (blood) cavity. The dorsal heart characteristic of protostomes has been retained, but the system of closed blood vessels has been replaced by an open circulatory system.

The shared anatomical features of arthropods form the obvious basis for considering them monophyletic [synapomorphies for the Arthropoda (Euarthropoda plus Onychophora) as listed by Weygoldt (117) are an alpha-chitin cuticle, mixocoel, dorsal blood vessel with paired ostia, pericardial sinus, complex brain, nephridia with sacculi, appendages with extrinsic and intrinsic muscles and terminal claws, centrolecithal eggs]. However, the interpretation has not been straightforward. Monophyly is supported by cladistic analyses of morphology (8, 117, 118). The onychophorans form the sister group of the euarthropods (117). Within the euarthropods, chelicerates is the sister clade to the mandibulates [crustaceans as a sister taxon to a myriapod plus hexapod clade (often referred to as unirames or tracheates)]. In the polyphyletic hypothesis, crustaceans, chelicerates, and unirames (including onychophorans) have separate origins from various hypothetical extinct annelid-like ancestors (72).

Onychophorans were recognized in the nineteenth century as sharing features of both arthropods and annelids, but we would now regard the annelid-like features as plesiomorphic. On the basis of such shared features as tracheae, onychophora were envisioned by Haeckel as direct ancestors of uniramians. He regarded the crustaceans as arising from a separate ancestor. Tiegs & Manton (107) pointed out that no matter how arthropods arose, some convergent evolution had to have occurred, and they suggested that arthropodization itself may have been a convergent feature.

Manton (71) developed a detailed view of arthropods as polyphyletic in origin based on the distinct locomotory structures and functions of the major living arthropod groups. She suggested that one type of locomotory system could not have given rise to another. She noted the distinct differences between the mandibles of crustaceans and uniramians, and held that they were functionally distinct and nonhomologous as jaws. Willmer (122) argued that if annelid-like worms were to evolve exoskeletons, the arthropodization would arise inevitably given the mechanical demands. Thus, convergence and polyphyly would be inevitable. Each arthropod group has a highly distinct pattern of appendages on the head. The patterns of the three living arthropod groups exhibit an immensely long evolutionary conservation through hundreds of millions of years. However, the tagmosis patterns of the Burgess Shale arthropods are in most cases quite different from those of the long-persisting major lineages (41, 72). Manton & Anderson (72) suggested that these highly divergent tagmosis patterns indicate a hidden and convergent evolution of "arthropods." Gould (41), on the other hand, suggested a common descent with a low level of developmental constraint among early arthropods. Tagmosis patterns thus later become fixed by rigid developmental rules.

Manton's arguments for polyphyly were weakened by her failure to consider synapomorphies and divergent evolution of characters after splitting of lineages (117). The fully evolved gnathobasic mandible of crustaceans and the whole limb mandible of uniramians may not be derivable from one another, but both could have descended divergently from a primitive arthropod head appendage. In the crustaceans, evolution of a gnathobasic mandible was favored, whereas in uniramians, the limb tips became the primary food handling elements. Weygoldt (117) suggested that the primitive unirames had biramous appendages. Unirames and chelicerates, as well as the few terrestrial crustaceans, have lost the outer parts of these appendages in evolving efficient terrestrial walking legs.

Other complex organ systems are consistent with monophyly. Paulus (82) concluded that the compound eye is a synapomorphy supporting arthropod monophy. The eyes of Cambrian trilobites underwent the same mode of accretive growth as was seen in development of the Drosophila eye, indicating a long conserved developmental mechanism (22, 86). The development of the central nervous systems of insects and crustaceans is similar (106, 119).

Anderson (3–5) suggested that fate maps and developmental features of the embryos could be used as indicators of the uniqueness or relatedness of arthropod groups. He concluded that arthropod development is too disparate for arthropods to be monophyletic. He united the onychophorans with myriapods, insects, and other minor groups that make up the uniramians because all exhibit a syncytial mode of cleavage with subsequent formation of a cellular blastoderm. The conceptual fate maps of their blastoderms are similar, and development of midgut, mesoderm, and ectoderm are similar, as is formation of somites. He suggested that uniramians and onychophorans have the greatest developmental similarity to annelids. Onychophorans and uniramians fate maps are consistent with an ancestry among the oligochaetes and leeches. Crustacean development is different. Those large eggs cleave syncytially and form a cellular blastoderm. Other crustaceans with smaller eggs exhibit complete cleavage, which Anderson (3–5) concluded represents the primitive mode of early development in crustaceans. The crustacean fate map is quite different from the uniramian/annelid fate map. Finally, chelicerate development and fate map is different from that of annelids or other arthropods. Weygoldt (117) criticized Anderson's analysis as typological. Emphasis was put on the unique features (apomorphies) of each group rather than on features shared with other groups. As with other proponents of arthropod polyphyly, characters shared by all arthropods were considered convergences. Proponents of arthropod polyphyly never established that any of the arthropod taxa are more closely related to any other metazoan taxon than they are to each other (117).

Polyphyly arguments depend on estimates of how much evolutionary change is possible. If two groups exhibit some difference, the argument is that transitions between features are constrained in such a way that a transition is forbidden. The difficulty with such statements is that they are generally unsupported by empirical data on how much evolutionary change is possible. For example, it has long been assumed that early development is highly constrained and thus resistant to evolutionary modification. Evolutionary conservation, however, says nothing per se about whether a feature can be modified. Recent studies of the evolution of larval stages show that quite conserved developmental features can be rapidly and radically modified (85, 125).

Arthropod Molecular Systematics

Arthropod molecular phylogenies have been inferred from partial 18S rRNA sequences (1, 111), complete 18S rDNA sequences (113, 123), 12S mitochondrial rDNA sequences (11), 18S rDNA and ubiquitin protein sequences (118). The questions addressed are arthropod monophyly, the arthropod sister group, the relationship among major arthropod groups, within major groups, and of some minor phyla thought to be related to the arthropods. All (11, 111, 113, 118, 123) except evolutionary parsimony (64) support arthropod monophyly. Only some analyses support the classic annelid-arthropod clade, the Articulata. The results of Wheeler et al (118), using a combined data set (607 bp of 18S rDNA, plus 550 bp of ubiquitin, plus 126 morphological characters), supported the Articulata, but their molecular data sets alone did not, nor did studies by Turbeville et al (111) with 18S rDNA and Ballard et al with 12S mitochondrial rDNA (11). The hypothesis of an annelid-arthropod clade has been questioned on morphological grounds (27) but supported by other analyses of morphological data sets (18, 117, 118). All trees support a monophyletic Chelicerata. The gene sequence trees support a mandibulate clade composed of insects (as representative unirames) and crustaceans. The myriapods are positioned as the sister group of insects in one analysis (118) but are found deep in the other trees, breaking up the unirames (11, 111). The difficulty may lie in inadequate sampling and problems of long branch length in milliped 18S rDNA. In counter to this, three myriapod 12S rDNAs were sampled, and myriapods again did not fall into a unirame clade with insects (11).

Pentastomids and Onychophorans

Two studies of arthropod molecular systematics include small phyla thought on anatomical or embryological grounds to be related to arthropods. A small phylum of parasitic "worms," called pentastomids that live in the nasal passages of vertebrates has been thought to be related to crustaceans, based on their embryology, sperm morphology, and nauplius-like larva. That hypothesis has been confirmed by 18S rRNA sequences (1). The relationships of the onychophora, which play such a central role in arguments about arthropod origins, have been controversial. They fall into the arthropods in the 12S mitochondrial rDNA tree (11) [but see (81) for possible artifacts], but as the sister group of all arthropods in the combined dataset tree (118). If onychophorans fall within the arthropods, it requires the hypothesis of loss of arthropodization by their lineage, and a different view of the position of the "lobopods" in the Cambrian radiation (see 19 for speculations on Cambrian lobopods).

Other molecular data also support the monophyly of arthropods. Intermediate filaments have been observed by electron microscopy and immunostaining with a broadly reactive antibody in all phyla examined (cnidarians, acoelomates, pseudocoelomates, lophophorates, deuterostomes, and protostomes), but these filaments were absent in all arthropods, where they are apparently replaced functionally by microtubules (13). Onychophorans were found to have peculiar filamentous structures and so could not be placed with the arthopods by this criterion. Consistent with their position as an arthropod offshoot, pentastomids lacked intermediate filaments. Loss of intermediate filaments and their functional replacement by a different class of intracellular fibers in all arthropod groups is consistent with other molecular data supporting the monophyly of arthropods (exclusive of the onychophorans).

In summary, current cladistic analyses of morphology and gene sequence data strongly support the monophyly of the arthropods. The hypothesis of an articulate clade comprising annelids as the sister group of arthropods is still in dispute. This is not a disagreement between molecules and morphology as both molecular and morphological analyses are inconclusive. The results inferred are sensitive to data sets used. The relationships among arthropod subphyla are also unresolved.

DEUTEROSTOMES

The second major classic eucoelomate clade, the deuterostomes, is a much smaller group than either the arthropods or the coelomate protostomes. The body plans of the deuterostome phyla are quite distinct from each other, and the group is linked primarily on the basis of shared morphological features [such as the presence of a hydropore (or pores) associated with the protocoel, and the anus forming near the site of the blastopore]. The unequivocal deuterostome taxa are urochordates, cephalochordates, vertebrates, hemichordates, and echinoderms. The chaetognathes and lophophorates have been associated more dubiously with the deuterostomes. The relationships among deuterostome phyla have been poorly defined. The application of molecular data will be of crucial importance to answer such basic questions as, which taxa belong to the deuterostomes, what is the pattern of relationships among the deuterostomes, and which taxon is the sister group of the chordates.

These questions have been addressed using 18S rDNA sequence data. The results with partial sequences support a monophyletic Deuterostomia only with evolutionary parsimony (64). Analyses of complete sequences from all major representatives support deuterostome monophyly (112, 114a). [We note that deuterostome monophyly is sensitive to the diploblastic outgroup utilized (JM Turbeville, RA Raff, unpublished)]. Lophophorates do not belong in the deuterostomes (36, 43). Chaetognaths or arrow worms have recently been investigated using 18S rDNA (105, 114a). These are dart-shaped planktonic carnivores that range in size from a few millimeters to 12 cm in length. The placement of chaetognaths among the deuterostomes has been justified on the bases of the mesoderm arising from the archenteron, their tripartite coelom, and the origin of their subepidermal muscles from mesoderm derived from the archenteron (18). It has been argued that a number of these traits are not so unambigous for chaetognaths (15, 122).

An 18S rDNA sequence of a chaetognath does not place them in a deuterostome clade (105, 114a). Instead they branch deep (with flatworms or molluscs) in the tree, possibly because of long branch lengths. In our unpublished analysis, which includes complete sequences of two nematodes, both long-branched taxa, the chaetognath forms the sister group to the nematodes. It is probably best to view their position based on 18S rDNA sequences with caution until more data are available.

Relationships Among Major Deuterostome Taxa

There are three unequivocal deuterostome phyla: hemichordates, chordates, and echinoderms. The orthodox hypothesis found in many textbooks (and the most parsimonious tree) places hemichordates as the sister group of the chordates (based primarily on the shared possession of gill slits and a dorsal hollow nerve cord). The second hypothesis places hemichordates as the sister group of the echinoderms on the basis of a shared heart/glomerular complex (94, 112). The third places the echinoderms as the sister group of the chordates. This corresponds to the hypothesis of Jefferies (58), who has interpreted the features of an extinct lower Paleozoic echinoderm group, the carpoids, as reflecting their position as ancestral chordates, which he calls calcichordates. These unusual unsymmetrical echinoderms were covered with large calcite plates and some (the stylophorans) bear a single arm. Jefferies suggested that the arm is homologous to the chordate tail, complete with notochord and nerve tube. According to this hypothesis, these animals display a reduction of structures on the right side. Both left/right asymmetry and the calcite skeleton are interpreted as synapomorphies of an echinoderm plus chordate clade. However, the calcite plates are typical of echinoderm skeletons (which would have had to have been lost by chordates), and there is little evidence that the tail is not an echinoderm feeding arm (79).

The hypothesis that echinoderms are the sister group of the chordates can be tested. The first attempt was made by Holland et al (54) using relatively short 18S rDNA sequences. They found weak support for hemichordates as the sister group of vertebrates. Turbeville et al (112) used complete 18S rDNA sequences and a variety of inference methods. The tree-length distribution test (53) suggests that the signal for inference of deuterostome phylogeny is weak in this molecule. Their analysis illustrates some of the difficulties that still plague high level molecular systematics of animal phyla. Neighbor-joining, Fitch-Margoliash, and maximum likelihood weakly supported a monophyletic Chordata. The maximum parsimony method did not support chordate monophyly: The urochordates (ascidians) fell outside of the rest of the chordates. The following pattern resulted: (outgroups ((echinoderms (hemichordate, urochordates))(cephalochordate, craniates))). Similarly, Wada & Satoh (114a) found weak support for an echinoderm-hemichordate clade by neighbor-joining, but they also found that ascidians fell out of the chordates and formed (albeit with very weak support) the sister group to the echinoderm-hemichordate branch. Morphological features strongly support inclusion of the urochordates in a chordate clade. These features include possession of a notochord, epichordal neural tube with associated embryonic neurenteric canal, endostyle, post-anal tail, equivalent fate maps, paired tail muscle cells, and vertebrate-like muscle actin genes. Given the strength of character support for a monophyletic Chordata, a contrary 18S rDNA-based inference is unacceptable without very powerful independent support. The problem may stem in part from a high rate of 18S rDNA evolution in the ascidian lineage. Combining the morphological and molecular data sets results in a monophyletic Chordata that is the sister group to hemichordate plus echinoderm clade (112). An 18S rDNA analysis (JM Turbeville, RA Raff, unpublished) utilizing a hemichordate as the outgroup is congruent with the favored morphological hypothesis (ascidians(cephalochordates, craniates)) (70, 95). However, in maximum parsimony analyses, placement of the urochordates into the chordates is sensitive to the outgroups used (112).

CONCLUSIONS

Thus far, most molecular data applied to phylum level relationships have come from gene sequences, especially from 18S rRNA and DNA. Other gene sequences are beginning to enter the database and will become increasingly important. Not all applications of gene sequence data have produced strong or consistent results. Molecular systematics has not yet been able to provide an adequate tree of phylum level relationships, but it has, as discussed above, provided convincing data relevant to such issues as the sister group of the Metazoa, arthropod monophyly, and the controversies over the relationships of nemertines, molluscs, and lophophorates. For the pseudocoelomate phyla, molecular data offer the best chance of finding phylogenetically informative characters. By analyzing sequences with multiple tree recovery algorithms and by applying appropriate tests for convergence, molecular sequences offer an enormous wealth of information for resolving many of the difficult problems in phylum-level metazoan relationships. With increasing methodological sophistication, and larger and more diverse data sets, molecular data will contribute further to our understanding of metazoan relationships and the structure of the Cambrian radiation. These contributions are particularly important given the dearth of morphological evidence for many of the most interesting problems in metazoan level phylogenetics.

Literature Cited

1. Abele LG, Kim W, Felgenhauer BE. 1989. Molecular evidence for inclusion of the phylum Pentastomida in the Crustacea. *Mol. Biol. Evol.* 6:685–91
2. Adoutte A, Philippe H. 1993. The major lines of metazoan evolution: Summary of traditional evidence and lessons from ribosomal RNA sequence analysis. In *Comparative Molecular Neurobiology,* ed. Y Pichon, pp. 1–30. Basel: Birkhäuser Verlag
3. Anderson DT. 1969. On the embryology of the cirripede crustaceans *Tetraclita rosea* (Krauss), *T. purpurascens* (Wood), *Chthamalus antennatus* (Darwin), *Chamaesipho columna* (Spengler), and some considerations of crustacean phylogenetic relationships. *Philos. Trans. R. Soc. B.* 256:183–235
4. Anderson DT. 1973. *Embryology and Phylogeny in Annelids and Arthropods.* Oxford: Pergamon. 495 pp.
5. Anderson DT. 1979. Embryos, fate maps, and the phylogeny of arthropods. In *Arthropod Phylogeny,* ed. AP Gupta, pp. 59–105. New York: Van Nostrand Reinhold
6. Archie JW. 1989. A randomization test for phylogenetic information in systematic data. *Syst. Zool.* 38:239–52
7. Ax P. 1985. The position of the Gnathostomulida and Platyhelminthes in the phylogenetic system of the Bilateria. In *The Origins and Relationships of Lower Invertebrates,* ed. S Conway Morris, JD George, R Gibson, HM Platt, pp. 168–80. Oxford: Clarendon

8. Ax P. 1989. Basic phylogenetic systematization of the Metazoa. In *The Hierarchy of Life,* ed. B Fernholm, K Bremer, H Jörnvall, pp. 229–45. Amsterdam: Elsevier Sci.
9. Backeljau T, Winnepenninckx B, De Bruyn L. 1993. Cladistic analysis of metazoan relationships: a reappraisal. *Cladistics* 9:167–81
10. Baldauf SL, Palmer JD. 1993. Animals and fungi are each others closest relatives—congruent evidence from multiple proteins. *Proc. Natl. Acad. Sci. USA* 90:11558–62
11. Ballard JWO, Olsen GJ, Faith DP, Odgers WA, Rowell DM, Atkinson PW. 1992. Evidence from 12S ribosomal RNA sequences that onychophorans are modified arthropods. *Science* 258:1345–48
12. Bartels J, Murtha MT, Ruddle FH. 1993. Multiple Hox/HOM-class homeoboxes in Platyhelminthes. *Mol. Phyl. Evol.* 2: 143–51
13. Bartnik E, Weber K. 1989. Widespread occurrence of intermediate filaments in invertebrates; common principles and aspects of diversion. *Eur. J. Cell Biol.* 50:17–33
14. Baverstock PR, Fielke R, Johnson AM, Bray RA, Beveridge I. 1991. Conflicting phylogenetic hypotheses for the parasitic platyhelminths tested by partial sequencing of 18S ribosomal RNA. *Intern. J. Parisitol.* 21:329–39
15. Bone Q, Kapp H, Pierrot-Bults AC. 1991. Introduction and relationships of

the group. In *The Biology of Chaetognaths*. ed. Q Bone, H Kapp, AC Pierrot-Bults, pp. 1–4. Oxford: Oxford Univ. Press

16. Bowring SA, Grotzinger JP, Isachsen CE, Knoll AH, Pelechaty SM, Kolosov P. 1993. Calibrating rates of early Cambrian evolution. *Science* 261:1293–98

17. Bremer K. 1988. The limits of amino-acid sequence data in angiosperm phylogenetic reconstruction. *Evolution* 42: 795–803

18. Brusca RC, Brusca GJ. 1990. *Invertebrates*. Sunderland: Sinauer. 922 pp.

19. Budd, G. 1993. A Cambrian gilled lobopod from Greenland. *Nature* 364:709–11

20. Christen R, Ratto A, Baroin A, Perasso R, Grell KG, Adoutte A. 1991. An analysis of the origin of metazoans, using comparisons of partial sequences of the 28S RNA, reveals an early emergence of triploblasts. *Eur. Mol. Biol. Org. J.* 10:499–503

21. Clark RB. 1979. Radiation of the Metazoa. In *The Origin of Major Invertebrate Groups*, ed. MR House, pp. 55–102. Oxford: Clarendon

22. Clarkson ENK. 1975. The evolution of the eye in trilobites. *Fossils Strata* 4:7–31

23. Conway Morris S. 1993. The fossil record and the early evolution of the Metazoa. *Nature* 361:219–25

24. Curry GB, Runnegar B. 1990. Amino acid sequences from lingulid and priapulid hemerythrins. *Geol. Soc. Am.* 22: A129 (Abstr.)

25. Donoghue MJ, Olmstead RG, Smith JF, Palmer JD. 1992. Phylogenetic relationships of Dipsacales based on rbcL sequences. *Ann. Missouri Bot. Gard.* 79: 333–345

26. Douglas SE, Murphy CA, Spencer DF, Gray MW. 1991. Cryptomonad algae are evolutionary chimaeras of two phylogenetically distinct unicellular eukaryotes. *Nature* 350:148–51

27. Eernisse DJ, Albert JS, Anderson FE. 1992. Annelida and Arthropoda are not sister taxa: A phylogenetic analysis of spiralian metazoan morphology. *Syst. Biol.* 41:305–30

28. Eernisse DJ, Kluge AG. 1993. Taxonomic congruence versus total evidence, and amniote phylogeny inferred from fossils, molecules, and morphology. *Mol. Biol. Evol.* 10:1170–95

29. Ehlers U. 1985. Phylogenetic relationships within the platyhelminthes. In *The Origins and Relationships of Lower Invertebrates*, ed. S Conway Morris, JD George, R Gibson, HM Platt, pp. 143–58. Oxford: Clarendon

30. Faith DP. 1991. Cladistic permutation tests for monophyly and nonmonophyly. *Syst. Zool.* 40:366–75

31. Faith DP, Cranston PS. 1991. Could a cladogram this short have arisen by chance alone?: On permutation tests for cladistic structure. *Cladistics* 7:1–28

32. Felsenstein J. 1978. Cases in which parsimony and compatibility methods will be positively misleading. *Syst. Zool.* 27: 401–10

33. Felsenstein J. 1985. Confidence limits on phylogenies: an approach using the bootstrap. *Evolution* 39:783–91

34. Felsenstein J. 1988. Phylogenies from molecular sequences: inference and reliability. *Annu. Rev. Genet.* 22:521–65

35. Felsenstein J, Kishino H. 1993. Is there something wrong with the bootstrap on phylogenies? A reply to Hillis and Bull. *Syst. Biol.* 42:193–200

36. Field KG, Olsen GJ, Lane DJ, Giovannoni SJ, Pace NR, et al. 1988. Molecular phylogeny of the animal kingdom. *Science* 239:748–53

37. Field KG, Turbeville JM, Raff RA, Best BA. 1990. Evolutionary relationships of phylum Cnidaria inferred from 18S rRNA sequence data. *Fourth Int. Congress Syst. Evol. Biol.* (Abstr.)

38. Freeman G, 1991. The bases for timing and regional specification during larval development in Phoronis. *Dev. Biol.* 147:157–173.

39. Gill LL, Hardman N, Chappell L, Qu LH, Nicoloso M, et al. 1988. Phylogeny of *Onchocerca volvulus* and related species deduced from rRNA sequence comparisons. *Mol. Biochem. Parasitol.* 28: 69–76

40. Grell K. 1982. Placozoa. In *Synopsis and Classification of Living Organisms*, ed. S Parker, 1:639. New York: McGraw-Hill

41. Gould SJ. 1989. *Wonderful Life*. New York: Norton. 347 pp.

42. Gouy M, Li WH. 1989. Molecular phylogeny of the kingdoms Animalia, Plantae, and Fungi. *Mol. Biol. Evol.* 6: 109–22

43. Halanych KM. 1993. The phylogenetic position of the lophophorates based on 18S ribosomal gene sequence data. *Am. Zool.* 33:288 (Abstr.)

44. Hasegawa M, Iida Y, Yano T, Takaiwa F, Iwabuchi M. 1985. Phylogenetic relationships among eukaryotic kingdoms inferred from ribosomal RNA sequences. *J. Mol. Evol.* 22:32–38

45. Hasegawa M, Hashimoto T, Adachi J, Iwabe N, Miyata T. 1993. Early branchings in the evolution of eukaryotes: Ancient divergence of entamoeba that lacks

mitochondria revealed by protein sequence data. *J. Mol. Evol.* 36:270–81

46. Hendricks L, De Baere R, Van Broeckhoven C, De Wachter R. 1988. Primary and secondary structure of the 18S ribosomal RNA of the insect species *Tenebrio molitor. FEBS Lett.* 232:115–20

47. Hendricks L, Van Broeckhoven C, Vandenberghe A, Van De Peer Y, De Wachter R. 1988. Primary and secondary structure of the 18S ribosomal RNA of the bird spider *Eurypelma californica* and evolutionary relationships among eukaryotic phyla. *Eur. J. Biochem.* 177:15–20

48. Hendriks L, DeBaere R, Van de Peer Y, Neefs J, Goris A. 1991. The evolutionary position of rhodophyte *Pophyra umbilicalis* and the basidiomycete *Leucosporidium scottii* among other eukaryotes as deduced from complete sequences of small ribosomal subunit RNA. *J. Mol. Evol.* 32:167–77

49. Hendy MD, Charleston MA. 1993. Hadamard conjugation: a versatile tool for modelling nucleotide sequence evolution. *N. Zeal. J. Bot.* 31:231–37

50. Hendy MD, Penny D. 1989. A framework for the quantitative study of evolutionary trees. *Syst. Zool.* 38:297–309

51. Hendy MD, Penny D, Steel MA. 1992. Discrete Fourier analysis for evolutionary trees. *Mathematical and Information Sciences Rep., Ser. B,* 92/2:1–13, Massey Univ., Palmerston North, New Zealand

52. Herzog M, Maroteaux L. 1989. Dinoflagellate 17S rRNA sequence inferred from the gene sequence: Evolutionary implications. *Proc. Natl. Acad. Sci. USA* 83:8644–48

53. Hillis DM, Huelsenbeck JP. 1992. Signal, noise, and reliability in molecular phylogenetic analyses. *J. Hered.* 83:189–95

54. Holland PWH, Hacker AM, Williams NA. 1991. A molecular analysis of the phylogenetic affinities of *Saccoglossus cambrensis* Brambell & Cole (Hemichordata). *Philos. Trans. R. Soc. Lond.* B332:185–89

55. Huelsenbeck JP, Hillis DM. 1993. Success of phylogenetic methods in the four-taxon case. *Syst. Biol.* 42:247–64

56. Hyman LH. 1940. *The Invertebrates: Protozoa through Ctenophora.* Vol. I. New York: McGraw-Hill

57. Hyman LH. 1951. *The Invertebrates: Platyhelminthes and Rhynchocoela. The Acoelomate Bilateria.* Vol. II. New York: McGraw-Hill

58. Jefferies RPS. 1986. *Ancestry of the Vertebrates.* London: Br. Mus. 376 pp.

59. Katayama T, Yamamoto M, Wada H, Satoh N. 1993. Phylogenetic position of acoel turbellarians inferred from partial 18S rDNA sequences. *Zool. Science* 10:529–536

60. Kimura M. 1981. Estimation of evolutionary distances between homologous nucleotide sequences. *Proc. Natl. Acad. Sci. USA* 78:454–458

61. Kojima S, Hashimoto T, Hasegawa M, Murata S, Ohta S, Seki H, et al. 1993. Close phylogenetic relationship between Vestimentifera (tube worms) and annelida revealed by the amino acid sequence of elongation factor-1a. *J. Mol. Evol.* 37:66–70

62. Kristensen RM. 1983. Loricifera, a new phylum with Aschelminthes characters from the meiobios. *Z. Zool. Syst. Evolutionsforsch.* 21:163–80

63. Lake JA. 1987. Rate-independent technique for analysis of nucleic acid sequences: Evolutionary parsimony. *Mol. Biol. Evol.* 4:167–91

64. Lake JA. 1990. Origin of the Metazoa. *Proc. Natl. Acad. Sci. USA* 87:763–66

65. Lake JA. 1991. The order of sequence alignment can bias the selection of tree topology. *Mol. Biol. Evol.* 8:378–85

66. Lake JA. 1994. *Proc. Natl. Acad. Sci. USA.* In press

67. Li W-H, Gouy M. 1991. Statistical methods for testing molecular phylogenies. In *Phylogenetic Analysis of DNA Sequences,* ed. MM Miyamoto, J Cracraft, pp. 249–77. New York: Oxford Univ. Press

68. Lockhart PJ, Penny D. 1993. The problem of GC content, evolutionary trees and the origins of Chl-a/b photosynthetic organelles: Are the prochlorophytes a eubacterial model for higher plant photosynthesis. *Photosynthesis Res.* 111:499–505

69. Lorenzen S. 1985. Phylogenetic aspects of pseudocoelomate evolution. In *The Origins and Relationships of Lower Invertebrates,* ed. S Conway Morris, JD George, R Gibson, HM Platt, pp. 210–23. Oxford: Clarendon

70. Maisey JG. 1986. Heads and tails: A chordate phylogeny. *Cladistics* 2:201–256.

71. Manton SM. 1977. *The Arthropods: Habits, Functional Morphology and Evolution.* Oxford: Oxford Univ. Press

72. Manton SM, Anderson DT. 1979. Polyphyly and the evolution of arthropods. In *The Origin of Major Invertebrate Groups,* ed. MR House, pp. 269–321. London: Academic

73. Marshall CR. 1992. Substitution bias, weighted parsimony, and amniote phylogeny as inferred from 18S rRNA sequences. *Mol. Biol. Evol.* 9:370–73

74. Morris PJ. 1993. The developmental role of the extracellular matrix suggests a monophyletic origin of the kingdom Animalia. *Evolution* 47:152–65

75. Nielsen C. 1991. The development of the brachiopod *Crania* (*Neocrania*) anomala (OF Fller) and its phylogenetic significance. *Acta zool., Stockh.* 72:7–28.

76. Olsen GJ. 1987. Earliest phylogenetic branchings: comparing rRNA-based evolutionary trees inferred with various techniques. *Cold Spring Harbor Symp. Quant. Biol.* 52:825–37

77. Ossowski I von, Hausner G, Loewen PC. 1993. Molecular evolutionary analysis based on the amino acid sequence of catalase. *J. Mol. Evol.* 37:71–76

78. Pace NR, Stahl DA, Lane DJ, Olsen GJ. 1985. Analyzing natural microbial populations by rRNA sequences. *Am. Soc. Microbiol. News* 51:4–12

79. Parsley RL. 1988. Feeding and respiratory strategies in Stylophora. In *Echinoderm Phylogeny and Evolutionary Biology,* ed. CRC Paul, AB Smith, pp. 347–61. Oxford: Clarendon

80. Patterson C. 1989. Phylogenetic relations of major groups: conclusions and prospects. In *The Hierarchy of Life,* ed. B Fernholm, K Bremer, H Jörnvall, pp. 471–88. Amsterdam: Elsevier Sci.

81. Patterson CD, Williams M, Humphries CJ. 1993. Congruence between molecular and morphological phylogenies. *Annu. Rev. Ecol. Syst.* 24:153–58

82. Paulus HF. 1979. Eye structure and the monophyly of the Arthropoda. In *Arthropod Phylogeny,* ed. AP Gupta, pp. 299–383. New York:Van Nostrand Reinhold

83. Penny D, Hendy MD, Steel MA. 1991. Testing the theory of descent. In *Phylogenetic Analysis of DNA Sequences,* ed. MM Miyamoto, J Cracraft, pp. 155–83. New York: Oxford Univ. Press

84. Penny D, Watson EE, Hickson RE, Lockhart, PJ. 1993. Some recent progress with methods for evolutionary trees. *New Zeal. J. Bot.* 31:275–88

85. Raff RA. 1994. Developmental mechanisms in the evolution of animal form: origins and evolvability of body plans. In *Early Life on Earth,* ed. S. Bengtson. In press

86. Ready DF. 1989. A multifaceted approach to neural development. *Trends Neural Sci.* 12:102–10

87. Rieger RM. 1976. Monociliated epidermal cells in Gastrotricha: Significance for concepts of early metazoan evolution. *Z. Zool. Syst. Evolut.-forsch.* 14:198–226

88. Rieger RM. 1985. The phylogentic status of the acoelomate organization within the Bilateria: histological perspective. In *The Origins and Relationships of the Lower Metazoa,* ed. S Conway Morris, JD George, HM Platt, R Gibson: pp. 101–22. Oxford: Clarendon

89. Rieger RM. 1986. Uber den Ursprung der Bilateria: Die beduetung der Ultrastruktur Forschung fur eines neues Verstehen der Metazoenevolution. *Verh. Dtsch. Zool. Gesellschaft* 79:31–50

90. Riutort M, Field KG, Turbeville JM, Raff RA, Baguna J. 1992. Enzyme electrophoresis, 18S rRNA sequences, and levels of phylogenetic resolution among several species of freshwater planarians (Platyhelminthes, Tricladida, Paludicola). *Can. J. Zool.* 70:1425–39

91. Runnegar B, Curry GB. 1992. Amino acid sequences of hemerythrins from Lingula and a priapulid worm and the evolution of oxygen transport in the Metazoa. *Abstr. 29th Int. Geological Congress, Kyoto*

92. Ruppert EE. 1982. Comparative ultrastructure of the gastrotrich pharynx and the evolution of myoepithelial foreguts in Aschelminthes. *Zoomorphology* 99:181–220

93. Ruppert EE. 1991. Introduction to aschelminth phyla: a consideration of mesoderm, body cavities, and cuticle. *Microscopic Anatomy of Invertebrates,* Vol. 4. *Aschelminthes,* ed. FW Harrison, EE Ruppert, pp. 1–17. New York: Wiley-Liss

94. Ruppert EE, Balser EJ. 1986. Nephridia in the larvae of hemichordates and echinoderms. *Biol. Bull.* 171:188–96

95. Schaeffer B. 1987. Deuterostome monophyly and phylogeny. *Evol. Biol.* 21:179–235

96. Scheltema AH. 1993. Aplacophora as progenetic aculiferans and the coelomic origin of mollusks as the sister taxon of Sipuncula. *Biol. Bull.* 184:57–78

97. Shenk MA, Steele RE. 1993. A molecular snapshot of the metazoan 'Eve.' *Trends Biochem. Sci.* 18:459–463.

98. Schram FR. 1991. Cladistic analysis of metazoan phyla and the placement of fossil problematica. In *The Early Evolution of Metazoa and Significance of Problematic Taxa,* ed. AM Simonetta, S Conway Morris, pp. 35–46. Cambridge: Cambridge Univ. Press

99. Smith AB. 1989. RNA sequence data in phylogenetic reconstruction: testing

the limits of its resolution. *Cladistics* 5:321–44

100. Sogin ML, Gunderson JH, Elwood HJ, Alonso RA, Peattie DA. 1989. Phylogenetic meaning of the kingdom concept: An unusual ribosomal RNA from *Giardia lamblia*. *Science* 243:75–77

101. Steel MA, Hendy MD, Penny D. 1992. Significance of the shortest tree. *J. Classif.* 9:71–90

102. Steel MA, Lockhart PJ, Penny D. 1993. Confidence in evolutionary trees from biological sequence data. *Nature* 364: 440–42

103. Sterrer W, Mainitz M, Rieger RM. 1985. Gnathostomulida: enigmatic as ever. In *The Origins and Relationships of Lower Invertebrates*, ed. S Conway Morris, JD George, R Gibson, HM Platt, pp. 181–99. Oxford: Clarendon

104. Swofford DL, Olsen GJ. 1990. Phylogeny reconstruction. In *Molecular Systematics*, ed. DM Hillis, C Moritz, pp. 411–501. Sunderland: Sinauer

105. Telford MJ, Holland PWH. 1993. The phylogenetic affinities of the chaetognaths: A molecular analysis. *Mol. Biol. Evol.* 10:660–76

106. Thomas JB, Bastiani MJ, Bate M, Goodman CS. 1984. From grasshopper to *Drosophila*: a common plan for neuronal development. *Nature* 310:203–7

107. Tiegs OW, Manton SM. 1958. The evolution of the Arthropoda. *Biol. Rev. Camb. Phil. Soc.* 33:255–337

108. Turbeville JM. 1986. An ultrastructural analysis of coelomogenesis in the hoplonemertine *Prosorhochmus americanus* and the polychaete *Magelona* sp. *J. Morphol.* 187:51–6

109. Turbeville JM. 1991. Nemertina. In *Microscopic Anatomy of Invertebrates*, Vol. 3: *Platyhelminthes and Nemertinea*, ed. FW Harrison, BJ Bogitsh, pp. 285–328. New York: Wiley-Liss

110. Turbeville JM, Field KG, Raff RA. 1992. Phylogenetic position of Phylum Nemertini, inferred from 18S rRNA sequences: Molecular data as a test of morphological character homology. *Mol. Biol. Evol.* 9:235–49

111. Turbeville JM, Pfeifer DM, Field KG, Raff RA. 1991. The phylogenetic status of arthropods, as inferred from 18SrRNA sequences. *Mol. Biol. Evol.* 8:669–686

112. Turbeville JM, Schulz JR, Raff RA. 1994. Deuterostome phylogeny and the sister group of the chordates: Evidence from molecules and morphology. *Mol. Biol. Evol.* In press

113. Van de Peer Y, Neefs JM, De Rijk P, De Wachter R. 1993. Reconstructing evolution from eukaryotic small-ribosomal-subunit RNA sequences: Calibration of the molecular clock. *J. Mol. Evol.* 37:221–32

114. Vossbrinck CR, Maddox JV, Friedman S, Debrunner-Vossbrinck BA, Woese CR. 1987. Ribosomal RNA sequence suggests microsporidia are extremely ancient eukaryotes. *Nature* 326:411–14

114a. Wada H, Satoh N. 1994. Details of the evolutionary history from invertebrates to vertebrates, as deduced from the sequences of 18S rDNA. *Proc. Natl. Acad. Sci. USA.* 91:1801–1804

115. Wainright PO, Hinkle G, Sogin ML, Stickel SK. 1993. Monophyletic origins of the metazoa: An evolutionary link with fungi. *Science* 260:340–42

116. West JG, Faith DP. 1990. Data, methods and assumptions in phylogenetic inference. *Aust. Syst. Bot.* 3:9–20

117. Weygoldt P. 1986. Arthropod interrelationships: the phylogenetic-systematic approach. *Z. zool. Syst. Evolut.-forsch.* 24:19–35

118. Wheeler WC, Cartwright P, Hayashi CY. 1993. Arthropod phylogeny: A combined approach. *Cladistics* 9:1–39

119. Whitington PM, Leach D, Sandeman R. 1993. Evolutionary change in neural development within the arthropods: axonogenesis in the embryos of two crustaceans. *Development* 118:449–61

120. Whittington HB. 1985. *The Burgess Shale*. New Haven: Yale Univ. Press. 151 pp.

121. Williams PL, Fitch WM. 1990. Phylogeny determination using dynamically weighted parsimony method. *Meth. Enzym.* 183:615–26

122. Willmer P. 1990. *Invertebrate Relationships. Patterns in Animal Evolution*. Cambridge: Cambridge Univ. Press

123. Winnepennickx B, Backeljau T, Van de Peer Y, De Wachter R. 1992. Structure of the small ribosomal subunit RNA of the pulmonate snail *Limicolaria kambeul*, and phylogenetic analysis of the Metazoa. *FEBS* 309:123–26

124. Deleted in proof

125. Wray GA Raff RA. 1991. The evolution of developmental strategy in marine invertebrates. *Trends Ecol. Evol.* 6:45–50

126. Zharkikh A, Li W-H. 1993. Inconsistency of the maximum-parsimony method: the case of five taxa with a molecular clock. *Syst. Biol.* 4:113–25

Annu. Rev. Ecol. Syst. 1994. 25:377–99

GENETICS AND ECOLOGY OF WHALES AND DOLPHINS[1]

A. Rus Hoelzel

Laboratory of Viral Carcinogenesis, Building 560, National Cancer Institute, Frederick, Maryland, 21702

KEY WORDS: marine mammals, cetaceans, population genetics, molecular genetics

Abstract

Cetacean species are widely distributed in the world's oceans, and their populations are impacted by human activity both directly through hunting and indirectly. The determination of genetic stock boundaries and genetic diversity within stocks is essential to the conservation of genetic diversity in these as in any other species, but cetaceans present special challenges. The definition of stock boundaries based on assumptions drawn from experience with terrestrial species has proven inadequate for cetaceans. For example, genetic stock boundaries do not necessarily correspond to patterns of apparent geographic isolation. Cetaceans are capable of migrating great distances, and in some cases breeding populations mix in seasonal feeding grounds up to thousands of miles from where they breed. Among those species that breed and forage within the same geographic range, intraspecific differences in feeding ecology can lead to the genetic differentiation of local populations. This review presents a summary of the available data on genetic diversity and differentiation within cetacean species, with an emphasis on the ecological context.

INTRODUCTION

Through adaptation to the marine environment, cetacean species have undergone extreme morphological evolution. In comparison with their terrestrial ancestors, cetaceans have been adapted to swimming by compression of the

---

[1]The US government has the right to retain a non-exclusive, royalty-free license in and to any copyright covering this paper.

forelimbs and expansion of the phalanges to form flippers. The hind limbs are gone or completely vestigial—reduced to a single pair of bones in some species. The cranial bones are perhaps the most modified, especially the mandibular and frontal bones, adapted to various specialized modes of feeding, and in odontocetes, to facilitate the acoustic imaging of the environment through echolocation. Numerous physiological adaptations of cetaceans allow, for example, prolonged apnea, excretion of excess salt, and deep diving. Cetaceans have evolved a high degree of mobility and versatility in the marine environment, and these characteristics are reflected in the genetic structure of cetacean populations.

Odontocetes (the toothed whales), which include all dolphin species, are adapted to feeding on highly mobile prey, primarily fish and cephalopods. Dolphins are relatively small and fusiform species, with extended rostrums and numerous monodont, conical teeth, adapted to grasping fast-moving prey. A remarkable aspect of dolphin behavior is the extreme level of sociality, perhaps associated with cooperative foraging behaviors (38). Most species are social to some extent, and many travel in large, stable schools. However, the closely related porpoises are less social and have spade-like teeth, perhaps adapted to pursuing smaller prey. Social cohesion can affect genetic differentiation through its effect on the age and range of dispersal and by the tendency for kin to associate in social groups.

Some dolphin and porpoise species inhabit nearshore waters, some are pelagic, and in some species both nearshore and offshore populations exist. The largest dolphin, the killer whale (*Orcinus orca*), preys on both fish and other marine mammals, which leads to differences in social behavior and habitat use (8, 23, 26). Several other dolphin species are also known to partition local habitat based on prey choice (see below). The beaked whales are more distantly related to other toothed whale species. They are typically but not exclusively solitary, and all species are pelagic. They feed at great depth, primarily on cephalopods. The largest odontocete, the sperm whale (*Physeter catadon*), also preys on cephalopods but travels in stable social groups.

Mystecetes (the baleen whales) are typically solitary, and some species are known to travel great distances between feeding and breeding grounds. The right whale (Eubalaena sp.) and bowhead whale (*Balaena mysticetus*) are adapted to feeding on small plankton, such as copopods, primarily in polar waters. The balaenopterids, which include the largest whale, the blue whale (*Balaenoptera musculus*), feed in both temperate and polar waters, primarily on fish and krill (a small crustacean). The gray whale (*Eschrichtius robustus*) is a benthic, nearshore feeder. Four species, the humpback whale (*Megaptera novaengliae*), the gray whale, the northern right whale (*Eubalaena glacialis*) and the southern right whale (*E. australis*) are known to migrate between specific breeding and feeding grounds on an annual cycle. The breeding behavior of other mysticetes is poorly understood, though most probably migrate between foraging and breeding areas.

In this review I describe what is known about the genetic structure of cetacean populations. The unusual dispersion and social behavior of some species makes the geographic identification of genetic stocks problematic, and the life history and behavior of other species, especially the beaked whales, are too poorly understood to permit an accurate interpretation of genetic data. In many cases the ecological context is central to the interpretation of behavior and genetic structure.

COLLECTION OF SAMPLES

The early genetic studies of cetacean species depended on the whaling industry to provide tissue samples. Many studies still use material from strandings, animals incidentally caught in fishing nets, and samples from aboriginal whaling and drive fisheries. The main problem with these sampling methods is that there is little control over which individuals or populations are sampled. The advantage is that samples can be obtained at relatively little cost. Dolphins held in captivity can be easily sampled for blood or a small skin biopsy. Stranded animals are most useful only if found fairly soon after death. A method involving capture and release was used quite effectively for sampling bottlenose dolphins (*Tursiops truncatus*) off the Florida coast (18), but the effort required is considerable, and the method is only practical for small, approachable species such as the bottlenose dolphin.

More recently the sampling method of choice has been biopsy darting (e.g. 33, 62). This allows the remote collection of skin samples from free-ranging animals as well as the design of a sampling strategy where specific individuals can be included in the analyses. The dart is typically a cylinder 0.5–1 cm in diameter, sharpened at the leading edge. A barbed shaft or barbs inside the cylinder retain the sample. The darts can be propelled at the tip of an arrow or airgun-dart or attached to the end of a pole. Samples collected at sea can be preserved for DNA analysis at ambient temperature in a salt/DMSO solution (3).

GENETIC ANALYSIS OF KINSHIP

Ever since the discovery of hypervariable genetic markers in the 1980s (see review 16), it has become possible to test with considerable resolution the variance in male and female reproductive success, and to measure kinship within and between social groups. With free-ranging cetacean species, however, it is often difficult to identify the relevant individuals to include in a genetic assessment. This problem has been overcome to a large extent by the visual recognition and resighting of individual animals over time. This has been accomplished using the photographic identification of distinctive coloration and markings. For example, individual humpback whales can be distinguished by the pattern of pigmentation on the underside of their flukes (32), and dolphins can often be identified by scars, marks on their dorsal fins, and patterns of coloration on their backs (8). Even with the success of this technique, however, only some species are accessible enough to permit the level of contact necessary. For example, to test paternity in species where partuition is not seasonal, known individuals must be followed at least through the period of gestation (11–14 months) to identify which males were in association with specific females at the time of conception. However, even without identifying specific potential fathers, it is possible to interpret the degree of polygyny, for example, through a comparative analysis of allelic variation in the adult male and offspring populations (see below).

Odontocetes

The most detailed study of the genetic structure of an odontocete social group to date is of whole pods captured during the Faroese pilot whale (*Globicephala melas*) drive fishery (1, 4). It is not known to what extent these pods are stable social groupings, or whether the sampling (capture) was inclusive of all relevant individuals, but the study provides a large data base on which to base future interpretations. Amos et al (1) used multilocus DNA fingerprints, and a range of internal bands they identified as a single locus, to test for paternity and kinship within pods. Five pods were sampled including 16 to 103 whales per pod. The two largest pods were compared, and genetic similarity was found to be greater within than between pods.

A later study using variation at six microsatellite loci reinforced this interpretation (4, 51). The study predicted that if kinship within groups is due to philopatry, then the alleles present in older cohorts should be overrepresented in the pod as a whole. An analysis in comparison with computer simulations seemed to bear this out and suggested that pods are matrifocal (4). The role of males in pilot whale pods remains elusive. Both studies taken together indicate that all males killed with the pod can be excluded from paternity in 33 of 34 mother/foetus pairs. Further, a test that compared hypothetical parent-offspring pairs (compatible in age and genotype) with that expected by chance suggested that there could be many more mother-offspring than father-offspring relationships among the whales captured within a pod. This suggests that reproductive age males either associate temporarily with matrifocal pods for mating or remain within their natal group, but mate during interactions with other pods. Amos et al (1) compared a small sample of adult males and females from within a pod (11 males and 16 females). Only animals within the age range of 18–35 years were included (determined by counting dental layers). A chi square test indicated that the allele distributions at the single locus system were significantly different ($p < 0.05$). However, a later analysis calculating likelihood ratios (based on allele frequencies in combined data from the minisatellite single locus system and a hypervariable microsatellite locus) suggested that adult males are more related to other whales in the pod in which they were captured, than to whales in the other pod (comparing the same two pods as before; 4).

One interpretation is that pilot whale males remain in their natal pods into maturity, but either temporarily or permanently join other pods to mate. All-male pods are occasionally seen (52), which may represent transitional groups moving between matrifocal pods. The apparent number of matings achieved by a given male was tested by comparing paternal alleles in the 34 foetuses with allele frequencies among adult males (4). There was little support for multiple matings, and no indication of extreme polygeny. Together these data indicate that pilot whale pods (which can include over 100 whales) are matrifocal, including related females and to some extent related males. Males disperse (at least in genetic terms), and variance in male reproductive success may be low, though understanding this aspect will require a more rigorous investigation. A long-term observational study would facilitate the interpretation of these data considerably. Female philopatry and male dispersal are typical of mammalian species, and one consequence is that the geographic pattern of genetic differentiation will be largely determined by the range of male dispersal. Such populations may be expected to show a high level of genetic structure in the matrilinially inherited mitochondrial DNA markers.

A similar social relationship has been determined by observational studies of known individual killer whales. Bigg et al (8) photographically identified and followed the movements and interactions of approximately 300 killer whales for 20 years. Among the pods of killer whales that prey on fish, both males and females apparently remain within their natal pods (which include approximately 5–50 whales) well into sexual maturity. No dispersal of single individuals could be verified. However, on several occasions large pods divided into smaller ones (there is some indication from the genetic data that this also happens in pilot whales—1). A genetic analysis using minisatellite DNA variation of a small number of captured or stranded whales from the killer whales in this study indicated little genetic differentiation between pods (27), consistent with expectations based on the observational studies. The social and ranging behavior among killer whales that prey on marine mammals is different from those that prey on fish (8, 23, 26). In this case, pods are smaller and lone males and all-female pods are relatively common. This type of social organization is more typical of serial polygeny and could lead to a less structured pattern of genetic variation at the population level.

Bottlenose dolphins inhabiting waters near Sarasota, Florida, also form stable, apparently matrifocal social groups (39). Genetic analysis using protein polymorphisms, mtDNA restriction fragment length polymorphism (RFLP) analysis, and chromosome markers suggested that kinship was greater within than between groups (18). There was also preliminary evidence that both males and females move between social groups. This was based on mtDNA halplotype analysis and the identification of a supernumerary chromosome in some individuals. Females in a given social group sometimes had different mtDNA haplotypes, indicating different maternal origins and suggesting the movement of females between groups. The chromosomal marker was seen in different groups and across three different mtDNA haplotypes, suggesting that it could have been distributed between groups by a male carrying the marker. The confirmation of each of these hypotheses will require further genetic analysis, but they are consistent with long-term observational studies based on the recognition of known individuals (e.g. 39).

Each of these three species, pilot whales, killer whales, and bottlenose dolphins, apparently form stable matrifocal groups, which is typical of mammalian social behavior. However, the degree of movement between groups, especially by males, apparently varies. Similar size groups are formed in many other dolphin species, but little is known about the composition or dynamics of these groups in most species. The role of males and females is most distinct in sperm whales (*Physeter catadon*), which form separate female/offspring and all male pods, and breeding apparently takes place during temporary associations.

Mysticetes

So far the humpback whale is the only baleen whale whose breeding behavior has been investigated using molecular techniques. Humpback whales migrate between feeding and breeding grounds annually. On the breeding grounds in Hawaiian waters, off Baja California, and near the Bahamas, humpback whales display with an elaborate song, which evolves over time but is shared by all whales on a given breeding ground each season (42, 44). It is thought that these 'singers' are males advertising to attract mates, and the sex of the singer has been verified in some cases by visual inspection (by divers). Females on the breeding grounds are often attended by whales known as 'escorts' who travel in association with cow/calf pairs or with a single 'focal' whale. There is frequent fighting between escorts, apparently for access to the female or focal whale. Molecular analysis has confirmed the assumption that escorts are usually males, and that the focal animal is usually a female (10). Sex was determined by collecting biopsy samples from individuals observed in these escort groups and analyzing the samples for the Y-chromosome–specific SRY gene (10). A male-specific segment of that gene was amplified using the polymerase chain reaction (41).

The cases where the molecular data contradicted the observational data were interesting. In 23% of the field identifications described as "positive" (N = 22), focal whales were male, not female. However, 95% of the whales identified as escorts were male as expected (N = 114). The all-male groups could be consorts, and positioning within these groups could reflect a dominance hierarchy. Samples are being collected to test the reproductive success of specific male escorts. A similar social structure appears to exist in right whales, where females are attended by large groups of males on the breeding grounds (43). An assessment of the degree of variance in male reproductive success would help the understanding of genetic structuring at higher levels. For example, if variance in male reproductive success were high, and movement between breeding grounds were rare, the rate of genetic differentiation between breeding stocks could be expected to be high (see below).

POPULATION LEVEL GENETIC DIVERSITY

Intrapopulation Diversity

ODONTOCETES   One consequence of social behavior in odontocetes could be a reduction in genetic variation within local populations. Low variation within local populations of killer whales (27) and pilot whales (1) could be examples of this (see above). In each case there is apparently a high degree of philopatry, but the extent of mixing between local populations is not yet known. A comparison of mtDNA control region sequences between killer whales from distant geographic regions did not reveal exceptionally high genetic distances for interpopulation comparisons, suggesting dispersal of females between contiguous or distant populations (27).

Demographic factors can also reduce variation in local populations. Many cetacean species have been overexploited either through hunting or indirectly, for example, through incidental entanglement in fishing nets. When a species or local population is reduced to a small enough population for a long enough period, then there will be a loss of genetic diversity (37). The severity of population bottleneck required to reduce variation depends on a number of factors including the initial level of variation prior to the bottleneck and demographic stochasticity during the recovery phase (see 30). A founder event has a similar effect, and this may explain the lack of genetic diversity observed in the Black Sea harbor porpoise (*Phocoena phocoena*; 49). Rosel (49) sequenced 394 bp from the 5′ end of the mtDNA control region from a total of 88 harbor porpoises representing populations in the North Pacific (N = 62), North Atlantic (N = 15), and Black Sea (N = 9). Nucleotide diversity among the North Pacific samples was 0.7%, 0.4% among the North Atlantic samples, and only 0.05% among the Black Sea samples. There were three haplotypes in the Black Sea, but each of two of them could be derived from the third by a single base pair change. This suggests a founder event where one haplotype survived and has since mutated to form two new haplotypes—though the sample size is small and further data are needed before it can be assumed that there are only three haplotypes. Change in this mtDNA region has been estimated to occur at 1–3% per million years (7, 29), which suggests a founder event 150,000 to 500,000 years ago.

MYSTICETES   There are some data to suggest founder events in mysticete species as well. Baker et al (7) investigated mtDNA control region variation in ten putative humpback whale populations. The level of variation was high in most geographic regions (with a nucleotype diversity of 0.75–0.95), but no variation was found among seven whales sampled off Hawaii, and five whales sampled off Southeast Alaska. An earlier study investigating mtDNA RFLP variation showed no variation among 20 samples from Southeast Alaska, and an order of magnitude lower variation among 16 samples from Hawaii, compared to 20 from central California and 28 from the Gulf of Maine (6). Studies tracing the movement of individual humpback whales based on photographic identification indicate that whales found feeding in Southeast Alaska in the summer are found in winter on the breeding grounds off Hawaii (13). Genetic analyses indicated that these populations are essentially identical (6, 7). Because mtDNA markers were used, this may reflect only the history and movement of females. As Baker and coworkers (7) point out, the whaling record does not indicate an emphasis on hunting in the central Pacific (which could have led to a bottleneck in this population). In fact nineteenth century records do not describe catches of humpback whales in Hawaiian waters at all, nor are baleen whales mentioned in the legends of native Hawaiians. Based on this they propose that the population that migrates between Hawaii and Alaska may have been colonized in this century. A small founding population could have limited the mtDNA diversity to a single haplotype.

Some species were hunted to levels low enough that a reduction in genetic diversity could be expected as a consequence. For example, the northern right whale was hunted extensively for centuries (since the tenth or eleventh century), and now after having been protected for about 50 years, only hundreds of whales remain. A preliminary analysis of mtDNA RFLP variation suggests limited genetic variation in this species compared to the southern right whale, which was not exploited as extensively. Relatively low levels of variation found for various markers in North Pacific minke whales (*Balaenoptera acutorostrata*; 28, 59) and the small-form Bryde's whale (*Balaenoptera edeni*; 60) may also reflect the demographic histories of these populations.

Sympatry and Habitat Division

ODONTOCETES   Killer whales can be grouped into two types of behavioral strategists, those feeding on marine mammals and those feeding on fish. In the nearshore waters off Vancouver Island, British Columbia, there are approximately 300 killer whales which have been studied intensively for the past 20 years (8). These whales, which can be individually distinguished, group into at least two populations. In one, the whales travel in stable social groups of 10–50 individuals and follow the seasonal salmon migration through the study area (20). The other population consists of smaller (1–8 whales), less stable social groups, that prey primarily on marine mammals. The coloration patterns of whales in these two populations can be statistically distinguished (5), and their geographic ranges overlap, though no interactions have been observed.

mtDNA sequence diversity has been compared between these populations, and with killer whales in the north and south Atlantic (27). Genetic distance between the marine-mammal-eating and other killer whales off Vancouver Island was as great as the genetic distance between either of the North Pacific populations and either of the north or south Atlantic populations. These populations were also compared for minisatellite DNA variation. The same population level distinctions were evident, and within population variation was very low, suggesting a high level of genetic uniformity within local killer whale populations (see above). The data suggest that sympatric, genetically differentiated populations coexist, but utilize the local habitat in different ways.

A similar situation may exist in other dolphin species. Nearshore and offshore forms of the bottlenose dolphin have been recognized based on morphology (see review in 34) and hematology (19), though it is not yet known to what extent these morphotypes may be genetically differentiated. Nearshore and offshore forms of the spotted dolphin (*Stenella attenuata*) can be distinguished by tooth and jaw structure (45). To some extent there is also a nearshore/offshore distinction in the distribution of two forms of common dolphin (*Delphinus delphus*; the 'short-beaked' and 'long-beaked' forms, Figure 1), though their ranges overlap (22). These morphotypes have been classified as members of the same species, and the morphological distinction is subtle enough to be very difficult to identify at sea.

Rosel et al (50) compared 328 bp of mtDNA control region sequence and 360 bp of mtDNA cytochrome *b* sequence between 8 short-beaked and 11 long-beaked forms of common dolphins from California. Short-beaked forms from the eastern tropical Pacific (N = 6) and the Black Sea (N = 4) were also analyzed. The two forms did not share any haplotypes for either genetic marker. The mtDNA control region net sequence divergence between short vs long-beaked common dolphins sampled in California was 1.78%, while the divergence between short-beaked dolphins from California, the eastern tropical

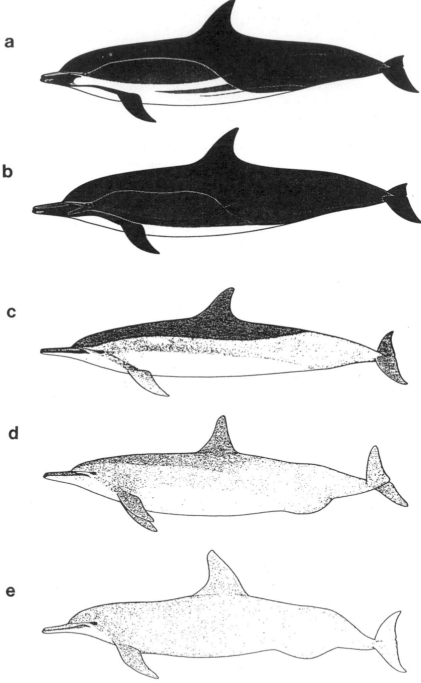

Pacific, and the Black Sea was 0.91%. The sympatric forms could represent a recent or seasonal admixture of two emerging or established species. However, isolation may also have been established or reinforced by differential habitat use. Differences in tooth and jaw structure could imply adaptations to different prey species (49).

Another extreme seems to exist among populations of spinner dolphins (*Stenella longirostris*). Four management stocks of spinner dolphins have been recognized in the eastern tropical Pacific, based on morphology (Figure 1) and geographical distribution. An eastern endemic form has light pigmentation on the dorsal surface, a forward angled dorsal fin, and a pronounced peduncle. The pantropical form is opposite in each respect. Between these two populations are two (northern and southern) designated stocks of an intermediate form, called the 'whitebelly' spinner, and recent studies have suggested that the two whitebelly stocks could represent a broad zone of hybridization or clinal integration between the pantropical and eastern forms (47). The ranges of the different forms overlap, and mixed schools are not uncommon (14).

Dizon et al (14) compared mtDNA RFLP data between 79 eastern, 45 whitebelly, and 11 pantropical spinners (though the latter were from the Timor Sea and may comprise a distinct race of pantropical spinner dolphins—46). Six six-base recognition site enzymes were used (out of 13 initially screened), each of which revealed high levels of variation within and between populations. Measures of pairwise genetic distance showed equally high within- and between-population diversity, comparing eastern and whitebelly forms. The Timor Sea pantropical spinners showed less within-population variation and unique haplotypes when compared with either of the other populations, but genetic distance was low comparing the Timor Sea population with either the eastern or whitebelly forms. Dizon et al (14) infer from these data significant genetic interchange at least between the eastern and whitebelly forms, despite the large differences in morphology.

MYSTICETES   Two forms of Bryde's whales exist near The Solomon Islands and Indonesia. Eight samples were collected and analyzed from what has been referred to as the 'small-form' Bryde's whale, though still classified as *Balaenoptera edeni* (60). They are about 2 m smaller than other Bryde's whales at maturity, but in other respects their morphology is very similar. Allozyme analysis at 45 loci indicated a Nei genetic distance (36) of D = 0.50 between the small-form and ordinary Bryde's whales, nearly as high as the genetic distance between Bryde's and fin (*Balaenoptera physalus*) or minke whales,

Figure 1   Morphotypes of common (after 49) and spinner dolphins (after 14): (a) short-beaked form of the common dolphin, (b) long-beaked form of the common dolphin, (c) the pantropical spinner dolphin, (d) the whitebelly spinner dolphin, and (e) the eastern spinner dolphin.

and much higher than the genetic distance between Bryde's and sei (*Balaenoptera borealis*) whales (D = 0.053—Ref. 60). The small-form Bryde's had a unique fixed gene at 10 loci, and polymorphism at only one out of 45 loci, which would seem to rule out the possibility that they represented an F1 hybrid between two of the other species in the genus. Together these results suggest that this could be a previously unknown species of Baleanopterid whale.

Migration and Temporal Patterns of Genetic Diversity

ODONTOCETES   The seasonal migrations of odontocetes are less well established than those of mysticetes. In some populations, dolphins are locally resident throughout the year (such as the bottlenose dolphins off the coast of Florida—39). Seasonal changes in prey (fish and cephalopod) distributions are more pronounced in temperate and polar regions, and this affects the seasonal movements of odontocetes inhabiting those regions. For example, killer whales that follow salmon migrations in the northeastern Pacific during the summer then move offshore during the winter, presumably to exploit a different resource (20). Beluga whales (*Delphinapterus leucas*) in Arctic and subarctic waters migrate between open waters in the winter and river estuaries and surrounding areas during the summer months, after the pack-ice has broken up.

JG Brown & JW Clayton (unpublished manuscript) compared 288 beluga whales from 14 different geographic areas from Alaska to the St. Lawrence seaway for mtDNA control region sequence variation. Earlier studies employing less variable markers (immunological markers and allozymes) did not show any distinction between beluga populations (35), though a preliminary mtDNA RFLP study (21) did show differentiation between Hudson Bay belugas and other stocks. Brown & Clayton found significant genetic differentiation between geographically distinct summering groups, which they attribute to maternal fidelity to summer migration areas. For example, populations on either side of the Hudson Bay were highly differentiated. The dominant haplotype in the west was not represented among whales sampled in the east, and these two populations both winter in the Hudson Strait, Davis Strait region southwest of Greenland (see Figure 2, JG Brown & JW Clayton, unpublished).

MYSTICETES   Many of the baleen whales are migratory, sometimes traveling thousands of kilometers between feeding and breeding grounds. This complicates the identification of boundaries between stocks from a management perspective, because most hunting takes place on feeding grounds, where in some cases reproductive stocks are known to mix (see 24 and below). Wada (58) investigated the allozyme difference between Korean and Japanese Sea

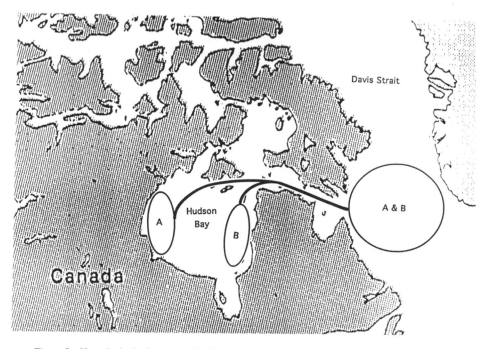

Figure 2 Hypothetical migration path of two genetically differentiated beluga whale populations that winter in the Davis Strait and spend summers in the Hudson Bay (after JG Brown & JW Clayton, unpublished).

minke whale populations and compared each with a third area to the north of Japan in the Okhotsk Sea. Minke whales are known to feed in this area and further north during spring and summer months. Wada found a clear distinction for the frequency of an *Adh-1* allele between Korean and Japanese samples, but an intermediate frequency was found in the Okhotsk Sea. Further, when the samples were examined by monthly catches, this distinction was seen only for April, suggesting that stocks from Korea and the Japanese Pacific coast were mixing in that part of the Okhotsk Sea in April. Possible biases related to the sex ratio in different catches or annual variation were tested and ruled out.

Minke whales feeding in the North Atlantic may also represent mixed assemblages of breeding stocks. Palsboll (40) investigated mtDNA RFLP variation among 80 minke whales from west Greenland, 9 from Norway, and 16 from Iceland. The 17 haplotypes revealed by use of 9 restriction enzymes could be classified into two groups. Estimated base pair substitutions between haplotypes within groups ranged from 0.4% to 0.9%, while comparisons be-

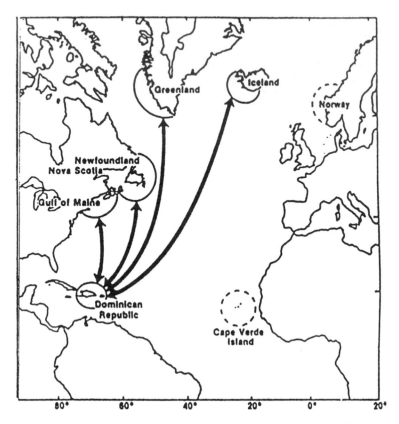

Figure 3 Proposed migration routes of humpback whales in the North Atlantic (after 6).

tween groups ranged from 1.9% to 2.5%. Both principal haplotypes were represented in all populations, and there was no significant difference in haplotype frequency. The simplest explanation is that these whales, taken on seasonal feeding grounds, represent two different breeding stocks. However, Danielsdottir et al (12) investigating allozyme variation also found genetic differentiation between these three populations (see below). Substantial population differentiation and low levels of variation within minke whale populations have been observed elsewhere for both mtDNA and nuclear markers (see below).

The opposite pattern, where whales from more than one feeding area converge on one breeding area, has been reported for humpback whales (Figure 3) based on the photographic identification of individuals (48) and on mtDNA RFLP and sequence data (6, 7). It is well established that females show considerable fidelity to feeding grounds (see 9), and this is reflected in the patterns of mtDNA diversity. However, individuals both male and female have been sighted on different feeding grounds in different years. On breeding grounds in the West Indies, Clapham et al (10) identified males within competitive groups (where several males apparently compete for access to one female) using the SRY y-chromosome-specific amplification described by Palsboll et al (41). They then matched identification photographs to determine that for 12 competitive groups, two or more of the males had been previously identified in different high latitude feeding areas (either West Greenland, Newfoundland, or the Gulf of Maine).

Geographic Isolation

ODONTOCETES   As described above, conspecific populations of dolphins that have differentiated genetically can coexist sympatrically, and this is correlated with habitat division in some cases. Divisions also exist between populations separated by geographic and less defined boundaries. I here review three examples.

Bottlenose dolphins are abundant in tropical and warm temperate waters worldwide. Morphological variation between geographically isolated populations, especially the Pacific vs the Atlantic, and local variants, such as nearshore vs offshore forms, has caused the species to be variously subdivided in the past, though they are all at present recognized as one species (see review in 34). Dowling & Brown (17) investigated mtDNA RFLP variation among bottlenose dolphins from the eastern North Pacific and western North Atlantic. North Atlantic samples were collected from the Gulf of Mexico and the eastern coast of the United States from Florida to North Carolina. Sample sizes were small for some locations, but clear differences were seen, especially between samples from the two oceans. Two methods were used, endlabelling restriction digested fragments of purified mtDNA, and probing blots of whole cell DNA with a mtDNA probe. By the former method, Atlantic vs Pacific populations were separated by an estimated 2.4% sequence divergence. Gulf of Mexico samples were separated from the other North Atlantic samples by an estimated 0.6% sequence divergence. By the latter method the difference between oceans was still the largest, but less pronounced, and the pairwise difference between some individuals in the Gulf of Mexico was greater. Taken together these data suggest some degree of isolation between the Pacific and Atlantic populations, and possibly two or more populations in the western North Atlantic.

Winan & Jones (61) investigated allozyme diversity among 360 Dall's porpoise (*Phocoenoides dalli*) collected from three contiguous locations in the Bering Sea and North Pacific Ocean. They analyzed 26 loci and found that 9 were polymorphic at the $P = 0.99$ level. Average heterozygosity was 0.058, which is within the range seen for other cetacean species. They divided their sample into three sets, north (N = 99), south (N = 200), and southwest (N = 61) of the Aleutian Islands. There were no significant differences in allele or genotype frequencies between the three areas. However, pooled data from the north and south areas were found to be out of Hardy-Weinberg equilibrium for three out of nine variable loci, showing a deficiency of heterozygotes. Because heterozygote deficiencies can result when two or more populations with differing allele frequencies are treated as one population (Wahlund effect), they subdivided samples and retested. Subsamples within the south area below 50°N to 52°N latitude then tested as being in Hardy-Weinberg equilibrium, while the northernmost subsample within that area was not. From this they estimate that at least two populations exist, and that they are separated north and south somewhere between 50°N and 52°N latitude, south of the Aleutian Islands. In this case there is apparently a natural geographic barrier at the Aleutian Islands; however, the likely population division is further south in open water.

Off the Pacific coast of Japan there are two forms of short-finned pilot whale (*Globicephala macrorhynchus*) which differ in morphology, the northern and southern forms. The northern form is larger and has a more distinct pattern of dorsal pigmentation (31). Wada (57) compared 204 northern form whales with 167 southern form whales for allozyme variability at 36 loci. He found very little variation, with only two variable loci in the northern form and five in the southern form. Average heterozygosity was 0.009 and 0.008 for the northern and southern forms, respectively, which is low compared to most other cetacean species (see 53, 60). For example Stenella sp. had an order of magnitude higher levels of average heterozygosity (53). At three out of the five variable loci, a contingency test for heterogeneity showed a significant difference in allele frequency between the two morphotypes, suggesting that they may represent genetically differentiated stocks. However, analysis using markers with greater resolution will be required to determine the degree of differentiation.

MYSTICETES   There are few obvious boundaries that might separate whales in the open ocean. In some cases, it has been assumed that the equator would serve as a boundary. For species such as the humpback whale that migrate between temperate and polar waters, the reproductive cycles for the populations on either side of the equator should be out of phase, and interbreeding restricted or prohibited. However, an individual humpback was sighted both in the Antarctic and in the northern hemisphere off the coast of Colombia (54). Further, mtDNA analyses that have compared Antarctic with North Pacific and North Atlantic humpback whales show that haplotypes are shared between oceans (7; PJ Palsboll, PJ Clapham, DK Mattila, F Larsen, R Sears, AR

Siegismund, J Sigurjonsson, O Vasquez, P Arctander, unpublished manuscript). Phylogenetic analyses showed three major clades when haplotypes from the North Atlantic and Antarctic are compared, one that includes whales sampled from Iceland, the western North Atlantic, and the Antarctic, one specific to Antarctic haplotypes, and one specific to the western North Atlantic (PJ Palsboll, PJ Clapham, DK Mattila, F Larsen, R Sears, AR Siegismund, J Sigurjonsson, O Vasquez, P Arctander, unpublished). Baker et al (7) included whales from the North Pacific and calculated the minimum number of migration events required to account for the current geographic distribution of haplotypes. They estimate that two were migrations from the southern oceans to the North Atlantic, two in the opposite direction, and two from the southern oceans to the North Pacific. Both studies indicate substantial divergence (4–5% sequence difference) between the three oceans.

The most extensive analysis of genetic variation within and between baleen whale populations to date is that conducted by Wada & Numachi (60). They investigated allozyme variation at up to 45 loci in fin, sei, Bryde's, and minke whales from 10 broad geographic areas, and they screened samples from a total of over 18,000 individuals.

Fin whale samples were compared between the Antarctic (samples collected over a very broad range from 10°E longitude to 60°W), the nearshore region off Spain, and from the north Pacific (from 160°E to 120°W). Of the four species studied by Wada & Numachi (60), the fin whale showed the least variation between populations. The Nei genetic distance from a comparison of 13 loci, over 400 individuals in the Antarctic and north Pacific, and 87 individuals from the Spanish coast, ranged from 0.0004–0.0007, too low to imply any distinction. The coefficient of genetic differentiation (GST) was estimated to be 0.071, implying that only 7% of the observed variation was due to variation between populations.

In a separate study, 283 fin whales caught off Iceland were compared at 40 allozyme loci with 46 fin whales caught off the coast of Spain (11). The genetic distance between these populations was 0.013, nearly two orders of magnitude greater than for the distances between the Spanish coast and either the north Pacific or the Antarctic. Danielsdottir et al (11) also describe differences between yearly catches of fin whales off Iceland. A comparison of 24 fin whales from the eastern coast of Canada, 24 from Iceland, and 19 from Norway at 32 allozyme loci indicated genetic differentiation between all three populations (12). Four out of five polymorphic loci tested showed significant differences in allele frequencies at the $p < 0.001$ level.

Sei whale samples were collected from the Antarctic and north Pacific over approximately the same range as for fin whales (60). Levels of polymorphism and heterozygosity were similar to those for fin whales, but the genetic distance comparing 13 loci between the Antarctic and north Pacific samples was higher

(0.0084). Neither the sei whale nor the fin whale samples showed heterogeneity in the Antarctic corresponding to the six management stocks defined by the International Whaling Commission (15).

Unlike the other members of the genus *Balaenoptera,* the Bryde's whale never migrates into polar waters. Its distribution is tropical and warm temperate. In the Pacific and Indian oceans, the International Whaling Commission recognizes nine management stocks. Wada & Numachi (60) compared Bryde's whales at 45 allozyme loci from four of these areas (from the western south Pacific, the Peruvian stock, the western north Pacific, and from either side of the south Indian Ocean). The sample size was greater than 100 individuals from all areas (over 1800 individuals from the western north Pacific). All comparisons showed low levels of genetic distance (ranging from D = <0.001–0.004). Although the genetic distance between the Indian Ocean and Pacific samples was low (D = 0.004), it was an order of magnitude higher than the comparison between the western and eastern sides of the south Pacific (Fiji vs Peru).

Due to the focus of interest for commercial whaling in the recent past, the minke whale has been investigated for genetic stock structure more thoroughly than other baleen whale species. Although the International Whaling Commission (IWC) imposed a moratorium on whaling in 1987, several whaling member countries feel that minke whale populations in the Antarctic, north Pacific, and north Atlantic are large enough to permit a resumption of whaling, and Iceland has recently quit the IWC in order to resume hunting. Danielsdottir et al (12) compared the population of minke whales off Norway (N = 118) with those off Iceland and West Greenland (N = 40) at 37 allozyme loci and found a Nei genetic distance of 0.023 comparing West Greenland and Norway, and 0.028 between Iceland and Norway. They conclude that the minke whales off Iceland and Norway represent separate breeding populations.

Wada & Numachi (60) compared 11,414 minke whales at 45 allozyme loci from a nearly circumpolar distribution in the Antarctic with 416 whales from the Japanese coastal fishery and 46 whales from the Korean coastal fishery. They found no heterogeneity between management areas in the Antarctic, though an earlier report based on a smaller sample size (56) had described a significant difference between allele distributions for two adjacent management areas. When they compared north Pacific with Antarctic samples, they found a genetic distance (D = 0.083) greater than that found between minke and sei whales (D = 0.053). Comparing samples from the Korean and Japanese coasts, they found a distance of D = 0.013.

Several DNA studies have made some of the same comparisons. Wada et al (60) isolated whole mtDNA from 79 Antarctic, 32 Korean coastal, 30 Japanese coastal, and one dwarf-form minke whale, and compared RFLP patterns using fourteen six-base recognition site restriction enzymes. Again, there was no distinction between Antarctic areas. However, consistent with the allozyme study, the genetic distance between north Pacific and Antarctic samples was high (an estimated 3.9% sequence difference). The dwarf-form minke sequence differed from that of the Antarctic minkes by 3.8%, and from that of the north Pacific minkes by 1.5%. Hoelzel & Dover (28) compared polymerase chain reaction–amplified control region mtDNA by RFLP analysis between Antarctic, north Pacific, and West Greenland populations. They found an estimated sequence difference of 5.1% between the Antarctic (N = 20) and north Pacific (N = 10) samples, and a similarly large difference between the north Atlantic (West Greenland, N = 10) samples and both north Pacific (9.0% difference) and Antarctic samples (3.4% difference). Other studies investigating satellite DNA (3) and minisatellite DNA (55) supported the same stock divisions.

CONCLUSIONS

In the oceans, the abundance and distribution of many species are influenced by such factors as currents and temperature gradients. Cold, highly oxygenated water is usually the most productive, and places where the water is cold throughout the water column, or where cold water at depth is brought up to the surface by upwelling, are some of the most productive in the world. At the same time, cetaceans are mammals and must maintain a high constant body temperature in a medium of high thermal conductivity. The long annual migrations of some cetacean species are apparently a compromise between the different requirements of maximizing feeding potential and breeding. This means that representatives of a given breeding population may, as in the case of humpback whales, be spread over great distances during the season when they congregate on feeding grounds. Alternatively, whales from two or more breeding populations may form a mixed assemblage on feeding grounds, as with minke whales in the North Pacific and North Atlantic. This may explain in part the high levels of polymorphism and lack of apparent population structure found for various species of baleen whale on feeding grounds in the Antarctic. If this is the case, then there are serious problems with modeling suitable catch quotas for whales in either the Antarctic or the North Sea (24).

In temperate and tropical waters, the distributional ecology of cetaceans is probably more closely tied with seasonal variations in the distributional ecology of their prey. Even within a season, there is considerable variation in the behavior and habitat of potential prey species. The specialist foraging of killer whales is an extreme example, where some killer whales consume marine mammals nearly exclusively, and others concentrate on fish prey. This resource division leads to differences in social (23, 26) and ranging behavior (8), and genetic differentiation has been described between sympatric populations of these two behavioral strategists (27). Some level of reproductive isolation may also result from other examples of local habitat division by dolphin species, especially where differences in morphology have been described (such as for nearshore vs offshore forms of several species). At the same time, pronounced differences in morphology may not correlate to patterns of genetic differentiation, as with the spinner dolphins in the eastern tropical Pacific.

Population boundaries reflect various factors including philopatry to natal breeding grounds and the sex and range of dispersing animals. As seen for humpback whales, there can be genetic migration between oceans and across the equator while considerable genetic differentiation is maintained between those populations (7). Some species such as the fin whale can show greater genetic differentiation between local populations than between populations that are geographically isolated by land mass or great distance. For several species with world-wide distributions, the degree of differentiation between oceans can be relatively small (such as with fin and sei whales), while others show considerable differentiation between populations in different oceans (such as minke and humpback whales). In other cases there can be an apparent boundary between genetically differentiated populations, where there is no obvious geographic boundary (as with Dall's porpoise in the North Pacific).

In general, a review of genetic diversity within cetacean species reveals the following four patterns that have important implications for conservation and management (25): (i) Apparent geographic boundaries between populations may not correspond to real genetic distinctions, and genetic population boundaries exist where there are no clear geographic boundaries. (ii) Feeding ground populations can represent mixed assemblages of breeding populations. (iii) The extent of morphological variation is not necessarily a good indicator of genetic distinction. And (iv) genetically differentiated sympatric populations exist, sometimes based on differential habitat use. The better we understand the relationship between these species and their complex marine environment, the more effectively we will be able to interpret observed patterns of genetic variation.

Literature Cited

1. Amos W, Barrett J, Dover GA. 1991. Breeding system and social structure in the Faroese pilot whale as revealed by DNA fingerprinting. *Rep. Int. Whaling Commiss. (Special issue)* 13:255–70

2. Amos W, Dover GA. 1991. The use of satellite DNA sequences in determining population differentiation in the minke whale. *Rep. Int. Whaling Commiss. (Special issue)* 13:235–44

3. Amos W, Hoelzel AR. 1991. Long-term preservation of whale skin for DNA analysis. *Rep. Int. Whaling Commiss.* (Special issue) 13:99–104

4. Amos W, Schlotterer C, Tautz D. 1993. Social structure of pilot whales revealed by analytical DNA profiling. *Science* 260:670–72

5. Baird RW, Stacey PJ. 1988. Variation in saddle patch pigmentation in populations of killer whales (*Orcinus orca*) from British Columbia, Alaska and Washington State. *Can. J. Zool.* 66: 2582–85

6. Baker CS, Palumbi SR, Lambertsen RH, Weinrich MT, Calambokidis J, O'Brien SJ. 1990. Influence of seasonal migration on geographic distribution of mitochondrial DNA haplotypes in humpback whales. *Nature* 344:238–40

7. Baker CS, Perry A, Bannister JL, Weinrich MT, Abernethy RB, et al. 1993. Abundant mitochondrial DNA variation and world-wide population structure in humpback whales. *Proc. Natl. Acad. Sci. USA* 90:8239–43

8. Bigg MA, Olesiuk PF, Ellis GM, Ford JKB, Balcomb KC. 1990. Social organization and genealogy of resident killer whales (*Orcinus orca*) in the coastal waters of British Columbia and Washington State. *Rep. Int. Whaling Commiss.* (Special issue) 12:383–406

9. Clapham PJ, Mayo CA. 1990. Reproduction of humpback whales observed in the Gulf of Maine. *Rep. Int. Whaling Commiss.* (Special issue) 12:171–75

10. Clapham PJ, Palsboll PJ, Mattila DK, Vasquez O. 1992. Composition and dynamics of humpback whale competitive groups in the West Indies. *Behaviour* 122:182–94

11. Danielsdottir AK, Duke EJ, Joyce P, Arnason AI. 1991. Preliminary studies on genetic variation at enzyme loci in fin whales (*Balaenoptera physalus*) and sei whales (*R. borealis*) from the north Atlantic. *Rep. Int. Whaling Commiss.* (Special issue) 13:115–24

12. Danielsdottir AK, Halldorsson SD, Arnason A. 1992. Genetic variation at enzyme loci in northeastern Atlantic minke whales. *Rep. Int. Whaling Commiss. (SC/44/NAB15)*

13. Darling JD, McSweeney DJ. 1985. Observations on the migrations of North Pacific humpback whales. *Can. J. Zool.* 63:308–14

14. Dizon AE, Southern SO, Perrin WF. 1991. Molecular analysis of mtDNA types in exploited populations of spinner dolphins (*Stenella longirostris*). *Rep. Int. Whaling Commiss.* (Special issue) 13: 183–202

15. Donovan GP. 1991. A review of IWC stock boundaries. *Rep. Int. Whaling Commiss.* (Special issue) 13:39–68

16. Dover GA. 1993. Evolution of genetic redundancy for advanced players. *Curr. Opin. Genet. Dev.* 3:902–10

17. Dowling TE, Brown WM. 1993. Population structure of the bottlenose dolphin as determined by restriction endonuclease analysis of mitochondrial DNA. *Mar. Mammal Sci.* 9:138–55

18. Duffield DA, Wells RS. 1991. The combined application of chromosome, protein and molecular data for the investigation of social unit structure and dynamics in *Tursiops truncatus. Rep. Int. Whaling Commiss.* (Special issue) 13: 155–70

19. Duffield DA, Ridgeway SH, Cornell LH. 1983. Hematology distinguishes coastal and offshore forms of bottlenose dolphins. *Can. J. Zool.* 61:930–33

20. Heimlich-Boran JR. 1986. Fishery correlations with the occurrence of killer whales in greater Puget Sound. In *Behavioural Biology of Killer Whales,* ed. B Kirkevold, JS Lockard, pp. 113–31. New York: Liss. 389 pp.

21. Helbig R, Boag PT, White BN. 1989. Stock identification of beluga whales using mitochondrial DNA markers: preliminary results. *Musk-Ox* 37:122–28

22. Heyning JE, Perrin WF. 1994. Two forms of common dolphin (genus *Delphinus*) from the eastern North Pacific; evidence for two species. *Contr. Sci. Los Angeles.* In press

23. Hoelzel AR. 1991. Killer whale predation on marine mammals at Punta Norte, Argentina; food sharing, provisioning and foraging strategy. *Behav. Ecol. Sociobiol.* 29:197–204

24. Hoelzel AR. 1991. Whaling in the dark. *Nature* 352:481

25. Hoelzel AR. 1992. Conservation genetics of whales and dolphins. *Mol. Ecol.* 1:119–25

26. Hoelzel AR. 1993. Foraging behavior and social group dynamics in Puget Sound killer whales. *Anim. Behav.* 45: 581–91

27. Hoelzel AR, Dover GA. 1991. Genetic differentiation between sympatric killer whale populations. *Heredity* 66:191–95

28. Hoelzel AR, Dover GA. 1991. Mitochondrial D-loop DNA variation within and between populations of the minke whale (*Balaenoptera acutorostrata*). *Rep. Int. Whaling Commiss.* (Special issue) 13:171–82

29. Hoelzel AR, Hancock J, Dover GA.

1991. Evolution of the cetacean mito-chondrial D-loop region. *Mol. Biol. Evol.* 8:475–93

30. Hoelzel AR, Halley J, O'Brien SJ, Campagna C, Arnbom T, et al. 1993. Elephant seal genetic variation and the use of simulation models to investigate historical population bottlenecks. *J. Hered.* 84:443–49

31. Kasuya T, Miyashita T, Kasamatsu F. 1988. Segregation of two forms of short-finned pilot whales off the Pacific coast of Japan. *Sci. Rep. Whales Res. Inst.* 39:77-90

32. Katona S, Baxter B, Brazier O, Kraus S, Perkins J, Whitehead H. 1979. Identification of humpback whales by fluke photographs. In *Behavior of Marine Animals*, Vol. 3, ed. HE Winn, BL Olla, pp. 33–44. New York: Plenum. 438 pp.

33. Lambertsen RH. 1987. Biopsy system for large whales and its use for cytogenetics. *J. Mammal.* 68:443–45

34. Leatherwood S, Reeves RR. 1982. Bottlenose dolphin (*Tursiops truncatus*) and other toothed cetaceans. In *Wild Mammals of North America*, ed. JA Chapman, GA Feldhamer, pp. 369–414. Baltimore MD: Johns Hopkins Univ. Press. 652 pp.

35. Lint DW, Clayton JW, Lillie WR, Postma L. 1990. Evolution and systematics of the beluga whale and other odontocetes: a molecular approach. *Can. Bull. Fish. Aquat. Sci.* 224:7–22

36. Nei M. 1972. Genetic distance between populations. *Am. Nat.* 106:283–92

37. Nei M, Maruyama T, Chakraborty R. 1975. The bottleneck effect and genetic variability in populations. *Evolution* 29:1–10

38. Norris KS, Dohl TP. 1980. The structure and function of cetacean schools. In *Cetacean Behavior*, ed. LM Herman, pp. 211–61. New York: Wiley. 460 pp.

39. Odell DK, Asper ED. 1990. Distributions and movements of freeze-branded bottlenose dolphins in the Indian and Banana rivers, Florida. In *The Bottlenose Dolphin*, ed. S Leatherwood, RR Reeves, pp. 515–40. San Diego, Calif: Academic. 653 pp.

40. Palsboll PJ. 1990. Preliminary results of restriction fragment length analysis of mitochondrial DNA in minke whales from the Davis Strait, northest and central Atlantic. *Rep. Int. Whaling Commiss. (SC/42/NHMi35)*

41. Palsboll PJ, Vader A, Bakke I, Raafat El-Gewely M. 1992. Determination of gender in cetaceans by the polymerase chain reaction. *Can J. Zool.* 70:2166–70

42. Payne K, Tyack P, Payne R. 1983. Pro-gressive changes in the songs of hump-back whales: a detailed analysis of two seasons in Hawaii. In *Communication and Behavior of Whales*, ed. R Payne, pp. 9–58. Boulder, Colo: Westview. 643 pp.

43. Payne R. 1986. Long-term behavioral studies of the southern right whale (*Eubalaena australis*). *Rep. Int. Whaling Commiss.* (Special issue) 10:161–67

44. Payne R, McVay S. 1971. Songs of humpback whales. *Science* 173:585–97

45. Perrin WF. 1975. Variation of spotted and spinner dolphins (genus *Stenella*) in the eastern tropical Pacific and Hawaii. *Bull. Scripps Inst. Ocean.* 21:1–206

46. Perrin WF, Miyazaki N, Kasuya T. 1989. A dwarf form of the spinner dolphin (*Stenella longirostris*) from Thailand. *Mar. Mammal Sci.* 5:213–27

47. Perrin WF, Scott MD, Walker GJ, Cass, VL. 1985. Review of geographical stocks of tropical dolphins (Stenella sp. and *Delphinus delphis*) in the eastern Pacific. *NOAA Tech. Rep. NMFS28* Unpublished. 28 pp.

48. Perry A, Baker CS, Herman LM. 1990. Population characteristics of individually identified humpback whales in the central and eastern north Pacific: a summary and critique. *Rep. Int. Whaling Commiss.* (Special issue) 12:307–18

49. Rosel PE. 1992. *Genetic population structure and systematic relationships of some small cetaceans inferred from mitochondrial DNA sequence variation.* PhD thesis. Univ. Calif., San Diego. 191 pp.

50. Rosel PE, Dizon AE, Heyning JE. 1994. Genetic analysis of sympatric morpho-types of common dolphins (genus *Delphinus*). *Mol. Biol. Evol.* In press

51. Schlotterer C, Amos W, Tautz D. 1991. Conservation of polymorphic simple sequence loci in cetacean species. *Nature* 354:63–65

52. Sergeant DE. 1962. The biology of the pilot or pothead whale (*Globicephala melaena*) in Newfoundland waters. *Bull. Fish. Res. Board Canada*, 132:1–184

53. Shimura E. Numachi K. 1987. Genetic variability and differentiation in the toothed whales. *Sci. Rep. Whales Res. Inst.* 38:141–63

54. Stone GS, Florez-Gonzalez L, Katona S. 1990. Whale migration record. *Nature* 346:705

55. van Pijlen I, Amos W, Dover GA. 1991. Multilocus DNA fingerprinting applied to population studies of the minke whale. *Rep. Int. Whaling Commiss.* (Special issue) 13:245–54

56. Wada S. 1982. Analysis of the biochem-

ical data by G-statistics. *Rep. Int. Whaling Commiss.* 32:707

57.   Wada S. 1988. Genetic differentiation between two forms of short-finned pilot whales off the Pacific coast of Japan. *Sci. Rep. Whales Res. Inst.* 39: 91–101

58.   Wada S. 1991. Genetic distinction between two minke whale stocks in the Okhotsk Sea coast of Japan. *Rep. Int. Whaling Commiss. (SC/43/Mi32)*

59.   Wada S, Kobayashi T, Numachi KI. 1991. Genetic variability and differentiation of mitochondrial DNA in minke whales. *Rep. Int. Whaling. Commiss.* (Special issue) 13:203–16

60.   Wada S, Numachi KI. 1991. Allozyme analyses of genetic differentiation among the populations and species of the Balaenoptera. *Rep. Int. Whaling Commiss.* (Special issue) 13:125–54

61.   Winans GA, Jones LL. 1988. Electrophoretic variability in Dall's porpoise in the North Pacific Ocean and Bering Sea. *J. Mammal.* 69:14–21

62.   Winn HE, Bischoff WL, Taruska AG. 1973. Cytological sexing of Cetacea. *Mar. Biol.* 23:343–46

Annu. Rev. Ecol. Syst. 1994. 25:401–22

FISHERIES ECOLOGY IN THE CONTEXT OF ECOLOGICAL AND EVOLUTIONARY THEORY

Kenneth T. Frank and William C. Leggett[1]

Marine Fish Division, Department of Fisheries and Oceans, Bedford Institute of Oceanography, PO Box 1006, Dartmouth, Nova Scotia B2Y 4A2, and [1]Department of Biology, McGill University, 1205 Avenue Dr. Penfield, Montreal, Quebec H3A 1B1

KEY WORDS: population regulation, density dependence, dispersal, life history theory

---

Abstract

This review examines the application of fisheries ecological data in the development and testing of ecological and evolutionary theory, and the use of such theory in the pursuit of fisheries science. The development of modern fisheries ecology is traced from the beginning of the twentieth century to illustrate the paradigm shifts that have influenced the discipline. The major influence of ecological theory on the development of fisheries ecology was the application of the theory of density dependence, developed in the context of studies of insect population biology. This led to the theoretical construct of a functional relationship between stock and recruitment—a central concept in fisheries ecology. Fisheries ecological data have not been used extensively in the testing of theory. Important exceptions include contributions to the development and testing of self-thinning, the role of density independence in population dynamics, metapopulation modelling and associated theory, and life history theory. The potential for the use of fisheries ecological data in the development and testing of ecological and evolutionary theory is great. The importance of a greater consideration of ecological and evolutionary theory in the development of fisheries ecology is advocated.

401

0066-4152/94/1120-0401$05.00

INTRODUCTION

The purpose of this review is twofold: to highlight areas where fisheries ecology has contributed directly or indirectly to the development and testing of ecological and evolutionary theory and vice versa; and, through the drawing of parallels between work in fisheries and elsewhere, to alert ecologists to the rich data base available for the elaboration and testing of theory. In so doing we remind fisheries ecologists of the importance of ecological and evolutionary theory to the development of their science.

We define *fisheries ecology* as that branch of science dealing with the study of factors influencing the abundance, distribution, and availability of commercially important species of fish. In fisheries ecology the unit of study is typically the population, and the time and space scales investigated are generally large (on the order of years and hundreds of kilometers). Because of its focus on the population as the unit of study, much of fisheries ecology might be viewed simply as applied population dynamics. We do not share this view. In fact, the study of the dynamic interaction between the fish and their biological and physical environment forms the very foundation of the discipline. Notwithstanding this reality, ecologists have made limited use of the unique data sets developed by fisheries scientists to test theory (145). Furthermore, fisheries ecologists have made only limited use of the extensive body of ecological theory in the development of their discipline. In contrast, the field of fish ecology, which is not treated in detail here, and which has typically focused on the individual and the species, has contributed significantly to the development and testing of ecological and evolutionary theory (176).

In this paper we draw largely on the literature dealing with temperate marine species which constitute the majority of the economically important and well-studied fisheries of the world. A considerable literature on freshwater fisheries exists. The fields of marine and freshwater fisheries ecology have followed largely independent paths with little cross-fertilization of ideas. Recent comparisons of the two systems have emphasized the differences between them. For example, Magnuson (102) has drawn the analogy of lakes as islands and oceans as continents, the former being regulated primarily by extinction and the latter by colonization processes.

FISHERIES ECOLOGY—A BRIEF REVIEW

Modern fisheries ecology, the systematic study of the dynamic interactions between the abundance and distribution of fishes, and the biotic and abiotic environments they occupy, was born in Norway at the beginning of the twentieth century (150). Throughout this century the central goal of the discipline has been to identify, and to develop the capacity to predict, the causes of interannual variation in the numbers of recruits to the population ("the recruitment problem"). Recruitment has several definitions. In the non-fisheries literature, recruitment is generally defined as the number of individuals belonging to a specific cohort that survive to enter the reproductive population; the number is determined from studies of life tables and survivorship curves (74, 75). In the fisheries literature, and especially in literature focused on the management of fisheries, recruitment has also been defined as the number of individuals belonging to a specific cohort that survive to harvestable size, which may or may not correspond in development to reproductive maturity. Other definitions are common. For example, reference is often made to the numbers of fish larvae that "recruit" to the juvenile stage. "Cohort" and "year-class" are used interchangeably.

Interannual variability in recruitment to marine fish populations is often enormous. The ratio of maximum to minimum recruitment has been summarized for several stocks (137). Stocks with the highest ratios included California sardine, Norwegian herring, West Greenland cod, and Georges Bank haddock, with values of 196, 130, 100, and 2700, respectively. The social and economic implications of such variation have been the principal motivation of fisheries ecological studies. In eastern Canada we are now witnessing unprecedented economic losses and social displacement as a consequence of the recent and dramatic decline in the abundance of Northern cod (129).

Throughout the eighteenth and nineteenth centuries, variations in the availability of commercially important marine fish species were considered to be the result of changes in their distributions relative to the location of "traditional" fishing locations (144, 150). This species-based migration model was shattered by the development and application of techniques that use the bony structures of fishes (scales, otoliths, vertebrae, spines) to estimate age. First developed by Dahl (36), the application of this technique to North Sea herring, whose fluctuations in abundance are legendary (29), led to the discovery that fluctuations in their abundance were the result of the infrequent appearance of dominant year-classes. The development of aging techniques, and Hjort's (68) use of the tool to demonstrate the importance of variation in cohort survival to the dynamics of marine fisheries, established the foundation for modern fisheries ecology. The discipline soon became focused on the population as the unit of study (144) and on the importance of the dynamics of age structure as the cause of, and the evidence for, changes in production and abundance. No other animal group has yielded such a rich body of data in this area.

The link between fish population dynamics and the dynamics of the biotic and abiotic environment was also established early in the century. Hjort (68) proposed that the success of individual cohorts was determined by the availability of food to the larvae at a critical period—the time of the transition of endogenous (yolk-sac) feeding to exogenous (planktivorous) feeding. Central to Hjort's "critical period" hypothesis is the assumption that recruitment success is determined during the larval stage, and that interannual variation in the animal's environment (in his view, food availability and the physical factors that influence that availability) is the primary determinant of variability in larval survival.

Hjort not only changed the direction of fisheries research with his application of aging techniques; his "critical period" hypothesis defined one of the two dominant foci of this research for more than 50 years. Fisheries ecology has since focused strongly on studies of the relationship between biotic/abiotic interactions during the early life stages of fishes and the resulting abundance of adults.

Hjort's hypothesis was generalized by Cushing (28). His "match-mismatch" hypothesis removed the restriction that mortality due to starvation leading to recruitment variability would be restricted to a singular "critical" stage in larval development. It was hypothesized that larval survival (and recruitment) would be linked to the strength of the temporal overlap in the production of larvae and the seasonal production cycle of the plankton. He also explicitly recognized the importance of abiotic factors as regulators of the timing and intensity of the seasonal plankton production cycle.

The development of methods for aging individual fishes, and for developing reliable age-structured data for whole populations, led to a second important focus of research in fisheries ecology—the study of the relationship between the abundance of the parent stock and the resulting yield of recruits to the population. Launched in 1954 by Ricker's now classic analysis (127), "stock-recruitment" studies dominated the research on adult stages of marine fishes from the 1950s through the 1980s and stimulated a great deal of research on biotic and abiotic determinants of abundance. Strongly influenced by the development of theories of density dependence as they applied to insect population dynamics (127, 144), stock-recruitment studies focused on the limits to population abundance and on analyses of the causes of variation about theoretically derived stock-recruit models. It was hoped that such relationships would allow identification of the long-term average stock size needed to maintain good recruitment, on average. Fisheries ecology thus quickly became focused on the design of strategies for dealing with variability about such long-term averages. Several variants of Ricker's basic model have been developed and applied (9, 27, 137, 140). These differ primarily in the source and intensity of the density-dependent controls assumed. Because the concept of self-regulation is common to all such models, equilibrium (123) and/or stationary (163) conditions have been assumed.

Stock-recruitment theory has had a profound influence on thinking and research on fish population dynamics and on the theory of harvesting (86, 137).

Central to this thinking was the widely accepted belief that density-dependent mechanisms would largely define the relationship between stock and recruitment, and that annual deviations from this expected relationship were likely to be the result of noncompensatory factors, the ultimate source of which may be abiotic. Thus, a knowledge of the underlying density-dependent model, together with a predictive understanding of noncompensatory forces leading to deviations from this model, could be expected to lead to reliable predictions of future abundance, and of the capacity of the population to sustain harvesting. Early life history studies and stock-recruitment studies merged in their focus on the causes of noncompensatory forces influencing recruitment.

One of the major theoretical outgrowths of stock-recruitment theory, and one of the major disappointments of applied fisheries science, was the concept of maximum sustained yield (MSY) (85). *Maximum sustained yield* is defined as the locus of maximum surplus reproduction (the population size at which the production of recruits maximally exceeded the number required to replace the parent stock). In theory, fish populations could be maintained at the level of parent stock required to achieve MSY through the harvesting of surplus recruits, thereby maximizing yield to the fishery. In practice, variance about even the most promising stock-recruit relationships was so high that application of the MSY concept was impossible. A large literature has since developed on how to replace the maximum sustained yield concept with the use of feedback and adaptive strategies for dealing with the high variability and uncertainty characteristic of population fluctuations in marine fishes (19, 103, 162).

Recently, stock-recruitment models, and those who seek to apply them, have come under increasing criticism. Welch (170) argues that such models have proven dangerous as a basis for fisheries management. Hall (60) has criticized the approach because of its "cake and eat it too" aspect, arguing that fisheries scientists have shown excessive willingness to impose theory on data rather than testing the null hypothesis that there is no relationship between stock and recruitment. In fact, fisheries biologists have tended to take constant recruitment (no relationship between stock and recruitment) as the null hypothesis (142). This is dangerous because the underlying assumption of such a hypothesis is that a stock capable of producing constant recruitment at any stock level cannot be collapsed by fishing (20, 142). However, in the absence of compensatory processes, the expected relation between recruitment and parent stock size would be direct proportionality, and this should, in principle, be the most appropriate null hypothesis (142). Indeed, the existence of compensatory processes can only be inferred from departures from proportionality in the relationship between stock and recruitment (10).

Peters (120) has noted that the inadequate fits of stock-recruitment models to the data are consistent with the difficulty of obtaining reliable data. Walters

& Ludwig (166) have also noted the difficulties involved in accurately measuring the key variables. This, Peters claims, has led fisheries scientists to accept poor fits rather than to challenge the underlying assumptions of the theory.

It should be noted, however, that the approach has not been barren, and that fisheries scientists have not been as oblivious to its shortcomings as the above critiques imply. Several researchers (3, 69, 90, 114, 156) have related the residuals from fitted stock-recruitment relationships to abiotic factors (e.g. temperature, river discharge, Ekman transport, sea level, sunlight) believed to influence survival during the egg and larval stages. Modelling recruitment using key factor analysis has also been undertaken (13, 40). Many investigators have used correlation techniques to explore the relationship between various biotic and abiotic variables and recruitment. Several promising relationships have been identified (26, 65, 95, 112, 146) and developed following detailed analyses of the dynamics of the production system, and/or of the biology of the species involved. In other cases where the relationships were identified primarily as a result of exploratory analyses, or where the potential regulators were identified intuitively, subsequent analyses using new data have frequently failed to sustain the initial models (43). These failures have led to repeated criticisms of the correlative approach and to calls for its abandonment (6, 58, 143, 164). Others have been less sceptical, noting that the causes of the model failures quoted in these critiques are numerous and correctable. Such causes include short data series, faulty intuition, faulty assumptions about the mode of operation of the environmental factors chosen, and incorrect averaging over the time and space scale of the independent variable (95, 135, 160).

As noted above, the work of Hjort and his colleagues led to a profound paradigm shift which dramatically diminished the perceived importance of fish movements as regulators of fishery yield. However, the new focus on the recruitment dynamics of populations, and the extensive application of mark-recapture methods originally developed by Petersen (121) to quantify changes in population abundance, soon provided a rich data base on the timing, speed, and extent of fish movements. Tagging studies soon revealed that many freshwater, anadromous, and marine fishes homed to natal spawning sites to reproduce (89). These findings reinforced the focus of fisheries ecological research on the population as the unit of study and on the importance of life cycle closure (62, 144) in fish population dynamics. The widespread evidence of homing, its apparent precision, and the early and uncritical acceptance of the hypothesis that such precision could only be explained by a "knowledge" of the location of home, and by true navigational ability on the part of individuals, caused research on fish migration quickly to become heavily focused on mechanisms of orientation and navigation in fishes. The validity of this hypothesis has since been challenged, and its potential negative effect on the advancement of a broader understanding of fish migrations has been high-lighted (91). Such studies did, however, contribute in a major way to the development and testing of theories of animal orientation and navigation (52, 108, 147).

Beginning in the early 1970s, studies of fish migrations as ecological phenomena (i.e. as distributional responses of various life stages to environmental factors such as temperature, salinity, tides, currents), and of their potential link to survival during the early life history and to reproductive requirements of populations, supplanted studies of orientation and navigation mechanisms as the focus of migration research. This transition was largely coincident with the realization that the timing and path of migrations may be less predictable than proponents of the navigation hypothesis had surmised (89).

Examples of such environmentally assisted/regulated migrations are numerous and dramatic. They include active and passive tidal stream transport of anadromous, catadromous, and marine fish larvae and juveniles (2, 91); temperature regulation of the seasonal distributions, timing, and path of migration (89, 91, 132, 133); and temperature-regulated occupation of larval "safe sites" (48, 50). The dynamics of these behaviorial responses to environmental signals also influence recruitment to adult populations, catch rates in mobile and fixed-gear fisheries (95, 134, 136, 157), and the evolution of life history strategies (92, 139).

Significant differences of view continue to surround the importance of early life history events in the dynamics of fish populations. However, research into the relative importance of density-dependent vs density-independent influences on parent stock relationships, and the role of biotic and abiotic factors as guides to fish migration, has generated a wealth of data on, and understanding of, the broader areas of ecology and evolution. It is to these questions that we now turn.

CONTRIBUTIONS TO ECOLOGY AND EVOLUTION

Gulland (59) observed that "fish have no direct terrestrial counterparts—a fox or lion does not start competing with mice." Fish begin life as one component of a complex plankton community with which they interact and compete directly during the most vulnerable life stage. As a consequence, the strongest parallels between fish and terrestrial organisms in the context of population processes are likely to be with plants, insects, and amphibians. Moreover, the applied focus of much of insect and plant ecology, resulting from the economic consequences of large fluctuations in the abundance and distribution of pest species, has created important parallels in the approach to the study of factors controlling population dynamics and distributions. Important similarities exist as well because of attempts to control and predict these processes.

Population Regulation

The difficulties encountered in applying stock-recruitment approaches to the management of fisheries, particularly in the short term, may be related to the failure to recognize or heed Gulland's (59) observation about the difference between fish and lions. In fact, notwithstanding the contributions of insect population dynamics to the development of stock-recruitment concepts, the approach taken in such studies has strongly paralleled the mammalian model, which has tended to collapse events occurring during several life stages into a single output.

A small number of researchers have suggested, or attempted to apply, a more phenological approach to recruitment prediction that incorporates the dynamics of individual life stages into the final prediction. Larkin et al (87) suggested that the life history stages be considered as successive stock recruitment relationships, with the output from one comprising the input to the next. Hempel (66) suggested that during the course of an individual fish's life it belongs to three or more populations (larval, juvenile, adult), each having its own dynamic, each phase being interconnected, and each having different requirements for growth and survival. More recent applications of this approach exist (61, 95, 116, 165). Unfortunately, the potential of this approach has not been fully explored in fisheries ecology, in part due to the difficulties involved in obtaining reliable census data on all individual life stages.

Collectively, these latter approaches are analogous to population regulation studies in plant, insect, and amphibian ecology whose models are built to separate the factors influencing survival and recruitment (63, 83, 173). Moreover, several of the fundamental hypotheses that have governed the development of thinking in fisheries ecology have direct parallels in the plant and insect literature and involve both density-dependent and density-independent factors.

DENSITY-DEPENDENT CONTROL   Several modelling studies have examined the role and timing of density-dependent forces as potential regulators of fish abundance. These have consistently concluded such feedback should be most intense, and should have the greatest influence on population numbers, during the larval and juvenile stages when densities are highest (33, 71, 72, 80, 141, 167). One of two outcomes has generally been assumed: (i) death as a direct result of competition for limited resources, and (ii) death as a secondary effect as, for example, through the influence of growth rate and/or condition on susceptibility to disease, predation, etc. Houde (71) argues that catastrophic environmental conditions need not occur to generate a hypothetical recruitment failure. Through modelling he has demonstrated that modest variation in the duration of the larval stage, induced by growth rate changes (8, 17) could, through its influence on cumulative mortality, lead to 100-fold variations in the numbers of larvae recruiting to the juvenile stage. Solid empirical support for the concept is, however, difficult to locate (8). The "stage duration" hypothesis is founded on the assumption that instantaneous mortality rates during the larval period are constant, or inversely related to larval size. However, numerous authors reported higher rates of predatory losses among larger and faster growing larvae (38, 51, 96, 117). Such growth-related differential mortality could counter, or even negate, the potential advantages of reduced stage duration (8).

Estimates of food consumption by larvae in the field indicate little effect on prey levels (30, 34, 78, 109, 122). This suggests that individual consumers cannot experience density-dependent reductions in growth or survival as a consequence of their impact on prey availability. Fortier & Harris (42) conclude that density dependence, if it occurs, is driven by grazing of the prey community by the full suite of zooplankton predators, and not as a consequence of the feeding activity of any single population. However, given the convergence of spawning times in temperate marine fish species and non-fish predatory components of the zooplankton, such combined predation could create food limitation and density dependence. This community-level density dependence, if it occurs, could mask the expected relationship between stock and recruitment.

Density dependent growth has, however, been repeatedly documented for fishes during the juvenile life stage (12, 35, 90, 168). Significant negative correlations between growth and population density of young fish have been assumed to represent intraspecific competition for food. Direct evidence in support of this claim, based on estimates of food supply or feeding rate, is generally lacking. Uncertain precision of the growth and density indices (15), sampling bias (35), and the tendency for increased dispersion of large year-classes (35, 44) suggests alternative interpretations are possible or that other mechanisms may be operating.

The density responses of some fish populations (e.g. stream-dwelling salmonids) are, however, consistent with the process of self-thinning frequently reported for terrestrial plant and animal communities. Compensatory changes in plant growth relative to density have a characteristic slope of $-3/2$ which has been termed the "self-thinning rule" (41, 57). For mobile animals, Grant (57) suggested several self-thinning lines are possible, with slopes ranging from -1 (carrying capacity constant), -1.33 (populations follow a constant metabolic rate rule, i.e. constant rate of energy input to the cohort) and -1.08 (assumes territory size limits density because territory size increases with body size). Variation in the elevation of self-thinning lines is hypothesized to result from interannual changes in carrying capacity. These concepts have not yet been extensively applied to the analysis of density-dependent responses in marine fish populations, even though the process of self-thinning was used by Beverton & Holt in deriving their theory of stock and recruitment. Recent evidence from demersal marine fishes suggests the existence of such processes (110). It is likely that the theory would be most applicable to fishes that exhibit limited movement and/or restricted ranges (but see 55).

DENSITY-INDEPENDENT FACTORS   Density-independent factors are widely believed to be important contributors to the high variability in recruitment evident in marine fish populations. Their influence is believed to operate primarily at the egg and larval stage. Mortality is extremely high during the early life stages then declines rapidly, resulting in survivorship curves approximating the Type III, positively skewed rectangular (37). For example, it is not uncommon for cod and haddock cohorts to decline by several orders of magnitude during the first six months of life (16).

The dominant hypothesis regarding the influence of density-independent factors on recruitment has been Cushing's "match-mismatch." This hypothesis rests on two fundamental assumptions: (i) fish in temperate waters spawn at a fixed time in the spring or fall, and (ii) abiotic factors (ocean climate) regulate the timing of the spring and autumn plankton blooms which provide the necessary food concentrations for growth and survival. Hence, interannual variation in larval survival is governed by the degree of overlap in time and/or space of the cycle of larval and food production. A full match is expected to produce an exceptional number of survivors and correspondingly high recruitment. A full mismatch, in contrast, is expected to result in recruitment failure. Well-known examples of exceptional year-classes include the legendary 1904 Atlantic herring year-class in Norway (144), the 1963 year-class of North Sea plaice (11), and the North Sea gadoid outburst during the 1960s (31, 67). The evidence accumulated to date is broadly consistent with the predictions of the hypothesis (14, 32, 94).

A related hypothesis, Lasker's "stable ocean" (88) argues that calm weather periods are essential to the development and maintenance of high-density food layers essential to successful first feeding in larval fishes. A test of this hypothesis (118, 119) revealed that while larval survival was related to the development of prey maxima, recruitment to the juvenile stage was unrelated to the occurrence of calm periods. This finding highlights the limitations of the single stage approach to recruitment analyses. It has recently been proposed that microscale turbulence, known to be generated by wind, upwelling, and tidal currents, facilitates successful feeding during the larval period (138). This hypothesis has been tested and extended (100, 101). Optimal turbulence levels vary with larval stages and exhibit a size-specific optima at intermediate turbulence levels. These new findings call into question the generality of the stable ocean hypothesis, and the dependence of the match-mismatch hypothesis on prey abundance alone.

The timing and success of critical life stage transitions have also been demonstrated to be under the influence of abiotic factors. These include arrested development and hatching synchrony (84, 161, 177), synchronous dispersal within populations and among co-occurring species in response to environmental signals (4, 47, 49, 50), timing of metamorphosis (8), and the onset of schooling (12, 46). Ultimately, recruitment will be the product of the survival at each life stage/transition as influenced by these non–density-dependent controls.

Direct parallels to these responses of fish populations to abiotic factors and the influence of such interactions on population abundance are found in seed germination (5, 53, 171) and dispersal (63, 76, 77, 171) and insect emergence (1, 82, 97, 175) and dispersal (21, 149, 151, 172). It is important to indicate exactly how some of these ecological studies parallel the wide variety of fisheries hypotheses identified. Reproductive timing in plants and insects is clearly related to maximizing the probability of offspring survival (53). Selective mortality, predator satiation, and weather cues have been demonstrated as causes of synchronous patterns of seed production and insect emergence (76, 97). Convergence in reproductive timing is also known to occur among sympatric species (82). Optimal responses to stochastic environmental events, such as seed dormancy and insect diapause, and the theoretical rules governing such life stage transitions (22) apply similarly to aquatic systems. At the community level, species coexistence may be facilitated by variable, asynchronous recruitment. This "storage effect" could act to maintain high species diversity and may be most applicable to trees and marine fishes whose population patterns of survivorship, fecundity, and life span are most appropriate for such a mechanism to operate (169).

Such responses are also consistent with the "safe site" concept (64) in which animals and plants selectively occupy habitats that favor enhanced survival in the face of stochastic variation in habitat suitability. This "match-mismatch" to the temporal or spatial occurrence of safe sites in aquatic systems is frequently cued to stochastic but reliable environmental signals such as temperature change (45, 47, 50, 81, 95, 104), turbulent mixing (18, 107), pressure fluctuations (106, 113, 126), rainfall (176), and cumulative degree days (124). Population outbreaks and dominant year-classes result from a full match of organism requirements for survival with their occurrence in space and time (25, 31, 95).

Population Dispersal

Population dynamic models applied to fish populations frequently ignore dispersive processes such as immigration and emigration (44, 99) because these populations are widely considered to be geographically isolated and self-reproducing. Emigration, whether active or passive, has commonly been viewed as the equivalent of mortality; emigrants are considered lost to the reproductive population. Numerous studies, focused on the egg and larval stages, have demonstrated the negative influence of emigration on recruitment (3, 105, 111, 115). This concept has recently been formalized by Sinclair (144) who coined the phrase "member-vagrant" to describe the importance of losses of eggs and larvae due directly to advection and diffusion to the abundance of local populations. This hypothesis also includes the negative effects of juvenile vagrants that fail to recruit to, and adults that stray from, the spawning population.

The vagrancy concept is not new to animal ecology. Wynne-Edwards (178) believed that dispersal acted as a safety valve providing immediate relief to the potentially negative effects of overpopulation. High population density relative to food supply is known to be associated with dispersal events, and theoretically derived rules governing such transitions have been devised (22). The potential gain from such dispersal is, however, strongly dependent on asynchrony in the food chain processes determining feeding success (70). It has been argued that temporal variation in food resources should select for increased dispersal, while spatial variation alone should select for reduced dispersal (79).

Recent analyses of dispersion phenomena in marine fishes have shown that profound demographic consequences can result. For example, some fish stocks are strongly coupled by dispersive processes (39, 44, 73). In these populations, dispersal of late larval and juvenile stages beyond the normal distributional range can result in mixing with adjacent stocks that may continue until the time of spawning. Return movements of mature adults to their birth site result in reciprocal fluctuations in the apparent abundance of the two stocks. Such fluctuations have previously been interpreted as changes in recruitment. This tendency for dispersion appears to be density dependent, i.e. strong year-classes in one population lead to dispersal to the second, thereby inflating the latter's apparent abundance while damping the realized population increase in the former (44). Such processes may contribute to the apparent synchrony of year-class strength in adjacent populations (23, 158).

In some circumstances receiving populations, generally associated with sink habitats, may be maintained exclusively by immigration from productive populations occupying source habitats (125). Such phenomena could explain both the apparent year-class synchrony in adjacent populations and the rapid decline/disappearance of one or more sink populations (and their subsequent reappearance) as a consequence of large-scale fluctuations in the productivity of the source. The possibility exists that the sudden and dramatic decline of cod throughout the North Atlantic may be the result of systematic overexploitation of spawning concentrations of a few key source populations that had historically maintained many sink populations through dispersal of surplus production. Atlantic and Pacific salmon production trends have been interpreted in a somewhat similar manner (98). These concepts are now captured in metapopulation models (56, 159). Such models deal with population regulation phenomena acting on large spatial scales, which are required for the analysis of the dynamics of marine fishes. They also generally assume that density-dependent emigration strongly influences the dynamics of both the receiving and the source populations. Recruitment dynamics may be seriously misinterpreted if such dispersive processes are ignored. The examples presented here of demographically coupled fish populations appear to represent some of the best empirical support in the ecological literature for metapopulation modelling and associated theory.

Life History Traits

Kikkawa (83), commenting on paradoxes in ecology, noted that no general theory can explain the regulation of animal numbers, but observed that "this does not mean that the regulation of animal numbers cannot be explained." He highlighted several ways of demonstrating that such a general theory of population regulation is unattainable. These included demonstrating that animals have different patterns of survival in different environments and showing that different species have reproductive strategies that span the so called r-k continuum. As a result, he argued, species with different life history strategies cannot be regulated by the same general mechanisms. It was suggested that to resolve this dilemma populations be classified according to the types of fluctuations they exhibit rather than the types of rules they should follow.

Clearly the dominant general theory governing research in fisheries ecology has been the theory of stock and recruitment. The difficulties experienced in applying this theory may derive from failure to recognize explicitly the distinct differences in evolved life history strategies exhibited by species and by populations within species. Kikkawa argues that classified groups might be characterized with respect to their physiological, genetic, and phylogenetic attributes, trophic relations, habitat requirements, and certain combinations of these and other properties. This implies an examination of recruitment variation from the perspective of the overall life history of the animal.

Modern life history theory—which owes much of its impetus to Cole's (24) simple question "why do animals reproduce more than once"—has provided an important framework for such analyses. Among the more general predictions of this theory are the expectation of a trade-off between energy allocation to reproduction vs that to survival, of the development of bet-hedging strategies (variable age at maturity, iteroparity) in stochastic environments, and of a coupling of life history traits and migration (131, 152–154). Perhaps more than in any other area of ecology and evolution, data developed by fisheries ecologists has played an important role in testing predictions of life history theory. We have selected some of the more significant applications of fisheries ecological data to the testing of the predictions of life history theory to highlight (92, 93, 130, 139, 174).

Schaffer & Elson (139) examined the life history traits of Atlantic salmon in relation to the energetics of migration and growth; they found, consistent with the predictions of theory, that (i) salmon spawning in long, fast-flowing rivers demanding high energy expenditure, exhibited longer feeding at sea, and delayed maturity in order to develop the energy reserves required for the migration; (ii) animals that exhibited rapid sea growth subsequent to the age of first possible reproduction delayed reproduction, thereby gaining the reproductive advantage associated with rapid growth and larger size (greater fecundity) at first reproduction, and (iii) high year-to-year variation in juvenile survival within a population was associated with high variability in age at first reproduction (bet hedging).

Populations of anadromous American shad, which home to their natal rivers and spawn in most major Atlantic coast rivers from New Brunswick to Florida, differ dramatically in their life history traits. Differences in the variability of thermal regimes in these rivers lead to increasing reproductive uncertainty with increasing latitude (92). Northern populations are characterized by delayed maturity, iteroparity, and reduced fecundity relative to southern populations which, at the extreme of the range, are semelparous, highly fecund, and mature at an early age. Average lifetime fecundity in all populations is equivalent in spite of the dramatic geographic cline in life history traits (93). These interpopulation differences have a genetic basis (7). For a related example from fish ecology, see (155).

Roff (130), who used the comparative approach to examine the life history constraints imposed on fishes by migration, found that migratory species were relatively large, matured late, and exhibited a faster growth rate and hence higher fecundity at age. Snyder (148) tested Roff's generalizations and found that migratory and nonmigratory populations of sticklebacks conformed to Roff's predictions, and that the differences were genetically based.

Winemiller & Rose (174) analyzed the life history patterns of 216 North American fish species. They report that fishes with relatively large body size exhibit a pattern of later maturity, high fecundity, small egg size, and reduced frequency of reproduction. Small fishes exhibited the inverse pattern. Species with large geographic ranges also tended to exhibit high fecundity. Anadromous fishes, consistent with Roff's (130) predictions, exhibited higher ages at maturation, faster adult growth rate, longer life spans, and larger egg sizes.

This broad conformity of life history traits of fishes with the predictions of life history theory, together with direct evidence in those species that have been most intensively studied, indicate that life history traits are adapted in ways that lead to a dampening of the effects of environmentally induced variability in offspring survival. The success of theory in predicting life history traits might be a useful starting point for a general theory of how population regulation arises from such traits in combination with other factors at the population level.

In general, fisheries ecologists have been slow to incorporate these findings into the study of the dynamics of populations. Sinclair (144), who has been a leader in this area, has correctly cautioned that uncritical interpretations of life history theory as applied to marine fishes may, in certain circumstances, lead to inappropriate conclusions as a consequence of failure to consider geographic constraints on the evolution of life history traits. Clearly life history theory is also subject to Kikkawa's criticism.

The application of life history concepts, and of the specific nature of life history adaptations in individual species/populations, to the understanding of reproductive success and recruitment in marine fishes has now begun (128, 144).

CONCLUSION

As stated at the outset, examples of the direct application of ecological or evolutionary theory to the study of fisheries ecology, and the application of fisheries ecological data to the testing of theory, are limited. In our view, the more significant examples of the use of fisheries data in the testing of theory to date include self-thinning, the important role of density independence in population dynamics, metapopulation modelling and associated theory, and life history theory. The theory of density dependence has been the dominant contribution of ecological theory to the development of ideas in fisheries ecology. The major outcome of this application has been the study of the functional relationship between stock and recruitment. The concept of self-regulation that underlies this relationship and is the foundation of population dynamics remains an enigma not only in fisheries (142) but in population biology in general (54). We conclude that a greater consideration of the implications of ecological and evolutionary theory could lead to significant advances in the understanding of processes of population regulation in marine fishes. On the other hand, the potential for the productive application of fisheries data in the development and testing of ecological and evolutionary theory is enormous.

ACKNOWLEDGMENTS

We wish to thank J Simon for help with the logistics of compiling the literature for this review. CT Marshall, KW Frank, and Drs. JE Carscadden and F Page made constructive comments on early drafts of the manuscript. We are particularly indebted to Dr. CJ Walters for his thorough and helpful review.

Any *Annual Review* chapter, as well as any article cited in an *Annual Review* chapter, may be purchased from the Annual Reviews Preprints and Reprints service. 1-800-347-8007; 415-259-5017; email: arpr@class.org

Literature Cited

1. Andrewartha HG. 1952. Diapause in relation to the ecology of insects. *Biol. Rev. Camb. Philos. Soc.* 27:50–107
2. Arnold GP. 1981. Movements of fish in relation to water currents. In *Animal Migration,* ed. DJ Aidley, pp. 55–79. Cambridge: Cambridge Univ. Press. 264 pp.
3. Bailey KM. 1981. Larval transport and recruitment of Pacific hake *Merluccius productus. Mar. Ecol. Prog. Ser.* 6:1–9
4. Bams RA. 1969. Adaptation in sockeye salmon associated with incubation in stream gravels. In *Symposium on Salmon and Trout in Streams,* ed. HR MacMillan, pp. 71–87. Vancouver: Univ. Br. Columbia. 388 pp.
5. Bazzaz FA. 1979. The physiological ecology of plant succession. *Annu. Rev. Ecol. Syst.* 10:351–71
6. Bell FH, Pruter AT. 1958. Climatic temperature changes and commercial yields of some marine fisheries. *J. Fish. Res. Board Can.* 15:625–83
7. Bentzen P, Leggett WC, Brown GG. 1993. Genetic relationships among the shads (Alosa) revealed by mitochondrial DNA analysis. *J. Fish. Biol.* 43:909–17
8. Bertram D. 1993. *Growth, development and mortality in metazoan early life histories with particular reference to marine flatfish.* PhD thesis. McGill Univ., Montreal. 219 pp.
9. Beverton RJH, Holt SJ. 1957. *On the Dynamics of Exploited Fish Populations. U.K. Ministry of Agriculture, Fisheries and Food, Fishery Investigations,* Ser. 2, Vol. 19. London: Her Majesty's Stationery Office. 533 pp.
10. Beverton RJH, Iles TC. 1992. Mortality rates of 0-group plaice (*Platessa platessa* L.), dab (*Limanda limanda* L.) and turbot (*Scophthalmus maximus* L.) in European waters. III. Density-dependence of mortality rates of 0-group plaice and some demographic implications. *Neth. J. Sea Res.* 29:61–79
11. Beverton RJH, Lee AJ. 1965. Hydrographic fluctuations in the North Atlantic Ocean and some biological consequences. In *The Biological Significance of Climatic Changes in Britain,* ed. CG Johnson, LP Smith, pp. 79–107. London, New York: Academic. 222 pp.
12. Blaxter JHS, Hunter JR. 1982. The biology of the clupeoid fishes. *Adv. Mar. Biol.* 20:3–194
13. Bradford MJ. 1992. Precision of recruitment predictions from early life stages of marine fishes. *Fish. Bull.* 90:439–53
14. Brander K, Hurley PCF. 1992. Distribution of early-stage Atlantic cod (*Gadus morhua*), haddock (*Melanogrammus aeglefinus*), and witch flounder (*Glyptocephalus cynoglossus*) eggs on the Scotian Shelf: a reappraisal of evidence on the coupling of cod spawning and plankton production. *Can. J. Fish. Aquat. Sci.* 49:238–51
15. Bromley PJ. 1989. Evidence for density-dependent growth in North Sea gadoids. *J. Fish. Biol.* 35 (Suppl. A): 117–23
16. Campana SE, Frank KT, Hurley PCF, Koeller PA, Page FH, Smith PC. 1989. Survival and abundance of young Atlantic cod (*Gadus morhua*) and haddock (*Melanogrammus aeglefinus*) as indicators of year-class strength. *Can. J. Fish. Aquat. Sci.* 46 (Suppl. 1):171–82
17. Chambers RC, Leggett WC. 1987. Size and age at metamorphosis in marine fishes: an analysis of laboratory-reared winter flounder (*Pseudopleuronectes americanus*) with a review of variation in other species. *Can. J. Fish. Aquat. Sci.* 44:1936–47
18. Checkley DM, Raman S, Maillet GL, Mason KM. 1988. Winter storm effects on the spawning and larval drift of a pelagic fish. *Nature* 335:346–48
19. Clark CW. 1976. *Mathematical Bioeconomics: The Optimal Management of Renewable Resources.* New York: Wiley & Sons. 352 pp.
20. Clark CW. 1985. *Bioeconomic Modelling and Fisheries Management.* New York: Wiley & Sons. 291 pp.

21. Clark WC, Jones DD, Holling CS. 1977. Patches, movements, and population dynamics in ecological systems: a terrestrial perspective. In *Spatial Pattern in Plankton Communities*, ed. J Steele, pp. 385–432. New York: Plenum. 470 pp.

22. Cohen D. 1967. Optimizing reproduction in a randomly varying environment when a correlation may exist between the conditions at the time a choice has to be made and the subsequent outcome. *J. Theor. Biol.* 16:1–14

23. Cohen EB, Mountain DG, O'Boyle RN. 1991. Local-scale versus large-scale factors affecting recruitment. *Can. J. Fish. Aquat. Sci.* 48:1003–6

24. Cole LC. 1954. The population consequences of life history phenomena. *Q. Rev. Biol.* 27:103–37

25. Crecco VA, Savoy T. 1984. Effects of fluctuations in hydrographic conditions on year-class strength of American shad (*Alosa sapidissima*) in the Connecticut River. *Can. J. Fish. Aquat. Sci.* 41:1216–23

26. Cury P, Roy C. 1989. Optimal environmental window and pelagic fish recruitment success in upwelling areas. *Can. J. Fish. Aquat. Sci.* 46:670–80

27. Cushing DH. 1973. Dependence of recruitment on parent stock. *J. Fish. Res. Board Can.* 30:1965–76

28. Cushing DH. 1978. Biological effects of climatic change. *Rapp. P.-v. Reun. Cons. Int. Explor. Mer* 173:107–16

29. Cushing DH. 1982. *Climate and Fisheries*. London: Academic. 373 pp.

30. Cushing DH. 1983. Are fish larvae too dilute to affect the density of their food organisms? *J. Plank. Res.* 5:847–54

31. Cushing DH. 1984. The gadoid outburst in the North Sea. *J. Cons. Int. Explor. Mer* 41:159–66

32. Cushing DH. 1990. Plankton production and year-class strength in fish populations: an update of the match/mismatch hypothesis. *Adv. Mar. Biol.* 26:249–93

33. Cushing DH, Harris JGK. 1973. Stock and recruitment and the problem of density dependence. *Rapp. P.-v. Reun. Cons. Int. Explor. Mer* 164:142–55

34. Daag MJ, Clarke ME, Hishiyama T, Smith SL. 1984. Production and standing stock of copepod nauplii, food items for larvae of the walleye pollock *Theragra chalcogramma* in the southeastern Bering Sea. *Mar. Ecol. Prog. Ser.* 19:7–16

35. Daan N, Bromley PJ, Hislop JRG, Nielsen NA. 1990. Ecology of North Sea fish. *Neth. J. Sea Res.* 26:343–86

36. Dahl K. 1907. The scales of herring as a means of determining age, growth and migration. *Rep. Norwegian Fish. Mar. Invest.* 2:1–36

37. Deevey ES. 1947. Life tables for natural populations of animals. *Q. Rev. Biol.* 22:283–314

38. DeLafontaine Y, Leggett WC. 1988. Predation by jellyfish on larval fish: an experimental evaluation employing in situ enclosures. *Can. J. Fish. Aquat. Sci.* 45:1173–90

39. Dickson RR, Brander KM. 1993. Effects of a changing windfield on cod stocks of the North Atlantic. *Fish. Oceanogr.* 2:3/4:124–53

40. Elliott JM. 1987. Population regulation in contrasting populations of trout *Salmo trutta* in two lake district streams. *J. Anim. Ecol.* 56:83–98

41. Elliott JM. 1993. The self-thinning rule applied to juvenile sea-trout, *Salmo trutta*. *J. Anim. Ecol.* 62:371–79

42. Fortier L, Harris RP. 1989. Optimal foraging and density-dependent competition in marine fish larvae. *Mar. Ecol. Prog. Ser.* 51:19–33

43. Frank KT. 1991. Predicting recruitment variation from year class specific vertebral counts: an analysis of the potential and a plan for verification. *Can. J. Fish. Aquat. Sci.* 48:1350–57

44. Frank KT. 1992. Demographic consequences of age-specific dispersal in marine fish populations. *Can. J. Fish. Aquat.* 49:2222–31

45. Frank KT, Carscadden JE. 1989. Factors affecting recruitment variability of capelin (*Mallotus villosus*) in the Northwest Atlantic. *J. Cons. Int. Explor. Mer* 45:146–64

46. Frank KT, Carscadden JE, Leggett WC. 1993. Causes of spatio-temporal variation in the patchiness of larval fish distributions: differential mortality or behaviour? *Fish. Oceanogr.* 2:3/4:114–23

47. Frank KT, Leggett WC. 1981. Wind regulation of emergence times and early larval survival in capelin (*Mallotus villosus*). *Can. J. Fish. Aquat. Sci.* 38:215–23

48. Frank KT, Leggett WC. 1982. Coastal water mass replacement: its effect on zooplankton dynamics and the predator-prey complex associated with larval capelin. *Can. J. Fish. Aquat. Sci.* 39:991–1003

49. Frank KT, Leggett WC. 1983. Multispecies larval fish associations: accident or adaptation? *Can. J. Fish. Aquat. Sci.*: 40:754–62

50. Frank KT, Leggett WC. 1983. Survival value of an opportunistic life-stage transition in capelin (*Mallotus villosus*). *Can. J. Fish. Aquat. Sci.* 40:1442–48

51. Fuiman LA. 1989. Vulnerability of Atlantic herring larvae to predation by yearling herring. *Mar. Ecol. Prog. Ser.* 51:291–99

52. Gauthreaux SA. 1980. *Animal Migration, Orientation and Navigation.* New York: Academic. 387 pp.

53. Giesel JT. 1976. Reproductive strategies as adaptations to life in temporally heterogeneous environments. *Annu. Rev. Ecol. Syst.* 7:57–79

54. Godfray HCJ, Hassell MP. 1992. Long time series reveal density dependence. *Nature* 359:673–74

55. Gordoa A, Duarte CM. 1991. Size-dependent spatial distribution of hake (*Merluccius capensis* and *Merluccius paradonus*) in Namibian waters. *Can. J. Fish. Aquat. Sci.* 48:2095–99

56. Gotelli NJ. 1991. Metapopulation models: the rescue effect, the propagule rain, and the core-satellite hypothesis. *Am. Nat.* 138:768–76

57. Grant JWA. 1993. Self-thinning in stream-dwelling salmonids. *Can. Spec. Publ. Fish. Aquat. Sci.* 118:99–102

58. Gulland JA. 1953. Correlations on fisheries hydrography. *J. Cons. Int. Explor. Mer* 18:351–53

59. Gulland JA. 1982. Why do fish numbers vary? *J. Theor. Biol.* 97:69–75

60. Hall CAS. 1988. An assessment of several of the historically most influential theoretical models used in ecology and of the data provided in their support. *Ecol. Model.* 43:5–31

61. Hankin DG. 1980. A multistage recruitment process in laboratory fish populations: implications for models of fish population dynamics. *Fish. Bull.* 78:555–78

62. Harden Jones FR. 1968. *Fish Migration.* New York: St. Martin's. 325 pp.

63. Harper JL. 1977. *Population Biology of Plants.* London: Academic. 892 pp.

64. Harper JL, Clatworthy JN, McNaughton IH, Sagar GR. 1961. The evolution and ecology of closely related species living in the same area. *Evolution* 15:209–27

65. Hayman RA, Tyler AV. 1980. Environment and cohort strength of Dover sole and English sole. *Trans. Am. Fish. Soc.* 109:54–70

66. Hempel G. 1965. On the importance of larval survival for the population dynamics of marine food fish. *Calif. Coop. Ocean. Fish. Invest.* 10:13–23

67. Hempel G. 1978. North Sea fisheries and fish stocks—a review of recent changes. *Rapp. P.-v. Reun. Cons. Int. Explor. Mer* 173:145–67

68. Hjort J. 1914. Fluctuations in the great fisheries of northern Europe viewed in the light of biological research. *Rapp. P.-v. Reun. Cons. Int. Explor. Mer* 20:1–228

69. Hollowed AB, Bailey KM. 1989. New perspectives on the relationship between recruitment of Pacific hake (*Merluccius productus*) and the ocean environment. *Can. Spec. Publ. Fish. Aquat. Sci.* 108:207–20

70. Horn HS. 1978. Optimal tactics of reproduction and life-history. In *Behavioural Ecology: an Evolutionary Approach,* ed. JR Krebs, NB Davies, pp. 411–29. Oxford: Blackwell Sci. 494 pp.

71. Houde ED. 1987. Fish early life dynamics and recruitment variability. *Am. Fish. Soc. Symp.* 2:17–29

72. Houde ED. 1989. Comparative growth, mortality, and energetics of marine fish larvae: temperature and implied latitudinal effects. *Fish. Bull.* 87:471–95

73. Hovgard H, Buch E. 1990. Fluctuations in the cod biomass off the West Greenland Sea ecosystem in relation to climate. In *Large Marine Ecosystems: Patterns, Processes, and Yields,* ed. K Sherman, LM Alexander, BD Gold, pp. 36–43. Washington, DC: Am. Assoc. Advanc. Sci. 242 pp.

74. Hutchinson GE. 1978. *An Introduction to Population Ecology.* New Haven, London: Yale Univ. Press. 260 pp.

75. Itô Y. 1980. *Comparative Ecology.* Cambridge: Cambridge Univ. Press. 436 pp.

76. Janzen DH. 1971. Seed predation by animals. *Annu. Rev. Ecol. Syst.* 2:465–92

77. Janzen DH. 1976. Why bamboos wait so long to flower. *Annu. Rev. Ecol. Syst.* 7:347–92

78. Jenkins GP. 1987. Comparative diets, prey selection, and predatory impact of co-occurring larvae of two flounder species. *J. Exp. Mar. Biol. Ecol.* 110:147–70

79. Johnson ML, Gaines MS. 1990. Evolution of dispersal: theoretical models and empirical tests using birds and mammals. *Annu. Rev. Ecol. Syst.* 21:449–80

80. Jones R. 1973. Density dependent regulation of the numbers of cod and haddock. *Rapp. P.-v. Reun. Cons. Int. Explor. Mer* 164:156–73

81. Kaartvedt S, Aksnes DL, Egge JK. 1987. Effect of light on the vertical distribution of *Pecten maximus* larvae. *Mar. Ecol. Prog. Ser.* 40:195–97

82. Karban R. 1982. Increased reproductive success at high densities and predator satiation for periodical cicadas. *Ecology* 63:321–28

83. Kikkawa J. 1977. Ecological paradoxes. *Aust. J. Ecol.* 2:121–36

84. Kingsford MJ. 1985. The demersal eggs and planktonic larvae of *Chromis dispilus* (Teleostei: Pomacentridae) in north-eastern New Zealand coastal waters. *N. Zeal. J. Mar. Freshw. Res.* 19: 429–38

85. Larkin PA. 1977. An epitaph for the concept of maximum sustained yield. *Trans. Am. Fish. Soc.* 106:1–11

86. Larkin PA. 1978. Fisheries management—an essay for ecologists. *Annu. Rev. Ecol. Syst.* 9:57–73

87. Larkin PA, Raleigh RF, Wilimovsky NJ. 1964. Some alternative premises for constructing theoretical reproduction curves. *J. Fish. Res. Board Can.* 21:477–84

88. Lasker R. 1975. Field criteria for survival of anchovy larvae: the relation between inshore chlorophyll maximum layers and successful first feeding fish. *Fish. Bull.* 73:453–62

89. Leggett WC. 1977. The ecology of fish migrations. *Annu. Rev. Ecol. Syst.* 8: 285–308

90. Leggett WC. 1977. Density-dependence, density-independence and recruitment in the American shad (*Alosa sapidissima*) population of the Connecticut River. In *Proc. Conf. Assessing the Effects of Power-Plant-Induced Mortality on Fish Populations*, ed. W. Van Winkle, pp. 3–17. New York: Pergamon. 380 pp.

91. Leggett WC. 1984. Fish migrations in coastal and estuarine environments: a call for new approaches to the study of an old problem. In *Mechanisms of Migration in Fishes*, ed. GP Arnold, JJ Dodson, WH Neil, pp. 159–78. New York: Plenum. 574 pp.

92. Leggett WC. 1985. The role of migrations in the life history evolution of fish. *Contr. Mar. Sci.* (Suppl.) 27:277–95

93. Leggett WC, Carscadden JE. 1978. Latitudinal variation in reproductive characteristics of American shad (*Alosa sapidissima*): evidence for population specific life history strategies in fish. *J. Fish. Res. Board Can.* 35:1469–78

94. Leggett WC, DeBlois E. 1994. Ho: recruitment in marine fishes is not regulated by starvation and predation in the egg and larval stages. *Neth. J. Sea Res.* In press

95. Leggett WC, Frank KT, Carscadden JE. 1984. Meteorological and hydrographic regulation of year-class strength in capelin (*Mallotus villosus*). *Can. J. Fish. Aquat. Sci.* 41:1193–201

96. Litvak MK, Leggett WC. 1992. Age and size selective predation on larval fishes: the bigger is better paradigm revisited. *Mar. Ecol. Prog. Ser.* 81:13–24

97. Lloyd M, Dybas HS. 1966. The periodical cicada problem. I. Population ecology. *Evolution* 20:133–49

98. Loftus KH. 1976. Science for Canada's fisheries rehabilitation need. *J. Fish. Res. Board Can.* 33:1822–57

99. MacCall AD. 1990 *Dynamic Geography of Marine Fish Populations*. Seattle: Univ. Wash. Press. 153 pp.

100. MacKenzie BR, Leggett WC. 1991. Quantifying the contribution of small scale turbulence to the encounter rates between larval fish and their zooplankton prey: the effects of wind and tide. *Mar. Ecol. Prog. Ser.* 73:149–60

101. MacKenzie BR, Miller TJ, Cyr S, Leggett WC. 1994. Evidence for a dome shaped relationship between turbulence and larval fish ingestion. *Limnol. Oceanogr.* In press

102. Magnuson JJ. 1988. Two worlds for fish recruitment: lakes and oceans. *Am. Fish. Soc. Symp.* 5:1–6

103. Mangel M. 1985. *Decision and Control in Uncertain Resource Systems*. New York: Academic. 255 pp.

104. Mann R. 1982. The seasonal cycle of gonadal development in *Arctica islandica* from the southern New England shelf. *Fish. Bull.* 80:315–26

105. Mann KH, Lazier JRN. 1991. *Dynamics of Marine Ecosystems*. Boston: Blackwell Sci. 466 pp.

106. Mann R, Wolf CC. 1983. Swimming behaviour of larvae of the ocean quahog *Arctica islandica* in response to pressure and temperature. *Mar. Ecol. Prog. Ser.* 13:211–18

107. Matlock GC. 1987. The role of hurricanes in determining year-class strength of red drum. *Contr. Mar. Sci.* 30:39–47

108. McCleave JD, Arnold GP, Dodson JJ, Neill WH. 1984. *Mechanisms of Migration in Fishes*. New York: Plenum. 574 pp.

109. Monteleone DM, Peterson WT. 1986. Feeding ecology of American sand lance *Ammodytes americanus* larvae from Long Island Sound. *Mar. Ecol. Prog. Ser.* 30:133–43

110. Myers RA, Cadigan NG. 1993. Density-dependent juvenile mortality in marine demersal fish. *Can. J. Fish. Aquat. Sci.* 50:1576–90

111. Myers RA, Drinkwater K. 1989. The influence of Gulf Stream warm core rings on recruitment of fish in the northwest Atlantic. *J. Mar. Res.* 47:635–56

112. Myers RA, Drinkwater KF, Barrowman NJ, Baird JW. 1993. Salinity and recruitment of Atlantic cod (*Gadus morhua*) in the Newfoundland region. *Can. J. Fish. Aquat. Sci.* 50:1599–609

113. Naidu S. 1970. Reproduction and breeding cycle of the giant scallop in Port au Port Bay, Newfoundland. *Can. J. Zool.* 48:1003–12

114. Nelson WR, Ingham M, Schaaf WE. 1977. Larval transport and year class strength of Atlantic menhaden, *Brevoortia tyrannus*. *Fish. Bull.* 75:23–41

115. Parrish RH, Nelson CS, Bakun A. 1981. Transport mechanisms and reproductive success of fishes in the California Current. *Biol. Oceanogr.* 1:175–203

116. Paulik GJ. 1973. Studies of the possible form of the stock-recruitment curve. *Rapp. P.-v. Reun. Cons. Int. Explor. Mer* 164:303–15

117. Pepin P, Shaers TH, DeLafontaine Y. 1992. The significance of body size to the interaction between a larval fish (*Mallotus villosus*) and a vertebrate predator (*Gasterosteus aculeatus*). *Mar. Ecol. Prog. Ser.* 81:1–12

118. Peterman RM, Bradford MJ. 1987. Wind speed and mortality rate of a marine fish, the northern anchovy (*Engraulis mordax*). *Science* 235:354–56

119. Peterman RM, Bradford MJ, Lo NCH, Methot RD. 1988. Contribution of early life stages to interannual variability in recruitment of northern anchovy (*Engraulis mordax*). *Can. J. Fish. Aquat. Sci.* 45:8–16

120. Peters RH. 1991. *A Critique for Ecology.* Cambridge: Cambridge Univ. Press. 366 pp.

121. Petersen CGJ. 1896. The early immigration of young plaice into the Limfjord from the German Sea. *Rep. Danish Biol. Sta.* 6:1–77

122. Peterson WT, Ausubel SJ. 1984. Diets and selective feeding by larvae of Atlantic mackerel *Scomber scombrus* on zooplankton. *Mar. Ecol. Prog. Ser.* 17:65–75

123. Pope JG. 1973. An investigation into the effects of variable rates of the exploitation of fishery resources. In *The Mathematical Theory of the Dynamics of Biological Populations,* ed. MS Bartlet, RW Hiorns, pp 23–34. London: Academic. 347 pp.

124. Potts GW, Wootton RJ. 1984. *Fish Reproduction: Tactics and Strategies.* London: Academic. 410 pp.

125. Pulliam HR. 1988. Sources, sinks, and population regulation. *Am. Nat.* 132:652–61

126. Qasim SZ, Rice AL, Knight-Jones EW. 1963. Sensitivity to pressure changes in teleosts lacking swim-bladders. *J. Mar. Biol. Assoc. India* 5:289–93

127. Ricker WE. 1954. Stock and recruitment. *J. Fish. Res. Board Can.* 11:559–623

128. Rijnsdorp AD, Daan N, van Beek FA, Heessen HJL. 1991. Reproductive variability in North Sea plaice, sole, and cod. *J. Cons. Int. Explor. Mer* 47:352–75

129. Rivard D, Maguire J-J. 1993. Reference points for fisheries management: the eastern Canadian experience. *Can. Spec. Publ. Fish. Aquat. Sci.* 120:31–57

130. Roff DA. 1988. The evolution of migration and some life history parameters in marine fishes. *Environ. Biol. Fish.* 22:133–46

131. Roff DA. 1992. *The Evolution of Life Histories: Theory and Analysis.* New York: Chapman & Hall. 535 pp.

132. Rose GA. 1993. Cod spawning on a migration highway in the north-west Atlantic. *Nature* 366:458–61

133. Rose GA, Leggett WC. 1989. Interactive effects of geophysically-forced sea temperatures and prey abundance on mesoscale coastal distributions of a marine predator, Atlantic cod (*Gadus morhua*). *Can. J. Fish. Aquat. Sci.* 46:1904–13

134. Rose GA, Leggett WC. 1989. Predicting variability in catch per effort in Atlantic cod (*Gadus morhua*) trap and gillnet fisheries. *J. Fish. Biol.* 35 (Suppl. A):155–61

135. Rose GA, Leggett WC. 1990. The importance of scale to predator-prey spatial correlations. *Ecology* 71:33–43

136. Rose GA, Leggett WC. 1991. Effects of biomass-range interactions on catchability of migratory demersal fish by mobile fisheries: an example of Atlantic cod (*Gadus morhua*). *Can. J. Fish. Aquat. Sci.* 48:843–48

137. Rothschild BJ. 1986. *Dynamics of Marine Fish Populations.* Cambridge: Harvard Univ. Press. 277 pp.

138. Rothschild BR, Osborn TR. 1988. Small-scale turbulence and plankton contact rates. *J. Plank. Res.* 10:465–74

139. Schaffer WM, Elson PF. 1975. The adaptive significance of variations in life history among local populations of Atlantic salmon in North America. *Ecology* 56:577–90

140. Schnute J. 1985. A general theory for analysis of catch and effort data. *Can. J. Fish. Aquat. Sci.* 42:414–29

141. Shepherd JG, Cushing DH. 1980. A mechanism for density dependent survival of larval fish as the basis of a stock-recruitment relationship. *J. Cons. Int. Explor. Mer* 39:160–67

142. Shepherd JG, Cushing DH. 1990. Regulation in fish populations: myth or mi-

rage? *Philos. Trans. R. Soc. Lond. B*: 1990: 29–42

143. Shepherd JG, Pope JG, Cousens RD. 1984. Variations in fish stocks and hypotheses concerning their links with climate. *Rapp. P.-v. Reun. Cons. Int. Explor. Mer* 185:255–67

144. Sinclair M. 1988. *Marine Populations: An Essay on Population Regulation and Speciation.* Seattle: Univ. Wash. Press. 252 pp.

145. Sinclair M, Maguire JJ, Koeller P, Scott JS. 1984. Trophic dynamics models in light of current resource inventory data and stock assessment results. *Rapp. P.-v. Reun. Cons. Int. Explor. Mer* 183:269–84

146. Skud BE. 1982. Dominance in fishes: the relation between environment and abundance. *Science* 216:144–49

147. Smith RJF. 1985. *The Control of Fish Migration.* Berlin: Springer-Verlag. 243 pp.

148. Snyder RJ. 1991. Migration and life histories of the threespine stickleback: evidence for adaptive variation in growth rate between populations. *Environ. Biol. Fish.* 31:381–88

149. Solbreck C. 1985. Insect migration strategies and population dynamics. *Contr. Mar. Sci.* 27(Suppl):641–62

150. Solemdal P, Sinclair M. 1989. Johan Hjort—founder of modern Norwegian fishery research and pioneer in recruitment thinking. *Rapp. P.-v. Reun. Cons. Int. Explor. Mer* 191:339–44

151. Southwood TRE. 1981. Ecological aspects of insect migration. In *Animal Migration,* ed. DJ Aidley, pp. 197–208. Cambridge: Cambridge Univ. Press. 264 pp.

152. Stearns SC. 1976. Life-history tactics: a review of the ideas. *Q. Rev. Biol.* 51:3–47

153. Stearns SC. 1977. The evolution of life history traits. *Annu. Rev. Ecol. Syst.* 8:145–71

154. Stearns SC. 1980. A new view of life-history evolution. *Oikos* 35:266–81

155. Stearns SC. 1983. A natural experiment in life-history evolution: field data on the introduction of mosquitofish (*Gambusia affinis*) to Hawaii. *Evolution* 37:601–17

156. Stocker M, Haist V, Fournier D. 1985. Environmental variation and recruitment of Pacific herring (*Clupea harengus pallasi*) in the Strait of Georgia. *Can. J. Fish. Aquat. Sci.* 42 (Suppl. 1):174–80

157. Taggart CT, Frank KT. 1987. Coastal upwelling and Oikopleura occurrence ("slub"): a model and potential applica-

tion to inshore fisheries. *Can. J. Fish. Aquat. Sci.* 44:1729–36

158. Thompson KR, Page FH. 1989. Detecting synchrony of recruitment using short, autocorrelated time series. *Can. J. Fish. Aquat. Sci.* 46:1831–38

159. Tuck GN, Possingham HP. 1994. Optimal harvesting strategies for a metapopulation. *Bull. Math. Biol.* 56: 107–27

160. Tyler AV. 1992. A context for recruitment correlations: why marine fisheries biologists should still look for them. *Fish. Oceanogr.* 1:97–107

161. Walker BW. 1952. A guide to the grunion. *Calif. Fish Game* 38:409–20

162. Walters CJ. 1986. *Adaptive Management of Renewable Resources.* New York: MacMillan. 374 pp.

163. Walters CJ. 1987. Nonstationarity of production relationships in exploited populations. *Can. J. Fish. Aquat. Sci.* 44(Suppl. 2):156–65

164. Walters CJ, Collie JS. 1988. Is research on environmental factors useful to fisheries management? *Can. J. Fish. Aquat. Sci.* 45:1848–54

165. Walters CJ, Juanes F. 1993. Recruitment limitation as a consequence of natural selection for use restricted feeding habitats and predation risk taking by juvenile fishes. *Can. J. Fish. Aquat. Sci.* 50: 2058–70

166. Walters CJ, Ludwig D. 1981. Effects of measurement errors on the assessment of stock-recruitment relationships. *Can. J. Fish. Aquat. Sci.* 38:704–10

167. Ware DM. 1975. Relation between egg size, growth, and natural mortality of larval fish. *J. Fish. Res. Board Can.* 32:2503–12

168. Ware DM. 1980. Bioenergetics of stock and recruitment. *Can. J. Fish. Aquat. Sci.* 37:1012–24

169. Warner RR, Chesson PL. 1985. Coexistence mediated by recruitment fluctuations: a field guide to the storage effect. *Am. Nat.* 125:769–87

170. Welch DW. 1986. Identifying the stock-recruitment relationship for age-structured populations using time-invariant matched linear filters. *Can. J. Fish. Aquat. Sci.* 43:108–23

171. White PS. 1979. Pattern, process, and natural disturbance in vegetation. *Bot. Rev.* 45:229–99

172. White TCR. 1976. Weather, food and plagues of locusts. *Oecologia* 22:119–34

173. Wilbur HM. 1980. Complex life cycles. *Annu. Rev. Ecol. Syst.* 11:67–93

174. Winemiller KO, Rose KA. 1992. Patterns of life-history diversification in

North American fishes: implications for population regulation. *Can J. Fish. Aquat. Sci.* 49:2196–218

175. Wolda H. 1988. Insect seasonality: why? *Annu. Rev. Ecol. Syst.* 19:1–18

176. Wooton RJ. 1990. *Ecology of Teleost Fishes.* London: Chapman & Hall. 404 pp.

177. Wourms JP. 1972. The developmental biology of annual fishes. III. Pre-embryonic and embryonic diapause of variable duration in the eggs of annual fishes. *J. Exp. Zool.* 182:389–414

178. Wynne-Edwards VC. 1962. *Animal Dispersion in Relation to Social Behaviour.* Edinburgh: Oliver & Boyd. 653 pp.

Annu. Rev. Ecol. Syst. 1994. 25:423–41

ECOLOGY AND EVOLUTION OF REPRODUCTION IN MILKWEEDS

Robert Wyatt

Institute of Ecology, University of Georgia, Athens, Georgia 30602

Steven B. Broyles

Department of Biological Sciences, State University of New York College at Cortland, Cortland, New York 13045

KEY WORDS: flowers, fruit-set, pollination, self-incompatibility, nectar

Abstract

Asclepiadaceae are the dicot counterparts to the Orchidaceae, which also transmit their pollen grains in large groups within pollinia. Unlike many terrestrial, nectar-producing orchids, however, milkweeds are characterized by low fruit-set, typically averaging 1–5%. Transfer of hundreds of pollen grains as a unit makes it possible to quantify pollinator activity and male and female reproductive success more directly and more easily in milkweeds than in plants with loose pollen grains. It also leads to the production of fruits whose seeds all share a single father, thus simplifying paternity analysis. Recent anatomical work has demonstrated that three of the five stigmatic chambers of milkweed flowers transmit pollen tubes to one of the two separate ovaries, whereas the other two chambers transmit only to the second ovary. Milkweed flowers are long-lived and produce copious nectar, which flows from nectaries within the stigmatic chambers to fill the hoods, which serve as reservoirs. Nectar also serves as the germination fluid for pollen grains, but concentrations above 30% inhibit germination. Most milkweeds are genetically self-incompatible and express an unusual late-acting form of ovarian rejection. Some weedy milkweeds, however, are self-compatible, and levels of self-insertion of pollinia are apparently high in these, as well as in self-incompatible, species. Early

423

0066-4152/94/1120-0423$05.00

attempts to explain the evolution of inflorescence size in milkweeds were hampered by failure to consider the genetic basis of the variation observed and by failure to determine the unit on which selection should act. Direct tests of the "pollen donation" hypothesis have cast doubt on the validity of the view that flower number and other floral traits evolved primarily to enhance male reproductive success. Milkweeds are pollinated by a diverse array of large bees, wasps, and butterflies, and these generalist pollinators effect extensive gene flow within and between populations, augmented by wind dispersal of comose seeds. Morphological and biochemical evidence support the view that limited, localized hybridization occurs between sympatric species of milkweeds.

INTRODUCTION TO MILKWEEDS AS A MODEL SYSTEM

The remarkably complex flowers of milkweeds (Asclepiadaceae) provide a number of unusual advantages to students of plant reproductive ecology. In many respects the Asclepiadaceae are the dicot equivalents of the Orchidaceae. Both families are unique within their respective classes in that their pollen grains cohere and transport as a unit, termed a "pollinium." Delivery of pollen in discrete packets has numerous reproductive consequences. It makes it possible to quantify the results of pollinator activity easily and directly: One need only count the numbers of pollinia removed from a flower or inserted into a flower to estimate the success of a hermaphroditic plant as a pollen donor (male function) or as a pollen receiver (female function). It also simplifies the counting of pollen units carried by the pollinators themselves. Coherence of the pollen into discrete packages makes it possible to use genetic markers to genotype individual pollinia following dispersal events. Moreover, delivery of hundreds of pollen grains to the stigma at one time ensures that more than enough grains are available to fertilize all of the ovules within an ovary. This results in consistently high seed numbers. Finally, this form of pollen delivery also leads to the production of fruits whose seeds all share a single father. This greatly simplifies paternity analysis, making the milkweed system ideal for studies focusing on sexual selection or gene flow.

Because milkweed pollinia must conform in size and shape to the stigmatic chambers into which they are inserted, there is the potential for mechanical barriers to interspecific hybridization. Such barriers to crossing are otherwise almost unknown in flowering plants. In *Asclepias* another unusual feature is the presence of a cucullate corona including hoods, which function as nectar reservoirs. This makes species of *Asclepias* ideal subjects for observations and experiments relating to nectar production, as the nectar can be easily removed and measured repeatedly. Moreover, the nectar is almost pure sucrose and represents the only reward for pollinating insects (the pollen being protected by the tough durable wall of the pollinium). Flowers of Asclepiadaceae also are unusual in consisting of two separate ovaries. This enables the study of resource competition not only between flowers but also within flowers. A final characteristic of the milkweed system is that most species appear to have an unusual form of late-acting self-incompatibility. Such systems appear to be rare in the angiosperms as a whole, and their functioning and genetic consequences are poorly understood.

Of course, there are also characteristics of milkweeds that make them undesirable subjects for some kinds of studies in reproductive ecology. One of these is the very low rate of fruit-set in natural populations (typically about 1% to 5%). Although fruit-set is higher from hand cross-pollinations, it never approaches 100%. This makes it difficult and time-consuming to carry out experiments designed to assess breeding systems, ovary competition, factors limiting fruit-set, etc. Another characteristic of milkweeds that might prove disadvantageous in some contexts is the fact that, unlike the case in many orchids, there is little specialization of different species for pollination by specific insects. For example, virtually any insect of appropriate size and behavior can effectively remove and insert pollinia of any species of *Asclepias*. Thus, the apparent high degree of coadaptation between some orchids and their species-specific pollinators does not exist in *Asclepias*.

STRUCTURAL AND FUNCTIONAL ASPECTS OF FLOWERS

Floral Morphology and Anatomy

Flowers of *Asclepias* consist of five showy, reflexed petals covering five smaller, usually green sepals (Figure 1A). Two separate, superior ovaries are united by their styles to form a gynostegium with five lateral stigmatic surfaces (Figure 1B, C). These surfaces are enclosed by the tightly abutting wings of adjacent anthers to produce five stigmatic chambers. From the bases of the five stamens extend the hoods, each of which usually contains an arching horn and which serve as reservoirs for the nectar secreted by nectaries located within the stigmatic chambers (25a). Together the hoods and horns comprise the corona (Figure 1A, B). There are five pollinaria (36), each of which consists of paired pollinia from adjacent anthers joined by translator arms to a corpusculum that sits just above the alar fissure, a narrow opening into the stigmatic chamber (Figure 1A, B, D).

Milkweed pollination is a two-stage process: (i) removal of a pollinarium occurs when a groove in the corpusculum catches on a bristle or other appendage of an insect and is forcibly pulled from the flower, and (ii) insertion is

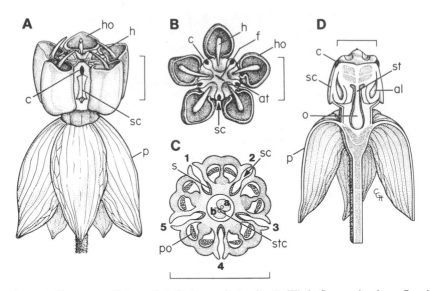

Figure 1 Flower morphology of *Asclepias amplexicaulis*. A. Whole flower, showing reflexed petals, corona of hoods and horns, and one surface of the gynostegium. B. Top view, showing location of the hoods and horns and entrances to the stigmatic chambers relative to the furrow in the gynostegium. C. Transverse section, showing the location of the stigmatic chambers relative to the styles. Chambers 1, 2, and 3 transmit pollen tubes to style a, whereas chambers 4 and 5 transmit pollen tubes to style b. D. Longitudinal section along the axis of the furrow, showing the location of the gynostegium relative to the style and ovary. In all figures, the scale delimits 3 mm. Abbreviations are: al, anther locule; at, anther tip; c, corpusculum; f, furrow; h, horn; ho, hood; o, ovary; p, petal; po, pollinium; s, stigmatic surface; sc, stigmatic chamber; st, style; and stc, stylar canal.

effected when a pollinium lodges in a stigmatic chamber (79). Following successful insertion, pollen tubes emerge from a thin-walled area along the convex surface of the pollinium (26, 66). Pollen tubes subsequently grow down the stylar canal and finally enter the ovary (17, 25).

Until recently, the relationship between the five stigmatic chambers and two subtending ovaries with respect to the pathway of pollen tube transmission was a mystery. It is now known that three adjacent stigmatic chambers transmit pollen tubes to one of the two separate ovaries, whereas the other two chambers transmit to the second ovary (58). These observations confirm Woodson's (77) expectations. Despite the obvious potential for pollen tubes to cross over at the point of fusion of the two styles, this was never observed in *A. amplexicaulis*. Out of hundreds of observations, Sparrow & Pearson (65) detected only one case of such "crossing-over" in *A. syriaca*. Morse (43), however, observed the phenomenon more commonly in natural populations of *A. syriaca* from Maine. He reported that nearly 2% of all successful hand pollinations with a single pollinium produced twin follicles, indicating fertilization of both ovaries by pollen tubes from a single stigmatic chamber.

Self-Incompatibility Systems

Early workers believed that all species of *Asclepias* are self-incompatible (17, 18, 28). Experimental crosses in *A. syriaca* by Moore (38) and Sparrow & Pearson (65) supported this view, and Woodson (77) arbitrarily discounted reports of successful self-pollination in this species (52, 67) and in *A. incarnata* (24). More recently, however, Kephart (30) and Kahn & Morse (29) reported low levels (< 5%) of self-compatibility in *A. syriaca.* Wyatt (78) reported that 2% of self-pollinations of *A. tuberosa* resulted in fruit-set, and Kephart (30) reported 29% success for self-pollinations of *A. incarnata.* Our recent work has uncovered two additional species that appear to be fully self-compatible: *A. curassavica,* in which 23.3% of self-pollinations were successful versus 25.9% of cross-pollinations; and *A. fruticosa,* in which 12.7% of self-pollinations set fruit versus 17.5% of cross-pollinations (R Wyatt, SB Broyles, unpublished data).

Experimental crosses in *A. perennis* and *A. texana* have revealed that these two species are completely self-incompatible (R Wyatt, AL Edwards, SR Lipow, CT Ivey, unpublished data), as is *A. verticillata* (30). *Asclepias subulata* also is largely self-incompatible: Only 1 of 99 self-pollinations resulted in fruit-set (R Wyatt, CT Ivey, SR Lipow, unpublished data). It appears, therefore, that those milkweeds investigated to date fall clearly into two distinct categories: (i) those that are largely or entirely self-incompatible (*A. exaltata, A. perennis, A. subulata, A. syriaca, A. texana, A. tuberosa,* and *A. verticillata*), and (ii) those that are largely or entirely self-compatible (*A. curassavica, A. fruticosa,* and *A. incarnata*).

It appears from the limited information available that self-incompatible species of *Asclepias* possess an unusual form of genetic self-incompatibility. Traditionally, two basic forms of self-incompatibility (SI) have been recognized: (i) sporophytic (SSI), in which pollen fails to germinate on an incompatible stigma; and (ii) gametophytic (GSI), in which incompatible pollen germinates, but pollen tube growth is arrested in the style. Recent reviews have argued, however, that other self-incompatibility systems exist that do not conform to the classically defined GSI and SSI models (57, 59). In many plants, it appears that SI may act very late, so that the incompatibility reaction occurs in the ovary. These ovarian self-incompatibility systems (OSI) can be categorized on the basis of whether pollen tube rejection occurs before or after the ovules are penetrated, whether syngamy occurs, and whether ovular inhibition is involved. Most of the details of OSI have not been worked out for any species of flowering plant, but it appears that OSI is typical of milkweeds (12,

29, 30, 65). It is intriguing that so inefficient a system exists in the morphologically derived genus *Asclepias.*

CAUSES OF LOW FRUIT-SET

Fruit-set is generally very low in natural populations of milkweeds, with averages ranging from 0.33% to 5.0% (84). Attempts to explain these low values have centered on two hypotheses: (i) resources to mature fruits are limiting, and (ii) insufficient numbers of compatible pollinia are reaching stigmatic chambers. Those favoring resource limitation have pointed out that pollinator activity is frequently high in milkweed populations and that abortion of apparently fertilized fruits is common (54, 55, 73, 75). Moreover, addition of inorganic fertilizer increased, whereas shading and leaf removal decreased, fruit-set in *A. syriaca* and *A. verticillata* (74). Application of fertilizer also increased fruit-set in *A. exaltata* (55). Chaplin & Walker (14) concluded that resources stored in the taproot of *A. quadrifolia* controlled flower and fruit production. All of these studies, however, are complicated to interpret because of possible side effects of the experimental treatments (especially addition of water) on nectar production (86), which, in turn, could have increased pollinator visitation. On the other hand, shading could have had a direct negative effect on pollinator activity.

Many populations of milkweeds undoubtedly receive high levels of pollination (13, 36, 60, 71). There is some question, however, as to how many of these apparently successful pollinium insertions represent self-pollinations or otherwise incompatible crosses. Hand pollinations increased fruit-set from natural levels of 0.33% to 14.8% in *A. tuberosa* (78, 82) and from 2.5% to 19.7% in *A. exaltata* (55). Morse & Fritz (44) argued that *A. syriaca* is pollen-limited, because they were able to double fruit-set by supplementing natural levels of pollination. Unfortunately, as with tests of the resource limitation hypothesis, all of these tests for pollen limitation are flawed in design (89). Wyatt (78, 80, 82, 83) has argued for a compromise position, noting that under different sets of circumstances, either pollination or resources can limit fruit-set in milkweeds.

NECTAR PRODUCTION

Compared to other plants, milkweeds have long-lived flowers that produce copious quantities of nectar. On average, individual flowers of *A. tuberosa* are reproductive for 7.4 + 0.34 (mean + standard deviation) days (82); of *A. exaltata,* 6.2 + 0.85 days (88); of *A. incarnata,* 4.87 + 1.66 or 3.87 + 1.04 days (early- versus late-season: Ref. 32); of *A. syriaca,* 5.18 + 1.24 or 5.30 + 0.82 days (32); and of *A. verticillata,* 6.30 + 2.02 or 5.14 + 1.18 days (32). Flowers that last 4 to 8 days are long-lived compared to most other flowering plants (53).

Cross-comparisons of nectar production data are complicated because different workers use different sampling methods (86). In *A. exaltata,* an average flower produced 63.5 µl of nectar over its 6-day life span, a net production considerably higher than that for *A. verticillata* (72), *A. quadrifolia* (49), *A. curassavica* (47, 81), or *A. syriaca* (40, 63, 71). In Southwick's (63) population of *A. syriaca,* for example, total nectar production ranged from 3.8 to 17.8 µl of nectar per flower. Similarly, the 23.6 mg of sucrose produced over 6 days by *A. exaltata* (88) greatly exceeds sugar production by *A. syriaca* (63, 64, 71), *A. quadrifolia* (49), or *A. verticillata* (72).

In *A. exaltata* (86, 88), *A. verticillata* (3, 71), and *A. syriaca* (63, 72), most nectar is secreted overnight. In *A. quadrifolia,* however, nectar production peaked in the morning and was very low at night. Nectar concentrations in *A. exaltata* and *A. syriaca* are typically low in the morning, averaging < 30%, but they increase steadily to 40–60% late in the afternoon (86, 88). These diurnal changes are strongly associated with temperature and relative humidity and are apparently caused by passive evaporation of the nectar within the open hoods (88). Similar increases in nectar concentration have been reported in *A. syriaca* (63, 71) and *A. verticillata* (72).

In *A. exaltata,* plants that produced more concentrated nectar had greater reproductive success (88). Nectar concentration was positively correlated with both the number of pollinia inserted per flower and the number of pollinaria removed per flower. Moreover, plants that produced concentrated nectar matured more fruits and had higher levels of fruit-set than plants that produced dilute nectar.

POLLEN VIABILITY AND GERMINATION

An unusual feature of milkweed reproduction is germination of the pollen in a nectar solution secreted within the stigmatic chamber. Among others, Shannon & Wyatt (61) reported that a 30% sucrose solution yielded highest germination of pollinia of *A. exaltata.* In *A. syriaca,* germination is inhibited by sucrose concentrations above 30% (20), and many failures of hand-pollinations may be due to inhibition of pollen germination by concentrated nectar within the stigmatic chamber (58, 84). Recently it has been shown that contamination of the germination fluid (nectar) by growth of microorganisms can inhibit germination (21, 22). Although these observations were made under laboratory conditions, these yeasts are natural contaminants of milkweed nectar and may be transmitted between flowers by insect pollinators.

Morse (40) speculated that the durable covering of milkweed pollinia should allow a long residence time on pollinators, thus enhancing pollen dispersal distances. This assumes that pollinia are resistant to desiccation and lose viability only slowly following their removal from flowers. In *A. syriaca,* pollinia appear to retain high germinability for at least 4 days under natural conditions (20). In vitro pollen germination experiments using *A. exaltata,* however, showed a loss of viability after 24 hr to about 50% of the original value (61). Pollinia of milkweed species, such as *A. syriaca,* that inhabit open sites may be more resistant to desiccation than are those of species such as *A. exaltata* that grow in moist forests and meadows.

It appears that pollen viability does not decline significantly merely as a function of flower age (42, 61). Stigma receptivity, however, decreased more than threefold over the five-day life span of flowers of *A. syriaca* (42). The possible repercussions of these effects are complicated because of the high degree of variation among plants in pollen germination and apparent fertilization in both *A. exaltata* (61) and *A. syriaca* (29).

LEVELS OF SELF-POLLINATION

Because most milkweeds are self-incompatible (see above), self-pollination (i.e. insertion of a pollinium into a stigmatic chamber of the same genetic individual) can be another cause of low fruit-set. This is an especially serious problem for milkweeds, as their flowers contain only five stigmatic chambers into which compatible pollinia can be inserted. Thus, "stigma clogging" by incompatible pollinia is very likely. Moreover, the late-acting incompatibility system of milkweeds opens up the possibility of ovules being preempted by fertilizations involving incompatible pollen tubes.

It has proved technically very difficult to estimate levels of self-pollination in milkweed populations. Conventional techniques for marking and following pollen dispersal (e.g. pollen-analogue dyes) do not work for milkweed pollinia. Pleasants & Ng (51) estimated levels of self-pollination in *A. syriaca* by comparing numbers of insertions in emasculated umbels to those in umbels with intact corpuscula. Over a range of umbel sizes, they calculated that 36% of inserted pollinia were self-insertions. There are problems, however, with the assumption that emasculation has only the effect of removing a source of self-pollen. Wyatt (79) has shown in similiar emasculation experiments on *A. tuberosa* that the presence of an intact corpusculum increases the likelihood of successful insertion. Thus, an alternative explanation of Pleasants & Ng's (51) result is simply that removal of pollinaria decreased the overall level of successful insertions. Using Pleasants & Ng's (51) interpretation, Wyatt's (79) data for *A. tuberosa* yield an estimate of 27.1% self-insertion.

After developing a technique for radioactively labelling pollinia (50), Pleasants (48) measured levels of self-insertion in a field plot of *A. syriaca*. Of 38 insertions into an umbel labelled with ^{14}C, 14 (37%) were from the labelled umbel itself. Moreover, most pollen dispersal occurred over short distances, suggesting that self-pollinations between stems of the same genetic individual might be very common for this species, which produces large clonal patches from gemmiferous roots. Effective levels of self-pollination are likely to be extremely high and could be a major factor limiting fruit-set.

Some recent studies suggest that self-inserted pollinia interfere with out-cross-pollinia and prevent them from entering the stigmatic chambers and penetrating the style. It is well-established that self-pollen germinates and penetrates ovaries as quickly as outcross-pollen in milkweed flowers (29). Competition among self- and outcross-pollen tubes reduces the number of ovules effectively fertilized by compatible sperm. Pollen competition studies in *A. exaltata* have demonstrated that self-pollen reduces the ability of cross-pollen to mature fruits by 49% when the self-pollination is performed simul-taneously with the cross-pollination, and by 81% when the self-pollination occurred 24 hr before the cross-pollination (12). In addition, fruit-set decreased 29% even when the self-pollination occurred 24 hr after the cross-pollination.

The loss in potential female reproductive success due to self-pollination may be very high. Seed-set in fruits maturing from flowers that were self-pollinated 24 hr prior to cross-pollination produced 37% fewer seeds. If we assume that self- and outcross-pollen compete in 30% of the ovaries and that self-pollina-tion within plants is approximately 66% (as determined for a self-incompatible milkweed like *A. syriaca:* 62), then self-pollination will reduce fruit production in about 20% of all ovaries that also receive compatible cross-pollinations. Thus, the impact of self-pollination on seed and fruit production is potentially great on flowers that may have also received compatible pollinia.

POLLEN DISPERSAL

Milkweed pollinia are dispersed by a diverse array of Hymenoptera and Lep-idoptera (31, 44, 71, 72, 77, 88). The distributions of interplant flight distances of insect pollinators are usually leptokurtic and skewed right, with most flights occurring over short distances (35). Interplant flight distances for bumblebees and large fritillary butterflies foraging on *A. exaltata* are similiar to observed patterns on other flowering plants. For example, nearly 80% of the interplant flight distances of large butterflies and bumblebees occurred over distances < 2 m, in natural and experimental populations of *A. exaltata* (6, 10). If pollen dispersal is correlated with pollinator flight distances, then effective pollen dispersal distances should also be leptokurtic and skewed toward longer dis-tances for milkweeds.

Pollinator flight distances may not reflect effective pollen dispersal distances within populations of milkweeds because of several unique features of milk-weed pollination. Following extraction of pollinaria from flowers, approximately 90 sec are required for the pollinarium to dry and reorient into a position that permits insertion. Insertion into stigmatic chambers is not possible during this time; thus, the opportunity for pollination on or near the pollen donor flower is decreased (55). In addition, slow turnover of pollinia on pollinators increases the probability of outcrossing and long-distance dispersal of pollen. Bumblebees foraging on *A. syriaca* picked up one pollinarium every 2–5 hr (40). Pollinia transported on bumblebee tarsi were retained approximately 6 hr, whereas pollinia deposited on mouthparts were retained longer than 24 hr (40). Pollinia typically are carried by insects on tarsi, rather than on mouthparts (37). Morse (39, 41) also suggested that carrying large numbers of pollinia reduces the foraging efficiency of bees and increases the probability of long-distance dispersal of pollen. These data suggest that pollinator flight distances are unlikely to provide realistic estimates of pollen dispersal.

Radioactive labelling of pollinia and paternity analysis of seeds have been used to measure effective pollen dispersal distances. Realized pollen dispersal distance determined from paternity exclusion analysis of seeds was three times greater than the mean pollen dispersal distance predicted from pollinator flight distances in populations of *A. exaltata* (6, 10). Mating was random with respect to interplant distances, and matings between neighboring plants were not significantly more common than matings between widely separated individuals. In contrast, by introducing a plant with radioactively labelled pollinia into a population of *A. syriaca,* Pleasants (48) found that 71% of the removed pollinia were inserted within 1 m of the labelled plant. Because *A. syriaca* forms extensive clones, many insertions will result in self-pollination of flowers on other ramets of the clone. It is unclear why pollen dispersal patterns should differ for these two milkweeds.

Differences in pollinators may account for part of the difference in pollen shadows for *A. exaltata* and *A. syriaca.* Pleasants's (48) study population of *A. syriaca* was pollinated by honeybees, whereas study populations of *A. exaltata* were pollinated by butterflies and bumblebees. Naturalists have observed that some native bumblebees (e.g. *Bombus griseocollis*) remove pollinaria and insert pollinia in other flowers far less frequently than do naturalized honeybees (*Apis mellifera*) (6, 37). More rapid turnover of pollinia on honeybees could have resulted in shorter pollen dispersal distances in Pleasants's (48) population of *A. syriaca.* Furthermore, apparent differences exist between lepidopterans and bumblebees in terms of pollination quality and efficiency (10, 44, 45). Thus, pollinators may differ in pollinium removal and deposition schedules on milkweeds, as they do in other flowering plants (69). Further investigations of the effects of pollinators on pollen dispersal in milkweed populations are warranted.

Slow pollinium turnover on large, strong-flying bees and butterflies may contribute to high levels of long-distance pollen dispersal between populations of milkweeds. Paternity exclusion analysis of seeds collected from natural populations of *A. exaltata* in Virginia showed that 11% to 50% of all seeds were sired by plants located outside the six populations that were examined (7, 10). These populations were isolated from other populations by 0.05–1.0 km. The correlation between levels of interpopulation pollen dispersal and isolation distance was statistically significant (Kendall's $t = -0.78$; $N = 7$; $P < 0.05$). Pollen-mediated gene dispersal reported for *A. exaltata* is among the highest reported for any insect-pollinated plant (23).

High levels of gene flow are likely to homogenize the gene pool among populations of milkweeds. Levels of genetic differentiation among populations of milkweeds are much lower than would be expected from studies of other outcrossing perennial herbs. For example, < 10% of total gene diversity (G_{ST} = 0.093) is found among populations of *A. exaltata* (11). Similarly, among-population diversity is low for the widespread *A. perennis* (G_{ST} = 0.082) and its rare sister species, *A. texana* (G_{ST} = 0.068: 19). In other outcrossing, animal-pollinated species, more of their genetic variation is typically partitioned among populations ($\overline{G}_{ST}$ = 0.197: 27). Clearly, insect dispersal of large pollinia and wind dispersal of comose seeds (46, 56) contribute to exceptionally high levels of gene flow and low levels of genetic differentiation among populations of milkweeds.

INTERSPECIFIC HYBRIDIZATION

Hybridization in milkweeds is rare, despite many opportunities for interspecific pollination in species that overlap in flowering phenology (31, 34) and habitat (34) and share many of the same generalist pollinators (30, 31, 32, 34, 37, 82). Many authors have attributed the paucity of hybridization in milkweeds to mechanical isolation brought about by a poor fit between pollinia and stigmatic chambers of different species. Recently, the effectiveness of this lock-and-key mechanism has been questioned (33, 34). High levels of interspecific insertions have been reported between several sympatric milkweeds that are not known to hybridize in nature. Mechanical isolation did, however, keep the large pollinia of *A. syriaca* from being inserted into stigmatic chambers of *A. incarnata* and *A. verticillata,* even though insertions of the small pollinia of *A. incarnata* and *A. verticillata* into *A. syriaca* were common (33).

Strong physiological barriers appear to be more important than mechanical barriers in preventing hybridization in milkweeds (30, 77). Foreign pollen germinates and penetrates ovules within the ovary, but seeds fail to develop (30). Even when some hybrid seeds develop, it has been speculated that these fruits may abort because they contain less than a full complement of seeds (85). This phenomenon might reinforce the mechanical isolation between two species that differ greatly in pollinium size. For example, pollinia of *A.*

incarnata would deliver approximately 99 pollen grains to stigmas of *A. syriaca* (R Wyatt, SB Broyles, unpublished data). Ovaries of *A. syriaca*, however, contain more than 200 ovules, and pollen of *A. incarnata* would at best fertilize only half of the ovules. These hybrid fruits with comparatively few seeds would therefore be more likely to abort than fruits with a full complement of seeds.

Hybrid sterility has been observed for artificially produced hybrids between *A. perennis* and *A. texana* (AL Edwards, CT Ivey, R Wyatt, unpublished data). This phenomenon was unexpected, given the relative ease with which interspecific crosses were performed and the vigorous germination and growth of the F_1 interspecific hybrids. Moreover, all species of *Asclepias* are isoploid, with $n = 11$ (77). Hybrids between *A. exaltata* and *A. purpurascens* also show reduced pollen viability relative to parental plants (SB Broyles, R Wyatt, unpublished data). Interestingly, no natural hybrids between these pairs of species have ever been observed. Present-day ranges of *A. perennis* and *A. texana* do not overlap, whereas those of *A. exaltata* and *A. purpurascens* overlap extensively.

Mechanical and physiological isolation is apparently lacking in the few documented cases of natural hybridization in milkweeds. Herbarium records of putative hybrids led Woodson (77) to list nine species pairs that he believed had hybridized in nature. Six of these pairs involved species that Woodson had assigned to different series or even subgenera. More recently, biochemical evidence has been used to document hybridization between *A. exaltata* and *A. quadrifolia* and between *A. purpurascens* and *A. syriaca* (87). Hybridization between *A. syriaca* and *A. speciosa* is supported by the production of artificial hybrids (67, 68), morphological analysis of hybrids (70, 77), and to a lesser extent by biochemical analysis of putative hybrids (1). Hybridization and introgression between *A. exaltata* and *A. syriaca* has been reported from several localities where the two occur sympatrically (34, 85, 87).

Hybridization and introgession between milkweed species appear to be limited. For example, in sympatric populations of *A. exaltata* and *A. syriaca,* fewer than 1% of the seeds produced on *A. exaltata* had been fertilized by pollen of *A. syriaca* (7). Hybridization between these species has been documented from areas associated with human disturbances or elevational gradients (85). These situations increase the likelihood of finding both *A. exaltata* (a forest species) and *A. syriaca* (a field/meadow species) in close proximity and in flower at the same time. Although hybridization is rare between these species, introgression of genes between these milkweeds may be greater than expected. Alleles diagnostic for *A. syriaca* have been found at low frequencies (<5%) in 22% of *A. exaltata* populations from the southern Appalachian Mountains. Therefore, even low levels of hybridization and introgression may provide a bridge for introducing novel genes into *A. exaltata*.

EVOLUTION OF INFLORESCENCE SIZE

The application of sexual selection theory to explain the evolution of inflorescence size (umbel size) in milkweeds has received considerable attention (2, 6, 8, 9, 13, 15, 51, 54, 55, 71–76, 80, 82). Milkweeds have drawn so much attention because it is easy, compared to plants with loose pollen grains, to estimate male reproductive success by counting the number of pollinaria removed from flowers. According to sexual selection theory, resource-limited fruit and seed production should drive the evolution of reproductive characters, such as inflorescence size, that enhance the probability of siring seeds on other plants (15, 16). Studies supporting this "pollen donation hypothesis" in milkweeds have demonstrated that (i) maximal fruit production is achieved on relatively small umbels, (ii) large umbels have more pollinaria removed than small umbels, and (iii) many more flowers are pollinated on most umbels than can set fruit.

Many of the underlying assumptions of the pollen donation hypothesis have not been critically evaluated in relation to milkweeds. The pollen donation hypothesis assumes that (i) fruit production is not limited by the quantity of pollen that flowers receive, (ii) umbel size is the target of selection, and (iii) male reproductive success is correlated with variation in umbel size. Fruit production may, however, be pollen-limited in many milkweed populations (see above). Simply counting pollinia received by flowers is not an accurate assessment of effective pollination, because pollinia might be improperly inserted (78, 79), pollinia might contain low-quality pollen grains (5), or pollinia might contain incompatible pollen (80). Moreover, a substantial percentage of flowers might not have received any pollinia (13, 88). Second, flower number per umbel is as variable within, as between, plants for many milkweed species (6, 14, 36, 60). In order for the pollen donation hypothesis to work, strong selection would need to target a few specific umbels from a diverse collection of small and large umbels on individual plants. It would, therefore, appear that variation in other inflorescence features, such as umbel number per plant, is more likely to affect reproductive success in milkweeds.

Recent tests have seriously challenged the pollen donation hypothesis in milkweeds. Using paternity exclusion analysis of seeds, Broyles & Wyatt (8, 9) demonstrated that mean umbel size was not significantly correlated with plant-level male or female reproductive success in a natural population of *A. exaltata*. To examine the effect of umbel size on reproductive success more closely, Broyles & Wyatt (13) performed a paternity analysis on seeds in an experimental population of *A. exaltata*, in which umbel size was manipulated by removing flowers. Plants with large umbels (18 flowers) attracted more pollinators and sired more seeds than did plants with small umbels (6 and 12 flowers), but they did not produce significantly more fruits than plants with small umbels. These observations are consistent with the pollen donation hypothesis and would offer strong support for the hypothesis if umbel size were a strong determinant of reproductive success of individuals. In both natural and experimental populations of *A. exaltata*, however, variation in male reproductive success was best explained by flower number per plant, not by flower number per umbel. Furthermore, male reproductive success did not increase at a faster rate than female reproductive success with respect to flower number per plant. Packaging of flowers into umbels and stems appears to be unimportant in determining male reproductive success in *A. exaltata*. On the other hand, umbel and stem number per plant explained most of the variation in female reproductive success. Thus, if natural selection functions to increase total reproductive success, then both male and female reproductive success will be maximized by addition of more inflorescences, rather than through increased investment in flowers on individual umbels.

Natural selection could, however, shape the evolution of umbel size if it were determined that developmental and/or architectural constraints limit the number of stems and umbels that plants produce. In this case, large umbels in milkweeds can maximize pollinator attraction and reproduction through both male and female functions. Pollinarium removal and pollinium insertion increase with inflorescence size (71, 78), even though increased insertion is not likely to result in greater fruit production. Female success, however, can increase on large umbels by selective maturation of fruits with many high-quality seeds. For example, in *A. speciosa*, fruits that contain fewer seeds or seeds that are growing more slowly are less likely to mature than fruits with many vigorous, fast-growing seeds (4). Moreover, pollen donors are known to differ in their ability to sire seeds from flowers within the same umbel (5). In *A. exaltata*, umbels that aborted several fruits generally contained more flowers, received more pollinia, and matured fruits with more seeds than did umbels that did not abort any fruits (9). The upper limit to inflorescence size in milkweeds may be set by the deleterious effects (loss of pollen and ovules) of increased self-pollination (80). Nevertheless, the evolution of inflorescence size probably represents a compromise among processes that simultaneously affect both male and female reproductive success.

It is unnecessary to invoke the "pollen donation hypothesis" to explain the existence of large floral displays and low fruit-set in milkweeds. The use of paternity analysis has permitted a more detailed examination of male reproductive success in natural populations than was previously possible. In both milkweeds and other hermaphroditic flowering plants, male success is generally quantified by counting pollinator visits and measuring pollen removal from flowers. Studies in natural populations of milkweeds have shown that the number of pollinaria removed, the usual estimator for male success, is more highly correlated with the number of seeds produced than with the number of seeds sired (8). The richness of paternity data will undoubtedly allow population biologists to examine similar processes in other flowering plants. As we learned from milkweed studies, the widespread application of sexual selection theory to explain the evolution of floral traits (2, 15, 16, 54) may be inappropriate for other hermaphroditic flowering plants as well.

CONCLUSIONS

Many unusual features of the milkweed reproductive system have contributed to the use of *Asclepias* as a model for studying various aspects of the ecology and evolution of plant reproduction. These include the delivery of pollen grains in discrete packets, accumulation of nectar in accessible reservoirs, use of nectar as the germination medium for pollen, production of two separate ovaries per flower, and possession of ovarian self-incompatibility. To some extent, all of these features have been capitalized upon by students of plant reproductive biology. Nevertheless, a great deal of potential has yet to be exploited by innovative and resourceful asclepiadologists.

We expect that paternity analysis in natural populations of milkweeds will enable workers to test various predictions from sexual selection theory regarding the selective forces driving the evolution of reproductive characters. In milkweeds it will be technically feasible to determine male and female reproductive success for all hermaphroditic plants in a population and then to relate these components of fitness to plant traits, such as inflorescence size. Paternity analysis should also permit detailed quantitative analysis of effective gene flow via pollen dispersal in natural populations. Moreover, by assessing pollen movement at several levels (e.g. electrophoresing single inserted pollinia), it will be possible to develop a very complete picture of the dynamics of pollen dispersal. When such analyses are carried out in the context of hybrid zones, the dynamics of interspecific pollen transfer will be revealed. In most plant species, this aspect of hybridization is often dealt with as a "black box."

We also predict a flurry of new studies using milkweeds to study details of nectar production and the foraging behavior of insects on flowers. Aside from plastic models, milkweed flowers appear to be among the most easily manipulated of flowers. It is surprising that no work has been done thus far involving manipulation of nectar rewards in milkweed flowers. Now that the relationship between the five stigmatic chambers and two ovaries of milkweed flowers is known, it is only a matter of time before experiments are carried out to assess the importance of ovary competition within, versus between, flowers. Finally, an area overripe for exploration at the present time is the nature and functioning of ovarian self-incompatibility in milkweeds. It is well-established that most milkweeds express a late-acting form of self-incompatibility, yet virtually nothing is known about how it functions, its phylogenetic distribution in the Apocynaceae/Asclepiadaceae clade, or its evolutionary origin and maintenance.

Clearly much remains to be done with this unusual and, in many ways, unique system for the study of plant reproduction.

ACKNOWLEDGMENTS

We thank a large number of coworkers who have collaborated with us on milkweed research over the past two decades, all of whom are identified in relevant publications. We also thank AL Edwards, CT Ivey, and SR Lipow for allowing us to cite unpublished data and for reading an earlier draft of this manuscript, DJ Futuyma for insightful comments, and CG Hahn for drawing the figure. Our research on milkweeds has been supported by fellowships and grants from the Whitehall Foundation, National Science Foundation, John Simon Guggenheim Foundation, Highlands Biological Station, and Mt. Lake Biological Station.

Any *Annual Review* chapter, as well as any article cited in an *Annual Review* chapter, may be purchased from the Annual Reviews Preprints and Reprints service. 1-800-347-8007; 415-259-5017; email: arpr@class.org

---

Literature Cited

1. Adams RP, Tomb AS, Price SC. 1987. Investigation of hybridization between *Asclepias speciosa* and *A. syriaca* using alkanes, fatty acids, and triterpenoids. *Biochem. Syst. Ecol.* 15:395–99
2. Bell G. 1985. On the function of flowers. *Proc. R. Soc. Lond. B.* 224:223–65
3. Bertin RI, Willson MF. 1980. Effectiveness of diurnal and nocturnal pollination of two milkweeds. *Can. J. Bot.* 58:1744–46
4. Bookman SS. 1983. Costs and benefits of flower abscission and fruit abortion in *Asclepias speciosa*. *Ecology* 64:264–73
5. Bookman SS. 1984. Evidence for selective fruit production in *Asclepias*. *Evolution* 38:72–86
6. Broyles SB. 1992. *Reproductive biology of poke milkweed*, Asclepias exaltata L.: *population structure, mating patterns, pollen dispersal, and the evolution of inflorescence size*. PhD thesis. Univ. Georgia, Athens. 206 pp.
7. Broyles SB, Schnabel A, Wyatt R. 1994. Evidence for long-distance pollen dispersal in milkweeds (*Asclepias exaltata*). *Evolution* In press
8. Broyles SB, Wyatt R. 1990. Paternity analysis in a natural population of *Asclepias exaltata*: multiple paternity, functional gender, and the "pollen donation hypothesis." *Evolution* 44:1454–68
9. Broyles SB, Wyatt R. 1990. Plant parenthood in milkweeds: a direct test of the pollen donation hypothesis. *Plant Species Biol.* 5:131–42
10. Broyles SB, Wyatt R. 1991. Effective pollen dispersal in a natural population of *Asclepias exaltata*: the influence of pollinator behavior, genetic similarity, and mating success. *Am. Nat.* 138:1239–49
11. Broyles SB, Wyatt R. 1993. Allozyme diversity and genetic structure in southern Appalachian populations of *Asclepias exaltata*. *Syst. Bot.* 18:18–30
12. Broyles SB, Wyatt R. 1993. The consequences of self-pollination in *Asclepias exaltata*, a self-incompatible milkweed. *Am. J. Bot.* 80:41–44
13. Broyles SB, Wyatt R. 1994. A reexamination of the "pollen donation hypothesis" in an experimental population of *Asclepias exaltata*. *Evolution*. In press
14. Chaplin SJ, Walker JL. 1982. Energetic constraints and adaptive significance of the floral display of a forest milkweed. *Ecology* 63:1857–70
15. Charnov EL. 1979. Simultaneous her-

maphroditism and sexual selection. *Proc. Natl. Acad. Sci. USA* 76:2480–84

16. Charnov EL. 1982. *The Theory of Sex Allocation.* Princeton, NJ: Princeton, Univ. Press

17. Corry TH. 1883. On the mode of development of the pollinium in *Asclepias cornuti* Decaisne. *Trans. Linn. Soc. London Ser. 2, Bot.* 75–84

18. Delpino F. 1865. Relazione sull 'apparechio della fecondazione nelle Asclepiadee. Torino: Published by author

19. Edwards AL, Wyatt R. 1994. Population genetics of the rare *Asclepias texana* and its widespread congener, *A. perennis. Syst. Bot.* 19:291–307

20. Eisikowitch D, Kevan PG, Fowle S, Thomas K. 1987. The significance of pollen longevity in *Asclepias syriaca* under natural conditions. *Pollen Spores* 29:121–28

21. Eisikowitch D, Kevan PG, Lachance MA. 1990. The nectar-inhabiting yeasts and their effect on pollen germination in common milkweed, *Asclepias syriaca* L. *Israel J. Bot.* 39:217–26

22. Eisikowitch D, Lachance MA, Evans PG, Willis S, Collins-Thompson DL. 1990. The effect of the natural assemblage of microorganisms and selected strains of the yeast *Metschnikowia reukauffi* in controlling the germination of pollen of the common milkweed *Asclepias syriaca. Can. J. Bot.* 68:1163–65

23. Ellstrand NC. 1992. Gene flow among seed plant populations. In *Population Genetics of Forest Trees,* ed. WT Adams, SH Strauss, DL Copes, AR Griffin, 6:241–56. Dordrecht, The Netherlands: Kluwer. 256 pp.

24. Fischer E. 1941. Der Anbau einer Faserund Bienen-futterpflanze. *Pflanzenbau* 17:212–18

25. Frye TC. 1902. A morphological study of certain Asclepiadaceae. *Bot. Gaz.* 34: 389–413

25a. Galil J, Zeroni M. 1965. Nectar system of *Asclepias curassavica. Bot. Gaz.* 126: 144–48

26. Galil J, Zeroni M. 1969. On the organization of the pollinium in *Asclepias curassavica. Bot. Gaz.* 130:1–4

27. Hamrick JL, Godt MJW. 1989. Allozyme diversity in plant species. In *Plant Population Genetics, Breeding, and Genetic Resources,* ed AHD Brown, MT Clegg, AL Kahler, BS Weir, pp. 44–64. Sunderland, Mass: Sinauer

28. Hildebrand R. 1866. Über die Befruchtung von *Asclepias cornuti. Bot. Zeit.* 24:376–78

29. Kahn AP, Morse DH. 1991. Pollinium

30. Kephart SR. 1981. Breeding systems in *Asclepias incarnata* L., *A. syriaca* L., and *A. verticillata* L. *Am. J. Bot.* 68:226–32

31. Kephart SR. 1983. The partitioning of pollinators among three species of *Asclepias. Ecology* 64:120–33

32. Kephart SR. 1987. Phenological variation in flowering and fruiting of *Asclepias. Am. Midl. Nat.* 118:64–76

33. Kephart SR, Heiser CB. 1980. Reproductive isolation in *Asclepias:* lock and key hypothesis reconsidered. *Evolution* 34:738–46

34. Kephart SR, Wyatt R, Parrella D. 1988. Hybridization in North American *Asclepias.* I. Morphological evidence. *Syst. Bot.* 13:456–73

35. Levin DA, Kerster HW. 1974. Gene flow in seed plants. *Ann. Missouri Bot. Gard.* 8:233–53

36. Lynch SP. 1977. The floral ecology of *Asclepias solanoana. Madroño* 24:159–77

37. Macior LW. 1965. Insect adaptation and behavior in *Asclepias* pollination. *Bull. Torrey Bot. Club* 2:114–26

38. Moore RJ. 1947. Investigations on rubber-bearing plants. V. Notes on the flower biology and pod yield of *Asclepias syriaca* L. *Can. Field Nat.* 61: 40–46

39. Morse DH. 1981. Modification of bumblebee foraging: the effect of milkweed pollinia. *Ecology* 62:89–97

40. Morse DH. 1982. The turnover of milkweed pollinia on bumblebees, and implications for outcrossing. *Oecologia* 53: 187–96

41. Morse DH. 1985. Costs in a milkweed-bumblebee mutualism. *Am. Nat.* 125: 903–5

42. Morse DH. 1987. Roles of pollen and ovary age in follicle production of the common milkweed *Asclepias syriaca. Am. J. Bot.* 74:851–56

43. Morse DH. 1993. The twinning of follicles by common milkweed *Asclepias syriaca. Am. Midl. Nat.* 130:56–61

44. Morse DH, Fritz RS. 1983. Contribution of diurnal and nocturnal insects to the pollination of common milkweed (*Asclepias syriaca* L.). *Oecologia* 60:190–97

45. Morse DH, Jennersten O. 1991. The quality of pollination by diurnal and nocturnal insects visiting common milkweed, *Asclepias syriaca. Am. Midl. Nat.* 125:18–28

46. Morse DH, Schmitt J. 1985. Propagule

size, dispersal ability, and seedling performance in *Asclepias syriaca*. *Oecologia* 67:372–79

47. Opler PA. 1983. Nectar production in a tropical ecosystem. In *The Biology of Nectaries*, ed. B. Bentley, T. Elias, pp. 30–79. New York: Columbia Univ. Press. 259 pp.

48. Pleasants JM. 1991. Evidence for short-distance dispersal of pollinia in *Asclepias syriaca* L. *Func. Ecol.* 5:75–82

49. Pleasants JM, Chaplin SJ. 1983. Nectar production rates of *Asclepias quadrifolia:* causes and consequences of individual variation. *Oecologia* 59:232–38

50. Pleasants JM, Horner HT, Ng G. 1990. A labelling technique to track dispersal of milkweed pollinia. *Func. Ecol.* 4: 823–27

51. Pleasants JM, Ng G. 1987. The relationship between inflorescence size and self-pollination in milkweed, *Asclepias syriaca. Iowa Acad. Sci.* (Abstr.)

52. Plotnikova T. 1938. An experiment in self-pollination of *Asclepias cornuti. Ukraine Acad. Sci., Inst. Bot. J. No. 26–27* (English summary)

53. Primack RB. 1985. Longevity of individual flowers. *Annu. Rev. Ecol. Syst.* 16:15–38

54. Queller DC. 1983. Sexual selection in a hermaphroditic plant. *Nature* 305:706–7

55. Queller DC. 1985. Proximate and ultimate causes of low fruit production in *Asclepias exaltata. Oikos* 44:373–81

56. Sacchi CF. 1987. Variability in dispersal ability of common milkweed, *Asclepias syriaca,* seeds. *Oikos* 49:191–98

57. Sage TL, Bertin RI, Williams EG. 1994. Ovarian and other late-acting self-incompatibility systems. *Sex. Pl. Reprod.* In press

58. Sage TL, Broyles SB, Wyatt R. 1990. The relationship between the five stigmatic chambers and the two ovaries of milkweed flowers: a three-dimensional assessment. *Israel J. Bot.* 39: 187–96

59. Seavey SR, Bawa KS. 1986. Late-acting self-incompatibility in angiosperms. *Bot. Rev.* 52:195–218

60. Shannon TR, Wyatt R. 1986. Reproductive biology of *Asclepias exaltata. Am. J. Bot.* 73:11–20

61. Shannon TR, Wyatt R. 1986. Pollen germinability of *Asclepias exaltata:* effects of flower age, drying time, and pollen source. *Syst. Bot.* 11:322–25

62. Shore JS. 1993. Pollination genetics of the common milkweed, *Asclepias syriaca* L. *Heredity* 70:101–8

63. Southwick EE. 1983. Nectar biology and nectar feeders of common milkweed, *Asclepias syriaca* L. *Bull. Torrey Bot. Club* 110:324–34

64. Southwick EE, Lopez GM, Sadwick SE. 1981. Nectar production, composition, energetics, and pollinator attractiveness in spring flowers of western New York. *Am. J. Bot.* 67:994–1002

65. Sparrow FK, Pearson NL. 1948. Pollen compatibilty in *Asclepias syriaca. J. Agric. Res.* 77:187–99

66. Sreedevi P, Namboodiri AN. 1982. The germination of pollinium and the organization of germ furrow in some members of Asclepiadaceae. *Can. J. Bot.* 60:166–72

67. Stevens OA. 1945. Cultivation of milkweed. *N. Dakota Agric. Exp. Sta. Bull.* 33:1–19

68. Stevens OA. 1945. *Asclepias syriaca* and *A. speciosa,* distribution and mass collections in North Dakota. *Am. Midl. Nat.* 34:368–74

69. Thomson JD, Thomson BA. 1992. Pollen presentation and viability schedules in animal-pollinated plants: consequences for reproductive success. In *Ecology and Evolution of Plant Reproduction,* ed R Wyatt, pp. 1–24. New York: Chapman & Hall. 397 pp.

70. Thomson JW, Wagner WH. 1978. Hybrid swarms of the prairie *Asclepias speciosa* with common *Asclepias syriaca. Ohio Biol. Surv. Notes* 15:264 (Abstr.)

71. Willson MF, Bertin RI. 1979. Flower-visitors, nectar production, and inflorescence size of *Asclepias syriaca. Can. J. Bot.* 57:1380–88

72. Willson MF, Bertin RI, Price PW. 1979. Nectar production and flower visitors of *Asclepias verticillata. Am. Midl. Nat.* 102:23–35

73. Willson MF, Price PW. 1977. The evolution of inflorescence size in *Asclepias* (Asclepiadaceae). *Evolution* 31:495–511

74. Willson MF, Price PW. 1980. Resource limitation of fruit and seed production in some *Asclepias* species. *Can. J. Bot.* 58:2229–33

75. Willson MF, Rathcke BJ. 1973. Adaptive design of the floral display in *Asclepias syriaca* L. *Am. Midl. Nat.* 92: 47–57

76. Wolfe LM. 1987. Inflorescence size and pollinaria removal in *Asclepias curassavica* and *Epidendrum radicans. Biotropica* 19:86–89

77. Woodson RE. 1954. The North Amerian species of *Asclepias* L. *Ann. Missouri Bot. Gard.* 41:1–211

78. Wyatt R. 1976. Pollination and fruit-set

in *Asclepias*: a reappraisal. *Am. J. Bot.* 63:845–51
79. Wyatt R. 1978. Experimental evidence concerning the role of the corpusculum in *Asclepias pollination. Syst. Bot.* 3: 313–21
80. Wyatt R. 1980. The reproductive biology of *Asclepias tuberosa*: I. Flower number, arrangement, and fruit-set. *New Phytol.* 85:119–31
81. Wyatt R. 1980. The impact of nectar-robbing ants on the pollination system of *Asclepias curassavica. Bull. Torrey Bot. Club* 107:24–28
82. Wyatt R. 1981. The reproductive biology of *Asclepias tuberosa.* II. Factors determining fruit-set. *New Phytol.* 88: 375–85
83. Wyatt R. 1982. Inflorescence architecture: how flower number, arrangement, and phenology affect pollination and fruit-set. *Am. J. Bot.* 69:587–96
84. Wyatt R, Broyles SB. 1990. Reproductive biology of milkweeds (*Asclepias*): recent advances. In *Biological Approaches and Evolutionary Trends in Plants*, ed. S Kawano, pp. 255–272. London: Academic. 417 pp.
85. Wyatt R, Broyles SB. 1992. Hybridization in North American *Asclepias.* III. Isozyme evidence. *Syst. Bot.* 17: 640–48
86. Wyatt R, Broyles SB, Derda GS. 1992. Environmental influences on nectar production in milkweeds (*Asclepias syriaca* and *A. exaltata*). *Am. J. Bot.* 79:636–42
87. Wyatt R, Hunt DM. 1991. Hybridization in North American *Asclepias.* II. Flavonoid evidence. *Syst. Bot.* 16:132–42
88. Wyatt R, Shannon TR. 1986. Nectar production and pollination of *Asclepias exaltata. Syst. Bot.* 11:326–34
89. Zimmerman M, Pyke GH. 1988. Reproduction in *Polemonium:* assessing the factors limiting seed set. *Am. Nat.* 131: 723–38

Annu. Rev. Ecol. Syst. 1994. 25:443–66

THE NATURE AND CONSEQUENCES OF INDIRECT EFFECTS IN ECOLOGICAL COMMUNITIES

J. Timothy Wootton

Department of Ecology and Evolution, University of Chicago, 1101 East 57th Street, Chicago, Illinois 60637

KEY WORDS   species interactions, competition, predation, mutualism, interaction, modifications

Abstract

Indirect effects occur when the impact of one species on another requires the presence of a third species. They can arise in two general ways: through linked chains of direct interactions, and when a species changes the interactions among species. Indirect effects have been uncovered largely by experimental studies that have monitored the response of many species and discovered "unexpected results," although some studies have looked for specific indirect effects predicted from simple models. The characteristics of such approaches make it likely that the many indirect effects remain uncovered, but the application of techniques such as path analysis may reduce this problem. Deterministic theory indicates that indirect effects should often be important, although stochastic models need exploration. Simulation models indicate that some indirect effects may stabilize multi-species assemblages. Five simple types of indirect effects have been regularly demonstrated in nature: exploitative competition, trophic cascades, apparent competition, indirect mutualism, and interaction modifications. Detailed experimental investigations of natural communities have yielded complicated effects. Indirect effects have the potential to affect evolutionary patterns, but empirical examples are limited. Future directions in the study of indirect effects include developing techniques to estimate interaction strength in dynamic models, deriving more efficient approaches to detecting indirect effects, evaluating the effectiveness of ap-

443

proaches such as path analysis, and investigating mechanisms in which a species changes how other species interact.

INTRODUCTION

Ecologists have long been interested in determining the role that biotic interactions play in ecosystems. Many workers have suggested that biotic interactions between pairs of species play a major role in shaping the ecological and evolutionary patterns in natural and human-impacted systems. If true, then one should expect that species will also indirectly affect other species when a species is involved in several strong pair-wise interactions. Finding indirect effects thus indicates that biotic interactions are important to ecological systems, and that community-level phenomena must be incorporated to understand and predict the dynamics of natural systems. Aside from the basic interest in this question, the existence of indirect effects can have important implications for applied problems. If indirect effects are common, then environmental impacts such as species introductions and the reduction or extinction of species can have widespread effects on the rest of the system. Furthermore, because indirect effects can offset or exacerbate direct effects of the manipulation, the task of predicting the consequences of particular environmental manipulations becomes extremely complicated (88, 101, 140). In this paper, I review the nature and occurrence of indirect effects in natural systems, discuss methodological approaches to their investigation, and suggest several directions for future research.

TERMINOLOGY

To define indirect effects, one must first define direct effects. I consider direct effects to arise as a result of a physical interaction between two species. Therefore direct effects would occur between a pair of species both in isolation and within multi-species communities of varying composition. Common examples include one species consuming, interfering with, or physically benefiting another. Logically then, indirect effects are those effects of one species on another that are not direct effects. The key feature of indirect effects is that they require the presence of intermediary species in order to arise. Thus, they are a property arising in multi-species assemblages.

In some instances, indirect effects have also been considered to arise through a change in a physical or chemical variable in the environment as well as through another species (116). For example, fish foraging activity may change the sedimentation regime of a stream, which in turn may influence invertebrates and algae (37, 94, 96). Similarly, a species may affect the availability of inorganic nutrients in the system, which in turn affects the population dynamics of producers (9, 41, 123, 131). The major feature in such examples is that the physical or chemical variable of interest acts in a dynamic manner, much as another species would. Little is gained in debating whether or not such effects should be considered "indirect," as long as the underlying mechanisms are recognized and the dynamics of the physical or chemical variable are accounted for when necessary. For the purposes of this paper, I consider indirect effects to be only those mediated through other species.

Indirect effects arise in ecological systems through two general mechanisms (Figure 1; 74, 133, 136). First, varying the abundance of one species can indirectly affect the abundance of another by changing the abundance of an intermediary species that interacts with them both—an interaction chain. For example, one species can reduce the abundance of its prey, thereby reducing the food base of other consumers of the prey (i.e. exploitative competition). Second, varying the abundance of one species can indirectly affect the abundance of a second species by changing the interaction between the second species and a third species, an interaction modification. Thus, these indirect effects arise because the interactions between pairs of species are not independent of other species. For example, increasing vegetation density may indirectly increase the abundance of a prey species by allowing it to hide from its predator, thereby reducing the intensity of the predator-prey interaction. Distinguishing between the two types of indirect effects is important because they have different implications for predicting system responses to a perturbation. Indirect effects arising from interaction chains can be predicted readily, given a knowledge of how species pairs directly interact. In contrast, indirect effects arising from interaction modifications can be uncovered by identifying instances where one species might affect the interaction between two others. At our present level of understanding, however, the quantitative consequences of interaction modifications can only be determined by experimental manipulations within the context of the community of interest; they cannot be predicted ahead of time.

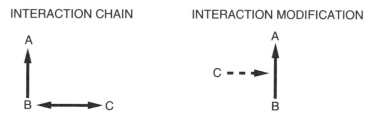

INTERACTION CHAIN   INTERACTION MODIFICATION

Figure 1   Two basic ways in which one species can indirectly affect another. Left, species C affects species A through a chain of direct interactions involving a change in the abundance of species B. Right, species C indirectly affects species A by modifying how species A interacts with species B. Adapted from (133).

The definition of indirect effects used here (see also 74, 116) is broader than the one advocated by Billick & Case (8), who suggest that it be applied only to interaction chains. Because "indirect" is the antonym to "direct," however, applying the term "indirect effects" only to a subset of the nondirect impacts of a species breeds confusion. Similarly, most discussion of interaction modifications has been carried out under the term "higher-order interactions." Much confusion has arisen in these studies, in part because higher-order interactions can also refer simply to nonlinear direct effects. Nonlinearities need not arise because one species changes how a species pair interacts (92). "Interaction modification" accurately describes the general mechanism by which this class of indirect effects arises, thereby drawing attention to the salient features to be investigated. Consequently empiricists need not be at the mercy of a mathematical description of a system in order to identify and study such indirect effects. Additional terminology (e.g. response, behavioral, morphological, and chemical indirect effects) has been proposed for specific mechanisms by which interactions are modified by other species (74, 116). This terminology can be useful in more precisely categorizing different types of indirect effects, but the recognition of interaction modifications as a general class of indirect effects is useful because they represent a fundamental difference in how such effects are modeled, yet provide a general mechanistic criterion to classify particular interactions identified from field observations.

THE DETECTION OF INDIRECT EFFECTS

Indirect effects have come to the attention of ecologists largely as a result of either experimental manipulations or large-scale (re-)introductions of species that have been placed in a multi-species context. Usually, they have been detected only when an experimental manipulation produces "unexpected results" on a target species (109). The reliance of ecologists on fortuitous results to identify indirect effects is a cause for concern, because such an approach is likely to give a biased or incomplete picture about the importance of indirect effects. This is particularly true when indirect effects have impacts in the same direction as those expected from more obvious direct interactions. Under such circumstances, the overall change observed between treatments usually is assumed to arise from the direct effect alone. This assumption is not always a safe or appropriate one to make. For example, removal experiments on birds that forage intertidally (132) have demonstrated a five-fold increase in the population of the limpet *Lottia digitalis,* consistent with a strong predator effect. A detailed examination of the results that accounted for changes in other members of the intertidal community showed, however, that the direct effects of predation actually caused only about a two-fold change in population size. The remaining change in population size of the limpet was attributable to negative indirect effects of birds on limpets arising from changes in the abundance of sessile species that comprise the habitat of the limpets.

Recently, indirect effects have been identified in studies that have looked explicitly for patterns of population changes predicted from a specified type of indirect effect (e.g. 14, 93, 102). Such studies still limit the identification of indirect effects, because predictions depend on the particular type of indirect effect under investigation. In these studies, other types of indirect effects continue to be detected only when results produce patterns that are unexpected both through the indirect interaction under study and through known direct interactions.

Aside from the disadvantages of current methodological protocols for uncovering indirect effects, an increasingly common statistical application would seem to bias against detecting indirect effects too. Multivariate analysis of variance (MANOVA) has commonly been applied to data sets involving multi-species response variables in order to control type I error rates (e.g. 77). Some ecologists may not appreciate that this technique factors out interdependencies between dependent variables. Such interdependencies are naturally present when indirect effects occur among the dependent variables, so this technique essentially excludes all variation in community structure due to indirect effects.

DEMONSTRATIONS OF INDIRECT EFFECTS IN NATURE

Despite the difficulty of identifying indirect effects, there is now a considerable amount of work showing that they can play an important role in natural systems. In this section, I review some of the common patterns of indirect effects found in field experiments or following the large-scale introduction or extinction of a species.

Exploitative Competition

Although often erroneously considered a type of direct interaction among species, exploitative competition is an indirect effect; one species indirectly reduces a second species by directly reducing the abundance of a shared resource (Figure 2a). This indirect effect has been the central focus of community-level studies for decades, and consequently there are numerous demonstrations of its importance in natural communities. Experimental studies of exploitative competition have been extensively reviewed by others (18, 21, 41, 103), so I do not do so here. In some cases, it is unclear whether the effects from experiments are the result of direct interference competition or indirect exploitative competition, because the mechanism of competition is often not investigated (120). It is important to emphasize that exploitative competition possesses no unique properties relative to other kinds of indirect effects. Thus, the extent to which exploitative competition has been demonstrated may pro-

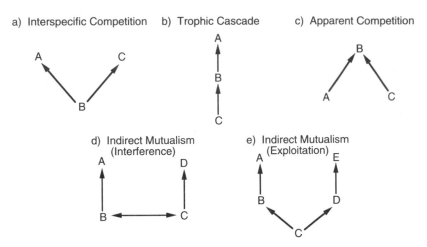

Figure 2 Five commonly investigated types of simple indirect effects. Horizontal arrows: interference competition (arrows show impacted species); vertical arrows: consumer-resource interactions (arrows determine direction of energy flow). Subfigures: (a) exploitative competition, (b) trophic cascade, (c) apparent competition, (d) indirect mutualism involving interference competition, (e) indirect mutualism involving exploitative competition.

vide an indication of the prevalence of other less-studied indirect effects. Furthermore, other types of indirect effects are just as likely as exploitative competition to have effects on the ecological and evolutionary dynamics of natural communities.

Trophic Cascades

Trophic cascades are indirect effects mediated through consumer-resource interactions (Figure 2b; 14, 82). Hairston et al's verbal theory of highly aggregated terrestrial food webs (45) clearly predicted their existence, and similar predictions have subsequently been made by many mathematical theories of food-chains (38, 78, 99, 112). Evidence from experiments and species invasions has now established that trophic cascades can occur in a variety of habitats.

In marine systems, the recovery of the sea otter (*Enhydra lutris*) in Alaska has been associated with large reductions in sea urchins (*Strongylocentrotus* spp.). In turn, reduced sea urchin grazing has led to increases in kelp cover, which has affected the structure of near-shore communities by changing both physical characteristics of the environment (e.g. water movement) and energy-flow patterns (via kelp detritus; 29, 30, 33, 34). Whelks (*Morula marginalba*), gulls (*Larus glaucescens*), and oystercatchers (*Haematopus bachmani*) in rocky intertidal systems also indirectly increase algal abundance by feeding on limpet grazers (*Lottia* spp.; 35, 132). In salt marshes, killifish (*Fundulus heteroclitus*) prey upon grass shrimp (*Palaemonetes pugio*), thereby indirectly enhancing the abundance of the anemone *Nematostella victensis* that the shrimp feed upon (56). Finally, human harvesting of grazing (*Fissurella* spp.) and of predatory gastropods (*Concholepas concholepas*) in Chile has led to increased abundances of algae (*Eridaea boryana*) and sessile invertebrates (*Perumytilus purpuratus*) that the gastropods consume (75, 76).

Examples of trophic cascades in freshwater systems are equally common. Experimental introductions of fish have caused reductions in the abundance of large zooplankton grazers and increases in phytoplankton biomass in the water column of lakes (14, 108), and small ponds (121). Similarly, removing fish has caused increases in snail grazers and declines in benthic algae in lakes (10, 68). Removing bass from Ozark streams has lead to an increase in grazing fish populations, which in turn has caused reductions in algae (95). In northern California rivers, however, removing steelhead has led to an increase in small predators, a decrease in grazers, and an increase in algae (93, 137).

Strong (117) has argued that, unlike aquatic habitats, trophic cascades are unimportant in terrestrial systems because of the compensatory effects of interspecific competition within trophic levels. Experimental manipulations of top consumers (particularly entire trophic levels), however, are much more difficult in terrestrial settings because the dominant species usually have much slower dynamics that operate over much larger spatial scales than those of aquatic systems. Thus, the relative scarcity of examples of trophic cascades from terrestrial systems may be a reflection of experimental limitations. Nevertheless, examples of trophic cascades do exist in terrestrial settings. In the East African savannah, the introduction of rinderpest caused reductions in grazers and browzers, leading to an increase in tree cover (110). On tropical islands, the removal of lizards caused an increase in phytophagus insects, and a subsequent increase in leaf damage (105, 114). Removal of spiders has caused an increase in grasshoppers and a decline in grassland vegetation (53). Exclusion of bird predators has caused increases in grazing insect abundance and a reduction in the biomass of oaks in Missouri (67). Along the Pacific coast of North America, the re-establishment of peregrine falcons (*Falco peregrinus*) has been associated with an increase in murres, cormorants, and oystercatchers, apparently as an indirect result of falcons feeding upon nest-robbing crows (84).

Apparent Competition

Apparent competition arises when two prey species share a common predator (Figure 2c; 50). In this case, an increase in one prey species may lead to an increase in the shared predator, causing a subsequent decline in the other prey species. Although examples are still scarce, some evidence of apparent competition exists in natural communities. In a subtidal marine community, increasing the abundance of bivalves leads to declines in the abundance of grazing gastropods because of increases in invertebrate predators (102). The introduction of the variegated leafhopper (*Erythroneura variabilis*) into the San Joaquin Valley of California causes reductions in the grape leafhopper (*Erythroneura elegantula*) by increasing the population of parasitic wasps (107). Increasing the abundance of the terrestrial isopod *Porcellio laevis* causes increases in the prevalence of an iridescent isopod virus, which in turn causes the reduction in a second isopod species, *Porcellio scaber* (44). More work needs to be done to determine the prevalence of this type of interaction in natural systems.

Indirect Mutualism and Commensalism

"Indirect mutualism" or "indirect commensalism," defined as indirect positive effects of one species on another (28, 71, 106, 122), can arise through a number of mechanisms. Typically, the indirect effect involves a consumer-resource interaction linked with either exploitative or interference competition (Figure 2d,e). In intertidal systems, adding starfish or snails indirectly increases the abundance of competitively inferior sessile species by reducing the abundance of mussels (*Mytilus*), the competitively dominant space-occupiers (66, 70, 80, 81). Intertidal bird predators indirectly increase the abundance of their snail prey (*Nucella* spp.) by consuming goose barnacles (*Pollicipes polymerus*), a competitively superior species to the snail's preferred prey, acorn barnacles (*Semibalanus cariosus*; 135). Birds also enhance the abundance of acorn barnacles by consuming limpets that "bulldoze" young individuals off the rocks while feeding (133). Grazing fish, molluscs, and crabs indirectly increase the abundance of crustose algae and diatoms by removing fleshy algae that shade and abrade shorter species (28, 72, 82). In some cases (28, 82) this shift in algal community structure has generated further indirect consequences by enhancing the abundance of grazers that specialize on crustose algae and diatoms. In the Gulf of California, adding predatory snails (*Acanthina angelica*) reduces the abundance of acorn barnacles (*Chthamalus anisopoma*), releasing algae (*Ralfsia* spp.) from competition for space, and enhancing the food supply of limpet grazers (*Lottia strongiana*). Likewise, limpets indirectly enhance snail abundance by reducing algal cover, thereby increasing the abundance of the snail's acorn barnacle prey through reduced competition for space (31, 32).

Similar patterns have been found in freshwater systems. For example, by feeding on competitively superior frog tadpoles, predatory salamanders indirectly increase the abundance of competitively inferior frog species in ponds (77, 127). Planktivorous fish, by preferentially feeding on large zooplankton, indirectly enhance the abundance of small zooplankton in lakes (11). Zooplankton prefer to feed upon green algae, thereby indirectly enhancing the abundance of blue-green algae in some lakes (59).

Several examples of indirect mutualisms and commensalisms exist in terrestrial systems. In the deserts of the southwestern United States, kangaroo rats (*Dipodomys* spp.) indirectly increase the abundance of small-seeded plant species by preferentially consuming large-seeded plants. As an indirect result, ants increase in abundance because they are only able to forage effectively on small seeds (26), and birds increase in local abundance apparently because after rodents reduce vegetation cover they can detect seeds more readily (119). In northern Europe, fieldfares (*Turdus pilarus*) aggressively defend their territories from avian predators, thereby indirectly enhancing the abundance of other avian species (111).

Indirect mutualisms and commensalisms may also play an important role in the dynamics of succession following disturbance (22). By removing rapidly invading, consumer-susceptible species, consumers may indirectly push the assemblage of sessile organisms of a system to a more consumer-resistant group of species by freeing them from preemptive interference competition. This mechanism of succession has been demonstrated frequently in experimentally manipulated marine systems. Intertidal snail and crab grazers remove early successional algal species, allowing grazer-resistant algae to become dominant (64, 65, 113). Subtidal starfish remove early successional species of bryzoans, allowing more resistant species to dominate (27). Intertidal bird and snail predators remove earlier successional blue mussels and goose barnacles, promoting the establishment of California mussels on rocky intertidal benches above tidal heights frequented by large starfish (130, 134). Evidence of consumer-driven succession also exists in lakes. By grazing on early-blooming algae, zooplankton shift the algal community over time to domination by cyanobacteria (100).

Because of the difficulties of performing experiments, there are few solid examples of consumer-driven succession in terrestrial settings. Ecologists have noticed patterns of late-succession plant species having higher levels of defenses against grazers, which suggests that this indirect effect may be important in terrestrial systems too (13, 16, 17, 40).

Interaction Modifications ("Higher-Order" Interactions)

As mentioned above (Figure 1, right), these indirect effects occur when one species modifies the interaction between two other species. Good examples of these indirect effects are limited because of difficulties in executing appropriate experiments and confusion about the mathematical criterion used in statistical tests (15, 92, 136). The strongest studies are those that can mechanistically identify how one species modifies the interactions between other pairs of species. Several mechanisms seem particularly likely to be important. Crypsis is one manifestation of an interaction modification. For example, sessile invertebrates indirectly affect limpet populations by making them harder for bird predators to discover, thereby changing the intensity of bird predation (73, 132, 133). Aquatic macrophytes may also reduce the intensity of predation by interfering with the detection of prey by fish predators (23, 125). Similarly, barnacles interfere with limpets grazing on algae (31, 36, 51).

Changes in behavior may also cause interaction modifications. For example, the activities of one predator species may flush out prey species, making them more susceptible to other predators. Alternatively, aggression among predators may reduce the consumption rate on the prey (129). Another possibility is that a predator is deterred from feeding in a patch by defended prey species, thereby reducing consumption of undefended prey species, an associational defense. For example, undefended plants perform better when associated with defended plants in the face of grazing (46, 90, 118). Some species may provide other species with defensive items, thereby indirectly reducing consumption rates. For example, the dinoflagellate *Protogonyaulax* spp. secretes a neurotoxin (the cause of paralytic shellfish poisoning or "red tide") that the butter clam *Saxidomus giganteus* can sequester in its body. The sequestered toxin deters gulls and other vertebrate predators from feeding on the clams (57). Furthermore, consumers may become satiated at high levels of prey availability. When several prey species are involved, a predator's consumption of one prey species may preclude the consumption of another prey species at the same time. Thus, a type II (49) functional response of a predator feeding on one prey (N1) when it also consumes another species (N2) should take the mathematical form:

$$c_1 N_1 / (1 + c_1 ht_1 N_1 + c_2 ht_2 N_2),$$

where c_x is the per-capita consumption rate of prey species x and ht_x represents the predator's handling time of an individual of species x. In this case, species 2 clearly reduces the predator's effect on species 1 (2). These mechanisms and others (1) indicate that interaction modifications may be common in natural systems, making community dynamics even more difficult to predict.

Overview

Indirect effects have commonly been observed in long-term experiments or observations of species introductions and deletions when ecologists have focused on the response of multiple members of the community, rather than on species pairs. For example, Menge's recent review of indirect effects in rocky intertidal communities (71) found exponential increases in the number of indirect effects detected with increases in the number of species considered. Indirect effects have been most commonly found in marine and freshwater systems but have also been detected in terrestrial systems, despite the above-mentioned logistical difficulties of doing so. Many indirect effects can be

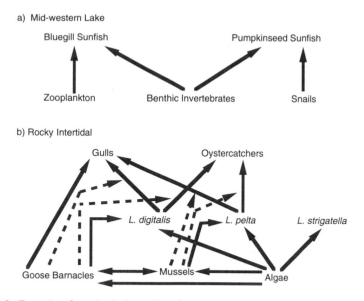

Figure 3 Examples of complex indirect effects in natural communities: (a) small lakes in Michigan (adapted from [79]), (b) a rocky intertidal community in Washington state (adapted from [132]). Vertical arrows: consumer-resource interactions, with arrow following the direction of energy flow. Arrows ending on horizontal: nonconsumptive direct interactions (interference competition, habitat preference) with arrows showing impacted species. Dashed arrows: modifications of consumer-resource interactions via crypsis.

ascribed to the simple types of interactions listed above (see also 71). However, there are often cases where the results do not fit such simple scenarios (71). For example, adding bluegill sunfish (*Lepomis gibbosus*) in Michigan lakes causes declines in zooplankton, declines in pumpkinseed sunfish (*Lepomis macrochirus*), and increases in snails through a combined effect of exploitative competition and apparent competition (Figure 3a; 79). Similarly, intertidal bird predators affect limpet grazers and algae through a morass of direct and indirect pathways (Figure 3b; 132). It is tempting to coin new terms to characterize such newly recognized patterns of interactions, but this type of enterprise would soon lead to a quagmire of new terminology. In characterizing indirect effects, it seems more appropriate simply to place them in the context of a functional web of interactions (71, 82) among species within a particular community.

In detailing the occurrence of indirect effects in the literature, it is also of interest to know whether direct effects are stronger than indirect effects. On the surface, one might expect direct effects to be stronger for a couple of reasons (106). First, because many indirect effects arise from chains of strong direct effects, strong direct effects are often required before even weak indirect effects can arise. Second, environmental variation may act independently on each species within an indirect pathway, thereby progressively weakening the strength of the effect relative to other sources of variation. Most arguments on the subject are verbal and deserve more rigorous theoretical investigation. Furthermore, the results of deterministic theory are at odds with these arguments: indirect effects in several theoretical studies can overwhelm direct effects in the absence of environmental variation (85, 140).

At present, empirical work on indirect effects seems too limited to merit strong generalizations on the relative importance of direct and indirect effects. In Schoener's literature analysis (106), selected direct effects tended to be stronger and to exhibit less variability than did indirect effects. This analysis, however, restricted its focus to direct effects that had been shown to be important, and it required assumptions in some cases about the whether an effect was predominantly direct or indirect when both could occur. The question may be better answered as techniques are developed to estimate simultaneously the strengths of various direct and indirect pathways that might affect a particular species, and when a better accounting of weak interactions is developed. For example, Menge (71) presented several possible approaches to assigning overall community variance due to direct versus indirect effects. In his analysis of a New England intertidal community, direct and indirect effects accounted for approximately equal amounts of community variation. Even if direct effects generally tend to be stronger than indirect effects, there are many examples where the direct effects of a manipulated species on a particular target species are much weaker than the indirect effects involved (e.g. 93, 132)

POSSIBLE EVOLUTIONARY CONSEQUENCES OF INDIRECT EFFECTS

Aside from the basic question of whether indirect effects exist, a related question of interest is whether indirect effects can play a role in evolutionary processes. Evidence bearing on this question is scarce, but several arguments can be made that indirect effects do affect the evolutionary trajectories of species. One would expect that if a species was negatively affected by another species, natural selection would favor the evolution of traits that either reduced the co-occurrence of the two species or reduced the impact of the indirect effects. This topic has been much discussed in relation to one indirect effect, exploitative competition. Arguments tracing back to Darwin (24) suggest that interspecific competition should lead to habitat separation between species or to character shifts that reduce competition between species pairs (12, 42). Although it is difficult to rigorously demonstrate such evolutionary changes in response to exploitative competition (20), there is some evidence that it occurs. For example, Grant (43) and his colleagues have mustered an enormous amount of observational support for the notion that competition for seeds has led to character divergence in Darwin's finches (*Geospiza* spp.). As mentioned above, there is nothing special about exploitative competition relative to other indirect effects, so one might predict similar evolutionary changes when a species is indirectly affected by another in a negative manner, regardless of the pathway.

Just as negative indirect effects should lead to traits favoring reduced coexistence or the minimization of the indirect effect, positive indirect effects should favor the evolution of traits that increase sympatry with the species involved and that maximize the effect of the indirect pathway. One possible example involves limpet habitat selection in the rocky intertidal communities of western North America (132). By making limpets more cryptic to bird predators, certain mussels and barnacles indirectly affect limpets by reducing the intensity of predation by birds. Corresponding with this indirect benefit, limpet species differentially select habitats that afford them the most protection from bird predation. Although the genetic basis of the behavior remains to be demonstrated, this scenario is consistent with positive indirect effects leading to the evolution of traits to maximize co-occurrence. There is some evidence that positive indirect effects may also affect morphological traits. For example, populations of the Caribbean tree lizard *Conocarpus erectus* have fewer trichomes on their leaves where high *Anolis* lizard densities reduce the abundance of grazing arthropods (104).

MECHANISMS PREVENTING INDIRECT EFFECTS

Although it is of great interest to know if and how indirect effects occur, it is also important to determine when and why indirect effects will not occur. Few studies have addressed the topic, although several causes have been proposed (106). The most obvious reason why indirect effects might not occur is because a system may be characterized by weak direct interactions between pairs of species. Even if direct effects are strong enough to be detected, they may be sufficiently weak that when linked together the effect of a perturbation becomes damped out. A second reason that indirect effects may not be important is if the system contains a great deal of environmental variation. For example, high rates of stress or disturbance might keep populations at sufficiently low levels that species do not interact strongly (5, 19). Similarly, if sufficient environmental noise affects the populations of each species in a chain of interactions, it may progressively swamp out the signal arising from the manipulation of a particular species at one end of the chain (106). Little theoretical work has been done on this topic (106), but in taking a path-analysis approach to analyzing indirect effects (described below), one can readily see how environmental stochasticity might weaken the importance of indirect effects.

A final reason that indirect effects might not be apparent is that other strong interactions may oppose the effect, yielding no net change. For example, birds that forage intertidally may have no net effect on some limpet species because positive indirect effects mediated through sessile habitat species may counteract the negative direct effects of predation (132). Under other circumstances, strong self-limitation may reduce interspecific effects (63, 135).

MATHEMATICAL ANALYSES AND PREDICTIONS OF INDIRECT EFFECTS

Theoretical Explorations

Theoretical analyses of multi-species models are critical to understanding and predicting the consequences of indirect effects; such analyses employ several different approaches. The most common approach uses dynamical models of multi-species systems. Because of the difficulty of analysis, these efforts have largely been restricted to linear equations taking the form of a standard Lotka-Volterra models. Methods have been developed to estimate indirect effects, based upon the per-capita effects of direct interactions among species pairs (58, 60, 101, 115). This approach has shown the potential for indirect effects to alter significantly the predicted relationships of species involved in either competition or consumer-resource interactions (25, 58, 60, 101, 106, 115, 122, 140). To date, this approach has not been particularly useful in predicting the consequences of changes to natural communities because of the difficulty in obtaining estimates of interaction coefficients in the field.

An early modification of this approach was loop analysis (61, 97). This technique, derived from the sort of linear models described above, allows qualitative predictions about how a system should respond to a perturbation, given that one knows the existence and the sign of interactions among the variables of interest. Because loop analysis does not require estimates of strengths of interaction among species, it has been more useful in making predictions. This approach is limited, however, to relatively simple community configurations; estimates of interaction strengths are required to make predictions in more complex webs.

The dynamic modeling approach has been extended to evaluate the consequences of adding nonlinear complexity. Work on models with nonlinear foraging effects (1–3) has shown a dizzying array of possible outcomes beyond those derived from linear models. Recent analyses have also started to incorporate spatial variation and higher-order terms representing interaction modifications (128). Although the complexity of the models precludes analytical results, simulation studies have come to the interesting conclusion that the higher-order terms promote the coexistence of species. If the result is general, then it would help answer the question of why natural communities can support so many species when simple mathematical models predict that they should be relatively unstable (69). Clearly more theoretical work is needed on these issues.

A second general theoretical approach has involved the analysis of networks of material flow through ecosystems (48, 85–87). In this approach, the amount of a material (e.g. carbon) residing in a particular compartment (e.g. within particular species, or within particular parts of the physical environment—sediments, for example) at a given point in time is divided fractionally between that which remains in the compartment and that which flows into other compartments at the next time interval. The method assumes that all compartments are at steady state, and that all flows (i.e. transition probabilities) are constant (48). Indirect effects are estimated by adding up all of the possible routes that a unit of material can travel from one compartment to another over an infinite time interval, compared to the magnitude of the direct exchange (48, 85–87). From these analyses, great importance has been placed upon the role of indirect effects in natural systems. This approach has spawned a lively controversy over methodological and interpretational issues (48, 62, 87, 91, 124). The network approach appears to have several drawbacks that may limit its utility in predicting the direct and indirect consequences of environmental impacts. The steady-state assumption would seem to make the method difficult to apply to a situation involving an environmental change (62), and it is hard to tell how nontrophic interactions (e.g. interference competition), which are known to be important components of indirect effects (e.g. 28, 64, 80, 81, 82, 113, 132, 134, 135), can be subsumed into this framework (62). Most importantly, the assumption that flows between compartments are independent of the states of the compartments seems particularly unrealistic. For example, most ecologists expect that higher numbers of predators should increase the total number of prey consumed, thereby increasing the flow from prey to predator compartments. Models taking the network approach have been parameterized for several systems and have been used in environmental impact assessments (87), but I know of no empirical tests of the approach. Experimental manipulations are clearly called for in these systems to determine the ability of this approach to predict the direct and indirect consequences of an impact on the environment.

Path Analysis

Another recent approach to the study of indirect effects is the application of path analysis (47, 138) to ecological systems. Path analysis is a statistical approach that estimates the degree to which changing a causal variable will affect a dependent variable through both direct and indirect pathways. Because this technique depends on the presence of variation in the system, the approach is strongest when applied to communities that are dynamic in space and time. Perhaps the best way to ensure this condition is to apply an experimental manipulation to the system and use the path analysis to follow the routes that the signal of the manipulation travels through other elements of the community (135).

Several kinds of information can be gleaned from a path analysis that make it a potentially important extension in the study of indirect effects. First, given several schemes of how variables are related to one another in a causal network, path analysis can be used to identify the most likely scenario (47, 135). This ability should promote a more frequent consideration of alternative causal hypotheses, particularly those that include multiple factors (98). Furthermore, because the technique deals with substantially more complex hypotheses than most statistical approaches, information on multiple variables is more likely to be obtained in empirical studies, thereby broadening our knowledge of possible impacts of a variable. Second, given a particular causal scheme, path analysis estimates the relative strength of direct and indirect pathways between pairs of variables. This represents a major improvement in the study of indirect effects, because the identification of important indirect effects is less tied to the discovery of "unanticipated results." Third, once a causal scheme and interaction strengths have been derived from path analysis, predictions can be made about the consequences of changing particular (previously unmanipulated) variables in the system. Obtaining such predictions of the direct and indirect consequences of an environmental impact has been and continues to be a basic goal of ecology.

The path-analysis approach has been applied in various forms to intertidal, freshwater, and terrestrial systems (6, 52, 126, 135). A key question is whether the technique works. Recently, experimental tests of path-analysis predictions have been carried out in a rocky intertidal community (135) and a steppe-boreal forest community (126). In both cases the predictions derived from path analysis have been supported, suggesting that the approach may indeed be useful. More applications and experimental tests of path analysis in different systems are clearly needed to evaluate its abilities to differentiate direct and indirect effects and predict their consequences.

There are several possible limitations of using path analysis that require further investigation. First, the results depend on the underlying causal scheme assumed. Therefore it is important for investigators to consider a variety of possible mechanisms by which an experimentally manipulated variable affects other variables of interest and to adjust their data collection accordingly. Second, the path analysis may do a poor job in systems near equilibrium, where little variation is available for the path analysis to work with. Third, although path analysis can incorporate reciprocal interactions (47, 139), its ability to do so has not been challenged to date in the ecological systems where it has been experimentally evaluated. Because path analysis is related to traditional linear regression techniques, which assume unidirectional causality, it is unclear whether it can adequately handle reciprocal effects. Fourth, interaction modifications are fairly tricky to handle in path analysis. They can be accommodated only if a specific higher-order functional form is assumed to describe the effect (e.g. the product of two causal variables). Finally, as the complexity of a causal model increases, sample sizes must increase too. This may limit the application of this approach in complex systems, unless those systems are organized to some extent into sub-modules of highly interacting species (82).

FUTURE DIRECTIONS

Evidence is now accumulating that indirect effects can play an important role in natural communities, indicating that they deserve further study. I foresee several directions to future research. First, experimental investigations of indirect effects will undoubtedly continue, given the rise of field experiments in ecology. It is critical that these experimental investigations monitor a variety of response variables and try to place these variables in the context of the entire community or ecosystem to better understand the mechanisms involved. Furthermore, techniques (e.g. path analysis) must be developed and applied to more accurately distinguish possible direct and indirect effects. Second, experiments will be of particular value if they are designed to test specific mechanistic models of community organization in order to evaluate predictive approaches in ecology. For example, several studies have experimentally investigated the degree to which simple food chain models predict the effects of adding or deleting top predators or varying productivity (14, 93, 121, 135, 137). Third, the development of theory for complex systems remains an important area of research. For example, Wilson's (128) simulations of systems incorporating higher-order interactions and spatial structure is one excellent starting point for further investigation into the consequences of complex interactions. Two other research directions seem particularly important and will likely lead to more extensive progress: the estimation of interaction strength in the field, and the determination and evaluation of mechanisms that modify interactions among species.

The development and testing of theory would be greatly facilitated by empirical measurements of interaction strength, because the predictions of many models change as interaction strength changes (97, 106, 140). An important initial consideration is what interaction strength means. To this end, it is critical to distinguish between interaction strength and effect strength. In theoretical treatments (69), interaction strength refers to the per-capita rates of

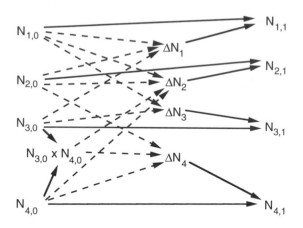

Figure 4 General structure of a path analysis designed to estimate interaction strength in dynamical models. $N_{y,0}$ represents abundance of species y at time $t = 0$; δN_y, represents the change in abundance from one census to the next ($t = 1$); $N_{y,1}$ represents abundance of species y at the subsequent census. Dashed arrows represent those terms yielding estimates of interaction strength to be incorporated into dynamical models. Arrows involving the term ($N_{3,0} \times N_{4,0}$) illustrate how an interaction modification can be incorporated into path analysis (here species 3 modifies the effect of species 4 on species 2).

change of one variable caused by the direct effects of another variable (e.g. the average effect of one predator individual on one prey individual). Effect strength is the total impact that one variable has on another (e.g. the effect of a predator species on a prey species). This latter quantity is measured most often in experimental manipulations. Effect strength differs from interaction strength in that a rare species with a high interaction strength may have the same total effect as a common species with a low interaction strength. Both quantities provide important information, but interaction strength is the quantity that belongs in models.

Several approaches seem possible to estimate interaction strength. One method is to manipulate the value of every variable in a system by a known amount and estimate the per-capita consequences on target species (7, 39, 83). In doing so, one must be careful to ensure that only direct effects are being measured. Although the experimental perturbation approach is powerful, it is logistically difficult, if not impossible, to apply to the entire system of interest. A second approach is to take observational data that directly reflect interaction strength in the models (i.e. per-capita rates). Many types of information have been used as indices of interaction strength (e.g. percent composition in a diet, energy flow rate), but frequently these measures correspond poorly to experimentally derived results (82). This situation is not surprising, because the units of the quantities being measured match the units neither of interaction strength nor of effect strength in the models. In contrast, measurements that have the proper units of interaction-and-effect strength can predict quite well the direct effects manipulations (130). A third approach is a synthetic framework that combines experimental manipulation, mathematical theory, and path analysis. A difficulty with previous applications of path analysis to ecological systems is that the path analyses were not structured in the same manner as the commonly used theoretical models. Most models examine rates of change in variables as a function of the states of other variables, whereas the path analyses to date have examined the state of one variable as a function of the states of other variables. Therefore the interaction strengths measured by path analysis and those required by the models are not equivalent. By defining the response variables in path analysis to match the structure of the models of interest (Figure 4), the two approaches could be linked, and the models could be more effectively developed. Furthermore, reciprocal interactions and nonlinearities are more easily handled in this format. A simplified version of this approach has been applied to communities of tide-pool sculpins, and the technique successfully predicted the consequences of experimental manipulations (89).

A final important area of future research on indirect effects is to identify the mechanisms by which interactions among species pairs are modified by other species. Past work addressing this question focused on quantitative evidence, but methodological, definitional, and theoretical difficulties have complicated the interpretations of this work (4, 8, 15, 92, 136). By considering instead how interaction modifications arise in natural systems, the identification of likely mechanisms should come more quickly, allowing novel experimental approaches to test their importance (e.g. 129), as well as focusing investigators' attention on the interactions to be studied most intensively.

CONCLUSION

In summary, the study of indirect effects remains an important area of investigation as ecologists worry about how abstract models can be while still yielding useful predictions about the consequences of an environmental change. If indirect effects are important, then models must account for community- and ecosystem-level phenomena in order to make useful predictions. Experimental demonstrations of indirect effects—both through chains of direct interactions and through one species modifying interactions between other species—are becoming increasingly common as experimentalists pay more attention to the responses of multiple variables. The relative importance of indirect effects remains uncertain because of methodological difficulties in determining their existence and strength, but recent applications of path analysis may improve the situation. Future progress is most likely to occur through the synthesis of experimental, statistical, and theoretical methods to derive estimates of interaction strength and the degree to which these strengths are modified by other ecological variables. Through this work, insight will be gained into whether the behavior of ecological systems can be decomposed into their component parts, and how predictable the consequences of human impacts are on the environment.

Any *Annual Review* chapter, as well as any article cited in an *Annual Review* chapter, may be purchased from the Annual Reviews Preprints and Reprints service. 1-800-347-8007; 415-259-5017; email: arpr@class.org

Literature Cited

1. Abrams PA. 1983. Arguments in favor of higher order interactions. *Am. Nat.* 121:887–91
2. Abrams PA. 1987. Indirect interactions between species that share a predator: varieties of indirect effects. See Ref. 54, pp. 38–54
3. Abrams PA. 1993. Indirect effects arising from optimal foraging. See Ref. 55, pp. 255–79
4. Adler FR, Morris WF. 1994. A general test for interaction modifications. *Ecology.* In press
5. Andewartha HG, Birch LC. 1954. *The Distribution and Abundance of Animals.* Chicago: Univ. Chicago Press
6. Arnold SJ. 1972. Species densities of predators and their prey. *Am. Nat.* 106:220–36
7. Bender EA, Case TJ, Gilpin ME. 1984. Perturbation experiments in community ecology: theory and practice. *Ecology* 65:1–13
8. Billick I, Case TJ. 1994. Higher order interactions in ecological communities: what are they and how can they be detected? *Ecology.* In press
9. Bosman AL, Du Toit JT, Hockey PAR, Branch GM. 1986. A field experiment demonstrating the influence of seabird guano on intertidal primary production. *Estuarine, Coastal Shelf Sci.* 23:283–94
10. Brönmark C, Klosiewski SP, Stein RA. 1992. Indirect effects of predation in a freshwater benthic food chain. *Ecology* 73:1662–74
11. Brooks JL, Dodson SI. 1965. Predation, body size, and composition of plankton. *Science* 150:28–35
12. Brown WL Jr, Wilson EO. 1956. Character displacement. *Syst. Zool.* 5:49–64
13. Bryant JP, Chapin FS III. 1986. Browsing-woody plant interactions during bo-

real forest plant succession. In *Forest Ecosystems in the Alaskan Taiga,* ed. K Van Cleve, FS Chapin III, PW Flanagan, LA Biereck, CT Dyrness, pp. 213–25. New York: Springer-Verlag
14. Carpenter SR, Kitchell JF, Hodgson JR. 1985. Cascading trophic interactions and lake productivity. *BioScience* 35:634–39
15. Case TJ, Bender EA. 1981. Testing for higher order interactions. *Am. Nat.* 118:920–29
16. Cates RG, Orians GH. 1975. Successional status and the palatability of plants to generalized herbivores. *Ecology* 56:410–18
17. Coley PD. 1983. Herbivory and defensive characteristics of tree species in a lowland tropical forest. *Ecol. Monogr.* 53:209–33
18. Colwell RK, Fuentes ER. 1975. Experimental studies of the niche. *Annu. Rev. Ecol. Syst.* 6:281–310
19. Connell JH. 1977. Diversity in tropical rainforests and coral reefs. *Science* 199:1302–10
20. Connell JH. 1980. Diversity and coevolution of competitors or the ghost of competition past. *Oikos* 35:131–38
21. Connell JH. 1983. On the prevalence and relative importance of interspecific competition: evidence from field experiments. *Am. Nat.* 122:661–96
22. Connell JH, Slatyer RO. 1977. Mechanisms of succession in natural communities and their role in community stability and organization. *Am. Nat.* 111:1119–44
23. Crowder LG, Cooper WE. 1982. Habitat structural complexity and the interaction between bluegills and their prey. *Ecology* 63:1802–13
24. Darwin CR. 1859. *The Origin of Species.* Reprinted 1976. New York: Macmillan

25. Davidson DW. 1980. Some consequences of diffuse competition in a desert ant community. *Am. Nat.* 116:92–105
26. Davidson DW, Inouye RS, Brown JH. 1984. Granivory in a desert ecosystem: experimental evidence for indirect facilitation of ants by rodents. *Ecology* 65:1780–86
26. Day RW, Osman RW. 1981. Predation by *Patiria miniata* (Asteriodea) on bryzoans: prey diversity may depend on the mechanism of succession. *Oecologia* 51:300–9
28. Dethier MN, Duggins DO. 1984. An "indirect commensalism" between marine herbivores and the importance of competitive hierarchies. *Am. Nat.* 124:205–19
29. Duggins DO. 1980. Kelp beds and sea otters: an experimental approach. *Ecology* 61:447–53
30. Duggins DO, Simenstad CA, Estes JA. 1989. Magnification of secondary production by kelp detritus in coastal marine ecosystems. *Science* 245:170–73
31. Dungan ML. 1986. Three-way interactions: barnacles, limpets, and algae in a Sonoran desert rocky intertidal zone. *Am. Nat.* 127:292–316
32. Dungan ML. 1987. Indirect mutualism: complementary effects of grazing and predation in a rocky intertidal community. See Ref. 54, pp. 188–200
33. Estes JA, Palmisano JF. 1974. Sea otters: their role in structuring benthic nearshore communities. *Science* 185:1058–60
34. Estes JA, Smith NS, Palmisano JF. 1978. Sea otter predation and community organization in the western Aleutian Islands, Alaska. *Ecology* 59:822–33
35. Fairweather PG. 1990. Is predation capable of interacting with other community processes on rocky reefs? *Aust. J. Ecol.* 15:453–64
36. Farrell TM. 1991. Models and mechanisms of succession: an example from a rocky intertidal community. *Ecol. Monogr.* 61:95–113
37. Flecker AS. 1992. Fish trophic guilds and the structure of a tropical stream: weak direct versus strong indirect effects. *Ecology* 73:927–40
38. Fretwell SD. The regulation of plant communities by the food chains exploiting them. *Perspect. Biol. Med.* 20:169–85
39. Gause GF. 1934. *The Struggle for Existence.* Reprinted 1964. New York: Hafner
40. Godfray HC Jr. 1985. The absolute abundance of leaf miners on plants of different successional stages. *Oikos* 45:17–25
41. Goldberg DE, Barton AM. 1992. Patterns and consequences of interspecific competition in natural communities: a review of field experiments with plants. *Am. Nat.* 139:771–801
42. Grant PR. 1972. Convergent and divergent character displacement. *Biol. J. Linn. Soc.* 4:39–68
43. Grant PR. 1986. *Ecology and Evolution of Darwin's Finches.* Princeton: Princeton Univ. Press
44. Grosholz ED. 1992. Interactions of intraspecific, interspecific, and apparent competition with host-pathogen population dynamics. *Ecology* 73:507–14
45. Hairston NG, Smith FE, Slobodkin LB. 1960. Community structure, population control, and competition. *Am. Nat.* 94:421–25
46. Hay ME. 1986. Associational defenses and the maintenance of species diversity: turning competitors into accomplices. *Am. Nat.* 128:617–41
47. Hayduk LA. 1987. *Structural Equation Modeling with LISREL.* Baltimore: Johns Hopkins Univ. Press
48. Higashi M, Patten BC. 1989. Dominance of indirect causality in ecosystems. *Am. Nat.* 133:288–302
49. Holling CS. 1965. The functional response of predators to prey density and its role in mimicry and population regulation. *Mem. Entomol. Soc. Can.* 45:1–60
50. Holt RD. 1977. Predation, apparent competition, and the structure of prey communities. *Theor. Popul. Biol.* 12:197–229
51. Johnson LE. 1992. Potential and peril of field experimentation: the use of copper to manipulate molluscan herbivores. *J. Exp. Mar. Biol. Ecol.* 160:251–62
52. Johnson ML, Huggins DG, DeNoylles F Jr. 1991. Ecosystem modeling with LISREL. *Ecol. Appl.* 1:383–98
53. Kajak A, Andrezejewska L, Wojcik Z. 1968. The role of spiders in the decrease of damage caused by Acridoidea on meadows—experimental investigations. *Ekol. Polska. Ser. A* 16:755–64
54. Kawanabe H, Cohen JE, Iwasaki K, ed. 1993. *Mutualism and Community Organization.* Oxford: Oxford Univ. Press
55. Kerfoot WC, Sih A, eds. 1987. *Predation: Direct and Indirect Impacts on Aquatic Communities.* Hanover: Univ. Press New Engl.
56. Kneib RT. 1988. Testing for indirect effects of predation in an intertidal soft-bottom community. *Ecology* 69:1795–805

57. Kvitek RG. 1991. Sequestered paralytic shellfish poisoning toxins mediate glaucous-winged gull predation on bivalve prey. *Auk* 108:381–92

58. Lawlor LR. 1979. Direct and indirect effects of n-species competition. *Oecologia* 43:355–64

59. Leibold MA. 1989. Resource edibility and the effects of predators and productivity on the outcome of trophic interactions. *Am. Nat.* 134:922–49

60. Levine SH. 1976. Competitive interactions in ecosystems. *Am. Nat.* 110:903–10

61. Levins R. 1975. Evolution in communities near equilibrium. In *Ecology and Evolution of Communities,* ed. ML Cody, JM Diamond, pp. 16–50. Cambridge: Harvard Univ. Press

62. Loehle C. 1990. Indirect effects: a critique and alternate methods. *Ecology* 71:2382–86

63. Lotka AJ. 1925. *Elements of Mathematical Biology.* Reprinted 1956. New York: Dover

64. Lubchenco J. 1978. Plant species diversity in a marine intertidal community: importance of herbivore food preference and algal competitive abilities. *Am. Nat.* 112:23–39

65. Lubchenco J. 1983. Littorina and Fucus: effects of herbivores, substratum heterogeneity, and plant escapes during succession. *Ecology* 64:1116–23

66. Lubchenco J, Menge BA. 1978. Community development and persistence in a low rocky intertidal zone. *Ecol. Monogr.* 48:67–94

67. Marquis RJ, Whelan CJ. 1994. Insectivorous birds increase growth of white oak through consumption of leaf-chewing insects. *Ecology.* In press

68. Martin TH, Crowder LB, Dumas CF, Burkholder JM. 1992. Indirect effects of fish on macrophytes in Bays Mountain Lake: evidence for a littoral trophic cascade. *Oecologia* 89:476–81

69. May RM. 1974. *Stability and Complexity in Model Ecosystems.* Princeton: Princeton Univ. Press

70. Menge BA. 1976. Organization of the New England rocky intertidal community: role of predation, competition, and environmental heterogeneity. *Ecol. Monogr.* 46:355–93

71. Menge BA. 1994. Indirect effects in marine rocky intertidal interaction webs: patterns and importance. *Ecol. Monogr.* In press

72. Menge BA, Lubchenco J, Ashkenas LR, Ramsey F. 1986. Experimental separation of effects of consumers on sessile prey in the low zone of a rocky shore in the Bay of Panama: direct and indirect consequences of food web complexity. *J. Exp. Mar. Biol. Ecol.* 100:225–69

73. Mercurio KS, Palmer AR, Lowell RB. 1985. Predator-mediated microhabitat partitioning by two species of visually cryptic, intertidal limpets. *Ecology* 66:1417–25

74. Miller TE, Kerfoot WC. 1987. Redefining indirect effects. See Ref. 54, pp. 33–37

75. Moreno CA, Lunecke KM, Lépez MI. 1986. The response of an intertidal *Concholepas concholepas* (Gastropoda) population to protection from Man in southern Chile, and the effects on benthic sessile assemblages. *Oikos* 46:359–64

76. Moreno CA, Sutherland JP, Jara JF. 1984. Man as a predator in the intertidal zone of southern Chile. *Oikos* 42:155–60

77. Morin PJ. 1983. Predation, competition, and the composition of larval anuran guilds. *Ecol. Monogr.* 53:119–38

78. Oksanen L, Fretwell SD, Arruda J, Niemela P. 1981. Exploitation ecosystems in gradients of primary productivity. *Am. Nat.* 118:240–61

79. Osenberg CW, Mittelbach GG, Wainwright PC. 1992. Two-stage life histories in fish: the interaction between juvenile competition and adult performance. *Ecology* 73:255–67

80. Paine RT. 1966. Food web complexity and species diversity. *Am. Nat.* 100:65–75

81. Paine RT. 1974. Intertidal community structure. Experimental studies on the relationship between a dominant competitor and its principal predator. *Oecologia* 14:93–120

82. Paine RT. 1980. Food webs: linkage, interaction strength and community infrastructure. *J. Anim. Ecol.* 49:667–85

83. Paine RT. 1992. Food-web analysis through field measurement of per capita interaction strength. *Nature* 355:73–35

84. Paine RT, Wootton JT, Boersma PD. 1990. Direct and indirect effects of peregrine falcon predation on seabird abundance. *Auk* 107:1–9

85. Patten BC. 1982. Environs: relativistic elementary particles for ecology. *Am. Nat.* 119:179–219

86. Patten BC. 1983. On the quantitative dominance of indirect effects in ecosystems. In *Analysis of Ecological Systems: State-of-the-Art in Ecological Modeling,* ed. WK Lauenroth, GV Skogerboe, pp. 27–37. Amsterdam: Elsevier

87. Patten BC. 1990. Environ theory and indirect effects: a reply to Loehle. *Ecology* 71:2386–93

88. Perfecto I. 1990. Indirect and direct effects in a tropical agroecosystem: the maize-pest-ant system in Nicaragua. *Ecology* 71:2125–34

89. Pfister CA. 1993. *The dynamics of fishes in intertidal pools.* PhD thesis. Univ. Wash., Seattle. 169 pp.

90. Pfister CA, Hay ME. 1988. Associational plant refuges: convergent patterns in marine and terrestrial communities result from different mechanisms. *Oecologia* 83:405–13

91. Pilette R. 1989. Evaluating direct and indirect effects in ecosystems. *Am. Nat.* 133:303–7

92. Pomerantz MJ. 1981. Do "higher order interactions" in competition systems really exist? *Am. Nat.* 117:583–91

93. Power ME. 1990. Effects of fish in river food webs. *Science* 250:811–14

94. Power ME. 1990. Resource enhancement by indirect effects of grazers: armored catfish, algae, and sediment. *Ecology* 71:897–904

95. Power ME, Matthews WJ, Stewart AJ. 1985. Grazing minnows, piscivorous bass, and stream algae: dynamics of a strong interaction. *Ecology* 66:1448–57

96. Pringle CM, Blake GA, Covich AP, Buzby KM, Finley A. 1993. Effects of omnivorous shrimp in a montane tropical stream: sediment removal, disturbance of sessile invertebrates and enhancement of understory algal biomass. *Oecologia* 93:1–11

97. Puccia CT, Levins R. 1985. *Qualitative Modeling of Complex Systems: An Introduction to Loop Analysis and Time Averaging.* Cambridge: Harvard Univ. Press

98. Quinn JF, Dunham AE. 1983. On hypothesis testing in ecology and evolution. *Am. Nat.* 122:602–17

99. Rosenzweig ML. 1973. Exploitation in three trophic levels. *Am. Nat.* 107:275–94

100. Sarnelle O. 1993. Herbivore effects on phytoplankton succession in a eutrophic lake. *Ecol. Monogr.* 63:129–49

101. Schaffer WM. 1981. Ecological abstraction: the consequences of reduced dimensionality in ecological models. *Ecol. Monogr.* 51:383–401

102. Schmitt RJ. 1987. Indirect interactions between prey: apparent competition, predator aggregation, and habitat segregation. *Ecology* 68:1887–97

103. Schoener TW. 1983. Field experiments on interspecific competition. *Am. Nat.* 122:240–85

104. Schoener TW. 1987. Leaf pubescence in buttonwood: community variation in a putative defense against defoliation. *Proc. Natl. Acad. Sci. USA* 84:7992–95

105. Schoener TW. 1989. Food webs from the small to the large. *Ecology* 70:1559–89

106. Schoener TW. 1993. On the relative importance of direct versus indirect effects in ecological communities. See Ref. 55, pp. 365–415

107. Settle WH, Wilson LT. 1990. Invasion by the variegated leafhopper and biotic interactions: parasitism, competition, and apparent competition. *Ecology* 71:1461–70

108. Shapiro J. 1979. The importance of trophic-level interactions to the abundance and species composition of algae in lakes. In *Hypertrophic Ecosystems,* ed. J Barica, L Mur, pp. 101–16. The Hague: Junk

109. Sih A, Crowley P, McPeek M, Petranka J, Strohmeier K. 1985. Predation, competition and prey communities: a review of field experiments. *Annu. Rev. Ecol. Syst.* 16:269–311

110. Sinclair A. 1979. Dynamics of the Serengeti ecosystem. In *Serengeti: Dynamics of an Ecosystem,* ed. A Sinclair, M. Norton-Griffiths, pp. 1–30. Chicago: Univ. Chicago Press

111. Slagsvold T. 1980. Habitat selection in birds: on the presence of other bird species with special regard to *Turdus pilaris. J. Anim. Ecol.* 49:523–36

112. Smith FE. 1969. Effects of enrichment in mathematical models. In *Eutrophication: Causes and Consequences,* pp. 631–45. Washington, DC: Natl. Acad. Press

113. Sousa WP. 1979. Experimental investigations of disturbance and ecological succession in a rocky intertidal community. *Ecol. Monogr.* 49:227–54

114. Spiller DA, Schoener TW. 1990. A terrestrial field experiment showing the impact of eliminating top predators on foliage damage. *Nature* 347:469–72

115. Stone L, Roberts A. 1991. Conditions for a species to gain advantage from the presence of competitors. *Ecology* 72:1964–72

116. Strauss SY. 1991. Indirect effects in community ecology: their definition, study and importance. *Trends Ecol. Evol.* 6:206–10

117. Strong DR. 1992. Are trophic cascades all wet? Differentiation and donor-control in speciose ecosystems. *Ecology* 73:747–55

118. Tahvanainen JO, Root RB. 1972. The influence of vegetational diversity on the population ecology of a specialized herbivore, *Phyllotreta cruciferaea* (Col-

eoptera: Chrysomelidae). *Oecologia* 10: 321–46

119. Thompson DB, Brown JH, Spencer WD. 1991. Indirect facilitation of granivorous birds by desert rodents: experimental evidence from foraging patterns. *Ecology* 72:852–63

120. Tilman D. 1987. The importance of the mechanisms of interspecific competition. *Am. Nat.* 129:769–74

121. Turner AM, Mittelbach GG. 1990. Predator avoidance and community structure: interactions among piscivores, planktivores, and plankton. *Ecology* 71:2241–54

122. Vandermeer J. 1980. Indirect mutualism: variations on a theme by Stephen Levine. *Am. Nat.* 116:441–48

123. Vanni MJ, Findlay DL. 1990. Trophic cascades and phytoplankton community structure. *Ecology* 71:921–37

124. Weigert RG, Kozlowski J. 1984. Indirect causality in ecosystems. *Am. Nat.* 124:293–98

125. Werner EE, Gilliam JF, Hall DJ, Mittelbach GG. 1983. An experimental test of the effects of predation risk on habitat use in fish. *Ecology* 64:1540–48

126. Wesser SD, Armbruster WS. 1991. Species distribution controls across a forest-steppe transition: a causal model and experimental tests. *Ecol. Monogr* 61:323–42

127. Wilbur HM, Fauth JE. 1990. Experimental aquatic food webs: interactions between two predators and two prey. *Am. Nat.* 135:176–204

128. Wilson DS. 1992. Complex interactions in metacommunities, with implications for biodiversity and higher levels of selection. *Ecology* 73:1984–2000

129. Wissinger S, McGrady J. 1993. Intraguild predation and competition between larval dragonflies: direct and indirect effects on shared prey. *Ecology* 74:207–18

130. Wootton JT. 1990. *Direct and indirect effects of bird predation and excretion on the spatial and temporal patterns of intertidal species.* PhD thesis. Univ. Wash., Seattle. 207 pp.

131. Wootton JT. 1991. Direct and indirect effects of nutrients on intertidal community structure: variable consequences of seabird guano. *J. Exp. Mar. Biol. Ecol.* 151:139–53

132. Wootton JT. 1992. Indirect effects, prey susceptibility, and habitat selection: impacts of birds on limpets and algae. *Ecology* 73:981–91

133. Wootton JT. 1993. Indirect effects and habitat use in an intertidal community: interaction chains and interaction modifications. *Am. Nat.* 141:71–89

134. Wootton JT. 1993. Size-dependent competition: effects on the dynamics versus the endpoint of mussel bed succession. *Ecology* 74:195–206

135. Wootton JT. 1994. Predicting direct and indirect effects: an integrated approach using experiments and path analysis. *Ecology* 75:151–65

136. Wootton JT. 1994. Putting the pieces together: testing the independence of interactions among organisms. *Ecology.* In press

137. Wootton JT, Power ME. 1993. Productivity, consumers and the structure of a river food chain. *Proc. Natl. Acad. Sci. USA* 90:1384–87

138. Wright S. 1934. The method of path coefficients. *Ann. Math. Stat.* 5:161–215

139. Wright S. 1960. The treatment of reciprocal interaction, with or without lag, in path analysis. *Biometrics* 16:423–45

140. Yodzis P. 1988. The indeterminacy of ecological interactions as perceived through perturbation experiments. *Ecology* 69:508–15

Annu. Rev. Ecol. Syst. 1994. 25:467–93

INTEGRATIVE APPROACHES TO EVOLUTIONARY ECOLOGY: *Anolis* Lizards as Model Systems

Jonathan B. Losos

Department of Biology, Campus Box 1137, Washington University, St. Louis, Missouri 63130-4899

KEY WORDS: competition, evolutionary ecology, phylogeny

Abstract

Two approaches characterize the study of evolutionary ecology. Prospective studies investigate how present-day ecological processes may lead to evolutionary change; retrospective studies ask how present-day ecological conditions can be understood as the outcome of historical events. I argue that the most appropriate test of an evolutionary ecological hypothesis requires an integration of these approaches. I illustrate this approach by examining the hypothesis that interspecific competition has been the driving force behind the evolutionary radiation of *Anolis* lizards in the Caribbean. This hypothesis is supported by four lines of evidence: 1. Anole communities are structured by competition; 2. Populations alter resource use in the presence of congeners; 3. Microevolutionary adaptation occurs in response to resource shifts; and 4. Macroevolutionary patterns are consistent with interspecific competition as the driving force behind anole adaptive radiation.

INTRODUCTION

A schism exists in the field of evolutionary ecology. To some, evolutionary ecology is prospective, inquiring how present-day ecological processes may lead to evolutionary change, whereas for others, the emphasis is retrospective, asking how present-day ecological conditions can be understood as the out-

467

come of historical events. This dichotomy can be seen by comparing Pianka's (89) ecologically oriented *Evolutionary Ecology* with Cockburn's considerably more evolutionary *Introduction to Evolutionary Ecology* (11) or by perusing the pages of the journal *Evolutionary Ecology.*

Although some authors suggest that "evolutionary ecology" should include both perspectives (e.g. 3, 101), most practitioners in the field usually consider only one. For example, I examined all papers in the first four volumes of the journal *Evolutionary Ecology* in addition to a haphazard survey of papers, books, and symposia with "evolutionary ecology" in the title. Of the papers surveyed, only 5.3% considered both aspects of evolutionary ecology. By contrast, 85.3% focused on prospective approaches and 9.3% dealt with historical determinants of currently observed ecological patterns.

Although both retrospective and prospective studies are interesting in their own right, each of these approaches is limited in scope. By their nature, historical studies concern patterns and their interpretation; experimentation is not possible. Thus, for example, one may make deductions about historical sequences or how the evolutionarily-acquired features of taxa limit or channel their ecology, but these inferences can be tested only in a correlational manner with historical data. On the other hand, although one can directly measure and experimentally verify the existence and magnitude of currently operative ecological processes, one cannot demonstrate either that these processes have operated in the past, or that the processes, extrapolated over time, are sufficient to explain evolutionary patterns.

An integrative approach, by contrast, can provide broad insight into the factors regulating biological diversity. The prospective and retrospective approaches are complementary. Historical approaches can suggest hypotheses about which processes have been important in shaping biological diversity. Prospective studies can verify the importance of these processes, at least in contemporary communities, and can examine whether they lead to microevolutionary change in the direction predicted by historical studies. Consider, for example, a hypothesis, addressed below, that interspecific competition has been the driving force behind adaptive radiation in a given lineage. This hypothesis entails three predictions: 1. Historical patterns implicate competition as a cause of evolutionary diversification; 2. interspecific competition is demonstrably important in present-day communities; and 3. such competition leads to microevolutionary change which, if extrapolated sufficiently over time, would produce observed macroevolutionary patterns. Confirmation of all three premises would strongly support the hypothesis that competition has been the primary force determining community structure and diversification.

One of the first attempts to integrate phylogenetic and prospective approaches to evolutionary ecology was a pioneering analysis of the evolution of communities of *Anolis* lizards in the Caribbean (141). In that paper, Williams argued that interspecific competition was the driving force behind the diversification of Caribbean anoles. The subsequent 22 years have seen both a tremendous amount of research on *Anolis* ecology and evolution and major conceptual advances in how these fields are studied.

The goals of this paper are two-fold. First, the body of the paper attempts to demonstrate the power of an integrative approach by focusing on the hypothesis that interspecific competition has been the major force guiding the community structure and evolution of Caribbean *Anolis* lizards. I address the three predictions detailed above by arguing that competition is a potent force structuring present-day anole communities, that anoles have evolved adaptations to allow resource partitioning, and that historical analyses implicate competition as a driving force in the anole adaptive radiation. Second, in the latter part of the paper, I return to a broader discussion of integrative approaches to evolutionary ecology and particularly address the question of how to interpret situations in which independent lines of investigation are inconsistent with respect to a given hypothesis.

BACKGROUND ON *ANOLIS*

Anolis is one of the largest vertebrate genera, with approximately 300 described species, half of which occur on Caribbean islands. Anoles are typically small, arboreal insectivores, but interspecific variation exists in size, habitat, and diet (see 130). Most of the small islands in the Caribbean contain 1–2 species of anoles, although islands that were part of considerably larger landmasses within the past 10,000 years, such as the Bahamas and satellite islands in the Greater Antilles, may have larger anole faunas (96, 123). Radiations on each of the Greater Antilles, however, have been more extensive, resulting in local communities with as many as ten sympatric species; total diversity exceeds 40 species on both Hispaniola and Cuba (142; Figure 1).

These radiations have produced essentially the same set of ecological types, termed "ecomorphs," on each island. An ecomorph is a group of "species with the same structural habitat/niche, similar in morphology and behavior, but not necessarily close phyletically" (141, p. 82). Quantitative morphological measurements indicate that members of each ecomorph type are truly convergent in morphology; species do indeed cluster in morphological space by ecomorph type rather than by phylogenetic affinity (69, 80). These differences in morphology correlate with differences in ecology and behavior (65, 67, 82, 83); functional studies indicate that differences in locomotor capabilities are responsible for these ecomorphological correlations (66, 75).

Ecomorphs are named for the microhabitat they normally utilize. Four ecomorph types (trunk-ground, trunk-crown, crown-giant, and twig) are common to all four Greater Antillean islands. The grass-bush ecomorph is present

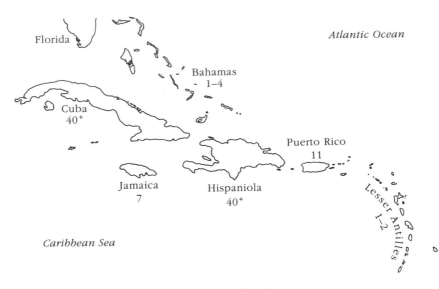

Figure 1   Caribbean islands and their anole species diversity.

on three of the islands but absent in Jamaica, whereas the trunk ecomorph is found only on Hispaniola and Cuba.

Higher-level phylogenetic relationships within anoline lizards are still controversial (5, 7, 33, 34, 37, 143), but reliable phylogenies exist for the radiations on Jamaica and Puerto Rico (see 69). Despite uncertainty about anoline relationships, the radiations on the islands clearly have been, for the most part, independent (142). Consequently, members of an ecomorph type are similar as a result of convergent evolution rather than recency of common ancestry.

INTERSPECIFIC INTERACTIONS AND THE EVOLUTION OF COMMUNITY STRUCTURE

Interspecific competition has often been invoked as the underlying driving force in adaptive radiation in general (22, 29, 114), and in the Caribbean anoline radiation in particular (69, 104, 141). Although particular hypotheses differ in details (69, 104, 141), all share a common scenario: Ecologically similar species compete strongly, creating strong selective pressure for species to diverge in resource use, thereby allowing coexistence. These competitive processes, extended over macroevolutionary time, are postulated to have produced the adaptive radiations observed today.

Four explicit predictions stem from the hypothesis that interspecific competition drives adaptive radiation:

1. Anole communities are structured by competition.
2. To minimize competitive pressures, populations alter their resource use as a function of which competitors are sympatric.
3. Microevolutionary (defined here as equivalent to within-species) changes in physiology, morphology, and behavior occur as populations adapt to alterations in resource use.
4. Macroevolutionary patterns are consistent with a scenario of adaptation to specialized niches in response to interspecific competition.

In the following, I illustrate that the evidence for these four propositions is strong.

Anole Communities Are Structured by Competition

SYMPATRIC ANOLES DIFFER IN RESOURCE USE   Differences in resource use among sympatric species have often been taken as evidence of ongoing and/or past competition (13, 124). *Anolis* species partition resources along three axes (93, 94, 98, 142): prey size, which is strongly correlated with body size (102, 115, 116, 125; but see 21); structural habitat, usually defined by measurements of perch height and diameter; and microclimate (i.e. thermal habitat). In the Greater Antilles, sympatric species invariably differ on at least one of these axes, with partitioning by structural habitat being most common (98, 116, 121; 126, 127). By contrast, in the Lesser Antilles, sympatric species invariably differ along at least two axes (106).

Variation in which resource axes are partitioned   The relative importance of these resource axes exhibits geographic and lineage-based differences. In the Lesser Antilles, species on two-species islands always differ greatly in body size, with one exception (118). However, the species differ in which the other resource axis is partitioned. In the northern Lesser Antilles, species (*bimaculatus* series) differ more in structural than in climatic microhabitat, whereas in the southern Lesser Antilles (*roquet* lineage), the reverse is true (106).

Similarly, in the Greater Antilles, where sympatric species partition resources on at least one axis, variation exists in which axis is partitioned by sympatric members of the same ecomorph [a null model confirms that sympatric members of the same ecomorph are more ecologically distinctive than expected by chance (118); exceptions occur at zones of contact between parapatric species (142)]. In contrast to the situation in the Lesser Antilles, however, the variation correlates with differences in structural habitat, rather than geography or lineage (127). The more terrestrial trunk-ground and grass-bush ecomorphs partition climatic microhabitats when two members of either type occur sympatrically. This partitioning is so strong that sympatric members of either ecomorph type often are almost allotopic, only overlapping at ecotones (e.g. forest edge—93, 107). These distributions are probably driven by physiological adaptation to different microhabitats (see Comparative Studies Among Closely Related Taxa), which explains why different species will occupy the same spot at different times of the day (119). Closely related members of these ecomorph types that use different climatic microhabitats differ only slightly in body size.

In contrast to these more terrestrial ecomorphs, sympatric members of the more arboreal trunk-crown, crown-giant, and twig ecomorphs differ substantially in body size (123, 142). Although sympatric trunk-crown anoles also differ in microclimate, these differences are considerably less marked than those displayed by trunk-ground and grass-bush anoles and do not prevent the species from occurring syntopically (119, 126, 127). Schoener & Schoener (127) proposed an explanation for the relationship between degree of arboreality and axis of resource partitioning.

Body size varies depending upon which congeners are sympatric   Resource partitioning is also evident when body size distributions are compared. In both the Lesser and Greater Antilles, when ecologically similar species are absent, anole species tend toward a similar body size, whereas communities composed of ecologically similar species often contain species diverging from this size. These observations suggest the possibility that an optimal body size may exist for anoles, but that competition in multi-species communities forces species to diverge from this optimum (117, 118).

Most (> 80%) of the anole populations occurring on one-species islands in both the Lesser and Greater Antilles (Greater Antillean one-species islands occur primarily in the Bahamas, Virgin Islands, and fringing islands in Cuba and Hispaniola) exhibit an adult body size that falls within a relatively narrow range (63, 117). Although it is possible that lizards of this size are more likely to be successful colonizers (see 117, 141), some of these islands were connected to larger mainlands historically and presumably originally contained more species; consequently, differential colonizing ability as a function of body size cannot be the sole explanation for the size of species on one-species islands. Rather, these data suggest that an optimal size appears to exist for one-species islands (117). The body size predicted to be optimal from a model incorporating energetic and physiological information corresponds closely with the size of species on these islands (85).

An optimal size may also exist for ecomorphs in multi-species communities in the Greater Antilles. On Cuba and Hispaniola, species that are similar in structural habitat and do not occur with other members of the same ecomorph type tend to be similar to each other in size (118).

However, the presence of sympatric congeners leads to deviations from the putatively optimal size both on two-species islands in the Lesser Antilles and in Greater Antillean communities containing two species of the same ecomorph type. In the Lesser Antilles, two-species islands are, with one exception, occupied by species greatly different in size (118). Null models indicate that these differences are greater than would be expected if species on two-species islands were randomly sampled from a source pool (68, 123). Similarly, in the Greater Antilles, syntopic members of the same ecomorph tend to differ greatly in size (118). As discussed above, this disparity in size occurs primarily among the more arboreal ecomorphs, whereas sympatric members of the more terrestrial ecomorphs are frequently of similar size and allotopic.

These deviations from the optimal size could result from evolutionary adjustments in sympatry or from differential colonization or extinction as a function of size similarity (8, 9, 68, 104). Distinguishing between these alternatives generally is difficult without an historical perspective (8, 68; see below), but in some cases, evolutionary adjustment seems probable. For example, *A. porcatus* is considerably smaller in central Cuba, where it is sympatric with the larger *A. allisoni,* than on the eastern and western thirds of the island, where the larger *A. allisoni* does not occur (121).

RESOURCE PARTITIONING DIMINISHES THE EFFECTS OF INTERSPECIFIC COMPETITION   The importance of resource partitioning in minimizing competition is demonstrated in two ways. First, experiments conducted in the Lesser Antilles demonstrated that two species with relatively minor differences in size and structural habitat competed strongly, whereas two species that partitioned resources along these axes had almost no effect on each other (86, 87, 105, 110). In the latter case, experimental habitat alterations that forced the species to overlap in structural habitat to a greater extent led to increased competitive effects on one of the species (110).

Second, the success of anole introductions (most accidental) is a function of how ecologically similar the introduced species is to the native species. Of 11 species introduced to islands on which no ecologically similar species occurred (ecological similarity defined in terms of overlap in the resource axes discussed above), 7 have achieved relatively large geographic ranges and none has become extinct. By contrast, of 12 species introduced to islands that already had an ecologically similar species, none has become widespread and 2 have become extinct (73).

Populations Alter Resource Use in the Presence of Congeners

HABITAT SHIFTS   For interspecific competition to have evolutionary consequences, the presence of competitors must lead to alterations in resource use.
A large body of evidence documents shifts in habitat use in anoles as a function
of sympatry with other species. Although anoles partition resources along three
axes, few studies have looked for competitive effects on diet or microclimatic
habitat use. One exception is a set of experimental studies in the Lesser
Antilles, in which ecologically similar species alter their diet and use of
microclimatic habitats when sympatric, but ecologically dissimilar species do
not (87, 110). Shifts in microclimatic habitat use were also detected in an
experimental removal in Florida (111).

In contrast to the paucity of studies examining effects on the diet and
microclimate resource axes, a number of studies document shifts in structural
habitat use. These studies, which include behavioral observations of species
interactions, experimental manipulations, and geographical and temporal comparisons, are summarized in Table 1. Two general results emerge from these
studies. First, in the presence of congeners, anoles tend to alter their habitat
use so as to minimize overlap (e.g. in the presence of a more terrestrial

Table 1 Studies demonstrating habitat shift in anoles in the presence of congeners.

Focal species	Locality	Context	Reference
opalinus	Jamaica	Observations made when lineatopus is present or absent	54
porcatus, sagrei	Cuba	Observations made when both or only one species present	12
sagrei	Florida	Experimental removal of cristatellus	111
gingivinus	St. Maarten	Experimental enclosures with and without wattsi pogus	87
wattsi wattsi, bima-culatus	St. Eustatius	Experimental enclosures with each species alone or both species together	87, 100
pulchellus	Puerto Rico	Experimental removal of crista-tellus	38
carolinensis, dis-tichus, grahami, sagrei	Caribbean islands	Comparisons of sites with or with-out other species, corrected for habitat availability	120
sagrei	Caribbean islands	Comparisons of sites with or with-out other species	62
cooki	Puerto Rico	Comparison of sites with or with-out cristatellus	55
conspersus	Grand Cayman	Comparison of habitat use before and after introduction of sagrei	73

congener, species become more arboreal). Second, when two species are sympatric, habitat shift is sometimes more apparent in the smaller of the two species.

Particularly instructive are the experimental studies by Roughgarden and colleagues in the Lesser Antilles in which different combinations of species were introduced into enclosed areas or onto a small island (86, 87, 105, 110; see 35 for a critique of the design of these experiments). On St. Maarten, the relatively ecologically similar species *A. wattsi pogus* and *A. gingivinus* both alter their habitat use when sympatric. By contrast, *A. bimaculatus* and *A. w. wattsi* on St. Eustatius are considerably more ecologically distinct, and only one of the species (the smaller *A. w. wattsi*) shifts its habitat use in sympatry. Not only is *A. bimaculatus* unaffected by the presence of *A. w. wattsi* on St. Eustatius, but the fitness effects on *A. w. wattsi* are also much less than the effects experienced by *A. w. pogus* on St. Maarten (86, 87, 110).

ECOLOGICAL RELEASE   If interspecific competition leads to partitioning of resources, then in the absence of competitors, a species may broaden its use of resources. This phenomenon, termed ecological release, has been widely sought among insular populations of various vertebrates, including anoles (reviewed in 122). Several studies looked for ecological release in climatic habitat use among Lesser Antillean anole species (44, 51, 97, 102, 108). Interpretation of such studies is problematic, however, because the ancestors of these populations may never have occurred in communities with other congeners; thus, no pre-ecological release comparison is possible (44, 62). Indeed, broad resource use may be a prerequisite for successful colonizers, rather than a result of post-colonization evolution (44).

By contrast, taxa clearly derived from ecologically specialized Greater Antillean species are potential candidates to exhibit ecological release and can be compared to close relatives in multi-species communities in the Greater Antilles. In several such species, ecological release in terms of climatic and structural habitat appears to occur (44, 58, 62; but see 26, 109).

Microevolutionary Adaptation Occurs in Response to Resource Shifts

EXPERIMENTAL AND ACCIDENTAL INTRODUCTIONS   Interspecific competition may lead not only to behavioral shifts in resource use, as described above, but also to changes in the selective environment, favoring microevolutionary adaptation to the new regime of resource use (reviewed in 135). No study to date has documented interspecific competition driving natural selection to produce microevolutionary change in anoles. However, one study attempted to show that populations transplanted into habitats greatly different from their native habitat experienced strong directional selection (76). Although the experimental design has been questioned (57, 60), this study does indicate that strong directional selection did occur in at least one population, and that population, coincidentally or not, was the one transplanted from the most different habitat (montane rainforest to xeric scrub).

Whether microevolutionary change occurs when populations use new habitats has also been investigated experimentally (JB Losos, KI Warheit, TW Schoener, unpublished). In this study it was also possible to make a priori predictions about the direction of adaptive morphological divergence. Populations of *A. sagrei* were introduced to extremely small islands in the Bahamas in 1977 (128). The vegetation available on these islands consisted of bushes and small trees, providing a habitat matrix of smaller-diameter supports than was available in the habitat of the source population. Based on the relationship between limb morphology and support diameter among species of Greater Antillean anoles (65, 82, 83, 91), one would predict that the populations should evolve relatively shorter limbs. Fifteen years after the introductions, the populations had diverged morphologically from the source population in the expected direction; further, the degree of divergence was correlated with how different was the island vegetationally on which a population occurred, relative to the source island. Studies are currently underway to assess the extent to which differences in relative limb length could be environmentally induced by differences in habitat.

Rapid microevolutionary change has also been documented in *A. leachi,* a large species that has increased to an even greater size since its introduction to Bermuda in the 1940s (92). In this case, the cause of the change is not clear, although one possibility is that the larger size of *A. leachi* evolved because the other species of *Anolis* on Bermuda (also introduced) are larger than the species with which *A. leachi* is sympatric on Barbuda and Antigua (70). However, the genetic basis for differences in body size in anoles still requires investigation (cf. 103).

COMPARATIVE STUDIES AMONG CLOSELY RELATED TAXA   The studies just discussed document rapid morphological change in populations experiencing new environments, but they do not clearly implicate the presence of competing species as the impetus for this change (although it is suggested for *A. leachi*). Such evidence is provided by comparisons among populations occurring sympatrically and allopatrically with other species. Such comparisons indicate that microevolutionary adaptation occurs in response to shifts in resource use on all three axes. Divergence along the prey-size axis is illustrated by several widespread species in which body size decreases as a function of the number of sympatric competitors (117). Adaptation to shifts in structural habitat occurs in Bahamian populations of *A. carolinensis* and *A. sagrei*. Both species differ

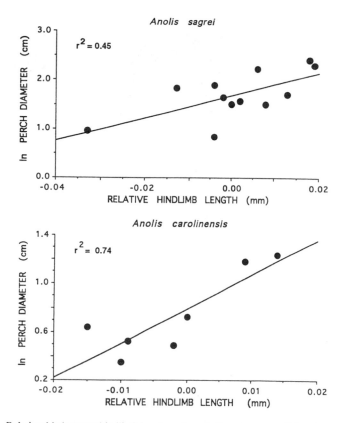

Figure 2   Relationship between hindlimb length and perch diameter among Bahamian populations of *A. carolinensis* (*top*) and *A. sagrei* (*bottom;* from 72).

in habitat use as a function of which competitors are present (120) and exhibit a relationship between relative limb length and structural habitat use (72; Figure 2). Finally, adaptive differences in thermal physiology are often exhibited by closely related taxa that partition resources along the microclimate axis (25; 42, 45, 48, 52, 107), although this result is not universal (40, 42, 43, 46; see also 78).

Macroevolutionary Patterns Are Consistent with Interspecific Competition as the Driving Force Behind Anole Adaptive Radiation

If interspecific competition has been the driving force behind the anole adaptive radiation in the Caribbean, then the preceding discussion suggests that two patterns should be evident in an historical analysis: First, adaptation for

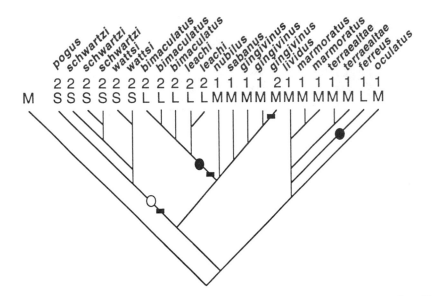

Figure 3 Evolution of body size in *Anolis* lizards of the northern Lesser Antilles (based on 68, 104). Numbers indicate the number of *Anolis* species on the island occupied by each taxon. Letters indicate body size (small, medium, or large). Circles represent major evolutionary changes in body size (*solid* = increase; *open* = decrease); bars represent the transition from an ancestor on a one-species island to a descendant on a two-species island. The statistical analysis in (68) used actual values rather than categorical variables. Perch height and body size are strongly correlated among these lizards (106); consequently, the evolution of perch height proceeded in an essentially identical manner. (Figure from 69 with permission)

resource specialization should evolve as additional species join a community. Second, based on the apparent prevalence of structural habitat partitioning in the Greater Antilles (see Sympatric Anoles Differ in Resource Use), evolutionary specialization should occur first along the structural habitat axis and only subsequently on the prey-size and microclimate axes. These hypotheses can be tested by using phylogenetic methods to reconstruct character evolution, assuming that a robust phylogenetic hypothesis exists for the group in question (4, 14, 36, 50, 59, 74).

The evidence is mixed concerning the prediction that the addition of species to a community leads to resource specialization in the Lesser Antilles. Supporting evidence comes from a phylogenetic analysis of the anole radiation in the northern Lesser Antilles, which indicates that divergence along two resource axes occurred when taxa came into sympatry (68; Figure 3). However, this result must be qualified owing to uncertainties about phylogenetic relationships and the existence of alternative evolutionary scenarios that require equally or only slightly less parsimonious reconstructions of character evolution (D Miles, A Dunham, unpublished; C Schneider, K de Queiroz, J Losos, unpublished). By contrast, analysis of the diversification of anoles in the southern Lesser Antilles does not indicate that evolutionary change in resource use was associated with the attainment of sympatry with competitors. For these taxa, ecological differences appear to have evolved in allopatry and have subsequently allowed coexistence of taxa (68).

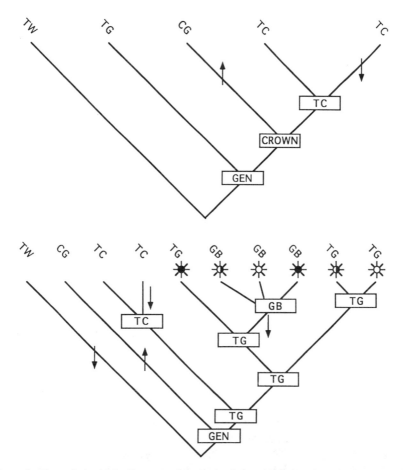

Figure 4   The evolution of *Anolis* community structure in Jamaica (*left*) and Puerto Rico (*right;* see 69 for methodological details). The names of the ecomorphs refer to the habitat they normally utilize (TW, twig; CG, crown-giant; TG, trunk-ground; TC, trunk-crown; GB, grass-bush). One ancestral taxon was reconstructed as inhabiting the crown, but not as belonging clearly to either of the crown ecomorphs (GEN, generalist). Arrows indicate the evolution of large or small size from the ancestral, intermediate body size. Circular symbols represent species that have specialized to distinctive climatic habitats (*open*—xeric, hot; *half-shaded*—lowland, moderate; *shaded*—montane, cool).

For Greater Antillean anoles, suitable phylogenies currently only exist for the Jamaican and Puerto Rican radiations (reviewed in 69). Comparison of macro-evolutionary patterns in these two radiations indicates not only that resource specialization has occurred as species are added to the community, but also that the two islands have followed an almost identical sequence in the evolution of ecomorphs (69; Figure 4). This unexpected finding suggests that a deterministic trajectory exists by which anole communities increase in species number and complexity. Further, this analysis suggests a proximate explanation for the absence of grass-bush anoles in Jamaica. Rather than occur- ring as a result of the lack of appropriate available resources, the absence of a grass-bush ecomorph on Jamaica may result simply because Jamaica has only reached the four-ecomorph stage, and the grass-bush anole is the fifth ecomorph to evolve in the community-evolution sequence. This explanation shifts inquiry from why the grass-bush ecomorph has not evolved to why Jamaica has not reached the five-ecomorph stage in the community-evolution sequence (69, 142).

Phylogenetic analysis also indicates that structural habitat specialization often precedes climatic or diet specialization (Figure 4). For example, on both Jamaica and Puerto Rico, prey-size divergence occurs within a clade of trunk-crown anoles. Further, on Puerto Rico, climatic specialization occurs subsequent to structural habitat specialization in both the trunk-ground and grass-bush ecomorphs. By contrast, the phylogenetic analysis indicates no cases in which climatic or diet specialization precedes structural habitat specialization (although several cases of substantial change in body size occur simultaneously with specialization to a particular microhabitat, such as the concordant evolution of small size and twig specialization in Puerto Rico). In addition, both phylogenies document that specialization for a particular microclimate evolved multiple times but that each ecomorph type evolved only once. Thus, it appears that a structural habitat niche, once filled, can be subdivided easily along microclimate lines but generally is resistant to evolutionary invasion by another lineage differentiating on the same island. [However, the successful invasion of Jamaica in recent years by the Cuban trunk-ground anole *A. sagrei* (140) indicates that other lineages may ecologically invade if they have already evolved the necessary specialization.]

Without a phylogeny for the anole radiations of Hispaniola and Cuba, the hypothesis that differentiation in structural habitat occurs prior to differentiation on other axes is difficult to test. Nonetheless, evolutionary inference is possible in cases in which closely related species (e.g. members of the same species complex) are all members of a single ecomorph type, yet have diverged in microclimate or prey size specialization. Specialization to different micro-climates has occurred among closely related members of the same ecomorph type in the trunk-ground (Hispaniola: 46, 129, 138; Cuba: 107) and, perhaps, grass-bush (Hispaniola: 41) ecomorphs, whereas prey-size specialization, as represented by body size, has occurred in the trunk-crown (Cuba: 121; Hispaniola: 139) and twig (Hispaniola: 142) ecomorphs. These examples indicate that microclimate and prey-size partitioning often occur subsequent to structural habitat specialization (assuming that the ancestor of these closely related groups of species was the same ecomorph as all of its descendants). However, in contrast to the Jamaican and Puerto Rican radiations, several of the ecomorphs may have evolved multiple times on Hispaniola and Cuba (5, 142), although without a robust phylogenetic hypothesis, this statement must be tentative. Further, most multiple origins of an ecomorph type appear to have occurred in the unusual montane faunas of these islands (142).

Competition, Predation, or Parasitism?

The data reviewed here make a compelling case that interspecific interactions not only are important in structuring extant anole communities in the Caribbean, but also have played an important role in shaping community evolution and adaptive radiation. Interspecific competition has generally been considered the process responsible for these patterns, but they could also be produced by other ecological processes. Recently, it has been suggested that predation or parasitism may be the driving force structuring anole communities. Although the importance, and even the incidence, of these processes has been neglected in studies of anole ecology, it is unlikely that either is a dominant structuring force in anole communities.

PREDATION   In recent years, theorists have suggested that patterns attributed to interspecific competition could also result from predation (31, 49, 53). Predation could exert its influence in two ways: either via the erstwhile competitors eating each other or via a predator that eats both species. In the anole communities of the Caribbean, both processes are unlikely to be important in shaping community structure.

Many species of *Anolis* have been reported to eat vertebrates occasionally, including other anoles (e.g. 19, 71, 95, 134). Most incidents of one *Anolis* eating another ("intra-guild predation"—90) involve an adult eating a juvenile or an adult of one species eating an individual of a considerably smaller species (19, 95). In some cases, such predation may affect habitat use of the prey species such as juvenile *A. aeneus* (134) and the small Hispaniolan *A. bahorucoensis* (19).

However, intraguild predation does not provide a general explanation for anole habitat partitioning for the simple reason that the species and size classes that interact most strongly are those least likely to prey on each other. Anoles are probably only capable of eating organisms considerably smaller than they are. Naganuma & Roughgarden (85) estimate the maximum prey size as one third of an anole's snout-vent length (based on 132; see also 134). Consequently, the effects of predation should be most strongly manifested in comparisons among different-sized taxa; adults of all but the smallest species should be immune to predation from other anoles, except perhaps from large crown-giants (71).

The data, however, indicate that the strongest interspecific effects on fitness parameters occur between similar-sized species, as noted above. Further, Schoener (120) documented that the strongest interspecific effects on habitat use were between relatively similar size-classes [e.g. females of a larger species and males of a smaller species or subadult males and adult females of similar-sized species; females are generally smaller than males in Caribbean species (117)], rather than between the size-classes most dissimilar in size and thus most likely to exhibit the effects of predation. Thus, intraguild predation cannot explain the large interspecific effects between similar-sized lizards, although it may explain habitat and behavioral shifts documented in females and juveniles of some species.

A second possibility is that sympatric anoles alter habitat use, not to minimize resource competition, but to avoid predators. It is possible that predators specialize, either evolutionarily or cognitively, to capture prey in particular habitats. Thus, by diverging in habitat use, anole species could, in theory, minimize predation pressure.

This is a difficult hypothesis to reject out-of-hand because so little is known about predation on anoles. Probably the most common anole predators in the Caribbean are certain birds (1, 81, 136, 137, 144, 146) and snakes (39) and perhaps sometimes mammals and invertebrates (136). On some islands, predation may be relatively intense (81, 136). Could the foraging behavior of these predators be responsible for patterns of anole resource partitioning and evolutionary diversification?

I suggest that this hypothesis is unlikely for several reasons:

1. Resources appear to be limiting for Caribbean anole populations. Supplementation experiments on anoles generally indicate that they are food-limited (reviewed in 32, 145).
2. Interspecific aggression sometimes occurs between ecologically similar species. In a number of instances, ecologically similar species exhibit interspecific territoriality and higher levels of interspecific agonistic behavior where they co-occur (references in 47).
3. Anoles respond in behavioral time to the presence of competitors. Several studies indicate that anoles alter their habitat use when individuals of other species are present (12, 54), which implicates interactions between the lizards themselves, rather than a mutual predator, as the most likely cause.
4. Experimental Studies of interspecific interactions Studies on St. Maarten indicated that, when one species was experimentally removed, the other species exhibited higher measures of body condition, prey intake, and egg

production. By contrast, on St. Eustatius, where two ecologically dissimilar species occur, removals had considerably less effect (87, 110).

A hypothesis that interspecific competition structures anole populations would predict each of these observations. Individual anoles are limited by food, defend resources against other species, and behaviorally alter their resource use when potential competitors are present. Further, the presence of competing species adversely affects anoles.

Of course, it is possible to formulate scenarios in which predation could produce the same results, but most such scenarios are fairly convoluted. For example, the presence of other anole species might attract more predators; thus, behavioral shifts could be a response to this increased risk of predation. Similarly, in the Lesser Antillean experiments, the removal of one species might reduce predation pressure, allowing the remaining species to alter resource use and exploit previously unavailable resources. Still, to explain these results, one must assume that *A. w. pogus* and *A. gingivinus* on St. Maarten share predators, but *A. w. wattsi* and *A. bimaculatus* on St. Eustatius do not. Many such scenarios seem implausible, but only experimental manipulations can demonstrate conclusively the role predators play in structuring anole communities.

PARASITISM   Differential susceptibility to parasites also has been invoked recently as an important determinant of community structure (e.g. 31, 131). On St. Maarten, *A. gingivinus* is more susceptible to malaria than *A. wattsi pogus*. The parasite is patchily distributed and, with one exception, *A. w. pogus* only occurs in areas in which malaria is found, whereas *A. gingivinus* occurs throughout the island. Thus, malarial infection may alter competitive relationships and allow coexistence of the two species (112). This finding is particularly significant because St. Maarten is the only island in the Lesser Antilles in which two relatively similar-sized species co-occur.

A similar hypothesis has been proposed for the anole fauna of Puerto Rico, in which malaria is prevalent in only one (*A. gundlachi*) of five species examined (113). Schall & Vogt (113) hypothesized that malaria mediates coexistence between *A. evermanni* and *A. gundlachi.* However, the situation in Puerto Rico does not parallel that in St. Maarten. In contrast to the species on St. Maarten, *A. evermanni* and *A. gundlachi* exhibit substantial differences in structural habitat use (64, 65, 93, 99, 127), foraging behavior (67), and time of activity (43). These differences are comparable to those displayed by other sympatric Greater Antillean anoles. Consequently, the coexistence of *A. evermanni* and *A. gundlachi* needs no special explanation. Indeed, in more open lowland habitats in Puerto Rico, *A. evermanni* is able to coexist with *A. cristatellus* (93, 127), a species quite similar to *A. gundlachi* in structural habitat and morphology, even though neither species is prone to malaria (113). Thus, at this point, the evidence does not indicate that susceptibility to malaria is an important determinant of community composition in the Greater Antilles.

DISCUSSION

Several lines of evidence implicate interspecific competition as an important force structuring communities of *Anolis* lizards in ecological and evolutionary time: sympatric species usually differ in resource use along one or more of three resource axes; when ecologically similar species cooccur, competitive effects on fitness occur; species alter their resource use in the presence of ecologically similar species; species adapt evolutionarily to shifts in resource use; and phylogenetic reconstructions are consistent with competition as a driving force behind the anole radiations. The congruence among these approaches provides strong support for the competition hypothesis.

Below, I first discuss integrative approaches to evolutionary ecology, and how one may interpret situations in which different lines of investigation provide inconsistent results, and then I discuss future directions in the study of anole communities.

Integrative Approaches to Evolutionary Ecology

Historical effects on the biological diversity of communities and regions are widely recognized (e.g. 6, 27, 100, 104). The cause of such effects, and how they may be studied, can be quite varied. In some cases, rare phenomena such as asteroid impacts, cycles of forest contraction and expansion, or the appearance of new taxa such as eutherian mammals, may produce large and long-standing effects on diversity. Because these are unique or infrequent events of large magnitude, they generally can only be studied using historical methods.

An alternative class of historical effects on diversity results from the evolutionary outcome of ongoing ecological processes. Such effects occur because ecological processes often have effects on fitness. If genetic variation exists for relevant traits among affected individuals, then fitness differentials may lead to microevolutionary change. Over time, these changes can be extrapolated to produce observed macroevolutionary patterns (e.g. 29).

To study this latter class of historical effects, both historical and ecological data are needed. In particular, an evolutionary ecological hypothesis may be corroborated if three premises are confirmed:

1. A given ecological process is an important determinant of community structure;
2. Selection studies indicate that the process leads to microevolutionary change.

3. Historical analyses yield macroevolutionary patterns similar to those that would have been predicted if observed microevolutionary processes were extrapolated over long periods of time.

These criteria obviously lead to a prescription for integrative studies of organismal diversity, incorporating phylogenetic and ecological studies. Confirmation of these three hypotheses would provide strong support to the premise that the process is and has been an important determinant of biological diversity. But what if one (or more) of these postulates is not confirmed? A number of possible explanations exist:

1. *Ecological processes may not have evolutionary effects.* Ecological processes could have important effects but not have any net evolutionary impact. This could result if no genetic variation existed for relevant traits, if all genetic variants were equally affected by the ecological process (i.e. no differential selective effect), or if the magnitude and direction of the selective effect varied through time such that the net effect over time was zero (e.g. 24).

2. *Microevolutionary change might not accumulate over time to produce macroevolutionary patterns.* The extent to which the processes governing macroevolutionary change are distinct from those operating within species is controversial (10, 28, 61). To the extent that speciation leads to large-scale changes of a fundamentally different nature than those that occur during anagenesis, then one may not be able to extrapolate from microevolutionary process to macroevolutionary pattern.

3. *The processes important in structuring extant communities may not be the same processes that have molded biological diversity historically.* The processes that regulate community structure probably depend on which taxa are present (88), but the composition of a community is constantly in flux as a result of immigration, extinction, speciation, and successional changes. In addition, environmental change may also lead to shifts in the relative importance of different ecological processes (56).

4. *Ecological or historical studies may be misleading.* Rare events of large impact may play a more important role in directing large-scale patterns than day-to-day processes (30, 133), but such events may rarely be observed in most short-term ecological studies. In a similar vein, historical analyses reconstruct past events relying on assumptions (e.g. parsimony) that may not always be correct. Thus, incongruence between ecological and historical analyses may result because one approach is simply mistaken about the importance of a particular process.

These possibilities underscore the advantages of taking an integrative approach to evolutionary ecology. By combining studies at several hierarchical scales, one can not only look for overall congruence between pattern and process, but one can also generate predictions at one level that can then be tested at another. Further, when analyses at different levels produce differing conclusions, additional testable hypotheses about the forces regulating biological diversity may be formulated. Because biological diversity ultimately reflects the interplay of ecological and evolutionary processes, only by considering both can we come to fully appreciate its genesis and maintenance.

Future Directions in Anole Evolutionary Ecology

For many years, community ecologists were preoccupied with documenting and comparing the structure of communities. However, observed ecological patterns often may result from a number of different processes (9, 15). In an effort to distinguish among competing hypotheses, ecologists in recent years have turned to either experimental or mechanistic approaches to understand community organization. Although both approaches have been used to some extent to understand anole communities, I suggest below several additional areas in which these approaches could prove fruitful.

INTERSPECIFIC INTERACTIONS   Given the tractability of anoles, it is surprising that they have not been used more in experimental studies. Examination of the relative importance of competition, predation, and parasites in mediating coexistence may prove particularly enlightening. Because these processes can produce very similar patterns, however, detailed mechanistic studies of interspecific interactions may be necessary to disentangle the relative importance of these processes. In addition, experimental studies could determine the extent to which resource partitioning is an evolutionarily fixed result of historical interactions as opposed to a behavioral response to ongoing interactions. Although a variety of data suggest that interactions occur between some ecomorphs (e.g. trunk-ground and trunk-crown ecomorphs), it would not be surprising to find that other ecomorphs (e.g. twig and crown-giant) do not currently compete.

On the other hand, the possible importance of other factors such as predation and parasitism is still uncertain. More detailed studies of community dynamics may yet yield surprises such as the possible importance of parasites in regulating anole interactions on St. Maarten (112). Examination of these questions may require a combined mechanistic/experimental approach to understand how the addition or removal of species affects other sympatric species.

RESOURCE PARTITIONING   Resource partitioning needs to be examined mechanistically at several levels. First, although anoles partition climatic and spatial resources, it is not clear how populations are limited by these resources. One possibility is that partitioning along these axes leads to differences in prey utilization. Surprisingly little information is available on whether sympatric anoles partition prey by taxon. A more detailed understanding of anole natural history and community dynamics, combined with experimental alterations of resource availability (including habitat), would be useful. In the one such study conducted to date, Rummel & Roughgarden (110) altered available perch heights and concluded that perch height and prey size are not independent resource axes in the Lesser Antilles.

Second, how do the specific adaptations of taxa allow them to use particular habitats more successfully? A correlation exists between morphology, habitat use, and behavior among anoles (65, 77, 82–84, 91). Further, differences in morphology lead to ecologically relevant differences in functional capacities such as running and jumping ability (65, 66, 75). These differences in performance also correlate with differences in behavior and habitat (65). What is needed now is an understanding of why different capabilities are required in different habitats. Why, for example, is greater sprinting speed correlated with the use of broad surfaces? Are lizards in these habitats farther from prey or cover? Do they face faster predators or prey? Is fast sprint speed not advantageous in cluttered habitats? Again, more detailed understanding of anole natural history is required.

EVOLUTIONARY CHANGE   Recent years have seen a proliferation of studies attempting to detect the operation of natural selection in nature (16). However, in the absence of information about how organisms interact with their environment, studies of selection can be difficult to interpret because a plausible explanation can be advanced for any result (16, 23). Selection studies are most informative when conducted on phenotypic traits for which the functional and behavioral significance is well understood. Well-studied taxa such as *Anolis* are thus obvious choices to understand the working of natural selection.

In addition, selection studies can be designed to test key components of hypotheses concerning anole adaptive radiation. A number of factors have been implicated as important in driving the anole radiation. Because humans have altered anole communities through both addition and subtraction, many species may currently experience directional selection. Selection studies can thus verify whether species using new habitats or species sympatric with new combinations of species actually do experience selective pressures to change in the hypothesized manner and whether, over several generations, microevolutionary change actually occurs. A robust understanding of the relationship between ecological process and microevolutionary response thus can test an important underlying hypothesis concerning anole adaptive radiation. Experimental introductions may also be used to test these hypotheses.

SPECIATION   Theories concerning anole radiation have focused on the cause of morphological differentiation among species but have paid little attention to how speciation actually occurs. Macro- and microvicariant models have been postulated as a prerequisite for the determination of allopatric speciation in the Greater Antilles (e.g. 142), but these models are usually speculative.

Most studies of evolutionary diversification draw a distinction between the population-genetic phenomena occurring during speciation and the subsequent ecological processes that affect survival and proliferation of new species. However, in anoles it is possible that speciation and ecology are intimately linked. Species-recognition in anoles relies upon visual signals, including the shape, color, and pattern of the dewlap and the pattern of head-bobbing (references in 20, 47). These features are also important in intraspecific communication. Further, the optimal visual signal may differ as a function of habitat (17, 18, 20). For example, a brightly colored dewlap may be most visible in a sun-lit habitat, whereas in dark forest, lighter colors may be more effective. Indeed, a relationship exists between dewlap color and habitat use among Greater Antillean anoles (20).

This raises the possibility that shifts in habitat use by an anole population may lead to shifts in signals to maximize intraspecific communication. Because the same signals are used for both intraspecific communication and species recognition, these changes may lead to the establishment of a new species-recognition signal. Hence, adaptation for increased communicative success may have the incidental effect of reproductively isolating a population from other populations in a species, thus causing speciation (e.g. 17, 79). Hence, the forces that promote ecological differentiation may also be responsible for the high rates of speciation in *Anolis*. This hypothesis can best be tested by combining historical and ecological approaches.

CONCLUSIONS

Biological diversity is the result of processes occurring presently and in the past. Understanding why some communities or regions are more diverse than others thus requires both studies of present-day ecology and inferences about processes operating in the past. Historical and ecological approaches are complementary: Each makes predictions that can be tested by reference to the other. The synergism resulting from the integration of the two is the most effective means by which we may understand the origin and structure of natural communities.

ACKNOWLEDGMENTS

For comments on previous versions of this paper, I thank M Butler, PE Hertz, RB Huey, DJ Irschick, JA Rodríguez-Robles, AS Rand, TW Schoener, OJ Sexton, and an anonymous reviewer. This work was supported by the National Science Foundation (DEB-9318642).

Literature Cited

1. Adolph S, Roughgarden J. 1983. Foraging by passerine birds and *Anolis* lizards on St. Eustatius (Neth. Antilles): implications for interclass competition, and predation. *Oecologia* 56:313–17

2. Andrews RM. 1979. Evolution of life histories: a comparison of *Anolis* lizards from matched island and mainland habitats. *Breviora* 454:1–51

3. Bradshaw AD. 1984. The importance of evolutionary ideas in ecology—and vice versa. In *Evolutionary Ecology,* ed. B Shorrocks, pp. 1–25. Oxford: Blackwell Sci.

4. Brooks DR, McLennan DA. 1991. *Phylogeny, Ecology, and Behavior: A Research Program in Comparative Biology.* Chicago: Univ. Chicago Press

5. Burnell KL, Hedges SB. 1990. Relationships of West Indian *Anolis* (Sauria: Iguanidae): an approach using slow-evolving protein loci. *Caribb. J. Sci.* 26:7–30

6. Cadle JE, Greene HW. 1993. Phylogenetic patterns, biogeography, and the ecological structure of neotropical snake assemblages. In *Species Diversity in Ecological Communities: Historical and Geographical Perspectives,* ed. RE Ricklefs, D Schluter, pp. 281–93. Chicago: Univ. Chicago Press

7. Cannatella DC, de Queiroz K. 1989. Phylogenetic systematics of the anoles: is a new taxonomy warranted? *Syst. Zool.* 38:57–68

8. Case TJ. 1983. Sympatry and size similarity in *Cnemidophorus.* In *Lizard Ecology: Studies of a Model Organism,* ed. RB Huey, ER Pianka, TW Schoener, pp. 297–325. Cambridge: Harvard Univ. Press

9. Case TJ, Sidell R. 1983. Pattern and chance in the structure of model and natural communities. *Evolution* 37:832–49

10. Charlesworth B, Lande R, Slatkin M. 1982. A neo-Darwinian commentary on macro-evolution. *Evolution* 36:474–98

11. Cockburn A. 1991. *An Introduction to Evolutionary Ecology.* Oxford: Blackwell Sci.

12. Collette BB. 1961. Correlations between ecology and morphology in anoline lizards from Havana, Cuba and southern Florida. *Bull. Mus. Comp. Zool.* 125:137–62

13. Colwell RK, Futuyma DJ. 1971. On the measurement of niche breadth and overlap. *Ecology* 52:567–76

14. Donoghue MJ. 1989. Phylogenies and the analysis of evolutionary sequences, with examples from seed plants. *Evolution* 43:1137–46

15. Drake JA. 1990. Communities as assembled structures: do rules govern pattern? *Trends Ecol. Evol.* 5:159–63

16. Endler JA. 1986. *Natural Selection in the Wild.* Princeton, NJ: Princeton Univ. Press

17. Endler JA. 1992. Signals, signal conditions, and the direction of evolution. *Am. Nat.* 139:S125–53

18. Endler JA. 1993. The color of light in forests and its implications. *Ecol. Monogr.* 63:1–27

19. Fitch HS, Henderson RW. 1987. Ecological and ethological parameters in *Anolis bahorucoensis,* a species having rudimentary development of the dewlap. *Amphibia-Reptilia* 8:69–80

20. Fleishman LJ. 1992. The influence of sensory system and the environment on motion patterns in the visual displays of anoline lizards and other vertebrates. *Am. Nat.* 139:S36–61

21. Floyd HG, Jenssen TA. 1983. Food habits of the Jamaican lizard, *Anolis opalinus:* resource partitioning and seasonal effects examined. *Copeia* 1983: 319–31

22. Futuyma DJ. 1986. *Evolutionary Biology.* Sunderland, MA: Sinauer. 2nd ed.

23. Garland T Jr, Losos JB. 1994. Ecological morphology of locomotor performance in squamate reptiles. In *Ecological Morphology: Integrative Organismal Biology,* ed. PC Wainwright, SM Reilly, pp. 240–302. Chicago: Univ. Chicago Press. In press

24. Gibbs HL, Grant PR. 1987. Oscillating selection on Darwin's finches. *Nature* 327:511–13

25. Gorman GC, Hillman S. 1977. Physiological basis for climatic niche partitioning in two species of Puerto Rican *Anolis* (Reptilia, Lacertilia, Iguanidae). *J. Herpetol.* 11:337–40

26. Gorman GC, Stamm B. 1975. The *Anolis* lizards of Mona, Redonda, and La

Blanquilla: chromosomes, relationships, and natural history notes. *J. Herpetol.* 9:197–205

27. Gorman OT. 1993. Evolutionary ecology and historical ecology: assembly, structure, and organization of stream fish communities. In *Systematics, Historical Ecology, and North American Freshwater Fishes*, ed. RL Mayden, pp. 659–88. Stanford, Calif: Stanford Univ. Press

28. Gould SJ. 1980. Is a new and general theory of evolution emerging? *Paleobiology* 6:119–30

29. Grant PR. 1986. *Ecology and Evolution of Darwin's Finches*. Princeton, NJ: Princeton Univ. Press

30. Grant PR. 1986. Interspecific competition in fluctuating environments. In *Community Ecology*, ed. J Diamond, TJ Case, pp. 173–91. New York: Harper & Row

31. Grosholz ED. 1992. Interactions of intraspecific, interspecific, and apparent competition with host-pathogen population dynamics. *Ecology* 73:507–14

32. Guyer C. 1988. Food supplementation in a tropical mainland anole, *Norops humilis:* effects on individuals. *Ecology* 69:362–69

33. Guyer C, Savage JM. 1986. Cladistic relationships among anoles (Sauria: Iguanidae). *Syst. Zool.* 35:509–31

34. Guyer C, Savage JM. 1992. Anole systematics revisited. *Syst. Biol.* 41:89–110

35. Hairston NG Jr. 1989. *Ecological Experiments: Purpose, Design, and Execution*. Cambridge: Cambridge Univ. Press

36. Harvey PH, Pagel MD. 1991. *The Comparative Method in Evolutionary Biology*. Oxford: Oxford Univ. Press

37. Hass CA, Hedges SB, Maxson LR. 1993. Molecular insights into the relationships and biogeography of West Indian anoline lizards. *Biochem. Syst. Ecol.* 21:97–114

38. Heatwole H. 1977. Habitat selection in reptiles. In *Biology of the Reptilia*, ed. C Gans, DW Tinkle, 7:137–55. New York: Academic

39. Henderson RW, Crother BI. 1989. Biogeographic patterns of predation in West Indian colubrid snakes. In *Biogeography of the West Indies: Past, Present, & Future*, ed. CA Wood, pp. 479–518. Gainesville, Fla: Sandhill Crane

40. Hertz PE. 1979. Sensitivity to high temperatures in three West Indian grass anoles (Sauria, Iguanidae), with a review of heat sensitivity in the genus *Anolis*. *Comp. Biochem. Phys. A* 63:217–22

41. Hertz PE. 1979. Comparative thermal biology of sympatric grass anoles (*Anolis semilineatus* and *A. olssoni*) in lowland Hispaniola (Reptilia, Lacertilia, Iguanidae). *J. Herpetol.* 13:329–33

42. Hertz PE. 1980. Responses to dehydration in *Anolis* lizards sampled along altitudinal transects. *Copeia* 1980:440–46

43. Hertz PE. 1981. Adaptation to altitude in two West Indian anoles (Reptilia: Iguanidae): field thermal biology and physiological ecology. *J. Zool.* 195:25–37

44. Hertz PE. 1983. Eurythermy and niche breadth in West Indian *Anolis* lizards: a reappraisal. In *Advances in Herpetology and Evolutionary Biology: Essays in Honor of Ernest E. Williams*, ed. AGJ Rhodin, K Miyata, pp. 472–83. Cambridge: Mus. Comp. Zool., Harvard Univ.

45. Hertz PE, Arce-Hernández A, Ramírez-Vázquez J, Tirado-Rivera W, Vázquez-Vives L. 1979. Geographical variation of heat sensitivity and water loss rates in the tropical lizard, *Anolis gundlachi*. *Comp. Biochem. Physiol. A* 62:947–53

46. Hertz PE, Huey RB. 1981. Compensation for altitudinal changes in the thermal environment by some *Anolis* lizards on Hispaniola. *Ecology* 62:515–21

47. Hess NE, Losos JB. 1991. Interspecific aggression between *Anolis cristatellus* and *A. gundlachi:* comparison of sympatric and allopatric populations. *J. Herpetol.* 25:256–59

48. Hillman SS, Gorman GC. 1977. Water loss, desiccation tolerance, and survival under desiccating conditions in 11 species of Caribbean *Anolis:* evolutionary and ecological implications. *Oecologia* 29:105–16

49. Holt RD. 1977. Predation, apparent competition and the structure of prey communities. *Theor. Popul. Biol.* 12:197–229

50. Huey RB. 1987. Phylogeny, history, and the comparative method. In *New Directions in Ecological Physiology*, ed. ME Feder, AF Bennett, W Burggren, RB Huey, pp. 76–98. Cambridge: Cambridge Univ. Press

51. Huey RB, Webster TP. 1975. Thermal biology of a solitary lizard: *Anolis marmoratus* of Guadeloupe, Lesser Antilles. *Ecology* 56:445–52

52. Huey RB, Webster TP. 1976. Thermal biology of *Anolis* lizards in a complex fauna: the cristatellus group on Puerto Rico. *Ecology* 57:985–94

53. Jeffries MJ, Lawton JH. 1984. Enemy-free space and the structure of ecological communities. *Biol. J. Linn. Soc.* 23:269–86

54. Jenssen TA. 1973. Shift in the structural habitat of *Anolis opalinus* due to congeneric competition. *Ecology* 54:863–69

55. Jenssen TA, Marcellini DL, Pague CA, Jenssen LA. 1984. Competitive interference between two Puerto Rican lizards, *Anolis cooki* and *Anolis cristatellus*. *Copeia* 1984:853–62

56. Kareiva PM, Kingsolver JG, Huey RB, eds. 1993. *Biotic Interactions and Global Change*. Sunderland, Mass: Sinauer

57. Katti M. 1992. Are *Anolis* lizards evolving? *Nature* 355:505–6

58. Laska AL. 1970. The structural niche of *Anolis scriptus* on Inagua. *Breviora* 349:1–6

59. Lauder GV. 1981. Form and function: structural analysis in evolutionary morphology. *Paleobiology* 7:430–42

60. Lawton JH, McArdle BH. 1992. Are *Anolis* lizards evolving? *Nature* 355: 506

61. Levinton J. 1988. *Genetics, Paleontology, and Macroevolution*. Cambridge: Cambridge Univ. Press

62. Lister BC. 1976. The nature of niche expansion in West Indian *Anolis* lizards I: ecological consequences of reduced competition. *Evolution* 30:659–76

63. Lister BC. 1976. The nature of niche expansion in West Indian *Anolis* lizards. II. Evolutionary components. *Evolution* 30:677–92

64. Lister BC. 1981. Seasonal niche relationships of rain forest anoles. *Ecology* 62:1548–60

65. Losos JB. 1990. Ecomorphology, performance capability, and scaling of West Indian *Anolis* lizards: an evolutionary analysis. *Ecol. Monogr.* 60:369–88

66. Losos JB. 1990. The evolution of form and function: morphology and locomotor performance in West Indian *Anolis* lizards. *Evolution* 44:1189–1203

67. Losos JB. 1990. Concordant evolution of locomotor behaviour, display rate, and morphology in *Anolis* lizards. *Anim. Behav.* 39:879–90

68. Losos JB. 1990. A phylogenetic analysis of character displacement in Caribbean *Anolis* lizards. *Evolution* 44:558–69

69. Losos JB. 1992. The evolution of convergent structure in Caribbean *Anolis* communities. *Syst. Biol.* 41:403–20

70. Losos JB. 1992. A critical comparison of the taxon-cycle and character-displacement models for size evolution of *Anolis* lizards in the Lesser Antilles. *Copeia* 1992:279–88

71. Losos JB, Andrews RM, Sexton OJ, Schuler AL. 1991. Behavior, ecology,

and locomotor performance of the giant anole, *Anolis frenatus*. *Caribb. J. Sci.* 27:173–79

72. Losos JB, Irschick DJ, Schoener TW. Adaptation and constraint in the evolution of specialization of Bahamian *Anolis* lizards. *Evolution*. In press

73. Losos JB, Marks JC, Schoener TW. 1993. Habitat use and ecological interactions of an introduced and a native species of *Anolis* lizard on Grand Cayman, with a review of the outcomes of anole introductions. *Oecologia* 95:525–32

74. Losos JB, Miles DB. 1994. Adaptation, constraint, and the comparative method: phylogenetic issues and methods. See Ref. 23, pp. 60–98

75. Losos JB, Sinervo B. 1989. The effect of morphology and perch diameter on sprint performance of *Anolis* lizards. *J. Exp. Biol.* 145:23–30

76. Malhotra A, Thorpe RS. 1991. Experimental detection of rapid evolutionary response in natural lizard populations. *Nature* 353:347–48

77. Malhotra A, Thorpe RS. 1991. Microgeographic variation in *Anolis oculatus*, on the island of Dominica, West Indies. *J. Evol. Biol.* 4:321–35

78. Malhotra A, Thorpe RS. 1993. An experimental field study of a eurytopic anole, *Anolis oculatus*. *J. Zool.* 229:163–70

79. Marchetti K. 1993. Dark habitats and bright birds illustrate the role of the environment in species divergence. *Nature* 362:149–52

80. Mayer GC. 1989. *Deterministic patterns of community structure in West Indian reptiles and amphibians*. PhD thesis. Harvard Univ., Cambridge, Mass.

81. McLaughlin JF, Roughgarden J. 1989. Avian predation on *Anolis* lizards in the northeastern Caribbean: an inter-island contrast. *Ecology* 70:617–28

82. Moermond TC. 1979. The influence of habitat structure on *Anolis* foraging behavior. *Behaviour* 70:147–67

83. Moermond TC. 1979. Habitat constraints on the behavior, morphology, and community structure of *Anolis* lizards. *Ecology* 60:152–64

84. Moermond TC. 1981. Prey-attack behavior of *Anolis* lizards. *Z. Tierpsychol.* 56:128–36

85. Naganuma KH, Roughgarden JD. 1990. Optimal body size in Lesser Antillean *Anolis* lizards—a mechanistic approach. *Ecol. Monogr.* 60:239–56

86. Pacala SW, Roughgarden J. 1982. Resource partitioning and interspecific competition in two two-species insular

Anolis lizard communities. *Science* 217: 444–46

87. Pacala SW, Roughgarden J. 1985. Population experiments with the *Anolis* lizards of St. Maarten and St. Eustatius. *Ecology* 66:129–41

88. Paine RT. 1966. Food web complexity and species diversity. *Am. Nat.* 100:65–75

89. Pianka ER. 1988. *Evolutionary Ecology.* New York: Harper & Row. 4th ed.

90. Polis GA, Myers CA, Holt RD. 1989. The ecology and evolution of intraguild predation: potential competitors that eat each other. *Annu. Rev. Ecol. Syst.* 20: 297–330

91. Pounds JA. 1988. Ecomorphology, locomotion, and microhabitat structure: patterns in a tropical mainland *Anolis* community. *Ecol. Monogr.* 58:299–320

92. Pregill G. 1986. Body size of insular lizards: a pattern of Holocene dwarfism. *Evolution* 40:997–1008

93. Rand AS. 1964. Ecological distribution in anoline lizards of Puerto Rico. *Ecology* 45:745–52

94. Rand AS. 1967. The ecological distribution of anoline lizards around Kingston, Jamaica. *Breviora* 272:1–18

95. Rand AS. 1967. Ecology and social organization in the iguanid lizard *Anolis lineatopus. Proc. US Natl. Mus.* 122:1–79

96. Rand AS. 1969. Competitive exclusion among anoles (Sauria: Iguanidae) on small islands in the West Indies. *Breviora* 319:1–16

97. Rand AS, Rand PJ. 1967. Field notes on *Anolis* lineatus in Curaçao. *Studies on the Fauna of Curaçao and Other Caribbean Islands* 24:112–17

98. Rand AS, Williams EE. 1969. The anoles of La Palma: aspects of their ecological relationships. *Breviora* 327:1–19

99. Reagan DP. 1992. Congeneric species distribution and abundance in a three-dimensional habitat: the rain forest anoles of Puerto Rico. *Copeia* 1992: 392–403

100. Ricklefs RE. 1987. Community diversity: relative roles of local and regional processes. *Science* 235:167–71

101. Rosenzweig ML. 1987. Evolutionary ecology. *Evol. Ecol.* 1:1–3

102. Roughgarden J. 1974. Niche width: biogeographic patterns among *Anolis* lizard populations. *Am. Nat.* 108:429–42

103. Roughgarden JD, Fuentes E. 1977. The environmental determinants of size in solitary populations of West Indian *Anolis* lizards. *Oikos* 29:44–51

104. Roughgarden J, Pacala S. 1989. Taxon cycle among *Anolis* lizard populations:

review of evidence. In *Speciation and its Consequences,* ed. D Otte, JA Endler, pp. 403–32. Sunderland, Mass: Sinauer

105. Roughgarden J, Pacala S, Rummel J. 1984. Strong present-day competition between the *Anolis* lizard populations of St. Maarten (Neth. Antilles). See Ref. 3, pp. 203–20

106. Roughgarden J, Porter W, Heckel D. 1981. Resource partitioning of space and its relationship to body temperature in *Anolis* lizard populations. *Oecologia* 50: 256–64

107. Ruibal R. 1961. Thermal relations of five species of tropical lizards. *Evolution* 15:98–111

108. Ruibal R, Philibosian R. 1970. Eurythermy and niche expansion in lizards. *Copeia* 1970:645–53

109. Ruibal R, Philibosian R. 1974. The population ecology of the lizard *Anolis acutus. Ecology* 55:525–37

110. Rummel JD, Roughgarden J. 1985. Effects of reduced perch-height separation on competition between two *Anolis* lizards. *Ecology* 66:430–44

111. Salzburg MA. 1984. *Anolis sagrei* and *Anolis cristatellus* in southern Florida: a case study in interspecific competition. *Ecology* 65:14–19

112. Schall JJ. 1992. Parasite-mediated competition in *Anolis* lizards. *Oecologia* 92: 58–64

113. Schall JJ, Vogt SP. 1993. Distribution of malaria in *Anolis* lizards of the Luquillo Forest, Puerto Rico: Implications for host community ecology. *Biotropica* 25:229–35

114. Schluter D, McPhail JD. 1992. Ecological character displacement and speciation in sticklebacks. *Am. Nat.* 140: 85–108

115. Schoener TW. 1967. The ecological significance of sexual dimorphism in size of the lizard *Anolis conspersus. Science* 155:474–78

116. Schoener TW. 1968. The *Anolis* lizards of Bimini: resource partitioning in a complex fauna. *Ecology* 49:704–26

117. Schoener TW. 1969. Size patterns in West Indian *Anolis* lizards. I. Size and species diversity. *Syst. Zool.* 18:386–401

118. Schoener TW. 1970. Size patterns in West Indian *Anolis* lizards. II. Correlations with the size of particular sympatric species—displacement and convergence. *Am. Nat.* 104:155–74

119. Schoener TW. 1970. Nonsynchronous spatial overlap of lizards in patchy habitats. *Ecology* 51:408–18

120. Schoener TW. 1975. Presence and absence of habitat shift in some widespread lizard species. *Ecol. Monogr.* 45:233–58

121. Schoener TW. 1977. Competition and the niche. In *Biology of the Reptilia*, ed. C Gans, DW Tinkle, 7:35–136. London: Academic

122. Schoener TW. 1986. Resource partitioning. In *Community Ecology: Pattern and Process*, ed. J Kikkawa, DJ Anderson, pp. 91–126. Melbourne: Blackwell Sci.

123. Schoener TW. 1988. Testing for nonrandomness in sizes and habitats of West Indian lizards: choice of species pool affects conclusions from null models. *Evol. Ecol.* 2:1–26

124. Schoener TW. 1989. The ecological niche. In *Ecological Concepts: the Contribution of Ecology to an Understanding of the Natural World*, ed. JM Cherrett, pp. 79–113. Oxford: Blackwell Sci.

125. Schoener TW, Gorman GC. 1968. Some niche differences in three Lesser-Antillean lizards of the genus *Anolis*. *Ecology* 49:819–30

126. Schoener TW, Schoener A. 1971. Structural habitats of West Indian *Anolis* lizards. I. Jamaican lowlands. *Breviora* 368:1–53

127. Schoener TW, Schoener A. 1971. Structural habitats of West Indian *Anolis* lizards. II. Puerto Rican uplands. *Breviora* 375:1–39

128. Schoener TW, Schoener A. 1983. The time to extinction of a colonizing propagule of lizards increases with island area. *Nature* 302:332–34

129. Schwartz A. 1979. A new species of cybotoid anole (Sauria, Iguanidae) from Hispaniola. *Breviora* 451:1–27

130. Schwartz A, Henderson RW. 1991. *Amphibians and Reptiles of the West Indies: Descriptions, Distributions, and Natural History*. Gainesville, Fla: Univ. Fla. Press

131. Settle WH, Wilson LT. 1990. Invasion by the variegated leafhopper and biotic interactions: parasitism, competition, and apparent competition. *Ecology* 71: 1461–70

132. Sexton OJ, Bauman J, Ortleb E. 1972. Seasonal food habits of *Anolis limifrons*. *Ecology* 53:182–86

133. Sousa WP. 1984. The role of disturbance in natural communities. *Annu. Rev. Ecol. Syst.* 15:353–91

134. Stamps JA. 1983. The relationship between ontogenetic habitat shifts, competition and predator avoidance in a juvenile lizard (*Anolis aeneus*). *Behav. Ecol. Sociobiol.* 12:19–33

135. Taper ML, Case TJ. 1992. Coevolution among competitors. *Oxford Surv. Evol. Biol.* 8:63–109

136. Waide RB, Reagan DP. 1983. Competition between West Indian anoles and birds. *Am. Nat.* 121:133–38

137. Wetmore A. 1916. Birds of Puerto Rico. *US Dep. Agric. Bull.* 326:1–140

138. Williams EE. 1963. *Anolis whitemanni*, new species from Hispaniola (Sauria, Iguanidae). *Breviora* 197:1–8

139. Williams EE. 1965. The species of Hispaniolan green anoles (Sauria, Iguanidae). *Breviora* 227:1–16

140. Williams EE. 1969. The ecology of colonization as seen in the zoogeography of anoline lizards on small islands. *Q. Rev. Biol.* 44:345–89

141. Williams EE. 1972. The origin of faunas. Evolution of lizard congeners in a complex island fauna: a trial analysis. *Evol. Biol.* 6:47–89

142. Williams EE. 1983. Ecomorphs, faunas, island size, and diverse end points in island radiations of *Anolis*. See Ref. 8, pp. 326–70

143. Williams EE. 1989. A critique of Guyer and Savage (1986): cladistic relationships among anoles (Sauria: Iguanidae): are the data available to reclassify the anoles? See Ref. 39, pp. 433–77

144. Wright SJ. 1981. Extinction-mediated competition: the *Anolis* lizards and insectivorous birds of the West Indies. *Am. Nat.* 117:181–92

145. Wright SJ, Kimsey R, Campbell CJ. 1984. Mortality rates of insular *Anolis* lizards: a systematic effect of island area? *Am. Nat.* 123:134–42

146. Wunderle JM Jr. 1981. Avian predation upon *Anolis* lizards on Grenada, West Indies. *Herpetologica* 37:104–8

Annu. Rev. Ecol. Syst. 1994. 25:495–520

THE ECOLOGICAL CONSEQUENCES OF SHARED NATURAL ENEMIES

R.D. Holt

Museum of Natural History, Department of Systematics and Ecology, University of Kansas, Lawrence, Kansas 66045

J.H. Lawton

NERC Centre for Population Biology, Imperial College, Silwood Park, Ascot SL5 7PY, United Kingdom

KEY WORDS: polyphagous enemies, apparent competition, enemy-free space, predation, indirect food webs, interactions, contingent theory

Abstract

When multiple victim species (e.g. prey, host) are attacked by one or more shared enemy species (e.g. predator, pathogen), the potential exists for apparent competition between victim populations. We review ideas on apparent competition (also called "competition for enemy-free space") and sketch illustrative examples. One puzzling aspect of this indirect interaction is the repeated rediscovery of the essential ideas. Apparent competition arises between focal and alternative prey populations because, in the long term, enemy abundance depends on total prey availability; by increasing enemy numbers, alternative prey intensify predation on focal prey. A frequent empirical finding, consistent with theory, is exclusion of victim species from local communities by resident enemies. Theory suggests victim-species coexistence depends on particular conditions. To understand fully the consequences of shared enemies requires a body of contingent theory, specifying the time-scale of the interactions (short- and long-term consequences of sharing enemies generally differ), the structure of the food-web encompassing the interactions, its spatial context, etc. The "core criterion" for a focal victim species to invade a community supporting a resident, polyphagous enemy is $r > aP$ (the invader's intrinsic rate of increase

495

0066-4162/94/1120-0495$05.00

should exceed attack rate times average enemy abundance). A growing body of data and observations test, and support, this prediction.

A GENERAL PROBLEM: CONTINGENT THEORY IN ECOLOGY

There is a growing recognition in ecology of the need for contingent theory—models tailored to particular systems—developed with strong links to synoptic views of the discipline as a whole (103, 154). This is particularly true for interactions between generalist natural enemies and their victims, a diverse class of interactions transcending taxa, habitats, and biomes. Here, the terms "enemy" and "victim" embrace, respectively: "true" predators, parasitoids, parasites, pathogens, or herbivores; and prey, hosts, or plants. For simplicity we often refer to "predators" and "prey," but unless otherwise stated our arguments apply broadly to all enemy-victim interactions.

What do the following miscellaneous interactions have in common—badgers and hedgehogs (34); sea urchins and the city of Los Angeles (144); rabbits and herbaceous vegetation (176, 177); ants, extrafloral nectaries, and caterpillars (87); Aleuts and sea otters (164); biological pest control by generalist natural enemies (77); and domestic cats and house sparrows (26)? The answer, which may not be obvious at first, is that all involve one (or more) "focal" victim species linked to other prey via a shared, generalist enemy or enemies: The thread tying these examples together is that in each, enemy impact upon particular victim species is profoundly influenced by the availability and productivity of alternative food or victims. Relationships between alternative prey mediated via shared predators are, we believe, a particularly pervasive and important class of indirect interactions (1, 2, 85, 110, 155, 168, 189a). Previous overviews of the role of predation in communities (25, 30, 42, 85, 111, 116, 162) have rarely highlighted the significance for prey species of sharing predators. The problems of coexistence and exclusion for prey species that share predators—due to an indirect interaction denoted by various names, including "apparent competition" (64) and "competition for enemy-free space" (80)—constitute the central focus of this review.

REDISCOVERING THE WHEEL

One of the most puzzling aspects of the phenomenon of shared enemies is the repeated rediscovery of its main ecological and evolutionary consequences by numerous authors, each apparently oblivious of earlier work. It is as if ecologists working on different taxa, in each decade, were continually to rediscover interspecific competition for resources! We have no idea why ecologists in general are so ignorant of the general importance of shared enemies and yet are so familiar with conventional, resource-based interspecific competition.

The essential ideas of enemy-free space and apparent competition are not new. Over 35 years ago, Williamson (185) first showed in a formal model that the consequences of two species sharing a natural enemy are, in general terms, identical to more conventional forms of interspecific competition for limiting resources. Holt (64) coined the term *apparent competition* to emphasize this formal similarity and, with collaborators (64, 66–72), developed a body of theory characterizing this interaction. Recently, Connell (28) and Reader (142) examined evidence for apparent competition between plants via shared herbivores. Jeffries & Lawton (80) listed 20 papers dealing with enemy-free space in general, conceptual terms. They summarize 81 papers, providing examples of ecological and evolutionary consequences, drawn from the marine, freshwater, terrestrial, and paleontological literature, and including organisms from microbes to mammals and phytoplankton to fish. Yet ecologists continue to rediscover one or more aspects of the general problem. Entomologists alone appear to have independently stumbled across the essential ideas at least 20 times (93). There are numerous discoveries of the basic phenomena in the parasite literature (47, 139, 140). Some authors (16, 17, 32, 38, 40, 41, 77, 86, 119, 123, 178) have clearly recognized the importance of alternative prey in defining the impact of polyphagous predators on focal prey in particular systems, without exploring broader implications. Several recent examples of rediscovery are elegant and important (6, 29, 35, 73, 90, 105, 106, 113, 119, 121, 143, 147, 159, 175). They suffer only from not being set in a more general, conceptual framework, and hence they contribute less to testing and developing theory than one might wish.

There are literally hundreds of examples we could use to illuminate how alternative prey for generalist predators influence the distribution and abundance of focal victim species; the papers cited here are illustrative, not exhaustive. To us, these examples tellingly highlight the need for the further development of contingent theories, in the first place tailored to the details of particular systems, then deliberately woven into a broader tapestry of thematically related contingent theories. Some such theories for shared generalist enemies are well-developed (e.g. 1, 2, 33, 50, 55, 58, 62, 64, 66–67, 69–72, 80, 93, 99a, 108, 122a, 153); others await explicit attention from theoreticians.

GENERAL FEATURES OF SHARED PREDATION

A logical series of questions about the effect of a generalist (polyphagous) predator on a focal prey type are: Does the predator cause that prey's local extinction? If not, by how much are prey numbers depressed below the carrying capacity set by other factors? Does predation regulate—or destabilize—the prey population? To answer these questions, one must consider three dimensions to the problem: (i) temporal scale; (ii) food web structure and feedbacks; and (iii) spatial scale. The reader will quickly come to appreciate the need for contingent theory!

Time Scales

The relative generation lengths of enemies and victims influence their linked dynamics (30, 32, 86). Observational time-scales strongly color interpretation (1, 2, 64, 70), and failure to consider time-scales can lead to misleading tests of theory (31). Feedback between predator and prey occurs when the rate of predation over some time-scale depends functionally upon prey abundance. At short time-scales, within a single predator generation, feedbacks are behavioral in nature (114) [e.g. predator satiation, switching, and aggregative responses (30, 57) and prey avoidance behaviors (100, 179)]. These short-term effects can be masked or exaggerated over multiple predator generations—the time-scale pertinent to understanding the role of predation in determining community structure—where the potential for predator numerical responses (30, 57) looms larger. At yet longer time-scales, phenotypic evolution within a given prey species will reflect the phenotypic spectrum characterizing all prey supporting shared predators; niche differentiation among prey may be driven by shared predation (19, 80, 93, 145, 160). Due to space limitation, we say little on evolution here.

SHORT TIME-SCALES   Within a single predator generation in a closed community, predator numbers are approximately fixed. The availability of alternative prey influences predator behavior, modulating the predation experienced by a focal prey via the predator's functional response; for instance, alternative prey may lower predation on a focal prey because of predator selectivity or satiation. One important consequence of labile predator behaviors is that generalist predators can contribute to density-dependent regulation of focal prey species (36, 40, 41, 55, 58) and at times can promote the coexistence of competing prey (190).

Such interactions have received considerable attention (29, 44, 48, 74, 143, 187). In short-term experiments with two species of amphibian tadpoles, survival of *Bufo americana* increased in mixed populations with *Rana palustris* when both were exposed to predation by *Anax junius* nymphs (Odonata) or by adult *Notophthalmus viridescens* salamanders. *Rana* survivorship was enhanced by *Bufo* in the presence of *Notophthalmus,* but not with *Anax* (181). Natural fluctuations in prey numbers produce comparable phenomena. On Santa Barbara island, predation by the barn owl (*Tyto alba*) extirpated burrowing owls (*Athene cunicularia*) when alternative prey (small mammals) declined (35), within a single barn owl generation.

These results are predicted by a traditional representation of a predator's short-term (functional) response to multiple prey-species—the generalized Holling "disk" equation (96, 118, 120). In this model, time spent handling one prey item precludes handling another. An increase in alternative prey relaxes predation upon a focal prey; conversely, decreases in alternative prey magnify predation upon focal prey. Thus, short-term interactions between alternative prey mediated by a shared predator can generate "apparent mutualism" (1, 2, 64, 70).

Various authors have explored a number of theoretical, but realistic, complications in short-term interactions between prey. For instance, the generalized disk equation does not incorporate adaptive foraging behavior by predators, or adaptive escape behavior by prey. If predators switch or forage optimally, apparent mutualism persists and may be reinforced (1, 2a, 50, 65). This and other indirect interactions [e.g. (+,−) called "apparent predation," "indirect antagonism" (74), or "contramensalism" (10)] may arise from shifts in prey behavior, or when predators violate canonical optimal foraging models (1, 65, 70, 74). Moreover, optimal patch use by predators can generate short-term, within-patch apparent competition between alternative prey (21, 70).

LONGER (TRANS-GENERATIONAL) TIME-SCALES   Predator abundance in a closed community in the long run is always a dependent variable, and in particular it depends on the availability and productivity of the entire prey base. The rate of predation experienced by a focal prey species can be profoundly influenced by the indirect, cumulative impact of alternative prey, sustaining predator populations at densities higher than allowed by the focal species alone. This dependency is often neglected in field studies of the community effects of polyphagous enemies. Understanding the assembly and persistence of species assemblages, we argue, requires a multi-generational perspective encompassing the full range of potential numerical responses.

Theoretical studies suggest that when predators are largely limited by prey availability (rather than by, say, nest-site availability), alternative prey should experience long-term, negative-negative interactions via shared predation (apparent competition), regardless of short-term contramensalism or apparent mutualisms due to predator satiation, adaptive foraging, or switching (2, 50, 64, 65, 71). These indirect interactions can lead to species exclusions. Ironically, the phenomenon is difficult to demonstrate by observing an established community, because the most severely affected species are absent from the system!

One approach is to compare local assemblages (6, 189). The species composition of *Enallagma* damselflies differs markedly but predictably between permanent lakes, depending upon whether the dominant polyphagous predators are fish (found only in some lakes), or larger Odonata (absent from lakes with fish). Different *Enallagma* species differ in vulnerability to these two predator types and are excluded from lakes with the "wrong" predators (109). (Apparent competition in this example is not just between *Enallagma* spp., but between *Enallagma* and the entire invertebrate prey complexes that support populations of fish or larger odonates). A very different example is provided by comparing fossil and extant marine communities. Certain isolated salt lakes lacking fish are dominated by benthic ophiuroids, and appear to provide a glimpse of Paleozoic marine communities, common in the fossil record, but now largely exterminated by the evolution of predatory fish in the Mesozoic (9).

Exclusion by shared enemies is best demonstrated by experiments, either controlled (3, 34, 82, 119, 121, 151, 152, 159) or more haphazard and accidental (20, 32, 86, 139). Exclusion by shared enemies is also seen dramatically in new, but transient, situations created by invasions, either of novel prey (90, 147, 157, 158), alien enemies (7, 11, 12, 23, 37, 73, 98, 113, 129–131, 150), or both (32, 133, 146). It is not the case that alien predators are somehow different from native predators (35), only that apparent competition to the point of exclusion can be readily observed during an invasion.

Exclusion by shared predators is not inevitable—just as species may compete for food, but coexist. Mechanisms promoting prey coexistence, despite shared predation, have been reviewed in detail elsewhere (64, 66, 69, 71). They include: (i) donor-controlled enemy dynamics; (ii) weak enemy numerical responses (i.e. limitation other than by prey, territoriality for instance); (iii) resource limitation for prey; (iv) resource and habitat partitioning by prey; (v) spatial and temporal refuges for prey; (vi) labile enemy behavior (e.g switching): and (vii) food-web effects (e.g. higher-order predators).

Food-Web Structures and Feedbacks

Sometimes indirect effects resulting from shared predation can be discerned species-by-species, and sometimes only diffusely. Published studies often conform to idealized food webs (Figure 1), although reality will often be more complex, encompassing both more types of prey, and more species of predator, than abstracted by the original author(s).

The simplest cases report a single enemy species and two distinct victim species (Figure 1a) (16, 35, 45, 48, 82, 127, 139, 158). The next most complicated cases (Figure 1b) focus on a single predator species attacking an entire suite of taxa, often with a focal prey (7, 13, 36, 46, 52, 79, 84, 98, 113, 126, 151, 164, 175). A yet more complicated scenario is for two or more distinct generalist predators to be supported by a pair of prey species (14), or by an entire suite of alternative prey, often with one prey selected for particular study (Figure 1c) (40, 41, 44, 55, 88, 91, 92, 101, 172, 178, 188). Finally, the focal prey "type" may be an ensemble of prey species, attacked by a complex of

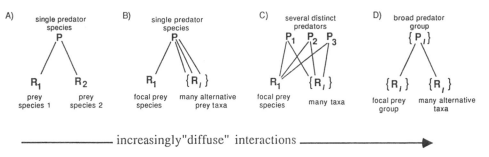

Figure 1 A series of food-web modules, representing increasingly complex and diffuse examples of natural enemies shared by two or more victim species (see text).

predators, treated as a group (Figure 1d) (15, 109). (Other food web effects, such as competition between prey, are discussed below.)

Over and above these relatively simple examples, many cases are difficult to classify, either because essential details are lacking, or because predator-prey interactions vary in complex ways in space and time. For instance, spatially varying assemblages of introduced mammals, sustained by a variety of native and introduced prey, pose major threats to breeding seabirds on sub-antarctic islands (81).

Spatial Scales

The nature of a predator's functional and numerical responses to a focal prey species depends upon the spatial relation of the predator, the focal prey, and alternative prey. In simple cases, the predator may encounter prey (both focal and alternatives) in a fine-grained fashion and have a numerical response to a weighted average of their abundances. This is a useful point of departure for theoretical models (64, 66, 169, 173, 174) and may adequately describe some empirical systems, such as "bottle" experiments (48, 91, 92) and relatively homogeneous field systems with mobile predators. Striped skunks, *Mephitis mephitis,* foraging for invertebrates in grassland, apparently encounter birds' nests at random, but nevertheless impose considerable mortality on several rare and endangered species, a process Vickery et al (175) called "incidental predation."

But often, enemies encounter focal and alternative victims in a coarse-grained fashion, through either time or space (136).

One can usefully distinguish three scenarios:

1. Enemies move from one habitat to another, and encounter different victim types in different habitats. For instance, Arctic foxes move into breeding areas of wading birds from adjacent habitats following lemming population declines and have a major effect on nesting success (172). Impacts on focal

victims depend upon when and how predators move in relation to their prey. In some circumstances, predator movement can cause high mortality to prey in low-productivity patches, because immigration boosts predator numbers above levels sustainable locally by those prey (56, 66, 68, 125, 141). For instance, polar bears, migrating on drift-ice in spring, directly or indirectly caused 60% mortality of brent goose eggs on Svalbard (104). Alternatively, predators may show aggregative responses (30, 57) to different prey in different habitats, generating switching (120) and reduced mortality for species in low density patches. It is less well known that an aggregative response can generate short-term apparent competition between alternative prey within a patch (70), exemplified by kangaroo rat, *Dipodomys merriami,* predation on experimental patches containing two sizes of seeds (21). Evolution of nest-site differences among coexisting bird species is favored by a qualitatively similar mechanism; in a given habitat, predators may be less successful when searching simultaneously for several different nest types than when concentrating on just one (105, 106).

2. Enemies may be stationary, but victims (focal, or other) move between habitats. Asymmetrical spatial flows reduce the potential for classical, density-dependent feedback between enemies and victims in any particular community. Two very different types of interaction are informative. Insectivorous birds are unable to attack below-ground larval stages of periodical cicadas, *Magicicada* spp.; accordingly, birds cannot show sustained numerical responses to cicadas between emergences and are swamped by the overwhelming numbers of adults that emerge synchronously into above-ground habitats (184). [Note that by emerging together (180), the three species of *Magicicada* are short-term apparent mutualists.] In contrast, sustained, predictable flows of food, which enter habitats from elsewhere at a rate independent of resident predators, have quite different effects. Such predators are donor controlled and may build up to large numbers, with major consequences for focal prey. Examples embrace a variety of unconventional flows of alternative foods, enemies, and focal victims: sewage outfalls that sustain sea urchin populations, reducing local communities of attached algae to the point of extinction (144); litterfall that contributes 20–50% to the diet of fallow deer, *Dama dama,* allowing the deer to all but obliterate plants in the forest understorey (12); and food provided by people for domestic cats, with serious consequences for bird and mammalian prey (26, 39; see also 136).

3. Echoing these spatial processes, there may be a partial or complete temporal (e.g. seasonal) separation of victims, supporting enemies that straddle seasons. The level of predation in one season may depend on the availability of prey in previous seasons. Examples of seasonal flows range from inter-continental migration of birds (with largely unexplored consequences for prey species), and within-continental migration of Odonata (187), to insect hosts

that live in physically separate habitats, coupled by shared, seasonally alternating parasitoids (128, 191). Differences in diel activity can generate asymmetrical relations among alternative prey. For example, Halle (54) reported
that a guild of diurnal avian predators attacked the vole *Microtus arvalis* at a
considerably higher rate than the mouse *Apodemus sylvaticus,* and suggested
that this reflects the partially diurnal habits of *Microtus.*

In brief, a full understanding of the dynamics of coexisting prey and the
potential for prey exclusion requires knowledge of both long- and short-term
predator responses to alternative prey, anchored in an appropriate spatial,
temporal, and community context. Good examples come from simple Arctic
and boreal ecosystems, where bird and mammal species show parallel (but
frequently asynchronous) cycles in numbers and breeding success, driven by
shared predators (22, 55, 88, 101, 172). Understanding these cycles requires
knowledge of long-term predator numerical responses, as well as shorter-term
aggregative and switching responses from declining prey (e.g. lemmings) to
alternative prey (e.g. nesting birds).

VICTIM EXCLUSION IN FEEDBACK AND NONFEEDBACK SYSTEMS

A simple but nonetheless illuminating criterion for exclusion of the focal prey
is:

$$0 < r < aP, \qquad\qquad\qquad 1.$$

where r is the average intrinsic rate of increase of the focal prey (when it is
rare enough to ignore intraspecific density-dependence), P is the average
abundance of resident predators, and a is the attack rate per predator, per prey
when the focal prey is rare (57, 64, 71, 93). If $r > 0$, the prey can cope with
the local environment and thus could invade in the absence of predation.
Exclusion occurs when $r < aP$.

In some circumstances, r increases with density at low densities (viz., Allee
effects), up to an r_{max} at some density, above which negative density-dependence is felt. Some prey species may be excluded from a community by
predation, yet able to persist if their initial abundance exceeds a certain
threshold level. This effect (which is compounded if predators have satiating
functional responses, reducing a at higher prey densities) can lead to alternative
stable states in prey community composition (64).

Examples of prey exclusion by polyphagous enemies have already been
given. Others include exclusion of the water boatman *Corixa expleta* from
low-salinity lakes by four species of parasitic water mites (116, 157a); elimination of *Tribolium castaneum* by the shared sporozoan parasite *Adelina
tribolii* in mixed cultures with *T. confusum* in Park's (127) classical experi-

ments (in the absence of *Adelina, T. castaneum* out-competes *T. confusum*); and the differential exclusion, or virtual exclusion, of three types of gall-forming nematine sawflies from arroyo willows by a guild of predatory ants (188).

The ambient density of predators, P, reflects the cumulative impact of the resident prey community that supports these predators, as well as many other aspects of community structure (see below). By definition, the magnitude of P is not governed by feedback with excluded prey! We therefore call predators that exclude prey "despotic."

Different assumptions about the relation between a despotic predator and alternative prey have interestingly different dynamic consequences:

1. Predators may be limited by nontrophic resources such as territories (40) or nest sites (77) (e.g. specially evolved "domatia" on host-plants for predatory ants—78). If so, the productivity of alternative prey may vary widely without influencing the propensity for exclusion of focal prey.

2. The enemy may be food-limited, but in a donor-controlled manner. This describes many systems in which the alternative food supply for the enemy is nonliving (e.g. 12)), such as nectar supplied by plants with extra-floral nectaries, attractive to predatory ants (78). These ants frequently (if not always—87) depress populations of one or more species of phytophagous insects exploiting the host plant, and in extreme cases exclude particular vulnerable species entirely (59, 97). Likewise, sea urchins are sustained at very high densities by Los Angeles sewage effluent and can therefore eliminate local populations of attached algae (144).

Donor control may also roughly describe systems in which domestic stock sustains infectious diseases (32), and those in which prey have absolute refuges (in space, time, body size, etc) from predation (61). Because heterogeneities of all sorts tend to generate "virtual" refuges from predation, touches of donor control may be common in many environments and may play an important role in maintaining stable, polyphagous, potentially despotic predator populations.

3. A vast literature, based largely on analyses of gut-contents and feces, shows that many vertebrate and invertebrate predators are not only polyphagous, but extremely so (186). In such cases, no particular prey species matters very much in determining P, which can often as a first approximation be treated as a constant rather than a dynamical variable responsive to that prey (58, 119). In contrast to case 1. above, predator abundance depends on the productivity of the full prey community. But in comparison with simpler community modules, indirect pairwise interactions between prey species will be difficult to discern (e.g. experimentally).

A Comment on "Ratio-Dependent Predation"

An alternative representation of predator-prey dynamics, ratio-dependent predation, has attracted recent attention; proponents argue that the functional response of a predator to its prey is determined not by prey density, R, but by the ratio of prey to predators, R/P (4, 8, 107).

In a simple ratio-dependent formulation, per capita prey mortality due to predation is independent of the number of predators. Hence if strict ratio-dependence were the rule, much of what we have said is wrong. However, the theory lacks unequivocal experimental support (149) and has been severely criticized for logical flaws (e.g. 2b). The numerous manipulative experiments showing an effect of predator abundance on prey mortality or density suggest that strict ratio-dependence is the exception, not the norm. Moreover, ratio-dependence may inadequately represent how predation affects the capacity of a victim population to increase when it is rare. The latter is what matters for community structure. We therefore accept Equation 1 as the most useful, general, albeit simple, working model for prey exclusion by resident, despotic predators.

REASONS FOR PREY EXCLUSION

The three components of the exclusion criterion (r, a, and P) correspond to (i) inherent physiological and life history properties of the prey in a given environment; (ii) behavioral attributes and environmental context of the predator-prey interaction (e.g. predator hunting efficiency, anti-predator escape strategies, availability of spatial refuges); and (iii) a summation of those factors that control predator numbers, particularly food web effects. We now "unpack" the biological complexities latent in the seemingly simple exclusion criterion; particular instances of exclusion (or coexistence) cannot be fully understood without considering all three components.

Exclusion Because of Low Prey r

If $r < 0$, the focal prey species is excluded from the local environment, irrespective of predation, unless sources provide continuous immigrants into sinks (68, 71, 141). If $r > 0$, but is low relative to mortality imposed by resident enemies, one tends to ascribe exclusion to predation. This is correct, but such exclusion also rests on the prey having a low r.

A prey species may have low r for many reasons, both autecological and synecological. Long generation length and/or low clutch size are obvious contributors. If the prey species is near the edge of its fundamental niche or close to its geographical limits, maximal growth rate may be low (95). For other species, resources may be inherently low in availability or quality, or depressed by interspecific competition.

These observations lead to a series of first-order expectations about prey species most at risk of exclusion from a local community by resident, polyphagous enemies. All else being equal, they are victims near the edge of their distributional ranges, specializing on scarce or low-quality resources, and overlapping in resource requirements with the resident prey community. These predictions appear not to have been tested experimentally [although the establishment of red-legged partridges, *Alectoris rufa,* in Britain in the face of heavy predation by foxes appears to have been favored by high fecundity (138)].

Exclusion Because of a High Attack Rate, a

Attack rates can be decomposed into a series of steps describing the time-course of a predator-prey encounter (63). Average attack rates will be reduced if victims (53a, 93, 161): (i) avoid enemies spatially or temporally; (ii) have behaviors or appearances that reduce detection and recognition; (iii) develop escape abilities; and (iv) resist attack. (Different, detailed considerations are pertinent to plant-herbivore, pathogen-host, and parasite-host interactions, without affecting the basic argument).

One simple conclusion is that if invading prey closely resemble resident prey in niche dimensions that determine vulnerability to predation, the invaders are likely to be excluded by resident enemies (80, 93). Only invaders with less vulnerable phenotypes should be able to enter the community (other things being equal, particularly r). Direct tests of these predictions are also lacking. But many aspects of the biology of coexisting hosts and parasitoids (93) and predators and prey (109, 189) are consistent with them. For example, phytophagous insect species in particular feeding niches are excluded entirely, or almost entirely, from host plants within the hunting ranges of predatory ants (46, 79, 188), seabirds nesting in different sites are differentially vulnerable to mink predation (7), and predation on bird nests increases with overlap in microhabitats exploited for nest placement (105).

Exclusion Because of High Predator Abundance, P

Obviously, a given invading prey is likely to be excluded if highly productive resident prey or nonliving food resources sustain a large population of resident predators (P). But the problem is more interesting than that. The invasion criterion for prey species i, $r_i > a_i P$, can be restated as:

$$r_i/a_i = P^*, \qquad\qquad 2.$$

where P^* is the greatest density of predators prey i can tolerate and still persist. If prey species i occurs alone with the predator, then P^* defines the maximum abundance of the predator sustainable by that prey. This leads to a simple criterion for dominance under shared predation: the species with the largest value for P^* tends to exclude alternative prey. This simple P^* rule characterizes the outcome of apparent competition in a diverse array of models, including classical Lotka-Volterra models (64, 66, 122a), host-parasitoid models (71), and some models with mixed exploitative and apparent competition (69). Microcosm studies (91, 122) have validated this criterion, but its efficacy in explaining dominance in field situations (e.g. 158) has not yet been tested experimentally. This "rule-of-thumb" for dominance may break down if key assumptions are violated, for instance, because of complex functional responses, or because predators themselves are attacked by higher-order enemies. Nonetheless, the P^*-rule provides a useful yardstick for gauging the likely importance of such complications.

When Prey Are Not Excluded

If the focal prey can invade (i.e. $r > aP$), many of the above considerations still pertain, but with the significant difference that various feedbacks may arise.

A focal prey may persist at an equilibrium density well below carrying capacity, because alternative prey support generalist enemies at high densities. Such a prey is unlikely to have a large effect on the remainder of its community. A particularly well-documented example is provided by Potts's (137) work on grey partridges, *Perdix perdix,* in which breeding densities are depressed by a complex of predators (e.g. foxes and crows) without significantly influencing predator numbers. Alternatively, invading focal prey can build up to high levels, allowing an increase in enemy numbers that depress and potentially threaten populations of resident, alternative victims. Two cases of invading phytophagous insects, a leafhopper *Erythroneura variabilis* in California (71, 158) and a gall wasp *Andricus quercuscalicis* in Britain (157), provide good examples. In the first case, increases in populations of a shared parasitoid, *Anargus epos,* appear to have caused marked reductions in resident *E. elegantula;* in the second, increases in gall inquilines potentially threaten native gall-formers.

Suppression, rather than exclusion, of focal prey populations by shared pathogens is also documented (32). Populations of another British gamebird, red grouse (*Lagopus lagopus*), are suppressed by louping-ill virus, maintained at high levels by alternative domesticated and wild hosts, including sheep, goats, hares, and deer. Models indicate that grouse alone cannot sustain the virus (32, 75, 76).

COMPLICATIONS OF COMMUNITY STRUCTURE

In natural communities, simple food-web modules are typically embedded in more complex webs, including higher-order predators, omnivory/intraguild predation (31, 132, 134), and predation upon competing prey (27, 69). It would take us too far afield to characterize in detail all these community-level complications, but it is useful to briefly consider a few.

Higher-Order Predators

By dampening or eliminating the long-term numerical response of intermediate predators to their prey (124), higher-order predators can weaken apparent competition between focal prey and alternative prey (64). Removal of higher-order predators allows intermediate predators to increase, often with devastating consequences for prey species lower in the trophic chain (24, 85, 164)—a phenomenon termed "mesopredator release" by Soulé et al (165). Prey species impacted by polyphagous predators undergoing mesopredator release are victims of apparent competition.

Analyzing the net impact of a guild of predators upon a focal prey becomes more complicated—and in interesting ways—given omnivory or intraguild predation. In the simplest case, higher-level predators not only attack intermediate predators (in this instance, the focal prey), they also share their resources (132, 134, 135). The combined problems of predation from, and interspecific competition with, higher-level predators make the persistence of intermediate predators more difficult (35, 134, 135, 156, 166, 167). In contrast, if focal prey experience predation from a predator guild, intraguild predation within that guild can relax the total impact of predation on the focal prey (34, 134). Again, we know of no experimental tests of these predictions.

Competition Among Prey

The traditional focus in community ecology has been on how predation modifies preexisting competitive interactions, permitting competitively incompatible species to coexist. This emphasis pervades both empirical (e.g. 99, 126) and theoretical (e.g. 27, 89, 102, 173, 190) studies. "Keystone species" were originally defined as predators with major impacts on competitively dominant prey (115, 126, 164). We by no means discount the potential importance of predation as a factor diversifying communities with strong competitive interactions. However, the notion that polyphagous predators characteristically promote prey coexistence [*Pisaster* providing the classical example—126)] is neither the simplest nor most general case. A single-minded emphasis on this effect may be a sociological factor predisposing ecologists to "rediscovering the wheel" of negative indirect interactions via shared predation.

If resident prey compete with focal prey, adding direct competition further hampers the persistence of focal prey. If resident prey compete among themselves, the number of predators supportable by those prey tends to be reduced, enhancing coexistence. Thus, the effect of prey competition upon focal prey exclusion depends on whether such competition acts among the alternative prey, or between them and the focal prey. Predation can also sharpen competitive interactions (53). For instance, if refuges are limited in supply, interspecific competition between victims may be intensified (61, 67, 74, 117).

Adding direct competition to predator-prey systems opens up a Pandora's box of dynamic possibilities, such as limit cycles and chaotic dynamics (49, 173, 174), essentially because of long feedback loops (predator A knocks down competitor B, which allows competitor C to increase, which cannot support predator A, which declines, allowing competitor B to increase again, and so on). The simple criterion for exclusion of focal prey (Equation 1) is rigorously true for constant P; more complicated criteria for invasion are needed for variable P (71). Simple criteria such as $r > aP$, however, help one gauge the significance of more refined criteria—yet another example of the conceptual utility of contingent theory.

In Community Ecology, It Gets Worse Before It Gets Better

The above remarks provide just a taste of the complexities inherent in almost any multi-species food web. A basic, humbling message emerging from theoretical studies of complex webs is that with many potential routes for indirect interactions between any species pair, it may be difficult to predict the effect of one species upon another (64, 189a, but see 110, 155).

Although it is important to recognize such complexities, it is equally important not to despair of simple explanations for broad ecological patterns. Contingent models of shared predators with pronounced numerical responses reveal that alternative prey can exclude or reduce the abundance of a focal prey. Without pretending to explain any particular example in all its details, this message rings loud and clear in many case studies.

SOME CASE HISTORIES

A few studies go beyond qualitative agreement with the predictions of contingent theory to more rigorous, quantitative tests.

Protist Assemblages

"Bottle experiments" (83) offer powerful opportunities to test the contingent theories of apparent competition. Lawler (91), for instance, grew two species of bacterivorous protists, *Chilomonas,* a flagellate (denoted C below), and *Tetrahymena* (T), a small ciliate, with a predatory protist, *Euplotes* (E), a larger ciliate, in assorted combinations. She demonstrated that in one predator–one prey treatments (E-C and E-T), E persisted for 40–60 generations with either C or T. Moreover, the two prey species coexisted in the absence of predation. She further noticed (personal communication) that compared with prey abundances in the absence of predation, E depressed T by around 50%, whereas C was depressed nearly two orders of magnitude; moreover, equilibrial abundance of E in the E-C treatment was much less than in the E-T treatments. These results suggest the predator has a much higher attack rate on C than on T, i.e. $a_C \gg a_T$, permitting over-exploitation of C.

Lawler further noted that T became very abundant in microcosms within two days after introduction into single-species cultures, while C did not become abundant until the fourth day, suggesting that $r_T \gg r_C$. Hence, $r_T/a_T > r_C/a_C$. Theoretically, the prey that supports the highest predator density (i.e. highest r/a, see above) should displace alternative prey. In other words, T should displace C in the face of shared predation by E.

This is exactly what happened. In experimental treatments beginning with E-C-T, C disappeared rapidly as E grew beyond the average abundance supported in the E-C treatments. This experiment appears to be the first demonstration of apparent competition in a controlled laboratory microcosm. It is completely consistent with the a priori theoretical expectations for prey dominance under apparent competition.

"Bottle experiments" have also been carried out by Nakajima & Kurihara (122). They used a continuous dialysis system with a bacterium (*Escherichia coli*) supporting a protozoan (*Tetrahymena thermophila*), to show that mutant bacterial clones with higher growth rates or lower attack rates successfully supplant resident clones by enhancing the predator's equilibrium density (generating increased predation upon the ancestral clones).

Habitat Partitioning

Habitat segregation between alternative prey is one potential outcome of shared predation (66). Schmitt (152) documented an example in a subtidal reef community and experimentally corroborated the role of shared predation.

This community has two distinct habitats: high-relief reefs, with surface texture providing refuges from predation; and cobble reefs, offering few refuges. A guild of mobile predators (e.g. lobster, octopus, whelk) attacks two functional groups of molluscan prey, sessile bivalves (the most common being *Chama arcana*) and mobile herbivorous snails (three species) (a "many-to-many" scenario; see Figure 1). Even though bivalves are preferred prey, they are common on high-relief reefs, where predators are also common but gastropods rare. Conversely, on cobble reefs gastropods are relatively common, but bivalves rare. The two functional prey groups use distinct resources and do not compete directly.

Experimental augmentation of bivalves in the cobble reef increased predators four-fold via aggregative numerical responses; this increased gastropod mortality and sharply reduced gastropod abundance. A correlative study using transplanted *Chama* on the cobble reef revealed that local predator abundance (and *Chama* mortality) tracked local gastropod abundance. Schmitt argued (personal communication) that even the low abundance of predators maintained by resident gastropods on the cobble reefs suffices to eliminate sessile bivalves there; reciprocally, the high abundance of predators maintained on high-relief reefs by partially protected bivalve populations keeps gastropods rare. Schmitt concluded that this habitat segregation reflects shared predation. Further field experiments are needed to validate this hypothesis fully.

One further example will suffice, incorporating several processes. Bergerud (16, 17) argued that on Newfoundland, large lynx populations (high P) are maintained by productive snowshoe hare populations in boreal habitats. Resulting intense lynx predation limits arctic hares to tundra—a habitat not occupied by snowshoe hares. Arctic and snowshoe hares introduced onto islands without predators coexist; in one case, later introducing lynx forced the arctic hare to extinction (112). Arctic hares have intrinsically low r's (60), particularly when restricted to boreal forest foodstuffs (14), making it unlikely that $r > aP$ in this habitat. Other predators, particularly red foxes, may be as important mortality agents for arctic hares as lynx (AT Bergerud, personal communication), implicating other alternative prey, such as voles, as determinants of arctic hare distribution. A final twist is that snowshoe hares are differentially highly vulnerable to avian predators in tundra habitats (14), which may in part explain the confinement of snowshoe hares to boreal forest.

APPLIED IMPLICATIONS OF SHARED PREDATION

Indirect interactions via shared predators are significant in many applied ecological problems. We touch on three.

Wildlife Disease Epidemiology

A well-known phenomenon in wildlife disease epidemiology is that certain hosts provide "reservoirs" for infectious disease agents, which can severely depress focal host populations (32, 116, 153). White-tailed deer sustain and can withstand a meningeal worm fatal to caribou. Hence, "caribou cannot be introduced to ranges where white-tailed deer have a high frequency of meningeal worm infections" (18, see also 72, 75, 76, 133, 146).

Biological Control

Aspects of apparent competition are important in biological pest control (71, 94). Howarth (73) summarizes several examples where polyphagous parasitoids, released as biological control agents, appear to have caused the local or regional extinction of the pest or of native, nontarget hosts. In the Florida Keys, the endemic cactus *Opuntia spinosissima* is threatened with imminent extinction by an introduced biocontrol agent, *Cactoblastis cactorum*, whose population is sustained by the target weed, an introduced cactus, *O. stricta* (163). Conversely, released biocontrol agents may fail to establish because of attack by indigenous, polyphagous enemies (51, 71, 90).

Biocontrol practitioners have long recognized the significance of several of the phenomena highlighted in this review as means of increasing predation and parasitism on economically important focal prey. These include the use of "supplementary resources" (e.g. pollen, nectar, and honeydew—management that boosts alternative "prey" numbers) to maintain high populations of potential control agents (43, 45, 77, 82, 178), and provisioning alternative nesting, feeding, or over-wintering sites to boost predator numbers, either within a crop, or to promote predator flows from a reservoir (77, 171).

Landscape Effects and Conservation

Alternative prey are significant causes in the declines of many endangered populations. Examples involving alien prey or predators were noted above. But native predators can also be important (175). Many forest landbirds are decreasing precipitously in forest fragments in the United States because of high parasitism by cowbirds and nest losses from avian and mammalian predators (108, 148, 182, 183). Large enemy populations are maintained by food outside forest fragments, including anthropogenic sources. Mesopredator release poses a significant conservation threat to prey populations via apparent competition in many parts of the world (137, 159, 165, 170).

Human impacts on landscapes may be exacerbated by polyphagous herbivores. In eastern and central United States, white-tailed deer have increased greatly due to forest clearing and deliberate game management, creating early to mid-successional vegetation. Deer move into mature forest stands, where their selective herbivory profoundly alters community composition (5). For instance, young age classes of hemlock, yew, and white cedar can be abundant within exclosures, yet absent outside.

FUTURE DIRECTIONS

Part of our original intent in sculpting this review was to provide a table, akin to that in Sih et al (162), cross-classifying a wide range of empirical studies of shared predation according to various criteria, then comparing the patterns in the table to basic theoretical expectations. We abandoned this goal, because only rarely do authors report all the pertinent information needed to relate empirical studies to theory. Ideally, the following desiderata would be provided:

1. Habitat
2. Number of enemy species, with Latin family and specific names (it is appalling how rarely the latter are provided outside the entomological literature)
3. Number of victim species, with Latin names
4. Knowledge of resource-based competition among victims
5. Nature of data (controlled experiment; quasi-experiment; observational)
6. Spatial scale (e.g. quadrat sizes)

7. Spatial context (open/closed to enemy/victim dispersal; landscape config-
uration)
8. Temporal scale of study (relative to enemy and victim generations)
9. Enemy response to focal victim populations (e.g. aggregative or reproductive response)
10. Impact on victims (e.g. none; apparent mutualism; victim exclusion).

This may appear a daunting list, but much can be gauged with reasonable
accuracy from basic natural history. Given the burgeoning state of the ecological literature, it is imperative that ecologists provide enough information to
link their published work with related empirical studies, and with general
theory, and in particular to identify those contingent theories that have explanatory power in particular systems.

The case studies and theory summarized in this review convince us that
indirect inhibitory interactions arising from shared enemies are a powerful,
dominant theme in natural communities, and a significant component in many
applied ecological problems. We have identified a number of theoretical predictions that have yet to be tested, and shared predation scenarios that have
yet to be examined theoretically. Much remains to be done.

ACKNOWLEDGMENTS

We thank Claire Challis, Wendy Wastell, Brenda Kaye, and Jennifer Wilson
for assistance in manuscript preparation, and Peter Abrams and Billy Schwei-
ger for perceptive comments on the manuscript. Preparation was supported
from the core grant to the NERC Centre for Population Biology, and by grants
from the US National Science Foundation and the British Ecological Society
to RD Holt. JH Lawton thanks the Konza Prairie for allowing him to share
ticks with bison; RD Holt thanks his lucky stars for walks in enemy-free space.
We thank numerous colleagues for preprints and advice. Apologies to the
authors of many excellent papers we had no room to cite.

Literature Cited

1. Abrams P. 1987. Indirect interactions
between species that share a predator:
varieties of indirect effects. See Ref. 85,
pp. 38–54
2. Abrams PA. 1987. On classifying interactions between populations. *Oecologia*
73:272–81
2a. Abrams PA. 1993. Indirect effects arising from optimal foraging. In *Mutualism
and Community Organization,* ed. H
Kawanabe, JE Cohen, K Iwasaki, pp.
255–79. Oxford: Oxford Univ. Press
2b. Abrams PA. 1994. The fallacies of
"ratio dependent predation." *Ecology.*
In press
3. Addicott JF. 1974. Predation and prey

community structure: an experimental study of the effect of mosquito larvae on the protozoan communities of pitcher plants. *Ecology* 55:475–92

4. Akçakaya HR. 1992. Population cycles of mammals: evidence for a ratio-dependent predation hypothesis. *Ecol. Monogr.* 62:119–42

5. Alverson WS, Waller DM, Solheim SL. 1988. Forests too deer: edge effects in northern Wisconsin. *Conserv. Biol.* 2:348–58

6. Anderson DJ. 1991. Apparent predator-limited distribution of Galápagos red-footed boobies *Sula sula. Ibis* 133:26–29

7. Andersson Å. 1992. Development of waterbird populations in the Bullerö archipelago off Stockholm after colonization by mink (Swedish, English summary). *Ornis Svecica* 2:107–18

8. Arditi R, Perrin N, Saiah H. 1991. Functional responses and heterogeneities: an experimental test with cladocerans. *Oikos* 60:69–75

9. Aronson RB, Sues H-D. 1987. The paleoecological significance of an anachronistic ophiuroid community. See Ref. 85, pp. 355–66

10. Arthur W, Mitchell P. 1989. A revised scheme for the classification of population interactions. *Oikos* 56:141–43

11. Atkinson IAE. 1985. The spread of commensal species of *Rattus* to oceanic islands and their effects on island avifaunas. In *Conservation of Island Birds,* ed. PJ Moors, pp. 35–81. *ICBP Tech. Publ. No. 3*

12. Atkinson IAE., Cameron EK. 1993. Human influences on the terrestrial biota and biotic communities of New Zealand. *Trends Ecol. Evol.* 8:447–51

13. Baillie S, Gooch S, Birkhead T. 1993. The effect of magpie predation on songbird populations. In *Britain's Birds in 1990–91: the Conservation and Monitoring Review,* ed. J Andrews, SP Carter, pp. 68–73. Thetford: Br. Trust for Ornithol., Joint Nature Conserv. Com.

14. Barta RM, Keith LB, Fitzgerald SM. 1989. Demography of sympatric arctic and snowshoe hare populations: an experimental assessment of interspecific competition. *Can. J. Zool.* 67:2762–75

15. Belovsky GE, Slade JB. 1993. The role of vertebrate and invertebrate predators in a grasshopper community. *Oikos* 68:193–201

16. Bergerud AT. 1967. The distribution and abundance of arctic hares in Newfoundland. *Can. Field Nat.* 81:242–48

17. Bergerud AT. 1983. Prey switching in a simple ecosystem. *Sci. Am.* 249:130–41

18. Bergerud AT, Mercer WE. 1989. Caribou introductions in eastern North America. *Wildl. Soc. Bull.* 17:111–20

19. Bernays E, Graham M. 1988. On the evolution of host specificity in phytophagous arthropods. *Ecology* 69:886–92

20. Black RW II, Hairston NG Jr. 1988. Predation driven changes in community structure. *Oecologia* 77:468–79

21. Brown JS, Mitchell WA. 1989. Diet selection on depletable resources. *Oikos* 54:33–43

22. Bulmer MG. 1975. Phase relations in the ten-year cycle. *J. Anim. Ecol.* 44:609–12

23. Burdon JJ. 1993. The role of parasites in plant populations and communities. In *Biodiversity and Ecosystem Function,* ed. E-D Schulze, HA Mooney, pp. 165–79. Berlin: Springer-Verlag

24. Carpenter SR, Kitchell JF, Hodgson JR, Cochran PA, Elser JJ, et al. 1987. Regulation of lake primary productivity by food web structure. *Ecology* 68:1863–76

25. Case TJ, Bolger DT. 1991. The role of introduced species in shaping the distribution and abundance of island reptiles. *Evol. Ecol.* 5:272–90

26. Churcher PB, Lawton JH. 1987. Predation by domestic cats in an English village. *J. Zool.* 212:439–55

27. Comins HN, Hassell MP. 1976. Predation in multi-prey communities. *J. Theor. Biol.* 62:93–114

28. Connell JH. 1990. Apparent versus "real" competition in plants. In *Perspectives on Plant Competition,* ed. JB Grace, D Tilman, pp. 445–74. New York: Academic

29. Copeland RS, Craig GB Jr. 1992. Interspecific competition, parasitism, and predation affect development of *Aedes hendersoni* and *A. triseriatus* (Diptera: Culicidae) in artificial treeholes. *Ann. Entomol. Soc. Am.* 85:154–63

30. Crawley MJ, ed. 1992. *Natural Enemies: The Population Biology of Predators, Parasites and Diseases.* Oxford: Blackwell Sci. 576 pp.

31. Diehl S. 1993. Relative consumer sizes and the strengths of direct and indirect interactions in omnivorous feeding relationships. *Oikos* 68:151–57

32. Dobson AP, Hudson PJ. 1986. Parasites, disease and the structure of ecological communities. *Trends Ecol. Evol.* 1:11–15

33. Dobson AP, May RM. 1991. Parasites, cuckoos, and avian population dynamics. See Ref. 129a, pp. 391–412

34. Doncaster CP. 1992. Testing the role of intraguild predation in regulating hedge-

hog populations. *Proc. R. Soc. Lond. B* 249:113–17
35. Drost CA, McCluskey RC. 1992. Extirpation of alternative prey during a small rodent crash. *Oecologia* 92:301–4
36. Dunn E. 1977. Predation by weasels (*Mustela nivalis*) on breeding tits (*Parus* spp.) in relation to the density of tits and rodents. *J. Anim. Ecol.* 46:633–52
37. Ebenhard T. 1988. Introduced birds and mammals and their ecological effects. *Swed. Wild. Res. - Viltrevy* 13(4):5–107
38. Elton C. 1927. *Animal Ecology.* London: Sidgwick & Jackson. 204 pp.
39. Elton CS. 1951. The use of cats in farm rat control. *Br. J. Anim. Behav.* 1:151–55
40. Erlinge S. 1987. Predation and noncyclicity in a microtine population in southern Sweden. *Oikos* 50:347–52
41. Erlinge S, Göransson G, Högstedt G, Jansson G, Liberg O, et al. 1984. Can vertebrate predators regulate their prey? *Am. Nat.* 123:125–33
42. Estes JA. 1994. Top-level carnivores and ecosystem effects: questions and approaches. In *Linking Species and Ecosystems*, ed. CG Jones, JH Lawton. New York: Chapman & Hall. In press
43. Evans EW. 1994. Indirect interactions among phytophagous insects: aphids, honeydew, and natural enemies. In *Individuals, Populations, and Patterns in Ecology*, ed. S Leather, N Mills, A Watt. Andover: Intercept. In press
44. Fauth JE, Resetarits WJ Jr. 1991. Interactions between the salamander *Siren intermedia* and the keystone predator *Notophthalmus viridescens*. *Ecology* 72:827–38
45. Flaherty DL. 1969. Ecosystem trophic complexity and Willamette mite, *Eotetranychus willamettei* Ewing (Acarina: Tetranychidae), densities. *Ecology* 50:911–15
46. Fowler SV, MacGarvin M. 1985. The impact of hairy wood ants, *Formica lugubris*, on the guild structure of herbivorous insects on birch, *Betula pubescens*. *J. Anim. Ecol.* 54:847–55
47. Freeland WJ. 1983. Parasites and the coexistence of animal host species. *Am. Nat.* 121:223–36
48. Gilbert JJ, MacIsaac HJ. 1989. The susceptibility of *Keratella cochlearis* to interference from small cladocerans. *Freshwater Biol.* 22:333–39
49. Gilpin ME. 1979. Spiral chaos in a predator-prey model. *Am. Nat.* 113:306–8
50. Gleeson SK, Wilson DS. 1986. Equilibrium diet: optimal foraging and prey coexistence. *Oikos* 46:139–44
51. Goeden RD, Louda SM. 1976. Biotic interference with insects imported for weed control. *Annu. Rev. Entomol.* 21:325–42
52. Gooch S, Baillie SR, Birkhead TR. 1991. Magpie *Pica pica* and songbird populations. Retrospective investigation of trends in population density and breeding success. *J. Appl. Ecol.* 28:1068–86
53. Grosholz ED. 1992. Interaction of intraspecific, interspecific, and apparent competition with host-pathogen population dynamics. *Ecology* 73:507–14
53a. Gross P. 1993. Insect behavioral and morphological defenses against parasitoids. *Annu. Rev. Entomol.* 38:251–73
54. Halle S. 1988. Avian predation upon a mixed community of common voles (*Microtus arvalis*) and wood mice (*Apodemus sylvaticus*). *Oecologia* 75:451–55
55. Hanski I, Hansson L, Henttonen H. 1991. Specialist predators, generalist predators, and the microtine rodent cycle. *J. Anim. Ecol.* 60:353–67
56. Hansson L. 1989. Predation in heterogeneous landscapes: how to evaluate total impact? *Oikos* 54:117–19
57. Hassell MP. 1978. *The Dynamics of Arthropod Predator-Prey Systems*. Princeton, NJ: Princeton Univ. Press. 237 pp.
58. Hassell MP, May RM. 1986. Generalist and specialist natural enemies in insect predator-prey interactions. *J. Anim. Ecol.* 55:923–40
59. Heads PA, Lawton JH. 1984. Bracken, ants and extrafloral nectaries. II. The effect of ants on the insect herbivores of bracken. *J. Anim. Ecol.* 53:1015–31
60. Hearn BJ, Keith LB, Rongstad OJ. 1987. Demography and ecology of the arctic hare (*Lepus arcticus*) in southwestern Newfoundland. *Can. J. Zool.* 65:852–61
61. Hixon MA, Menge BA. 1991. Species diversity: prey refuges modify the interactive effects of predation and competition. *Theoret. Pop. Biol.* 39:178–200
62. Hochberg ME, Clarke RT, Elmes GW, Thomas JA. 1994. Population dynamic consequences of direct and indirect interactions involving a large blue butterfly and its plant and red ant hosts. *J. Anim. Ecol.* 63:375–91
63. Holling CS. 1965. The functional response of predators to prey densities and its role in mimicry and population regulation. *Mem. Entomol. Soc. Can.* 45:5–60
64. Holt RD. 1977. Predation, apparent competition, and the structure of prey

communities. *Theoret. Pop. Biol.* 12: 197–229
65. Holt RD. 1983. Optimal foraging and the form of the predator isocline. *Am. Nat.* 122:521–41
66. Holt RD. 1984. Spatial heterogeneity, indirect interactions, and the coexistence of prey species. *Am. Nat.* 124:377–406
67. Holt RD. 1987. Prey communities in patchy environments. *Oikos* 50:276–90
68. Holt RD. 1993. Ecology at the mesoscale: the influence of regional processes on local communities. In *Species Diversity in Ecological Communities: Historical and Geographical Perspectives,* ed. RE Ricklefs, D Schluter, pp. 77–88. Chicago: Univ. Chicago Press
69. Holt RD, Grover J, Tilman D. 1994. Simple rules for interspecific dominance in systems with exploitative and apparent competition. *Am. Nat.* 144:741–71
70. Holt RD, Kotler BP. 1987. Short-term apparent competition. *Am. Nat.* 130: 412–30
71. Holt RD, Lawton JH. 1993. Apparent competition and enemy-free space in insect host-parasitoid communities. *Am. Nat.* 142:623–45
72. Holt RD, Pickering J. 1985. Infectious disease and species coexistence: a model of Lotka-Volterra form. *Am. Nat.* 126:196–211
73. Howarth FG. 1991. Environmental impacts of classical biological control. *Annu. Rev. Entomol.* 36:485–509
74. Huang C, Sih A. 1990. Experimental studies on behaviorally mediated, indirect interactions through a shared predator. *Ecology* 71:1515–22
75. Hudson P. 1992. *Grouse in Space and Time: The Population Biology of a Managed Gamebird.* Fordingbridge, UK: Game Conservancy. 224 pp.
76. Hudson PJ, Dobson AP. 1991. Control of parasites in natural populations: nematode and virus infections of red grouse. See Ref. 129a, pp. 413–32
77. Huffaker CB, Messenger PS, eds. 1976. *Theory and Practice of Biological Control.* New York: Academic
78. Huxley CR, Cutler DF, eds. 1991. *Ant-Plant Interactions.* Oxford: Oxford Univ. Press. 601 pp.
79. Ito F, Higashi S. 1991. Variance of ant effects on the different life forms of moth caterpillars. *J. Anim. Ecol.* 60:327–34
80. Jeffries MJ, Lawton JH. 1984. Enemy free space and the structure of ecological communities. *Biol. J. Linn. Soc.* 23:269–86
81. Jouventin P, Weimerskirch H. 1991. Changes in the population size and demography of southern seabirds: management implications. See Ref. 129a, pp. 297–314
82. Karban R, Hougen-Eitzman D, English-Loeb G. 1994. Predator-mediated apparent competition between two herbivores that feed on grapevines. *Oecologia.* 97: 508–11
83. Kareiva P. 1989. Renewing the dialogue between theory and experiments in population ecology. In *Perspectives in Ecological Theory,* ed. J Roughgarden, RM May, SA Levin, pp. 68–88. Princeton, NJ: Princeton Univ. Press
84. Kenward RE, Parish T, Robertson PA. 1992. Are tree species mixtures too good for grey squirrels? In *The Ecology of Mixed-Species Stands of Trees,* ed. MGR Cannell, DC Malcolm, PA Robertson, pp. 243–53. Oxford: Blackwell Sci.
85. Kerfoot WC, Sih A, eds. 1987. *Predation: Direct and Indirect Impacts on Aquatic Communities.* Hanover, NH: Univ. Press New Engl.
86. Knowlton N. 1992. Thresholds and multiple stable states in coral reef community dynamics. *Am. Zool.* 32:674–82
87. Koptur S. 1991. Extrafloral nectaries of herbs and trees: modelling the interaction with ants and parasitoids. In *Ant-Plant Interactions,* ed. CR Huxley, DF Cutler, pp. 213–30. Oxford: Oxford Univ. Press
88. Korpimäki E, Huhtala K, Sulkava S. 1990. Does the year-to-year variation in the diet of eagle and Ural owls support the alternative prey hypothesis? *Oikos* 58:47–54
89. Kretzschmar MR, Nisbet RM, McCauley E. 1993. A predator-prey model for zooplankton growing on competing algal populations. *Theoret. Pop. Biol.* 44:32–66
90. Lake PS, O'Dowd DJ. 1991. Red crabs in rain forest, Christmas Island: biotic resistance to invasion by an exotic snail. *Oikos* 62:25–29
91. Lawler SP. 1993. Direct and indirect effects in microcosm communities of protists. *Oecologia* 93:184–90
92. Lawler SP. 1993. Species richness, species composition and population dynamics of protists in experimental microcosms. *J. Anim. Ecol.* 62:711–19
93. Lawton JH. 1986. The effect of parasitoids on phytophagous insect communities. In *Insect Parasitoids. 13th Symposium of the Royal Entomological Society, London,* ed. J Waage, D Greathead, pp. 265–87. London: Academic
94. Lawton JH. 1990. Biological control of plants: a review of generalisations, rules

and principles using insects as agents. In *Alternatives to Chemical Control of Weeds: Proceedings of an Int. Conf., Rotorua, New Zealand,* ed. C Bassett, LJ Whitehouse, JA Zabkiewicz, pp. 3–17. Rotorua: Ministry of For., For. Res. Inst. Bull. 155

95. Lawton JH. 1993. Range, population abundance and conservation. *Trends Ecol. Evol.* 8:409–13

96. Lawton JH, Beddington JR, Bonser R. 1974. Switching in invertebrate predators. In *Ecological Stability,* ed. MB Usher, MH Williamson, pp. 141–58. London: Chapman & Hall

97. Lawton JH, Heads PA. 1984. Bracken, ants and extrafloral nectaries I. The components of the system. *J. Anim. Ecol.* 53:995–1014

98. Lawton JH, Woodroffe GL. 1991. Habitat and the distribution of water voles: why are there gaps in a species' range? *J. Anim. Ecol.* 60:79–91

99. Leibold MA. 1991. Trophic interactions and habitat segregation between competing *Daphnia* species. *Oecologia* 86:510–20

99a. Leibold MA. 1994. A graphical model of keystone predators in food webs: effects of productivity on abundance, incidence, and diversity patterns in communities. MS

100. Lima SL, Dill LM. 1990. Behavioral decisions made under the risk of predation: a review and prospectus. *Can. J. Zool.* 68:619–40

101. Lindstrom E, Angelstam P, Widen P, Andren H. 1987. Do predators synchronize vole and grouse fluctuations? *Oikos* 48:121–24

102. Louda SM, Keeler KH, Holt RD. 1990. Herbivore influences on plant performance and competitive interactions. In *Perspectives on Plant Competition,* ed. JB Grace, D Tilman, pp. 415–44. New York: Academic

103. MacArthur RH. 1972. *Geographical Ecology: Patterns in the Distribution of Species.* New York: Harper & Row. 269 pp.

104. Madsen J, Bregnballe T, Mehlum F. 1989. Study of the breeding ecology and behaviour of the Svalbard population of light-bellied brent goose *Branta bernicla hrota. Polar Res.* 7:1–21

105. Martin TE. 1988. On the advantage of being different: nest predation and the coexistence of bird species. *Proc. Natl. Acad. Sci. USA* 85:2196–99

106. Martin TE. 1988. Processes organizing open-nesting bird assemblages: competition or nest predation? *Evol. Ecol.* 2:37–50

107. Matson P, Berryman A. 1992. Ratio-dependent predator-prey theory. *Ecology* 73:1529–66

108. May RM, Robinson SK. 1985. Population dynamics of avian brood parasites. *Am. Nat.* 126:475–94

109. McPeek MA. 1990. Determination of species composition in the *Enallagma* damselfly assemblages of permanent lakes. *Ecology* 71:83–98

110. Menge BA. 1995. Indirect effects in marine rocky intertidal interaction webs: patterns and importance. *Ecol. Monogr.* In press

111. Menge BA, Sutherland JP. 1987. Community regulation: variation in disturbance, competition, and predation in relation to environmental stress and recruitment. *Am. Nat.* 130:730–57

112. Mercer WE, Hearn BJ, Findlay C. 1981. Arctic hare populations in insular Newfoundland. In *Proc. World Lagomorph Conf.,* ed. K. Myers, CD MacInnes, pp. 450–68

113. Miller DJ. 1989. Introduction and extinction of fish in the African Great Lakes. *Trends Ecol. Evol.* 4:56–59

114. Miller TE, Kerfoot WC. 1987. Redefining indirect effects. See Ref. 85, pp. 33–37

115. Mills LSM, Soulé ME, Doak DF. 1993. The keystone-species concept in ecology and conservation. *BioScience* 43:219–24

116. Minchella DJ, Scott ME. 1991. Parasitism: a cryptic determinant of animal community structure. *Trends Ecol. Evol.* 6:250–54

117. Mittlebach GG, Chesson PL. 1987. Predation risk: indirect effects on fish populations. See Ref. 85, pp. 315–32

118. Murdoch WW. 1973. The functional response of predators. *J. Appl. Ecol.* 10:335–42

119. Murdoch WW, Bence J. 1987. General predators and unstable prey populations. See Ref. 85, pp. 17–30

120. Murdoch WW, Oaten A. 1975. Predation and population stability. *Adv. Ecol. Res.* 9:1–25

121. Naeem S. 1988. Predator-prey interactions and community structure: chironomids, mosquitoes and copepods in *Heliconia imbricata* (Musaceae). *Oecologia* 77:202–9

122. Nakajima T, Kurihara Y. 1994. Evolutionary changes of ecological traits of bacterial populations through predator-mediated competition. I. Experimental analysis. *Oikos.* In press

122a. Nakajima T, Kurihara Y. 1994. Evolutionary changes of ecological traits of bacterial populations through predator-

mediated competition. II. Theoretical considerations. *Oikos.* In press
123. Newton I. 1991. Concluding remarks. See Ref. 129a, pp. 637–54
124. Oksanen L, Fretwell SD, Arruda J, Niemalä P. 1981. Exploitation ecosystems in gradients of primary productivity. *Am. Nat.* 118:240–62
125. Oksanen T. 1990. Exploitation ecosystems in heterogeneous habitat complexes. *Evol. Ecol.* 4:220–34
126. Paine RT. 1966. Food web complexity and species diversity. *Am. Nat.* 100:65–75
127. Park T. 1948. Experimental studies on interspecific competition. I. Competition between populations of the flour beetles, *Tribolium confusum* Duval and *Tribolium castaneum* Herbst. *Ecol. Monogr.* 18:267–307
128. Parnell JR. 1964. The parasite complex of two seed beetles *Bruchius ater* (Marsham) (Coleoptera: Bruchidae) and *Apion fuscirostre* Fabricius (Coleoptera: Curculionidae). *Trans. R. Entomol. Soc., London* 116:73–88
129. Pech RP, Sinclair ARE, Newsome AE. 1994. Predation models for primary and secondary prey species. *Aust. Wildlife Res.* In press
129a. Perrins, CM, Lebreton J-D, Hirons GJM, eds. 1991. *Bird Population Studies: Relevance to Conservation and Management.* Oxford: Oxford Univ. Press
130. Pimm SL. 1987. The snake that ate Guam. *Trends Ecol. Evol.* 2:293–95
131. Pimm SL. 1991. *The Balance of Nature?* Chicago: Univ. Chicago Press. 434 pp.
132. Pimm SL, Lawton JH. 1978. On feeding on more than one trophic level. *Nature* 275:542–44
133. Plowright W. 1982. The effects of rinderpest and rinderpest control in east Africa. *Symp. Zool. Soc. Lond.* 50:1–28
134. Polis GA, Holt RD. 1992. Intraguild predation: the dynamics of complex trophic interactions. *Trends Ecol. Evol.* 7:151–55
135. Polis GA, Myers CA, Holt RD. 1989. The ecology and evolution of intraguild predation: potential competitors that eat each other. *Annu. Rev. Ecol. Syst.* 20: 297–330
136. Polis GA, Holt RD, Menge BA, Winemiller K. 1995. Time, space, and life history: influences on food webs. In *Food Webs: Integration of Patterns and Dynamics,* ed. GA Polis, K Winemiller. London: Chapman & Hall. In press
137. Potts GR. 1986. *The Partridge: Pesticides, Predation and Conservation.* London: Collins. 274 pp.
138. Potts GR. 1993. Red-legged partridge.

In *The New Atlas of Breeding Birds in Britain and Ireland:1988–1991,* ed. DW Gibbons, JB Reid, RA Chapman, pp. 134–35. London: Poyser, TAD
139. Price PW, Westoby M, Rice B. 1988. Parasite-mediated competition; some predictions and tests. *Am. Nat.* 131:544–55
140. Price PW, Westoby M, Rice B, Atsatt PR, Fritz RS, et al. 1986. Parasite mediation in ecological interactions. *Annu. Rev. Ecol. Syst.* 17:487–505
141. Pulliam HR. 1988. Sources, sinks, and population regulation. *Am. Nat.* 132: 652–61
142. Reader RJ. 1992. Herbivory as a confounding factor in an experiment measuring competition among plants. *Ecology* 73:373–76
143. Resetarits WJ Jr. 1991. Ecological interactions among predators in experimental stream communities. *Ecology* 72: 1782–93
144. Ricklefs RE. 1979. *Ecology.* New York: Chiron Press
145. Ricklefs RE, O'Rourke K. 1975. Aspect diversity in moths: a temperate-tropical comparison. *Evolution* 29:313–24
146. van Ripper C III, van Ripper SG, Goff ML, Laird M. 1986. The epizootiology and ecological significance of malaria in Hawaiian land birds. *Ecol. Monogr.* 56:327–44
147. Robinson JV, Wellborn GA. 1988. Ecological resistance to the invasion of a freshwater clam, *Corbicula fluminea:* fish predation effects. *Oecologia* 77: 445–52
148. Robinson SK. 1992. Population dynamics of breeding neotropical migrants in a fragmented Illinois landscape. In *Ecology and Conservation in Neotropical Migrant Landbirds,* ed. JM Hagen III, DW Johnston, pp. 408–18. Washington, DC: Smithsonian Inst. Press
149. Ruxton GD, Gurney WS C. 1992. The interpretation of tests for ratio-dependence. *Oikos* 65:334–35
150. Savidge JA. 1987. Extinction of an island forest avifauna by an introduced snake. *Ecology* 68:660–68
151. Schlumprecht H. 1989. Dispersal of the thistle gallfly *Urophora cardui* and its endoparasitoid *Eurytoma serratulae* (Hymenoptera: Eurytomidae). *Ecol. Entomol.* 14:341–48
152. Schmitt RJ. 1987. Indirect interactions between prey: apparent competition, predator aggregation, and habitat segregation. *Ecology* 68:1887–97
153. Schmitz OJ, Nudds TD. 1994. Parasite-mediated competition in deer and moose: how strong is the effect of me-

ningeal worm on moose? *Ecological Appl.* 4:91–103

154. Schoener TW. 1986. Overview: Kinds of ecological communities—ecology becomes pluralistic. In *Community Ecology*, ed. J. Diamond, TJ Case, pp. 467–79. New York: Harper & Row

155. Schoener TW. 1993. On the relative importance of direct versus indirect effects in ecological communities. In *Mutualism and Community Organization*, ed. H Kawanaba, J Cohen, pp. 365–408. Oxford: Oxford Univ. Press

156. Schoener TW, Spiller DA. 1987. Effects of lizards on spider populations: manipulative reconstruction of a natural experiment. *Science* 236:949–52

157. Schönrogge K, Stone GN, Crawley MJ. 1994. Spatial and temporal variation in guild structure: Parasitoids and inquilines of *Andricus quercuscalicis* Burgsd. (Hymenoptera: Cynipidae) in its native and alien ranges. *Oikos.* In press

157a. Scudder GGE. 1983. A review of factors governing the distribution of two closely related corixids in the saline lakes of British Columbia. *Hydrobiologia* 105:143–54

158. Settle WH, Wilson LT. 1990. Invasion by the variegated leafhopper and biotic interactions: parasitism, competition, and apparent competition. *Ecology* 71:1461–70

159. Sieving KE. 1992. Nest predation and differential insular extinction among selected forest birds of central Panama. *Ecology* 73:2310–28

160. Sih A. 1985. Evolution, predator avoidance, and unsuccessful predation. *Am. Nat.* 125:153–57

161. Sih A. 1987. Predators and prey lifestyles: an evolutionary and ecological overview. See Ref. 85, pp. 203–24

162. Sih A, Crowley P, McPeek M, Petranka J, Strohmeier K. 1985. Predation, competition and prey communities: a review of field experiments. *Annu. Rev. Ecol. Syst.* 16:269–311

163. Simberloff D. 1992. Conservation of pristine habitats and unintended effects of biological control. In *Selection Criteria and Ecological Consequences of Importing Natural Enemies*, ed. WC Kauffman, JR Nechols, pp. 103–17. Landham, MD: Entomol. Soc. Am.

164. Simenstad CA, Estes JA, Kenyon KW. 1978. Aleuts, sea otters, and alternate stable-state communities. *Science* 200:403–11

165. Soulé ME, Bolger DT, Alberts AC, Wright J, Sorice M, et al. 1988. Reconstructed dynamics of rapid extinctions of chaparral-requiring birds in urban habitat islands. *Conserv. Biol.* 2:75–92

166. Spiller DA, Schoener TW. 1988. An experimental study of the effect of lizards on web-spider communities. *Ecol. Monogr.* 58:57–77

167. Spiller DA, Schoener TW. 1990. Lizards reduce food consumption by spiders: mechanisms and consequences. *Oecologia* 83:150–61

168. Strauss SY. 1991. Indirect effects in community ecology: their definition, study and importance. *Trends Ecol. Evol.* 6:206–10

169. Takehuchi Y, Adachi N. 1983. Existence and bifurcation of stable equilibrium in two-prey, one-predator communities. *Bull. Math. Biol.* 45:877–900

170. Terborgh J, Winter B. 1980. Some causes of extinction. In *Conservation Biology: An Evolutionary-Ecological Perspective*, ed. ME Soulé, BA Wilcox, pp. 119–33. Sunderland, Mass: Sinauer

171. Thomas MB, Wratten SD, Sotherton NW. 1991. Creation of 'island' habitats in farmland to manipulate populations of beneficial arthropods: predator densities and emigration. *J. Appl. Ecol.* 28:906–17

172. Underhill LG, Prys-Jones RP, Syroechkovski EE Jr, Groen NM, Karpov V, et al. 1993. Breeding of waders (Charadrii) and brent geese *Branta bernicla bernicla* at Pronchishcheva Lake, northeast Taimyr, Russia, in a peak and a decreasing lemming year. *Ibis* 135:277–92

173. Vance RR. 1978. Predation and resource partitioning in one predator-two prey model communities. *Am. Nat.* 112:797-813

174. Vandermeer J. 1991. Contributions to the global analysis of 3-D Lotka-Volterra equations: dynamic boundedness and indirect interactions in the case of one predator and two prey. *J. Theor. Biol.* 148:545–61

175. Vickery PD, Hunter ML Jr, Wells JV. 1992. Evidence of incidental nest predation and its effects on nests of threatened grassland birds. *Oikos* 63:281–88

176. Watt AS. 1981. A comparison of grazed and ungrazed Grassland A in East Anglian Breckland. *J. Ecol.* 69:499–508

177. Watt AS. 1981. Further observations on the effects of excluding rabbits from Grassland A in East Anglian Breckland: The pattern of change and factors affecting it (1936–1973). *J. Ecol.* 69:509–36

178. Way MJ, Khoo KC. 1992. Role of ants in pest management. *Annu. Rev. Entomol.* 37:479–503

179. Werner EE. 1992. Individual behavior and higher-order species interactions. *Am. Nat. (Supplement)* 140:5–32
180. White J. 1980. Resource partitioning by ovipositing cicadas. *Am. Nat.* 115:1–28
181. Wilbur HM, Fauth JE. 1990. Experimental aquatic food webs: interactions between two predators and two prey. *Am. Nat.* 135:176–204
182. Wilcove DS. 1985. Nest predation in forest tracts and the decline of migratory songbirds. *Ecology* 66:1211–14
183. Wilcove DS, McLellan CH, Dobson AP. 1986. Habitat fragmentation in the temperate zone. In *Conservation Biology. The Science of Scarcity and Diversity,* ed. ME Soulé, pp. 37–56. Sunderland, Mass: Sinauer
184. Williams KS, Smith KG, Stephen FM. 1993. Emergence of 13-yr periodical cicada (Cicadidae: Magicicada): phenology, mortality, and predator satiation. *Ecology* 74:1143–52
185. Williamson MH. 1957. An elementary theory of interspecific competition. *Nature* 180:422–25
186. Wise DH. 1993. *Spiders in Ecological Webs.* Cambridge: Cambridge Univ. Press. 328 pp.

187. Wissinger S, McGrady J. 1993. Intraguild predation and competition between larval dragonflies: direct and indirect effects on shared prey. *Ecology* 74:207–18
188. Woodman RL, Price PW. 1992. Differential larval predation by ants can influence willow sawfly community structure. *Ecology* 73:1028–37
189. Woodward BD. 1983. Predator-prey interactions and breeding-pond use of temporary-pond species in a desert anuran community. *Ecology* 64:1549–55
189a. Wootton JT. 1994. The nature and consequences of indirect effects in ecological communities. *Annu. Rev. Ecol. Syst.* In press
190. Yodzis P. 1986. Competition, mortality, and community structure. In *Community Ecology,* ed. J Diamond, TJ Case, pp. 480–91. New York: Harper & Row
191. Zwölfer H. 1961. A comparative analysis of the parasite complexes of the European fir budworm, *Choristoneura murinana* (Hub.) and the North American spruce budworm *C. fumiferana* (Clem.). *CIBC Tech. Bull.* 1:1–162

NOTE ADDED IN PROOF

We have become aware of two excellent examples of apparent competition between plants due to shared herbivory, both highlighting spatial consequences: MA Parker & RB Root, 1981. Insect herbivores limit habitat distribution of a native composite, *Machearanthera canescens Ecology* 62:1390–92; CD Thomas, 1986. Butterfly larvae reduce host plant survival in vicinity of alternative host species. *Oecologia* 70:113–17. A broad overview of herbivore effects on plant communities is provided by N Huntly, 1991. Herbivores and the dynamics of communities and ecosystems *Annu. Rev. Ecol. Syst.* 22:477–503. Finally, poaching on African buffalo (a grazer) in the Serengeti has depressed lion numbers, leading to marked increase in impala (a browser)—an inadvertent experiment indicating the existence of apparent competition via shared lion predation: ARE Sinclair, 1995. Population limitation of resident herbivores. In *Serengeti II: Research, Management and Conservation of an Ecosystem,* ed. ARE Sinclair and P Arcese. Chicago: Univ. Chicago Press. In press. We would welcome being told of additional pertinent examples.

Annu. Rev. Ecol. Syst. 1994. 25:521–46

DIVERSIFICATION IN AN ARID WORLD: The Mesembryanthemaceae

H-D. Ihlenfeldt

Institut für Allgemeine Botanik und Botanischer Garten, Universität Hamburg, D-22609 Hamburg, Germany

KEY WORDS: evolution and habitats, adaptation, homeotic genes

Abstract

Mesembryanthemaceae (ca. 2000 species, 116 genera) is the dominating family in the succulent flora of the Succulent Karoo Region (southern Africa), to which the vast majority of species of this family are restricted. The success of this family is due to an extremely broad spectrum of highly specialized life forms and life strategies. The family must have developed a tempo of evolution that is probably unrivaled within angiosperms, since the mega-niche Succulent Karoo opened up only 5 million years BCE at most.

The remarkable biodiversity of this family is the result of a complex inter-action between the availability of numerous diverse niches, often very small in extent, a climate characterized by steep gradients, and a strong genetic drift caused by a very low rate of gene exchange among populations and frequent breakdown of populations. It is suggested that a special genetic potentiality plays a crucial role in the evolutionary processes of this family. This potenti-ality allows fast and easy rearrangement of high-ranking regulatory "homeotic" genes. This process is strongly canalized by the shortage of resources, i.e. changes in one place in the life cycle usually force changes in other places. Perhaps by repeated changes in homeotic genes, this family has succeeded in evolving numerous novel and diverse complex life forms and life cycles in spite of a very limited set of subordinate ontogenetic programs.

INTRODUCTION

Arid habitats are believed to exhibit a reduced or poor plant biodiversity compared, for instance, with tropical rainforests. This reduced biodiversity

521

0066-4162/94/1120-0521$05.00

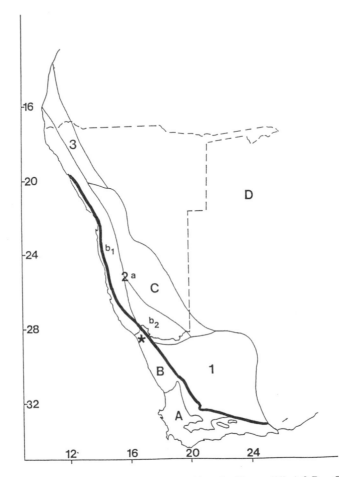

Figure 1 Phytogeographical division of southern Africa after Jürgens (41). A & B — Greater Cape Flora, A — Cape Floristic Region, B — Succulent Karoo Region. C — Nama Karoo Region. * — location of the Numees Valley (see text).

applies both to the spectrum of life forms and to species density, i.e. the number of species per unit of area. Moreover, most of the biomass present in those habitats is said to be concentrated in a few species only.

Although this appears to be true of most arid regions of the world, there are astonishing exceptions. The most outstanding exception according to our present knowledge is the so-called Succulent Karoo Region (area B in Figure 1) along the west coast of southern Africa. This is a summer-dry area affected by the cold Benguela Current, with an annual precipitation between 10 and

300 mm. The vegetation of this region is the counterpart of the succulent vegetation of Baja California or northern Chile. Phytogeographically this area had been included in the Kingdom of Palaeotropis (58, 59), but a recent analysis based on more than 1700 taxa (41, 42) revealed that it forms part of the Cape Floral Kingdom (now Greater Cape Flora, see 41).

To give an idea of the climate and the richness of the flora, results are cited of the extensive study by Jürgens (38) of the Numees Valley located in the center of the Succulent Karoo Region, about 60 km from the coast (see Figure 1; climatic data see 57): Mean annual precipitation is ca. 60 mm per year with fluctuations between 18 mm and 117 mm, and most of the rain between March and October (winter rainfall area); air temperatures from 4 to 42°C; relative air humidity dropping to only 10% but rising quite regularly to nearly 100% during the night; causing fog formation on about 80 days a year, and the humidity of the fog can be utilized by the plants to a certain degree (57). Within an area of only 1.3 km^2 331 species of angiosperms (updated figure, personal communication) were found growing in 17 distinct communities. This species density approaches figures given for tropical rainforests of western Africa (47).

THE MESEMBRYANTHEMACEAE

Judged from the number of species, two families are dominant in the Numees area, the Asteraceae and the Mesembryanthemaceae, each representing about 15% of the total flora. Almost nothing is known about the evolutionary strategies of the Asteraceae of this region. As to Mesembryanthemaceae, however, extensive studies into distribution patterns, population structure, the anatomy, morphology, and ecophysiology of the family during the last 25 years and a number of recent taxonomic revisions now allow us to outline the evolutionary processes and strategies of this successful family.

The Mesembryanthemaceae (sometimes included in Aizoaceae s.l.; 3, 22) is one of the three great families of succulent plants, each comprising about 2000 species: the Cactaceae (stem succulents, nearly completely restricted to the New World), the Crassulaceae (leaf succulents, occurring in both the New and Old World), and the Mesembryanthemaceae (leaf succulents, originally restricted to the Old World). Approximately 90% of the last family are native of the winter rainfall area of southern Africa (Succulent Karoo), which comprises ca. 200,000 km^2. Less than 2% of the species naturally occur outside the area shown in Figure 2 (East and North Africa, Madagascar, Australia, New Zealand and Chile). All species occurring outside the African continent (including Arabia) are seashore plants, which probably reached their present habitats by long-distance dispersal by seabirds (6).

Compared with other families of succulent plants, the range of life forms

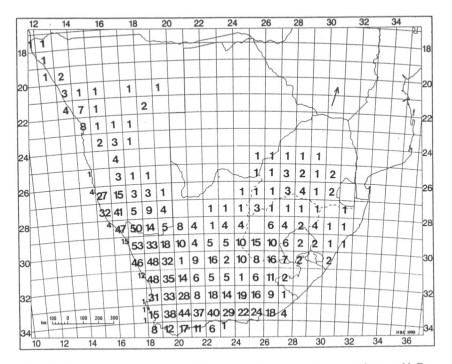

Figure 2 Density of genera/subgenera of Mesembryanthemaceae in a square degree grid. From Jürgens (41).

within this family is unrivaled: true annuals (*Dorotheanthus*); short-lived perennials (*Mesembryanthemum* species); trailing woody plants (*Cephalophyllum*); upright semishrubs (a very widespread type), shrubs and small trees up to 3 m high (*Ruschia*); highly succulent compact forms (*Conophytum, Lithops*); forms sometimes permanently ("window plants," *Fenestraria*) or periodically (*Conophytum* species) sunken in the ground; some not rarely extremely miniaturized (*Oophytum, Diplosoma, Conophytum* species); stem succulents with deciduous succulent leaves (*Psilocaulon, Monilaria*); and even geophytes with a subterranean tuber (*Phyllobolus* species).

Superimposed are two quite diverse basic investment strategies (30, 32, 57): one adapted to surviving even prolonged droughts as adult plants (with the drawback of low output from photosynthesis and consequent slow growth), and another one adapted to periods of precipitation above average (with the advantage of fast growth but with the drawback of the risk of not surviving prolonged droughts as adult plants; these plants have to rely on survival by seeds). Plants representing these two investment strategies successfully coexist in the same habitat—one of the reasons for the high degree of diversity of the

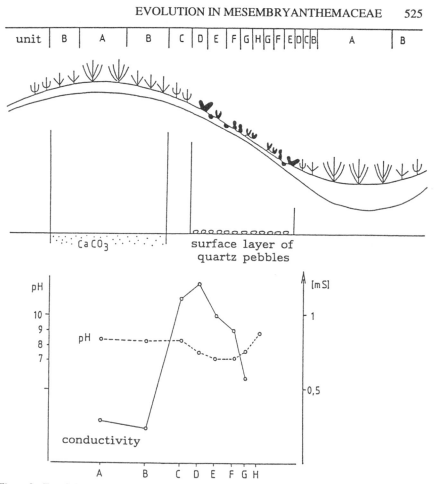

Figure 3   Top: Schematic longitudinal section of a hill with a zonation of distinct plant communities (A–H) dominated by members of the Mesembryanthemaceae. The varying depth of a catena of fine textured soil above the bedrock, the occurrence of free limestone in the soil (CaCO₃), and a surface layer of quartz pebbles are indicated. Bottom: pH and conductivity in the center of the units A–H. (Adapted from Jürgens—38).

Mesembryanthemaceae in relation to their area of distribution (see below). Occurrence of two strategy types in one habitat can cause dramatic fluctuations in species composition and distribution of biomass among the species, according to climatic cycles typical of this region (30).

   In recent years several attempts have been made to develop a system and an improved terminology for the description of life and growth forms in Mesembryanthemaceae and succulents in general (24, 30, 38–40, 57).

   Of the three great families of succulent plants, the Mesembryanthemaceae are furthermore unrivaled with regard to diversity in relation to their area of distribution. Up to 53 genera or subgenera (not species!; see Figure 2) may occur in a single degree square (ca 100 × 100 km). Over vast areas of the Succulent Karoo Regions, this family is the dominant plant group; in whole series of plant formations, this family may represent more than 50% of the species and up to 90% of the biomass (38; Figure 3). This fact is the more astonishing since the ecological mega-niche (arid winter rainfall area with moderate temperatures) now dominated by this family, opened up only 5 million years BCE at best (1, 8, 9), i.e. this family developed a tempo of evolution that is probably unique within the angiosperms.

A comprehensive overview of the morphology of the Mesembryanthemaceae has recently been compiled by Hartmann (19) as well as a list of accepted genera (116 genera; 22). These two papers also provide a nearly complete list of the relevant taxonomic literature; author's names of taxa are therefore omitted in the present paper. Ample data on the ecology and the physiological properties of this family can be found in the recently published book of von Willert et al (57), which also provides a complete overview of the relevant literature on these aspects. The present article, therefore, focuses on the evolutionary processes and strategies of this family. An attempt is made to explain the extreme diversity of this family, apparently the basis of its great success, which must have been achieved in a remarkably short period of time.

Any attempt to explain the extreme biodiversity and high tempo of evolution in this family has to take into account:

*the taxonomic structure of the family and patterns of distribution of the taxa,
*the environment in which the family evolved, and
*the genetic loads of the family.

TAXONOMIC STRUCTURE AND PATTERNS OF DISTRIBUTION

It is a well-known trait of the Mesembranthemaceae that there exist on the one hand well-defined and isolated taxa with often very limited distribution, sometimes comprising only a few populations (e.g. genus level: *Oophytum*—28, *Caryotophora*—44; species level: *Dorotheanthus booysenii*—36) and on the other hand vast complexes of closely related taxa that are difficult to delimit; the latter ones usually have much larger distribution areas (e.g. genus level: *Lithops*, whose distribution area extends from Transvaal through the Cape Province as far as northern Namibia—7). This remarkable phenomenon is independent of the taxonomic level in question, i.e. it can be observed at any taxonomic level between population and subfamily. A widespread and taxonomically well-defined genus may be composed both of a number of species difficult to delimit and with wide distributions and of a few well-defined species with very limited distribution (e.g. *Conophytum*—12), or vice versa (e.g. *Mitrophyllum*—48). These findings may be interpreted as that the family is composed of taxa of different age, with the well-defined and isolated taxa (species and especially genera) being relicts of ancient evolutionary levels. In the case of taxa above species level, this, however, does not imply that they are no longer able to start a new phase of adaptive radiation (*Monilaria*—35). A different interpretation is presented at the end of this article.

Another widespread trait of the family is the reticulate distribution of characters considered to be taxonomically important. Again this is true of all taxonomic levels. This trait strongly impedes cladistic analyses that attempt to elucidate the phylogeny of a group, and in many cases it causes problems in the delimitation of natural taxonomic units, especially on the genus level (e.g. subtribe Dorotheanthinae—36; *Ihlenfeldtia-Vanheerdea-Tanquana* complex—20, 21). On the level of taxa above genus, this phenomenon is even more apparent: natural groups of closely related genera, deduced from the presence of synapomorphies, are surrounded by "satellite" genera, clearly more distantly related, but with distinct affinities to other "clusters" of genera.

The taxonomic situation described above especially affects the genus concept applied in this family. Originally all species of this family were united in a single genus, *Mesembryanthemum* L. With the growing number of known species, however, it became necessary to split this genus, and several attempts using mainly life-form characters were made. At the beginning of the 1920s, NE Brown (5) discovered that the complicated fruits of the family rendered a variety of distinct characters that could be used to delimit taxonomic units, especially on the genus level. Nevertheless, habit and life form continued to be an important character complex. In the end the complicated taxonomic structure of the family led to a genus concept which has been criticized as being very narrow. Of the contemporary authors, only Bittrich (overview of the subfamily Mesembryanthemoideae—2) and Hammer (monograph of *Conophytum*—12) adopted a broader genus concept and lumped together some of the established genera. Most of the active workers in this family, however, still prefer a narrow concept. This is a kind of compromise. On the one hand the narrow genus concept generates a considerable number of genera with only few species, even quite a number of monotypic genera, but on the other hand there are still genera left that have more than 200 species. The advantage of the current concept is that within a genus life forms are usually identical or at least very similar. In many cases, however, genera grouped together mainly on the basis of fruit structure (19, 22) often exhibit very diverse life forms, e.g. the "*Leipoldtia* group" (22); another example is discussed later in this article.

The current genus concept has the great advantage that it is possible to deal with the evolutionary processes within the family in two sections: One is the population-species level (which substantially means speciation), and the second one is the genus level (in many instances this means evolutionary changes in life form and life cycle).

THE ENVIRONMENT

The Succulent Karoo Region is characterized by a great variety of geological formations, a rugged geomorphology with steep reliefs, and a great variety of soils, which are generally poor in nutrients but rich in other ions (37). Steep gradients in ion content and in pH within short distances are typical (37, 38), and in many places a high content of heavy metal ions is obvious.

Annual precipitation decreases from about 300 mm in the south to nearly zero in the north. These values, however, do not reflect the amount of water actually available in a given site. Geomorphology (slopes, altitude, exposition, rain shadow) and soil conditions greatly modify the amount of utilizable water. Rock crevices may collect water from a considerable area of bare rock, and a soil with a high content of clay combined with a surface of densely packed quartz pebbles may strongly impede penetration of the water, which runs off quickly (38). Moreover, a pronounced gradient in temperature (increasing) and air humidity (decreasing) occurs from west (coast) to east. Due to the cold Benguela Current, temperatures usually are moderate, but occasional hot "bergwinds" falling down the escarpment of the central African plateau may cause a drastic rise of the air temperature and a dramatic drop of the relative humidity (down to only 10%; 57). The Benguela Current, too, is the source of high relative air humidity, especially during the night when fog formation is not rare. Dew may contribute considerably to the water budget of certain species, e.g. in the genus *Mitrophyllum* most species are restricted to the slopes with the highest amount of fog precipitation (48). Summarizing, one can state that the Succulent Karoo Region is characterized by a very large range of quite diverse ecological niches, often very small in extent and forming a sort of mosaic. This is reflected by the size of areas inhabited by distinct plant communities (38) and the size of populations in highly specialized taxa of this family as well. Areas between 30 and 100 m^2 are quite common, and distances between neighboring populations may reach several kilometers (compare e.g. dot maps in 7, 12, 15, 16, 35).

THE GENETIC LOADS

Pollination System and Breeding System

The flowers of the Mesembryanthemaceae, usually exhibiting numerous large and showy petals (of staminodial origin) and nectaries, are adapted to insect pollination. Most flowers admit a wide range of pollinators; nevertheless there are also specialized types (psychophily, phalaenophily; 19, 45, 56). Whether in some rare cases—where flowers are small and numerous, petals reduced and pollen copious—wind pollination does occur in the natural habitats (2, 10) still is questionable. Most species are protandrous (19, 23) and in addition self-sterile (45, 46, 50, 53). A general exception are annuals and some short-lived species, which are self-fertile and perform self-pollination toward the end of anthesis, or they may even be cleistogamic (30, 36, 52). Flowering time is well synchronized within the populations, and the period of flowering usually short. Furthermore, opening of the flowers is usually restricted to a certain period of the day (often around midday, hence the name "midday-flowers" for this family). Self-sterility and "mass flowering" greatly promote gene exchange within a population. Most of the documented pollinators, however, do not fly far (46, 55), so that gene exchange among populations by pollen transport is much restricted.

Experiments with cultivated plants have revealed that genetic barriers between related species (of the same genus or of closely related genera) may be weak or even completely missing (13, 46). Even species of genera that are considered to be only remotely related can be hybridized successfully (13). This is in contradiction to observations in the field, where hybrids are extremely rare (13, 46). Up to now, only a few instances of successful introgression between neighboring populations have been reported (7, 12, 13, 31, 35). In natural habitats, species are usually stabilized by spatial separation and/or different flowering times and/or different flower structure (46), and sometimes by different levels of ploidy (up to 30% of the species of genera examined may be tetra- or hexaploid; 19). Species of the same genus that occur truly sympatrically usually are the most distantly related; such species cannot even be hybridized artificially (e.g. *Lithops*—7).

Dispersal System

The dissemination system of the Mesembryanthemaceae is well adapted to their arid habitats (overview 18, 19, 29). Seeds usually mature toward the end of the vegetation period but are retained in capsules that remain closed. Thereby the seeds are protected against predators and severe environmental conditions during the dry period. The capsules open as soon as they have been soaked by water; in the natural habitats, the seeds are ejected when raindrops hit the open capsule. Thus the seeds are released only when conditions for fast germination are favorable. As soon as the rain stops and the capsules desiccate, they close again. Capsules may last for many years, and the hydrochastic opening mechanism may function for long periods. The distances across which the seeds are ejected rarely exceed 1 m. One can characterize the dispersal system as a defensive one, aimed at assuring that the progeny stays in the (often small) ecological niche and replaces the mother plant in case of death, or that it conquers a site next to the mother plant which might become available by the death of an individual of a competing species (29). The patchwise distribution of Mesembryanthemaceae species in many plant associations is a striking phenomenon. In rare cases, long-distance transport of seeds or detached capsules ("tumble fruits"; 17) or parts of them (mericarps; 29, 33) is achieved by the wind. Dispersal of seeds or detached capsules by running water appears to be possible, but there are very few distribution patterns that hint at this possibility (48). In any case, long-distance dispersal is a rare event in this family.

Gene exchange among populations can be achieved by either pollen or seed transfer. It is obvious that the highly adapted dispersal system of the Mesembryanthemaceae also strongly impedes gene export from a population. To summarize, in this family, gene exchange among populations should be a rare event. This presumption is substantiated by cases of intraspecific differences in flower colors, which is not at all rare. Within a population, flower color is usually uniform; obviously a strong selection pressure urges toward uniformity. Populations that are only 200 m apart may, however, exhibit different colors. Frequent gene exchange among these populations should easily be detected by the occurrence of individuals with aberrant flower colors. Field observations show that also in these cases individuals with deviating flower colors are rare indeed.

EVOLUTIONARY PROCESSES ON THE POPULATION-SPECIES LEVEL

These processes are discussed by means of two genera that well illustrate the basic patterns encountered in this family. Generally, the Mesembryanthemaceae provide good examples of what has been called "diversifying natural selection," a process "that enhances the adaptedness of populations that live in heterogeneous environments" (11). Prerequisites are the availability of numerous diverse ecological niches (already described above) and a pronounced tendency toward polymorphism of the organisms involved. The latter precondition is well known of the Mesembryanthemaceae. The reasons are numerous: the much restricted gene flow among the populations (see above); the often small size of the ecological niches (see above), which in part is due to extreme ecological specialization. This in turn promotes an usually low number of individuals per population, further decreased by the prevailing "contracted" vegetation caused by the low amount of precipitation, which additionally limits the number of individuals per population in a given site. Further, long-term climatic cycles regularly cause complete or nearly complete breakdown of populations, which have to be reestablished from a few surviving individuals or even from the seed bank in the soil. All these factors promote genetic drift.

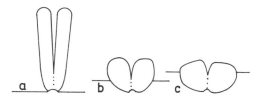

Figure 4 Morphology of a single leaf pair in *Argyroderma;* the horizontal line indicates top of soil. (a) *A. fissum,* the putative ancestor of the genus; (b) *A. crateriforme* and (c) *A. delaetii,* advanced members of the eco-cline shown in Figure 5. (Adapted from Hartmann—16).

In the past, in many cases this has led to "overdescription" of "species" (e.g. *Conophytum,* ca. 450 names for 84 specific taxa after revision—12).

The genus *Argyroderma* provides the first example. This genus is specialized to a certain type of soil, a kind of sandy clay with a surface layer of quartz pebbles. Only *A. fissum* is adapted to sandy soils. The whole distribution area of the genus is small (ca. 2500 km²; Figure 5). The putative ancestor (14, 16), *A. fissum,* is found all over this area. In spite of its comparatively wide distribution, this taxon is remarkably uniform (14, 16). Within the genus, 10 further taxa can be delimited to which the rank of a species or subspecies has been attributed (16). All these taxa share a common feature: The leaves are enlarged and tend to be hemispherical in shape (Figure 4); thus the sole pair of opposite leaves of every stem approaches more or less perfectly the shape of a sphere, which, due to a more favorable surface-volume relation, reduces water losses by cuticular transpiration. Furthermore the advanced types tend to grow more or less deeply sunken in the ground, and the degree of ramification (i.e. leaf pairs per individual plant) decreases from more than 50 ramets per individual to unbranched plants. Enlarged water-storing capacity per leaf pair, plus reduced losses and reduced water demand per individual, enables the advanced types to exist in more arid niches. Most of these advanced taxa are allopatric, in some cases with some overlapping of the individual distribution areas (Figure 5). Truly sympatric occurrence of two species, however, is extremely rare, and in the few examples (*A. pearsonii* and *A. delaetii*—16) the two species involved are not closely related and furthermore are isolated by different flowering times.

The derived taxa evolved along different lineages from the ancestor (16). *A. congregatum* (1 in Figure 5), *A. crateriforme* (2 in Figure 5) and *A. delaetii* (3 in Figure 5) can be considered as members of a former eco-cline that runs from west to east, where the habitats are more arid due to lower frequencies of fog events and higher temperatures. The increasing aridity of the sites is reflected by the fact that the number of competing species greatly decreases; *A. delaetii* often forms pure stands (14). This eco-cline is in the state of final

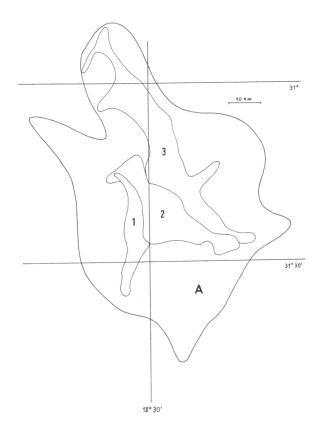

Figure 5 Distribution of *Argyroderma fissum,* the putative ancestor of the genus (A, outer contour), and the the former eco-cline of the advanced species *A. congregatum* (1), *A. crateriforme* (2), and *A. delaetii* (3). Compare text. (Adapted from Hartmann—16).

disruption, i.e. there are only few populations left that show transitional characters between the adjacent, already distinct taxa. Of course, one cannot exclude with certainty the possibility that the few transitional populations may be secondary due to introgression. To summarize, in this example the development of advanced taxa adapted to more arid habitats did not lead to an extension of the original distribution area of the genus; rather, the number of sites inhabited by members of this genus increased considerably.

The genus *Monilaria* provides another example (see Figure 6). Three species, *M. moniliformis* (1 in Figure 6), *M. chrysoleuca* (2 in Figure 6), and *M. scutata* (3 in Figure 6) are closely related. Their stems are thickened and can store considerable amounts of water. In principle, they are allopatric. Judged from certain anatomical characters relevant for protection of stored water (35),

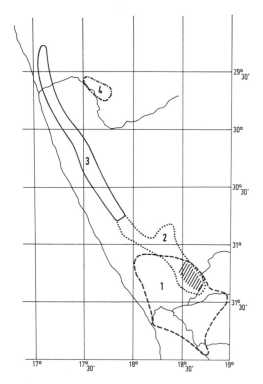

Figure 6 Distribution of the genus *Monilaria. M. moniliformis* (1), *M. chrysoleuca* (2) and *M. scutata* (3) form an eco-cline following the decrease of precipitation from south to north. *M. obconica* (4) is an isolated tetraploid highland species. Hatched area: the putative ancestor of the genus, *M. pisiformis;* compare text. (Adapted from Ihlenfeldt & Jörgensen—35).

they form an eco-cline following the decrease of annual precipitation from south to north. The distribution areas of *M. moniliformis* (1) and *M. chrysoleuca* (2) show considerable overlap. Nevertheless, in this zone most of the populations are clearly separable; only a few exhibit intermediate characters. In this overlapping zone another species, *M. pisiformis* (hatched area), occurs. This species is easily discernible as the stems are much thinner and less succulent. Again most of the populations are clearly separable; only a very few can be classified as intermediate between *M. chrysoleuca* and *M. pisiformis*. The low number of intermediate populations is quite remarkable because all species flower at the same time of the year. This observation again proves the low rate of gene flow among the populations in the family. The area of a fifth, also thick-stemmed, species, *M. obconica* (4, in Figure 6), is clearly separated from the other species. Moreover, the ecology of the habitats is quite different: *M.*

obconica grows at an elevation of ca. 1000 m, whereas the remaining species are lowland species. Because it is tetraploid, *M. obconica* is also genetically isolated from the other species, which are diploids.

One must conclude from comparison with related genera that a thick water-storing stem represents an advanced character. Consequently, the comparatively thin-stemmed *M. pisiformis* should represent the ancestral state or even the ancestor itself. If one assumes that the present distribution area of *M. pisiformis* still represents the original area, one must conclude that in *Monilaria* (different from the example *Argyroderma*) the development of advanced taxa with an enlarged water storing capacity led to a considerable extension of the original distribution area. This hypothesis is corroborated by a phenomenon one could call gene erosion. In the area of *M. pisiformis* (where two other species also occur) there is, independent from the taxa, a high diversity in certain characters such as color of the flowers and epidermis structure of the leaves, though within a given population uniformity prevails. Proceeding southward or northward from this center, this variety gradually disappears (35).

Summarizing, one can state that in this family speciation follows well-known patterns, though speciation is modified and greatly enhanced by certain traits of the genetic loads and the environment. The processes dealt with hitherto, however, do not really explain the extreme diversity of this family. The crucial question is: Which evolutionary processes made possible the development of so many diverse life forms? Moreover, there is broad evidence that comparable advanced life forms evolved several times independently in different lineages of the family. Thus one has to look for processes that generate novel life forms or transform one life form to another.

EVOLUTIONARY PROCESSES ON THE GENUS LEVEL

An example that provides deep insight into these kinds of processes is the subtribe Mitrophyllinae, comprising six well-defined genera and two odd isolated members, which could be considered as monotypic genera. The subtribe

---

Figure 7 Growth forms in the subtribe Mitrophyllinae. (a) *Mitrophyllum mitratum*—the plant body is composed of vegetative short shoots at the base of the plant and slender reproductive long shoots. (b) *Monilaria scutata*—the succulent stems are covered by black scale-like structures, each of which originates from the sclerified connate base of the second leaf pair of the previous season. (c) *Dicrocaulon brevifolium*—a low erect-growing dwarf shrub whose framework is composed of long shoots bearing numerous lateral short shoots; this species can develop more than two leaf pairs per season. (d) *Meyerophytum meyeri*—a cushion-like dwarf shrub, also composed of long shoots and short shoots. (e) *"Monilaria" globosa*—a compact highly succulent plant, composed of short shoots only. (f) *Oophytum nanum*—a miniature plant, composed of short shoots only; the emerging globose first leaf pairs of the season are still covered by the papery remnants of the last pair of previous season. For details compare text and note the different scales.

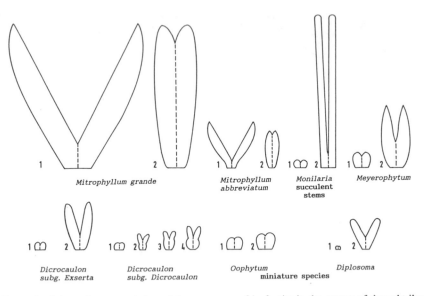

Mitrophyllum grande   *Mitrophyllum   Monilaria   Meyerophytum*
                             *abbreviatum   succulent
                                                 stems*

Dicrocaulon   Dicrocaulon   Oophytum   Diplosoma
subg. Exserta   subg. Dicrocaulon miniature species

Figure 8 Schematic representation of the sequences of leaf pairs in the genera of the subtribe Mitrophyllinae. Figures indicate the position of the individual leaf pair within the sequence of a single season. All figures at same scale; compare text.

inhabits a narrow strip of nearly 400 km in extent along the west coast of the Succulent Karoo Region. Annual precipitation ranges from about 250 mm in the south to less than 100 mm in the north. The life forms are extremely diverse (Figure 7): shrubs up to 1 m high (*Mitrophyllum*); cushion-like dwarf shrubs (*Meyerophytum, Dicrocaulon*); sparsely branched subshrubs with thick water storing succulent stems, up to 20 cm high (*Monilaria*); compact highly succulent plants with enlarged leaves (*"Monilaria" globosa;* this species has to be excluded from the genus *Monilaria,*—35), and compact miniature plants, between 0.5 and 3 cm in size (*Diplosoma* and *Oophytum*).

That these genera are closely related in spite of their heterogeneity in life forms is well established (26). Three of the relevant characters of this group have to be explained here:

*All members exhibit heterophylly, i.e. the leaf pairs developed within one season differ in their morphology (shape and size, see Figure 8). Usually only two pairs of leaves are developed on each stem per season.

*Each stem can only develop a single flower that terminates the annual shoot. A flowering stem is continued in the next growing season by one or more lateral shoots.

*All members belong to the peculiar group termed "recycling succulents" (30,

32), so far known only from this family. Toward the end of the vegetation period, all available water reserves are concentrated in the last pair of leaves. From the beginning of the dry season onwards, all utilizable water from this last (second) leaf pair is translocated into another pair of leaves, which gradually develops inside this second pair. Finally, the remnants of this pair cover the new pair as a papery sheath. At the beginning of the next rainy season, the new pair reaches its final size by uptake of water. Breaking the sheath, it emerges and starts functioning.

When one tries a cladistic analysis of this group, it turns out there is only a limited number of—in part alternative—components or standardized units that concern all fields of the life cycles of the genera: vegetative phase (of juvenile and adult individuals as well), flowering, dispersal of seeds, and establishment of seedlings. Moreover, most of the alternative units are of quantitative nature, e.g. large leaves vs small leaves, large flowers vs small flowers, long internodes vs short internodes. The available components are interwoven in a complicated manner to form a variety of complex life forms and life strategies (see Table 1).

Available space suffices to discuss in some detail only two sets of units that concern the vegetative phase. Comments point out some of the ways this bears on other fields of the life cycles.

The first set of units is a very simple one and not at all a special one. The internodes between two subsequent leaf pairs can remain unextended (resulting in a so-called short shoot), or they may stretch (resulting in a long shoot). During their juvenile stage, all species of the Mitrophyllinae grow according to the short-shoot mode. In the adult stage, there are two possibilities:

A. The short shoot mode may be retained for the whole lifespan: This is the case in "Monilaria" globosa and in the genera *Diplosoma* and *Oophytum,* two genera of miniature species.
B. At least some stems develop into long shoots. In this case there are three different options:

1. Only a few reproductive (i.e. flower-bearing) stems grow according to the long-shoot mode, whereas the majority of the stems at the base of the plant continue to grow according to the short-shoot mode (*Mitrophyllum*). The short shoots at the base of the plant may develop into reproductive stems in case the original reproductive stems have been eaten by herbivores or have died off due to exhaustion. In other words, the short shoots at the base of the plant form a sort of reserve from which the plant can regenerate. It is obvious that this strategy limits the number of flowers that can be formed per season and per individual, and this again is the reason why the flowers (and the capsules that develop from these flowers) are usually large. The

Table 1 Advantages and disadvantages of components

Advantages	Disadvantages
A. Large leaves	
Favorable surface-volume relation	High investments per leaf
↦ low water losses by cuticular transpiration during dry season	↦ additional costs (reinforcement of stems)
	↦ sparse branching limits number of flowers per season
↦ high water storing capacity	↳ loss of even a single leaf means considerable loss of biomass
↦ flowering before beginning of rainy season possible	High thermal stress reduces number of possible habitats
↳ capsules exposed to predators for only very short period	
B. Small leaves	
Low investments per leaf	Unfavorable surface-volume relation
↦ loss of a single leaf does not mean considerable loss of biomass	↦ high water losses by cuticular transpiration during dry season
↦ formation of additional leaves possible in good seasons	
↦ additional output from photosynthesis	↦ flowering only towards the end of rainy season possible when pollinators are rare
↳ formation of additional flowers possible	↳ change in pollination system
C. Succulent stems	
Low water losses due to periderm	High investments necessary
↦ only low investments for water storing leaves necessary	↦ only very limited number of stems possible
↦ good storing capacity for carbohydrates	↦ limits number of flowers per season
↳ flowering even after bad previous season possible	↦ large flowers and large capsules
	↳ loss of even a single stem means high loss of biomass
D. Miniaturizing	
Use of shade cast by pebbles	Very low output from photosynthesis
↦ use of condensed water possible	↦ only very limited number of flowers possible
↳ only low investments for water storing structures necessary	↳ flowers small

position of the flowers on long shoots well above the ground certainly enhances their visibility and makes pollination by the rare insects safer. On the other hand, in case the reproductive elongated erect stems are eaten by herbivores, the plant will be unable to produce flowers at all until some of the short shoots at the base of the plant have developed into reproductive long shoots.

2. After a compact juvenile stage, the framework of the plant body is composed of long shoots that bear lateral short shoots. Depending on the ratio between these two types, this results in a more or less cushion-like growth habit (*Dicrocaulon, Meyerophytum*). The enlarged number of stems confers considerable advantages: the plant is less vulnerable to damage caused by herbivores, especially because the water reserves are distributed over many stems (see below). Owing to the enlarged number of stems, there is an option for numerous, though smaller, flowers per season. Because they are produced as single terminal flowers, all of them can be opened at the same time, covering the plant with a mass of flowers, thus compensating for the smaller size and reduced visibility of a single flower. Moreover, a large number of flowers diminishes the pollination risk, and a large number of small capsules with comparatively few seeds per capsule reduces the risk of loss by predators. This option is implemented by some members of the genus *Dicrocaulon*. The remaining species and the genus *Meyerophytum* do not make use of this option; they produce a smaller number of large flowers.

3. Very peculiar and unique is a third mode: Of the two internodes formed every season, one stretches and the other remains unextended (a kind of mixture of the two basic modes). This enables the plant to form water-storing succulent stems at reasonable cost, i.e. investment of carbohydrates (*Monilaria*). Nevertheless, investments are still high compared with normal stems. On the other hand, this strategy saves high investments for water storing capacities in the leaves; however, only a limited number of stems of equal structure can be developed. This in turn limits the number of flowers per season, resulting in comparatively large flowers, from which large capsules with numerous seeds develop.

The second set of units concerns leaf morphology and is tightly connected with the problem of the water storing capacity. Two opposite leaves of the same type always form a pair. In principle, three leaf types are discernible (see Figure 9):

*The first type (I) is a very short, nearly hemispherical leaf; the two opposite leaves are laterally connate for nearly their complete length. This is the type that most Mesembryanthemaceae exhibit as cotyledons.

*A second type (III) is represented by a pair of long narrow leaves, connate

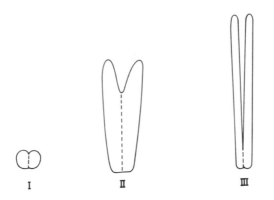

Figure 9 The three leaf types in the subtribe Mitrophyllinae exhibited by adult plants. I. A pair of hemispherical leaves, nearly completely connate (representing the type of the cotelydons of most Mesembryanthemaceae). II. Intermediate type, occasionally also occurring in other, not closely related genera of the family. III. A long pair, only connate at the base (normal foliage type of adult plants in Mesembryanthemaceae).

only at the very base. This is the normal foliage type that most Mesembryanthemaceae exhibit as adult individuals.

*The Mitrophyllinae (and some other members of the family) manifest a third type (II), which is intermediate: long leaves, but connate for a considerable portion of their total length, half the length or even more.

Most adults of the Mitrophyllinae develop only two pairs of leaves every season. Consequently, they can make use of only two of the three leaf types, and the special sequence of leaf types is used to delimit the genera (compare Figure 9).

The relevant type that defines the water storing capacity during the dry season is the second leaf pair of the season, because it is inside the connate part of this pair where the recycling of the utilizable water takes place. The length of the connate portion limits the maximum size of the newly developing leaf pair, and by this, the maximum storing capacity per stem during the dry season. *Mitrophyllum,* by using the intermediate type with an extraordinarily long connate portion as the second pair, has by far the highest storing capacity, which enables this genus to live in the most arid part of the distribution area of the subtribe.

Things are complicated by the fact that each leaf type exists in two versions: a much enlarged version and a small version, sometimes occurring even within the same genus (*Mitrophyllum;* see Figure 8).

As shown in Table 1, in principle, large leaves are more efficient for water storing due to their favorable volume-surface relation which reduces water losses by cuticular transpiration. However, they demand high investments of carbohydrates and cause additional costs for reinforcement of stems. Together with the limited availability of water to fill up these stores, this limits the number of leaf pairs per individual. Large leaves furthermore convey higher thermal stress to the plants and higher risk of loss of biomass by herbivores.

There are alternatives (Table 1): replacement of quality by quantity (B, *Meyerophytum, Dicrocaulon*), i.e. numerous small leaves; replacement by stem succulence (C, *Monilaria*); and miniaturizing of the whole plant body (D), thus being able to make use of shade cast by quartz pebbles and to exploit water that condenses on these pebbles during the night; this renders high investments for water storing structures during the dry season unnecessary (*Oophytum, Diplosoma*).

The arrows in Table 1 are intended to give an idea of the network of relations or interdependencies among the characters when they are reviewed under functional aspects and with special regard to the very limited resources of these plants.

INTERPRETATION

A detailed analysis of the inventory of characters in this group of genera reveals that characters from all fields of the life cycle are interdependent. The network of interdependencies is implemented by the very limited availability of resources, especially by the low output from photosynthesis which is much affected by almost permanent water stress. In an evolutionary context, this means that evolution is strongly canalized: changes in one place (e.g. additional investments for larger water storing capacity or an alternative basic strategy such as miniaturizing) force changes (e.g. economizing) in other places of the life cycle. In spite of the low number of available components, this group has been able to evolve an astonishing number of extremely diverse life cycles. This not only is typical of the group described but also applies to many other groups of this family. The broad array of often highly specialized life forms is the basis for the ability of the Mesembryanthemaceae to conquer a very wide range of diverse and partly extremely difficult ecological niches, an ability which in turn is the basis of their great success in the Succulent Karoo Flora.

The nature of the evolutionary processes that in Mesembryanthemaceae transform one life form to another becomes clear as soon as one takes into consideration the ontogenetic processes involved in the example described above:

*Plants are characterized as open systems; their development can be described as a repeated production of modules, such as internodes, leaves, etc. The change from one module to another is induced by the expression of a particular high-ranking gene. As soon as a decision of this kind is taken, another set of lower ranking genes becomes active for the realization of the particular characters of the module in question.

*The sequence of different leaf types and their position within the individual ontogeny and within the annual growth cycle is controlled by high-ranking genes that selectively activate the three available leaf subprograms composed of sets of cooperating genes that determine the outer form of the leaf.

*Each leaf starts as a very small protrusion at the apex. Final size may be determined by how long the genes are active.

*Each internode formed at the apex starts as a very short piece of stem. Whether it remains short or stretches depends on how long the genes that cause elongation growth are active.

*Each stem starts as a very thin stem. Final diameter may depend on how long genes that control the activity of the meristems involved are active. Prolonged activity of the meristem that builds up the cortex of the stem can lead to an increased amount of parenchyma capable of contributing to the water-storing capacity of the plant (i.e. development of a succulent stem).

Genes described above, i.e. genes affecting early decisions of the developmental program of an organism, have been called homeotic genes. Normally this level of high-ranking regulatory genes appears to be very well stabilized and to be little affected by selection pressures; this is the basis of homology (43). The first to assume that changes in the ontogenetic program might be involved in the phylogeny of this family was Schwantes (51, 54), who speculated that highly succulent unbranched and stemless forms such as *Lithops* and *Lapidaria* ("Sphaeroidea") might be juvenile stages of related genera with shrubby growth form, capable of flowering ("neoteny" or subterminal ontogenetic abbreviation—25). Ihlenfeldt (25, 26, 27) pointed out that fundamental changes in the whole ontogeny of the plant by rearrangement of ontogenetic subprograms apparently play an important role in the evolution of the Mesembryanthemaceae. Many instances have been described since.

At present there is a broad discussion on the significance of homeotic genes with respect to phenomena such as parallelism, homology, analogy, and homoiosis, and on consequences for the reconstruction of phylogeny (for overview, see 43; "process morphology", see 49). The outstanding ability of the Mesembryanthemaceae to evolve diverse life forms on the basis of a limited set of subordinate ontogenetic programs can be attributed to their ability to produce novel arrangements on the level of high-ranking regulatory genes, and to their ability to integrate various alternative subprograms from all parts of the life cycle to form novel complex life strategies. Apparently this is made possible by an unusual genetic potentiality that allows easy and fast rearrangement of homeotic genes.

THE ENIGMA OF THE ISOLATED TAXA WITH VERY LIMITED DISTRIBUTION

As has already been mentioned, in the subtribe Mitrophyllinae, there are two well-defined homogeneous species, each of which would deserve the rank of a monotypic genus: *Mitrophyllum roseum* and *"Monilaria" globosa*. Similar instances are known from other groups of the family. Of the two species mentioned, the first one is known only from two populations on a single hill, whereas the latter is quite common in an area of approximately 10×20 km. In a very unusual way *Mitrophyllum roseum* combines a number of characters of certain species of the genus *Dicrocaulon* (growth form, more than two pairs of leaves per season, rose flower color) with characters of the genus *Mitrophyllum* (leaf sequence). *"Monilaria" globosa* combines characters of *Meyerophytum* (leaf sequence) with characters of certain species of *Mitrophyllum* (leaf size, compact growth form of the juvenile stages). There are no characters that could be interpreted as primitive for the family (4), so that there is no basis for the assumption that these taxa might represent relicts of an ancient level of evolution within the Mitrophyllinae. All characters, which mainly belong to the group of "alternative" characters (see above), do occur in the closely related genera. One could, therefore, assume that both taxa are hybrids that by chance were produced not long ago and that succeeded in establishing themselves in one of the numerous available ecological niches. In fact, the putative parents do occur in the region, though in different niches. These two putative hybrids, however, do not split as one would expect and as has been reported from natural hybrid populations (7). That failure to split could be due to amphidiploidy can be excluded: Both species are diploids (34, and H-D Ihlenfeldt, unpublished data).

One can speculate that these taxa arose either by nonsexual horizontal gene transfer, recently discussed in literature (e.g. 43), or by reactivation of silenced genes and suppression of previously active genes in the common gene pool of this group (and possibly of the whole family) due to changes in the arrangement and activity of high-ranking homeotic genes. As has already been pointed out, minor changes on this level can force many additional changes ("canalization") due to external constraints caused by the shortage of resources. These novel types described above may form the starting point of a new event of adaptive radiation after successful establishment (*"Monilaria" globosa,* or these types may become extinct within a short time due to failure to reach and colonize additional suitable habitats (*Mitrophyllum roseum*); the latter species already exhibits signs of reduced fertility (48).

Acknowledgments

The studies in the Mesembryanthemaceae would not have been possible without the generous financial support provided by the Deutsche Forschungsgemeinschaft, to which the author is especially grateful. He is indebted to the numerous members of the Mesembryanthemaceae study group in Hamburg, who made important contributions to the knowledge of this family. The author is especially grateful to Hans-Helmut Poppendieck, Heidrun Hartmann, Norbert Jürgens, Volker Bittrich, Michael Struck, Maike Gerbaulet, Michael Dehn, Sigrid Liede, Ingeborg Niesler, and Stefan Rust. The author is also indebted to Dieter von Willert and his working group (Bayreuth/Münster) and Benno Eller (Zürich, Switzerland) who studied the ecophysiology of this family. Furthermore, he thanks Steven Hammer (Belen, New Mexico) who commented critically on this paper and contributed unpublished field data.

Any *Annual Review* chapter, as well as any article cited in an *Annual Review* chapter,
may be purchased from the Annual Reviews Preprints and Reprints service.
1-800-347-8007; 415-259-5017; email: arpr@class.org

Literature Cited

1. Axelrod DI, Raven PH. 1978. Late Cretaceous and Tertiary vegetation history of Africa. In *Biogeography and Ecology of Southern Africa*, ed. MJA Werger, pp. 77–130. The Hague: Junk

2. Bittrich V. 1986. Untersuchungen zu Merkmalsbestand, Gliederung und Abgrenzung der Unterfamilie Mesembryanthemoideae (Mesembryanthemaceae Fenzl). *Mitt. Inst. Allg. Bot. Hamburg* 21:5–116

3. Bittrich V, Hartmann H. 1988. The Aizoaceae—a new approach. *Bot. J. Linn. Soc.* 97:239–54

4. Bittrich V, Struck M. 1989. What is primitive in Mesembryanthemaceae? An analysis of evolutionary polarity of character states. *S. Afr. J. Bot.* 55:321–31

5. Brown NE. 1921. Mesembryanthemum and some genera separated from it. *Gard. Chron.* 1921:151–52

6. Carlquist S. 1983. Intercontinental dispersal. *Sonderb. Naturwiss. Ver. Hamburg* 7:37–47

7. Cole DT. 1988. *Lithops. Flowering Stones*. Randburg: Acorn

8. Deacon HJ. 1983. Another look at the Pleistocene climates of South Africa. *S. Afr. J. Sci.* 79:325–28

9. Deacon HJ. 1983. An introduction to the Fynbos region, time scales and palaeoenvironments. *S. Afr. Nat. Sci. Progr. Rep.* 75:1–20

10. Dehn M. 1992. Untersuchungen zum Verwandtschaftskreis der Ruschiinae (Mesembryanthemaceae Fenzl). *Mitt. Inst. Allg. Bot. Hamburg* 24:91–198

11. Dobzhansky T, Ayala FJ, Stebbins GL, Valentine JW. 1977. *Evolution*. San Francisco: Freeman

12. Hammer S. 1993. *The Genus Conophytum*. Pretoria: Succulent Plant Publ.

13. Hammer S, Liede S. 1990. Natural and artificial hybrids in Mesembryanthemaceae. *S. Afr. J. Bot.* 56:356–62

14. Hartmann H. 1975. Speciation in Mesembryanthemaceae. *Boissiera* 24:255–61

15. Hartmann H. 1976. Monographie der Gattung *Odontophorus* N.E.Br. (Mesembryanthemaceae Fenzl). *Bot. Jahrb. Syst.* 97:161–225

16. Hartmann H. 1978. Monographie der Gattung *Argyroderma* N.E.Br. (Mesembryanthemaceae Fenzl). *Mitt. Inst. Allg. Bot. Hamburg* 15:121–235

17. Hartmann H. 1982. Monographien der Subtribus Leipoldtiinae. III. Monographie der Gattung *Fenestraria* (Mesembryanthemaceae). *Bot. Jahrb. Syst.* 103:145–83

18. Hartmann H. 1988. Fruit types in

Mesembryanthema. *Beitr. Biol. Pflanz.* 63:313–49

19. Hartmann H. 1991. Mesembryanthema. *Contrib. Bolus Herb.* 13:75–157
20. Hartmann H. 1992. *Ihlenfeldtia,* a new genus of Mesembryanthema (Aizoaceae). *Bot. Jahrb. Syst.* 114:29–50
21. Hartmann H. 1992. Revision of the genus *Vanheerdea* L. Bolus ex H.E.K. Hartmann. *Bradleya* 10:5–16
22. Hartmann H. 1993. Aizoaceae. In *The Families and Genera of Vascular Plants,* Vol. II, ed. K Kubitzki, JG Rohwer, V Bittrich, pp. 37–69. Berlin: Springer
23. Hartmann H, Dehn M. 1987. Monographien der Leipoldtiinae. VII. Monographie der Gattung *Cheiridopsis* (Mesembryanthemaceae). *Bot. Jarb. Syst.* 108:567–663
24. Hartmann H, Jürgens N. 1989. Wuchsformen bei Mesembryanthemen. *9th Symp. Anat. Morphol. Syst. Wien* p. 19 (Abstr.)
25. Ihlenfeldt H-D. 1971. Über ontogenetische Abbreviationen und Zeitkorrelationsänderungen und ihre Bedeutung für Morphologie und Systematik. *Ber. Dtsch. Bot. Ges.* 84:91–107
26. Ihlenfeldt H-D. 1971. Zur Morphologie und Taxonomie der Mitrophyllinae Schwantes (Mesembryanthemaceae). *Ber. Dtsch. Bot. Ges.* 84:655–60
27. Ihlenfeldt H-D. 1975. Some trends in the evolution of the Mesembryanthemaceae. *Boissiera* 24:249–54
28. Ihlenfeldt H-D. 1978. Morphologie und Taxonomie der Gattung *Oophytum* N.E.Br. (Mesembryanthemaceae). *Bot. Jahrb. Syst.* 99:303–28
29. Ihlenfeldt H-D. 1983. Dispersal of Mesembryanthemaceae in arid habitats. *Sonderb. Naturwiss. Ver. Hamburg* 7:381–90
30. Ihlenfeldt H-D. 1985. Lebensformen und Überlebensstrategien bei Sukkulenten. *Ber. Dtsch. Bot. Ges.* 98:409–23
31. Ihlenfeldt H-D. 1988. Taxonomic problems in the genus *Dorotheanthus* (Mesembryanthemaceae). *Monogr. Syst. Bot. Missouri Bot. Gard.* 25:581–86
32. Ihlenfeldt H-D. 1989. Life strategies of succulent desert plants. *Excelsa* 14:75–83
33. Ihlenfeldt H-D, Gerbaulet M. 1990. Untersuchungen zum Merkmalsbestand und zur Taxonomie der Gattungen *Apatesia* N.E.Br., *Carpanthea* N.E.Br., *Conicosia* N.E.Br., *Herrea* Schwantes und *Hymenogyne* Haw. (Mesembryanthemaceae). *Bot. Jahrb. Syst.* 111:457–98

34. Ihlenfeldt H-D, Hartmann H, Poppendieck H-H. 1978. Chromosomenzahlen der Mitrophyllinae Schwantes (Mesembryanthemaceae). *Mitt. Inst. Allg. Bot. Hamburg* 16:171–81
35. Ihlenfeldt H-D, Jörgensen S. 1973. Morphologie und Taxonomie der Gattung *Monilaria* (Schwantes) Schwantes s. str. (Mesembryanthemaceae). *Mitt. Staatsinst. Allg. Bot. Hamburg* 14:49–94
36. Ihlenfeldt H-D, Struck M. 1986. Morphologie und Taxonomie der Dorotheanthinae Schwantes (Mesembryanthemaceae). *Beitr. Biol. Pflanz.* 61:411–53
37. Jähnig U, Jürgens N. 1993. Standorteigenschaften arider Böden der Namib. *Mitt. Dtsch. Bodenkundl. Ges.* 72:947–50
38. Jürgens N. 1986. Untersuchungen zur Ökologie sukkulenter Pflanzen des südlichen Afrika. *Mitt. Inst. Allg. Bot. Hamburg* 21:139–365
39. Jürgens N. 1987. Growth forms in succulents. *XIV Int. Bot. Congr. Berlin Symp.* 4th Symposium, 18th contribution (Abstr.)
40. Jürgens N. 1990. A life form concept including anatomical characters, adapted for the description of succulent plants. *Mitt. Inst. Allg. Bot. Hamburg* 23a:321–41
41. Jürgens N. 1991. A new approach to the Namib Region I: Phytogeographic subdivision. *Vegetatio* 97:21–38
42. Jürgens N. 1992. *Vegetation der Namib. Ein Beitrag zur ökologischen Gliederung und Florengeschichte einer afrikanischen Trockengebietsflora.* Habilitationsschrift. Univ. Hamburg
43. Kubitzki K, v. Sengbusch P, Poppendieck H-H. 1991. Parallelism, its evolutionary origin and systematic significance. *Aliso* 13:191–206
44. Leistner OA. 1958. A new genus of Mesembryanthemaceae. In *Notes on Mesembryanthemum and Allied Genera III,* ed. HML Bolus, pp. 289–91. Cape Town: Univ. Cape Town
45. Liede S, Hammer S. 1990. Aspects of floral structure and phenology in the genus *Conophytum* (Mesembryanthemaceae). *Plant Syst. Evol.* 172:229–40
46. Liede S, Hammer S, Whitehead V. 1991. Observations on pollination and hybridization in the genus *Conophytum* (Mesembryanthemaceae). *Bradleya* 9:93–99
47. Mosango M, Lejoly J. 1990: La forêt dense humide à *Piptadeniastrum africanum* et *Celtis mildbraedii* des en-

virons de Kisangani (Zaïre). *Mitt. Inst. Allg. Bot. Hamburg* 23b:853–70

48. Poppendieck H-H. 1976. Untersuchungen zur Morphologie und Taxonomie der Gattung *Mitrophyllum* Schwantes s. lat. *Bot. Jahrb. Syst.* 97: 339–413

49. Sattler R. 1992. Process morphology: structural dynamics in development and evolution. *Can. J. Bot.* 70:708–14

50. Schwantes G. 1916. Zur Biologie der Befruchtung bei Sukkulenten. *Monatsschr. Kakteenkunde* 26:34–43

51. Schwantes G. 1921. Zur Stammesgeschichte der Sphäroidea. *Monatsschr. Kakteenkunde* 31:23–29, 34–38, 165–73

52. Schwantes G. 1950. *Micropterum* Schwant. *Kakteen Sukk.* 1(2):4–5, 1(3):6–8

53. Schwantes G. 1953. The cultivation of the Mesembryanthemaceae. *Cact. Succ. J. Gr. Br. Suppl.*, pp. 1–36

54. Schwantes G. 1957. *Flowering Stones and Mid-Day Flowers*. London: Ernest Benn

55. Struck M. 1992: Pollination ecology in the arid winter rainfall region of southern Africa: a case study. *Mitt. Inst. Allg. Bot. Hamburg* 24:61–90

56. Vogel S. 1954. Blütenbiologische Typen als Elemente der Sippengliederung. In *Botanische Studien,* Vol. 1, ed. W Troll, H v. Guttenberg. Jena: Fischer

57. von Willert DJ, Eller BM, Werger MJA, Brinckmann E, Ihlenfeldt H-D. 1992. *Life Strategies of Succulents in Deserts with Special Reference to the Namib Desert.* Cambridge: Cambridge Univ. Press

58. Werger MJA. 1978. Biogeographical division of southern Africa. In *Biogeography and Ecology of Southern Africa,* ed. MJA Werger, pp. 145–70. The Hague: Junk

59. Werger MJA. 1978. The Karoo-Namib Region. In *Biogeography and Ecology of Southern Africa,* ed. MJA Werger, pp. 231–99. The Hague: Junk

Annu. Rev. Ecol. Syst. 1994. 25:547–72

GENETIC DIVERGENCE, REPRODUCTIVE ISOLATION, AND MARINE SPECIATION

Stephen R. Palumbi

Department of Zoology and Kewalo Marine Laboratory, University of Hawaii, Honolulu, Hawaii 96822

KEY WORDS: allopatric speciation, dispersal, molecular evolution, mate recognition, gamete incompatibility

Abstract

In marine species, high dispersal is often associated with only mild genetic differentiation over large spatial scales. Despite this generalization, there are numerous reasons for the accumulation of genetic differences between large, semi-isolated marine populations. A suite of well-known evolutionary mechanisms can operate within and between populations to result in genetic divergence, and these mechanisms may well be augmented by newly discovered genetic processes.

This variety of mechanisms for genetic divergence is paralleled by great diversity in the types of reproductive isolation shown by recently diverged marine species. Differences in spawning time, mate recognition, environmental tolerance, and gamete compatibility have all been implicated in marine speciation events. There is substantial evidence for rapid evolution of reproductive isolation in strictly allopatric populations (e.g. across the Isthmus of Panama). Evidence for the action of selection in increasing reproductive isolation in sympatric populations is fragmentary.

Although a great deal of information is available on population genetics, reproductive isolation, and cryptic or sibling species in marine environments, the influence of particular genetic changes on reproductive isolation is poorly understood for marine (or terrestrial) taxa. For a few systems, like the co-evolution of gamete recognition proteins, changes in a small number of genes may give rise to reproductive isolation. Such studies show how a focus on the physiology, ecology, or sensory biology of reproductive isolation can help uncover the

547

genetic changes associated with speciation and can also help provide a link between the genetics of population divergence and the speciation process.

INTRODUCTION

The formation of species has long represented one of the most central, yet also one of the most elusive, subjects in evolutionary biology. Darwin (28) sought out the mechanisms and implications of natural selection in order to explain the origins of species. Later, both Dobzhansky (29) and Mayr (88) would use speciation as a pivot around which to spin their divergent yet complementary views of the evolutionary process. They called their works *Genetics and the Origin of Species* and *Systematics and the Origin of Species,* perhaps to emphasize that they were using genetics and systematics primarily to advance understanding of the speciation process (45).

As a result of these efforts, and the series of papers that developed and used the new synthesis, a basic model of speciation arose. Now termed *allopatric speciation,* the basic scenario is familiar to virtually all evolutionary biologists: A large, continuous population is broken up into smaller units by extrinsic barriers; genetic exchange between these separated populations ceases, and genetic divergence takes place between them; the build-up of genetic differences leads to intrinsic barriers to reproduction. If the separated populations (now separate species) reconnect with one another through the breakdown of the original extrinsic barriers, they will remain reproductively isolated and selection for increased reproductive isolation may occur (30).

Much of the early evidence for this process was based on discovery of species groups at the range of stages predicted by the above scenario (88). Some species have broad distributions, often with local variants. Other species are easily divided into allopatric subspecies whose taxonomic rank is debated. In other cases, two similar but slightly different species inhabit the same region, yet are distinguished by mating preferences or habitat differences that limit hybridization between them.

Even though Mayr (89) could identify this series in marine species, there have been relatively few attempts to examine patterns and processes of speciation in marine habitats. This is unfortunate because marine species often represent a serious challenge to the idea of allopatric speciation, especially in marine taxa with high fecundity and larvae that can disperse long distances. These life history traits result in species that have large geographic ranges, high population sizes, and high rates of gene flow between distant localities.

Such attributes might be expected to limit the division of a species' range into allopatric populations. Few absolute barriers to gene flow exist in oceans, and as a result, even widely separated regions may be connected genetically. Furthermore, marine populations tend to be large, which can slow genetic divergence between populations. Population genetics has shown that many species with these life history traits have little genetic population structure and appear to act as large, panmictic units (101). For these species, allopatric speciation events may be infrequent and slow (89).

Yet, speciation in these taxa is common enough that marine species with these life history traits dominate important marine groups like echinoderms (33) and fish (17, 58). Furthermore, some types of marine habitats like coral reefs and the soft sediments of the deep sea have a huge number of species (46, 47, 74, 113, 149), some of which appear to be closely related (71, 101). Thus, the generalization that speciation must be rare in marine taxa with high dispersal appears to be incorrect.

In fact, a number of factors affect the chance of speciation through allopatric mechanisms in the sea. Like most useful generalizations, the process of allopatric speciation as described above includes a wide range of exceptions. What mechanisms are there that might enhance population subdivision and promote genetic divergence in species with high dispersal? How does reproductive isolation evolve in recently diverged species? What aspects of marine speciation have attracted the most research, and where are the future opportunities? To answer some of these questions (at least partly), and to arrange these topics in a manageable way, I have separated them into (i) opportunities for population subdivision, (ii) mechanisms of genetic differentiation, and (iii) reproductive isolation in closely related species. Together, these sections highlight the success of research into marine speciation, but they point out the existence of a major gap in our understanding.

OPPORTUNITIES FOR POPULATION SUBDIVISION

Population genetic studies of marine species have shown that, especially along continental margins, high dispersal potential is often associated with only mild genetic differentiation over large scales (101). These results suggest high levels of gene flow between populations, but there may often be limits to the actual dispersal of marine species with high dispersal potential (122). These limits vary widely with species, habitat, local ocean conditions, and recent history, and they may create ample opportunity for genetic divergence. Although such limits may seldom create absolute barriers to gene flow, they may often limit gene flow in some directions or at some times. Thus, partially isolated populations may occur quite commonly in marine systems. Throughout this section, the main focus is on mechanisms by which marine species with high dispersal may become at least partially isolated. The goal is to summarize ways in which these populations can diverge genetically despite their potential for gene exchange. Species with low dispersal often show interesting and unexpected biogeographic patterns (e.g. 63) or remarkable levels of genetic distinction over mere meters (138a), but in general it is no mystery how genetic barriers in low dispersal species arise (49).

Invisible Barriers

Even if larvae were simply passive planktonic particles, drifting helplessly in ocean currents (5, but see next section), gene flow across the world's oceans would be neither continuous nor random. The physics of a liquid ocean on a spinning globe, heated differentially at the poles and the equator, will always provide complex oceanic circulation (124). Today, these patterns include a prevailing westward-flowing equatorial current and two large circulation centers in the northern and southern hemispheres in both the Pacific and Atlantic Oceans. Schopf (124) suggested that these basic patterns also occurred in the past, and that biogeographic boundaries—the defining limits of biogeographic provinces—are typically set by these physical forces (see also 61, 133).

If most planktonic dispersal follows these currents, then movement from one circulation center to the others might be infrequent. Data on the distribution and abundance of fish (60), planktonic copepods (90), and other zooplankton (87) show that even the open ocean is a fragmented habitat. Across a large geographic scale, species composition of planktonic communities may be determined by currents such as gyres and mesoscale eddies (122). Although few data exist on the influence of these currents on species formation, gene flow across the oceans is probably constrained and directed by such circulation patterns.

Smaller geographic features also influence oceanic circulation, and probably gene flow as well. On the east coast of North America, Cape Hatteras and Cape Cod define biogeographic boundaries set by near-shore currents and a steep temperature gradient (124). Along this coast, genetic variation seems to be over a far shorter geographic scale than those predicted by gene flow estimates based on larval biology and current patterns (1, 11, 108, 120). Similarly, on the west coast of North America, Point Conception is a focus for the range endpoints of many species (61, 143). The Indonesian Archipelago is also a biogeographic indicator, separating Indian Ocean from Malayan provinces (124, 143). Several studies have shown that this complex of islands represents a barrier to gene flow within species (8) as well as separating closely related species (91).

A different type of pattern has been seen in the central Pacific (67). Here, the fish and gastropods of the islands of the Pacific tectonic plate are sometimes very different from those of archipelagoes on other plates: across a tectonic boundary, archipelagoes sometimes have very different faunas. Springer (131) suggested that the fish species tend to remain on archipelagoes of a particular plate, despite the potential for dispersal across plate boundaries (123), and that "plate effects" have built up over a long time (see also 66). The generality of this pattern is not clear, however, and further research is warranted.

Isolation by Distance

Oceanic currents are sometimes able to carry larvae far from their parents (121–123). For example, populations of spiny lobsters in Bermuda seem to be dependent on long distance larval transport along the Gulf Stream (52). However, there may be a limit to gene flow even in species with larvae that can disperse long distances (144, 145). Although long-lived larvae may drift for many months (114, 121), *successful* transport over long distances may be rare (62). Larvae that disperse over long distances may have a greater chance of wafting into unfavorable environments than do larvae that disperse short distances. This is coupled with a diffusion effect: The density of larvae thins with increasing distance from the center of larval production so that settlement events per available area decline with distance from the source of propagules. Lecithotrophic larvae can also be constrained by energy supply; long periods in the plankton consume energy stores, leaving little metabolic reserve for metamorphosis (114; planktotrophic larvae may not always have these limits—95).

Geographic patterns of genetic variation of marine fish and invertebrates suggest that isolation by distance occurs, but only over the largest geographic scales. Isolated islands in the Pacific Ocean, like the Hawaiian and Society Islands, appear to harbor populations with reduced genetic variation (98, 103, 150). These reductions are probably due to two physical factors. First, these islands are a long distance from neighboring archipelagoes. Second, equatorial currents flow westward toward the center of the Indo-West Pacific, and so both Hawaii and the Society Islands are "upstream" from the rest of the Indo-West Pacific. When the equatorial current breaks down, or when large water masses move from west to east across the Pacific during El Niño years (153), this dispersal barrier may disappear (115).

Isolation by distance effects may be weakest in species that inhabit continental margins, where extreme populations are connected through intermediate, stepping-stone populations. We have found that Atlantic and Pacific populations of the sea urchins *Strongylocentrotus droebacheinsis* and *S. pallidus* can be very similar genetically (102, 104). This pattern can change for populations on different sides of an ocean basin where no intermediate populations exist. For example, littorinid snails with planktotrophic larvae have little genetic divergence along the east coast of North America but are very divergent on opposite sides of the Atlantic (9, 10).

Behavioral Limits to Dispersal

The physical barriers discussed above can play an important role in limiting gene flow and creating genetic structure within oceanic populations even if larvae are passive planktonic particles. However, additional aspects of marine life histories can lead to limited genetic dispersal. Burton & Feldman (19) showed that genetic differences in marine organisms can occur on a geographic scale that is much less than that predicted by their dispersal potential. For some species, dispersal occurs at a stage during which the individual can control its movements. For example, fresh water eels spawn in marine habitats, and their larvae migrate from spawning grounds to continental river mouths (2). American and European populations of eels both breed in the Sargasso Sea, but adult populations are genetically distinct (2). This suggests that these larval fish can control the direction of their migration from the joint breeding ground to the rivers inhabited by adults. Larger marine animals, like turtles and whales, have long been known to be capable of this type of migration, and genetic structure in these species is on a geographic scale far smaller than their potential range of movement (4, 14).

However, small larvae and adults may also have some control over their dispersal. Burton & Feldman (19) showed that the intertidal copepod *Tigriopus californicus* showed strong genetic differences over just a few kilometers of coastline. One explanation for this pattern is that juveniles and adults may have behavioral adaptations that prevent their being swept off the rocky outcrops that they inhabit. Such behavioral nuances are known for the amphipod *Gammarus zaddachi,* which migrates in and out of estuaries by rising into the water only during those seasonal tidal currents that will take individuals seaward in winter and upstream in the spring (57). Crustacean larvae are known to regulate their depth in a complex way that may allow retention in estuaries (27) or return them to coastal habitats after initial transport offshore (107). Few, if any, genetic differences have been attributed to these larval behavioral abilities (100, but see 92), but only a small number of species have been examined.

Selection

As shown by several well-known studies in marine systems, gene flow may be curtailed by selection as well as by limited dispersal. In the mussel *Mytilus edulis,* estuarine habitats of Long Island Sound are colonized regularly by migrants from oceanic, coastal zones. However, strong selection at a leucine amino-peptidase locus alters gene frequencies of settlers in the Sound, creating a strong genetic cline (53, 75). In the salt marsh killifish, *Fundulus heteroclitus,* selection at one of the lactate dehydrogenase (LDH) loci appears to create a strong cline in gene frequencies along the steep temperature gradient of the east coast of North America (reviewed in 108). Temperature and allozyme properties combine in these fish to create differences in development rate, swimming endurance, oxygen transport, and patterns of gene expression (108). A cline in mitochondrial haplotypes also parallels the LDH cline, and these concordant patterns suggest a dual role for phylogenetic history and natural selection in the divergence of southern and northern populations of this fish (11).

Recent History

One of the most surprising marine genetic patterns was discovered in the widespread oyster *Crassostrea virginica*. Despite a larval dispersal stage in this species that lasts for several weeks, Reeb & Avise (111) demonstrated a strong genetic break midway along the east coast of Florida. Populations north and south of this break differed in mitochondrial DNA sequences by about 3% despite the lack of an obvious barrier to genetic exchange. Populations spanning this break have broadly similar patterns of allozyme variation, a result that had been interpreted as evidence for widespread gene flow (18). Karl & Avise (65) showed that patterns of nuclear DNA differentiation match the mtDNA patterns, not the allozyme patterns, and they suggested that balancing selection is responsible for the allozyme similarities. Reeb & Avise invoked history to explain these varied genetic patterns: populations of estuarine species like *C. virginica* may have been isolated during periods of low sea level in the Pleistocene when large coastal estuaries drained. Thus, the genetic pattern we see today may be far from equilibrium, and it reflects neither contemporary genetic exchange nor the larval dispersal potential of this species.

Unique historical events may have been instrumental in the speciation of stone crabs in the Gulf of Mexico. Western and eastern Gulf populations of *Menippe mercenaria* were probably separated during periods of low sea level during the Pliocene or Pleistocene. Today, two species exist allopatrically in the southeastern United States (12). There is a broad hybrid zone where these species meet in the Gulf of Mexico (13), but there also appears to be a second region where allozyme frequencies are intermediate between species. This second region is on the Atlantic coast of Florida, close to the mouth of the Sewanee Strait, a temporary seaway that connected the Gulf and the Atlantic during periods of high sea level in the Miocene and Pliocene (12). A combination of genetic and geological data suggests that the brief existence of this seaway injected genes from the western Gulf species deep into the range of the eastern Gulf/Atlantic species. Although this injection occurred long ago, the genetic signature of the event persists despite the potential for long distance gene flow in this species (12, 13).

The tropical Pacific ocean has been a backdrop for a great deal of faunistic change in the Pleistocene. Although sea surface temperatures probably did not change much during glacial cycles (24), sea levels changed repeatedly by up to 150 m (105). During sea level regressions, shallow back reefs and lagoons dried out. Higher sea level may have drowned some fringing reefs. Associated with these changes have been many local extinctions and recolonizations by the marine fauna of isolated reefs (48, 76, 105). For example, the cone snail

Conus kahiko is found commonly in the fossil record of Hawaii until about 100,000 years ago, when it disappeared and was replaced by the morphologically similar *Conus chaldaeus* (76).

Recent evidence from two species groups suggests that the Pleistocene may have been a period of rapid speciation. Sibling species of *Echinometra* sea urchins arose and spread throughout the Pacific over the past 0.5–2 million years (103). Likewise, sibling species of butterfly fish in at least two subgenera of *Chaetodon* differentiated from their Indian Ocean counterparts during the past million or so years (91). In the latter case, concordant patterns of species differentiation based on molecular phylogenies strongly suggest that divergence was affected by extrinsic factors such as dispersal barriers during sea level fluctuations (91).

Some taxa have probably been affected more strongly than others by the flush-fill cycle in the Pacific. Soft-sediment (e.g. lagoon-inhabiting) bivalves have low species richness on isolated archipelagoes where such habitats were severely reduced by low sea level. This may explain a previously uncovered but poorly understood pattern of lower bivalve endemicity on isolated islands (66).

Cronin & Ikeya (27a) regard cycles of local extinction followed by recolonization as opportunities for speciation. Their analysis of arctic and temperate ostracods suggests that these opportunities only seldom result in new species. However, there have been a large number of opportunities for speciation during the past 2.5 million years, and as a result, speciation has occurred in 15% to 30% of ostracods during this time period.

MECHANISMS OF GENETIC DIFFERENTIATION

Genetic Differentiation in Large Populations

The types of genetic changes that occur during speciation have fueled debate for many years. A great deal of attention has been focused on small populations derived by colonization of a novel habitat. These founder events (88) can lead to rapid genetic changes that have been described as genetic revolutions (21, 22) or genetic transiliences (138). Such changes are thought to alter substantially the genetic architecture of a population, allowing rapid accumulation of a large number of genetic differences that can then lead to reproductive isolation.

In addition to these genomic reconstructions, normal genetic variants may accumulate more quickly in small than large populations. Under several reasonable models of molecular evolution, most mutations are slightly deleterious. Kimura (69) showed that this type of mutation could drift in a small population as if it were neutral, rising to fixation with about the same probability as a strictly neutral change. By contrast, in large populations, in which drift is minor, even slightly deleterious mutations will be eliminated by natural selection. Kimura's analysis shows that as population size decreases, the fraction of "nearly neutral" mutations increases. The result is that the overall rate of molecular evolution may increase for small populations as compared to large populations.

It is unlikely that evolutionary models that rely on very small population sizes will explain a large fraction of speciation events among marine organisms with the potential for long-distance dispersal. This is because populations that become allopatrically or parapatrically separated from one another (by some of the mechanisms reviewed above) are likely to be large in extent and in population size. Furthermore, multiple invasions of a new habitat (like an island) are much more likely for marine organisms with long distance dispersal than for gravid female flies, birds, lizards, etc. As a result, the genetic differentiation of allopatric marine populations has been thought to be a slow process, requiring many millions of years to accomplish (89, 117, 131).

Although many efforts have been made to identify and explain major genetic changes during founder events (see 22 for review), other workers have argued that the well-known genetic processes of mutation and selection may be the most powerful forces creating reproductive isolation (5, 6). When selection acts, gene frequencies can shift quickly, even in large populations. Thus, a shifting selective regime can generate large genetic differences very quickly, even between large populations that are not completely isolated. Given the extensive geographic ranges of many marine species, it is not difficult to imagine environmental gradients that impose differential selection in different areas (108). In fact, these types of environmental gradients have produced some of the best-known examples of selection acting on individual allozyme loci (see above). Thus, speciation can result from the shifting adaptive landscape envisioned by Barton & Charlesworth (7), as populations throughout an extensive geographic range adjust to local selective pressures.

Newly Discovered Mechanisms of Genetic Divergence

Our view of the acrobatics of the genome during divergence has changed substantially since the allopatric model was proposed. Molecular tools have revealed a host of evolutionary mechanisms that might contribute to the divergence of genomes in large and small populations. These mechanisms may act along with selection in large populations to promote genetic differentiation of semi-isolated marine populations.

Transposable elements exist in the genomes of virtually all taxa (36, 51), including marine groups like sea urchins (130). Transposons are short stretches of DNA capable of directing their own replication and insertion through either a DNA or an RNA intermediate. They disrupt genome function by inserting into otherwise functional genes and can greatly increase mutation rate (136). Yet, although they may reduce fitness, transposable elements can spread rapidly through even a large population (42). For instance, natural populations of *Drosophila melanogaster* throughout the world may have been invaded by transposable "P" elements within a period of 20–30 years (118).

Rose & Doolittle (116) suggested that invasion of allopatric populations by different transposable elements may greatly reduce the fitness of hybrids between populations. This is because the mechanisms that limit the copy number of a particular transposable element in a genome may disappear in hybrids (34), allowing rampant transposition and an increase in mutation rate. Rose & Doolittle could not find an obvious case of species formation by invasion of transposons, but the clear demonstration of hybrid dysgenesis in *Drosophila* shows how the basic mechanism can operate (68, 118).

One of the hallmarks of transposable elements is that they exist in multiple copies throughout the genome. Other gene regions, however, also occur as multiple copies. Even though they do not transpose, they often show extraordinary evolutionary dynamics. For example, the nuclear ribosomal genes are typically found in a long tandem array containing hundreds of copies of this gene cluster (reviewed in 54). Although ribosomal genes tend to be variable between species, the multiple gene clusters within the array tend to be identical to one another. If simple mutation and Mendelian inheritance were the only genetic processes occurring in these clusters, we would expect to find a great deal of variation between gene clusters on a chromosome, perhaps even more than we find between species. However, in general, the tandem clusters of ribosomal genes are remarkably similar.

The process that homogenizes multiple copies of a DNA segment within a population has been called concerted evolution and has been documented for a number of multi-gene families (55). Two mechanisms operate during concerted evolution. Unequal crossing-over changes the number of tandem DNA segments on two homologous chromosomes. Through stochastic processes, this gain and loss of segments will result in extinction of some segments and eventual fixation of one type (31). Hillis et al (55) also showed that biased gene conversion operated in tandem arrays of ribosomal gene clusters. In gene conversion, sequences on one chromosome are used to change the sequence of homologous regions of the second chromosome. Biased gene conversion is the preferential replacement of one type of sequence with another. Dover (31, 32) has pointed out that this mechanism could result in the rapid sweep of a particular sequence through a large population. Termed *molecular drive,* this rapid shift in the properties of a genome could play a role in rapid genetic divergence of large populations during speciation (31). Shapiro (126) lists a suite of genetic mechanisms that might contribute to the reorganization of whole genomes during evolution.

None of these mechanisms (gene conversion, concerted evolution, molecular drive, hybrid dysgenesis, etc) has been strongly implicated in particular speciation events (116), and it has been argued that such mechanisms are unnecessary to explain most cases of speciation (6, 7). Yet, modern genetic research continues to uncover mechanisms, like these, that can substantially remold genomes. Furthermore, some of these changes can spread through populations in a nonmendelian way. As a result, the genetic divergence of populations through mutation, selection, and drift can perhaps be augmented by other types of genetic change. For our purposes, it is enough to point out that these mechanisms operate well in large populations, and that there are a plethora of possible mechanisms for generating large genomic differences in relatively short periods of time.

REPRODUCTIVE ISOLATION IN CLOSELY RELATED SPECIES

The formation of species requires the evolution of reproductive isolation (7, 25, 71, 88). If allopatric populations are brought back together, and no barrier to reproduction exists, then whatever genetic differences had accumulated between isolates will be shared throughout the rejoined population. As a result, understanding marine speciation requires an understanding of reproductive isolation between species. The most illuminating examples are likely to be those in which the isolating mechanisms act between two recently derived species. In these cases we are more likely to be examining changes that occurred during speciation (although it is usually impossible to prove this in any given case).

Mechanisms of Reproductive Isolation

In general, reproductive barriers are classified into pre-zygotic and post-zygotic categories (25). For marine species, post-zygotic isolation is seldom studied because of the difficulty of raising offspring through complex life cycles and through long generation times. However, pre-zygotic mechanisms of reproductive isolation are well studied and fall into several broad types.

MATE PREFERENCE   In terrestrial taxa, mate preferences are known to vary between closely related species (e.g. 30), and this form of reproductive isolation is receiving more attention in marine systems. Snell & Hawkinson (128) found mating preferences among sympatric and allopatric populations of the estuarine rotifer *Branchionus plicatilis,* possibly because of species-specific reaction to a diffusable mating signal (41). Male fiddler crabs (genus *Uca*) engage in elaborate courtship displays in which the single large claw is waved and rapped on the substrate. Although morphological differences between species are often slight, the waving and rapping components of courtship often differ significantly (119). Other crustaceans such as stomatopods (110), amphipods (132), and isopods (129) also have complex behavioral mechanisms or chemical detection abilities that may isolate sibling species. The behavioral component of assortative mating is the most important in maintaining isolation among several sympatric species of the isopod genus *Jaera* (129). These differences probably arose during the Pleistocene diversification of this genus (129).

The large claw of alpheid shrimp is used in aggression between males or between females of the same species or between males and females of different species. Species separated by the Isthmus of Panama have quickly become reproductively isolated: Male-female pairs from different species are behaviorally incompatible (73). Although these pairs have been allopatrically separated by a land-barrier, there are also sympatric shrimp species that appear to be behaviorally isolated in very similar ways. Thus, the mechanism of reproductive isolation so clearly seen across the Isthmus of Panama appears to operate within ocean basins as well.

Weinberg et al (147) showed that this type of behavioral change could be detected on a very small geographic scale. In the polychaete genus *Nereis,* males and females react territorially to members of the same sex but form mated pairs after intersexual encounters. Populations of *N. acuminata* from the Atlantic and Pacific coasts of North America showed strong aggression toward each other when a male and female from opposite coasts were paired (147). Surprisingly, east coast populations separated by only 110 km also showed a significant degree of aggression. The common infaunal polychaete *Capitella* is composed of several cryptic species that are reproductively isolated even when they occur sympatrically (46).

Fish can also show strong behavioral control over mate choice. In the tropical genus *Hypoplectrus* (the hamlets), sibling species are defined on the basis of color pattern differences: Few ecological or morphological distinctions can be found among sympatric species (35). Field observations show that spawning is almost exclusively (~ 95%) between individuals of the same color pattern (35). Work within other species has also shown that females can distinguish males on the basis of their color pattern and that they choose mates using species-specific rules (146). This degree of color discrimination is not always observed, however. Among butterfly fish of the genus *Chaetodon,* sibling species are distinguished by discrete color pattern differences. However, mating occurs randomly between species along a narrow hybrid zone in the Indo-West Pacific (91). In this genus, sibling species are largely allopatric as opposed to the largely sympatric distribution of behaviorally isolated hamlets (35).

HABITAT SPECIALIZATION   Reproductive isolation can also be associated with habitat specialization. Recently diverged Baltic Sea species of the amphipod

Gammarus have developed marked differences in salinity tolerance that prevent their hybridization (77). A group of hydroid species that inhabits the shells used by hermit crabs shows strict habitat specialization: Different hydroid species use the shells of different hermit crab species (20). Coral species in the genus *Montastrea* appear to segregate on the basis of depth and light levels (72, 74). Knowlton & Jackson (72) discuss other examples from coral reefs of niche use differentiation among sibling species (see also 71). Species of the isopod *Jaera* (129) show slight habitat segregation, but this mechanism of isolation is thought to be less important than the behavioral isolation discussed above.

SPAWNING SYNCHRONY   Many marine species spawn eggs and sperm into the water column or lay demersal eggs that are fertilized externally. For sedentary invertebrates, fertilization success is a strong function of proximity to another spawning individual (84, 106). As a result, selection for spawning synchrony may occur in these species, and closely related species can be isolated by changes in the timing of spawning. Among three sympatric sea cucumber species in the genus *Holothuria* on the Great Barrier Reef, two show strong, seasonal patterns of spawning (50). In the tropical Pacific, the sea urchin *Diadema savignyi* spawns at full moon. A broadly sympatric species, *D. setosum,* spawns at full moon in some localities but at new moon in others. Where spawning overlaps, hybrids between the two species are common (JS Pearse, personal communication). Species in this genus separated by the rise of the Isthmus of Panama have also diverged in spawning time (81, 82). Examples of sympatric species that show differences in the timing of spawning come from hermit crabs (112), bivalves (15, 109), sponges (63a), coral reef fish (39), and gastropods (140). Knowlton (71) lists 26 examples of spawning asynchrony in cryptic or sibling marine species.

However, differences in the timing of spawning are not ubiquitous among sympatric marine species (3, 50, 71). Hundreds of coral species spawn together on the Great Barrier Reef during a few nights in the summer (3). In temperate habitats, numerous species spawn in the spring, sometimes during mass spawning events (106), perhaps because spawning time is constrained by seasonal availability of planktonic food (56). As a result, other mechanisms of reproductive isolation probably exist to limit cross-fertilization among gametes of different species spawned at the same time.

FERTILIZATION   Fertilization is easily studied in many marine species, and a great deal has been discovered about fertilization mechanisms in these taxa. By contrast, there are relatively few studies of fertilization patterns between closely related species. Nevertheless, the data available suggest a number of generalizations.

Some species pairs fertilize readily in the laboratory when their gametes are mixed together. The sea stars *Asterias forbesi* and *A. vulgaris* occur over a narrow sympatric zone along the northeast coast of North America. There is only slight differentiation in spawning season for these species, and sperm and eggs of both species can cross-hybridize (125). Sea urchins in several genera can also cross with one another (but see below) (83, 103, 134, 135). Certain kelp species distributed in the north and south Atlantic can cross-fertilize (although normal offspring are not always produced—139).

Complete fertilization in hybrid crosses is not the most common result, however. Instead, species that can cross-fertilize often do so incompletely or unidirectionally. That is, the eggs of one species will be receptive to the sperm from the second, but the reverse crosses fail (83, 135, 141). Of the three "successful" crosses performed by Buss & Yund (20) between species in the hydroid genus *Hydractinia,* two showed asymmetric success. Rotifer mating preferences show the same pattern (128). These patterns are remarkably similar to the mate choice asymmetries in insects discussed by Coyne & Orr (26). An interesting but unanswered question is why such similar patterns emerge from biological mechanisms as different as marine fertilization and insect mate choice.

In some taxa, certain species' eggs tend to be "choosier" than others. For example, eggs of the sea urchin *Strongylocentrotus droebachiensis* can be cross-fertilized to a greater degree (134) than the eggs of congeners (they are also more easily fertilized at low concentrations of conspecific sperm; see 84). Eggs of the sea urchin *Colobocentrotus atratus,* which occurs only in intertidal areas with high wave action, also show high cross-fertilizability (16). Again there is an analogy to the literature on mate choice in insects. Species differ in the receptivity of females to heterospecific males. Changes in this receptivity have been hypothesized to be important to rapid species formation (64).

In a few known cases, fertilization barriers are reciprocal and strong. Buss & Yund (20) recorded 6 out of 9 crosses between hydroid species that resulted in less than 5% developing eggs, although in this case it has not been conclusively shown that fertilization failure (as opposed to developmental failure) was the cause of these patterns. Sibling species of the serpulid polychaete *Spirobranchus* show strong reciprocal fertilization barriers (86), producing about 5% developing eggs in interspecific crosses. Crosses between four species of abalone showed low fertilization unless sperm concentrations were 100 times normal. Even under these conditions, only 10–30% of the eggs were fertilized (on average), except in one cross (and in only one direction) which produced 96% fertilization (80). Among Hawaiian sea urchins in the genus *Echinometra,* we have shown that there are strong reciprocal barriers to fertilization (93, 103). This result has been observed for the two species in this genus on Guam (93), although the species complexes on Okinawa and in the Caribbean show some cases of asymmetric gamete compatibility (83, 141).

Selection for Reproductive Isolation?

A classic problem in speciation research is distinguishing those changes that occur during speciation from those that occur afterwards as a result of reproductive isolation. For example, do mate choice differences arise by random drift between partially or completely isolated populations? Or is there selection for reduced mating between species that are already developmentally incompatible because of evolutionary changes at other loci (88)? Work on species separated by the Isthmus of Panama has shown that reproductive isolation at the fertilization (83) and mate recognition levels (73) can arise without contact between newly formed species. These changes, especially incomplete, asymmetric barriers to fertilization, are probably due to random drift of these characters in isolated populations (83). They cannot be due to selection for reproductive isolation because these species have been geographically isolated since their initial separation in the Pliocene (82). Interestingly, the patterns of mate recognition and fertilization failure seen across Panama are also seen between sympatric species in the Caribbean, the eastern Pacific, and the tropical Pacific (73, 83, 103, 141). This suggests that reproductive divergence without reinforcing selection has occurred in at least some of these cases, or that the signatures of selection and random divergence are remarkably similar.

Some of the strongest evidence for the operation of selection in reducing hybridization comes from comparison of allopatric and sympatric species in the same species group (26), but this type of evidence is rare in marine taxa. Snell & Hawkinson (128) showed that sympatric populations of *Brachionis* rotifers had stronger mating discrimination than did allopatric populations; this is consistent with the idea of selective reinforcement of reproductive barriers. Isopod species exhibit a slight pattern in this direction—the average hybridization rate for females from sympatric populations is roughly half that of allopatric species—but males show no geographic effects (Ref. 129, Table 1). Sympatric species of alphaeid shrimp are more different morphologically than are allopatric species with the same degree of genetic divergence (N Knowlton, personal communication), suggesting that selection is acting differently in sympatric vs allopatric comparisons.

Examples in which an allopatric/sympatric difference is not seen strongly include urchins (83) and butterfly fish (91). In the latter case, species with overlapping distributions show less discrimination in mate choice experiments than do species that are allopatric. These divergent results suggest that selection for increased or decreased reproductive isolation may occur in marine systems, but it is not ubiquitous, and significant isolation can evolve over reasonably short time periods without it.

GENETICS OF REPRODUCTIVE ISOLATION—A MISSING LINK

The preceding pages highlight the large amount of information on mechanisms of genetic divergence of marine populations. Likewise, there have been many studies of the ways in which recently diverged marine species have become reproductively isolated. But a large gap remains between these two types of information. We know why genetic change might take place, but not how these changes affect reproductive isolation. We know what types of physiological, ecological, or sensory changes give rise to reproductive isolation, but not which genetic changes have produced them. The link between genetics and reproductive isolation is largely missing.

Recently, interest has increased in genetic divergence of particular loci that are strongly involved in reproductive isolation and species recognition (25). For some systems, it has been possible to examine the evolution of proteins that are involved in creating barriers to gene flow. For example, two genes are involved in the control of reproductive season in some insects (137). Coyne (25) lists other examples in which only a few loci affect reproductive isolation; but even for terrestrial taxa, the list is very short.

Mechanisms of Reproductive Isolation and the Evolution of Recognition

Part of the reason for this lack of understanding is that studies of reproductive isolation are phenomenological: They describe the interactions of individuals within and between species in order to detect the phenomenon of reproductive isolation. This approach is sufficient to understand the nature of species differentiation, but it does not explain the mechanisms of reproductive isolation and leaves open the question of how those mechanisms evolved. Reproductive isolation can involve many different molecular, physiological, or sensory systems (see above for just a few examples), and so it is difficult to generalize from one isolation mechanism to another.

A few aspects of reproductive isolation, however, can be investigated at all levels (molecular, physiological, sensory) and in practically all taxa. One such aspect is recognition, the means by which individuals recognize members of the same species and distinguish them from other individuals in the environment. Most modes of reproductive isolation involve some component of recognition, except for strict allopatry (e.g. the Isthmus of Panama) or strong habitat selection (77).

Surprisingly, very little is known about the recognition process in marine species. Some fish are thought to use color patterns as a mating cue (35, 91). Limb vibrations appear to be part of the courtship process in some crustaceans (110, 119, 129). A diffusable pheromone can induce mating behavior in rotifers (128). Bioluminescent ostracods may broadcast their identity with patterns of bioluminescence (J Morin, personal communication). Sperm attraction to eggs has been documented in a large number of invertebrate taxa (see for example 94a). In none of these systems has the genetics of recognition been determined, in part because of the potentially complex nature of genetic control over some of these recognition processes.

Evolutionary studies of recognition that have been performed to date have focused on simple interactions that are amenable to genetic analysis. Examples include the *per* locus effects on mate signaling in *Drosophila* (148), the S-allele system that mediates self-incompatibility in plants (23), the mating type loci in protozoa (94), and loci governing fertilization in marine invertebrates (101, 142).

Although these studies involve only a few simple examples, they are derived from a range of taxa. Thus, it is interesting that there are so many similarities at the genetic level. Comparisons of amino acid sequences for proteins involved in gamete or mating type discrimination reveal a general pattern of large differences among species (23, 79, 94, 148) or among alleles within species (23, 93). These differences often exceed those predicted on the basis of silent changes in these proteins, and so they appear to reflect the action of some type of selection for variation sensu Hughes & Nei (59, 23, 79). The nature of these selective forces is well understood for self-incompatibility in plants, because in these systems there is strong selection for heterozygosity (151). By contrast, the existence of similar selective forces is unclear for marine invertebrates.

An additional problem is that reproductive isolation is unlikely to be caused by large differences at a single genetic locus. Dobzhansky (29), Mayr (88), and many other evolutionary biologists have pointed out that if reproductive isolation is caused by a single, large, dominant mutation at a single locus, then the individuals possessing that mutation cannot breed and so have zero fitness. A recessive mutation might drift in a population until it is expressed in many individuals at the same time, but this mechanism is likely to operate well only in small populations. Orr (99) discussed the impact of maternal inheritance on this result.

However, recognition loci seldom act alone. In most cases there are both signals and signal receptors, and these are likely to be produced through the action of different loci (self-incompatibility systems of plants and protists are major exceptions in that they involve only a single locus—23, 94). Where recognition occurs because of the interaction of at least two loci, mathematical models have shown how polymorphisms can be maintained within populations (70) and lead to reproductive divergence (78, 152). In general, polymorphisms are maintained if individuals with a particular allele at one locus prefer mates with a particular allele at the other locus. If there are multiple combinations of these matched alleles, they can be stable within populations (152; SR Palumbi, submitted). This type of model, developed to understand sexual selection, has seldom been applied to marine systems. As a result, the application of these results to concrete examples of speciation of marine organisms is lacking.

However, marine species have provided some of the best mechanistic views of the recognition process. This is because many marine invertebrates spawn eggs and sperm into the water. In these taxa, complex pre-mating behavioral differences are limited to the interactions of short-lived gametes, avoiding the complex behavioral genetics that might dominate reproductive isolation in many vertebrates or arthropods. For species that spawn at the same time (e.g. the hundreds of coral species that mass spawn in the Pacific—3), or for temperate invertebrates that are slaved to strong seasonal reproduction (106), such gamete interactions may determine levels of hybridization between species, or determine fertilization success within species.

Two gamete recognition systems have received the most attention. In abalone, sperm penetrate the outer chorion layer through the action of a protein called lysin (142). Although the mechanisms of lysin action are obscure, the protein has been purified and its secondary structure determined (127). Lysins act efficiently only on the chorion coatings of their own species (142), and fertilizations between abalone species are low except at high sperm concentration (80). Amino acid sequences of lysins from several species show high ratios of replacement to silent site changes (79) as discussed above. Furthermore, the areas of high amino acid replacement occur along a part of the protein that appears, from the crystalline structure, to play an important functional role (127).

The other well-studied system is the fertilization mechanism of sea urchins. In this class, a sperm protein called bindin attaches sperm to the vitelline coat of eggs and promotes egg-sperm fusion (37, 44). Bindin is expressed only in sperm where it occurs in a tightly packed vesicle. After the acrosome reaction, bindin coats the outside of the acrosomal process. The mature bindin shows a central area of high amino acid conservation: 95% of the amino acids are conserved between urchins separated by 150–200 million years. By contrast, the flanking regions both show large sequence differences between species (40, 43, 96)

A series of detailed experiments has failed to show a simple relationship between bindin sequence and attachment of bindin to the egg receptors. Although small pieces of the bindin protein, synthesized as peptides, can show species specificity (97), there is no single substitution that alters overall bindin-egg interactions (85). Between very closely related sea urchins that show gamete incompatibility, we have shown multiple amino acid substitutions, many clustered in a short region of the protein (101), as well as a suite of insertion/deletion events. Interestingly, many bindin alleles occur in the species we have studied, and these alleles differ from one another in ways that are qualitatively similar to the differences we have seen between species (high amino acid substitution, plus rampant insertion/deletion events) (E Metz, SR Palumbi, in preparation). These results suggest that there is a continuum of bindin function and that differences in amino acid sequence accumulate to give rise to more and more differentiated sperm-egg binding properties.

Of course, the egg receptor plays an equally critical role in this process. Recently, the gene for the sea urchin egg receptor has been isolated and sequenced (38). Although comparative sequence data are not yet available for a large number of species, preliminary results suggest that the extracellular component of the egg receptor is highly variable between species (38). The ability to analyze both sperm attachment and egg receptor proteins in sea urchins makes this system particularly interesting in the analysis of gamete interactions.

Directions for Future Research

One of the largest gaps in our knowledge about speciation remains the link between genetic divergence and mechanisms of reproductive isolation (25). Even in systems amenable to formal genetics, like *Drosophila,* an understanding of the genetics of speciation is only slowly emerging (25).

Unfortunately, for many species it is not possible to perform the genetic miracles that are commonplace in a *Drosophila* laboratory. Marine species are especially difficult to raise because of long generation times, complex life cycles, or obscure mating requirements. In terrestrial systems, these limitations are sometimes overcome by using the natural laboratories of hybrid zones to illuminate the genetic nature of species boundaries. Such studies are rare in marine systems (see 12, 91) but may be profitably used in the future.

A complementary approach is suggested by recent successes in understanding the evolutionary genetics of gamete recognition proteins. Here, a physiological process (gamete binding and fusion) was explored with the full power of the modern molecular toolbox, and the genes responsible for the phenomenon were isolated. Without formal genetics, these studies have shown the importance of particular modes of molecular evolution to the evolution of species recognition.

For some marine taxa, this approach is especially appealing because the simple mating dynamics of free-spawning invertebrates eliminates many of the complexities of mate choice in behaviorally complex vertebrates or arthropods. These simple mating cues (e.g. the mating pheromone of rotifers—128 or simple gamete recognition processes—104) allow the possibility of unraveling reproductive isolation at the genetic level.

Perhaps osmoregulatory differences in amphipods, or chemosensory systems in rotifers, or visual pigment differences in fish could be understood in terms of the gene products that create the physiological, ecological, or sensory differences responsible for currently recognized patterns of reproductive isolation. Although such research is technically difficult and may not uncover *all* the genes responsible for reproductive isolation, this approach can serve as a strong alternative to the study of the genetics of reproductive isolation.

CONCLUSIONS

Although examples of genetic homogeneity over large distances are common in marine systems, there are also many examples of population structure in marine species with high dispersal potential. Such species probably do not "see" the ocean as a single, undifferentiated habitat, either because of environmental differences among localities or because of a large number of physical mechanisms known to produce at least partial isolation between populations. Genetic divergence within these partially isolated gene pools is probably not as slow as thought originally. Various mechanisms exist to generate genetic differences between large isolated populations. Some of these mechanisms include evolutionary processes that have only recently been recognized, whereas others include fairly standard applications of selection theory. Finally, history has played a strong role in the development of marine biogeographic patterns. Cycles of sea level rise-and-fall during the Pleistocene have affected near-shore marine communities, and these cycles were probably exacerbated by the steepening of latitudinal thermal gradients. As a result, even populations that are well connected today by gene flow may have been isolated in the very recent past.

The link between genetic divergence of populations and reproductive isolation is poorly known for marine (or terrestrial) species. How do genetic changes lead to the physiological, ecological, or sensory differences that define sibling species? How do they create reproductive isolation? What are the mechanisms by which species recognition evolves? Studies of gamete recognition show how a focus on the mechanisms of reproductive isolation can lead to the discovery of the genes responsible for species recognition.

This suggests a general approach to speciation research that is based on investigations of the physiological, ecological, and sensory differences that give rise to species recognition and perhaps to reproductive isolation. Such investigations would lead to increased understanding of the underlying genetic mechanisms by which recognition evolves within and between species, and they provide important evidence to help fill major gaps in our understanding of speciation.

ACKNOWLEDGMENTS

I thank T Duda, N Knowlton, WO McMillan, G Roderick, S Romano, R Strathmann, G Vermeij, and an anonymous reviewer for comments on the manuscript. Supported by grants from the National Science Foundation.

Literature Cited

1. Avise JC. 1992. Molecular population structure and biogeographic history of a regional fauna: a case history with lessons for conservation biology. *Oikos* 63:62–76

2. Avise JC, Helfman GS, Saunders NC, Hales LS. 1986. Mitochondrial DNA differentiation in North Atlantic eels: Population genetic consequences of an unusual life history pattern. *Proc. Natl. Acad. Sci. USA* 83:4350–54

3. Babcock RC, Bull GD, Harrison PL, Heyward AJ, Oliver JK, et al. 1986. Synchronous spawning of 105 scleractinian coral species on the Great Barrier Reef. *Mar. Biol.* 90:379–94

4. Baker CS, Perry A, Abernethy B, Alling A, Bannister J, et al. 1993. Abundant mitochondrial DNA variation and world-wide population structure in humpback whales. *Proc. Natl. Acad. Sci. USA* 90:8239–43

5. Banse K. 1986. Vertical distribution and horizontal transport of planktonic larvae of echinoderms and benthic polychaetes in an open coastal sea. *Bull. Mar. Sci.* 39:162–75

6. Barton NH. 1989. Founder effect speciation. In *Speciation and Its Consequences*, ed. D Otte, JA Endler, pp. 229–56. Sunderland, Mass: Sinauer. 679 pp.

7. Barton NH, Charlesworth B. 1984. Genetic revolutions, founder effects and speciation. *Annu. Rev. Ecol. Syst.* 15:133–64

8. Benzie JA, Stoddart JA. 1992. Genetic structure of crown-of-thorns starfish (*Acanthaster planci*) in Australia. *Mar. Biol.* 112:631–39

9. Berger EM. 1973. Gene-enzyme varia-tion in three sympatric species of *Littorina. Biol. Bull.* 145:83–90

10. Berger EM. 1977. Gene-enzyme variation in three sympatric species of *Littorina*. II. The Roscoff population with a note on the origin of North American *Littorina. Biol. Bull.* 153:255–64

11. Bernardi G, Sordino P, Powers DA. 1993. Concordant mitochondrial and nuclear DNA phylogenies for populations of the teleost fish *Fundulus heteroclitus. Proc. Natl. Acad. Sci. USA* 90:9271–74

12. Bert TM. 1986. Speciation in western Atlantic stone crabs (genus *Menippe*): the role of geological processes and climatic events in the formation and distribution of species. *Mar. Biol.* 93:157–70

13. Bert TM, Harrison RG. 1988. Hybridization in western Atlantic stone crabs (genus *Menippe*): evolutionary history and ecological context influence species interactions. *Evolution* 42:528–44

14. Bowen BW, Meylan AB, Avise JC. 1989. An odyssey of the green sea turtle: Ascension Island revisited. *Proc. Natl. Acad. Sci USA* 86:573–76

15. Boyden CR. 1971. A comparative study of the reproductive cycles of the cockles *Cerastoderma edule* and *C. glaucum. J. Mar. Biol. Assoc. UK* 51:605–22

16. Branham JM. 1972. Comparative fertility of gametes from six species of sea urchins. *Biol. Bull.* 142:385–96

17. Brothers EB, Williams DM, Sale PF. 1983. Length of larval life in twelve families of fishes at "One Tree Lagoon", Great Barrier Reef, Australia. *Mar. Biol.* 76:319–24

18. Buroker NE. 1983. Population genetics of the American oyster *Crassostrea*

virginica along the Atlantic coast and the Gulf of Mexico. *Mar. Biol.* 75:99–112

19. Burton RS, Feldman MW. 1982. Population genetics of coastal and estuarine invertebrates: does larval behavior influence population structure? In *Estuarine Comparisons*, ed. VS Kennedy, pp. 537–51. New York: Academic. 709 pp.

20. Buss LW, Yund PO. 1989. A sibling species group of *Hydractinia* in the northeastern United States. *J. Mar. Biol. Assoc. UK* 69:857–74

21. Carson HL. 1982. Speciation as a major reorganization of polygenic balances. In *Mechanisms of Speciation*, ed. C Barigozzi, pp. 411–33. New York: Liss. 546 pp.

22. Carson HL, Templeton AR. 1984. Genetic revolutions in relation to species phenomena: The founding of new populations. *Annu. Rev. Ecol. Syst.* 15:97–131

23. Clark AG, Kao T-H. 1991. Excess non-synonymous substitution at shared polymorphic sites among self-incompatibility alleles of Solanaceae. *Proc. Natl. Acad. Sci. USA* 88:9823–27

24. CLIMAP Project. 1976. The surface of the ice age earth. *Science* 191:1131–37

25. Coyne JA. 1992. Genetics and speciation. *Nature* 355:511–15

26. Coyne JA, Orr HA. 1989. Patterns of speciation in *Drosophila*. *Evolution* 43:362–81

27. Cronin TW, Forward RB. 1986. Vertical migration cycles of crab larvae and their role in larval dispersal. *Bull. Mar. Sci.* 39:192–201

27a. Cronin TW, Ikeya N. 1990. Tectonic events and climatic change: Opportunities for speciation in cenozoic marine ostracoda. In *Causes of Evolution: A Paleontological Perspective*, ed. RM Ross, WD Allmon. Chicago: Univ. Chicago Press. 479 pp.

28. Darwin CR. 1872. *On the Origin of Species by Natural Selection*. New York: Appleton. 5th ed. 447 pp.

29. Dobzhansky TH. 1937. *Genetics and the Origin of Species*. Reprinted 1982. New York: Columbia Univ. Press. 364 pp.

30. Dobzhansky TH. 1970. *Genetics of the Evolutionary Process*. New York: Columbia Univ. Press. 505 pp.

31. Dover G. 1982. Molecular drive, a cohesive mode of species evolution. *Nature* 299:111–17

32. Dover G. 1989. Linkage disequilibrium and molecular drive in the rDNA gene family. *Genetics* 122:249–52

33. Emlet RB, McEdward LR, Strathmann RR. 1987. Echinoderm larval ecology viewed from the egg. *Echinoderm Stud.* 2:55–136

34. Engels WR. 1981. Hybrid dysgenesis in *Drosophila* and the stochastic loss hypothesis. *Cold Spring Harbor Symp. Quant. Biol.* 45:561–65

35. Fischer EA. 1980. Speciation in the Hamlets (*Hypoplectrus*: Serranidae)—a continuing enigma. *Copeia* 1980:649–59

36. Flavell AJ. 1992. *Ty1-copia* group retrotransposons and the evolution of retroelements in the eukaryotes. *Genetics* 86:203–14

37. Foltz KR, Lennarz WJ. 1993. The molecular basis of sea urchin gamete interactions at the egg plasma membrane. *Dev. Biol.* 158:46–61

38. Foltz KR, Partin JS, Lennarz WJ. 1993. Sea urchin egg receptor for sperm: sequence similarity of binding domain and hsp70. *Science* 259:1421–25

39. Foster SA. 1987. Diel and lunar patterns of reproduction in the Caribbean and Pacific segeant major damselfishes *Abudefduf saxatilis* and *A. troschelii*. *Mar. Biol.* 95:333–43

40. Gao B, Klein LE, Britten RJ, Davidson EH. 1986. Sequence of mRNA coding for bindin, a species-specific sea urchin sperm protein required for fertilization. *Proc. Natl. Acad. Sci. USA* 83:8634–38

41. Gilbert JJ. 1963. Contact chemoreceptors, mating behavior and sexual isolation in the rotifer genus *Brachionus*. *J. Exp. Biol.* 40:625–41

42. Ginzburg LR, Bingham PM, Voo S. 1984. On the theory of speciation induced by transposable elements. *Genetics* 107:331–41

43. Glabe CG, Clark D. 1991. The sequence of the *Arbacia punctulata* bindin cDNA and implications for the structural basis of species-specific sperm adhesion and fertilization. *Dev. Biol.* 143:282–88

44. Glabe CG, Vacquier VD. 1977. Species-specific agglutination of eggs by bindin isolated from sea urchin sperm. *Nature* 267:836–38

45. Gould SJ. 1982. Introduction. See Ref. 19

46. Grassle JF, Maciolek NJ. 1992. Deep-sea species richness: regional and local diversity estimates from quantitative bottom samples. *Am. Nat.* 139:313–41

47. Grassle JP, Grassle JF. 1976. Sibling species in the marine pollution indicator *Capitella* (Polychaeta). *Science* 192:567–69

48. Grigg RW. 1988. Paleooceanography of coral reefs in the Hawaii-Emporer chain. *Science* 240:1737–43

49. Hansen TA. 1983. Modes of larval de-

velopment and rates of speciation in early tertiary neogastropods. *Science* 220:501–2

50. Harriot VJ. 1985. Reproductive biology of three congeneric sea cucumber species, *Holothuria atra, H. impatiens,* and *H. edulis* at Heron Island, Great Barrier Reef. *Aust. J. Mar. Freshwater Res.* 36:51–57

51. Hartl DL, Lozovskaya ER, Lawrence JG. 1992. Nonautonomous transposable elements in prokaryotes and eukaryotes. *Genetica* 86:47–53

52. Hateley JG, Sleeter TD. 1993. A biochemical genetic investigation of spiny lobster (*Panulirus argus*) stock replenshment in Bermuda. *Bull. Mar. Sci.* 52:993–1006

53. Hilbish TJ, Koehn RK. 1985. The physiological basis of natural selection at the LAP locus. *Evolution* 39:1302–17

54. Hillis DM, Dixon MT. 1991. Ribosomal DNA: Molecular evolution and phylogenetic inference. *Q. Rev. Biol.* 66:411–53

55. Hillis DM, Moritz C, Porter CA, Baker RJ. 1991. Evidence for biased gene conversion in concerted evolution of ribosomal DNA. *Science* 251:308–10

56. Himmelman JH. 1979. Factors regulating the reproductive cycles of two Northeast Pacific chitons *Tonicella lineata* and *T. insignis. Mar. Biol.* 50: 215–25

57. Hough AR, Naylor E. 1992. Biological and physical aspects of migration in the estuarine amphipod *Gammarus zaddachi. Mar. Biol.* 112:437–43

58. Hourighan TF, Reese ES. 1987. Midocean isolation and the evolution of Hawaiian reef fishes. *Trends Ecol. Evol.* 2:187–91

59. Hughes AL, Nei M. 1988. Pattern of nucleotide substitution at major histocompatibility complex class I loci reveals over dominant selection. *Nature* 335:167–70

60. Incze LS, Ortner PB, Schumacher JD. 1990. Microzooplankton, vertical mixing and advection in a larval fish patch. *J. Plankton Res.* 12:365–79

61. Jablonski D, Flessa K, Valentine JW. 1985. Biogeography and paleobiology. *Paleobiology* 11:75–90

62. Jackson JBC. 1986. Modes of dispersal of clonal benthic invertebrates: Consequences for species' distributions and genetic structure of local populations. *Bull. Mar. Sci.* 39:588–606

63. Johannesson K. 1988. The paradox of Rockall: Why is a brooding gastropod (*Littorina saxatilis*) more widespread than one having a planktonic larval dispersal stage (*L. littorea*)? *Mar. Biol.* 99:507–13

63a. Johnson MF. 1978. Studies of the reproductive cycles of the calcareous sponges *Clathrina cortiacea* and *C. blanca. Mar. Biol.* 50:73–79

64. Kaneshiro KY. 1980. Sexual solation, speciation and the direction of evolution. *Evolution* 34:437–44

65. Karl SA, Avise JC. 1992. Balancing selection at allozyme loci in oysters: Implications from Nuclear RFLPs. *Science* 256:100–2

66. Kay EA. 1983. Patterns of speciation in the Indo-West Pacific. In *Biogeography of the Tropical Pacific,* ed. FJ Radovsky, PH Raven, SH Sohmer, 72:15–31. Honolulu: Bishop Mus. Press. 221 pp.

67. Kay EA, Palumbi SR. 1987. Endemism and evolution in Hawaiian marine invertebrates. *Trends Ecol. Evol.* 2:183–87

68. Kidwell M. 1982. Hybrid dygenesis in *Drososphila melanogaster.* A syndrome of aberrant traits inducing mutation, sterility and male recombination. *Genetics* 86:813–33

69. Kimura M. 1979. Model of effectively neutral mutations in which selective constraint is incorporated. *Proc. Natl. Acad. Sci. USA* 76:3440–44

70. Kirkpatrick M. 1982. Sexual selection and the evolution of female choice. *Evolution* 36:1–12

71. Knowlton N. 1993. Sibling species in the sea. *Annu. Rev. Ecol. Syst.* 24:189–216

72. Knowlton N, Jackson JBC. 1994. New taxonomy and niche partitioning on coral reefs: jack of all trades or master of some? *Trends Ecol. Evol.* 9:7–9

73. Knowlton N, Weigt LA, Solorzano LA, Mills DK, Bermingham E. 1993. Divergence in proteins, mitochondrial DNA, and reproductive compatibility across the Isthmus of Panama. *Science* 260: 1629–32

74. Knowlton N, Weil E, Weigt LA, Guzman HM. 1992. Sibling species in *Montastraea annularis,* coral bleaching, and the coral climate record. *Science* 255:330–33

75. Koehn RK, Newell RI, Immerman F. 1980. Maintenance of an aminopeptidase allele frequency cline by natural selection. *Proc. Natl. Acad. Sci. USA* 77:5385–89

76. Kohn AJ. 1981. *Conus kahiko,* a new Pleistocene gastropod from Oahu, Hawaii. *J. Paleontol.* 54:534–41

77. Kolding S. 1985. Genetic adaptation to local habitats and speciation process within the genus *Gammarus* (Amphipoda: Crustacea). *Mar. Biol.* 89: 249–55

78. Lande R. 1981. Models of speciation by sexual selection on polygenic traits. *Proc. Natl. Acad. Sci. USA* 78:3721–25

79. Lee YH, Vacquier VD. 1992. The divergence of species-specific abalone sperm lysins is promoted by positive Darwinian selection. *Biol. Bull.* 182:97–104

80. Leighton DL, Lewis CA. 1982. Experimental hybridization in abalone. *Int. J. Inverbr. Reprod. Dev.* 5:273–82

81. Lessios HA. 1981. Reproductive periodicity of the echinoids *Diadema* and *Echinometra* on the two coasts of Panama. *J. Exp. Mar. Biol. Ecol.* 50:47–61

82. Lessios HA. 1984. Possible prezygotic reproductive isolation in sea urchins separated by the Isthmus of Panama. *Evolution* 38:1122–48

83. Lessios HA, Cunningham CW. 1990. Gametic incompatibility between species of the sea urchin *Echinometra* on the two sides of the Isthmus of Panama. *Evolution* 44:933–41

84. Levitan DR. 1993. The importance of sperm limitation to the evolution of egg size in marine invertebrates. *Am. Nat.* 141:517–36

85. Lopez A, Miraglia SJ, Glabe CG. 1993. Structure/function analysis of the sea urchin sperm adhesive protein bindin. *Dev. Biol.* 156:24–33

86. Marsden J. 1992. Reproductive isolation in two forms of the serpulid polychaete, *Spirobranchus polycerus* (Schmarda) in Barbados. *Bull. Mar. Sci.* 51:14–18

87. Maynard NG, 1976. The relationship between diatoms in the surface sediments of the Atlantic Ocean and the biological and physical oceanography of overlying waters. *Paleobiology* 2:91–121

88. Mayr E. 1942. *Systematics and the Origin of Species.* Reprinted 1982. New York: Columbia Univ. Press. 334 pp.

89. Mayr E. 1954. Geographic speciation in tropical echinoids. *Evolution* 8:1–18

90. McGowan JA, Walker PW. 1985. Dominance and diversity maintenance in an oceanic ecosystem. *Ecol. Monogr.* 55:113–18

91. McMillan WO. 1994. *Speciation, species boundaries, and the population biology of Indo-West Pacific butterflyfishes (Cheatodontidae).* PhD thesis. Dep. Zool., Univ. Hawaii, Honolulu, HI

92. McMillen-Jackson AL, Bert TM, Steele P. 1994. Population genetics of the blue crab (*Callinectes sapidus* Rathbun): modest populational structuring in a background of high gene flow. *Mar. Biol.* 118:53–65

93. Metz EC, Kane RE, Yanagimachi H, Palumbi SR. 1994. Fertilization between closely related sea urchins is blocked by incompatibilities during sperm-egg attachment and early stages of fusion. *Biol. Bull.* In press

94. Miceli C, La Terza A, Bradshaw RA, Luporini P. 1992. Identification and structural characterization of cDNA clone encoding a membrane-bound form of the polypeptide pheromone Er-1 in the ciliate protozoan *Euplotes raikovi*. *Proc. Natl. Acad. Sci. USA* 89:1988–92

94a. Miller RL. 1985. Demonstration of sperm chemotaxis in echinodermata: Asteroidea, Holothuroidea, Ophiuroidea. *J. Exp. Zool.* 234:383–414

95. Miller SE, Hadfield MG. 1990. Developmental arrest during larval life extends life span in a marine mollusc. *Science* 248:356–58

96. Minor J, Gao B, Davidson E. 1989. The molecular biology of bindin. In *The Molecular Biology of Fertilization*, ed. H Schatten, G Schatten, pp. 73–88. San Diego: Academic. 384 pp.

97. Minor JE, Britten RJ, Davidson EH. 1993. Species-specific inhibition of fertilization by a peptide derived from the sperm protein bindin. *Mol. Biol. Cell* 4:375–87

98. Nishida M. Lucas JS. 1988. Genetic differences between geographic populations of the crown-of-thorns starfish throughout the Pacific region. *Mar. Biol.* 98:359–68

99. Orr HA. 1991. Is single-gene speciation possible? *Evolution* 45:764–69

100. Ovenden JR, Brasher DJ, White RW. 1992. Mitochondrial DNA analyses of the red rock lobster *Jasus edwardsii* supports an apparent absence of population subdivision throughout Australasia. *Mar. Biol.* 112:319–26

101. Palumbi SR. 1992. Marine speciation on a small planet. *Trends Ecol. Evol.* 7:114–18

102. Palumbi SR, Kessing B. 1991. Population biology of the trans-arctic exchange: mtDNA sequence similarity between Pacific and Atlantic sea urchins. *Evolution* 45:1790–805

103. Palumbi SR, Metz E. 1991. Strong reproductive isolation between closely related tropical sea urchins (genus *Echinometra*). *Mol. Biol. Evol.* 8:227–39

104. Palumbi SR, Wilson AC. 1989. Mitochondrial DNA diversity in the sea urchins *Strongylocentrotus purpuratus* and *S. droebachiensis. Evolution* 44:403–15

105. Paulay G. 1990. Effects of late Cenozoic sea-level fluctuations on the bivalve fau-

nas of tropical oceanic islands. *Paleobiology* 16:415–34

106. Pennington JT. 1985. The ecology of fertilization of echinoid eggs: The consequences of sperm dilution, adult aggregation, and synchronous spawning. *Biol. Bull.* 169:417–30

107. Phillips BF, McWIlliam PS. 1986. The pelagic phase of spiny lobster development. *Can. J. Fish. Aquat. Sci.* 43:2153–63

108. Powers DA. 1987. A multidisciplinary approach to the study of genetic variation within species. In *New Directions in Physiological Ecology,* ed. ME Feder, A Bennet, W Burggren, RB Huey, pp. 38–70. Cambridge, UK: Univ. Cambridge Press. 364 pp.

109. Rae JG. 1978. Reproduction in two sympatric species of *Macoma* (Bivalvia). *Biol. Bull.* 155:207–29

110. Reaka ML, Manning RB. 1981. The behavior of stomatopod crustacea and its relationship to rates of evolution. *J. Crustacean Biol.* 1:309–27

111. Reeb CA, Avise JC. 1990. A genetic discontinuity in a continuously distributed species: mitochondrial DNA in the American oyster, *Crassostrea virginica.* *Genetics* 124:397–406

112. Reese ES. 1968. Annual breeding seasons of three sympatric species of hermit crabs, with a discussion of the factors controlling breeding. *J. Exp. Mar. Biol. Ecol.* 2:308–18

113. Rex MA, Stuart CT, Hesler RR, Allen JA, Sanders HL, Wilson GDF. 1993. Global scale latitudinal patterns of species diversity in the deep-sea benthos. *Nature* 365:636–39

114. Richmond RH. 1987. Energetics, competency, and long distance dispersal of planula larvae of the coral *Pocillipora damicornis.* *Mar. Biol.* 93:527–33

115. Richmond RH. 1990. The effects of the El Niño/Southern oscillation on the dispersal of corals and other marine organisms. In *Global Ecological Consequences of the 1982–83 El Niño-Southern Oscillation,* ed. PW Glynn, pp. 127–40. Amsterdam: Elsevier. 563 pp.

116. Rose MR, Doolittle F. 1983. Molecular biological mechanisms of speciation. *Science* 220:157–62

117. Rosenblatt RH. 1963. Some aspects of speciation in marine shore fishes. In *Speciation in the Sea,* pp. 171–80. *Syst. Assoc. Special Publ. 5*

118. Rubin GM, Kidwell MG, Bingham PM. 1982. The molecular basis of P-M hybrid dysgenesis: The nature of induced mutations. *Cell* 29:987–94

119. Salmon M, Ferris SD, Johnston D, Hyatt G, Whitt GS. 1979. Behavioral and biochemical evidence for species distinctiveness in the fiddler crabs, *Uca speciosa* and *U. spinicarpa.* *Evolution* 33:182–91

120. Saunders NC, Kessler LG, Avise JC. 1985. Genetic variation and geographic differentiation in mitochondrial DNA of the horseshoe crab, *Limulus polyphemus.* *Genetics* 112:613–27

121. Scheltema RS. 1971. Larval dispersal as a means of genetic exchange between geographically separated populations of shoal-water benthic marine gastropods. *Biol. Bull.* 140:284–322

122. Scheltema RS. 1986. On dispersal and planktonic larvae of benthic invertebrates: An eclectic overview and summary of problems. *Bull. Mar. Sci.* 39:290–322

123. Scheltema RS, Williams IP. 1983. Long distance dispersal of planktonic larvae and the biogeography and evolution of some Polynesian and Western Pacific molluscs. *Bull. Mar. Sci.* 33:545–65

124. Schopf TJM. 1979. The role of biogeographic provinces in regulating marine faunal diversity through geologic time. In *Historical Biogeography, Plate Tectonics, and the Changing Environment,* ed. J Gray, AJ Boucot. pp. 449–57. Corvallis: Oregon State Univ. Press. 500 pp.

125. Schopf TJM, Murphey LS. 1973. Protein polymorphism of the hybridizing seastars *Asterias forbesi* and *Asterias vulgaris* and implications for their evolution. *Biol. Bull.* 145:589–97

126. Shapiro JA. 1992. Natural genetic engineering in evolution. *Genetica* 86:99–111

127. Shaw A, McRee DE, Vacquier VD, Stout CD. 1993. The crystal structure of lysin, a fertilization protein. *Science* 262:1864–67

128. Snell TW, Hawkinson CA. 1983. Behavioral reproductive isolation among populations of the rotifer *Branchionus plicatilis.* *Evolution* 37:1294–305

129. Solignac M. 1981. Isolating mechanisms and modalities of speciation in the *Jaera albifrons* species complex (Crustacea, Isopoda). *Syst. Zool.* 30:387–405

130. Springer M, Davidson EH, Britten RJ. 1991. Retroviral-like element in a marine invertebrate. *Proc. Natl. Acad. Sci. USA* 88:8401–4

131. Springer V. 1982. Pacific plate biogeography with special reference to shorefishes. *Smithson. Contrib. Zool.* 367:1–182

132. Stanhope MJ, Connelly MM, Hatwick B. 1992. Evolution of a crustacean com-

munication channel: behavioral and ecological genetic evidence for a habitat-modified, race specific pheromone. *J. Chem. Ecol.* 18:1871–87

133. Stanley SM. 1986. Population size, extinction, and speciation: the fission effect in Neogene Bivalvia. *Paleobiology* 12:89–110

134. Strathmann M. 1987. *Reproduction and Development of Marine Invertebrates of the Northern Pacific Coast.* Seattle: Univ. Wash. Press. 670 pp.

135. Strathmann RR. 1981. On the barriers to hybridization between *Strongylocentrotus droebachiensis* (O.F. Muller) and *S. pallidus* (G.O. Sars). *J. Exp. Mar. Biol. Ecol.* 55:39–47

136. Syvanen M. 1984. The evolutionary implications of mobile genetic elements. *Annu. Rev. Genet.* 18:271–93

137. Tauber CA, Tauber MJ, Necholo JR. 1977. Two genes control seasonal isolation in sibling species. *Science* 197:592–93

138. Templeton AR. 1982. Mechanisms of speciation—a population genetic approach. *Annu. Rev. Ecol. Syst.* 12:23–48

138a. Todd CD, Lambert WJ. 1993. *Population genetics of intertidal nudibranch molluscs: Does a planktonic larva confer dispersal?* Presented at Larval Ecol. Meet., Port Jefferson, NY

139. tom Dieck (Bartsch) I, de Oliveira EC. 1993. The section Digitatae of the genus *Laminaria* (Phaeophyta) in the northern and southern Atlantic: crossing experiments and temperature responses. *Mar. Biol.* 115:151–60

140. Tutschulte T, Connell JH. 1981. Reproductive biology of three species of abalones (Haliotis) in southern California. *Veliger* 23:195–206

141. Uehara T, Shingaki M. 1984. Studies on the fertilization and development in the two types of *Echinometra mathaei* from Okinawa. *Zool. Sci.* 1:1008

142. Vacquier VD, Carner KR, Stout CD. 1990. Species specific sequences of abalone lysin, the sperm protein that creates a hole in the egg envelope. *Proc. Natl. Acad. Sci. USA* 87:5792–96

143. Valentine JW. 1973. *Evolutionary Paleoecology of the Marine Biosphere.* Englewood Cliffs, NJ: Prentice-Hall

144. Vermeij GJ. 1987. *Evolution and Escalation: An Ecological History of Life.* Princeton, NJ: Princeton Univ. Press. 527 pp.

145. Vermeij GJ. 1987. The dispersal barrier in the tropical Pacific: implications for molluscan speciation and extinction. *Evolution* 41:1046–58

146. Warner RR, Schultz ET. 1992. Sexual selection and male characteristics in the bluehead wrasse, *Thalassoma bifasciatum*: Mating site acquisition, mating site defense, and female choice. *Evolution* 46:1421–42

147. Weinberg JR, Starczak VR, Mueller C, Pesch GC, Lindsay SM. 1990. Divergence between populations of a monogamous polychaete with male parental care: premating isolation and chromosome variation. *Mar. Biol.* 107:205–13

148. Wheeler DA, Kyriacou CP, Greenacre ML, Yu Q, Rutila JE, et al. 1991. Molecular transfer of a species-specific behavior from *Drosophila simulans* to *Drosophila melanogaster*. *Science* 251:1082–85

149. Wilson GD, Hesler RR. 1987. Speciation in the deep sea. *Annu. Rev. Ecol. Syst.* 18:185–207

150. Winans GA. 1980. Geographic variation in the milkfish *Chanos chanos*. I. Biochemical evidence. *Evolution* 34:558–74

151. Wright S. 1939. The distribution of self-fertility alleles in populations. *Genetics* 24:538–52

152. Wu C-I. 1986. A stochastic simulation study on speciation by sexual selection. *Evolution* 39:66–82

153. Wyrtki K. 1985. Sea level fluctuations in the Pacific during the 1982–83 El Niño. *Geophys. Res. Lett.* 12:125–28

Annu. Rev. Ecol. Syst. 1994. 25:573–600

ADAPTATION AND CONSTRAINT IN THE COMPLEX LIFE CYCLES OF ANIMALS

Nancy A. Moran

Department of Ecology and Evolutionary Biology, University of Arizona, Tucson, Arizona 85721

KEY WORDS: adaptation, developmental constraint, metamorphosis

Abstract

Life cycles that incorporate discrete, morphologically distinct phases predominate among animals. One explanation for the abundance and long-term persistence of complex life cycles is that they represent adaptive mechanisms for decoupling the developmental processes that underlie morphological traits of alternative phases, thereby allowing phases to respond independently to different selective forces. Another explanation is that complex life cycles persist due to developmental constraints acting on particular phases; in particular, larvae may represent the conservation of traits of early development while adult traits evolve more freely in response to selection. A role of adaptive decoupling is generally supported by comparative data on the relative extent of morphological evolution of larvae and adults and on the frequency of elimination of phases from complex life cycles. Adaptive decoupling could result from developmental compartmentalization, and within such developmental systems, heterochrony could be a route to the deletion of a life cycle phase. The extent to which complex life cycles represent adaptive mechanisms for severing developmental linkages could be elucidated both by quantitative genetic studies comparing levels of genetic correlation among phase-specific traits and by molecular developmental studies of gene expression in alternative phases.

INTRODUCTION

The lives of many animals unfold as discrete phases that exhibit contrasting morphological, physiological, behavioral, or ecological attributes. These life cycles have been lumped as a class and called *complex life cycles* (henceforth CLCs). The phenomena that are considered to belong to this class are immensely varied. Still, segmentation of the life history can be discerned as a pervasive trait; it has evolved over and over in a wide diversity of taxa and has persisted for long periods of time. Many of the pivotal events in the diversification of life have been associated with the origin, persistence, or loss of CLCs. For example, one of the most species-rich of all clades, the holometabolous insects, arose in association with metamorphosis and divergence between the larval and adult stages. Clearly, a general understanding of the bases of patterns of organic diversity depends on an explanation for why so many life histories are organized into discrete phases joined by abrupt transitions rather than as single phases that are static or continuously changing. Just as important for this understanding is a grasp of how organizing the life history into discrete phases affects patterns of evolutionary change within each phase.

The literature on the evolution of CLCs is divided into two, largely discrete, bodies. Ecologically oriented biologists, beginning with Istock (43) and continuing to the present (e.g. 24, 34, 44, 79, 108–110, 113), have emphasized the view that complex life cycles are adaptations to garner resources, particularly food, in two or more discretely different ways. Meanwhile, developmental biologists have an even longer record of attention to the problems of CLCs and their evolutionary bases. Since before Darwin, zoologists have been concerned with the relative degree of conservation of traits of different life cycle stages (e.g. 1, 4, 22, 31, 33, 74, 89–94, 116–118). At issue is the degree to which attributes of one phase can undergo adaptive evolution without correlated, presumably largely negative, effects on the other stage: To what extent are larvae and adults free to evolve independently of one another? These authors have viewed the malleability of development under selection as a major determinant of life cycle evolution. At the same time, they frequently have invoked selection based in ecological factors as a force in the evolution of larval or adult phenotypes (e.g. 46, 63, 97, 105). A particular concern for these developmentally oriented animal biologists has been the question of whether early stages of ontogeny are more conserved than late stages: The answer will have clear implications for the evolution of CLCs. Because addressing such issues requires evolutionary reconstructions that specify the direction of evolution, this set of authors has devoted relatively more attention to the phylogenetic context of CLCs.

Although all admit that both developmental and ecological factors are crucial in understanding CLCs, few authors have yet considered the roles of both simultaneously in addressing the questions of how and why CLCs originate, why they sometimes revert to single-phase life cycles, and why they sometimes persist for extremely long periods. Recent developments in each of the literatures on CLCs hint that some merging of the two perspectives is imminent, though the bridges remain to be developed. This paper is an attempt to encourage the broader perspective, one that should ultimately help to provide answers to the questions of why most animals have CLCs and how this fact affects their evolution and ecology. Although my focus is primarily on animals, many of the issues are relevant to other eukaryotes.

DEFINING COMPLEX LIFE CYCLES

The succession of two or more discrete phases qualifies a life cycle as complex. Despite this seemingly straightforward criterion, exact definitions of complex life cycles are elusive. Transformations within CLCs typically involve many traits, including ecological as well as morphological, physiological, and behavioral ones, and the life cycles that exhibit dramatic transformations are extremely varied. Which traits should show discrete differences among phases so that a life cycle may be designated a CLC?

Morphological Versus Ecological Criteria

Even in attempting to define complex life cycles, we are confronted by the gulf between ecological approaches and the developmental and phylogenetic approaches to the evolution of CLCs. Ecologically oriented authors (e.g. 34, 43, 44, 79, 108, 109, 113) have considered CLCs to be those life cycles in which alternative phases occupy different niches; they have emphasized the use of different food resources by alternative phases. This ecological criterion implies, at the least, that each phase has a niche and that phases can be compared in terms of overlap in resource use. But many organisms with two discrete life cycle phases fail to meet this criterion. Often, alternative phases appear to have different functions; for example, one may locate a food source whereas the other actually feeds. In such cases, which include marine invertebrates with nonfeeding larvae (45, 46, 63, 97, 89, 91, 118) and insects with nonfeeding adults (11), comparison of resources used by alternative phases is meaningless, and the niche can only be defined in terms of the life cycle as a whole. Conversely, many animals show dramatic shifts in resource use at a particular size or age (e.g. 108, 110, 111), but their life cycles are not generally considered to be complex unless the shift is accompanied by discrete morphological change.

In this review, I adopt a definition of complex life cycles similar to that used by developmentally oriented evolutionary biologists (e.g. 3, 25, 74, 90,

107, 116, 118): *a life cycle is considered to be complex if it contains two or more postembryonic phases differing discretely in morphology.* These can be either phases in an individual's development (the usual focus) or distinct generations in a multiple-generation life cycle. That intermediate cases exist, in which the transition between phases is less abrupt (examples in 103), is not unexpected. Lineages have evolved from having simple to complex life cycles and back, and as a result, depending on developmental mechanisms, we might expect a continuum between simple and complex life cycles.

An ecological definition of complex life cycles automatically links theory on CLCs to processes of survival and reproduction, that is, to the machinery of selection. However, for the goal of approaching CLCs with the broadest possible perspective, a morphological definition has the advantage that it does not bias our attention toward a particular subset of the forces that could underlie CLCs. Also, a morphology-based definition coincides with the most widespread and longstanding criterion for designating a life cycle as complex. Selection based in ecological conditions is one likely contributor to the evolution of morphologically distinct phases, but other forces, both adaptive and not, could also play a role. Although the morphological definition is relatively hypothesis-free, adopting it does not imply that growth and survivorship, particularly adaptation for alternative means of feeding, are not primary in the evolution of CLCs.

Single-Generation vs Multiple-Generation Complex Life Cycles

Most emphasis has been given to single-generation complex life cycles, in which the same individual progresses from a larval stage to an adult stage via metamorphosis. Multiple-generation CLCs, which incorporate a succession of two or more types of morphologically distinct individuals, have evolved in a wide variety of organisms, often in association with fluctuating environmental conditions. Examples include multivoltine insect species that produce alternative phenotypes at different points in a seasonal cycle (40, 84), aphids and some animal parasites that produce distinct generations using alternative host taxa (8, 13, 40, 64, 67), and aphids, cynipid wasps, rotifers, daphnids, and others that alternate between sexual and parthenogenetic reproduction (7, 9, 38, 49, 67, 93). Multiple-generation complex life cycles are widespread in eukaryotes other than animals, as in marine algal species that show two forms associated with seasonal changes in predation patterns (56), and in host-alternating rust fungi (81). Although most of this review is concerned with single generation CLCs, a comparison with multiple-generation cycles offers a useful opportunity to separate constraints arising from physiological continuity of alternative phenotypes from constraints arising from sharing a genome.

THE UBIQUITY OF COMPLEX LIFE CYCLES IN ANIMALS

As a number of authors have pointed out, complex life cycles are prevalent (summaries in 47, 59, 103, 109). The majority of multicellular animals undergo CLCs, and this claim is true whether organisms are tallied as individuals, as species, or as phyla. The claim holds whether we adopt the ecological definition based on separate niches of alternative phases, used by Istock and other ecologically oriented authors (24, 34, 43, 44, 109, 113), or a definition based on morphological transformation. Within the Metazoa, CLCs abound in marine, terrestrial, freshwater, and parasitic groups (91, 103, 109). In many instances, CLCs are clearly extremely ancient. In the classic cases of amphibians and holometabolous insects, phylogenetic distributions of CLCs combined with fossil evidence indicate persistence for over two hundred million years in each case (51, 105). The CLCs of marine invertebrates are even more ancient. In about half of metazoan phyla, and possibly in the Metazoa as a whole, the primitive life cycle is a CLC with a larva that undergoes metamorphosis to become a distinct adult (39, 47, 91, 94, 103, 116). In a number of cases, larval types of different phyla appear to be homologous (e.g. the Annelida and Mollusca, Echinodermata and Hemichordata) (15, 47, 91). Apparent homology of larvae of different phyla, in combination with fossil-based dates for adults and, in a few cases, larvae, imply that CLCs must date to near the time of the Cambrian radiation of metazoan phyla, roughly 550 million years ago (91). These ancient larval types are retained in many extant species of numerous phyla.

The high proportion of animal species that show CLCs is largely due to the success of CLC groups and not to a high frequency of origins of CLCs. Thus, the origin of separate larval and adult phases, separated by metamorphosis, has occurred once in an ancestor of all the Holometabola (39, 50) and probably only once in a common ancestor of the Amphibia (23, 94). The CLCs of marine invertebrates appear to have arisen infrequently as well; in general, evidence supports a single origin for all members of one or more phyla (47). Thus, CLCs abound due to the success of the rather small set of lineages in which they have originated rather than to a high rate of origin from simple life cycles. An evolutionary explanation for why CLCs are so frequent should provide reasons for their persistence within lineages and for the success of clades descending from CLC ancestors, but not for a high rate of origin.

Multiple-generation CLCs also appear sometimes to be ancient and may have arisen rather rarely. For example, alternation between sexual and parthenogenetic phases arose once in the common ancestor of all Aphidoidea, during the Jurassic or earlier (67). Morphologically distinct generations that are associated with alternative host plant taxa date to the early Eocene in some aphids (65) and have arisen only a few times within aphids (64, 67). The host alternating cycles of rusts appear to have evolved once in a common ancestor prior to the spread of angiosperms; a Carboniferous origin has been proposed (81).

Though many CLCs have persisted for very long periods of time, CLC lineages often have reduced or completely eliminated one life cycle phase, as summarized later in this paper. Istock (43) suggested that these instances reflect the evolutionary instability that was a major conclusion of his analysis (outlined below). However, the ubiquity and antiquity of CLCs among modern species suggests that any tendency to life cycle reduction is insufficient to counter the success, in terms of diversification rates, persistence and abundance, of lineages with CLCs.

LEVELS OF SELECTION AND COMPLEX LIFE CYCLES

The ecological perspective that originated with Istock (43) emphasized the contributions of alternative life cycle phases to population growth. This emphasis has persisted. For example, Werner (108) states that "the salient ecological feature of complex life cycles is the shift in niche that occurs at metamorphosis. The evolution of such life histories ultimately must be related to the causes and consequences of this ontogenetic niche shift in niche." Most ecologically oriented models have focussed on delineating the environmental conditions under which lineages with two-phase life cycles will persist in competition with lineages with single-phase life cycles.

Ecological models, which compare success of competing species or populations (e.g. 34, 43), are appropriate for addressing whether a lineage with a CLC can invade or persist in a community of species with simple life cycles. However, evolutionary models, which focus on relative fitnesses within populations, are required in order to determine whether CLC genotypes will spread in a population that shows heritable variation for life cycle type. The fact that CLCs can have interesting implications for ecological processes does not imply that ecological processes underlie the evolutionary origin and persistence of discrete life cycle phases within populations. Haefner & Edson (34) explicitly recognized the limitations of population models for CLCs, including their own and others: "Our concern has been with community composition, not with individual fitness (the proper context of evolutionary discussions). Consequently our results bear only peripherally on questions of the evolution of CLCs." To the extent that selective forces acting on genes or individuals are not identical to those at the population level, the outcome of selection on life cycle traits may not act to maximize population growth and may even hinder it. To cite an example possibly relevant to the evolution of CLCs, selection can favor dispersal rates higher than those optimal at the population level (19, 35). Because sedentary individuals concentrate their competitive interactions on genetic relatives, alleles that promote dispersal compete less among themselves. Thus, they can spread even where the level of ecological competition between individuals is the same for all genotypes, as in completely saturated environments. Applying this example to CLCs, selection could sometimes favor the retention of phases with dispersal capabilities even when such retention has zero or negative effect on overall population growth.

Another category of traits for which selection may differ at the gene and population levels is the set of traits involved in sexual selection. In a CLC, one life cycle phase may confer an individual mating advantage that has no effect or negative effect on population growth. Among cases in which adult features may be retained due to sexual selection are holometabolous insect lineages in which males retain wings and associated adult morphology whereas females are wingless and larviform. From the perspective of overall population growth, wings might be expected to be most beneficial in females, who can use them to locate new resource patches for development of larval progeny. The frequent occurrence of species with larviform, wingless females and winged males suggests that wing retention in males is favored in the context of maximizing numbers of matings rather than habitat location. Examples are found in Lepidoptera (e.g. all Psychidae, some Tortricidae, Pyralidae, Lymantriidae, Geometridae), Diptera (e.g. some Sciaridae), Strepsiptera (all), Coleoptera (e.g. some Lampyridae and related families, some Dermestidae, some Rhipiphoridae), and Hymenoptera (e.g. all Mutillidae, some Tiphiidae) (11).

Arguments that invoke selection at higher levels should specify mechanisms involving properties of the higher level (41, 45, 101). A number of authors have argued for a role of species-level selection in the evolution of CLCs, based on paleontological evidence for rates of cladogenesis and species extinction, both species level properties (37, 45, 98, 95). For example, Jablonski (45) showed that Late Cretaceous gastropod species with dispersive, feeding larvae speciate less but persist longer and have larger geographic ranges than do species with reduced, nonfeeding larvae that do not disperse as widely. The claim that loss of larval feeding affects rates of speciation and extinction within clades was justified in the context of a specific argument involving the capacities for dispersal of individual larvae, the resulting amount of gene flow among localities, and the associated probability of reproductive isolation among populations of lineages with and without larval feeding. Parts of the argument have been supported further by genetic analysis from modern populations of sea urchin species with feeding and nonfeeding larvae (62).

WHEN DO COMPLEX LIFE CYCLES EVOLVE AND PERSIST?

Explanations for the origin and persistence of complex life cycles can be grouped into three categories. The first is the ecological view that originated with Istock (43): Complex life cycles are mechanisms for adaptive switching between alternative means of garnering resources. A number of authors (34, 85, 105, 109, 113) have retained this view of CLCs but have varied one or more of Istock's original assumptions. This category has received the most explicit attention in the recent literature, and so I give only brief coverage. The second explanation resembles the first in that CLCs are viewed as adaptations for producing discrete phenotypes suited for different functions. However, the major fitness contribution of one or the other phase involves some aspect of fitness other than nutrient acquisition and survival, for example, mate location, habitat selection, or dispersal (e.g. 14). This view is also an obvious extension of quantitative genetics approaches to deconstructing fitness into different components (52, 86).

The third category of explanation contrasts with the first two: complex life cycles are not adaptive, but result from developmental constraints. Under this view, certain phases within a CLC persist not due to selection but because they are required as part of inflexible developmental pathways underlying adaptive phenotypes of other parts of the life cycle. Under one such view, the ancient larval types that characterize whole phyla show simplified morphology determined primarily by developmental requirements (17a). Such larvae are distinguished from "adaptive" larvae such as those of Amphibia and holometabolous insects, which are acknowledged to possess attributes selected in the context of particular larval functions. Under another variant of the developmental constraint explanation, characteristics of larval phases are as vestigial traits reflecting past selection, similar to the gills and tail expressed in the early embryonic stages of humans. Finally, there is rather widespread acceptance of the idea that early stages of ontogeny are evolutionarily conservative due to developmental constraints, and the associated implication that larvae themselves may be retained as a result of such constraints (summary in 31).

Complex Life Cycles as Adaptations for Resource Exploitation

In Istock's model (43), alternative phases occupy separate ecological niches. The overall level of growth achieved by populations that shift between discrete ecological niches is compared to that achieved by populations confined to a single ecological niche. The contribution of each life cycle phase to overall population growth depends on the balance between growth, or nutrient acquisition, and mortality. Thus, the "resources" relevant to Istock's model and to other ecologically based explanations for the persistence of CLCs (e.g. 109, 113) include both food and factors affecting survivorship. Istock showed that a CLC could invade a community consisting of populations confined to single niches. However, he reasoned further that, should one phase of a CLC have a more advantageous balance of mortality and growth, selection within the CLC population should favor expanding that phase at the expense of the other, less productive, phase. By this argument, stability of the CLC exists only under approximate equivalence of contributions of each phase. Such stability was expected to be fleeting because any ecological or evolutionary change altering the balance in contributions between the two phases would result in selection for eliminating the less productive phase entirely. Istock brought attention to the potentially destabilizing effect of adaptive evolution within phases: Selection will act continually to optimize the growth-mortality balance within each phase, and the resulting adaptations for greater efficiency in one phase are unlikely to be matched by parallel increases in efficiency of the other phase. The persistence of CLCs becomes even less tenable under Istock's framework if one allows that short-term environmental fluctuations will affect profitability of alternative niches. For example, extrinsic changes affecting larval food supply or mortality factors generally will not be countered by similar changes in the adult niche, thus upsetting any balance in contributions of the larval and adult phases and generating selection for elimination of the less productive phase.

Istock's (43) conclusion that complex life cycles are evolutionarily unstable, sometimes called "Istock's Dilemma," was unsettling. As summarized above, and as Istock himself noted, complex life cycles frequently do persist, suggesting that one or more of the conditions of his argument often must not pertain. Much of the literature on the evolution of CLCs can be interpreted as explorations of the consequences of certain likely violations of the original Istock assumptions, although only the ecologically oriented papers are explicitly framed in this way. Two of Istock's ecological assumptions have been questioned: the constancy of resources through the life cycle (85, 105, 113), and the equivalence of different life cycle stages in terms of potential for resource exploitation (24, 109); these are considered next.

TRANSIENCE OF RESOURCE TYPE   Although Istock's primary model (43) did allow for changing availability of one or the other resource used by alternative phases, his conclusion regarding CLC instability was based on constancy of availability of resources used by alternative phases. Selection should act to expand a more productive phase at the expense of a less productive phase, but expansion of a phase may be limited by ephemerality of its niche. In such cases, selection may favor retention of a less productive phase as long as it adds to overall growth and reproduction. Wassersug (85, 104, 105) argued that the tadpole stage of anurans was an adaptation for exploiting transient resources, particularly those arising from seasonal peaks in primary productivity in aquatic habitats. Wilbur (113) extended this view to organisms with CLCs in general, pointing out that many organisms, including many with CLCs, exploit constantly changing environments. Due to changes in ecological conditions during the course of an individual's life, not all options for acquiring food are equally available at all points in a life cycle. A particularly pervasive source of fluctuations in resource availability within a particular niche is seasonality. Metamorphosis and a discrete shift in phenotype may be favored by selection if, by changing its phenotype, an organism is able to shift to a new niche when an original one has disappeared or deteriorated. If the shift in phenotype enables continued growth and reproduction, it may be favored by selection even if the novel phase is less productive, in the Istock sense of the balance of mortality and growth, than the ancestral phase. The only requirement is that inclusion of the novel phase improves the overall fitness achieved in the life cycle.

SIZE-DEPENDENT SHIFTS IN RESOURCE UTILIZATION   Werner (109) pointed out an additional, very general, reason that organisms often might switch between discrete alternative morphologies as a means of maximizing food acquisition throughout ontogeny. In almost all life cycles, individuals grow from small to large. Animals increase in size by as much as three to five orders of magnitude during their development (72). Such extreme changes in body size will drastically affect the profitability of particular ways of exploiting the environment (e.g. 72, 111). In other words, growth could enforce an automatic shift in niche even in the absence of changes in environmental conditions external to the organisms. This shift in mode of nutrient acquisition may well correspond to a shift in optimal body plan. If optimal body plan changes continuously with size, then selection would favor gradual change in shape during development. Alternatively, optimal body plans for large and small individuals may differ discretely, with intermediates showing relatively low fitness in any habitat. In such cases, a radical shift in body plan, through metamorphosis, might be favored by selection.

Werner (109) considered how the expected growth rate of a population is affected by switching ecological niches. Ebenman (24) also focussed on the possibility that a discrete shift in morphology allows optimization of foraging at different body sizes; however, his model is an explicitly evolutionary one, focussed at the level of individual fitness. He explored how selection can favor CLCs as mechanisms enabling phases to evolve independently. Thus, Ebenman maintained the emphasis of Istock, Wilbur, and Werner on demographic parameters such as growth, fecundity, and survivorship, but his model is based on maximizing fitness at the individual level.

Ecological changes resulting from growth (24, 109) or from fluctuations in external environmental conditions (105, 114) are only a subset of the possible bases for selection for independence of different life cycle stages. Decoupling the evolution of various stages could be favored as a means of optimizing performance at conflicting tasks other than those involved in nutrient acquisition. This possibility is considered next.

Alternative Phases and Alternative Fitness Tasks

Istock (43) did not emphasize the distinction between losing a phase entirely and reducing a phase to a brief, nonfeeding stage. Within his demographic and ecological framework, these outcomes differ little. Still, the biological consequences of the two outcomes are radically different: Imagine how all aspects of the biology of mayflies would be affected should they lose their winged imago stage. Mayflies are but one example of the large set of taxa with complex life cycles in which one phase, either larval or adult, neither feeds nor grows. Nonfeeding adults abound in many orders of insects and especially in the Holometabola (e.g. 11, 69); nonfeeding larvae are common in various marine invertebrate phyla (45–47, 63, 88, 91). Why are these apparently ecologically useless phases retained in these life cycles?

In the second category of explanation, CLCs are adaptations for decoupling phenotypes expressed at different life cycle stages, allowing independence of adaptations for conflicting tasks. Under this view, life cycles evolve in the context of multiple, potentially conflicting, components of fitness, such as those resulting from the separate effects of sexual and natural selection (52) or the effects of selection in multiple discrete environments (68, 82, 99, 100). Particular selective agents may act primarily at certain stages of development, causing selection on a particular trait to reverse itself during the course of an individual's development. Overall fitness may be increased if phenotypes at one stage of development diverge from phenotypes at other stages. However, traits of different stages may be causally linked during the course of development, generating genetic correlations among stages. Such linkages may prevent stages from responding independently to selection (1, 16, 17, 20, 78, 86). Separation of development into largely independent sets of processes, with each set corresponding to a distinct segment of the life cycle, could permit traits of each stage to evolve to states that are more nearly optimal for the particular subset of selective forces acting at that stage. If developmental linkages between stages are low, independent adaptation of phases might lead to different fitness tasks being assigned to different life cycle stages.

How might a life cycle phase contribute to fitness other than through food acquisition and growth? Clearly there are many possibilities: Fitness may depend on finding or attracting mates, providing for young, surviving through harsh environmental conditions, surviving predation, etc. To the extent that different fitness-related activities can be allocated to separate stages in the life cycle, a CLC may represent adaptation for maximizing performance at a number of these tasks while simultaneously avoiding the constraints of design that prevent a single phenotype from performing multiple tasks efficiently. This idea of CLCs as a mechanism for efficiently performing more than one task is at the core of Werner's (109) and Ebenman's (24) proposals. However, both authors emphasized acquisition of food; they did not explore the possibility that the primary function of one or the other life stage may consist of some other task important to fitness. That fitness depends on more than feeding at a maximal rate is obvious to most evolutionary biologists, including population biologists, field researchers focussing on particular taxa, systematists, and evolutionary geneticists. Nonetheless, the point has been underappreciated in the ecological literature on the evolutionary maintenance of CLCs, perhaps due to the predominance of an ecological, population-level perspective.

The possibility that CLCs are adaptations for division of labor among developmental stages is illustrated by abundant cases in which phases appear to be specialized for particular components of fitness and in which combining tasks into a single phase would seem to have necessarily detrimental effects on fitness due to design constraints. Thus, in many animals, fitness of males depends on traits that are not expressed in the larval stage: Success in mate-finding may depend on agility, mobility, strength, brightness, ornamentation, or vocalization. These abilities may be incompatible with larval design. For example, male frogs call mates using specialized mouth structures, and design constraints may preclude transfer of these structures to the tadpole stage, which is characterized by mouthparts specialized for suspension feeding (105).

Female reproduction also may depend on traits not expressed in larvae. Consider what a female butterfly must accomplish to achieve any fitness greater than zero. It must eat and grow as a larva. Following metamorphosis, it may have to learn which leaf shapes correspond to favorable larval hosts at different times of the year and then locate, identify, and oviposit preferentially on those individual plants that will permit larval progeny to succeed (e.g. 70). Success at this second set of tasks depends on an array of structures and behaviors only present in the adult: these include wings and antennae in addition to the actual reproductive structures. As the only stage possessing wings, adult insects have the potential to contribute uniquely to fitness. Bryant (14) presented a model showing that retention of a mobile adult phase able to select favorable larval habitat could be adaptive even if the adult itself does not feed.

Dispersal is another major function that is often confined to a single phase of a CLC. In amphibians and insects, usually only adults disperse. In marine invertebrates with sessile adults, larvae commonly are considered as a specialized dispersive phase (e.g. 63, 83, 97), although Strathmann (91) argues against this interpretation.

Organisms with simple life cycles also must achieve multiple tasks in order to achieve even minimal fitness. When will selection favor dividing tasks among distinct life cycle phases and when will it favor their completion by a single phase? Critical factors include the strength of the trade-off between efficiencies at alternative tasks, the differential in abilities of small and large individuals to perform each task, and the functionality of morphologically intermediate phenotypes. Size-dependent foraging modes (24) are one category among the diverse set of situations in which different life cycle stages are selected to have different morphologies. Size differences between stages could underlie other causes of differences in selection on alternative phases. For example, depending on the mechanics of dispersal, selection might favor adaptations for dispersal at either small or large sizes. In other cases, division of tasks between life cycle stages appears not to be size related. In insects other than Ephemeroptera, only adults are winged. As a consequence, the adult stage is usually the stage specialized for dispersal as well as other activities, such as habitat selection, that require much mobility.

Ebenman (24) explicitly restricts his theory to situations in which alternative phases occurred within the ontogeny of a single individual. The benefits of decoupling phenotypic traits of different stages are also likely to be central to the origin and persistence of multiple-generation CLCs, in which alternative phenotypes are displayed by different individuals (68, 84). Multiple-generation CLCs illustrate that size differences need not always underlie the divergence among discrete phases of a CLC: Individuals of different generations all begin as a single cell of roughly similar size but nonetheless develop adaptations suited for different tasks.

Developmental Constraints

The third and final factor that may act to maintain complex life cycles is developmental constraint. Werner (109) and Ebenman (24) viewed complex life cycles as adaptations that arise in the context of inescapable trade-offs affecting performance at particular tasks. However, they considered only a subset of the possible kinds of constraint by focussing on constraints of design but not constraints of development (6, 61). Constraints of design would be present even if prior commitment of lineages to particular developmental processes presented no obstacle to the evolution of new, adaptive ontogenetic patterns.

Lineages are committed to developmental programs comprised of a complex interplay of processes involving many loci and precise timing (1, 4). These interlinked processes can constrain responses to selection, effectively halting the evolution of particular traits (52). Thus, traits of one stage may persist due to selection on developmentally linked traits expressed during other stages (16, 17). Several models have been proposed as descriptions of how selection acts on developmentally correlated characters (e.g. 16, 17, 20, 78, 86). Morphological evolution, reduction, or elimination of a developmental stage may sometimes be impossible, at least without negative effects on phenotypes of other stages. As a result, genetic variation for traits of the constrained stage is likely to be lacking or highly correlated with variation in traits of other stages that are important to fitness. In the context of CLCs, constraint due to developmental linkage could preserve traits that are maladaptive in one life cycle phase while other phases are free to respond to selection.

That early developmental stages are evolutionarily more conservative than late developmental stages has been accepted as a tendency if not a law by many biologists, including Darwin (28, 31, 33, 89). The idea was developed by von Baer who stated that early stages of development display characters in common with other similar species and that late stages of development display features that are more distinctive for a particular species (1828, as cited in 22, 31, 80). Acceptance has been based in part on observations of actual ontogenetic sequences; for example, vertebrate embryos show comparatively little variation among related species, indicating that they are relatively conserved. In addition, the acceptance of a trend toward conservation of early stages is grounded in certain assumptions about development (89), namely, that traits of early developmental stages are causal steps in the production of traits of late developmental stages (e.g. 22, 60). One variant of this assumption holds that the causal linkages of development unfold in a tree-like fashion, with early events affecting several later events (e.g. 86). In general, these consequences are expected to consist of random perturbations that decrease fitness of the later stage.

Under the view that developmental constraint is a major factor underlying the abundance of CLCs, larvae and larval attributes are retained as necessary steps on the way to production of the adult. An extreme version of this view would imply that larvae reflect ancient, outmoded adaptation, whereas adults have evolved in response to modern selective regimes. One rendering approaches Haeckel's (33) biogenetic law: the idea that ontogeny mostly evolves through terminal addition of ontogenetic stages and that larval traits correspond to adult traits of ancestors and cannot be eliminated (31). Depending on the nature of developmental linkages, the converse is also possible: Features of the adult may be retained as correlated effects of larval traits that are favored by selection.

Genetic correlations among stage-specific traits, as estimated in quantitative genetic studies, are not reliable indicators of the long-term limits on trait evolution. Genetic correlations themselves may evolve, in response to selection or other forces. Indeed, under the adaptive explanations for CLCs, metamorphosis is seen as the product of selection for decreased genetic correlations between traits of alternative phases. A deeper understanding of the limits to trait independence is likely to come from detailed studies of developmental and genetic mechanisms rather than from quantitative genetic experiments.

TESTING IDEAS FOR THE EVOLUTION OF COMPLEX LIFE CYCLES

The first two categories of explanation above are similar in viewing complex life cycles as adaptations for incorporating more than one distinct phenotype into a life cycle. They differ only in the proposed function of the divergence of phase-specific phenotypes: Do phases represent alternative adaptations for acquiring food, or do they reflect the segregation of other kinds of fitness tasks? The problem of determining the fitness contributions of particular phases falls within the domain of population biology; it consists of quantifying components of fitness within the two stages and examining the fitness consequences of variation in phase-specific features (e.g. 26). The more fundamental distinction, and the one that is my focus, is between the first two categories of explanation and the final one: To what extent do CLCs originate and persist as adaptations, and to what extent do they represent the outcomes of developmental constraints that prevent the elimination of one or the other phase?

The most general and plausible adaptive explanation of CLCs is that they reflect selection for decoupling traits of different stages of development. This idea, henceforth the "adaptive decoupling" hypothesis, combines both of the adaptive hypotheses outlined above. It can be contrasted with the hypothesis that CLCs are maintained as a result of developmental constraints acting to preserve certain developmental stages. Both possibilities are consistent with the view that developmental interdependencies constitute a pervasive force constraining the evolution of morphological change (1, 2, 4). To the extent that developmental constraints underlie CLCs, CLCs are the maladaptive consequence of such constraints; to the extent that adaptive decoupling underlies CLCs, they are adaptive responses to widespread selection to curtail such interdependencies. The adaptive decoupling explanation does not preclude the continued importance of developmental constraints in organisms that have CLCs; it only requires that metamorphosis decreases the importance of such constraints.

Predictions in several arenas depend on the roles of developmental constraint and adaptive decoupling as factors underlying the evolution of CLCs. These include (i) the phylogenetic patterns of variation in larval and adult characters, (ii) the ecological context of variations in phase-specific traits, (iii) the developmental and genetic mechanisms underlying metamorphosis and differentiation of phases, and (iv) patterns of genetic correlation between traits of different phases.

Phylogenetic Patterns in Variation of Phase-Specific Traits and Their Ecological Contexts

The adaptive decoupling hypothesis predicts that each phase can respond to selection independently of the other. In single-generation complex life cycles, this implies that larval characteristics should sometimes be more conserved and sometimes more labile than adult traits, depending on the history of selection on larval and adult phases. Adaptive divergence of larvae among species is expected in clades in which larvae have been subjected to new selective forces as a result of changed ecological conditions. Thus the adaptive decoupling hypothesis predicts that, comparing among lineages, larvae sometimes vary more than adults and that this variation is adaptive. In addition, the hypothesis predicts that the larval stage will sometimes be reduced or eliminated with no major consequences for the adult. DeBeer (22) used the term *caenogenesis* to refer to situations in which larvae had undergone adaptive morphological evolution independently from adults. He noted some cases of caenogenesis as evidence against the universality of von Baer's "law" that early developmental stages will be more conservative (see also 8a, 31).

In the most prominent groups of organisms with CLCs, including amphibians, marine invertebrates, and insects, larvae are sometimes dramatically altered, reduced, or eliminated entirely with no major consequences for adult traits.

AMPHIBIANS   A larval stage and CLC is presumed to have occurred in the common ancestor of all modern amphibians (94). However, evolutionary lability of the larval stage is evident within the three main groups of modern amphibians—salamanders, anurans, and caecilians. In anurans, tadpole traits can show adaptive variation with no apparent consequences for adults. For example, comparing among anuran species, tadpoles have undergone widespread adaptive evolution in feeding morphology (106). In some spadefoot toad species, larvae can adopt either of two discrete feeding morphologies but still produce the same adult morphology at metamorphosis (12, 73). Similarly, in some salamanders, alternative larval morphs, termed *typical* and *cannibalistic,* are produced as an adaptive response to different ecological conditions (18), indicating that larval morphology can vary adaptively without disrupting adult function. In the marbled newt, *Tritura marmoratus,* larvae appear to have undergone adaptive evolution in limb morphology that is not paralleled in adults (10). Although these observations suggest independence of larvae and adults, some constraining influence of developmental linkages between life cycle phases is suggested by observations of adult morphology in salamander lineages that have lost free-living larval stages (102). In lineages lacking larvae, adults appear to have increased specialization in morphological traits.

Variation in larval traits might be considered weak evidence against developmental constraints, because constraints may enforce conservation of certain larval features while others are free to respond to selection. Stronger evidence comes from the abbreviation or elimination of the larval stage, a transition that has occurred repeatedly in association with the production of large eggs and/or metamorphosis at small body size (25, 102, 105). All three orders of amphibians include lineages that have eliminated a free-living larval stage by telescoping larval development into the egg or mother (23). Within anurans, elimination of a free-living tadpole appears to have occurred in at least 20 lineages (105). These losses in anurans appear to have occurred in response to selection imposed by ecological factors, particularly those associated with relatively aseasonal tropical environments that lack predictable algal blooms in temporary ponds. In salamanders, loss of the larva appears to be an adaptation for eliminating dependence on aquatic environments.

In contrast to anurans, from which the adult (frog) stage has never been eliminated (105), a variety of salamander lineages have lost the original terrestrial adult phase, existing facultatively or obligately as aquatic larvae able to reproduce (23, 31, 114). Instances of loss of the adult phase appear to have occurred in response to similar conditions, specifically prolonged persistence of larval habitat (114).

MARINE INVERTEBRATES Marine larvae appear to have undergone widespread adaptive evolution in response to selective agents specific to particular environments, and this adaptive evolution encompasses abbreviation of the larval phase as well as modification of morphology. Larval types of marine invertebrates can be divided into categories as follows: feeding free-living; nonfeeding free-living; direct developing (juvenile hatches from egg); and viviparous (larval development occurs within mother who produces juvenile directly) (63). Some of the possible transitions between these categories occur readily within particular groups, as evidenced by variation in larval type among closely related lineages. More than one category may occur within the same genus, as in certain gastropods (e.g. 115), or within the same species, as in certain polychaetes (55).

Morphological evolution of early developmental stages can exceed that of adult stages in echinoderms (90). In echinoids, larval characters show a high degree of evolutionary lability, and diversification of larval morphology has occurred independently of changes in adult morphology (116, 117).

The adaptive decoupling hypothesis is further supported by the widespread reduction of the larval phase. Reduction or complete loss of a free-living, feeding (planktotrophic) larval phase has occurred repeatedly within several phyla, including echinoderms and molluscs (46, 47, 63, 88, 90, 91, 97, 118). In echinoids alone, the transition from feeding to nonfeeding larvae has occurred at least 14 times (25a). Loss of feeding by the larval stage in echinoids is followed by convergent morphological evolution in different lineages, suggesting that morphology of larvae, whether feeding or nonfeeding, is shaped by selection (90, 116). The shift to nonfeeding larvae is not linked to morphological change of adults (90).

As in anurans, lability of larval features is not without limit in marine invertebrates. For example, in echinoderms, once larval feeding has been eliminated, its reacquisition appears to be rare or completely absent (88). In addition, many larval characters of echinoderms have been retained since the early Paleozoic, the time of the common ancestor of the major echinoderm classes (90, 116).

INSECTS   Cases are known in which larvae of the same or closely related insect species show dramatic morphological variation while the corresponding adults are nearly identical. For example, caterpillars of the moth *Nemoria arizonaria* develop as either of two morphs depending on the time during the season (32). The differences, which involve the structure of the integument as well as head and jaw morphology and coloration, are adaptive. Individuals hatching in spring mimic the spring catkins of their host plants; those hatching in summer, after the catkins have fallen, mimic twigs of the hosts. Other groups within the Holometabola also show evidence of greater rates of adaptive morphological evolution in the larval stage than in the adult stage (22).

Larvae are rarely eliminated from insect life cycles, possibly because (except in Ephemeroptera) winged adults cannot molt and thus only larvae show substantial growth. Despite this limitation, free-living larval forms have been reduced or eliminated in species of several families (e.g. 27, 69). Elimination of the adult stage or reduction of adult features is somewhat more common, having evolved numerous times in gall midges (87), as well as in some beetles (e.g. 96), fleas, strepsipterans, and others (11).

OTHER SINGLE GENERATION CLCS   Other instances of adaptive evolution or loss of one developmental stage with no major consequences for other stages include a variety of host-alternating animal parasites. These have undergone both independent adaptation of stages occurring in different host taxa and elimination of hosts and corresponding phases (8, 13, 48). Astigmatid mites show adaptive evolution of immature stages possessing specialized structures for dispersal; in some species, alternative types of nymphs are produced depending on environmental conditions, with the same adult morphology being produced from either (42).

The above sampling of cases suggests that adaptive morphological evolution of larvae is widespread and that larval evolution often can occur largely independently of adult morphology. Stronger evidence against the idea that larval stages are retained due to developmental constraints comes from the fact that, in most groups with CLCs, a free-living larval stage has been repeatedly reduced or eliminated entirely. The proportion of CLCs that have been studied sufficiently to evaluate the role of adaptive decoupling vs developmental constraint is small and likely to remain so. Examples such as those listed can be used to build a case for adaptive decoupling often playing a role in the evolution of complex life cycles. At the same time, a significant role of developmental constraint is indicated by the facts that CLCs arise relatively rarely and are mostly extremely ancient (e.g. 47, 91, 94).

Genetic Correlations Between Traits of Different Phases

The adaptive decoupling hypothesis for complex life cycles predicts that genetic correlations between traits of different life cycle phases should be less than genetic correlations between traits of the same phase. Although these are some of the most direct predictions of the adaptive decoupling hypothesis, studies designed to test these predictions have yet to be carried out for either single-generation or multiple-generation CLCs (cf 66). The implementation and interpretation of such studies will be complicated. The selection of study traits, given the divergence between phenotypes of alternative life cycles, will require careful consideration. Also, since correlations might be expected to be higher between traits of temporally proximal stages in any kind of life cycle, correlations between phases of a CLC might be compared to correlations that are similarly temporally spaced within simple life cycles.

Developmental Mechanisms Underlying Production of Discrete Life Cycle Phases.

Instances of major changes in larval morphology and of reduction or loss of larval stages demonstrate that it is often possible to alter a developmental route without changing an endpoint, supporting the adaptive decoupling hypothesis. What is the developmental evidence that complex life cycles and metamorphosis promote independence of larval and adult traits? Dependence of adult traits on larval ones is expected when larval features are an intrinsic part of the causal sequences of events that produce adult structures (3). Studies of the developmental origins of adult traits can indicate the instances in which such causal links are present.

COMPARTMENTALIZATION   In at least some CLCs, the causal mechanisms involved in the development of larval and adult structures are largely independent. This independence stems from the phenomenon of compartmentalization: adult structures formed from cell lineages that remain undifferentiated during the larval phase (3, 53, 54). In contrast, in cases in which larval structures are modified to form adult structures, called remodeling, greater linkage between larval and adult features is expected. According to Alberch's (3) view, development tends to be irreversible: Only very limited dedifferentiation and remodeling of existing tissues is possible. This view is consistent with results from anurans (25).

In compartmentalized developmental systems, larval and adult cells are apportioned early in development. The two sets of cells adopt separate developmental programs, each unfolding according to its own schedule in response to specific, system-wide cues, usually hormonal (3, 53). Larval programs are accelerated relative to adult programs. Cell death is an important property of compartmentalized systems; it is the primary mechanism by which larval structures disappear at metamorphosis. Rather than being reformed into adult structures, cells involved in larval structures die and are replaced by expansion and differentiation of other cell lineages.

Compartmentalization is widespread at least as a partial basis for the divergence of the alternative phases of CLCs. In anurans, the coiled gut of the larva degenerates through programmed cell death at metamorphosis and is replaced by the short adult gut which forms from undifferentiated cells (29). In salamanders, compartmentalization rather than remodeling appears to underlie the production of the distinct larval and adult epibranchial elements (3). The most well-studied CLC is that of *Drosophila,* in which compartmentalization is extreme: Most larval cells die at metamorphosis, and adult structures are formed by cell lineages that have remained undifferentiated as the imaginal discs (5, 30, 53, 54). Similar processes, though sometimes not as extreme, occur at metamorphosis in other Holometabola (5).

One view of the evolutionary origins of metamorphosis is that it arose from a more gradual developmental pathway through concentration of morphogenesis at a particular time. This gradual route has been suggested for anurans (105, 107) and for insects (50). Similarly, Alberch (3) suggests that remodelling is the primitive state for the CLC of salamanders but that compartmentalization has evolved secondarily in some structures where it permits greater evolutionary independence and a more abrupt morphological transition.

MOLECULAR GENETICS   Perhaps the most dramatic advances in the understanding of constraints on life cycle evolution will come from developmental molecular genetics. Knowledge of how particular loci function in alternative phases of a CLC promises to illuminate the mechanisms causing interdependencies of phases. Progress in this arena will be greatly enhanced by the fact that intensive molecular genetic studies have focussed on a metazoan clade that has CLCs, namely *Drosophila* (53).

HETEROCHRONY AND PHASE REDUCTION   Within a CLC that is underlain by a compartmentalized developmental system, the elaboration or reduction of either the adult or the larval phase might be effected through heterochrony, a change in the relative timing of events during development (4, 22, 31, 59, 77). This effect of heterochrony is most likely in highly compartmentalized developmental systems in which later events in development are not dependent on earlier events (1, 2). In such systems, expression of either larval or adult traits can be moved forward, delayed, or eliminated without preventing expression of traits of the other phase: Adult features are not causally dependent on events of larval development. If the only adult features moved forward are those conferring reproductive capabilities, with other adult features being eliminated, the adult phase of the life cycle may be deleted, resulting in paedomorphosis (31). In contrast, if the developmental processes that underlie most aspects of adult morphology are accelerated, with larval trait expression simultaneously reduced or eliminated, then the larval phase of the life cycle will be lost. Various authors have suggested a role of heterochrony in the repeated evolutionary reduction of larval or adult phases of several phyla of marine invertebrates, amphibians, and insects (21, 31, 36, 57–59, 71, 90–92, 118).

A promising approach to a better understanding of life cycle simplification through loss or reduction of a phase is comparison of patterns of gene expression in species with larvae to related species that have omitted larvae (75, 76). Direct development of the sea urchin, *Heliocidaris erythrogramma,* is achieved through elimination of part of the larval pattern of gene expression (71). During development, some cell lineages skip larval patterns of gene expression, as observed in related species with larvae. Instead, adult functions of these cells are accelerated. Thus, elimination of the larval stage has been achieved through heterochrony.

In summary, compartmentalization may provide the developmental lability that is central to the evolutionary origin, persistence, and loss of CLCs. First, it may be the central adaptation that allows the evolutionary origin of a CLC: The abrupt differences between larval and adult morphologies that are the defining characteristics of a CLC may not be possible through remodelling alone. Second, compartmentalization facilitates the subsequent independent evolution of larval and adult traits. This independence potentially allows each phase to adapt independently to changing selective pressures and thus increases the likelihood that the CLC itself will confer a fitness advantage relative to a single-phase life cycle. Finally, in the event that reduction or elimination of a phase is selectively advantageous, such a transition can be readily achieved in a highly compartmentalized developmental system, through heterochrony or a change in the relative rates of processes underlying larval and adult features. Thus, the general effect of compartmentalization is to make CLCs more responsive to selection. As Strathmann (90) pointed out, studies that relate the degree of independence of adult and larval morphological evolution to the degree of developmental compartmentalization are so far lacking. Such studies would be extremely useful in evaluating the role of compartmentalization in facilitating phase-specific adaptation.

Expectations for Multiple-Generation Complex Life Cycles

What is the expectation for levels of developmental linkage in multiple-generation complex life cycles? In these life cycles, each phase develops independently from a single cell stage. Thus, the task of metamorphosis, to construct a wholly new phenotype from a preexisting one, is circumvented: The body of one phase does not have to be transformed into the body of the other phase. Since the cell lineages that produce the structures of alternative phases are entirely separate, multiple-generation CLCs may be considered to be extreme cases of developmental compartmentalization, with minimum levels of developmental linkages among traits of alternative phases. However, the early stages of development are often identical for alternative phenotypes, with divergence occurring only after development proceeds to a particular stage that is sensitive to environmental influence (68, 84). Selection for adaptive matching of phenotype to environmental conditions may favor delaying the divergence of developmental pathways, because environmental cues may be more informative later in development (59, 68). For example, in some wing polyphenic insects, commitment to one or the other developmental pathway occurs near the end of the first larval instar, in response to cues associated with habitat quality (examples in 59). In such cases, any change in early development may be expected to affect both phases, constraining their independent evolution. In addition, the independent evolution of phases in a multiple generation CLC may be constrained by the fact that many of the same loci are expressed in alternative morphs even after developmental pathways have diverged (e.g. 66).

Where reduction of a CLC to a single phase is selectively advantageous, it might be expected to occur more readily in multiple-generation CLCs. Because each phase has some capacity to acquire resources and reproduce, simple deletion of a phase is a potential route to life cycle reduction. Such reduction appears to have occurred repeatedly in most groups that show multiple-generation CLCs (57–59, 67, 68, 112). In contrast, in single generation CLCs, loss of a phase may require combining incompatible traits and functions; for example, larval design could prohibit the incorporation of traits necessary for seeking mates and producing progeny (e.g. 105). In some instances, however, developmental and design constraints could prevent the loss of a phase from a multiple-generation CLC. If activities important to fitness are divided between the alternative generations, then reduction to a single-phase life cycle through simple elimination of a generation may be accompanied by major costs. In many multiple-generation CLCs, dispersal, sexual reproduction, overwintering, or other functions are accomplished by a generation separate from those that achieve the greatest reproductive increases. I have argued previously that in aphids, which show some of the most elaborate multiple-generation CLCs in animals, the evolution of seasonal alternation between host plant taxa is largely due to constraints imposed by the generation that feeds in early spring; lineages that manage to eliminate this generation show single-host life cycles (64, 67).

SUMMARY AND CONCLUSIONS

Life cycles with morphologically distinct phases, or complex life cycles, are ubiquitous among animals; they occur in the majority of phyla, they have evolved many times independently, and they have often persisted for extremely long periods. Evolutionary studies of CLCs can mostly be divided into two sets, depending on whether they adopt an ecological or a developmental emphasis. So far these have remained almost completely separate; however, recent developments within each offer possible insight into the processes that affect the evolution of morphologically distinct life cycle phases. One general hypothesis for the abundance of CLCs is that they are responses to widespread selection for developmental independence of different stages of the life cycle. Under this view, metamorphosis and its underlying developmental mechanisms permit adaptive evolution within a life cycle stage, without correlated negative effects on traits of alternative phases. Independently adapted life cycle phases may be advantageous whenever selective forces vary in the course of the life cycle. Among possible bases for such divergence in selection at different stages are (i) optimal structural designs for food acquisition that vary with body size, (ii) different life cycle stages encountering different ecological conditions, and (iii) different components of fitness, such as abilities to garner resources, abilities to attract mates, or abilities to disperse to new habitat, that are concentrated within particular life cycle stages. A final hypothesis for the evolution of CLCs is that they result from developmental constraints. In particular, such constraints can cause traits of early stages to be preserved due to developmental linkages with traits of later stages.

Ecological, comparative, developmental, and genetic studies are relevant for elucidating the roles of adaptive decoupling and developmental constraint in the evolution of complex life cycles. The adaptive decoupling hypothesis, which is consistent with the idea that developmental linkages are a major constraining force in evolution, is supported by instances of adaptive morphological variation among larvae of related species and by cases of reduction or elimination of larvae from CLCs. Developmental compartmentalization, with separate cell lineages underlying larval and adult traits, appears to be a primary mechanism for the decoupling of life cycle phases; the most well-studied example is in the holometabolous insects, in which the imaginal discs remain undifferentiated in larva and produce adult structures during metamorphosis. If larval and adult traits are compartmentalized, reduction or elimination of one phase can occur relatively readily through heterochrony. Multiple-generation CLCs circumvent the necessity of metamorphosis, because distinct phases arise from separate zygotes. Genetic correlations between traits of the alternative phases are nonetheless likely, because early development is often identical for alternative phenotypes and since many of the same loci are expressed even after developmental pathways diverge. Loss of a phase from a multiple-generation life cycle is expected to occur more readily because all phases have the ability to grow and reproduce.

Further insight concerning the roles of constraint vs adaptive decoupling in the evolution of CLCs could be achieved through quantitative genetic studies of correlations between traits of the same and different phases, phylogenetic studies of variation in animal clades with CLCs, and through developmental and molecular genetic studies of the causal links between traits of different stages.

ACKNOWLEDGMENTS

I thank Carol von Dohlen and Diana Wheeler for stimulating discussions of complex life cycle evolution. Judie Bronstein, Goggy Davidowitz, Alan De-Queiroz, Conrad Istock, David Stern, Diana Wheeler, and a reviewer gave many insightful comments on an earlier draft of the manuscript. My research on complex life cycles of aphids has been supported by the National Science Foundation (BSR-8806068, DEB-9210386, DEB-9306495).

Any *Annual Review* chapter, as well as any article cited in an *Annual Review* chapter, may be purchased from the Annual Reviews Preprints and Reprints service. 1-800-347-8007; 415-259-5017; email: arpr@class.org

---

Literature Cited

1. Alberch P. 1980. Ontogenesis and morphological diversification. *Am. Zool.* 20: 653–67
2. Alberch P. 1982. Developmental constraints in evolutionary processes. In *Evolution and Development,* ed. JT Bonner, 313–32. Berlin: Springer-Verlag. 356 pp.
3. Alberch P. 1987. Evolution of the developmental process: irreversibility and redundancy in amphibian metamorphosis. In *Development as an Evolutionary Process,* ed. R Raff, E Raff, pp. 23–46. New York: Liss. 329 pp.
4. Alberch P, Gould SJ, Oster GF, Wake DB. 1979. Size and shape in ontogeny and phylogeny. *Paleobiology* 5:296–317
5. Anderson DT. 1972. The development of holometabolous insects. In *Developmental Systems: Insects,* Vol. 2, ed. SJ Counce, H Waddington, pp. 166–242. London: Academic. 615 pp.
6. Arnold SJ. 1992. Constraints on phenotypic evolution. *Am. Nat.* 140S:85–107
7. Askew RR. 1984. The biology of gall wasps. In *Biology of Gall Insects,* ed. TN Ananthakrishnan, pp. 223–59. London: Edward Arnold. 362 pp.
8. Baer JG. 1971. *Animal Parasites.* London: World Univ. Library. 256 pp.
8a. Ballard WW. 1981. Morphogenetic movements and fate maps of vertebrates. *Am. Zool.* 21:391–99
9. Birky CW, Gilbert JJ. 1971. Parthenogenesis in rotifers: the control of sexual and asexual reproduction. *Am. Zool.* 11: 245–57
10. Blanco MJ, Alberch P. 1992. Caenogenesis, developmental variability, and evolution in the carpus and tarsus of the marbled newt *Triturus marmoratus. Evolution* 46:677–87
11. Borror DJ, Triplehorn CA, Johnson NF. 1989. *An Introduction to the Study of Insects.* New York: Harcourt Brace. 875 pp.
12. Bragg AN, Bragg WN. 1958. Variations in the mouth parts in tadpoles of *Scaphiopus bombifrons* Cope (Amphibia: Salientia). *Southwest. Nat.* 3:55–69
13. Brooks DR, McClellan DA. 1993. *Parascript: Parasites and the Language of Evolution.* Washington: Smithsonian Inst. 429 pp.
14. Bryant EH. 1969. A system favoring the evolution of holometabolous development. *Ann. Entomol. Soc. Am.* 62:1087–91
15. Buchsbaum R, Buchsbaum M, Pearse J, Pearse V. 1987. *Animals Without Backbones.* Chicago: Univ. Chicago Press. 572 pp. 3rd ed.
16. Cheverud JM. 1984. Quantitative genetics and developmental constraints on evolution by selection. *J. Theor. Biol.* 110:155–71

17. Cheverud JM, Rutledge JJ, Atchley WR. 1983. Quantitative genetics of development: genetic correlations among age-specific trait values and the evolution of ontogeny. *Evolution* 37:895–905

17a. Cohen J, Massey BD. 1982. Larvae and the origins of major phyla. *Biol. J. Linn. Soc.* 19:321–28

18. Collins JP, Cheek JE. 1983. Effect of food and density on development of typical and cannibalistic salamander larvae in *Amystoma tigrinum nebulosum. Am. Zool.* 23:77–84

19. Comins HN, Hamilton WD, May RA. 1980. Evolutionarily stable dispersal strategies. *J. Theor. Biol.* 82:205–30

20. Crespi BJ. 1990. Measuring the effect of natural selection on phenotypic interaction systems. *Am. Nat.* 135:32–47

21. Cunningham CW, Buss L. 1993. Molecular evidence for multiple episodes of paedomorphosis in the family Hydractiniidae. *Biochem. Syst. Ecol.* 21:57–69

22. DeBeer G. 1958. *Embryos and Ancestors.* Oxford: Clarendon. 159 pp.

23. Dent JN. 1968. Survey of amphibian metamorphosis. In *Metamorphosis: A Problem in Developmental Biology,* ed. W Etkin, LI Gilbert, pp. 271–311. New York: Appleton-Century-Crofts. 578 pp.

24. Ebenman B. 1992. Evolution in organisms that change their niches during the life cycle. *Am. Nat.* 139:990–1021

25. Elinson RP. 1987. Change in developmental patterns: embryos of amphibians with large eggs. In *Development as an Evolutionary Process,* ed. R Raff, E Raff, pp. 1–21. New York: Liss. 329 pp.

25a. Emlet RB. 1991. Functional constraints on the evolution of larval forms of marine invertebrates: experimental and comparative evidence. *Am. Zool.* 31:707–25

26. Endler JA. 1986. *Natural Selection in the Wild.* Princeton: Princeton Univ. Press. 336 pp.

27. Ferrar P. 1987. A guide to the breeding habits and immature stages of *Diptera Cyclorrhapha.* Leiden: EJ Brill. 907 pp.

28. Fink WL. 1982. The conceptual relationship between ontogeny and phylogeny. *Paleobiology* 8:254–64

29. Fox H. 1981. Cytological and morphological changes during amphibian metamorphosis. In *Metamorphosis,* ed. LI Gilbert, E Frieden, pp. 327–62. New York: Plenum. 578 pp. 2nd ed.

30. Gehring WJ, Nüthiger R. 1972. The imaginal discs of *Drosophila.* In *Developmental Systems: Insects,* Vol. 2, ed.

SJ Counce, H Waddington pp. 211–90. London: Academic. 615 pp.

31. Gould SJ. 1977. *Ontogeny and Phylogeny.* Cambridge, Mass: Belknap. 501 pp.

32. Greene E. 1989. A diet-induced developmental polymorphism in a caterpillar. *Science* 243:643–46

33. Haeckel E. 1866. *Generelle Morphologie der Organismen.* Berlin: Reimer. 574 pp.

34. Haefner JW, Edson JL. 1984. Community invasion by complex life cycles. *J. Theor. Biol.* 108:377–404

35. Hamilton WD, May RA. 1977. Dispersal in stable habitats. *Nature* 269:578–81

36. Hanken J, Wake DB. 1993. Miniaturization of body size: organismal consequences and evolutionary significance. *Annu. Rev. Ecol. Syst.* 24:501–19

37. Hanson T. 1980. Influence of larval dispersal and geographic distribution on species longevity in neogastropods. *Paleobiology* 6:193–207

38. Hebert PDN. 1978. The population biology of *Daphnia* (Crustacea, Daphnidae). *Biol. Rev.* 53:387–408

39. Highnam, KC, 1981. A survey of invertebrate metamorphosis. In *Metamorphosis,* ed. LI Gilbert, E Frieden, pp. 43–73. New York: Plenum. 578 pp. 2nd ed.

40. Hille Ris Lambers D. 1966. Polymorphism in Aphididae. *Annu. Rev. Entomol.* 11:47–78

41. Hoffman A. 1984. Species selection. *Evol. Biol.* 18:1–20

42. Houck MA, O'Connor BM. 1990. Ecological and evolutionary significance of phoresy in the Astigmata. *Annu. Rev. Entomol.* 36:611–36

43. Istock CA. 1967. The evolution of complex life cycle phenomena: an ecological perspective. *Evolution* 21:592–605

44. Istock CA. 1984. Boundaries to life history variation and evolution. In *A New Ecology: Novel Approaches to Interactive Systems,* ed. PW Price, CN Slobochikoff, WS Gaud, pp. 143–68. New York: Wiley. 515 pp.

45. Jablonski D. 1986. Larval ecology and macroevolution in marine invertebrates. *Bull. Mar. Sci.* 39:565–87

46. Jablonski D, Lutz RA. 1983. Larval ecology of marine benthic invertebrates: paleobiological implications. *Biol. Rev.* 58:21–89

47. Jègersten G. 1972. *Evolution of the Metazoan Life Cycle.* New York: Academic. 282 pp.

48. Kennedy CR. 1965. The life cycle of *Archigetes limnodrili* (Yamaguti) (Cestoda: Caryophyllaeidae) and its de-

velopment in the invertebrate host. *Parasitology* 55:427–37

49. Kinsey AC. 1920. Phylogeny of cynipid genera and biological characteristics. *Bull. Am. Mus. Nat. Hist.* 42:357–402

50. Kukalova-Peck J. 1978. Origin and evolution of insect wings and their relationship to metamorphosis, as documented by the fossil record. *J. Morphol.* 156:53–126

51. Labandiera CC, Sepkoski JJ. 1993. Insect diversity in the fossil record. *Science* 261:3111–15

52. Lande R, Arnold SJ. 1983. The measurement of selection on correlated characters. *Evolution* 37:1210–26

53. Lawrence PA. 1992. *The Making of a Fly: The Genetics of Animal Design.* Oxford: Blackwell. 228 pp.

54. Lawrence PA, Morata G. 1976. The compartment hypothesis. In *Insect Development,* ed. PA Lawrence. New York: John Wiley & Sons. 230 pp.

55. Levin LA, Zhu J, Creed E. 1991. The genetic basis of life-history characters in a polychaete exhibiting planktotrophy and lecithotrophy. *Evolution* 45:380–97

56. Lubchenco J, Cubit J. 1980. Heteromorphic life histories of certain marine algae as adaptations to variations in herbivory. *Ecology* 61:676–87

57. Matsuda R. 1979. Abnormal metamorphosis and arthropod evolution. In *Arthropod Phylogeny,* ed. AP Gupta, pp. 137–256. New York: VanNostrand-Reinhold. 762 pp.

58. Matsuda R. 1982. The evolutionary process in talitrid amphipods and salamanders in changing environments, with a discussion of "genetic assimilation" and some other evolutionary concepts. *Am. Nat.* 60:733–49

59. Matsuda R. 1987. *Animal Evolution in Changing Environments.* New York: John Wiley & Sons. 355 pp.

60. Maynard Smith J. 1983. The genetics of stasis and puctuation. *Annu. Rev. Genet.* 7:11–25

61. McKitrick MC. 1993. Phylogenetic constraint in evolutionary theory: has it any explanatory power? *Annu. Rev. Ecol. Syst.* 24:307–30

62. McMillan WO, Raff RA. 1992. Population genetic consequences of developmental evolution in sea urchins (genus *Heliocidaris*). *Evolution* 46:1299–312.

63. Mileikovsky SA. 1971. Types of larval development in marine bottom invertebrates, their distribution and ecological significance: a re-evaluation. *Mar. Biol.* 10:193–213

64. Moran NA. 1988. The evolution of host alternation in aphids: evidence that spe-

cialization is a dead end. *Am. Nat.* 132:681–706

65. Moran NA. 1989. A 48-million-year-old aphid-host plant association and complex life cycle: biogeographic evidence. *Science* 245:173–75

66. Moran NA. 1991. Phenotype fixation and genotypic diversity in the life cycle of the aphid, *Pemphigus betae. Evolution* 45:957–70

67. Moran NA. 1992. The evolution of life cycles in aphids. *Annu. Rev. Entomol.* 37:321–48

68. Moran NA. 1992. The evolutionary maintenance of alternative phenotypes. *Am. Nat.* 139:971–89

69. Oldroyd H. 1964. *The Natural History of Flies.* New York: WW Norton. 324 pp.

70. Papaj DR, Rausher MD. 1987. Genetic differences and phenotypic plasticity as causes of variation in oviposition preference in *Battus philenor. Oecologia* 74:24–30

71. Parks AL, Parr BA, Chin J-E, Leaf DS, Raff RA. 1988. Molecular analysis of heterochronic changes in the evolution of direct developing sea urchins. *J. Evol. Biol.* 1:27–44

72. Peters RH. 1983. *The Ecological Implications of Body Size.* Cambridge: Cambridge Univ. Press. 329 pp.

73. Pfennig D. 1990. The adaptive significance of an environmentally-cued developmental switch in an anuran tadpole. *Oecologia* 85:101–7

74. Raff RA. 1987. Constraint, flexibility and phylogenetic history in the evolution of direct development in sea urchins. *Devel. Biol.* 119:6–19

75. Raff RA, 1992. Direct developing sea urchins and the evolutionary reorganization of early development. *Bioessays* 14:211–18

76. Raff RA. 1992. Evolution of developmental decisions and morphogenesis—the view from two camps. *Development* 1992S:15–22

77. Raff RA, Wray GA. 1989. Heterochrony: developmental mechanisms and evolutionary results. *J. Evol. Biol.* 2:409–34

78. Riska, B. 1986. Some models for development, growth, and morphometric correlation. *Evolution* 40:1303–11

79. Rowe L, Ludwig, D. 1991. Size and timing of metamorphosis in complex life cycles: time constraints and variation. *Ecology* 72:413–27

80. Russell ES. 1916. *Form and Function: A Contribution to the History of Animal Morphology.* Chicago: Univ. Chicago Press. 383 pp. Reprinted 1982

81. Savile, DBO 1976. Evolution of the rust fungi (Uredinales) as reflected by their ecological problems. *Evol. Biol.* 9:137–207

82. Scheiner SM. 1993. Genetics and evolution of phenotypic plasticity. *Annu. Rev. Ecol. Syst.* 24:35–68

83. Scheltema RS. 1971. Larval dispersal as a means of genetic exchange between geographically separated populations of shallow-water benthic marine gastropods. *Biol. Bull.* 140:284–322

84. Shapiro AM. 1976. Seasonal polyphenism. *Evol. Biol.* 9:259–333

85. Slade NA, Wassersug RJ. 1975. On the evolution of complex life cycles. *Evolution* 29:568–71

86. Slatkin M. 1987. Quantitative genetics of heterochrony. *Evolution* 41:799–811

87. Steffan WA. 1973. Polymorphism in *Plastosciara perniciosa. Science* 182: 1265–66.

88. Strathmann RR. 1978. The evolution and loss of feeding larval stages of marine invertebrates. *Evolution* 32:894–906

89. Strathmann RR. 1988. Larvae, phylogeny, and von Baer's law. In *Echinoderm Phylogeny and Evolutionary Biology,* ed. CRC Paul, AB Smith, pp. 53–68. Oxford: Clarendon. 373 pp.

90. Strathmann RR. 1992. Heterochronic developmental plasticity in larval sea urchins and its implications for evolution of non-feeding larvae. *Evolution* 46: 972–86

91. Strathmann RR. 1993. Hypotheses on the origin of marine larvae. *Annu. Rev. Ecol. Syst.* 24:89–117

92. Strathmann RR. 1993. Abundance of food affects relative size of larval and postlarval structures of a molluscan veliger. *Biol. Bull.* 185:232–39

93. Suomalainen E, Saura A, Lokki J. 1987. *Cytology and Evolution in Parthenogenesis.* Boca Raton, FL: CRC. 216 pp.

94. Szarski H. 1957. The origin of the larva and metamorphosis in Amphibia. *Am. Nat.* 91:283–301

95. Taylor PD. 1988. Major radiation of cheilostome bryozoans: triggered by the evolution of a new larval type? *J. Histor. Biol.* 1:45–64

96. Taylor V. 1978. A winged elite in a subcortical beetle as a model for a prototermite. *Nature* 276:89–91

97. Thorson G. 1950. Reproductive and larval ecology of marine bottom invertebrates. *Biol. Rev.* 25:1–45

98. Valentine JW, Jablonski D. 1983. Larval adaptations and patterns of brachiopod diversity in space and time. *Evolution* 37:1052–61

99. Via S. 1987. Genetic constraints on the evolution of phenotypic plasticity. In *Genetic Constraints on Adaptive Evolution,* ed. V Loeschcke, pp. 47–71. Berlin: Springer-Verlag. 188 pp.

100. Via S, Lande R. 1985. Genotype-environment interaction and the evolution of phenotypic plasticity. *Evolution* 39:505–22

101. Vrba ES, Eldridge N. 1984. Individuals, hierarchies and processes: towards a more complete evolutionary theory. *Paleobiology* 10:146–71

102. Wake DB. 1982. Functional and developmental constraints and opportunities in the evolution of feeding systems in urodeles. In *Environmental Adaptation and Evolution: A Theoretical and Empirical Approach,* ed. D Mossakowski, G Roth, pp. 51–66. Stuttgart: Fischer-Verlag. 302 pp.

103. Wald G. 1981. Metamorphosis: an overview. In *Metamorphosis* ed. LI Gilbert, E Frieden, pp. 1–42. New York: Plenum. 578 pp. 2nd ed.

104. Wassersug RJ. 1974. The evolution of anuran life cycles. *Science* 185:377–78

105. Wassersug RJ. 1975. The adaptive significance of the tadpole stage with comments on the maintenance of complex life cycles in amphibians. *Am. Zool.* 15:405–17

106. Wassersug RJ. 1980. Internal oral features of larvae from eight anuran families: functional, systematic, evolu- tionary, and ecological considerations. *Univ. Kansas Mus. Zool. Misc. Publ.* 68:1–146

107. Wassersug RJ, Hoff K. 1982. Developmental changes in the orientation of the anuran jaw suspension: A preliminary exploration into the evolution of anuran metamorphosis. *Evol. Biol.* 15:223–45

108. Werner EE. 1986. Amphibian metamorphosis: growth rate, predation risk, and the optimal size at transformation. *Am. Nat.* 128:319–41

109. Werner EE. 1988. Size, scaling, and the evolution of complex life cycles. In *Size-Structured Populations,* ed. B Ebenman, L Persson, pp. 60–81. Berlin: Springer-Verlag. 284 pp.

110. Werner EE. 1992. Individual behavior and higher order species interactions. *Am. Nat.* 140S:5–32

111. Werner EE, Gilliam JF. 1984. The ontogenetic niche and species interactions in size-structured populations. *Annu. Rev. Ecol. Syst.* 15:393–425

112. West-Eberhard MJ. 1989. Phenotypic plasticity and the origins of diversity. *Annu. Rev. Ecol. Syst.* 20:249–78

113. Wilbur HM. 1980. Complex life cycles. *Annu. Rev. Ecol. Syst.* 11:67–93

114. Wilbur HM, Collins JP. 1973. Ecological aspects of amphibian metamorphosis. *Science* 182:305–14
115. Woodward BB. 1909. Darwinism and malacology. *Proc. Malac. Soc. Lond.* 8:272–86
116. Wray GA. 1992. The evolution of larval morphology during the post paleozoic radiation of echinoids. *Paleobiology* 18: 258–87
117. Wray GA. 1992. Rates of evolution in developmental processes. *Am. Zool.* 32:123–34
118. Wray GA, Raff RA. 1991. The evolution of developmental strategy in marine invertebrates. *Trends Ecol. Evol.* 6:45–50

Annu. Rev. Ecol. Syst. 1994. 25:601–28

MALE PARENTAL BEHAVIOR IN BIRDS

Ellen D. Ketterson and Val Nolan, Jr.

Department of Biology and Center for the Integrative Study of Animal Behavior, Indiana University, Bloomington, Indiana 47405

KEY WORDS   paternal care, mating systems, sexual dimorphism, male removal, hormones, phenotypic engineering

Abstract

Male parental care is rare in most groups of animals but common in birds. Among birds considerable variation exists in the form of care provided their young; recent developments in several areas may help to explain this variation. These include (i) improvements in the comparative method, particularly the use of phylogenies to investigate questions about the evolution (origin) of behavior; (ii) field studies designed to determine the fitness benefits of care and the selective factors that maintain it; and (iii) investigations of the mechanisms of care, especially its neural and hormonal bases. The greatest challenge to functional explanations of male care has come from the revelation that males frequently provide care for offspring that are not genetically their own. Resolving this apparent paradox will require more accurate accounting of the costs and the benefits of care, and greater attention to the form of care provided by males. Data support the following working hypothesis: In taxa in which males incubate, as opposed to providing other forms of care, sexual selection may be less intense and males may resemble females more in appearance and physiology. Males may also be more likely to be the genetic sires of the offspring they care for, perhaps because of the greater cost to females of multiple matings or perhaps because male incubation is incompatible with sustained sexual behavior. Ultimately, an integrated consideration of history, function, mechanism, and development should reveal the interplay of factors that have led to male parental care in birds and that account for its maintenance in various forms today.

601

0066-4162/94/1120-0601$5.00

INTRODUCTION

With the exception of brood-parasitic species, which lay their eggs in the nests of other birds, virtually all birds provide some form of parental care after eggs are laid, and parental care by males is more extensive in birds than in any other vertebrate class (9, 38, 43, 70). Sometimes the care provided by males is indirect, as when they help build nests (65% of avian subfamilies—70) or feed the female during egg-laying or incubation (48% of avian subfamilies—70), but what clearly distinguishes male birds is the extent to which they give direct attention to eggs or young. In most avian subfamilies, according to Silver et al (70), males incubate eggs (68%) and care for young, either by feeding them (71%) or by escorting them to feeding sites (73%). While male care after egg-laying or parturition has been described in fish, amphibians, and mammals, it is decidedly rarer than in birds (9, 23, 25, 63).

Our particular interest in this subject stems from 12 years of study of male parental behavior in the dark-eyed junco (*Junco hyemalis*), a species of songbird in which males feed and protect offspring but leave most other forms of parental behavior to females. As behavioral ecologists, we tested whether male care was essential to reproduction by removing males at the time the eggs hatched and comparing the reproductive success of male-aided and unaided females (100, 101, 102). We next manipulated testosterone in male juncos, a treatment that simultaneously enhances courtship and suppresses parental behavior (38, 41). Our objective was to determine whether females, which had to increase parental effort to compensate for reduced male investment, based their subsequent choice of mates on the level of care previously provided by males.

We review the results of these studies below, but one initial finding of interest concerned the behavior of the males that took over the territories of those we removed. These replacement males courted the females and frequently paired with them for subsequent breeding attempts. Typically, replacement males provided only limited care for their predecessors' young, but a few behaved like fathers and brought food (40, 100). We became curious about the stimuli that elicited care and hence about the nature of the mechanisms governing care and how they might evolve (38). Our findings that testosterone suppressed male feeding (41, 71) while enhancing attractiveness to females (D Enstrom, ED Ketterson, V Nolan, Jr, unpublished data) suggested that this hormone might be pivotal to the physiology underlying variation in patterns of mating and parental care (38).

We concluded that a full understanding of the evolution of male parental care would require all the approaches recommended by Niko Tinbergen (74): historical, functional, mechanistic, and developmental. One of our objectives in this article is to develop the case for a multi-level analysis of parental care.

Another objective is to report particular advances in the study of male parental care in birds as an example of general progress in the area of behavioral ecology. The key questions we address are: (i) How does current understanding of avian parental behavior differ from previous understanding (e.g. 15–20 years ago)? And (ii) what areas of research were not foreseen about two decades ago? We begin with an overview of the changes that have occurred, which we do not attempt to document in detail, then narrow the focus to topics to be addressed in greater detail.

Then and Now

Some 20 years ago, behavioral ecologists interested in reproduction focused on mating systems, parental care, and sexual dimorphism in secondary sex characteristics (size, structures, or coloration). Perhaps the central question was why male parental care tended to co-occur with monogamy and sexual monomorphism, whereas reduced male care, or its absence, was correlated with polygyny and with sexual dimorphism (43, 77). Those most interested in the evolution of polygyny made within-species comparisons of the reproductive success of primary (first-mated) and secondary females of polygynous males (7, 54, 76). The expectation was that secondary females would receive less male help than they would receive if mated monogamously, but that nevertheless they would be as successful as females mated to males that attracted only one mate. Thus, the polygyny threshold model (54, 76) predicted that territories held by polygynists would provide resources sufficient to compensate for diluted male care. Those most interested in monogamy assumed that the parental care provided by males in monogamous species was essential, or at least very important, for successful reproduction (43, 99), and some began field studies to test that hypothesis (28, 62). Also underway were long-term assessments of variance in male lifetime reproductive success; males were marked and counted as were the females they paired with and the offspring produced (10). Variance in reproductive success was expected to be greater among males than among females and to be correlated with mating system; sexual selection, it was thought, would be more intense in polygynous species, accounting for their greater sexual dimorphism (10, 57, 68).

Preconceptions may have inhibited other lines of inquiry. For example, adaptation and mechanism were often studied independently, probably in the belief that "how" and "why" questions represented different levels of analysis. Adaptation was assumed, and the challenge was to identify function; mechanisms would evolve to enable function. Less attention was paid to the origin of parental behavior, as opposed to its current utility, perhaps because any hypothetical reconstruction would remain controversial as long as systematists could not agree on which taxa were primitive and which were advanced.

Rather, attention was focused on seeking ecological correlates of interspecific variation in mating systems, parental care, and dimorphism.

Behavioral ecologists are still interested in these issues, but their approaches to such studies have changed significantly. Methodological breakthroughs permit us to measure genetic relatedness with greater certainty at many levels (e.g. parent-offspring, species complexes). The discoveries that extra-pair paternity (EPP) is frequent in pair-bonded birds and that females of some species lay eggs in the nests of other females (conspecific brood parasitism, CBP) (4, 29, 84) have had enormous impact on the way we think about mating systems, male care, and sexual dimorphism. In fact, the existence of EPP and CBP have called into question our ability even to categorize mating systems and have cast doubt on past efforts to measure reproductive success (24). Terms like *monogamy* and *polygyny,* formerly entrenched in the literature, are now qualified as "apparent monogamy" or "social monogamy" when they are contrasted with alternatives such as "social or overt polygyny."

Extra-pair paternity also raised the unforeseen question of how male parental care evolved if the cared-for individuals frequently are not related to the care giver (4, 50, 57, 85, 88, 106). This has generated new interest in assessing the fitness consequences of male parental care in order to determine just how costly or beneficial it is (3, 16, 49, 81, 100, 102). EPP also raises the possibility that the variance in reproductive success among apparently monogamous male birds is greater than was once thought, but, given the existence of CBP, not necessarily greater than in females. Much research has focused on mate choice by females and the implications of choice for sexual selection and sexual dimorphism (summarized in 4).

Improved techniques for measuring relatedness have also acted synergistically with improvements in the comparative method (33) and in methods for building phylogenies (21), fostering greater confidence in hypotheses about adaptation (5, 53). The better our phylogenies, the greater our ability to reconstruct the order of evolution of behavioral alternatives and thus to ascertain the relative importance of ancestry and ecology in accounting for interspecific variation. In the near future we can expect a spate of studies relating behavior to molecular-based phylogenies.

Understanding of the ultimate causes of parental behavior, mating systems, and sexual dimorphism has also been enhanced by the study of mechanism. The circulating level of testosterone, a hormone long known to orchestrate sexual differentiation and thus to account for sexual dimorphism, has now been shown to correlate with mating systems and parental behavior (92–94). We may be on the verge of identifying the proximate mechanisms (i.e. changes in organismal attributes) that accompany loss or origin of social monogamy, male parental care, and monomorphism. Knowledge of mechanisms, particularly developmental mechanisms, is also important if we are to describe limits to adaptation, i.e. constraints, and also to account for the plasticity that apparently allows males to adjust their investment in mating and parental effort. Such adjustments are likely to depend on the relative frequency of individuals that behave in alternative ways, emphasizing the importance now attached to frequency dependence. Some such alternative adjustments may be made early in the individual's life, as we are just beginning to appreciate the role of early environment in shaping variation in reproductive tactics (36, 67, 80a, 97).

Clearly, we cannot develop all these points in a brief review. One theme we explore is that further progress in understanding the evolution of male care is likely to require both careful specification of the kind of care under consideration (i.e. the nature of the behavior), particularly whether the care is incubation of eggs, and also study of the mechanisms controlling care. Labeling of any behavior that appears beneficial to offspring simply as parental care may obscure important differences and impede integration of knowledge about history, function, mechanism, and development.

MALE PARENTAL CARE: THE COMPARATIVE APPROACH TO ITS EVOLUTION

Ecological and Life History Approaches to the Evolution of Male Parental Care

Lack (43) is probably most closely associated with the view that the environment as a collection of currently acting selective factors is the overriding determinant of avian life history traits. Although this approach is now seen by most as overly simplistic, Lack (43) was able to document and explain, to an extraordinary degree, the considerable variation in avian mating systems and parental care. An example of this approach is the speculation that in some shorebirds, classical polyandry—in which females have multiple mates and are the brighter sex and males incubate—may have evolved in the high arctic environment because time available for reproduction is so brief (19, 20, 44, 61). If females and males originally shared incubation, and if then, in response to the need for haste, females began to lay one clutch to be incubated by males and another for themselves, such a system would be only one step away from the system in which females lay clutches sequentially for more than one male and cease to incubate eggs themselves (20). This explanation is intuitively very appealing, but how might it be tested?

To render the study of the evolution of parental care more objective, Silver et al (70) used canonical correlation to describe the taxonomic distribution of avian parental behavior among 237 avian subfamilies. Five kinds of male parental care (feeding female, building nest, incubating, feeding young, escorting young) were coded along with 29 life-history traits and ecological attributes (e.g. mode of development of young, degree of sexual dimorphism, preference in diet, selection of habitat). Four pairs of canonical correlates were identified, the first accounting for 29% of the variation in male care, and several widely held adaptive hypotheses that had not yet been subjected to rigorous testing were supported.

Male parental care was found to be more common when the mating system is apparently monogamous and the habitat terrestrial. It is also more common when young are altricial as opposed to precocial; interestingly, mode of development also affects the kind of care provided. Males with altricial young are likely to feed the female and the young, but not to incubate eggs. Conversely, males with precocial young are more likely to incubate or to escort young to feeding sites, but not to feed the female or young. This seems reasonable from a mechanistic perspective, because the motor patterns of feeding are similar whether the recipient is a male's mate or his offspring, whereas the mechanisms underlying incubation are likely to prove quite different (see below).

Silver et al (70) concluded that biparental care was probably ancestral to altriciality and that, when the latter evolved, the already heavily invested female required help in gathering food for the dependent young. Thus, they reasoned, male feeding of young may have come later and replaced male help with incubation.

Sexual Monomorphism, Speciation, and the Form of Male Parental Care

In a recent study, Pierotti & Annnett (60) reported that intra-generic hybrids are more common in taxa with male care than in those without male care. They suggested as a reason that females that rely on care may be more attuned to cues that predict it and correspondingly less attuned to exaggerated traits such as body size or plumage color (e.g. 37, 90, 105). If females whose mates provide care are less particular about male appearance when selecting a mate, speciation may be less likely and hybridization more likely (cf West-Eberhard—83).

To support their thesis, Pierotti & Annett (60) examined avian families or subfamilies (46 nonpasserines, 48 passerines) and classified them as [socially] monogamous and monomorphic (MM, 55 taxa), [socially] monogamous and dimorphic (MD, 24 taxa), or [overtly] polygynous and dimorphic (PD, 9 taxa). The MM group contained fewer species per family than the MD and PD groups, and the percentage of species that hybridized was greatest in the MM group (34%), intermediate in the DM group (17%), and least in the PD group (9%). Because male parental care is characteristic of socially monogamous species, they concluded that speciation is less likely when the sexes are similar and males provide care.

Especially striking was a difference in the kind of parental behavior exhibited: Among passerines, the parental behavior most closely associated with sexual monomorphism was male incubation (60, data from 77). Male nest-building was common in both the monomorphic and dimorphic taxa (94% and 71%), as was feeding of nestlings (100% and 80%); but while males incubated in 79% of the monomorphic taxa, they did so in only 25% of the dimorphic taxa. Recall the finding by Silver et al (70) that male birds tend either to incubate or to feed but not to do both. Pierotti & Annett added that incubation tends to co-occur with monomorphism, while nest-building and feeding of young do not. How might this come about?

One suggestion is that male incubation may be the most confining of male parental behaviors because it consumes time and cannot be postponed. Further, incubation may require physiological changes that interfere with sexual behavior. In either case incubation may come at the cost of lost mating opportunities, reducing variability in reproductive success among males. If true, and if we accept the generalization that sexual dimorphism (including dichromatism) usually reflects the intensity of sexual selection, then collectively these observations suggest that sexual selection may be less intense in taxa in which males incubate, hypothetically because males allocate more effort to parental care and less effort to mating. If we now add that females belonging to taxa with male incubation may base their choice of mate on attributes that correlate with parental behavior rather than appearance (and especially if natural selection independently favors cryptic appearance in incubating males), then the observed co-variation among male incubation, monomorphism, and social monogamy may be better explained. Conversely, dimorphism and polygyny may be more likely to evolve in lineages in which males do not incubate but may feed, especially if feeding does not greatly interfere—in terms of time or underlying physiology—with attracting multiple mates (overt polygyny) or, in socially monogamous species, with extra-pair mating opportunities.

These arguments could be countered by noting the existence of extra-pair paternity in monomorphic species in which males incubate (e.g. 51; L Whittingham, unpublished data on house martins) or by asserting that even in monomorphic species females prefer males of higher "quality" (4). Confirmation or rejection will require more data on the incidence of EPP and on variance in male reproductive success (the only reliable basis for defining mating system) in species in which males do and do not incubate. Also needed are more data comparing variance in success between males and females. However, to repeat, one key to unraveling the linkage among sexual dimorphism, criteria used by females in mate choice, and male parental care is to refine the classification of species according to kind of care males provide.

Advances in the Comparative Approach

The integration of phylogenetic systematics and ecology has led to fundamental improvements in the comparative method (5, 11, 33) and may represent the most important breakthrough in evolutionary biology of the past two decades (53). Critics of comparative method studies carried out prior to this integration argue that statistical analyses like those used by Silver et al (70) and Pierotti & Annett (60) assume independence of data points, whereas because related taxa are inextricably linked by their common ancestry, they cannot be treated as independent. In this view, it is simply not helpful to know how many taxa exhibit a trait; rather we must know how many times a trait has arisen and how many times it has been lost (48). To know these things, we need an accurate determination of each lineage's history (48).

Phylogenetic analysis offers testable hypotheses about sequence of trait evolution, ease of reversal (strength of constraints), and rates of evolution. When applied to parental care, it offers objective criteria for determining whether males exhibit care because care is more adaptive than any alternative route to reproductive success or because ancestry dictates it. A fine example of this approach is that of Winkler & Sheldon (98), who mapped the mode of nest construction of swallows onto a molecular phylogeny and found this trait to be very conservative. Among 17 species exhibiting five diverse modes of nest construction, each mode apparently arose only once. Further, what would seem intuitively to be the simplest form, adoption of a cavity already made by another species, is the most recently evolved (Figure 1).

McKitrick (48) applied phylogenetic techniques to the evolution of male parental care in 60 avian taxa and confronted all the difficulties that currently face cladists: multiple and barely distinguishable trees, trees whose form varies depending upon which taxa are analyzed and upon assumptions regarding the likelihood of reversals of traits, incomplete behavioral data, etc. Nevertheless, her work is very provocative, and, because the methods are rigorous and clearly stated, it will be possible to adjust conclusions as more data become available and methods of analyzing trees progress.

Of the 60 taxa considered by McKitrick (48), 52 exhibited some form of biparental care. Her phylogenies were based on both anatomical (69 characters) and behavioral (15 characters) data, and her outgroup was a reptilian ancestor with no post-hatching care. When she used feeding of young as the index of parental care, her trees showed male care (male-only or biparental) as the ancestral state in birds. Thereafter, male feeding was lost only twice: once in the lineage that led to ducks and grouse and once in the hummingbird genus *Eulampis,* which clusters with birds of prey.

Interestingly, and unlike Silver et al (70), McKitrick (48) found no association between male care and altriciality. Of the two losses of male feeding,

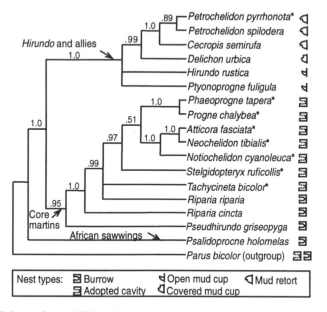

Figure 1 Phylogenetic tree (50% majority rule consensus, based on DNA hybridization) of 17 swallows and a titmouse that maps the origin of different modes of nest construction. English names for the taxa are, from top to bottom, cliff swallow, South African cliff swallow, rufous-chested swallow, house martin, barn swallow, rock martin, brown-chested martin, gray-breasted martin, white-banded swallow, white-thighed swallow, blue-and-white swallow, southern rough-winged swallow, tree swallow, sand martin, banded martin, gray-rumped swallow, black sawwing, and tufted titmouse (from Winkler & Sheldon—98).

one was associated with the evolution of precocial young and one was not. McKitrick's analysis does support the speculation that male-only incubation tends to be preceded by biparental incubation (19): male-only care originated five times, and ratites (e.g. rheas, kiwis) were the only one of these groups not preceded by an ancestor in which both sexes incubated. The stability of male incubation in the lineages described by McKitrick (48) and the conservatism of nest construction in swallows (98) both suggest that various forms of parental care may exhibit considerable inertia, i.e. be difficult to modify once they have evolved.

Unfortunately, much of the active field work on parental care in birds is done with songbirds (passerines), and McKitrick's anatomical data do not readily distinguish among members of this large group. The blackbirds (Icterinae), for example, should prove a fascinating subfamily for study because mating systems vary greatly and care in some species is biparental, in others female-only, and in still others (brood parasites) absent. Finally, some icterines are sexually dimorphic is size but not color, while others are similar in size but are dichromatic (43; D Enstrom, unpublished data). Even greater diversity among shorebirds (which are not passerines) would make them another fascinating group for study (20, 43, 44, 61).

In sum, biparental care is probably ancestral in birds. Mode of development may or may not have a determining effect on the kind of parental behavior expressed by males, although it seems likely that it does. Groups in which males incubate may be more likely to be socially monogamous and monomorphic, and they may also be less likely to speciate. They may also be less variable in their reproductive success and under less intense sexual selection. The degree of lability of all these traits and their order of evolution are not yet known, but there is reason to be optimistic that questions like these will be answered.

MALE PARENTAL CARE: THE FUNCTIONAL APPROACH

Turning to a cost-benefit approach, we ask whether male care, in the species in which it exists, has been shown to be the option most beneficial to males, i.e. is it currently maintained by natural selection? Such questions require a theoretical framework (85, 88, 96), the most enduring treatment of which has been provided by Maynard Smith in his classic paper, "Parental investment: a prospective analysis" (46).

Maynard Smith (46) assessed conditions that should favor the evolution of care and also asked when it should be biparental and when uniparental. The obvious consideration is the extent to which care improves survival of offspring, but other crucial considerations are (i) the relative benefits of providing care instead of deserting one's mate to pursue alternative reproductive options, and (ii) the response the mate might make to its partner's decision to stay or desert. If, for example, parental care improves survival of young but only one parent is sufficient, the sexes are in conflict: Which should stay and which desert? According to Maynard Smith (46), the answer lies in differences between the sexes in the factors that determine fecundity. Males must weigh the benefits of care against the costs of lost opportunities for additional matings. Females must weigh the benefits of care against any increase in fecundity that might come from withholding care, e.g. the ability to produce a larger current clutch or to produce a future clutch more promptly.

Most interesting was Maynard Smith's insight (46) into the evolutionary interdependence of these effects. Whether males gain more from extra matings than they would from parental care, in a population in which care is left to the female, will depend on the extent to which female-only care decreases female fecundity. If female-only care lengthens the time required to rear young, then males seeking extra matings will find fewer fertile females to mate with. Similarly, if female-only care results in smaller clutches, then the payoff of extra matings to males is reduced. On the other hand, if male care leads to laying of larger clutches by females, then male care may be favored. In short, the optimal solution for one sex depends on the choice made by the other sex.

According to Maynard Smith (46), biparental care in birds should prevail over mate desertion because, in general, reproductive success when two parents attend young is greater and because the male's chances of finding additional mating partners is low. Using his notation, biparental care is maintained because

$$P_2 > P_1 + pP_1,$$

where P_2 is productivity with two parents, P_1 is productivity with one parent, and p is the males' probability of finding an additional mate.

Tests of the Fitness Consequences of Male Parental Care

Can this theory be tested? A value may be placed on male care by measuring any change in reproductive success that occurs when care is eliminated or reduced. Several approaches have been used, including removal of the male leaving the female to rear her young alone (3, 16, 100) as well as manipulations that reduce male care. Care can be reduced by hormonal treatment (35, 38, 41, 71), by psychological means (e.g. decreasing the male's perception of his paternity—13), or by making flight more difficult, thereby increasing the energetic cost of care (72, 103). The other approach is to manipulate the environment, e.g. by varying the operational sex ratio (73), the reproductive value of the brood (106), or the need for male help (87, 105). In the following subsection, we focus on removal of the male.

With or without such manipulations, any accounting of benefits of care must consider the extent to which it facilitates the production of genetic offspring, and thus the accounting must be corrected for the occurrence of extra-pair fertilizations (EPF) both by the focal male and by other males at his expense. Using molecular techniques, workers are rapidly accumulating information on the frequency of extra-pair fertilizations and addressing the question whether certainty of paternity is related to level of paternal care (4, 51, 85, 88). The approaches have been theoretical (85, 88, 104), comparative (51), and experimental (13, 89).

Male Removal Studies

Male removal was pioneered by Gowaty (28) and Richmond (62) and summarized by Wolf et al (100), Bart & Tornes (3), Dunn & Hannon (16), and

Clutton-Brock (9). The technique assumes that the benefits of male care can be assessed by measuring any decline in female reproductive success that occurs when females are unaided. Benefits of male care can be direct or indirect and also immediate or delayed. Direct benefits are improved reproductive success, attributable to the male's attentiveness to the offspring themselves. Indirect benefits are derivative, through the female and her well being, which, of course, is relevant to the male only insofar as female condition affects male fitness. An example of a direct benefit would be an increase in numbers of fledglings produced when the male feeds or protects nestlings; an indirect benefit would be a difference in the female's ability to produce more young as a result of the male's contribution to care. Immediate benefits are those associated with the breeding attempt during which the male was removed, again including numbers of fledglings produced. Delayed benefits bear on future reproduction, assuming that the male and female remain paired, and include time elapsed before the female makes her next breeding attempt, the size of the next clutch, or the probability of her surviving to breed again. Delayed effects can be direct, e.g. if caring for young affects male survival, or indirect, e.g. if an unaided female is less likely to survive and females are the limiting sex.

Most male removal studies have reported only the numbers of fledglings produced by aided and unaided females, but a few have followed young to independence (e.g. 100). That is, they have reported only direct, immediate effects. Some have shown a significant decline in success when the male is absent (e.g. 65), but others have not (e.g. 28, 45). When males incubate, they may be essential to successful reproduction (16, 20), but males that merely guard the young or provide food are more likely to have either no detectable effect on success or only an incremental effect (3, 16, 100). Among passerines, only female black-billed magpies failed to produce any young in the absence of male help, and in this species males incubate (16).

In their summary of male removal studies in passerines, Wolf et al (100) speculated that male help was rarely essential to nest success, i.e. to the production of at least one fledgling, and that it appeared to be more beneficial in cavity nesters where broods are often larger, or at high latitudes, where weather may constrain the ability of females to feed when poikilothermic nestlings require brooding. Bart & Tornes's review (3) included both passerine and nonpasserine species, and, as interpreted by Webster (81), reported that unaided females produced 77% as many fledglings as did aided females. Bart & Tornes concluded that male parental behavior may be maintained by selection as long as it is at least occasionally beneficial and of relatively low cost to the male (see 18). Dunn & Hannon (17) showed that the effects of male removal may be apparent only when food is limiting.

Can male removal studies address Maynard Smith's prediction that in order

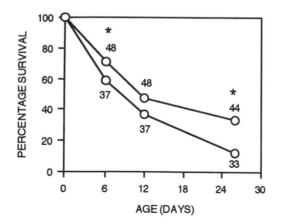

Figure 2 Survivorship of young dark-eyed juncos raised from hatching to independence by one (lower line) or two (upper line) parents. Measurements are mean percentage of broods surviving to three ages: mid-way through the nestling phase (day 6), fledging (day 12), and independence (day 26), from Wolf et al (100, asterisks indicate one-tailed *p* < 0.05, Mann-Whitney U).

for biparental care to evolve, the productivity of two parents should be greater than that of one? In juncos that had broods of four at hatching (the modal size), unaided and aided females did not differ in numbers of young fledged (Figure 2). However, unaided females reared an average 1.0 young to independence (i.e. of parental care), as compared to 2.6 for pairs (100, Figure 2). Thus, two parents do rear more young than one, and if a male were to desert his brood without caring for them, he would need to acquire an average of three females to perform as well as a care-giving male. And what of the other term in Maynard Smith's inequality, the probability of obtaining additional mating opportunities? We return to this subject below but emphasize here that in juncos, male parental care provides immediate and direct benefits.

Birkhead & Moller (4, pp. 241–42) asserted that "evidence from male removal studies conducted so far should not be used at all since the value of male care can be evaluated only if widowed females do not suffer any reduction in survival or future reproduction" and "survival and reproduction costs [of male removal] are experienced [by the female] at some later stage." We disagree with this position for two reasons. First, it cannot be assumed that females experience costs at some later stage; later effects, if any, on females should be measured to determine whether they are important. Second, increased female effort in compensation for the loss of male help is not necessarily detrimental to male fitness.

We looked for delayed and indirect effects in the junco (102) and found that unaided females lost significantly more mass while tending nestlings alone (10.4% vs 8.6%; here and below, the mean or percentage for unaided females is stated first). However, we found no significant differences in the percentage of females that renested after our manipulation (100% vs 91%), in the interval between broods (19.3 vs 15.9 days; the later brood apparently sired by males that replaced removed males), or in the size of the clutch in the female's next nesting attempt (3.5 vs 3.7 eggs) (all $p > 0.1$) (102). We also found no treatment difference in the proportion of females that returned to breed the following season (58% vs 45%) or in recruitment to the breeding population of young reared by aided and unaided females (15% vs 15%) (100; ED Ketterson, V Nolan, Jr, L Wolf, unpublished data; see also 28, 45).

We grant the low power of these comparisons because the sample sizes were small. Thus the conclusions are tentative. Nevertheless, they suggest that delayed costs of unaided reproduction to female juncos are not large. The reason for this is probably that fledglings cared for by unaided females were more likely to die before independence; after such deaths, the parental effort required of unaided females was reduced. If this is correct, then, contrary to the position of Birkhead & Moller, the cost to the male and female junco of lost male help is immediate and up front; it is the production of fewer independent young from the current brood. One observation (ED Ketterson, V Nolan, Jr, L Wolf, unpublished data), however, does suggest an important potential delayed effect on the fitness of males, if they were to force females to rear young alone. Although unaided females were as likely to survive (i.e. be present next year; see above), they were not as likely as aided females to return to the previous year's breeding territory (48% vs 75%). Because some male juncos fail to obtain females (the adult sex ratio favors males) and surviving pair members usually re-mate in successive years, this tendency of unaided females to switch territories would increase the probability that nonhelping males would remain unmated next year. However, in this situation the agent favoring male parental care could be female mate choice (sexual selection), not the cost of reproduction to unaided females (102).

In sum, we argue that male removal studies, like those in which male care is experimentally reduced (13, 35, 38, 41, 71, 72, 103), can provide a useful approach to quantifying the utility (fitness benefits) of male parental care. Rather than discourage their use, we suggest that they employ standardized methods and be performed on species in which samples can be large enough to allow confidence in the conclusions. Immediate and delayed, as well as direct and indirect, effects need to be quantified. In addition, more studies on nonpasserines are needed. Perhaps most importantly, more emphasis needs to be placed on the kind of care males provide. For example, is male care not only beneficial but essential when males incubate? When males simply guard their nests, what is the benefit of guarding (45, 105, 107; K Yasukawa, WE Richmond, unpublished data)?

Paternity and Male Parental Care

A critical consideration when quantifying the benefits of male care is the male's relatedness to the young in question (29, reviews in 4, 84). Despite earlier theoretical work to the contrary (57, 82), most students of avian parental behavior assumed that male care could evolve only if the beneficiaries were the genetic offspring of the caregivers. Newer theoretical treatments identify conditions in which relatedness and level of care are expected to co-vary (85, 88, 104) and also those in which they are not (85, 88).

Empirical studies are addressing the question both comparatively and experimentally (13, 50, 51, 89). Moller & Birkhead (51) reported, in interspecific comparisons, that the proportion of feeding trips by males can be used to predict the proportion of young sired with their social mates. Their conclusion has been seriously challenged by Dale (12), and resolution of this question must await more extensive data and data of higher quality. On the intraspecific level, Davies (13) has outlined an experimental approach to determining the effect of perceived level of paternity on paternal care: He systematically removed males for short periods of time during the female's presumed fertile period and then returned the males and documented their subsequent behavior. Male dunnocks are more likely to care for young if they have access to the female at the time she is presumably fertile (15).

One question of particular interest is whether the frequency of extra-pair fertilizations will vary with the form of male care provided, and the prediction made above was that EPP will be lower when males incubate (e.g. in black vultures and oystercatchers, 14, 34). Moller & Birkhead reviewed (51) 50 species in which both percentage of young sired by extra-pair males in focal nests and percentage of incubation contributed by the males associated with those nests were reported. Among the 38 passerines reviewed, males of three species incubated. We calculated that for those males the average percentage of young sired by extra-pair males was 5.2%; this compares to 19.1% extra-pair young among the 35 passerines whose males did not incubate (statistical tests are not appropriate). Among the nonpasserine species, five exhibited no paternal care at all. Of those with male care, six showed male incubation and one did not. For these seven species, the respective percentages of young sired by extra-pair males were 2.3% and 5.4%. Thus, EPP appears low in nonpasserines with male care but, consistently with the prediction, was lowest in species in which males incubate. We tried to compare sister taxa to see whether the percentage of time spent incubating was negatively related to EPP, but for various reasons (e.g. no sister taxon available for comparison) only four comparisons were possible. In all four, EPP was lower in the taxon with the higher percentage of male incubation. Obviously, a firmer conclusion that EPP is lower when males incubate will require more data.

Many people working in this area are fascinated with the question of why females mate with more than one male. The question is beyond the scope of this review, but one area of overlap between multiple mating by females and male parental care relates to a theme that we have already raised repeatedly: the relationship among kind of male care provided, monomorphism, and the basis for female mate choice. Regardless of the importance of variation among males—in their quality, their genes, or their resources—it seems likely to us that the fundamental reason that females mate multiply is to ensure the acquisition of sufficient sperm to fertilize their ova. This hypothesis has been considered and rejected several times (4, 84, but see 80b) but has not been adequately tested. If incubating males in sexually monomorphic species are less variable in their sperm production than males in more sexually dimorphic species whose males do not incubate, then females in monomorphic species may have less to gain from multiple mating (i.e. less need of fertilization insurance) and possibly more to lose (potential loss of male parental care). Have we any reason to expect less variation in sperm production in monomorphic males that incubate? To our knowledge, no one has looked directly, but we have speculated that one cost of extreme sexual dimorphism may be impaired fertility in some individuals owing to possible inhibitory effects of high testosterone levels on sperm production (ED Ketterson, V Nolan, Jr, in preparation; T Kast, ED Ketterson, V Nolan, Jr, unpublished data). Here again, more data are needed.

Relative Costs and Benefits of Mating Effort and Parental Effort: Phenotypic Engineering

One of our research objectives is to determine whether male parental behavior is currently maintained by natural selection, and for that we need a quantitative accounting of both detrimental and beneficial effects of male care on male fitness. Traditionally, these questions have been explored by comparisons of individuals in various natural categories (e.g. older vs younger males when these differ in amount of care provided), but an obvious drawback with this approach is that it is not experimental. One simple experimental approach is to alter the environment and ask how paternal behavior responds to food enhancement or to a change in brood size. But, as Lack said (43, p. 8), "It is easy to change the number of eggs or young in a nest, but no one has yet found how to make a monogamous species polygynous or a solitary species colonial...." Thus it was a breakthrough when Wingfield (91) reported that treatment with testosterone could induce polygyny in species that were ordinarily monogamous. We followed this lead, in an approach we call "phenotypic engineering," by using testosterone implants to alter the physiology and behavior of dark-eyed juncos (summarized in 38).

We found that experimental males (T-males) and controls (C-males) differed in a number of traits that would be expected to influence reproductive fitness. T-males sang more (41), were more attractive to females (D Enstrom, ED Ketterson, V Nolan, Jr, unpublished data), and ranged over wider areas (8). Concomitantly they fed young less frequently (41) and were slower to detect, hence to defend against, a model predator placed near the nest (M Cawthorn, D Morris, V Nolan Jr, ED Ketterson, and CR Chandler, unpublished data). The question then was the effects of these differences on the relative fitness of T-males.

We are now attempting to quantify all components of reproductive success for both phenotypes to determine whether individuals that deviate from the norm are at a disadvantage. Starting with the traditional measure of reproductive success, *apparent success,* which is the number of young produced by the female(s) mated (pair-bonded) to the focal male, we have found virtually no difference between T- and C-males (38, 41). *Realized success* consists of apparent success with social mates minus paternity losses to extra-pair fertilizations by other males (cf 24 for a different use of "realized success"), and this requires knowledge of the genetic relationship between male and offspring. On this measure, T-males perform slightly but not significantly less well than controls, and we are now asking whether these losses might be because T-males range more widely and leave their mates unprotected (PG Parker, ED Ketterson, V Nolan, Jr, SA Raouff, C Ziegenfus, T. Peare, and CR Chandler, unpublished data). The greater challenge comes in quantifying *covert success,* which consists of all the focal male's fertilizations of females other than his social mate. On this measure, T-males perform somewhat better than controls to date, perhaps again because of greater time spent off territory (8) or greater attractiveness (D Enstrom, ED Ketterson, Val Nolan, Jr, C. Ziegenfus, unpublished data), but here, too, the differences are not significant (PG Parker, ED Ketterson, V Nolan, Jr, SA Raouff, C Ziegenfus, T Peare, and CR Chandler, unpublished data). *Actual reproductive success* is the sum of realized and covert success, and on this measure, T- males and C-males are virtually identical (PG Parker, ED Ketterson, V Nolan Jr, SA Raouff, C Ziegenfus, T. Peare, and CR Chandler, unpublished data).

Returning to Maynard Smith (46) and his formula predicting patterns of parental behavior, we ask, "Does the model fit the junco?" Male parental care increases the production of independent young (Figure 2, 100). But evidently males could decrease their contribution to feeding nestlings by 50%, as they do when they are treated with testosterone, and suffer no fitness loss in apparent reproductive success (41). However, the physiological changes required to reduce the level of care might have consequences for other components of fitness such as survival or increased losses to EPP (39, 52; PG Parker, ED Ketterson, V Nolan, Jr, SA Raouff, C Ziegenfus, T Peare, and CR Chandler, unpublished data). This study is still in progress.

Some workers might prefer to test Maynard Smith's model by altering the environment, not the animal. One way to vary the availability of fertile females is through simulated nest predation, i.e. destruction of nests on neighboring territories. In many open-nesting passerines, nest predation can be very high, at least in some years (e.g. > 80% in juncos), and until late in the breeding season the effect of predation is usually to cause the female to become fertile again as she begins a new attempt to reproduce. She is then a candidate for EPFs. It would be interesting to know whether high predator densities influence the time males allocate to guarding their nests as opposed to seeking extra mating opportunities. (Note that high nest predation also decreases the likelihood that nests will succeed in producing young, and this lowers the probable value of any EPFs.) We expect more studies that address the responses of males to natural or staged variation in the environmental contingencies that may raise or lower the value of male care. Male red-winged blackbirds, for example, change their contribution to parental care in response to changes in brood size, nestling age, and food abundance (86, 87, 106, see also 73; K Yasukawa, WE Richmond, unpublished data).

If we find that males are effective at adjusting their allocation of effort to parenting and mating, depending upon environmental contingencies, a next step will be to determine the cues and physiological adjustments required to make the necessary behavioral shifts. To answer questions like these, more needs to be known of the mechanisms underlying male parental behavior.

MALE PARENTAL CARE: THE MECHANISTIC APPROACH

Phylogenies may reveal the order of evolution of life-history traits, and studies of function aim to reveal why one or another of a set of alternative behaviors is maintained under natural selection. But neither of these approaches can reveal what happens in organismal terms when paternal care evolves from non-care, or vice-versa. Full understanding of the evolution of a behavior pattern requires that we know what physiological changes in the animal permit or accompany the behavioral change. Such changes might include altered response to the stimuli that elicit care, alterations in the processing and integrating of potentially conflicting sensory input, and altered output in the cascade of neural and neurosecretory events that influence target tissues and affect internal state and behavior. Obviously, evolution of one pattern of male parental care from another necessarily involves modification in this whole system.

As an example, modifications that might be required in the evolution of male incubation (from no-incubation) could include a shift in response to eggs as stimuli to sit on, coordination of male and female bouts of sitting, alteration of schedule in male appetite so as to facilitate feeding in bouts rather than ad lib, and, in some cases, development of a brood patch for effective transfer of heat to eggs. At least equally important might be suppression of responsiveness to competing stimuli such as other males or fertile females.

Comparisons with Male Parental Care in Mammals

The systems of vertebrate parental behavior that are best understood at the level of mechanisms are mammalian (for review see 64). However, because male parental care is relatively rare in mammals, the focus has been on maternal behavior (e.g. 42, 59). More recently, researchers have addressed mechanisms of male parental care (6, 30–32, 98a, 108).

Despite the obvious differences in avian and mammalian parental behavior, e.g. birds do not lactate and mammals do not incubate, one common element is the formation of parental bonds with young. Moreover, there is reason to think that the mechanisms underlying these attachments in a variety of vertebrate taxa may be related to secretions of the neurohypophysis (59, 69, 98a). Pursuit of other parallels between birds and mammals might also prove profitable. Gubernick et al (32) studied the medial preoptic area (MPOA) of the hypothalamus, a region known to be important to the control of maternal behavior in mammals and one that is usually sexually dimorphic. In the California mouse, males provide parental care and the MPOA is sexually dimorphic only until the first litter is reared, after which the sexes become similar in this nucleus. Application of comparable methods to birds might reveal the neural basis for interspecific sex differences in the expression of parental care (69).

Circulating Hormones and Male Parental Care in Birds

Study of mechanisms of parental care in birds has focused on circulating hormones, by addressing the effects that hormones have on behavior as well as the effects that behavior and the environment have on hormones. Major contributions have been made by studying the inhibitory effect of testosterone on male parental behavior (summarized by Wingfield et al in 94) and investigating the role of prolactin in avian incubation (summarized by Goldsmith— 26, 27). The approach has been comparative, as workers have elected to study free-living species believed to be of special interest because of some aspect of their life history. For example, in cooperative breeders, juvenile birds often delay maturation and participate in care of siblings or unrelated young, and researchers have asked whether their hormone profiles suggest delayed reproductive maturation (47, 66, 79, 95). Similarly, Dufty (15) compared the hormone profile of a brood parasite, which provides no care, to those of caregiving species. For role-reversed species, in which females court males and males incubate (22, 56), the question has been this: Are the hormone profiles of the sexes the reverse of those of the much more numerous species in which only females incubate or in which incubation is biparental (26, 27, 40)?

Testosterone and Male Parental Care

As a rule, the onset of breeding in male birds is associated with a rise in gonadotropins followed by a rise in testosterone (1). Later, when males begin to behave parentally, testosterone levels decline (1). Wingfield and his associates have amassed information on seasonal profiles of testosterone from a variety of birds (93, 94). They have also documented the endocrine response of males to females and to male intruders (actual or simulated) onto their territories (92, 93), and following the lead of Silverin (71), they and we have manipulated testosterone levels in the field and measured the consequences for male parental behavior, mating behavior, and fitness (35, 91, summaries in 38, 78).

Interesting generalizations have emerged from interspecific comparisons of testosterone profiles (92, 94). One is that most species show a peak in testosterone (T) early in the breeding season, but the duration of this peak varies. For some species, it is brief; for others, it lasts nearly the entire breeding season. Another is that species differ in their hormonal responsiveness to intruders or other challenges; some show little response, while others exhibit rapid elevation of plasma T following a territorial intrusion. This variability can be related to mating system and parental care. First, the duration of the seasonal peak is longer in overtly polygynous species than in socially monogamous ones (93). Second, experimental treatment with T suppresses feeding of offspring in species in which males feed young (35, 38, 41, 71) and disrupts incubation in species in which males incubate (55). It can also induce polygyny in normally monogamous sparrows (91). Third, the species that are most responsive to intruders are those that provide the most parental care (92).

To examine the links among hormones, mating system, and parental care in a sample of 20 species, Wingfield et al (94) correlated the height of the breeding season peak of testosterone with an index of aggressiveness and parental care. High values of the ratio represented aggressive species that provide little parental care, low values the opposite. When the height of the seasonal peak was plotted against aggression/parental care (Figure 3), species in which males provide care showed the greatest seasonal variation in hormone levels, suggesting that testosterone secretion is suppressed except for brief periods. Males not providing care appear to live in a state of readiness to behave aggressively, and that state evidently interferes with the tendency to behave parentally. The states are loosely associated with traditional mating systems (Figure 3). Further progress will require better measures of aggressiveness, clearer distinctions regarding the kind of male care, data on variance

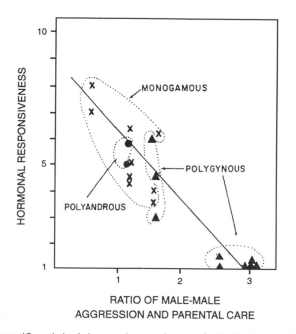

Figure 3 Interspecific variation in hormonal responsiveness of males (i.e. height of breeding season peak of testosterone) in relation to male-male aggression and parental care. High, sharp seasonal peaks in testosterone are associated with low levels of aggression and high parental care and also with social monogamy. In aggressive species with little parental care, the seasonal testosterone peak is less pronounced and the mating system tends toward overt polygyny, from Wingfield et al (94).

in actual male reproductive success, and data from species with different phyletic histories.

Less attention has been paid to the fact that socially monogamous, care-giving males may be less rigid in their allocation to parental and mating effort and accordingly more responsive hormonally to changing environmental cues. If so, the mixed reproductive strategy first delineated by Trivers (75) may have a mechanistic counterpart. We have begun an investigation of the coordinated responses of male juncos to changed opportunities for mating or parenting. We measure their testosterone levels, song rate, sperm density, and feeding of nestlings when they are confronted with enhanced broods or fertile females. We anticipate that other workers ask similar kinds of questions in other species.

The key general point, however, is that testosterone—a compound that we ordinarily associate with male courtship and sexual behavior, aggression, and sexual dimorphism—may lie at the core of the trade-off between mating effort and parental effort. A useful, if simple, analogy may be to think of testosterone as a compound that can be turned up or down like a dial regulating the volume of sound. When testosterone is blaring, male and female roles diverge; when it is muted, the sexes look and behave more alike.

Prolactin and Male Parental Care

For prolactin, the generalization is that onset of incubation is associated with a decline in gonadal steroids and a rise in circulating levels of prolactin (1). In the few species in which only males incubate, prolactin is higher in males than females (22, 56). When both sexes incubate, both have elevated prolactin, and when only the female incubates, usually only she shows elevated prolactin (review in 27). There are exceptions to these generalizations. In cooperative breeders, helping birds also have elevated prolactin, even though they do not incubate (79; C Vleck, unpublished data on Mexican jays). Similarly, although they do not incubate, male European starlings in North America, and brown-headed cowbirds of both sexes, elevate prolactin during the breeding season (1). These examples serve as reminders that hormones have multiple effects and that evolution of hormonal regulation need not proceed simply by altering secretion (27). Tissue responsiveness to hormones can also show evolutionary change (22). Indeed, Ball et al (2) report that the brain tissue of cowbirds binds less radioactive prolactin than the same tissue in parental species, an indicator of lowered brain tissue response to prolactin.

A longstanding debate has centered on whether prolactin causes incubation or simply supports it once it has been initiated (26, 27, 64). Future developments regarding the role of prolactin are likely to come from experimental field studies involving the administration of vasoactive intestinal peptide (VIP), a releasing hormone for prolactin (C Vleck, personal communication). In particular, we might learn whether the physiology of incubation, if it can be experimentally induced, conflicts with the physiology of mating.

MALE PARENTAL CARE: THE DEVELOPMENTAL APPROACH

We expect important future discoveries in the area of development. Much individual variation in behavior or morphology, once assumed to be genetic, is now thought to reflect age or condition. In lizards, Hews et al (36) have shown that the expression of alternative reproductive strategies in males can be induced by applying testosterone early in life. In mammals (house mice), vom Saal (e.g. 80a) has shown that a female's aggressiveness and sexual attractiveness as an adult are influenced naturally by whether her intra-uterine neighbors are male or female. Until recently, similar effects were not widely known in birds. Schwabl (67) reported that adult female canaries and zebra finches incorporate testosterone into their eggs and that the amount incorporated varies with order of laying the eggs. Last-laid eggs have higher levels of testosterone, regardless of the sex of the offspring, and young from last-laid eggs achieved higher social status in flocks of juveniles (67; see 97 for a summary of possible research questions raised by these findings). Applying the concept of environmentally induced variability in phenotypic expression to male parental care, we might predict that males hatched in years when food is abundant would develop to focus on achieving fitness via extra-pair fertilizations, whereas those hatched when food is scarce and the value of male parental care greater might behave more parentally as adults. This assumes, of course, some environmental carry-over from year to year and is meant merely to illustrate the kinds of effects that may be discovered.

SUMMARY

In sum, males of most avian taxa that help care for offspring tend to be what we now call socially monogamous and, in contrast to overtly polygynous males, to be more likely to resemble females in external appearance and morphology. That these traits often present themselves as a complex presumably reflects their common evolutionary history. It may also indicate common causation, for example, by hormonal mechanisms, some of which act early in development. Improvements in the comparative method, including the phylogenetic approach, may reveal not only the conditions that are ancestral and the lineages that have proved to be more plastic but also the ecological correlates of male parental care. Using a functional approach and molecular techniques for determining relatedness, the effort to discover which components of fitness are most affected by male care can be expected to continue. Future studies focusing on proximate causation and development will illuminate the links between genes and behavior through which evolution proceeds.

What may have emerged from our attempt at a unifying consideration of history, function, and causation, beyond confirmation of the traditional view that male birds that do care-giving also tend to be sexually monomorphic and socially monogamous, is the view that incubation is the form of care most closely linked to monomorphism and that extra-pair paternity may be less common in groups with male incubation. If this proves true, and if variance in male reproductive success in groups in which males incubate proves to be lower than in groups without male care or groups with other forms of care, we will then need to determine why. One cause may be the greater dependence of females on male care when it is in the form of incubation; another might be lower variability among males that incubate in the amount of sperm they offer to ensure fertilization. These effects could increase the costs to females of multiple mating while decreasing the benefits. As a result, when selecting a mate, females in species in which males incubate may be more attentive to cues that predict paternal behavior and less attuned to exaggerated traits. A predicted outcome, beyond monomorphism, is less highly developed species-isolating mechanisms and more local adaptation (60). At the level of mechanism, ontogenetic, seasonal, and short-term variation in testosterone may lie at the core of this complex of traits, including male allocation to mating and various forms of parental care. We think of this as the integrative approach to male parental care in birds and hope that it will prove fruitful.

ACKNOWLEDGMENTS

Our sincere thanks to D Enstrom, D Monk, L Whittingham, K Yasukawa, and an anonymous reviewer for critical readings and helpful suggestions. We also thank the National Science Foundation and Indiana University for financial assistance and express our heartfelt appreciation to our students and assistants and to our colleagues at Indiana University and Mountain Lake Biological Station.

Literature Cited

1. Ball GF. 1992. Endocrine mechanisms and the evolution of avian parental care. *Proc. XXth Int. Ornithol. Congr.*, pp. 984–91. Christchurch, New Zealand

2. Ball GF, Dufty AM, Goldsmith AR, Buntin JP. 1988. Autoradiographic localization of brain prolactin receptors in parental and non-parental songbird species. *Soc. Neurosci. Abstr.* 14:88

3. Bart J, Tornes A. 1989. Importance of monogamous male birds in determining reproductive success. *Behav. Ecol. Sociobiol.* 24:109–16

4. Birkhead TR, Moller AP. 1992. *Sperm Competition in Birds*. London: Academic. 282 pp.

5. Brooks DR, McLennen DA. 1991. *Phylogeny, Ecology and Behavior*. Chicago, IL: Univ. Chicago Press. 434 pp.

6. Brown E. 1993. Hormonal and experiential factors influencing parental behaviour in male rodents: an integrative approach. *Behav. Processes* 30:1–28

7. Carey MD, Nolan V Jr. 1979. Population dynamics of indigo buntings and the evolution of avian polygyny. *Evolution* 33:1180–92

8. Chandler CR, Ketterson ED, Nolan V Jr, Ziegenfus C. 1994. Effects of testosterone on spatial activity in free-ranging male dark-eyed juncos, *Junco hyemalis*. *Anim. Behav.* In press

9. Clutton-Brock TH. 1991. *The Evolution of Parental Care*. Princeton: Princeton Univ. Press. 331 pp.

10. Clutton-Brock TH, ed. 1988. *Reproductive Success: Studies of Individual Variation in Contrasting Breeding Systems*. Chicago, IL: Univ. Chicago Press

11. Clutton-Brock TH, Harvey PH. 1979. Comparison and adaptation. *Proc. R. Soc. Lond. B* 205:547–65

12. Dale J. 1994. Comment on Moller and Birkhead (1993): Does certainty of paternity covary with paternal care? *Anim. Behav.* In press

13. Davies NB. 1992. *Dunnock Behaviour and Social Evolution*. Oxford: Oxford Univ. Press. 272 pp.

14. Decker MD, Parker PG, Minchela DJ, Rabenold KN. 1993. Monogamy in black vultures: genetic evidence from DNA fingerprinting. *Behav. Ecol.* 4:29–35

15. Dufty, AM Jr., Goldsmith AR, Wingfield JC. 1987. Prolactin secretion in a brood parasite, the brown-headed cow-

bird, *Molothrus ater. J. Zool. Lond.* 212: 669–75

16. Dunn PO, Hannon SJ. 1989. Evidence of obligate male parental care in black-billed magpies. *Auk* 106:635–44

17. Dunn PO, Hannon SJ. 1992. Effects of food abundance and male parental care on reproductive success and monogamy in tree swallows. *Auk* 109:488–99

18. Dykstra CR, Karasov WH. 1993. Nesting energetics of house wrens (*Troglodytes aedon*) in relation to maximal rates of energy flow. *Auk* 110:481–91

19. Emlen ST, Oring LW. 1977. Ecology, sexual selection, and the evolution of mating systems. *Science* 197:215–23

20. Erckmann WJ. 1983. The evolution of polyandry in shorebirds: an evaluation of hypotheses. In *Social Behavior of Female Vertebrates,* ed. SK Wasser, pp. 113–68. New York: Academic

21. Felsenstein J. 1985. Phylogenies and the comparative method. *Am. Nat.* 125:1–15

22. Fivizzani AJ, Oring LW, El Halwani ME, Schlinger BA. 1990. Hormonal basis of male parental care and female intersexual competition in sex-role reversed birds. In *Endocrinology of Birds: Molecular to Behavioural,* ed. M Wada et al, pp. 273–86. Tokyo: Japan Sci. Press/Berlin: Springer-Verlag

23. Francis, CM, Anthony, EL, Brunton JA, Kunz, TH. 1994. Lactation in male fruit bats. *Nature* 367:691–92

24. Gibbs HL, Weatherhead PJ, Boag PT, White BN, Tabak LM, Hoysak DJ. 1990. Realized reproductive success of polygynous red-winged blackbirds revealed by DNA markers. *Science* 250: 1394–97

25. Gittleman JL. 1981. The phylogeny of parental care in fishes. *Anim. Behav.* 29:936–41

26. Goldsmith AR. 1983. Prolactin in avian reproductive cycles. In *Hormones and Behaviour in Higher Vertebrates,* ed. J. Balthazart, E Prove, R Gilles, pp. 375–87. Berlin: Springer-Verlag

27. Goldsmith AR. 1992. Prolactin and avian reproductive strategies. *Proc. XXth Int. Ornithol. Congr.,* pp. 2063–71. Christchurch, New Zealand

28. Gowaty PA. 1983. Male parental care and apparent monogamy among eastern bluebirds (*Sialia sialis*). *Am. Nat.* 112: 144–57

29. Gowaty PA, Karlin AA. 1984. Multiple maternity and paternity in single broods of apparently monogamous eastern bluebirds (*Sialia sialis*). *Behav. Ecol. Sociobiol.* 15:91–95

30. Gubernick DJ, Nelson RJ. 1989. Prolactin and paternal behavior in the bipa-

rental california mouse, *Peromyscus californicus. Horm. & Behav.* 23:203–10

31. Gubernick DJ, Schneider KA, Jeanotte LA. 1994. Individual differences in the mechanisms underlying the onset and maintenance of paternal behavior and the inhibition of infanticide in the monogamous biparental California mouse, *Peromyscus californicus. Behav. Ecol. Sociobiol.* 34:225–31

32. Gubernick DJ, Sengelaub DR, Kurz EM. 1993. A neuroanatomical correlate of paternal and maternal behavior in the biparental california mouse (*Peromyscus californicus*). *Behav. Neurosci.* 107: 194–201

33. Harvey PH, Pagel MD. 1991. *The Comparative Method in Evolutionary Biology.* Oxford: Oxford Univ. Press. 239 pp.

34. Heg D, Ens BJ, Burke T, Jenkins L, Kruijt JP. 1993. Why does the typically monogamous oystercatcher (*Haematopus ostralegus*) engage in extra-pair copulations? *Behaviour* 126:257–89

35. Hegner RE, Wingfield JC. 1987. Effects of experimental manipulation of testosterone levels on parental investment and breeding success in male house sparrows. *Auk* 104:462–69

36. Hews DK, Knapp R, Moore, MC. 1994. Early exposure to androgens affects adult expression of alternative male types in tree lizards. *Horm. & Behav.* 28:96–115

37. Hoi-Leitner M, Nechtelberger H, Dittami J. 1993. The relationship between individual differences in male song frequency and parental care in blackcaps. *Behaviour* 126:1–12

38. Ketterson ED, Nolan V Jr. 1992. Hormones and life histories: an integrative approach. *Am. Nat.* 140: S33–62

39. Ketterson ED, Nolan V Jr., Wolf L, Dufty AM Jr., Ball GF, Johnsen TS. 1991. Testosterone and avian life histories: effect of experimentally elevated testosterone on corticosterone, body mass, and annual survivorship of male dark-eyed juncos (*Junco hyemalis*). *Horm. & Behav.* 25:489–503

40. Ketterson ED, Nolan V Jr., Wolf L, Goldsmith A. 1990. Effect of sex, stage of reproduction, season, and mate removal on prolactin in dark-eyed juncos. *Condor* 92:922–30

41. Ketterson ED, Nolan V Jr., Wolf L, Ziegenfus C. 1992. Testosterone and avian life histories: effects of experimentally elevated testosterone on behavior and correlates of fitness in the dark-eyed junco (*Junco hyemalis*). *Am. Nat.* 140:980–99

42. Krasnagor NA, Bridges RS, eds. 1990. *Mammalian Parenting: Biochemical, Neurobiological, and Behavioral Determinants.* New York: Oxford Univ. Press

43. Lack D. 1968. *Ecological Adaptations for Breeding in Birds.* London: Methuen. 409 pp.

44. Ligon JD. 1994. The role of phylogenetic history in the evolution of contemporary avian mating systems and parental care systems. *Curr. Ornithol.* 10:1–46

45. Martin K, Cooke F. 1987. Bi-parental care in willow ptarmigan: a luxury? *Anim. Behav.* 35:369–79

46. Maynard Smith J. 1977. Parental investment: a prospective analysis. *Anim. Behav.* 25:1–9

47. Mays N, Vleck CM, Dawson J. 1991. Plasma luteinizing hormone, steroid hormones, behavioral role, and nest stage in cooperatively breeding Harris' Hawks (*Parabuteo unicinctus*). *Auk* 108: 619–37

48. McKitrick MC. 1993. Phylogenetic analysis of avian parental care. *Auk* 109: 828–46

49. Mock DW, Fujioka M. 1990. Monogamy and long-term pair bonding in vertebrates. *Trends Ecol. Evol.* 5:39–43

50. Moller AP. 1988. Paternity and paternal care in the swallow, *Hirundo rustica.* *Anim. Behav.* 36:996–1005

51. Moller AP, Birkhead TR. 1993. Certainty of paternity covaries with paternal care in birds. *Behav. Ecol. Sociobiol.* 33:261–68

52. Nolan, V Jr., Ketterson ED, Ziegenfus C, Cullen DP, Chandler CR. 1992. Testosterone and avian life histories: effects of experimentally elevated testosterone on prebasic molt and survival in male dark-eyed juncos. *Condor* 94:364–70

53. Nylin S. 1991. The phylogenetic approach to ecology [book review]. *Evolution* 45:1731–33.

54. Orians GH. 1969. On the evolution of mating systems in birds and mammals. *Am. Nat.* 103:589–603

55. Oring LW, Fivizzani AJ, El Halawani ME. 1989. Testosterone-induced inhibition of incubation in the spotted sandpiper (*Actitus mecularia [sic]*). *Horm. & Behav.* 23:412–13

56. Oring LW, Fivizzani AJ, El Halawani ME, Goldsmith A. 1986. Seasonal changes in prolactin and luteinizing hormone in the polyandrous spotted sandpiper, *Actitis macularia. Gen. Comp. Endocrin.* 62:394–403

57. Parker GA. 1984. Sperm competition and the evolution of animal mating strategies. In *Sperm Competition and the Evolution of Animal Mating Systems,* ed. RL Smith, pp. 1–60. Orlando, FL: Academic

58. Payne RB. 1984. Sexual selection, lek and arena behavior, and sexual size dimorphism in birds. *Ornithol. Monogr.* 33:1–52

59. Pedersen CA, Caldwell JD, Jirikowski GF, Insel TR, eds. 1992. *Oxytocin in Maternal, Sexual, and Social Behaviors.* New York: Ann. NY Acad. Sci. Vol. 652. 492 pp.

60. Pierotti R, Annett CA. 1993. Hybridization and male parental investment in birds. *Condor* 95:670–79

61. Pitelka FA, Holmes RT, MacLean SF. 1974. Ecology and evolution of social organization in Arctic sandpipers. *Am. Zool.* 14:185–204

62. Richmond A. 1978. *An experimental study of advantages of monogamy in the cardinal.* PhD thesis, Indiana Univ., Bloomington, IN

63. Ridley M. 1978. Paternal care. *Anim. Behav.* 26:904–32

64. Rosenblatt JS. 1992. Hormone-behavior relations in the regulation of parental behavior. In *Behavioral Endocrinology,* ed. JB Becker, MS Breedlove, D Crews, pp. 219–59. Cambridge MA: MIT Press

65. Sasvari L. 1986. Reproductive effort of widowed birds. *J. Anim. Ecol.* 55:553–64

66. Schoech SJ, Mumme RL, Moore MC. 1991. Reproductive endocrinology and mechanisms of breeding inhibition in cooperatively breeding Florida scrub jays (*Aphelocoma c. coerulescens*). *Condor* 93:354–64

67. Schwabl H. 1993. Yolk is a source of maternal testosterone for developing birds. *Proc. Natl. Acad. Sci. USA* 90: 11446–50

68. Selander RK. 1972. Sexual selection and dimorphism in birds. In *Sexual Selection and the Descent of Man,* ed. B Campbell, Ch. 8. London: Heinemann

69. Silver R. 1990. Avian behavioral endocrinology: status and prospects. In *Endocrinology of Birds, Molecular to Behavioral,* ed. M Wada, S Ishii, CG Scanes, pp. 261–72. Berlin:Springer-Verlag

70. Silver R, Andrews H, Ball GF. 1985. Parental care in an ecological perspective: a quantitative analysis of avian subfamilies. *Am. Zool.* 25:823–40

71. Silverin B. 1980. Effects of long acting testosterone treatment on free-living pied flycatchers, *Ficedula hypoleuca,* during the breeding period. *Anim. Behav.* 28:906- 12

72. Slagsvold T, Lifjeld JT. 1989. Ultimate adjustment of clutch size to parental feeding capacity in a passerine bird. *Ecology* 69:1918–22

73. Smith HG, Montgomerie R. 1992. Male incubation in barn swallows: the influence of nest temperature and sexual selection. *Condor* 94:750–59

74. Tinbergen N. 1963. On aims and methods of ethology. *Z. Tierpsychology* 20:410–33

75. Trivers RL. 1972. Parental investment and sexual selection. In *Sexual Selection and the Descent of Man*, ed. B. Campbell, pp. 136–79. London: Heinemann

76. Vehrencamp SL, Bradbury JW. 1984. Mating systems and ecology. In *Behavioural Ecology*, ed. JR Krebs, NB Davies. pp. 251–78. Oxford: Blackwell Sci. 2nd ed.

77. Verner J, Willson MF. 1969. Mating systems, sexual dimorphism and the role of male North American passerine birds in the nesting cycle. *Ornithol. Monogr.* 9:1–76

78. Vleck CM, Dobrott SJ. 1993. Testosterone, antiandrogen, and alloparental behavior in bobwhite quail foster fathers. *Horm. & Behav.* 27:92–107

79. Vleck CM, Mays NA, Dawson JW, Goldsmith A. 1991. Hormonal correlates of parental and helping behavior in the cooperatively breeding Harris' Hawk, *Parabuteo unicinctus. Auk* 108:638–48

80a. Vom Saal FS. 1983. Models of early hormonal effects on intrasex aggression in mice. In *Hormones and Aggressive Behavior*, ed. Svare, BB. pp. 197–222. New York: Plenum

80b. Wagner RH. 1992. The pursuit of extra-pair copulations by monogamous female razorbills: how do females benefit. *Behav. Ecol. Sociobiol.* 29:455–64.

81. Webster MS. 1991. Male parental care and polygyny in birds. *Am. Nat.* 137:274–80

82. Werren JH, Gross MR, Shine R. 1980. Paternity and the evolution of male parental care. *J. Theoret. Biol.* 82:619–31

83. West-Eberhard MJ. 1983. Sexual selection, social competition, and speciation. *Q. Rev. Biol.* 58:155–83

84. Westneat DF, Sherman PW, Morton ML. 1989. The ecology and evolution of extra-pair copulations in birds. In *Current Ornithology*, ed. DM Power, 7:331–69. New York: Plenum

85. Westneat DF, Sherman PW. 1993. Parentage and the evolution of parental behavior. *Behav. Ecol.* 4:66–77

86. Whittingham LA, Robertson RJ. 1993. Nestling hunger and parental care in red-winged blackbirds. *Auk* 110:240–46

87. Whittingham LA, Robertson RJ. 1994. Food availability, parental care and male mating success in red-winged blackbirds (*Agelaius phoeniceus*). *J. Anim. Ecol.* 63:139–50

88. Whittingham LA, Taylor PD, Robertson RJ. 1992. Confidence of paternity and male parental care. *Am. Nat.* 139:1115–25

89. Whittingham LA, Dunn PO, Robertson RJ. 1993. Confidence of paternity and male parental care: an experimental study in tree swallows. *Anim. Behav.* 46:139–39

90. Wiggins DA, Morris RD. 1986. Criteria for female choice of mates: courtship feeding and paternal care in the common tern. *Am. Nat.* 128:126–29

91. Wingfield JC. 1984. Androgens and mating systems: testosterone-induced polygyny in normally monogamous birds. *Auk* 101:665–71

92. Wingfield JC. 1992. Mating sytems and hormone-behavior interactions. *Proc. XXth Int. Ornithol. Congr.*, pp. 2055–62. Christchurch, New Zealand

93. Wingfield JC, Ball GF, Dufty AM Jr., Hegner RE, Ramenofsky M. 1987. Testosterone and aggression in birds. *Am. Sci.* 75:602–08

94. Wingfield JC, Hegner RE, Dufty AM Jr., Ball GF. 1990. The "challenge" hypothesis: theoretical implications for patterns of testosterone secretion, mating systems, and breeding strategies. *Am. Nat.* 136:829–46

95. Wingfield JC, Hegner RE, Lewis DM. 1991. Circulating levels of luteinizing hormone and steroid hormones in relation to social status in the cooperatively breeding white-browed sparrow weaver, *Ploceopasser mahali. J. Zool. Lond.* 225:43–58

96. Winkler DW. 1987. A general model for parental care. *Am. Nat.* 130:526–43

97. Winkler DW. 1993. Commentary: Testosterone in egg yolks: an ornithologist's perspective. *Proc. Natl. Acad. Sci. USA* 90:1139–41

98. Winkler DW, Sheldon FH. 1993. Evolution of nest construction in swallows (Hirundinidae): a molecular phylogenetic perspective. *Proc. Natl. Acad. Sci. USA* 90:5705–07

98a. Winslow JT, Hastings N, Carter SC, Harbaugh CR. 1993. A role for central vasopressin in pair bonding in monogamous prairie voles. *Nature* 365:545–48

99. Wittenberger JF. 1979. The evolution of mating systems in birds and mammals. In *Handbook of Behavioral Neurobiology*, Vol. 3, *Social Behavior and Communication*, ed. P Marler, J. Van-

denbergh, pp. 271–349. New York: Plenum

100. Wolf L, Ketterson ED, Nolan V Jr. 1988. Paternal influence on growth and survival of dark-eyed junco young: do parental males benefit? *Anim. Behav.* 36: 1601–18

101. Wolf L, Ketterson ED, Nolan V Jr. 1990. Behavioural response of female dark-eyed juncos to the experimental removal of their mates: implications for the evolution of male parental care. *Anim. Behav.* 39:125–34

102. Wolf L, Ketterson ED, Nolan V Jr. 1991. Female condition and delayed benefits to males that provide parental care: a removal study. *Auk* 108:371–80

103. Wright J, Cuthill I. 1989. Manipulation of sex differences in parental care. *Behav. Ecol. Sociobiol.* 25:171–81

104. Xia X. 1992. Uncertainty of paternity can select against paternal care. *Am. Nat.* 139:1126–29

105. Yasukawa K, Knight RL, Skagen SK. 1987. Is courtship intensity a signal of male parental care in red-winged blackbirds (*Agelaius phoeniceus*)? *Auk* 104: 628–34

106. Yasukawa K, Leanza F, King CD. 1993. An observational and brood-exchange study of paternal provisioning in the red-winged blackbird, *Agelaius phoeniceus. Behav. Ecol.* 4:78–82

107. Yasukawa K, Whittenberger LK, Nielsen TA. 1992. Anti-predator vigilance in the red-winged blackbird, *Agelaius phoeniceus:* do males act as sentinels? *Anim. Behav.* 43:961–69

108. Yogman MW. 1990. Male parental behavior in humans and nonhuman primates. In *Mammalian Parenting,* ed. NA Krasnegor, RS Bridges, pp 461–481. New York: Oxford Univ. Press

Annu. Rev. Ecol. Syst. 1994. 25:629–60

RELATIONSHIPS AMONG MAXIMUM STOMATAL CONDUCTANCE, ECOSYSTEM SURFACE CONDUCTANCE, CARBON ASSIMILATION RATE, AND PLANT NITROGEN NUTRITION: A Global Ecology Scaling Exercise

E.-Detlef Schulze

Lehrstuhl Pflanzenökologie, Universität Bayreuth, D-95440, Bayreuth, Germany

Francis M. Kelliher

Manaaki Whenua - Landcare Research, PO Box 31-011, Christchurch, New Zealand

Christian Körner

Botanisches Institut, Universität Basel, Schönbeinstrasse 6, CH-4056 Basel, Switzerland

Jon Lloyd

RSBS-Plant Environmental Biology Group, The Australian National University, GPO Box 475, Canberra, ACT 2601, Australia

Ray Leuning

CSIRO, Centre for Environmental Mechanics, GPO Box 820, Canberra, ACT 2601, Australia

KEY WORDS: leaf-specific area and nitrogen concentration, stomatal and canopy and surface conductance, soil evaporation, ecosystem CO_2 assimilation, leaf area index, global distribution of assimilation and evaporation

629

Abstract

This review provides a theoretical framework and global maps for relations between nitrogen-(N)-nutrition and stomatal conductance, g_s, at the leaf scale and fluxes of water vapor and carbon dioxide at the canopy scale. This theory defines the boundaries for observed rates of maximum surface conductance, G_{smax}, and its relation to leaf area index, Λ, within a range of observed maximum stomatal conductances, g_{smax}. Soil evaporation compensates for the reduced contribution of plants to total ecosystem water loss at $\Lambda < 4$. Thus, G_{smax} is fairly independent of changes in Λ for a broad range of vegetation types. The variation of G_{smax} within these boundaries can be explained by effects of plant nutrition on stomatal conductance via effects on assimilation.

Relations are established for the main global vegetation types among (i) maximum stomatal conductance and leaf nitrogen concentrations with a slope of 0.3 mm s^{-1} per mg N g^{-1}, (ii) maximum surface conductance and stomatal conductance with a slope of 3 mm s^{-1} in G per mm s^{-1} in g, and (iii) maximum surface CO$_2$ uptake and surface conductance with a slope of 1 μmol m^{-2} s^{-1} in A per mm s^{-1} in G. Based on the distribution of leaf nitrogen in different vegetation types, predictions are made for maximum surface conductance and assimilation of carbon dioxide at a global scale. The review provides a basis for modeling and predicting feedforward and feedback effects between terrestrial vegetation and global climate.

INTRODUCTION

Predictions of global climate change in response to anthropogenic carbon dioxide and trace gas emissions and pandemic changes in land use require a new scale in ecology, beyond the organismic and regional to the global scale (56). Mooney (42) thus advocates the development of "global ecology." This underlies the urgent need for understanding the interaction of feedforward effects of terrestrial ecosystem processes on global climate and feedback effects of global climate on ecosystem biology (16).

The role of terrestrial vegetation in surface-atmosphere exchange has gained increasing attention because of the potential importance of these ecosystems as a sink for atmospheric carbon (2, 11, 67). In addition, vapor loss from land surfaces, governed by leaf area, by stomatal and aerodynamic conductances of plant canopies, and by the contribution of evaporation from soils, affects numerous terrestrial processes ranging from the biogeochemical cycling of elements (59) to the development of climate itself (61). In this review, we present mathematical theory connecting the distribution of leaf nitrogen (N) to maximum canopy conductance for evaporation (G_{cmax}) via maximum stomatal conductance (g_{smax}). We then develop an expression for maximum ecosystem surface (plant canopy and soil) conductance for evaporation (G_{smax}) in terms of G_{cmax} (25). The theory demonstrates the linear nature of relationships between N, A, and G_{cmax}. The additional underlying problem of comparing species with vastly different leaf structures and longevity is addressed by investigating leaf nitrogen concentrations and contents in relation to specific leaf areas.

Given the overwhelming global variation in structure and physiology of plants, our approach in the final analysis is not mechanistic in the sense that all steps of mass and energy transfer between leaf and atmosphere are simulated. Recognizing also the limitations dictated by the availability of field data, we correlate empirical field data on the basis of known physiological processes, and we use functional relationships at the biome level. We begin our analysis by compiling data on the distribution of leaf nitrogen concentration, specific leaf area and leaf area index of major vegetation types of the world. A relation between leaf N and g_{smax} is determined on the basis of available data sets, followed by the scaling of g_{smax} to G_{smax} and of G_{smax} to maximum carbon dioxide (CO_2) assimilation by vegetated ecosystems under optimal environmental conditions (A_{smax}). Finally, organ (leaf) and ecosystem data are extrapolated to the global scale including leaf N concentrations, g_{smax}, G_{smax}, and A_{smax}.

This exercise in global ecology required us to overcome a number of problems related to the available information from field studies. Ecologists and micrometeorologists have generally confined their CO_2 and water vapor flux measurements to the scale of the leaf or the canopy, and these studies rarely considered plant nutrition. Similarly, few plant nutrient studies included flux measurements. Added to this myopia of separate scientific disciplines, we note that "modern" publications are commonly condensed to statistical synopses of levels of significance for averages. It has become acceptable to draw conclusions without presenting the supporting data. Nevertheless, data for our purpose were available by combining information from a variety of sources at the level of and across broad vegetation types. Future global ecology scaling exercises would benefit from multidisciplinary studies of vegetation, in which complete data sets are documented.

THE LINKAGE BETWEEN LEAF AND CANOPY AND ECOSYSTEM-SCALE PROCESSES

Carbon Assimilation Rate and Leaf and Canopy Conductance for Evaporation

The relation between leaf-level photosynthetic CO_2 assimilation rate (A) and stomatal conductance for CO_2 (g_{sc}) is well understood (e. g. 14). Thus, we may write:

$$A = (c_a - c_i) \, g_{sc} \qquad\qquad\qquad 1.$$

where c_a and c_i are the atmospheric and leaf internal CO_2 concentrations, respectively. Stomatal conductance for water vapor within the boundary layer is $g_s = 1.6 \, g_{sc}$ where the constant is the ratio of the molecular diffusivities for water vapor and CO_2. There is a wealth of information indicating that the ratio A/g_s is conservative for a broad range of conditions. Especially if we inspect photosynthetic capacity (30, 57), the ratio c_i/c_a is maintained within narrow bounds by normally functioning leaves (34). Thus, we expect close relation between maximum leaf assimilation rate and maximum stomatal conductance.

Leaf photosynthesis has been successfully linked to leaf nitrogen concentration (N) and intercepted radiation for sun and shade leaves within a canopy (see Appendix 3, 15, 19, 60). Despite the fact that macro- and micro nutrients are important for plant performance (37), leaf nitrogen and photosynthetic capacity are also closely correlated for a wide variety of species (57), although the underlying mechanism is more complicated than for the comparison of sun and shade leaves because of inherent species-specific differences in leaf structure and longevity. We assume that the leaf photosynthetic rate at light saturation A_{max} is a linear function of N, because of its role in the photosynthetic enzyme Rubisco:

$$A_{max} = \alpha_N(N - N_t) \qquad\qquad\qquad 2.$$

where N_t is a threshold below which there is no photosynthesis, α_N is a constant, and N does not reach a level whereby A_{max} becomes saturated. N_t is mainly determined by leaf structure and longevity (50). This problem is discussed in a separate section below.

Water Vapor Transfer Between Terrestrial Ecosystems and Atmosphere

The exchange of CO_2 and water vapor between terrestrial ecosystems and the atmosphere occurs from both plant canopies and soil. It is thus necessary to define an ecosystem surface conductance for evaporation (G_s) that describes the combined transfers because the canopy conductance G_c accounts for only the plant canopy. The total evaporation rate (E) from a vegetated ecosystem is the sum of contributions from the canopy (E_c) and the soil (E_s):

$$E = E_c + E_s \qquad\qquad\qquad 3.$$

The canopy contribution may be described with the Penman-Monteith equation, using the single-layer or "big-leaf" approximation (49):

$$\lambda E_c = \frac{\varepsilon \, R_{ac} + \lambda \, \rho \, D \, G_a}{\varepsilon + 1 + G_a/G_c} \qquad\qquad\qquad 4.$$

Here R_{ac} is the flux density of available energy intercepted by the canopy (net radiation less energy into thermal storage); $D = q_{sat}(T) - q$ is the air saturation deficit at a reference height above the surface; T is temperature (°C) and q specific humidity at the reference height (g m^{-3}); $q_{sat}(T)$ the saturation specific humidity; $\varepsilon = (\lambda/c)\, dq_{sat} / dT$ the rate of change of the latent heat content of saturated air with change in sensible heat content ($\varepsilon = 2.477$ at 10°C); ρ air density (1.240 kg m^{-3} at 10°C); λ latent heat of vaporization of water (2477 kJ kg^{-1}); and c_p the isobaric specific heat of air (1.27 at 10°C). The bulk aerodynamic conductance, G_a, is assumed the same for heat and water vapor transfer, and G_c is the bulk canopy conductance, pertaining to evaporation from the canopy alone. The gas constant for water vapor (0.462 m^3 kPa kg^{-1} K^{-1} at 10°C) must be included in Equation 4 for D expressed in kPa (26, 31).

The usual way of obtaining an ecosystem (vegetation canopy and soil) surface conductance (G_s) from field measurements is to invert the Penman-Monteith equation for the entire surface, not just the canopy (25). This is Equation 4 with E_c replaced by E, R_{ac} by the available energy flux density R_a $= R_{ac} + R_{ag}$ for the entire surface (R_{ag} being the available energy at the ground close to the soil surface), and G_c by G_g. To assess how G_g differs from G_c because of the contribution of evaporation from the ground, E_g, we assume that it occurs at the equilibrium rate (7, 47):

$$\lambda E_g = \varepsilon\, R_{ag} / (\varepsilon + 1) \qquad\qquad 5.$$

from which we define the conductance of the ground surface, G_g, as:

$$G_g = \frac{\varepsilon\, R_{ag}}{\lambda\, \rho\, (\varepsilon + 1)\, D_{soil}} \qquad\qquad 5a.$$

where D_{soil} is the vapor pressure deficit above the soil surface.

This is a plausible estimate for soil evaporation in conditions of adequate soil water supply, when soil evaporation is determined meteorologically rather than by the diffusion of water in the soil (45) or through the litter layer. Consequently it is a reasonable estimate for use in the present maximum conductance analysis. The energy available at the soil surface R_{ag} can be estimated as $R_{ag} = \tau\, R_a$ where $\tau = e(-\kappa_a\Lambda)$ is the fraction of available energy transmitted downward through leaf area index Λ. The coefficient κ_a, for the attenuation of available energy, is on average 0.6 for forests and herbaceous vegetation (22, 38). Λ does not account for light attenuation by branches and stems in forests, which may reach a projected area of 1. This should be considered when calculating Q.

Combining Equations 3 to 5 gives an expression for G_s in terms of G_c and other canopy properties:

$$G_s = G_c \left[\frac{1 + G_a/(\varepsilon G_i) + \tau\, G_a/[(\varepsilon + 1)G_c]}{1 + G_a/(\varepsilon\, G_i) - \tau} \right] \qquad\qquad 6.$$

where

$$G_i = R_a/(\lambda \rho D) \qquad\qquad 7.$$

is a simplified expression for the isothermal conductance for water vapor from the ground surfaces, G_g (40). This shows that $G_s \rightarrow G_c$ as $\Lambda \rightarrow \infty$, because the canopy transmission τ and the soil evaporation both tend to zero. At smaller Λ, G_s exceeds G_c because of the effect of soil evaporation.

Ecophysiology has accumulated very detailed information of how g_s responds to changing light, temperature, and air humidity (48, 55, 57), and these response functions have been successfully incorporated into gas exchange models (68) using multiple-constraint functions (21, 58):

$$g_s = g_{smax}\, f_Q(Q)\, f_D(D)\, f_T(T)\, f_M(M) \qquad\qquad 8.$$

where the linear or nonlinear functions f_x (between 0 and 1) account for the constraints on g_{smax} imposed by light (Q), air saturation deficit (D), temperature (T), and soil moisture (M). However, this can only be achieved if the level is known at which plants operate under optimal conditions, i. e. g_{smax}, which may differ between plants by a factor > 10 (29) and vary with leaf development and soil conditions (30). Therefore we focus on defining g_{smax} and G_{smax} in the following section.

Leaf Area Index and Maximum Conductances at Stomatal, Canopy, and Ecosystem Surface Scales

Following Kelliher et al (24), we can use the theory expressed by Equation 6 to examine the effect of Λ and g_{smax} on maximum values of G_c (G_{cmax}) and G_s (G_{smax}) (Figure 1). At high Λ, modeled G_{cmax} and G_{smax} converge because there is little available energy for soil evaporation beneath a dense plant canopy. However, the magnitude of modeled G_{cmax} and G_{smax} varies directly with the input value of g_{smax}. As expected, G_{cmax} declines markedly with decreasing Λ, and it ultimately reaches the origin. Decreasing radiation absorption by the plant canopy means increasing available energy for evaporation from the wet soil surface, and G_{smax} departs significantly from G_{cmax} at $\Lambda < 4$. For low values of g_{smax}, G_{smax} may even increase at decreasing Λ. The importance of soil evaporation is most appreciated in agriculture (18), and the contribution of forest floor soil evaporation was significant in *Nothofagus* and *Larix* forests (26 and unpublished data).

The measured G_{smax} data of a wide range of vegetation types are confined in Figure 1 by two (solid) lines, defined by modeled G_{smax} calculated using the upper and lower range of measured g_{smax} (29). The only exceptional data are the maximum $G_{smax} = 50$ mm s^{-1} for wheat (1, 44), and the minimum $G_{smax} = 5$ mm s^{-1} for tundra vegetation (13, 17). It is possible that the wheat was influenced by advection and the tundra by a dry surface layer of lichen and

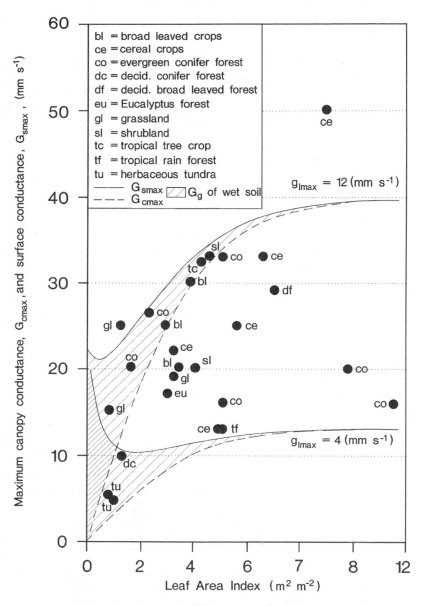

Figure 1 The relationship between measured (symbols) and modeled (lines) maximum surface conductance (G_{smax}) of herbaceous, woody, and agricultural crop vegetation and one-sided canopy leaf area index (Λ) after Kelliher et al (25). Relationships between modelled G_{smax} and Λ are given by solid lines for two values of maximum stomatal conductance ($g_{cmax} = 4$ and 12 mm s^{-1}). Relationships between modeled maximum canopy conductance G_{cmax} and Λ are given by dashed lines for the same values of g_{smax}. The shaded area between G_{smax} and G_{cmax} indicates the contribution of soil evaporation, which becomes significant below $\Lambda = 4$ (after Kelliher et al, 25).

moss (68). Essentially then, the G_{smax} model defines the boundaries within which, under optimal conditions, natural transfer of water vapor will occur from terrestrial vegetated ecosystems to the atmosphere. We expect that G_{smax} will converge with G_{cmax} if the soil surface is dry or insulated by dry litter or plant debris. In the following we explore further causes for variation within these limits.

CONDUCTANCES FOR EVAPORATION, CARBON ASSIMILATION RATE, AND PLANT NUTRITION

To link photosynthetic capacity and stomatal conductance to plant nutrition, we must leave the realm of environmental physics and enter the complex domain of plant metabolism. The understanding of plant metabolic processes is largely phenomenological (e.g. 54, 65).

Unfortunately, there are few data sets combining plant nutrition, maximum leaf and ecosystem evaporation, conductances, and carbon assimilation rate measurements for the same site, an exception being the Maruia *Nothofagus* forest site in New Zealand (20, 26, 31). Because of this limitation, we confine our investigation to the level of and among vegetation types defined in Table 1. Our analysis is based on the hypothesis that the maximum evaporation conductance and carbon assimilation rates are determined by plant nutrition (15). Based on plant physiology, the primary relationship should be between assimilation rate and plant nutrition because of the direct link between nitrogen nutrition and CO_2 fixation. A second logical step would be to develop a relation between assimilation rate and evaporation conductance. Thus, we use leaf nitrogen data to define the nutrient status of a particular vegetation type. However, because of the existing data we decided to use leaf nitrogen concentration to initially estimate g_{smax}, which in turn becomes a predictor of G_{smax} and A_{smax}.

Leaf N concentration (mg (g dry weight)$^{-1}$) increases with specific leaf area (SLA, one-sided leaf area per dry weight, $m^2 kg^{-1}$) in a comparison of about 140 woody and herbaceous species representing all global vegetation types (Figure 2, Appendix 2), and including an additional 21 species of different leaf structures as studied by Field & Mooney (15), and data by Körner (28). High N concentrations are found in plants with thin leaves, and generally such plants also have the highest growth rates (32). N content per unit one-sided leaf area appears to be fairly constant for a broad range of leaf types (solid line in Figure 2: 1.57 g N m^{-2}) including agricultural crops, natural grasses, deciduous trees, and tropical rain forest trees. The data range is between 1 and 4 g N m^{-2}. In conifers N content is higher than predicted by the regression line (3.02 mg N m^{-2}). This may in part reflect the problem of relating needle metabolic processes to one-sided rather than total leaf area. Also broad leaves

Table 1 Average values of nitrogen (N) concentration, specific leaf area (SLA), maximum stomatal conductance (g_{max}), maximum surface conductance (G_{max}), canopy assimilation (A_{max}), and leaf area index of different vegetation types, arranged in alphabetical order. (Source: N-concentration and SLA; see Appendix 1. g_{max} (57), G_{max} (48), A_{cmax}: Table 2, LAI (97). n.d.: no data)

Vegetation type	Nitrogen concentration (mg g^{-1})			Specific leaf area (m^2 kg^{-1})			Max. stomatal conductance (mm s^{-1})			Max. surface conductance (mm s^{-1})			Max. canopy assimilation (μmol m^{-2}s^{-1})			Leaf area index (m^{2}m^{-2})		
	av.	s.e	n	av.	s.e	n	av.	s.e	n	av.	s.e	n	av.	s.e	n	av.	s.e	n
Broadleaved crops	38.4	1.8	9	23.6	1.7	9	12.2	1.0	6	32.9	5.4	5	35.0		1	3.4	0.3	3
Cereals	33.6	3.2	7	25.3	1.9	7	11.0	1.0	5	33.8	5.2	4	33.5	4.2	4	6.1	31.2	4
Deciduous conifers	20.7	1.7	5	11.3	1.4	6	3.8	1.1	3	10.0		1	6.0		1	1.5		1
Evergreen conifers	11.0	0.6	14	4.1	0.4	7	5.5	0.5	26	20.6	1.9	11	22.5	3.3	4	7.4	2.4	2
Monsoonal forest	11.4	0.9	13	4.3	0.6	13	3.5	0.2	5	n.d.			n.d.			n.d.		
Sclerophyllous shrubland	11.4	0.6	42	6.9	0.7	7	4.8	0.5	35	22.0	2.0	2	23		1	4.2	0.3	2
Temperate deciduous trees	19.6	2.7	5	11.5	2.4	5	4.6	0.4	22	20.7	3.7	3	21.5	1.5	2	5.6	0.5	12
Temperate deciduous fruit trees	23.8	2.4	6	10.1	2.4	6	n.d.			n.d.			n.d.			n.d.		
Temperate evergreen broadleaf	13.4	1.0	5	5.7	0.8	5	5.1	1.3	6	14.0		1	16		1	6.4	1.1	5
Temperate grassland	25.5	1.6	28	16.9	1.3	28	8.2	1.8	5	20.7	3.5	5	21.5	3.5	2	1.8	0.5	9
Tropical deciduous forest	27.1	1.0	47	14.1	2.3	10	6.6	0.6	4	n.d.			n.d.			n.d.		
Tropical fruits & plantation	13.6	4.6	21	6.8	0.2	21	n.d.			33.0		1	41		1	n.d.		
Tropical grassland	10.7	4.0	7	n.d.			n.d.			4.		1	12		1	n.d.		
Tropical rainforest	16.5	1.6	210	9.9	1.4	210	5.0	0.8	17	12.5	0.5	3	27.6	6.6	3	6.3	0.5	3
Tundra	20.5	1.1	12	n.d.			5.9	0.7	14	4.7	0.4	2	4.7	0.4	.2	0.7	0.2	3

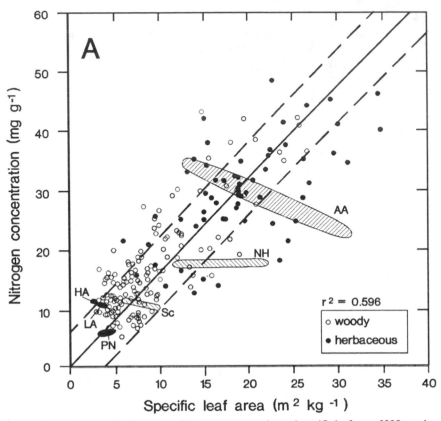

Figure 2 (A) Relationship between leaf nitrogen concentration and specific leaf area of 235 woody and herbaceous species. Data are listed in Appendix 1 with 21 additional species from Field & Mooney (15). Thick solid line indicates a regression through the origin (y = 1. 57 x, s.e. of y:6.32; r^2 = 0.60), dashed lines indicate the standard error of the y-estimate. The shaded areas indicate the range of data measured by Körner (28). (B) (On facingn page) Leaf nitrogen content (g m^{-2}) of leaves with different leaf structure as related to mass per unit leaf area (g m^{-2}). The shaded area shows the range of data as measured by Körner (28) which are also depicted in Figure 2A. Thin solid lines indicate different N-concentrations (mg g^{-1}).

of evergreen temperate trees (e.g. for *Eucalyptus*) and deciduous fruit plantations reach high N content per unit leaf area. However, we note that woody plants with higher nitrogen content do not necessarily have higher g_{smax} compared with otherwise similar woody plants (Table 1, Figure 2). This needs closer inspection at the species level (30).

The close relation between N concentration (mg g^{-1}) and specific leaf area (m^2 kg^{-1}) is surprising at first glance. From physiological experiments using the same species and from field observations inspecting the same lifeform, we know that nitrogen concentration would remain constant or decrease rather than increase with increasing specific leaf area (hatched areas in Figure 2; 28).

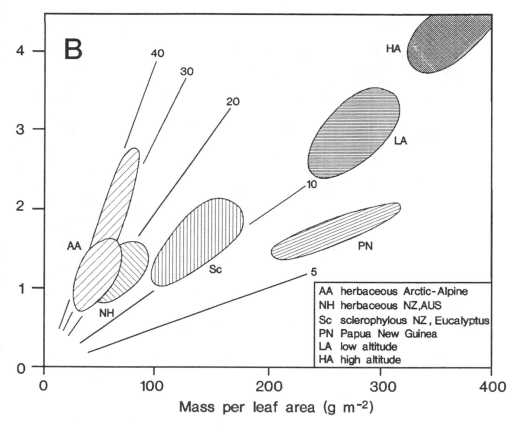

B

AA herbaceous Arctic-Alpine
NH herbaceous NZ,AUS
Sc sclerophylous NZ, Eucalyptus
PN Papua New Guinea
LA low altitude
HA high altitude

Mass per leaf area (g m^{-2})

In these cases N content (g m^{-2}) is linearly related to mass per unit leaf area (formerly termed specific leaf weight) (g m^{-2}), but the response is different for each type of leaf structure. In contrast to Figure 2B, and including a large range of plant species and leaf structures, thick leaves e.g. of tropical trees may contain the same amount of N per unit leaf area as herbaceous plants, but tropical tree leaves have lower N concentration in all leaf compartments. Apparently, plants have developed different strategies of distributing N in leaves at rather similar N content per area and these strategies appear to be related to leaf longevity (9, 50). In a species comparison mainly the C content seems to vary depending on leaf-longevity because structures and defense are required in long-lived leaves, which will lead to a dilution of N if supply is limited. The broad range of data points in this survey and in the study of Field & Mooney (15) indicates that the hatched areas in the study of Körner (28) exemplify responses of specific situations, but there is a continuum of the relation between nitrogen and specific leaf area along the whole scale.

At the present time no general model connects N content, specific leaf area, and leaf longevity. Stitt and Schulze (54, 65) explain the observed correlation between N concentration and specific leaf area (SLA) by interactions between assimilation and growth. High N supply not only results in high N concentrations in Rubisco, it also affects the sink strength of growing regions, including direct interactions between N concentration and extension growth. Under these conditions thin leaves with low C/N ratio are produced (33). In contrast, under conditions of reduced growth, assimilates accumulate. In a species comparison, conditions of reduced growth lead also to thicker leaves, which in turn bind N in non-photosynthetic tissues or require N for plant protection of longer-lived leaves. The C/N ratio increases as N content remains constant and N concentration decreases. Thus, the correlation between relative growth rate and specific leaf area (32) may be interpreted to be the result and not the cause of a regulatory process. With respect to the present analysis of G_s and A_s, the concentration of the CO_2-fixing enzyme is important for photosynthesis (66), and this is why we related conductances to N concentration rather than to N content per area. The emerging pattern of a linear response between N concentration and specific leaf area (SLA) is useful for the present analysis. It is not intended as a complete interpretation of all the underlying processes.

Within the same habitat or on the same soil type, plants have high and low N concentrations in thin and thick leaves, and short and long life spans (52). Thus, biodiversity and evolution of plant species become important if an extrapolation from soil fertility to plant nutrition is anticipated. There is no universal relationship between soil and plant nutrient status. Disturbance, regeneration, and other community processes will determine which plant type is being established, and the plant species will determine the N concentration in leaves at given supply by variation in SLA; this will affect G_{smax} and A_{smax}.

Based on our analysis of leaf N, a direct linear relationship between maximum stomatal conductance (g_{smax}) and leaf nitrogen concentration in different vegetations emerges (Figure 3A). In contrast g_{smax} was not related to N content per unit leaf area. The relation shown in Figure 3A would have consequences for land use changes presently occurring throughout the world, because g_{smax} reaches the greatest values for anthropogenic vegetations, mainly agricultural crops. We are aware of the fact that the regression in Figure 3A is determined mainly by the difference between natural and anthropogenic vegetation. In fact, Körner (29) concluded that variation of g_{smax} in woody plants is remarkably conservative. This observation reflects one underlying problem of our analysis, namely, that herbaceous communities were not classified into smaller groups, while woody communities were subdivided into several categories. We also expect that the non-normal distribution of data points would disappear if individual measurements and not averages were compared. This is demonstrated by Figure 3C (see below) where assimilation rates and surface conductances are compared for the same study sites. In this case there is no separation into clusters. Woody as well

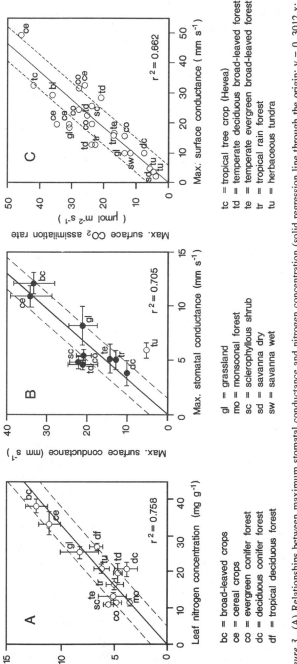

Figure 3 (A) Relationships between maximum stomatal conductance and nitrogen concentration (solid regression line through the origin: y = 0. 3012 x; dashed line: s. e. of y: 1.358, r²:0.758; Table 1). (B) Maximum surface conductance and stomatal conductance (solid regression line through the origin: y = 2.996 x; dashed line: s.e. of y: 4.495, r²: 0.705, Table 1), and (C) maximum surface assimilation and maximum surface conductance (solid regression line through the origin: y = 1.048 x; dashed line: s. e. of y: 6.445, r²: 0.684, Table 2).

as herbaceous communities cover the full ranges of assimilation and conductance just as much as natural and cultivated vegetation. The even distribution of data points in Figure 3C is the main justification for assuming a linear relation, also for the relation between stomatal conductance and leaf nitrogen. In Figure 3A, the deviation of sclerophyllous and conifers from the regression line may additionally result from the problem of using one-sided leaf area and from averaging across a diverse range of species (30). The N concentrations of temperate deciduous forests are probably underestimated (see Appendix 1). The 22 deciduous forest sites of the International Biological Programme (IBP) had a leaf N concentration of 24.3 mg g^{-1}. (The authors became aware of citation 23 after this review was completed.)

As suggested by the theory (Equation 6 and Figure 1), we expect a relationship between leaf-scale g_{smax} and ecosystem-scale G_{smax}. In fact, g_{smax} and G_{smax} of most global vegetation types generally fall within very narrow limits of a significant linear regression (Figure 3B). The slope of 3 suggests that there is a critical "active" leaf area index, below which soil evaporation contributes significantly to total ecosystem evaporation (Figure 1). The only significant exception in the relationship is tundra with a much lower G_{smax} than expected from g_{smax}. This may relate to dry surface layers of mosses and lichens affecting G_{smax} but not g_{smax}. Tropical deciduous forests and sclerophyllous shrubs tend to be higher than the average of other vegetation types with more closed canopies. Also temperate deciduous forests were underestimated by the regression in Figure 2B. However, if we predict G_{smax} from N measured in the IBP study sites (23) and neglect an overestimation of g_{smax} and use only the values calculated by the regression, we derive exactly the measured value of G_{smax} for temperate deciduous forest.

The next step in our analysis relates G_{smax} to A_{smax}, which follows from the conservative nature of leaf-scale A/g_s (43). At the scale of an ecosystsem, micrometeorological measurements of G_{smax} and A_{smax} include a variety of plant species and soils. Nevertheless a highly significant relationship exists between G_{smax} (mm s^{-1}) and A_{smax} (μmol m^{-2} s^{-1}) (Figure 3C; Table 2), which does not separate categories of woody vs non-woody vegetation as in Figure 3A and 3B, and shows overlapping variation for agricultural and non-agricultural ecosystems. This encourages us to propose that the approach used in Figure 3A and B is justified. The highest value of A_{smax} is found in a closed-canopy field of wheat (1), and the lowest in dry savanna (A Miranda, H Miranda, personal communication). The C_4 plant corn reached higher values (factor 1.6) than predicted by the regression line. Also tropical forests have higher rates of CO_2 assimilation in relation to conductance than was predicted by the regression line. This may be interpreted to suggest that more levels of leaf area contribute to the flux of CO_2 than of water vapor in dense canopies.

Table 2 Measurements of maximum surface conductance (G_{smax}, mm s^{-1}) and maximum surface CO_2 flux (A_{smax}, μmol m^{-2}s^{-1}) in different vegetation types. The data were taken from daily corses or from plots of assimilation versus conductance. We are grateful to Dr. Grace (UK), Dr. Miranda (Brazil) and Drs. Fitzgerald (USA) for supplying us unpublished data on rainforest (a) and savanna (b) and temperate deciduous forest (c). The data on Larix and *Pinus radiata* are unpublished data of Kelliher and Hollinger (d)

Vegetation type	Location	G_{smax} (mm s^{-1})	A_{smax} (μmol m^{-2}s^{-1})	Source G_{smax}	Source A_{smax}
Triticum aestivum (wheat)	Boardman (USA)	50	45	Baldocchi (1)	Baldocchi (1)
Glycine max (soybean)	Mead (USA)	30	35	Baldocchi et al (3)	Baldocchi et al (4)
Hordeum vulgare (barley)	Rothamstad (UK)	33	25	Monteith (39)	Biscoe et al (5)
Zea mays (corn)	Ontario (Canada)	20	30	McGinn & King (36)	Desjardins (8)
Zea mays (corn)	Boardman (USA)	20	34	Baldocchi (1)	Baldocchi (1)
Hevea brasiliensis (rubber)	Abidjan (Nigeria)	33	41	Monteney et al (41)	Monteney et al (41)
Temperate deciduous forest	Maruia (NZ)	29	20	Hollinger et al (20)	Hollinger et al (20)
Temperate deciduous forest	Oak Ridge (USA)	13	23	Verma et al (70)	Baldocchi et al (3)
Temperate deciduous forest	Havard (USA)	22	25	Fitzgerald (c)	Wofsy et al (72)
Pinus radiata	Haupapa (NZ)	22	27	Kelliher (d)	Kelliher (c)
Pinus sylvestris	Thetford (UK)	25	27	Stewart (63)	Jarvis et al (22)
Picea sitchensis	Fetteresso (UK)	20	23	Jarvis et al (22)	Jarvis et al (22)
Pseudotsuga menziesii	Dunsmuir (Canada)	16	13	Price & Black (46)	Price & Black (46)
Larix gmelinii	Yakutsk (Russia)	10	7	Kelliher et al (d)	Hollinger et al (d)
Mediterranean sclerophyll	Florence (Italy)	24	23	Valentini et al (69)	Valentini et al (69)
Eucalyptus maculata	NSW (Australia)	14	16	Dunin et al (10)	Leuning et al (35)
Tropical Rainforest	Ducke (Brazil)	13	22	Shuttleworth (62)	Fan (12)
Tropical Rainforest	Jaru (Brazil)	16	17	Grace (a)	Grace (a)
Tallgrass prairie	FIFE (USA)	19	30	Steward & Verma (64)	Kim & Verma (27)
Mixed Prairie	Matador (Canada)	10	13	Ripley & Redman (51)	Ripley & Redman (51)
Savanna wet (grass & trees)	Aguas (Brazil)	10	12	Miranda (b)	Miranda (b)
Savanna dry (trees only)	Aguas (Brazil)	3	4	Miranda (b)	Miranda (b)
Tundra	Alaska (USA)	5	5	Fitzjerrald & Moore (17)	Coyne & Kelly (6)
Tundra	Alaska (USA)	4	4	Fan et al. (13)	Fan et al (13)

GLOBAL DISTRIBUTION OF LEAF NITROGEN CONTENT, MAXIMUM STOMATAL CONDUCTANCE AND ECOSYSTEM SURFACE CONDUCTANCE, AND CARBON DIOXIDE ASSIMILATION RATE

Leaf nitrogen concentrations of different vegetation types (Table 1) and the relations shown in Figure 3 were the basis for estimation of g_{smax}, G_{smax}, and A_{smax} at a global scale. Vegetation distribution came from global maps of land-use types (71). Some corrections for deviation from the general relationships were made (see Appendix 2) for coniferous forest, which consistently had higher stomatal and surface conductances than predicted from needle N concentration. Surface conductance of temperate deciduous forests was also adjusted (see above). For C_4 crops we allowed for a higher assimilation than predicted from surface conductance (factor 1.6), while tropical grasslands do not seem to deviate from the general relation. Tundra vegetation had consistently lower gas exchange than would have been expected from leaf nitrogen. In these cases the average measured data (Table 1) were used instead of the values predicted by the regression model.

The global maps of leaf nitrogen, conductances, and assimilation (Figure 4, see p. 660ff) give a new perspective of plant processes at a global scale. Leaf nitrogen shows a broad band of high concentrations in the mid latitudes of Europe reaching into Asia, while most other regions of the globe, especially in the southern hemisphere, show very similar and low leaf nitrogen values. Regions of high leaf nitrogen concentration also include the agricultural regions of North America, India, and East Asia. Considering present changes in land use, we anticipate major changes in N distribution in future decades.

As a consequence of the global distribution of leaf nitrogen, stomatal (Figure 4B) and surface conductance (Figure 4C) show high values in the northern mid latitudes. The deciduous coniferous forest of Siberia contribute to higher conductances in Asia than in boreal North America. Generally the southern hemisphere remains at low levels of surface conductance. Only the drought deciduous forests of South America obtain G_{smax} values similar to those of the deciduous conifers of Siberia. It is apparent that there is relatively little variation of g_{smax} and G_{smax} in natural woody vegetation. They operate at very low N and thus G_{smax}, in contrast to anthropogenic conditions where a major increase in the range of g_{smax} and G_{smax} occurs.

The differences across vegetation types are further exaggerated when inspecting A_{smax} (Figure 4D). The broad band of high A_{smax} across Europe, East Asia, and North America illustrates the main agricultural and industrial regions.

The differences between the northern and southern hemisphere are even more pronounced if latitudinal averages are calculated (Figure 5A). Neglecting the arctic region, we observe a highly significant increase of average leaf N

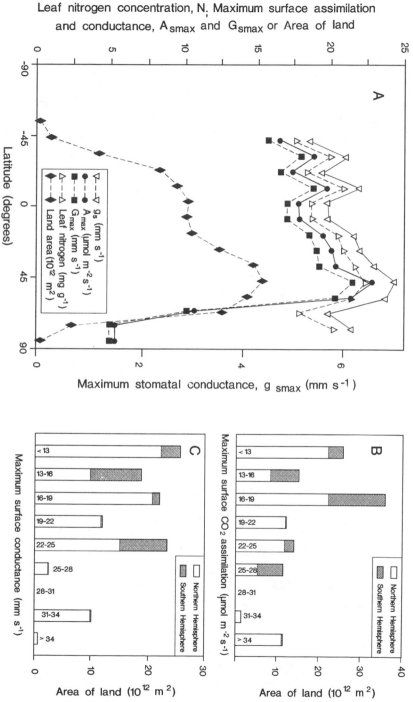

Figure 5 (A) Calculated 10° averages of leaf nitrogen concentrations (N), maximum surface assimilation (A_{smax}), maximum surface conductance (G_{smax}), maximum stomatal conductance (g_{smax}), and area of global land surface. Distribution of land cover by different conductance (B) and assimilation (C) classes for the northern and southern hemisphere.

concentration, g_{smax}, G_{smax}, and A_{smax} from southern to northern latitudes. A maximum is reached between 40° and 60° north. If different classes of surface conductance and assimilation rate are shown for the total land area regions of high A_{smax} and G_{smax} are not found in the southern hemisphere. An exception is 300 km^2 of *Pinus radiata* plantation forest in New Zealand, Chile, and Australia, which has a much higher conductance and assimilation than the native plant cover (Table 2). However, high A_{smax} and G_{smax} occurs over 10% and 20% of the northern continents, respectively. This coincides with the estimated global area of agriculture. We expect that this will feed back on tropospheric CO_2 concentration and contribute to the larger northern than southern hemisphere seasonal variations (56). The distribution of A_{smax} differs from predictions by Woodward & Smith (73), who showed global distributions of leaf photosynthesis based on the highest gas exchange measurements by single species. Their global map predicted highest rates in tropical climates due to C_4 photosynthesis. Since our predictions are based on actual flux measurements covering all major vegetation types of the globe, we think that the map of A_{smax} presented in this review may be closer to actual land-surface fluxes.

The present scaling exercise allows quantification of possible anthropogenic effects on global gas exchange through land use. This will be important for predicting future effects of terrestrial ecosystems on global climate, because more and more forest vegetation is being replaced by herbaceous crops or successional stages.

THE PLANT CARBON AND WATER BALANCE

The present study concentrates on the investigation of maximum rates, and it is beyond the scope of this review to also make predictions of seasonal carbon and water balances. Kelliher et al (25) have reviewed relations between surface conductance and environmental variables for coniferous forests and grasslands. It emerges that coniferous forest are more sensitive to changes in radiation and air saturation deficits than grasslands, but less sensitive to soil drought (54). Conifers appear to close their stomata at lower soil water content than do herbaceous plants. In addition, woody vegetations are generally deeper rooted and thus access a larger soil and water volume than do herbaceous species. Apparently, the main variable that determines seasonal carbon balance is the total length of the growing season. Not only for arid regions, but also for many sub-humid and humid vegetation types, the length of the growing season will depend on the functional rooting depth; *Acacia albida* in Africa or *Allocasuarina dicansiana* in Central Australia are explicit examples. In agricultural crops rooting depth is in many regions confined to ploughing depth, and plant water loss may exceed soil water storage even in humid climates. In these cases growth of crops may depend totally on rain inputs during the growing season. In contrast, deeper rooted trees will be less dependent on rainfalls in the growing season unless rooting depth has been altered by anthropogenic influences such as acid rain (53).

The present study suggests that the carbon and water balance will be strongly dependent on g_{smax} and its relation to nutrition as well as the length of the growing season, which is dependent on rooting depth.

CONCLUSIONS

1. Plant nutrition is a major determinant of maximum values of stomatal conductance, ecosystem surface conductance, and ecosystem CO_2 assimilation rate. The main effect of plant biodiversity on the investigated processes is through changes in specific leaf area and leaf longevity, and associated effects on N concentration, which in turn will feed back on canopy gas exchange. A better general understanding of N in leaves and its relation to photosynthesis, storage, and structure is needed.

2. Maximum stomatal conductance appears to be linearly related to leaf nitrogen concentration (mg g^{-1}) for broad categories of vegetation types with variable leaf longevity. There is no relationship with nitrogen content per unit leaf area (mg m^{-2}). Within vegetation types (woody vs herbaceous vs crops), the relation between maximum stomatal conductance and nutrition is relatively conservative.

3. There is a linear relationship between maximum stomatal and ecosystem surface conductances. The slope of 3 is supported by a model based on micrometeorological theory. Below a leaf area index of about 3, wet soil evaporation contributes significantly to total ecosystem evaporation.

4. A linear relationship exists between maximum ecosystem surface conductance for evaporation and maximum ecosystem carbon dioxide assimilation rate for a large range of vegetation types, including woody and herbaceous as well as natural and cultivated vegetations. This suggests existing knowledge of leaf-level processes can be scaled up to the ecosystem.

5. There are some exceptional vegetation types. In evergreen coniferous forest, maximum stomatal conductance was higher than expected from the leaf nitrogen concentration. This suggests that needle morphology is not well represented on a one-sided basis. It is not entirely clear why tundra has maximum ecosystem surface conductance and carbon dioxide assimilation rate much less than expected from the leaf nitrogen concentration. Maximum ecosystem surface conductance and carbon dioxide assimilation rates are higher than average for the C4 agricultural crop corn.

6. Expressing present knowledge, a map of terrestrial land cover makes it clear that the northern hemisphere has higher maximum evaporation conductances and carbon assimilation rates not only because of its greater land area but also because of higher latitudinal average concentrations of leaf

nitrogen. This is in part a natural phenomenon due to the large proportion of deciduous vegetation in the northern and mainly evergreen vegetation in the southern hemisphere. A second reason is anthropogenic: a large area of nitrogen-fertilized crop land in the northern hemisphere and industrial N depositions. The highest categories of assimilation and conductance (as defined in this study) were not present in the southern hemisphere, based on available land-surface description.

7. There are of course some limitations in this first global ecology scaling exercise of maximum evaporation conductances and carbon assimilation rates. The existing data, especially the paucity of simultaneous plant nutrition and flux measurements, confines us to the use of broad vegetation types as the level of integration which inevitably introduced errors. In addition, the present land-cover database for global land-cover classes used in our analyses does not conform exactly with the botany of vegetation classification. We believe that there is much scope for improvement in the measurement data base and in the land-cover classification of global vegetation.

ACKNOWLEDGMENTS

This study was initiated by discussions of the Scientific Committee of the IGBP core project GCTE (Global Change and Terrestrial Ecosystems). This study was supported by the German Bundesministerium fur Forschung und Technologie (contract 0339476A), which supported one of the authors as guest scientist. The study was also supported by the German Humboldt Foundation and the CSIRO/ANU Biology Center which allowed the senior author to work at Canberra. This study would not have been possible without the library facilities at CSIRO. The senior author is especially grateful for the helpful support of the staff of the Black Mountain library in finding even exotic literature. FM Kelliher thanks the New Zealand Foundation for Research, Science and Technology for its continued support of international atmospheric research. We thank Jim Randerson (Carnagie Institution, Stanford) for help with the final maps.

Appendix 1 Nitrogen concentrations (mg g^{-1}) and specific leaf areas (m^2 kg^{-1}) of different plant species and vegetation types. Average data are presented in cases where a large number of species had been investigated, or if only average data were published. The data are arranged according to the land-cover classes used for mapping, and within each category species are arranged in alphabetical order. We are grateful to Dr. Agnus, CSIRO Canberra, and Dr. Keating, CSIRO Brisbane, who supplied unpublished data on rice (*a*) and on wheat and sugracane (*b*) respectively. Also the data by Schulze on cashew, mango, spruce, *Betula, Eucalyptus,* and *Fagus* (*c*) were not previously published. (*d*) We became aware of the review by Kannah & Ulrich (23) on temperate deciduous forests after this review was completed. The N-concentrations of this vegetation type are underestimated.

Plant species	N		SLA	Source
Broadleaved crops				
Glycine max (USA)	46.0		34.7	(104)
Glycine max (USA)	44.0		26.7	(75)
Gossypium hirsutum (Australia)	29.4		20.0	(126)
Helianthus annuus (Spain)	41.9		15.1	(76)
Phaseolus vulgaris (Peru)	41.3		24.2	(100)
Phaseolus vulgaris (Guatemala)	37.5		24.4	(100)
Phaseolus vulgaris (Mexico)	36.6		22.6	(100)
Phaseolus vulgaris (El Salvador)	35.7		22.5	(100)
Phaseolus vulgaris (Argentina)	33.4		22.1	(100)
Average (s.e.)	38.4	(1.8)		
Cereals				
Oryza sativa (Australia)	27.8		16.6	(a)
Oryza sativa (Japan)	24.2		23.0	(105)
Triticum aestivum (Australia)	48.3		23.0	(b)
Triticum aestivum (Australia)	41.0		30.2	(81)
Triticum aestivum (Australia)	34.3		31.3	(91)
Triticum aestivum (Australia)	31.0		27.0	(81)
Zea mays (Australia)	28.5		26.3	(91)
Average (s.e.)	33.6	(1.9)		
Deciduous conifers				
Larix decidua (Germany)	26.3		9.9	(102)
Larix gmelinii (old trees, Russia)	15.6		6.8	(117)
Larix gmelinii (young trees, Russia)	21.9		10.8	(117)
Larix laricina (Canada)	20.1		15.8	(101)
Larix occidentalis (USA)	20.0		14.7	(90)
Average (s.e.)	20.7	(1.7)		
Evergreen conifers				
Picea abies (Germany)	11.9		4.9	(109)
Picea abies (Germany)	13.7		3.9	(102)
Picea abies (NZ)	9.6		2.8	(117)
Picea abies (Sweden)	8.2			(c)
Picea abies (Denmark)	11.3			(c)
Picea abies (France)	10.4			(c)
Picea glauca (USA)	8.7			(c)
Picea mariana (Canada)	10.0		4.1	(101)
Pinus contorta (USA)	9.5		3.2	(90)

Appendix 1 *(Continued)*

Plant species	N		SLA	Source
Pinus pinea (France)	11.3			(c)
Pinus sylvestris (Sweden)	9.5			(c)
Pinus sylvestris (Germany)	16.6		3.3	(102)
Average (s.e.)	11.0	(0.6)		
Monsoonal forest				
Eucalyptus 10 species (Australia)	11.8		4.9	(c)
Metrosideros polymorpha (USA)	6.3		3.8	(89)
Metrosideros polymorpha (USA)	7.1		4.2	(125)
Average (s.e.)	11.4	(0.9)		
Sclerophyllous scrub				
Chaparral 39 species (USA)	10.4			(15)
Finebos 33 species (RSA)	12.9			(98)
Arbutus menziesii (USA)	9.4		7.6	(88)
Heteromeles arbutifolia (USA)	8.8		4.9	(88)
Leucadendron xanthoconus (USA)	9.6		9.1	(82)
Prunus ilicifolia (USA)	15.4		4.6	(88)
Rhamnus californica (USA)	12.3		8.5	(88)
Umbellularia californica (USA)	12.5		6.1	(88)
Average (s.e.)	11.4	(0.6)		
Temperate deciduous trees				
Quercus lobata (USA)	27.0		8.3	(113)
Acer negundo (USA)	29.5		14.6	(83)
Liriodendron tulipifera (USA)	15.4		16.0	(107)
Quercus robur (Austria)	18.5		12.0	(59)
Betula pubescens (Norway)	13.9			(c)
Fagus sylvatica (France)	14.0			(c)
Betula pubescens (Sweden)	18.6		12.5	(123)
4 sites Hubbard Brook (USA)	24.2			(d)
4 sites *Fagus sylvatica* (Germany)	25.1			(d)
15 IBP sites (Russia)	21.3			(d)
Average (s.e.)	19.6	(2.7)	see (d)	
Temperate deciduous fruit trees				
Juglans regia (USA)	22.5		12.5	(94)
Juglans regia (USA)	22.6		9.5	(84)
Juglans regia (USA)	30.0		12.0	(95)
Pyrus communis (USA)	14.2		5.2	(116)
Malus domestica (USA)	29.7		15.6	(87)
Malus domestica (Canada)	23.9		6.2	(85)
Average (s.e.)	23.8	(2.4)		
Temperate grassland				
Lolium perenne (Netherlands)	25.0		12.5	(106)
Anthoxanthum odoratum (Netherlands)	25.0		17.5	(106)
Rumex acetosa (Netherlands)	40.0		35.0	(106)
Plantago lanceolata (Netherlands)	25.0		15.0	(106)
Succisa pratensis (Netherlands)	15.0		15.0	(106)

Agrostis vinealis (Netherlands)	25.1	12.5	(78)
Corynephorus canescens (Netherlands)	28.6	16.0	(78)
Lolium perenne (UK)	35.0	14.0	(115)
Lysimachia vulgaris (Netherlands)	21.5	6.0	(19)
Carex diandra (Netherlands)	20.8	8.8	(74)
Carex rostrata (Netherlands)	16.5	12.5	(74)
Carex lasiocarpa (Netherlands)	15.8	7.4	(74)
Carex acutiformis (Netherlands)	12.8	13.9	(74)
Ranunculus hirtus (NZ)	14.0	16.7	(97)
Ranunculus repens (NZ)	21.4	24.6	(97)
Ranunculus lappaceus (NZ)	15.1	15.0	(97)
Ranunculus enysii (NZ)	13.1	10.6	(97)
Ranunculus pachyrrhizus (NZ)	17.6	9.6	(97)
Ranunculus sericophyllus (NZ)	25.5	9.6	(97)
Achillea millefolium (Switzerland)	38.8	19.3	(9)
Ranunculus acris (Switzerland)	29.3	15.3	(9)
Potentilla anserina (Switzerland)	30.9	19.1	(9)
Trifolium repens (Switzerland)	45.0	29.5	(9)
Carum carvi (Switzerland)	35.0	26.4	(9)
Chrysanthemum leucanthemum (Switzerland)	34.6	19.4	(9)
Taraxacum officinale (Switzerland)	36.0	34.4	(9)
Geum rivale (Switzerland)	29.6	29.1	(9)
Primula elatior (Switzerland)	18.2	23.6	(9)
Average (s.e.)	25.5 (1.6)		

Temperate evergreen broadleaf tree

Eucalyptus blakleyi (Australia)	15.1	4.9	(99)
Eucalyptus pauciflora (Australia)	11.6	3.1	(96)
Quercus agrifolia (USA)	14.0	5.9	(113)
Nothofagus menziesii (NZ)	14.6	6.2	(96)
Griselinia littoralis (NZ)	10.7	8.2	(96)
Eucalyptus delegatensis (Austr.)	11.5		(127)
Eucalyptus pauciflora (Australia)	12.0		(127)
Eucalyptus div. spec. (Australia)	19.3		(127)
Average (s.e.)	13.5 (1.0)		

Temperate evergreen crop

Citrus paradisii (Spain)	25.2	9.5	(124)

Tropical deciduous forest

Boscia albitrunca (Botswana)	35.3		(86)
Grevia flava (Botswana)	27.6		(86)
Acacia tortilis (Botswana)	26.5		(86)
Acacia luederitzii (Botswana)	31.6		(86)
Commiphora pyracanthoides (Botswana)	23.1		(86)
Humboldtiella arborea (Venezuela)	31.5	26.8	(121)
Mansoa verrucifera (Venezuela)	27.1	17.5	(121)
Lonchocarpus dipteroneurus (Venezuela)	41.3	24.2	(121)
Beureria cumanensis (Venezuela)	28.7	11.6	(121)
Pithecellobium dulce (Venezuela)	25.8	15.1	(121)
Pithecellobium ligustricum (Venezuela)	34.9	15.4	(121)
Memora sp. (Venezuela)	20.5	7.8	(121)

Appendix 1 *(Continued)*

Plant species	N		SLA	Source
Capparis verrucosa (Venezuela)	31.5		8.9	(121)
Capparis aristiguetae (Venezuela)	22.1		6.9	(121)
Morisonia americana (Venezuela)	26.1		6.8	(121)
Acacia (15 species Namibia)	27.8			(102)
Non-Acacia (16 species Namibia)	24.5			(102)
Average (s.e.)	27.1	(1.0)		
Tropical fruit plantations				
Mangifera indica (Australia)	12.8		6.5	(c)
Anacardium occidentale (Australia)	14.4		8.3	(c)
Average	13.6	(4.6)		
Tropical grassland				
C4 grasses 5 species (Mali)	5.0			(110)
C3 annuals 10 species (Mali)	10.0			(110)
Cynodon dactylon (Botswana)	28.6			(86)
Eragrostis rigidor (Botswana)	8.9			(86)
Tragus berterionianus (Botswana)	17.0			(86)
Panicum maximum (Botswana)	21.28			(86)
Average (s.e.)	10.7	(4.0)		
Sugarcane				
Saccharum officinarum (Australia)	12.0		9.0	(b)
Tropical Rainforest				
23 Tree species (Brazil)	15.0		10.0	(114)
68 Tree species (Mexico)	16.8		13.4	(112)
40 Tree species (Australia)	16.2		9.9	(77)
22 Tree species (Sri Lanka)	16.7		9.4	(92)
74 Tree species (Australia)	18.9		19.1	(122)
10 Tree species (Venezuela)	28.9		19.6	(108)
4 Tree species (Venezuela	17.9		7.6	(121)
8 Mixed forest sp. (Venezuela)	12.7		7.4	(103)
5 Tall Catinga sp. (Venezuela)	11.6		7.6	(103)
9 Low Catinga sp. (Venezuela)	7.4		4.7	(103)
9 Tree species (Cameroon)	19.0			(93)
Average (s.e.)	16.5	(1.6)		
Tundra				
Gramineae (6 species, USA)	18.6			(80)
Calamagrostis canadensis (USA)	23.8			(79)
Carex aquatilis (USA)	18.5			(79)
Agrostis tenuis (UK)	22.1			(111)
Festuca ovina (UK)	13.2			(111)
Anthoxanthum odoratum (UK)	26.9			(111)
Carex div. spec. & herbs (UK)	20.4			(111)
Average (s.e.)	20.5	(1.1)		

References to Appendix 1 (numbering continues from main reference section):
74. Aerts R et al. 1992. *J. Ecol.* 80:653–64

75. Allen LH et al. 1988. *Crop Sci.* 28:84 – 94
76. Andrane A et al. 1993. *Plant Soil* 149:175 – 84
77. Basset Y. 1991. *Oecologia* 87:388 – 93
78. Boot RGA et al. 1992. *Physiologia Plant.* 86:152 – 60
79. Chapin FS III. 1989. *Ecology* 70:269 – 72
80. Chapin FS III. 1978. *Ecol Stud.* 29:483 – 507
81. Condon AG et al. 1992. *Aust. J. Agric. Res.* 43:935 – 47
82. Davis GW et al. 1992. *S. Afr. J. Bot.* 58:56 – 62
83. Dawson TE et al. 1993. *Ecology* 74:798 – 815
84. Erez A et al. 1985. *J. Plant Nutrit.* 8:103 – 15
85. Erf JA et al. 1989. *J. Am. Soc. Hort. Sci.* 114:191 – 96
86. Ernst WHO et al. 1989. J Proctor (ed.) Blackwell 97 – 120
87. Ferree DC et al. 1988. *J. Am. Soc. Hort. Sci.* 113:699 – 703
88. Field CB et al. 1993. *Oecologia* 60:384 – 89
89. Gerrish G. 1992. *Pac. Sci.* 46:315 – 24
90. Gower ST. 1989. *Tree Physiol.* 5:1 – 11
91. Hocking PJ et al. 1991. *Aust. J. Plant Physiol.* 18:339 – 56
92. Jayasekera R. 1992. *Vegetatio* 98:73 – 81
93. Kazda M et al. 1992. F. Hallé (ed) Paris pp 217 – 30
94. Klein I et al. 1991. *HortScience* 26:183 – 85
95. Klein I et al. 1991. *J. Plant Nutrit.* 14:463 – 84
96. Körner CH. 1986. *Oecologia* 69:577 – 88
97. Körner CH et al. 1885. *Oecologia* 66:443 – 55
98. Kruger FJ. 1987. *NATO ASI Ser.* 15:415 – 28
99. Landsberg J. 1990. *Aust. J. Ecol.* 15:73 – 87
100. Lynch J et al. 1992. *Crop Sci.* 32:633 – 40
101. MacDonald SE et al. 1990. *Can. J. For. Res.* 20:995 – 1000
102. Matyssek R. 1985. PhD. Thesis, Bayreuth. 224 pp.
103. Medina E et al. 1989. In J Proctor (ed.) pp. 217–40. Blackwell
104. Mulchi CL et al. 1992. *Agric. Ecosystems Environ.* 38:107 – 18
105. Murayama S et al. 1987. *Jpn. J. Crop Res.* 56:198 – 203
106. Olff H, 1992. *Oecologia* 89:412 – 21
107. O'Neill EG et al. 1987. *Plant & Soil* 104:3 – 11
108. Olivares E et al. 1992. *J. Veg. Sci.* 3:383 – 92
109. Oren R et al. 1988. *Oecologia* 77:151 – 62
110. Penning de Vries FWT (ed.) *PUDOC*, pp 304 – 45
111. Perkins DF et al. 1978. *Ecol. Stud.* 27:303 – 33
112. Pompa J et al. 1992. *Oikos* 63:207 – 14
113. Puttick GM. 1986. *Oecologia* 68:589 – 94
114. Reich PB et al. 1991. *Oecologia* 86:16 – 24
115. Ryle GJA. 1992. *J. Exp. Bot.* 43:811 – 18
116. Sanchez EE et al. 1990. *J. Am. Soc. Hort. Sci.* 115:934 – 37
117. Schulze E-D et al. 1984. *Plant Cell Environ.* 7:293 – 99
118. Schulze E-D et al. 1990. *Oecologia* 82:158 – 61
119. Schulze E-D et al. 1991 *Oecologia* 88:451 – 55
120. Schulze E-D et al. 1994. *Can. J. For. Res.* In press
121. Sobrado MA. 1991. *Func. Ecol.* 5:608 – 16
122. Stewart GR. 1990. *Oecologia* 82:544 – 51
123. Sveinbjörnsson B et al. 1992. *Func. Ecol.* 6:213 – 20
124. Syvertson JP. 1984. *J. Am. Soc. Hort. Sci.* 109:807 – 12
125. Vitousek PM et al. 1988. *Oecologia* 77:565 – 70
126. Wong S-C. 1990. *Photosynthesis Res.* 23:171 – 80
127. Woods PV. 1983. *Aust. J. Ecol.* 8:287 – 99

Appendix 2 Global land-cover classes of Wilson & Sellers (71) and the predicted levels of N-concentration (mg g^{-1}), maximum stomatal conductance (g_{smax}, mm s^{-1}), maximum surface conductance (G_{smax}, mm s^{-1}) and maximum canopy CO$_2$ assimilation (A_{smax}, μmol m^{-2}s^{-1}). Vegetation types with unknown characteristics are indicated by the number 100. N-concentrations are according to Appendix 1. g_{smax}, G_{smax}, and A_{smax} were predicted from N-concentrations (Figure 3: $g_{smax} = 0.3012\,N$, $G_{smax} = 2.996\,g_{smax}$, $A_{smax} = 1.05\,G_{smax}$), unless otherwise stated.

Type	No.	N	g_{smax}	G_{smax}	A_{smax}	Remark
Rice	4	27.8	8.4	25.1	26.3	
Needleleaf evergreen forests	10,11	11.0	5.5	20.6	21.6	g_{smax} & G_{smax} = Tab.1
Mixed forst	12,13	15.8	4.6	13.8	14.5	av.conifer & deciduous
Evergreen broadleaf forests	14,19	13.4	4.0	12.1	12.7	
Evergreen broadleaf crop	15	25.2	7.6	22.7	23.9	Citrus
Evergreen broadleaf shrub	16	10.4	3.1	9.4	9.9	
Deciduous needleleaf forests	17,18	20.7	3.8	11.4	12.0	
Deciduous broadleaf forests	20,21	19.6	5.9	20.7	21.7	G_{smax} = Tab. 1
Deciduous tree crop	22	23.8	7.2	21.5	22.6	
Drought deciduous forests	23,24,25,26,27,28	27.1	8.2	24.5	25.7	
Temperate grasslands	30,31	25.2	7.7	23.0	24.2	
Tropical savanna	32,34,35,37,39,71,73	18.9	5.7	17.1	17.9	av. 23&36
Tropical pasture	33	17.1	5.2	15.4	16.2	*Panicum maximum*
Semiarid rough grazing	36	10.7	3.2	9.7	10.1	trop. grasses
Arable cropland	40,41	36.0	10.8	32.5	34.1	av. cereals & broadleaf
Sugarcane	43	12.0	3.6	10.8	20.0	A_{smax} *1.6 (Fig. 1)
Maize	44	28.5	8.6	25.7	33.0	A_{smax} *1.6 (Fig. 1)
Gossypium	45	29.4	8.9	26.5	27.9	
Irrigated cropland	47,48	38.4	11.6	34.7	36.4	
Tropical Rainforest	50	16.5	5.0	14.9	15.6	
Tropical tree crop	46,49,51	13.6	4.1	12.3	12.9	
Tropical broadleaf forests	52	19.2	5.8	17.3	18.2	Trop decid & Monsoon
Tundra, bog, marsh	2,61,62	20.5	6.2	5.0	5.3	g_{smax} & G_{smax} = Tab.1
Water, ice, desert, urban, Mangrove	0,1,3,5,70,80	100	100	100	100	

Appendix 3
NITROGEN CONCENTRATION AND CANOPY PHOTOSYNTHESIS

A simplified mathematical theory links photosynthesis of a leaf canopy to leaf nitrogen concentration (N) and intercepted radiation. Leaf photosynthetic rate at light saturation A_{max} is a linear function of N because of its function in the photosynthetic enzyme:

$$A_{smax} = a_N (N - N_t)$$ A1.

where N_t is a threshold below which there is no photosynthesis and which represents N bound to leaf structures, α_N is a constant and N does not reach a level whereby A_{max} becomes saturated. The photosynthetic carbon assimilation rate (A) of an individual leaf exposed to absorbed irradiance Q_l is represented by the non-rectangular hyperbola:

$$\theta A^2 - (\alpha Q_l + A_{max}) + \alpha Q_l A_{max} = 0$$ A2.

where α is the quantum efficiency and θ determines the shape of the hyperbola (limiting cases are $\theta = 0$ for rectangular hyperbola and $\theta = 1$ for Blackman response, 69). A_{max} is dependent on N (103), α and θ are independent of N (60).

Depending on the position of individual leaves, A will vary because Q_l decreases with depth within the canopy and because N (and hence A_{max}) also varies. In a horizontally homogenous canopy the vertical variation of Q_l and N is described by analogous exponential functions of cumulative leaf area index (ξ) measured from the top of the canopy (38,19):

$$Q_l (x) = k_Q Q_o \exp(-k_Q \xi)$$ A3.

and

$$N(\xi) = (N_o - N_l) \exp(-k_N \xi) + N_l$$ A4.

where k_Q is the extinction coefficients for Q and k_N the distribution coefficient for N. Irradiance at the top of the canopy is Q_o, and N_o is the corresponding leaf nitrogen concentration. Radiation $Q_l = dQ/d\xi$, where Q is incident radiation. The leaf area index $\Lambda = \int d\xi$ and total canopy nitrogen is $N_c = \int N(\xi) d\xi$. We assume that the nitrogen concentration of the uppermost leaves is the same for all values of Λ and thus N_c increases with Λ, which may not be the case in reality if N is limited.

Dependence of A on vertical position in the canopy ($A(\xi)$) can now be calculated by substituting Equations A1, A3, and A4 into (A2). Integration over all leaves in the canopy results in canopy assimilation $A_c = \int A(\xi) d\xi$. This is a maximum assimilation rate, limited only by incident light for a given amount of available nitrogen. Although numerical methods are required to solve for A_c in terms of six parameters (a_N, N_l, α, θ, k_Q, and k_N), considerable insight into the behaviour of A_c comes from limiting cases of $\theta = 0$ (rectangular hyperbola) and $\theta = 1$ (Blackman function) for the light response curve of individual leaves in the canopy. For $\theta = 0$, Equation A2 reduces to:

$$A = \frac{A_{maxo}}{1 + \chi_o} \qquad \text{A5.}$$

where A_{maxo} is the light-saturated photosynthesis rate for a leaf with nitrogen concentiation N_o and $X_o = A_{maxo}/[\alpha\, Q_o])$ is the ratio of A_{maxo} and potential photosynthesis at light intensity Q_o. Substituting of Equations A3 and A4 into A5 yields:

$$A(x) = -\frac{A_{maxo}\exp(-k_Q\xi)}{\exp[-(k_Q - k_N)\xi + \chi_o} \qquad \text{A6.}$$

Plants should distribute their available N_o so as to maximize canopy assimilation rate (15, 19). A_c will be maximal for a fixed quantity of canopy N when the partial derivative $\partial A/\partial N$ is constant throughout the canopy such that the benefit to photosynthesis of a small increase in N of one leaf is matched by an equal loss from another. Unless this occurs, A_c may be increased by the redistribution of N (19). $\partial A/\partial N$ is at a maximum when nitrogen is distributed in the canopy in proportion to the photosynthetically active radiation; this occurs when $k_Q = k_N$. The integral of Equation A6 over the total leaf area index (Λ_c) then reduces to the simpler form:

$$A_c = A_{maxo}\int_0^\Lambda \frac{\exp(-k_Q\xi)d\xi}{1 + \chi_o} = \frac{A_{maxo}\,(1 - \exp(-k_Q\Lambda))}{k_Q(1 + \chi_o)} \qquad \text{A7.}$$

There is no analytical solution to the integral of Equation A6 when $k_Q \neq k_N$ and numerical methods are required to calculate A_c. A correct comparison of the effects of variation in k_N on A_c must ensure that the total amount of N in the canopy is the same for both optimal ($k_O = k_N$) and sub-optimal N distributions ($k_Q \neq k_N$). Total canopy N is given by the integral of Equation A4:

$$N_c = \int_0^\Lambda N(\xi)\,d\xi = \frac{(N_{x1}-N_1)}{k_N}(1 - \exp(-k_N\Lambda)) + \Lambda N_1 \qquad \text{A8.}$$

where N_{xl} represents maximum N concentrations for leaves at the top of the canopy when N is distributed sub-optimally. A similar equation may be written for an optimal N distribution by substituting k_Q for k_N in Equation A8 and using N_{vo} for N concentration of the uppermost leaves. Equating N_c for both cases and utilizing Equation 1 yields the ratio of maximum leaf-level assimilation rates for the two cases, viz

$$\frac{A_{maxl}}{A_{maxo}} = \frac{k_N\,(1 - \exp(k_Q\Lambda))}{k_Q\,(1 - \exp(k_N\Lambda))} \qquad \text{A9.}$$

where A_{maxl} is the maximum assimilation rate to be used when $k_Q \neq k_N$ for calculating: (1) $A(\xi)$ with Equation A6, instead of A_{maxo}, and (2) $X_l = A_{maxl}/(k_Q \, \alpha \, Q_o)$ instead of X_o.

The N distribution coefficient k_N can be varied to examine the effect of nitrogen distribution on calculated canopy rate. There is little effect of k_N on A_c until $\Lambda > 2$, and a maximal 10–20% reduction from the optimum thereafter occurs for both light response functions when $k_N = 0$. Modelled A_c is reduced by less than 6% from the optimum when $k_N = 0.25$ and $\Lambda = 6$. With the expectation of a direct correlation between irradiance on leaves and their N concentrations, we suggest that k_N will usually be greater than 0.25. Consequently, one could assume an optimal N distribution and use this approach to model canopy assimilation rate.

Literature Cited (see also References in Appendix 1)

1. Baldocchi DD. 1994. A comparative study of mass and energy exchange rates over a closed C3 (wheat) and open C4 (corn) crop: I. The partitioning of available energy into latent and sensible heat exchange. *Agric. For. Met.* 67:191–220

2. Baldocchi DD, Hicks BB, Meyers TD. 1988. Measuring biosphere-atmosphere exchanges of biologically related gases with micrometeorological methods. *Ecology* 69:1331–40

3. Baldocchi DD, Hicks BB, Camara P. 1987. A canopy stomatal resistance model for gaseous deposition to vegetated surfaces. *Atmos. Environ.* 21:91–101

4. Baldocchi DD, Verma SB, Rosenberg NJ. 1981. Environmental effects of CO_2 flux and CO_2-water flux ratio of alfalfa. *Agric. Met.* 24:175–84

5. Biscoe PV, Scott RK, Monteith JL. 1995. Barley and its environment. III. Carbon budget of the stand. *J. Appl. Ecol.* 12:269–93

6. Coyne PI, Kelley JJ. 1975. CO_2 exchange over the Alaskan arctic tundra: meteorological assessment by an aerodynamic method. *J. Appl. Ecol.* 12:587–611

7. Denmead OT, McIlroy IC. 1970. Measurements of non-potential evaporation from wheat. *Agric. Met.* 7:283–302

8. Desjardins RL, Brach EJ, Alvo P, Schuepp PH. 1982. Aircraft monitoring of surface carbon dioxide exchange. *Science* 216:733–35

9. Diemer M, Körner Ch, Prock S. 1992. Leaf life span in wild perennial herbaceous plants: A survey and attempts at a functional interpretation. *Oecologia* 89:10–16

10. Dunin FX, McIlroy IC, O'Loughlin EM. 1985. A lysimeter characterization of evaporation by eucalypt forest and its representatives for the local environment. In *The Forest-Atmosphere Interaction*, ed. BA Hutchinson, BB Hicks, pp. 271–91. Lancaster: Reidel

11. Eamus D, Jarvis PG. 1989. The direct effects of increase in the global atmospheric CO_2 concentration on natural and commercial temperate trees and forests. *Adv. Ecol. Res.* 19:1–55

12. Fan S-M, Wofsy SC, Bakwin PS, Jacob DJ, Fitzjarrald DR. 1990. Atmosphere-biosphere exchange of CO_2 and O_3 in the central Amazon forest. *J. Geophys. Res.* 95(D10):16851–64

13. Fan SM, Wofsy SC, Bakwin PS, Jacob DJ, Anderson SM, et al. 1992. Micrometeorological measurements of CH_4, and CO_2, exchange between the atmosphere and subarctic tundra. *J. Geophys. Res.* 97:16627–43

14. Farquhar GD, Sharkey TD. 1982. Stomatal conductance and photosynthesis. *Annu. Rev. Plant Phys.* 33:317–47

15. Field CB, Mooney HA. 1986. The photosynthesis-nitrogen relationship in wild plants. In *On the Economy of Plant Form and Function*, ed. TJ Givnish, pp. 25–56. Cambridge: Cambridge Univ. Press

16. Field CB, Chapin FS III, Matson PA, Mooney HA. 1992. Responses of terrestrial ecosystems to the changing atmosphere: A resource-based approach. *Annu. Rev. Ecol. Syst.* 23:201–35

17. Fitzjarrald DR, Moore KE. 1992. Turbulent transport over tundra. *J. Geophys. Res.-Atmos.* 97DI5:16,717–29

18. Greenwood EAN, Turner NC, Schulze E-D, Watson GD, Venn NR. 1992. Ground water management through increased water use by lupin crops. *J. Hydrol.* 134:1–11

19. Hirose T, Werger MJA, Pons TL, VanRheenen JWA. 1988. Canopy structure and leaf nitrogen distribution in a stand of *Lysimachia vulgaris* L. as influenced by stand density. *Oecologia* 77: 145–50

20. Hollinger DY, Kelliher FM, Byers JN, Hunt JE, McSeveny TM, Weir PL. 1994. Carbon dioxide exchange between an undisturbed old-growth temperate forest and the atmosphere. *Ecology* 75:134–50

21. Jarvis PG. 1976. The interpretation of the variations in leaf water potential and stomatal conductance found in canopies in the field. *Philos. Trans. R. Soc. Lond. Ser. B* 273:593–610

22. Jarvis PG, James GB, Landsberg JJ. 1976. Coniferous Forest. In *Vegetation and the Atmosphere*, Vol 2, ed. JL Monteith, pp. 171–240. Case Studies. London: Academic

23. Khanna PK, Ulrich B. 1991. Ecochemistry of temperate deciduous forests. In *Temperate Deciduous Forests*, ed. E Röhrig, B Ulrich. *Ecosystems of the World* 7:12–163

24. Kelliher FM, Leuning R, Raupach MR, Schulze E-D. 1994. Maximum conductance for evaporation of global vegetation types. *Agric. For. Met.* In press

25. Kelliher FM, Leuning R, Schulze E-D. 1993. Evaporation and canopy characteristics of coniferous forest and grassland. *Oecologia* 95:153–63

26. Kelliher FM, Köstner BMM, Hollinger DY, Byers JN, Hunt JE, et al. 1992. Evaporation, xylem sap flow, and tree transpiration in a New Zealand broadleaved forest. *Agric. For. Met.* 62:53–73

27. Kim J, Verma SB. 1990. Components of surface energy balance in a temperate grassland ecosystem. *Bound. Layer Met.* 51:401–17

28. Körner CH. 1989. The nutritional status of plants from high altitudes. *Oecologia* 81:379–91

29. Körner CH. 1994. Leaf diffusive conductances in the major vegetation types of the globe. See Ref. 56

30. Körner CH, Scheel JA, Bauer H. 1979. Maximum leaf diffusive conductance in vascular plants. *Photosynthetica* 13:45–82

31. Köstner BMM, Schulze E-D, Kelliher FM, Hollinger DY, Byers JN, et al. 1992. Transpiration and canopy conductance in a pristine broad-leaved forest of *Nothofagus:* An analysis of xylem flow and eddy correlation measurements. *Oecologia* 91:350–59

32. Lambers H, Poorter H. 1992. Inherent variation in growth rate between higher plants: A search for physiological causes and ecological consequences. *Adv. Ecol. Res.* 23:188–261

33. Lauerer M, Saftic D, Quick WP, Labate C, Fichtner K, et al. 1993. Decreased ribulose-1,5-bisphosphate carboxilaseoxygenase in transgenic tobacco transformed with "antisense" rbcs. VI. Effects on photosynthesis in plants grown at different irradiance. *Planta* 190:322–45

34. Leuning R. 1994. A critical appraisal of a combined stomatal-photosynthesis model for C3 plants. *Plant Cell Environ.* In press

35. Leuning R, Attiwell PM. 1978. Mass, heat and momentum exchange between a mature Eucalyptus forest and the atmosphere. *Agric. Met.* 19:215–41

36. McGinn SM, King KM. 1990. Simultaneous measurements of heat, water vapour and CO_2 fluxes above alfalfa and maize. *Agric. For. Met.* 49:331–49

37. Marschner H. 1986. *Mineral Nutrition of Higher Plants*. London: Academic. 674 pp.

38. Monsi M, Saeki T. 1953. Über den Lichtfaktor in den Pflanzengesellschaften und seine Bedeutung fur die Stoffproduktion. *Jpn. J. Bot.* 14:22–52

39. Monteith JL. 1965. Evaporation and environment. *Symp. Soc. Exp. Ecol.* 19: 205–34

40. Monteith JL, Unsworth MH. 1990. *Principles of Environmental Physics*. London: Edward Arnold. 291 pp. 2nd ed.

41. Monteney BA, Barbier JM, Bernos CM. 1985. Determination of the energy exchange of a forest-type culture: *Hevea brasiliensis*. In *The Forest-Atmosphere*

Interaction, ed. BA Hutchinson, BB Hicks, pp. 211–33. Dordrecht: Reidel

42. Mooney HA. 1993. *The Development of Global Ecology. 15. International Biological Congress, Yokohama Japan.* Special Lecture Abstr.

43. Mooney HA, Field CB. 1989. Photosynthesis and plant productivity—scaling to the biospere. In *Photosynthesis,* ed. WR Briggs, pp. 19–44. New York: Alan R. Liss

44. Perrier A, Kateriji N, Gosse G, Itier B. 1980. Etude in situ de l'évaporation réele d'une culture de blé. *Agric. For. Met.* 21:295–311

45. Philip JR. 1957 Evaporation, and moisture and heat fields in the soil. *J. Meteorol.* 14:354–66

46. Price DT, Black TA. 1990. Effects of short-term variation in weather on diurnal canopy CO_2 flux and evapotranspiration of a juvenile douglas-fir stand. *Agric. For. Met.* 50:120–59

47. Priestley CHB, Taylor RJ. 1972. On the assessment of surface heat flux and evaporation using large-scale parameters. *Month. Weather Rev.* 100: 81–92

48. Raschke K. 1979. Movements of stomata. *Encyclopedia of Plant Physiology NS* 7:383–441

49. Raupach MR, Finnigan JJ. 1988. Single-layer evapotranspiration models are incorrect but useful, whereas multilayer models are correct but useless. *Aust. J. Plant Physiol.* 15:715–26

50. Reich PB. 1993. Reconciling apparent discrepancies among studies relating life span, structure and function of leaves in contrasting plant life forms and climates: "the blind men and the elephant retold." *Func. Ecol.* 7:721–25

51. Ripley RA, Redmann RE. 1976. In *Vegetation and the Atmosphere,* ed. JL Monteith, pp. 351–98. London: Academic

52. Schulze E-D. 1982. Plant life forms as related to plant carbon, water, and nutrient relations. *Encyclopedia Plant Physiol.* Berlin, Heidelberg: Springer Verlag. 12B:615–76

53. Schulze E-D. 1989. Air pollution and forest decline in a spruce (*Pice abies*) forest. *Science* 244:776–83

54. Schulze E-D, ed. 1994. *Flux Control in Biological Systems.* San Diego: Academic. 487 pp.

55. Schulze E-D. 1994. The regulation of plant transpiration: Interactions of feedforward, feedback and futile cycles. See Ref. 54, pp. 203–37

56. Schulze E-D, Caldwell MM. 1994. *Ecophysiology of Photosynthesis. Ecol.*

Stud. 100. Berlin, Heidelberg: Springer Verlag. 576 pp.

57. Schulze E-D, Hall AE. 1982 Stomatal control of water loss. *Ency. Plant Physiol. N.S.* 12B:181–230

58. Schulze E-D, Lange OL, Evenari M, Kappen L, Buschbom U. 1976. An empirical model of net photosynthesis for the desert plant *Hammada scoparia* (Pomel) Iljin. I. Description and test of the model. *Oecologia* 32:355–72

59. Schulze E-D, Zwölfer H. 1987. Potentials and limitations of Ecosystem Analysis: Synthesis. *Ecol. Stud.* 61: 416–23

60. Sellers PJ, Berry JA, Collatz GJ, Field CB, Hall FG. 1992. *Remote Sens. Environ.* 42:187–216

61. Shukla J, Mintz Y. 1982. Influence of land-surface evapotranspiration on the earth's climate. *Science* 215:1498–1501

62. Shuttleworth WJ. 1988. Evaporation from Amazonian rainforest. *Proc. R. Soc. Lond. Ser. B* 233:321–46

63. Stewart JB. 1988. Modelling surface conductance of pine forest. *Agric. For. Met.* 43:19–35

64. Stewart JB, Verma SB. 1992. Comparison of surface fluxes and conductances at two contrasting sites within the Fife area. *J. Geophys. Res.* 97:8623–28

65. Stitt M, Schulze E-D. 1994. Plant growth, storage, and resource allocation: from flux control in a metabolic chain to the whole-plant level. See Ref. 54, pp. 57–118

66. Stitt M. 1994. Flux control at the level of the pathway: studies with mutants and transgenic plants having a decreased activity of enzymes involved in photosynthetic partitioning. See Ref. 54, pp. 13–36

67. Tans PP, Fung IY, Takahashi T. 1990. Observational constraints on the global atmospheric CO_2 budget. *Science* 247: 1431–38

68. Tenhunen JD, Siegwolf RA, Oberbauer SF. 1994. Effects of phenology, physiology, and gradients in community composition, structure, and microclimate on tundra ecosystem CO_2 exchange. See Ref. 56, 100:431–59

69. Valentini R, Scarascia Mugnozza GE, De Angelis P, Bimbi R. 1991. An experimental test of the eddy correlation technique over a Mediterranean macchia canopy. *Plant, Cell Environ.* 14: 987–94

70. Verma SB, Baldocchi DD, Anderson DE, Matt DR, Clement RJ. 1986. Eddy fluxes of CO_2 water vapour and sensible heat over a deciduous forest. *Bound. Layer Meteorol.* 36:71–91

71. Wilson MF, Henderson-Sellers A. 1985. A global archive of land cover and soils data for use in general circulation climate models. *J. Climatol.* 5:119–43

72. Wofsy SC, Goulden ML, Munger JW, Fan SM, Bakwin PS, et al. 1993. Net exchange of CO_2 in a mid-latitude forest. *Science* 260:1314–17

73. Woodward Fl, Smith TM. 1994. Predictions and measurements of the maximum photosynthetic rate, A_{max}, at a global scale. See Ref. 56, 100:491–508

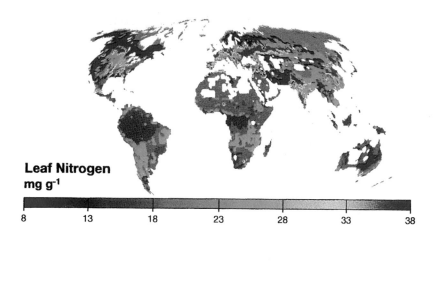

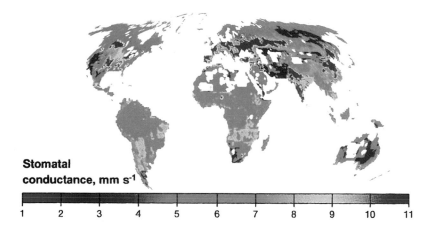

Fig. 4: Global map of **(A)** leaf nitrogen concentrations (N). Also shown is a color legend. **(B)** maximum stomatal conductance (g_{max}, mm s^{-1}), **(C)** maximum surface conductance (Gmax, mm s^{-1}), and **(D)** maximum canopy assimilation (A_{max}, μmol m^{-2}s^{-1}). The maps were prepared on the basis of values assigned to land-cover classes (Appendix 2) of primary and secondary vegetation types of Wilson and Henderson-Sellers (72). These were plotted on a 1x1 degree grid scale. White color indicates areas with insufficient information.

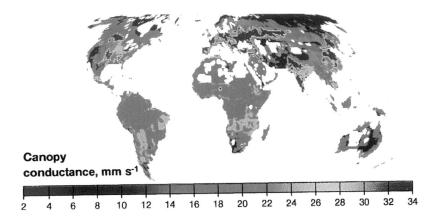

Canopy conductance, mm s⁻¹

2 4 6 8 10 12 14 16 18 20 22 24 26 28 30 32 34

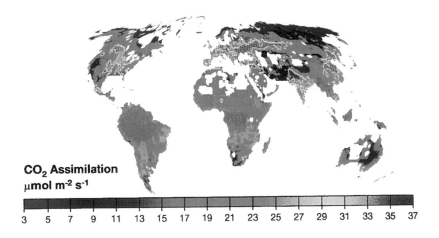

CO₂ Assimilation μmol m⁻² s⁻¹

3 5 7 9 11 13 15 17 19 21 23 25 27 29 31 33 35 37

SUBJECT INDEX

A

Abiotic factor
  fish populations and, 411
  seasonal plankton production
    cycle and, 404
  seed dispersal and, 263, 267,
    268–71
  seed loss and, 277
  seed/microsite interactions
    and, 274–75
Abrams, P. A., 227–28, 229, 230
Acacia albida, 646
Acanthaster planci, 240
Acanthina angelica, 450
Acanthiodiaptomus
  specific P content, 10
Acanthocephalan, 360–62
Acartia tonsa
  growth rate, 13–14
Acid rain, 647
Acoelomate, 359
Acorn barnacle, 450
Acoustic insect
  graded aggressive signals, 118
  inhibitory resetting, 105
  synchrony and alternation,
    108–9
Acoustic signaling
  aggression and, 117
  "beacon" effect, 111
  sexual advertisement and, 98
Acoustic spree, 119
Acquired immunodeficiency syn-
    drome (AIDS), 128, 143,
    154–56
Acquisition, 4
Acridid grasshopper
  call timing, 106, 118
  dawn and dusk choruses, 99
  rhythmic alternation, 112
Acris
  geographic variation in calls,
    299–300
Acris crepitans
  clinal variation in call fre-
    quency, 316
  dominant vocalization fre-
    quency, 298
Acris gryllus
  clinal variation in call fre-
    quency, 316
Acropora, 237–56
  biogeographic consistency, 250
  biogeographic diversity, 250–
    51

breeding compatibilities, 244–
    46
  gene flow, 247–49
  mating incompatibilities, 243–
    44
  patterns of distribution, 251–52
  phylogenetic studies, 253–54
  species boundaries, 241–47
Acropora brueggemanni, 251
Acropora cuneata, 240
  taxonomic distinctiveness, 248
Acropora donei, 254
Acropora formosa, 248
Acropora millepora
  fertilization in reciprocal
    crosses, 242
  natural hybrids, 243
Acropora nasuta, 249
Acropora nobilis, 248
Acropora palifera, 240
  taxonomic distinctiveness, 248
Acropora pulchra
  natural hybrids, 243
Acropora pulchra/millepora cross
  karyotypic analysis, 247
Acropora selago, 251
Acropora tenuis, 254
Adaptation, 32
  complex life cycles and, 580–82
  cost-benefit approach, 227–29
  enemy-related, 220
Adaptive decoupling, 573, 588
Adaptive radiation, 527
  interspecific competition and,
    470, 477–81
Adelina triboli, 503
Adoutte, A., 358
Advertisement call, 293
  acoustic properties
    female selectivity and, 303
    maximum broadcast area, 296
African green monkey
  immunodeficiency virus, 131
Aggression
  acoustic signaling and, 117
Ahlgren, G., 9
AIDS
  See Acquired immunodefici-
    ency syndrome
Aizoaceae, 523
Alberch, P., 591–92
Alectoris rufa, 506
Alfalfa
  leghemoglobins, 340
Algae
  biochemical and elemental
    composition, 4

C:N:P ratios
  in situ patterns, 7–8
  toxicity, 14
Algal biomass
  biomolecules, 3
Algal cell
  nutrient content, 5
Algal-herbivore interface, 1–23
Algal nutrient limitation, 5–9
Allocasuarina dicansiana, 646
Allopatric speciation, 548
Allopolyploidy, 316
Allozyme electrophoresis, 248
Allozyme heterozygosity
  fitness and, 57–58
Allozyme polymorphism, 52
Allozyme variation, 53
  population size and, 48
Alpheid shrimp, 558
Alternation
  competitive, 116–17
  as epiphenomenon, 108–9
  mechanism, 107–8
  phase delay, 105–7
  rhythmic, 112–14
  signal interactions and, 99
Amazon molly, 71–72
Ambystoma
  multiple hybrid events, 75
Ambystoma tigrinum
  growth rate
    allozyme genotype and, 57
Amino acids
  herbivore growth and, 8
  pelagic herbivore diet and, 3
  replacement substitutions, 149
Amos, W., 380
Amphibian
  complex life cycles, 577, 588–89
  male parental care, 602
  unisexual, 72
Amphipod, 558
  osmoregulatory differences, 566
Anargus epos, 507
Anax junius, 498
Andersen, T., 17
Anderson, D. T., 365–66
Andrewartha, H. G., 168
Andricus quercuscalicis, 507
Anemone, 449
Angiosperm
  chloroplast DNA
    inheritance, 47
Animal phyla
  Cambrian radiation
    DNA sequencing and, 351–
      71

Protist assemblage
  apparent competition and, 509–
    10
Protocoel, 368
Protogonyaulax, 452
Protozoan, 3
  herbivore diets and, 3
Pseudacris crucifer
  advertisement calls, 297
Pseudacris streckeri
  call timing, 107
Pseudacris triseriata feriarum
  call structure, 316
Pseudocoelomate, 360–62
Pseudogamy paradox, 78–79
Pseudogynodioecy, 240
Pseudotsuga menziesii
  seed zones, 54
Psilocaulon, 524
Psychophysics
  competitive signaling and,
    114–16
Pterophylla camellifolia
  call timing, 106
Pteroptyx cribellata
  phase delay synchrony, 103
Pteroptyx malaccae
  perfect synchrony, 109
Pteroptyx tener
  perfect synchrony, 109
Pumpkinseed sunfish, 453

Q

Quaking aspen
  growth rate
    allozyme genotype and, 57
Quantitative variation, 36–37
  nature, 38–40
  potential for strong effects, 37–
    38
Quasispecies theory
  HIV and, 145–47
Quinn, J. F., 176

R

Raff, R. A., 351–71
Rainbow trout
  oxygen consumption
    allozyme heterozygosity
      and, 57
Rain forest
  seed dispersal, 282–83
Rana catesbeiana
  advertisement calls, 296
Rana esculenta
  call pulse rates, 316
  hemiclonal genomes
    deleterious recessive muta-
      tions, 89
  hybridogenesis, 73
  hypoxic stress and, 80

  multiple hybrid events, 75
  sexual selection and, 314
Rana palustris, 498
Rana pipiens
  brainstem oscillators, 295
Rand, A. S., 315
Randomly amplified polymor-
    phic DNA inheritance, 46,
    49
Rare clone advantage, 79
Raven, P. H., 220
Reaction-diffusion theory
  metapopulations and, 177–78
Reader, R. J., 497
Recognition
  reproductive isolation and,
    562–65
Recolonization, 169–70
  founder effect and, 172
Recruitment, 403
Redfield ratio, 4, 6
Red fox, 511
Red grouse, 507
Red-legged partridge, 506
Red Queen hypothesis, 86, 220,
    224–27
Red tide, 452
Red-winged blackbird
  parental care, 618
Reeb, C. A., 553
Reeve, E. C. R., 40
Refield, A. C., 4
Reinfelder, J. R., 16
Release call, 294–95
Rendel, J. M., 33
Renicola roscovita, 205
Renicolid, 205
Reproduction
  nonrecombinant, 72–74
  parthenogenetic, 576
  sexual, 576
  unisexual
    sperm limitation and, 78
Reproductive ecology
  milkweed, 423–38
Reproductive isolation, 547, 557–
    62
  genetics, 562–66
  mechanisms, 557–61
Reptile
  squamate, 72
Resource heterogeneity model
  unisexuality and, 85–86
Resource partitioning
  evolutionary ecology and, 486–
    87
  interspecific competition and,
    473
Resource utilization
  complex life cycles and, 582
Restriction fragment length poly-
    morphism (RFLP), 46, 48–
    49, 54, 382, 387, 391

Retrovirus
  animal, 130
  genes, 129
  oncogenic, 129
  replication, 144
Reverse transcriptase
  hypermutation and, 145
Reverse transcription, 129
REV protein, 129
REV response element (RRE),
    129
RFLP
  See Restriction fragment
    length polymorphism
Rhee, G.-Y., 6
Rhesus macaque
  immunodeficiency virus, 130–
    32
Rhizobia
  phylogeny, 335
Rhizobium, 335
Rhynocoel, 359
Rhythm generation
  signal interactions and, 100–10
Rhythmic alternation, 112–14
Rhythmic synchrony, 110–12
Rhythm preservation hypothesis,
    110–11
Richards, D. G., 299
Richmond, A., 611
Ricker, W. E., 404
Rico-Hesse, R., 137–38
Rieseberg, L. H., 59
Rigby, M., 153
Right whale, 378
Rinderpest, 449
RNA polymerase II
  HIV replication and, 129
RNA sequencing, 356
RNA virus
  mutation and, 144–47
Robertson, F. W., 40
Robertson, A., 40
Robson, E. M., 193
Rocky intertidal community
  path analysis and, 458
Rodent
  seed dispersal and, 273, 282
Roff, D. A., 414
Rose, K. A., 414
Rose, M. R., 556
Rosel, P. E., 383, 385
Rosenzweig, M. L., 225, 227
Rotifer, 3, 360–62, 557, 560, 576
  chemosensory systems, 566
  mating behavior, 563
Roughgarden, J., 475, 481, 487
RRE
  See REV response element
Rubisco, 632
Ruditapes decussatus
  allozyme genotype viability, 57
Rummel, J. D., 487

CUMULATIVE INDEXES

CONTRIBUTING AUTHORS, VOLUMES 21–25

681

CUMULATIVE INDEXES

CHAPTER TITLES, VOLUMES 21–25

VOLUME 23 (1992)
SPECIAL SECTION ON GLOBAL ENVIRONMENTAL CHANGE

ANNUAL REVIEWS

a nonprofit scientific publisher
4139 El Camino Way
P.O. Box 10139
Palo Alto, CA 94303-0139 • USA

Annual Reviews publications may be ordered directly from our office; through booksellers and subscription agents, worldwide; and through participating professional societies. **Prices are subject to change without notice. We do not ship on approval.**

- **Individuals:** Prepayment required on new accounts. in US dollars, checks drawn on a US bank.

- **Institutional Buyers:** Include purchase order. Calif. Corp. #161041 • ARI Fed. I.D. #94-1156476

- **Students / Recent Graduates:** $10.00 discount from retail price, per volume. *Requirements:* 1. be a degree candidate at, or a graduate within the past three years from, an accredited institution; 2. present proof of status (photocopy of your student I.D. or proof of date of graduation); 3. Order direct from Annual Reviews; 4. prepay. This discount **does not** apply to standing orders, *Index on Diskette*, Special Publications, ARPR, or institutional buyers.

- **Professional Society Members:** Many Societies offer *Annual Reviews* to members at reduced rates. Check with your society or contact our office for a list of participating societies.

- **California orders** add applicable sales tax. • **Canadian orders** add 7% GST. Registration #R 121 449-029.

- **Postage paid** by Annual Reviews (4th class bookrate/surface mail). UPS ground service is available at $2.00 extra per book within the contiguous 48 states only. UPS air service or US airmail is available to any location at actual cost. UPS requires a street address. P.O. Box, APO, FPO, not acceptable.

- **Standing Orders:** Set up a standing order and the new volume in series is sent automatically each year upon publication. Each year you can save 10% by prepayment of prerelease invoices sent 90 days prior to the publication date. Cancellation may be made at any time.

- **Prepublication Orders:** Advance orders may be placed for any volume and will be charged to your account upon receipt. Volumes not yet published will be shipped during month of publication indicated.

N O T E	For copies of individual articles from any *Annual Review*, or copies of any article cited in an *Annual Review*, call **Annual Reviews Preprints and Reprints (ARPR)** toll free 1-800-347-8007 (fax toll free 1-800-347-8008) from the USA or Canada. From elsewhere call 1-415-259-5017.

ANNUAL REVIEWS SERIES *Volumes not listed are no longer in print*	**Prices, postpaid, per volume. USA/other countries**	Regular Order Please send Volume(s):	Standing Order Begin with Volume:
❑ *Annual Review of* **ANTHROPOLOGY**			
Vols. 1-20 (1972-91)....................................$41 / $46			
Vols. 21-22 (1992-93)....................................$44 / $49			
Vol. 23 (avail. Oct. 1994).... ....$47 / $52		Vol(s). _____	Vol. _____
❑ *Annual Review of* **ASTRONOMY AND ASTROPHYSICS**			
Vols. 1, 5-14, 16-29 (1963, 67-76, 78-91) ....$53 / $58			
Vols. 30-31 (1992-93)....................................$57 / $62			
Vol. 32 (avail. Sept. 1994)...................$60 / $65		Vol(s). _____	Vol. _____
❑ *Annual Review of* **BIOCHEMISTRY**			
Vols. 31-34, 36-60 (1962-65,67-91)....................$41 / $47			
Vols. 61-62 (1992-93)....................................$46 / $52			
Vol. 63 (avail. July 1994).....................$49 / $55		Vol(s). _____	Vol. _____
❑ *Annual Review of* **BIOPHYSICS AND BIOMOLECULAR STRUCTURE**			
Vols. 1-20 (1972-91)....................................$55 / $60			
Vols. 21-22 (1992-93)....................................$59 / $64			
Vol. 23 (avail. June 1994)...................$62 / $67		Vol(s). _____	Vol. _____